Sculptor Bill Hopen's work "Mortality" is a monument to over two thousand men who perished of silicosis while drilling the Hawk's Nest tunnel in West Virginia in the 1930s.

The artist's work contrasts the tunnel worker's outward appearance of great physical strength with his inward feeling of weakness and impending death, tragically caused by inhaled silica.

The sculpture is located in the grounds of the Tamarack, Beckley, West Virginia.

Photo credit: Jurgin Lorenzen
Art credit: Sculptor Bill Hopen of Sutton, West Virginia

Occupational Lung Disease

Occupational Lung Disease

An International Perspective

EDITED BY

Daniel E. Banks

Professor and Chief
Section of Pulmonary and Critical Care Medicine
West Virginia University School of Medicine
Morgantown
USA

AND

John E. Parker

Adjunct Professor of Medicine
West Virginia University School of Medicine and,
Chief, Examination Processing Branch
National Institute for Occupational Safety and Health
Morgantown
USA

CHAPMAN & HALL MEDICAL

London · Weinheim · New York · Tokyo · Melbourne · Madras

Published by
Chapman & Hall, an imprint of Lippincott–Raven Publishers Inc., 2–6 Boundary Row, London SE1 8HN, UK

Lippincott–Raven Publishers, 227 East Washington Square, Philadelphia, PA 19106–3780, USA

Typeset in 9/10½ Times by Genesis Typesetting, Laser Quay, Rochester, Kent

Printed by St Edmundsbury Press, Bury St Edmunds, Suffolk

ISBN 0 412 73630 6

A catalogue record for this book is available from the British Library

Library of Congress Catalog Card Number: 98–70535

Contents

Color plates appear between pages 12 and 13

Contributors vii
Preface xi

Part One: Introduction to occupational lung disease

1 The world-wide problem of occupational lung disease
Daniel E. Banks 3

2 Pulmonary function and disability
N. LeRoy Lapp 17

3 Identification, monitoring and control of dust exposures
Frank J. Hearl 35

4 Defining workplace exposures to vapors and gases
Roy J. Rando 53

5 Airways obstruction and occupational inorganic dust exposure
Mei-lin Wang and Daniel E. Banks 69

6 Workers' compensation for occupational lung disease: German–US parallels
Claus Piekarski, Elizabeth A. Jennison and John E. Parker 83

7 Occupational health, occupational illness: tuberculosis, silicosis and HIV on the South African mines
Brian Williams, Catherine Campbell, Nokuzola Mgogi and Immo Kleinschmidt 95

8 Occupational lung diseases in Brazil
Eduardo Algranti 105

Part Two: Pneumoconioses

9 The diagnosis of pneumoconiosis and novel therapies
Jeffrey L. Wilt, John E. Parker and Daniel E. Banks 119

10 Mechanisms of pneumoconiosis
Ken Donaldson 139

11 Lung diseases in coal workers
John E. Parker and Daniel E. Banks 161

12 Coal workers' pneumoconiosis in developing nations
Changqi Zou and Guo-An Shen 183

13 Quartz and silicosis
William G.B. Graham 191

14 Silicosis in developing countries
Surinder K. Jindal and Jagdeep Whig 213

15 Asbestos-related lung diseases
Raymond Bégin 219

16 Changing patterns of asbestos-related pulmonary diseases
Colin R. Woolf 239

17 Asbestos-related lung diseases in Russia
Favel M. Kogan 247

18 A national program to reduce the risk of asbestos-related disease in Finland
Antti Tossavainen and Matti S. Huuskonen 255

19 Man-made vitreous fibers and respiratory health effects
Jacques Ameille, J.C. Pairon and P. De Vuyst 263

20 Lung disease from metal exposure
Benoit Nemery 279

Part Three: Occupational asthma, respiratory illness due to dust exposures, and organic dust diseases

21 Assessment of airway responsiveness relevant to occupational exposures
Susan M. Tarlo, Neil Alexis and Frances Silverman 309

22 Mechanisms of occupational asthma
Chris Stenton, Gavin Spickett and David Hendrick 327

23 Occupational asthma due to low-molecular weight compounds
P. Maestrelli, M. Saetta, C.E. Mapp, L.M. Fabbri, J.E. Parker and D.E. Banks 345

24 Isocyanate-induced asthma
Cristina E. Mapp 375

25 Occupational asthma due to high-molecular weight agents
Jeremy R. Beach, Francis Thien and E. Hadyn Walters 381

26 Urban epidemic asthma
Josep M. Antó and Jordi Sunyer 401

27 The effects of cotton dust on the lung
Ragnar Rylander 409

28 Byssinosis in India and the developing world
Rohini V. Chowgule 421

29 Hypersensitivity pneumonitis
John E. Salvaggio, Ann E. Vockroth and Manuel Lopez 435

30 Acute toxic chemical exposures: mechanisms and outcomes
Paul Cullinan and A.J. Newman Taylor 453

Part Four: Pulmonary carcinogenesis

31 Lung carcinogenesis: molecular and cellular mechanisms
Heather H. Nelson, Xi Wang and Karl T. Kelsey 467

32 Occupationally induced pulmonary malignancy (excluding asbestos-related cancers)
David C. Christiani 485

Index 499

Contributors

Dr Neil Alexis
Division of Respirology
University of Toronto
223 College Street
Toronto, Ontario, M5T 1R4
Canada

Dr Eduardo Algranti
Division of Occupational Health
Fundacentro
Rua Capote Valente, 710
05409–002-São Paulo SP
Brazil

Dr J. Ameille
Consultation de pathologie professionnelle
Hôpital Raymond Poincaré
Université Paris V
Paris
France

Dr Josep M. Antó
Respiratory and Environmental Health Research Unit
Institut Municipal d'Investigació Mèdica
C Doctor Aiguadar 80
E-08003 Barcelona
Spain

Dr Raymond Bégin
Service de Pneumologie
CHU de Sherbrooke
3001–12e avenue Nord
Sherbrooke, Quebec, J1H 5N4
Canada

Dr Daniel E. Banks
Professor and Chief
Section of Pulmonary and Critical Care Medicine
West Virginia University School of Medicine
Morgantown, WV, 26506–9166
USA

Dr Jeremy R. Beach
Senior Clinical Lecturer
Institute of Occupational Health
University of Birmingham
Birmingham B15 2TT
UK

Dr Catherine Campbell
Division of Social Psychology
London School of Economics and Political Science
Houghton Street
London WC2A 2AE
UK

Dr Rohini Chowgule
Pulmonary Physician
Associate Professor of Medicine
Bombay Hospital Institute of Medical Sciences
Bombay University
Bombay Hospital MRC Building
1st Floor, New Marine Lines
Bombay 400 020
India

Dr David C. Christiani
Professor of Occupational Medicine and Epidemiology
Dept of Environmental Health
Harvard School of Public Health,
and Professor of Medicine
Harvard Medical School
Boston MA 02115
USA

Dr Paul Cullinan
Occupational and Environmental Medicine
Royal Brompton National Heart and Lung Hospital
Sydney Street
London SW3 6NP
UK

Professor Ken Donaldson
Professor and Reader in Pathobiology
Napier University
Department of Biological Sciences
10 Colinton Road
Edinburgh EH10 5DT
UK

Professor Leonardo M. Fabbri
University of Ferrara
Disease Center on Asthma and COPD
Via Savorlarola 9
44100 Ferrara
Italy

Dr William G.B. Graham
420 Lake Road
Box 2180
Charlotte VT 05445
USA

Frank J. Hearl
Division of Respiratory Disease Studies
Centres for Disease Control and Prevention
National Institute for Occupational Safety and Health
1095 Willowdale Road
Morgantown, West Virginia 26505–2888
USA

Dr David Hendrick
Departments of Respiratory Medicine and Immunology
Newcastle General Hospital
Westgate Road
Newcastle upon Tyne NE4 6BE
UK

Dr Matti S. Huuskonen
Uusimaa Regional Institute of Occupational Health
Arinatie 3 A
00370 Helsinki
Finland

Dr Elizabeth A. Jennison
Excel Corporate Care
Middletown
Ohio
USA

Dr Surinder K. Jindal
Additional Professor and Head
Department of Pulmonary Medicine
Postgraduate Institute of Pulmonary Medicine and Research
Chandigarh-160012
India

Dr Karl T. Kelsey
Occupational Health Program
Department of Environmental Health
Harvard School of Public Health
665 Huntington Avenue
Boston MA 02115
USA

Dr Immo Kleinschmeidt
Epidemiology Research Unit
Medical Bureau for Occupational Diseases
PO Box 30606
Braamfontein 2017
South Africa

Prof. Favel M. Kogan
Medical Research Center for Prophylaxis and Health Protection in Industrial Workers
30, Popov Street
Ekaterinburg 620014
Russia

Professor N. LeRoy Lapp
Emeritus Professor
Section of Pulmonary and Critical Care Medicine
West Virginia University Health Sciences
P.O. 9166
Morgantown, WV, 26505–9166
USA

Dr Manuel Lopez
Section of Allergy and Immunology
Tulane University Medical Centre
1700 Perdido Street
New Orleans
Louisianna 70112
USA

Dr P. Maestrelli
Institute of Occupational Medicine
University of Padova
Via Giustiniani, 2
35128 Padova
Italy

Dr Cristina E. Mapp
Institute of Occupational Medicine
University of Padova
Via Giustiniani, 2
35128 Padova
Italy

Dr Nokuzola Mgogi
Epidemiology Research Unit
Medical Bureau for Occupational Diseases
PO Box 30606
Braamfontein 2017
South Africa

Dr Heather H. Nelson
Occupational Health Program
Department of Environmental Health
Harvard School of Public Health
665 Huntington Avenue
Boston MA 02115
USA

Dr Benoit Nemery
Katholieke Universiteit Leuven
Laboratorium voor Pneumologie
Herestraat 49
B-3000 Leuven
Belgium

Professor Anthony Newman Taylor
Occupational and Environmental Medicine
Royal Brompton National Heart and Lung Hospital
Sydney Street
London SW3 6NP
UK

Dr J.C. Pairon
Service de pneumonologie et de pathologie professionnelle and Inserm Unité 139
Centre hospitalier Intercommunal
Créteil
France

Dr John E. Parker
Adjunct Professor of Medicine
West Virginia University School of Medicine, and Chief, Examination Processing Branch
National Institute for Occupational Safety and Health
Morgantown, WV, 26505
USA

Professor Dr Claus Piekarski
Institut fur Arbeitswissenschaften der RAG Aktiengesellschaft
Wengeplatz 1
Postfach 18 01 62
D-44369 Dortmund 18
Germany

Dr Roy J. Rando
Associate Professor of Public Health
Department of Environmental Health Sciences
Tulane University School of Public Health
New Orleans, LA, 70112
USA

Dr Ragnar Rylander
Department of Environmental Medicine
Gothenburg University
Box 414
405 30 Gothenburg
Sweden

Dr M. Saetta
Institute of Occupational Medicine
University of Padova
Via Giustiniani, 2
35728 Padova
Italy

Dr John E. Salvaggio
Section of Allergy and Immunology
Tulane University Medical Centre
1700 Perdido Street
New Orleans
Louisianna 70112
USA

Professor Guo-An Shen
Professor and Head
Department of Occupational Medicine
Sichuan Institute of Labour Hygiene and Occupational Disease
19 Dian Xin Road
Chengdu 610041
People's Republic of China

Dr Frances Silverman
Division of Respirology
University of Toronto
223 College Street
Toronto, Ontario, M5T 1R4
Canada

Dr Gavin Spickett
Regional Department of Immunology
Royal Victoria Infirmary
Queen Victoria Road
Newcastle upon Tyne NE1 4LP
UK

Dr Chris Stenton
Department of Respiratory Medicine
Royal Victoria Infirmary
Queen Victoria Road
Newcastle upon Tyne NE1 4LP
UK

Dr Jordi Sunyer
Respiratory and Environmental Health Research Unit
Institut Municipal d'Investigació Mèdica
C Doctor Aiguadar 80
E-08003 Barcelona
Spain

Dr Susan M. Tarlo
Division of Respirology
University of Toronto
223 College Street
Toronto, Ontario, M5T 1R4
Canada

Dr Francis Thien
Senior Lecturer
Department of Respiratory Medicine
Monash Medical School
Alfred Hospital
Prahran, Victoria,
Australia

Dr Antti Tossavainen
Institute of Occupational Health
Topeliuksenkatu 41 A
SF-00250 Helsinki
Finland

Dr Ann E. Vockroth
Section of Allergy and Immunology
Tulane University Medical Centre
1700 Perdido Street
New Orleans
Louisianna 70112
USA

Professor P. De Vuyst
Chest Department
Hôpital Erasme
Université libre de Bruxelles
Brussels
Belgium

Professor E. Hadyn Walters
Professor of Medicine
Director of Respiratory Medicine
Monash Medical School
Alfred Hospital
Prahran, Victoria,
Australia

Dr Mei-Lin Wang
Research Assistant Professor of Medicine
Section of Pulmonary and Critical Care Medicine
West Virginia University School of Medicine
Morgantown, WV, 26506–9166
USA

Dr Xi Wang
Occupational Health Program
Department of Environmental Health
Harvard School of Public Health
665 Huntington Avenue
Boston MA 02115
USA

Dr Jagdeep Whig
Professor and Head
Chest Diseases and Tuberculosis Unit
Dayanand Medical College and Hospital
PO Box 285
Ludhiana 141001
India

Dr Brian Williams
Epidemiology Research Unit
Medical Bureau for Occupational Diseases
PO Box 30606
Braamfontein 2017
South Africa

Dr Jeffrey Wilt
Assistant Clinical Professor of Medicine
Section of Pulmonary and Critical Care Medicine
Michigan State University
Blodgett Memorial Medical Centre
Grand Rapids, MI 49506
USA

Dr Colin R. Woolf
Former Director
Division of Respirology
Toronto General Hospital
University of Toronto Department of Medicine
1166 Bay Street, Suite 1405
Toronto, Ontario M5S 2X8
Canada

Professor Changqi Zou
Director and Professor
Institute of Occupational Medicine
Chinese Academy of Preventive Medicine
WHO Collaborating Center for Occupational Health
29 Nan Wei Road
Beijing, 100050
People's Republic of China

Preface

Diseases of the lung caused by workplace exposures have been recognized for centuries. Despite substantial knowledge about the agents responsible for these diseases, as well as control measures to reduce exposures, tragically these diseases persist throughout both the developed and developing world.

In this textbook, we have reviewed the current knowledge base for the recognition, pathogenesis, treatment and prevention of many of the recognized occupational lung diseases. We have tried not only to present the information in didactic format, but to bring this to life by providing a global perspective. Our interest in providing this international perspective makes us realize that although our cultures may be very different, the problems experienced within them are very similar, with physicians and public health professionals struggling to progress from the recognition of these illnesses to their prevention.

Although much of this book addresses the traditional pneumoconioses related to asbestos, silica, and coal mine dust exposures, as well as how to understand the workplace environments, we have also expanded these issues to look at the pathogenesis of these diseases and their prevalence in the developing world. We have attempted to bring these issues up to date by building on the work of our Chinese colleagues and changing our perspective on these illnesses by viewing them as having potential therapeutic options. Further, we have reviewed some of the social challenges that interact with occupational lung diseases.

In addition to byssinosis and hypersensitivity pneumonitis, the more recently recognized relationships of asthma and the workplace and the development of cancer associated with workplace exposures are also discussed. These disorders are currently challenging the traditional dust-related diseases as the most prevalent occupational lung disease in many nations.

Further special problems have been presented. These include the country-wide effort in Finland to reduce the burden of asbestos-related diseases, the occurrence of epidemics of asthma in different cities around the world, the difficulties that the population of Brazil faces as it changes from an agrarian society to an urban-based community, and what may be the most complex and worrisome problem in the study of occupational lung disease, i.e. the collision between silicosis, tuberculosis and HIV infection in South Africa which is exacting a painful toll.

The conception and maturation of this book has spanned several years. It builds upon the collective experience of numerous international scientists and physicians. We are proud of this effort and we recognize the competence of the authors, as well as thank them for their hard work. Each is an expert in his or her field of occupational lung disease and each brings a very unique approach to this work. Their dedication and commitment, as well as their care and compassion, for those afflicted with lung diseases caused or aggravated by work is quite palpable on these pages. Additionally, their knowledge that active attention to exposure control by occupational safety and health professionals, and by employers and workers, is crucial to the prevention of these disorders, is reflected in each chapter. We have been both privileged and humbled by the lifetime contribution of the many scientists whose work is referenced and who have shaped this textbook.

We must also recognize the competence of our colleagues at Chapman & Hall, Publishers. Joanna Koster, Ph.D., the editor, has shepherded this effort through the three years that it has taken to get to this point. In addition, Anne Crabb, Eleanor Gillam and, most recently and most importantly, Jo Pillai have worked with us on a near-daily basis to accomplish this task and finish this job. Without their competence, this book would still be a dream. We appreciate their help.

We sincerely thank these experts for their contributions and hope that others will find this text helpful in the noble pursuit of the recognition, treatment and prevention of occupational lung diseases.

As we have attempted to put the polishing touches on this text, we feel compelled to close our effort by paraphrasing a quote commonly attributed to Winston Churchill, 'You never really finish a book, you just finally abandon it.'

Finally, on a more personal note, Sheena Banks, Christine Parker, the five Banks and two Parker children, have supported this effort by accepting the common exchange of family time for time committed to this book for the past several years. They have been and are marvelous.

Daniel E. Banks

John E. Parker

Morgantown, West Virginia

March, 1998

Part One Introduction to occupational lung disease

1 The world-wide problem of occupational lung disease

Daniel E. Banks

The critics say further that mining is a perilous occupation to pursue, because the miners are sometimes killed by the pestilential air which they breathe; sometimes their lungs rot away; sometimes the men perish by being crushed in masses of rock; sometimes, falling from the ladders into the shafts, they break their arms, legs, or necks; and it is added there is no compensation which should be thought great enough to equalize the extreme dangers to safety and life. These occurrences, I confess, are of exceeding gravity, and moreover, fraught with terror and peril, so that I should consider that the metals should not be dug up at all, if such things were to happen very frequently to the miners, or if they could not safely guard against such risks by any means. Who would not prefer to live rather than to possess all things, even the metals? For he who thus perishes possesses nothing, but relinquishes all to his heirs. But since things like this rarely happen, and only in so far as workmen are careless, they do not deter miners from carrying on their trade any more than it would deter a carpenter from his ...

(Agricola, 1555).

INTRODUCTION

This quote from Agricola helps us to understand the widely held perspective of worker safety and health in his time. First, there can be no question that the risks of injury and lung illness were realized to be a part of the miner's life, even in the 16th century. Yet, the statement that the health risks associated with mining are comparable with the risk a carpenter might take in falling from a ladder appears to be confusing. Unless such work was somehow much more dangerous than it is thought of today, this quotation clearly underestimates what we historically know about the health risks in mining (Meiklejohn, 1951, 1952a,). Similarly, although few of us know much about the ways mining was performed in the 16th century, it is difficult for us today to understand the perspective that health impairment in a miner, when such events occur, would be dismissed as occurring infrequently and being attributable to the miner's carelessness.

Such a viewpoint apparently reflected the 'cost of doing business' in mining and defined the manner in which worker health was perceived in the 16th century. The approach of our age is much different. Over time, a series of changes in the social structure of a number of cultures has resulted in a dramatically different perspective. Governments of a great number of countries have implemented formal policies of workers' compensation, financially supported by the employer (with costs based on the frequency and severity of claims presented by their workers), yet mandating that representatives of the government be responsible for deciding whether a worker's ill health is the result of work. Interestingly, the policies of a government and the judgment of the government's representatives, and not the clinical features of a worker's illness nor the

Occupational Lung Disease. Edited by Daniel E. Banks and John E. Parker. Published in 1998 by Chapman & Hall.
ISBN 0 412 73630 6

degree of impairment, may influence whether disability payments are awarded (Liss *et al.*, 1997), or influence the policy made regarding the awarding of benefits (Ellison and Yeung, 1995).

When one addresses the continuing development of occupational lung disease in many countries around the world, several concerns are soon apparent. First, it is commonly recognized that there are a large number of workers exposed to potentially harmful agents in the workplace, yet, defining the scope of the problem in terms of the number of workers at risk or the number adversely affected is next to impossible for several reasons. Not only is there inadequate record-keeping, but not infrequently, the relationship between exposures at work and disease is underestimated. Inadequate statistics are the norm for most of the bodies which keep such records in the great number of countries. In just a few countries is there sufficient emphasis in maintaining accurate reports of disease to allow physicians to be confident that the data accurately represent the severity of the problem. Second, there are workers at risk in both developed and developing countries, and epidemics of a variety of occupational illnesses have been well documented in many different societies (Figure 1.1 and Plates 1–4). Although it appears that the number of cases of disease in a working population are diminished by aggressive enforcement of effective regulatory policies in societies where such approaches exist, it remains unclear whether epidemics of respirable disease in isolated workplaces can ever be controlled. As an example, we are unaware of any society which has rid itself of silicosis (Table 1.1). In the US, most investigators remain convinced that the number of Vermont granite shed workers (perhaps the most well-reported population of US workers with a continuing risk for silicosis in this century) who develop silicosis will remain minimal (Graham, *et al.*, 1991), if it occurs at all, yet many of us recognize that US workers employed in surface mining, or in the processing of silica flour, will continue to develop silicosis (Banks *et al.*, 1981, 1983). Third, although occupational pulmonary illnesses are thought to be essentially preventable, they continue to occur. Conveying the public health perspective that dust-related diseases are dangerous to a worker's respiratory health can be difficult, particularly when the period from first exposure to the development of disease may be 20 or more years. Furthermore, the frequently suggested solution of wearing personal respiratory protective devices throughout the workday is an unrealistic expectation. The increase in the work of breathing, the discomfort, the poor-fit sometimes attributable to facial hair, and the inability to speak and adequately communicate with one's fellow worker is almost too much to ask of any worker. Finally, the

Figure 1.1 One of the most tragic epidemics of silicosis has been reported among the pit diggers of north-eastern Brazil. In this area of the world where rainfall is minimal and drought appears to be the norm, the lack of water supply to the rural population has led dwellers to manually excavate pits in search of underground water. Briefly, two workers take turns hand-digging water wells which extend approximately 20 meters into the earth and are approximately 1.5 metres in circumference. The digging is done in a dusty, poorly ventilated space. Digging is mainly done with a hammer, chisel, or shovel, although explosive charges may be needed when the rock cannot be manually excavated. The rock is sandstone which, in places, appoximates pure crystalline silica. Measured dust concentrations exceed several hundred-fold the permissible allowable exposures. Of 366 men employed, 26.8% had category 0/1 or greater pneumoconiosis. Of these silicotics, 83% had a strong reaction to PPD. In a separate cohort of 86 silicotics followed for 3 years, 30 (35%) died of respiratory failure. None of these had active tuberculosis. The illustration shows men lowering the digger by a rope and also removing the dirt and fractured rock as needed. (Photographs courtesy of Marcia Alcantra Holandra, MD.)

recognition of the work-relatedness of the disease is not always appreciated. As an example, the development of asthma in a worker with isocyanate exposure can be very confusing when there is no clear understanding of the series of events which have resulted in the worker becoming sensitized to these agents.

How should the effectiveness of prevention of occupational lung disease in a society be measured? The answer is unclear and attempting to compare the occupational lung health of workers across different societies has never been satisfactorily considered. Numerous concepts make the comparisons difficult and often quite 'unfair', as societies are so different around

Table 1.1 Examples of epidemics of silicosis reported in the 1990s

Authors and year	*Country*	*Description*
Holandra *et al.* (1995)	Brazil	366 of 1050 pit diggers were screened with chest radiographs for silicosis. 121/366 had positive films. Silica exposures were attributable to manual pit digging.
Saiyed *et al.* (1995)	India	44 of 292 pottery workers developed silicosis. Of these, 16 developed tuberculosis and another 28 without silicosis developed tuberculosis.
Neukirch *et al.* (1994)	France	29 of 301 men employed as pottery plant workers developed silicosis. Interestingly, the level of lung function was independent of the presence or absence of silicosis.
Cavarini *et al.* (1995)	Italy	231 new cases of silicosis were recognized among 2480 actively employed ceramic workers in central Italy employed from 1974–1987 followed for an average of 8 years.
Partanen *et al.* (1995)	Finland	Although the population at risk is not reported, 667 Finns have developed silicosis from 1964–1992. The most dangerous jobs, in terms of the greatest number of reported cases, were casting and founding and work in the glass, ceramics, and building products industry.
Meijers (1996)	Netherlands	A retrospective study of 1794 Dutch ceramic workers revealed 124 cases of simple silicosis.
Rosenman *et al.* (1996)	US	Of 1072 current workers employed in an automotive factory for at least 5 years from 1969 to 1991, 60 had pneumoconiosis. The risk increased with years of employment, with 12% of workers employed >30 years having silicosis.
Provencher (1997)	Canada	Over a 13-month period beginning in 1992, 25 cases of silicosis were recognized by the Quebec Workers' Compensation Board. Six workers were under 40 years of age and five of these were sandblasters.

the world. First, comparing the number of respiratory illnesses due to acute chemical exposures may not be a good measure because such excessive exposures are likely to be relatively infrequent. For that reason, it would seem that the effects of chronic exposures and their resultant illnesses would be more representative of worker health. Second, the comparison of the health of workers employed with sophisticated technologic processes would be limited, as these processes would be most frequently recognized in more developed countries. Therefore, it might be more appropriate to compare chronic dust exposures (likely to be an exposure common to many countries or societies) than chronically occurring chemical exposures. Finally, if one were to address the effects of dust exposure, what dust would be relatively most common? Exposure to silica may be the most appropriate, as silica exposure appears to be common to nearly every type of mining and ore processing. The result of inhaling excessive numbers of silica particles of respirable size results in a chronic, progressive illness with defining characteristics which can arguably be used as an example for the adequacy of preventive measures used in workplaces throughout different societies (Figure 1.2).

Figure 1.2 This Indian worker has remained clean while 'renewing' a grinding wheel, applying a mixture of molasses and sand on a belt and allowing this to adhere to the grinding wheel. When this has dried, the wheel will serve as an effective abrasive surface.

Perhaps the most notorious failures in preventing occupational lung disease have occurred in reported epidemics of silicosis in the American workplace (Figure 1.3). These outbreaks have caused the premature death of many American workers in this century (Betts, 1900; Russell *et al.*, 1929, Russell, 1941; Hosey *et al.*, 1957; Banks, *et al.*, 1981, 1983; Suratt *et al.*,

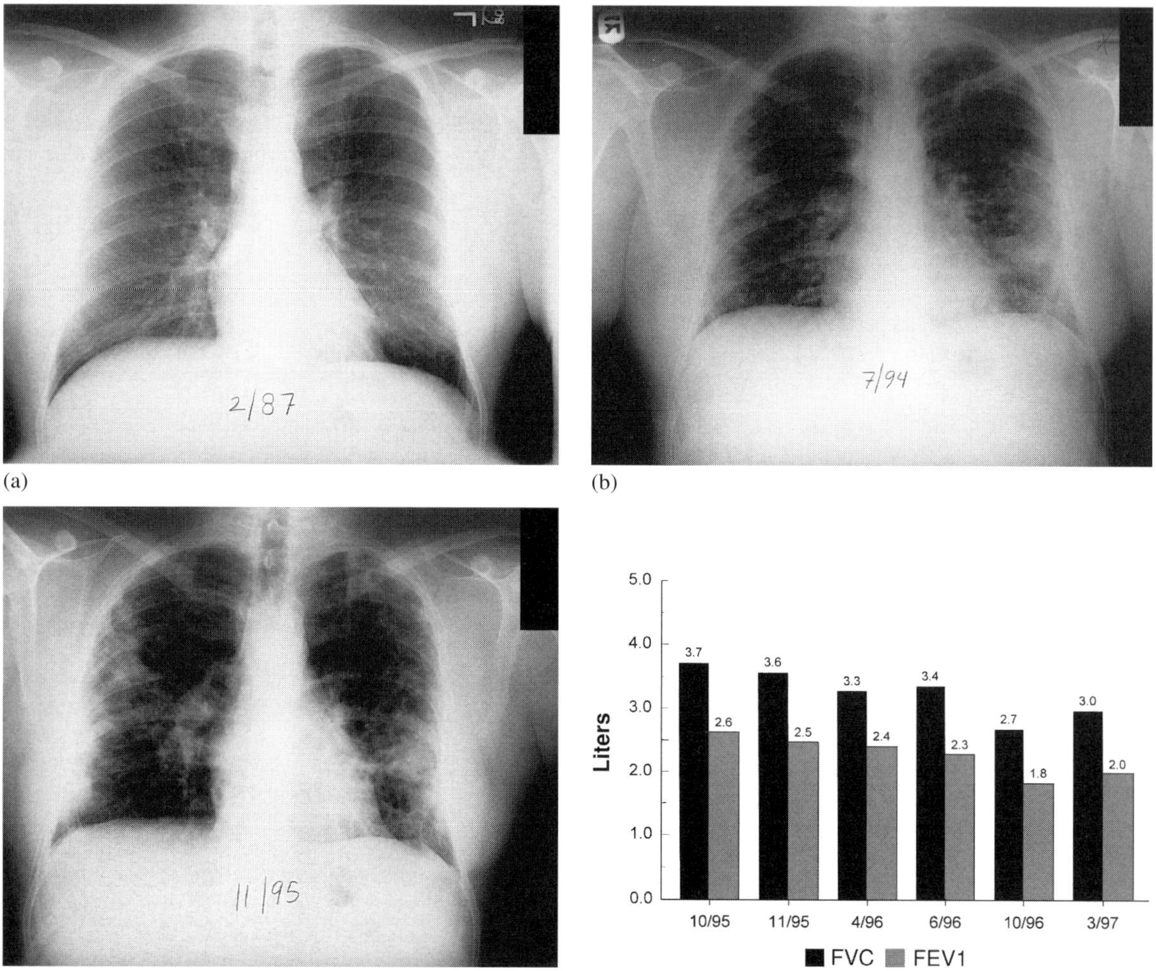

(a) (b) (c) (d)

Figure 1.3 These chest radiographs are from a 43-year-old man employed as an underground miner in southern West Virginia. Since 1975, he had worked for 13 years as a continuous miner/operator and in 1993, he operated a continuous mining machine for several months, drilling through hard rock in order to connect two coal seams. This was a very dusty job, where he commented that he 'could not see one foot in front of his face because of the extensive amount of dust'. The man was dyspneic at rest, and had a morning cough. He had no history of cardiac illness and his physical examination was unremarkable, including lungs which were quiet to auscultation. The 1987 radiograph (a) showed a background of minimal small rounded opacities. In 1994 (b), the radiograph was much different, with apparent consolidated areas in the lower zones. These features were thought to have developed following the inhalation of silica dust in the continuous mining of hard rock. Progression was evident in the 1995 radiograph (c). Overall, it appears that these radiographs reflect the development of very aggressive silicosis which occurred on a background of minimal coal workers' pneumoconiosis. A plot of lung function over time showed (d) a decline in FVC of 560 ml/yr and an FEV_1 decline of 480 ml/yr. This rate of decline approximates ten times the expected rate of yearly decline in lung function of a healthy worker.

1977; Fleming *et al.*, 1990). An author would be remiss if the worst recognized outbreak of silicosis, which occurred during the construction of the Gauley Bridge tunnel in West Virginia in 1930 and 1931, were not described. In this episode, more than 400 of the estimated 2000 men engaged in rock drilling died, and about 1500 contracted silicosis and were eventually disabled (Corn, 1980). Specific reasons why each of these epidemics have occurred are not immediately apparent, but a number of forces appear to be common to many. These include, but are certainly not limited to, the relatively long latency periods which may mask the relationship between exposure and development of silicosis, inadequate (or even careless) understanding of the relationship between dust exposure and disease, development of power tools with the potential to generate massive amounts of respirable-size dust particles, the association of silicosis with infectious diseases (specifically mycobacterial infection) (Sherson and Lander, 1990; Cowie, 1994), and respiratory protection which was initially thought effective, but later shown to be inadequate. On a more conceptual level, perhaps social change associated with world-wide industrialization and the growth of very remunerative international trade has played a role.

Importantly, in each report, regardless of whether the outbreak dated from the turn of the century or occurred in the 1980s, a common theme has been the lack of intervention by those knowledgeable about the health effects of fibrogenic dust exposure. In earlier days, the worker depended on the owner and supervisor, individuals with knowledge of the processes, to protect them (Glindmeyer and Hammad, 1988). More recently, state and federal regulatory agencies have identified and implemented 'safe' levels of exposure which have been shown to reasonably prevent the development of silicosis (Graham *et al.*, 1991). Even so, the disease has continued. Ironically, a report from England the same year that Graham described the absence of silicosis in Vermont granite workers detailed aggressive silicosis in stonemasons, showing how severe silicosis can be in the same type of industry when adequate respiratory protection is not employed (Seaton *et al.*, 1991).

In some mines and quarries, compliance with the permissible exposure level has been achieved by instituting wetting techniques and improving ventilation. Alternatively, in industries such as sandblasting, silica mining, and rock drilling, workers remain potentially exposed to many times the acceptable limit of respirable free silica (Figure 1.4). In the United States, particles much less toxic than silica have been available for blasting for a number of years (McKay *et al.*, 1980; Stettler *et al.*, 1988), and the National Institute for Occupational Safety and Health (1974) recommended substitution of non-silica-containing particles in abrasive blasting in 1974. A great variety of materials (corundum, glass beads, pumice, sawdust, slags, steel grit and shot and walnut shells) are commercially available. Yet, inaction on this issue by state and federal leadership has clearly added to the number of cases.

Figure 1.4 Long wall mining is the most mechanized and most productive means of underground coal mining. The number of underground mines employing long wall mining techniques continues to increase in the US (and now is used in approximately 100 mines, accounting for approximately 30% of US coal production). Long wall mining produces up to 40 tons of coal per minute, and in some instances, nearly 4000 tons per work shift. In this process, the continuous miner cuts out a long wall of coal which may be 600 to 800 feet wide and 5000 to 12 000 feet long. The figure shows the formation of the long wall. Clearly visible in this photograph are the coal face and the hydraulically powered roof supports. At the far end of the photograph and poorly visible is the cutting machine or shearer with water spraying to minimize the dust generation. A conveyor belt which runs beneath the shearer the length of the long wall captures the newly excavated coal.

SURVEILLANCE AND RECOGNITION OF THE SCOPE OF THE PROBLEM

Although reports of silicosis outbreaks in the US have appeared at relatively frequent intervals in the US pulmonary and occupational literature, updates of a governmental bulletin have generated a series of snapshots regarding the incidence and prevalence of occupational pulmonary diseases in the US over the two past decades. This has helped the community of occupational physicians in the US focus on the 'big picture' of these illnesses nation-wide. This bulletin, the *Work-related lung diseases surveillance report (1996)*, presents the data clearly, often in a helpful graphic format, and describes information which has helped us to better understand trends regarding occupational lung disease in America. A series of relevant conclusions are summarized in Table 1.2

There can be several emphases when one is dealing with the prevention of occupational lung diseases. Primary prevention, the backbone of prevention of all workplace disease, is best achieved by eliminating exposures. This is most effective and requires initiatives on several fronts, yet it seems to most often require an 'epidemic' of occupational illness in the workplace for this approach to be implemented. From a social perspective, the development, legislation and implementation of scientifically correct regulations is important. Engineering controls are usually necessary

Table 1.2 Trends in occupational lung disease in the US

Asbestosis
From 1968–1992, the number of deaths attributable to asbestosis have risen dramatically. In 1968, less than 1 American in 1 000 000 died with asbestosis. In the most recent year, nearly 6 per 1 000 000 deaths are due to this disease. Perhaps the 'good news' is that the number has remained relatively unchanged for the three most recent years of the survey. In addition, the mean age of death was 60 years in 1968. In 1992, the mean age of those dying with asbestosis was 74 years.

In 1991, the jobs at greatest risk for dying from asbestosis include plumbers and pipefitters, electricians, insulation workers, laborers and carpenters The most dangerous industry to work in is ship building and repair.

With an occasional exception, the greatest number of deaths occurred in the states bordering the oceans, or along the Gulf coast. These are sites where shipyard work has been most prevalent.

Coal workers' pneumoconiosis (CWP)
In contrast to asbestosis, from 1968–1992, the number of deaths attributable to CWP have declined dramatically. In 1968, approximately 12 Americans in 1 000 000 died with CWP; this peaked at 18 in 1972, and, in 1991, the most recent year, 8 per 1 000 000 died of this disease. Perhaps the continuing fall in the number of US coal miners has played an important part in this rate of decline as there is little evidence that the dust levels have declined in the past 10 years. On average, CWP appears to affect survival relatively less than asbestosis. The mean age of death for CWP was 72 years in 1968 and 78 years in 1992.

Silicosis
The number of workers dying with silicosis has declined dramatically. In 1968, 12 per million died with silicosis. In 1991, the number approximated 2 per million. In most instances, the mortality from silicosis parallels the mortality from CWP. The mean age of death from silicosis was 72 years in 1968. In 1992, the mean age of death was 75 years. Unlike asbestosis, mortality among blacks exceeds that among whites when the figures are race-adjusted.

Byssinosis
The number of deaths from byssinosis has varied from 0.6 to 0.12 per million per year for the years 1979 to 1991, with no clear pattern of increase or decrease in prevalence. Using death certificate data, 12 Americans died of byssinosis in the US in 1991, and 15 in 1992.

to meet the ambient air quality standards mandated for many agents. A 'corporate culture' which recognizes that it is in the corporation's best interest to address problems associated with excessive exposures before the recognition of disease in their workers must be present to have any success in preventing these illness. Secondary prevention is frequently initiated only after primary prevention has failed. This is achieved through the development of an organized surveillance program which allows for the early detection of disease. These initiatives can be very successful but they are frequently very costly and require a strong commitment by both management and labor. Tertiary prevention is the prevention of disability and complications of existing disease by improved means of diagnosis and effective treatment strategies. Such a process plays a small role in occupationally related lung diseases. Typically, once dust-induced lung disease has occurred, or a worker has developed occupational asthma to agents in the workplace, the worker is best advised to leave the workplace. Although symptomatic therapy for these illnesses is commonly offered and may provide some relief, the physician is powerless when it comes to affecting the natural history of asbestosis or silicosis, or to allow a worker with asthma attributable to isocyanate sensitivity to return to the workplace.

Based on death certificate information, the *Work-related lung diseases surveillance report* shows a generalized downward trend in US mortality rates from 1968 to 1992 in total deaths due to fibrogenic dust exposure. There was a 60% decline in deaths attributable to coal workers' pneumoconiosis (CWP), a 15% decline in deaths due to silicosis, and 18% fewer deaths from inhalation of agents such as aluminum, beryllium and heavy metals. Although 100 890 US workers have been reported to have died from pneumoconiosis over the past 24 years, the number has declined yearly.

Yet, even though the number of US death certificates from 1968–1992 with mention of silicosis and CWP are clearly declining, interpretation of this information is difficult. The most important problem in understanding these data is the absence of a denominator. The number of workers with exposure to these dusts (which certainly has changed over time, and, at least in the US, has apparently lessened) is unavailable. Information is expressed as deaths from either of these illnesses per million deaths, rather than the more informative deaths per million exposed. Therefore, it is not clear whether the percentage developing disease is less. This latter conclusion may be relevant in view of available coal mine exposure data (Figure 1.5), which shows that the percentage of coal dust samples exceeding the permissible exposure limit has remained relatively steady over the past decade. Second, both silicosis and CWP typically have a long latency time, and death certificate data from 1992, for example, represent work exposures during the 1950s, 1960s, 1970s and 1980s. Excessive exposures are not likely to be recognized in mortality statistics for approximately 30–40 years, making it difficult to assess the impact of current dust measurements.

Despite the attractive presentation of the above data, Rosenman and colleagues (1997) cast doubt on the

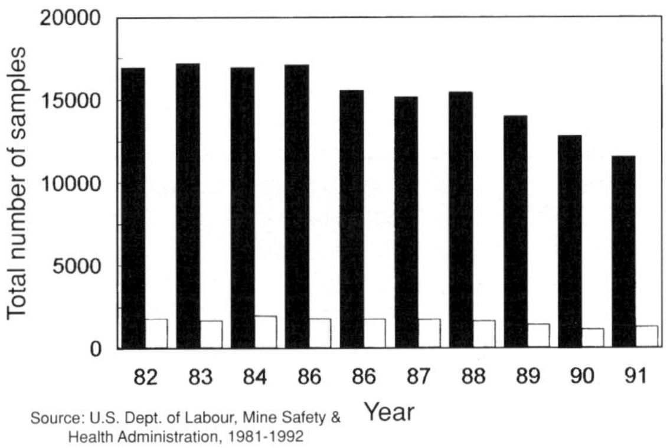

Figure 1.5 These data reflect the number of dust measurements in coal mines and the number of measurements which exceed the permissible exposure limit. (Supplied by the Mine Safety and Health Administration, a branch of the US Department of Labor; the data describe samples collected from 1981 to 1992.) Although the number of samples have decreased over time, the ratio of the number of samples which exceed the allowable exposure limit to the total number of samples collected remains unchanged. This leads to concerns about the usefulness of data from death certificates as an accurate means of identifying the number of deaths attributable to CWP.

usefulness of death certificate data. Their work leads one to be quite concerned that death certificate data significantly underestimate the prevalence of this chronic occupational illness in the population. They presented a case series of silicotics reported to the Michigan state health department from 1985 to 1995. Cases were identified through a series of sources. Of 557 recognized during this time, 12% were identified through physician referral, 63% by hospital records, 10% by death certificates, 15% by workers' compensation data, and less than 1% by index case follow-up. In a New Jersey database (Stanbury *et al.*, 1995), the families of 177 of 329 cases of silicosis recognized during the years 1979 to 1992, were interviewed to determine the number applying for workers' compensation benefits. Of these 177, 122 (69%) never filed, leading to the recognition that the database of workers' compensation claims is not an accurate reflection of the prevalence of silicosis. Furthermore, the authors correctly point out that failure to compensate these patients for medical care undercuts the effects of economic incentives for employers and employees to introduce and follow safe work practices.

Another approach to addressing the prevalence of pneumoconiosis has been employed by investigators of the National Institute for Occupational Safety and Health. Data from the National Study of Coal Workers' Pneumoconiosis, a program which provides

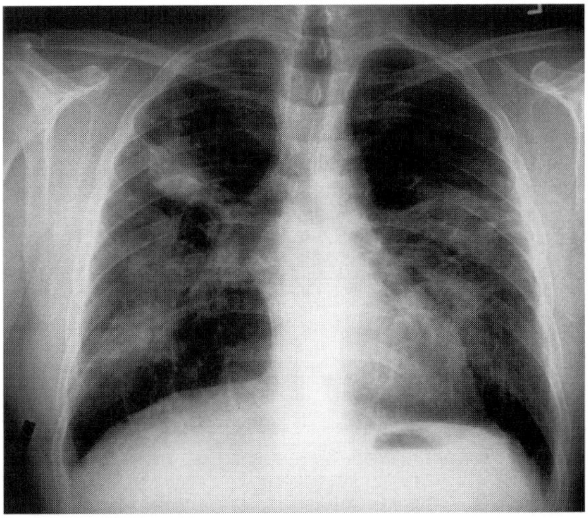

Figure 1.6 (a) Investigations undertaken in a worker with exposure to different dusts in a tool and die plant. This 49-year-old man had been employed for 18 years as a die setter with exposure to a variety of very fine powders of metals and their alloys used to make different dies. In this process, dust is poured into the molds and compressed using between 100 and 800 tons of pressure. This generates heat and fuses the material into the desired shape. It then undergoes sintering (grinding) and further heat treating. Approximately 15 years ago the man was evaluated for sarcoidosis, as a chest radiograph had been interpreted to show hilar areas which were described as being prominent bilaterally in association with distinct small rounded opacities in the parenchyma. Laboratory investigations initially performed to support a diagnosis of sarcoidosis were negative and a biopsy of a right supraclavicular node was non-diagnostic. He had only a brief smoking history in the 1970s. Over the year or two before this evaluation he had commented on increasing dyspnea, which has progressed to shortness of breath on dressing. He described the workplace as very dusty with a large amount of this dust throughout the work area. Chest examination was not remarkable. Interestingly, none of his fellow workers had respiratory impairment and he was unaware of any co-workers who had recognized dust disease in association with employment in this shop. At the time of this assessment, an angiotensin converting enzyme level was mildly elevated. Spirometry showed an FEV_1 of 0.96 litres (33% predicted), an FVC of 3.02 (66% predicted) and an FEV_1/FVC of 38%. Post-bronchodilator inhalation, the FEV_1 increased 17%. The total lung capacity was 4.44 litres (67% predicted), with the residual volume : total lung capacity ratio being 34% (predicted 27%). The chest radiograph (a) shows elevation of the right hilum with basilar emphysema but without mediastinal or hilar adenopathy. There are conglomerate mass lesions in the right upper and lower zones and left mid zone, with a parenchymal pattern of perfusion category 2/2 of 'R' size small opacities and 'C' size large opacities. A follow-up radiograph one year later (not shown) showed progression, primarily on the left. Analysis of dust particles revealed some to contain only carbon (three of these had small amounts of silica); others contained iron, copper and zinc; some were essentially pure nickel; and others had iron, nickel, copper and zinc in varying combinations. (*continued overleaf*)

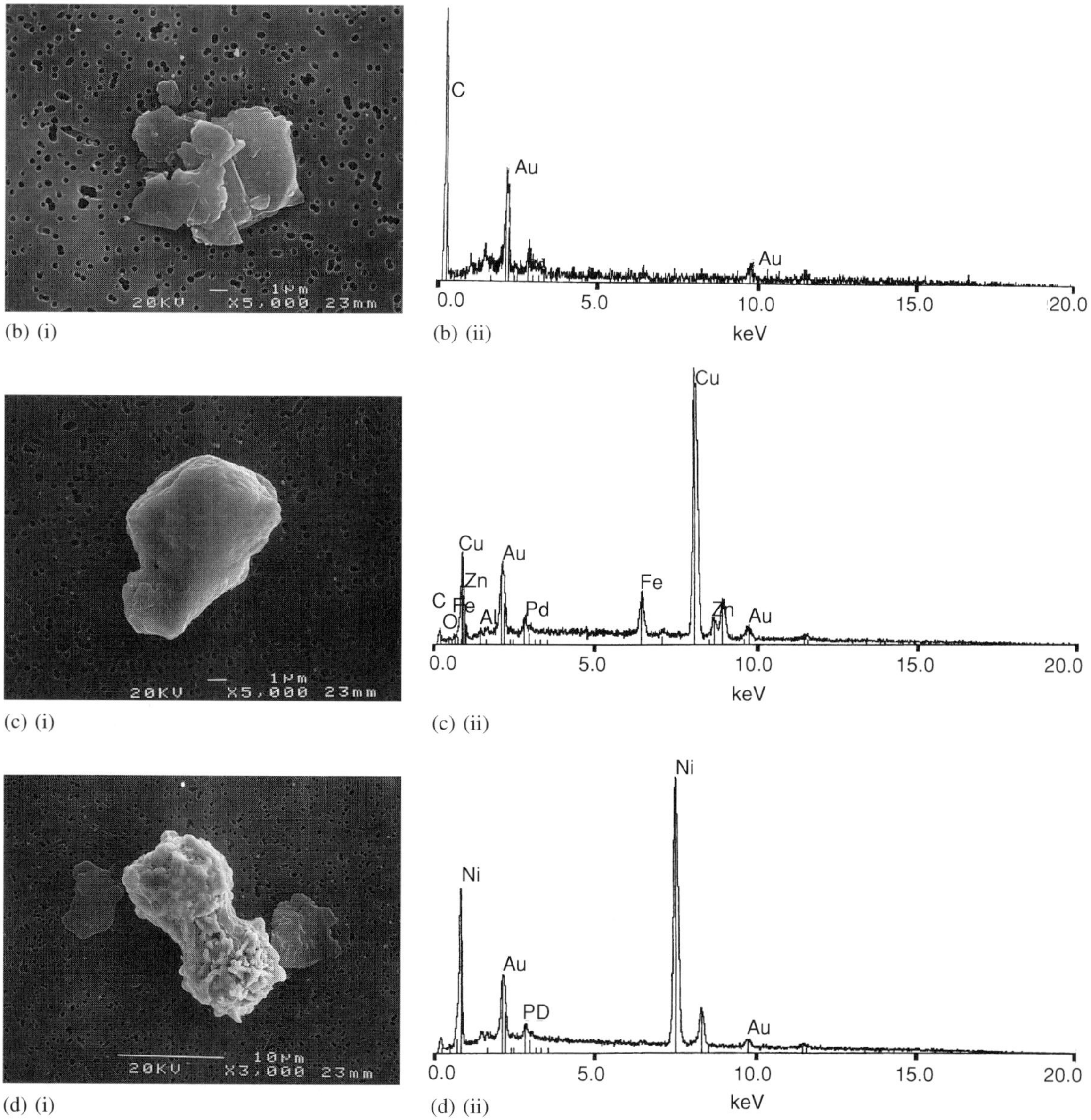

Figure 1.6 (b–d) Photomicrographs of three particles with the corresponding X-ray diffraction pattern. The element Au (gold) in each of the displays is a marker to standardize the information. The likely diagnosis appears to be carbon pneumoconiosis; however, the question of a previous diagnosis of sarcoidosis has confused the issue. The worker refused an open lung biopsy. (Dust analysis courtesy of Val Vallyathan, PhD.)

the prevalence of CWP in the population of miners which participate, have recently been presented in summary form (Attfield and Castellan, 1992; Attfield and Althouse, 1992). Briefly, although the data are complicated by X-ray reader variation, there appears to be a reduction in the prevalence and incidence of CWP in current miners who serially participated in this study. The most important concern, however, is whether the work is representative of the miners' respiratory health (Figure 1.6). For example, in 1986, the US Bureau of Mines estimated that there were 78 000 coal miners in the country. Yet, only 5131 miners participated in

Round 3 of the National Study of Coal Workers' Pneumoconiosis which was ongoing at that time. Validation that the miners who participated were representative of the mining population at large was not presented, leading to concerns regarding the adequacy of the sample.

The Canadian experience of silicosis surveillance in active miners appears to be different, although it is not clear that the data clearly describe the total number of workers at risk. Between 1979 and 1992, 68 701 miners had been screened by chest radiographs at least once, and 211 had been found to have silicosis. All had greater than 14 years of exposure. Of these 211, 14.4% were copper, gold or silver miners, 17.4% were uranium miners, and 10% were foundry workers (Finkelstein, 1995).

Finally, there is considerable temptation to use dust sample measurements as a surrogate for identified cases of silicosis. Reports which describe dust exposures often provide a wealth of information, but extrapolating this information to workers with silicosis may or may not be reasonable. For example, Freeman and Grossman (1995) pointed out that in 52% of the 255 US industries where respirable quartz measurements were made by Occupational Safety and Health Administration representatives from 1980 to 1992, the mean silica exposure was below the permissible limit, yet understanding the sampling strategy at each workplace, the adequacy of the sample size, the relationship of the worker within the work area to the site of sampling, and whether the measurements are comparable from one industry to another is not possible. Similarly, in a large survey (4839 samples) addressing the level of total dust and respirable silica dust in airborne samples collected in 20 Chinese mines and nine pottery factories from 1951 to 1987, there was a significant decline in total dust and silica content over this period. Conclusions which correlate the excessive dust exposures to the prevalence of pneumoconiosis were not made. Yet, one must accept that these data are very worrisome and that the use of such surveillance data provides evidence for the need for primary prevention activity before disease is documented.

Regrettably, there are examples that a decline in mortality attributable to dust disease which was reported in the US has not occurred world-wide. In South Africa, where more than 600 000 men were employed in mining in 1992 (approximately 90% Blacks) (Central Statistical Survey, 1992) for example, Murray *et al.* (1996) reviewed autopsy data from 16 454 black gold miners dying from trauma (mean age of death at the beginning of the study was 32.8 years). The prevalence of silicosis increased from 9.3% to 12.8% from 1975 to 1991, and the number with severe silicosis increased from 0.3% to 2.0%, and occurred in workers with a mean employment duration of just 4.8 years. Tuberculosis was present in 2.5% of the population. Furthermore, perhaps one-quarter of the ex-miners who return to their homeland from the South African mines return with previously undiagnosed pneumoconiosis and a history of tuberculosis (Steen *et al.*, 1997).

Starzynski *et al.* (1996) reported similar findings in Polish workers. Although denominator data are again lacking, there appears to be a significant increase in the number of workers with CWP in the Polish coal fields (Figure 1.7). On average, the men had been exposed to coal dust for 20.9 years during their employment.

In summary, epidemiologic data on the incidence and prevalence of pneumoconiosis are very often unreliable, and there is evidence that these illnesses are likely to be underestimated, at least in the US. Basing conclusions about the prevalence of pneumoconiosis on death certificate data alone is insufficient to identify the impact of work-related respirable disease on a community. Death certificates do not appear to be the most accurate measure of the number of workers dying with pneumoconiosis, although this approach to occupational mortality may, however, provide comparable data within a population which can be serially addressed and reflect on-going trends. Over the next decade, the continued decline in the number of death certificates showing pneumoconiosis

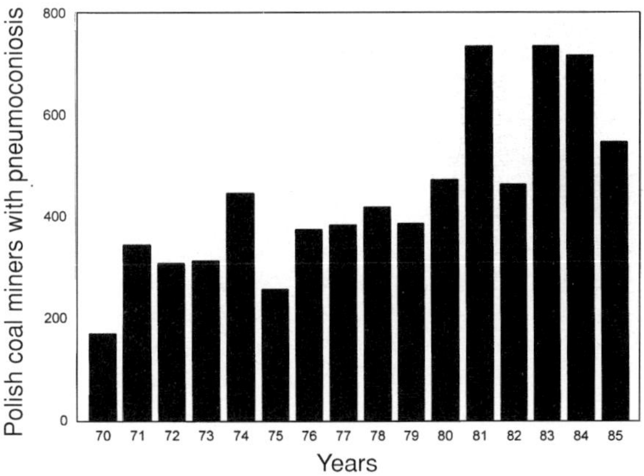

Figure 1.7 Data showing a gradual increase in the number of coal miners recognized to have pneumoconiosis by year for the years 1970 to 1985. The cohort was made up of male workers who were diagnosed in the out-patient departments of Occupational Medicine Clinics and reported to the National Register of Occupational Diseases. It is not clear whether the increase in the number of men diagnosed to have this illness reflects the implementation of a more effective system to recognize disease, a larger population at risk, changes in exposure to those at risk, or other factors which may result in changes in these data.

may confirm the conclusion of a decreased number of workers dying with these diseases as suggested by some epidemiologic studies. Yet, with the recognition of the inadequacy of death certificate data, the challenge is to keep accurate records regarding the development of cases of occupational lung disease. A combined approach, led by physician reporting of illnesses as they are recognized, and including reports from workers' compensation bureaus and hospital discharges, appears to be needed to define accurately the impact of these illnesses on worker health. These expensive and time-consuming requirements for the accurate recording of cases of occupationally related lung disease make it unlikely that accurate statistical analyses will be available in most countries.

Figure 1.8 An Italian furniture worker handling wood recently painted with isocyanate-based paints. The experience of isocyanate-related asthma in Italy has been well described by Drs Mapp and Fabbri and their colleagues from the Universities of Padua and Ferrara in an extensive series of articles. The recognition of this illness in this population has stimulated the management and labor representatives in these industries to address the methods of work to minimize exposures and decrease the number of workers who become sensitized. This has been very effective with a significant decrease in the number of workers with isocyanate-induced asthma. (Photo courtesy of Leonardo Fabbri, MD.)

IS ALL OCCUPATIONAL LUNG DISEASE PREVENTABLE?

There are more than 140 agents recognized to cause occupational asthma (Butcher and Banks, 1992). Asthma following exposure to a large number of these agents has only been recognized over the past 50 years, and these illnesses are the direct result of advances in technology. Of all the agents which can induce asthma in the workplace, isocyanate exposure is perhaps the best recognized (Figure 1.8). Development of asthma in association with exposure to isocyanate has been a consistent feature associated with the use of these agents since their inception (Fuchs and Valade, 1951). Isocyanate exposure has also been reported to be the most common cause of occupational asthma in Quebec and Ontario (Malo, 1990; Tarlo and Broder, 1991; Tarlo *et al.*, 1995).

The recognition that isocyanates are particularly strong sensitizers has been identified by United Kingdom chest and occupational physicians who reported newly identified cases of work-related respiratory disease. The work-related and occupational respiratory disease (SWORD) project calculated the incidence of occupational asthma by region, sex, occupation and age group (Meredith, 1993). This organized approach to reporting data increased the estimate of the number of new cases to about three times of what was previously thought. The group with the greatest chance to develop asthma was spray painting (658 cases of asthma per million workers), a group who typically have exposure to isocyanates in their daily work, and a group which comprised 22% of the 1085 reported cases of occupational asthma.

Isocyanates are formed from the mixture of amines with phosgene. A primary aliphatic or aromatic amine is dissolved in a solvent such as xylene or monochlorobenzene and mixed with phosgene dissolved in the same solvent for several hours at high heat. At the end of the reaction, fractions of isocyanate, reclaimable phosgene, hydrochloric acid, excess solvent and waste residues are the result. Toluene diisocyanate (TDI), one of a number of isocyanates, is a volatile compound of low molecular weight that exists in the 2,4 and 2,6 isomeric forms. It has been, until recently, the most commonly produced isocyanate. Urethanes are formed by a chemical reaction between polyglycols or polyols and diisocyanates. In a controlled chemical reaction, this reactivity allows isocyanate molecules to bind polyol molecules to form a polymeric mass. When water is added to these reacting chemicals, carbon dioxide is released, causing the mixture to carbonate or foam. By adding catalysts such as tin and various amines the chemical reactions are enhanced.

A primary feature of the isocyanates as a class of chemicals is their extreme chemical reactivity. Although this reactivity is a major asset to their use as chemical agents, it is also responsible for their adverse interactions in the lung and sensitization with the development of asthma. Inhalational exposure to isocyanate vapors can occur in the production of the chemical, during its transport, or in association with the production of polyurethane foams and coatings.

Plates 1 and 2 provide a historical background for the manufacture of asbestos cement products. They date from the 1970s and show work at the Tapanila plant in Finland. In contrast to the way that asbestos is handled today, the absence of respiratory protective equipment is noticeable. Asbestos exposures are associated not only with the direct handling and cutting of the asbestos cement product, but also with the general dustiness of the workplace. Note the film of dust on the floor and on the workers' aprons. (Photos courtesy of M.S. Huuskonen.)

Plate 3 In China as in other countries, addressing the generation of fibrogenic dusts in industrial processes typically requires the installation of effective engineering controls. The Plate shows a packing machine (momentarily not working) without local exhaust ventilation. The result is excessive exposures with dust deposited everywhere. An impinger sampler on the worker's belt with the sampler located in the breathing zones showed dust concentrations to range between 72 and 188 mg/m^3.

Plate 4 shows an increasingly effective system whereby a local exhaust hood was located as near as possible to the dust source generated in the cutting of moulding in a glazed tile factory. In this instance, the dust concentration was less than 2 mg/m^3. (Photograph courtesy of Professor Yuying Liu.)

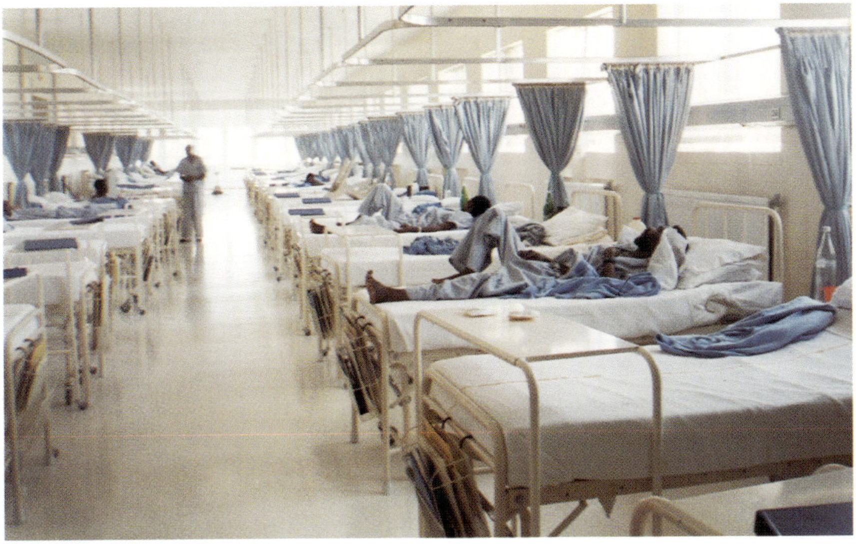

Plate 5 A ward for mine workers with tuberculosis at a local mine hospital.

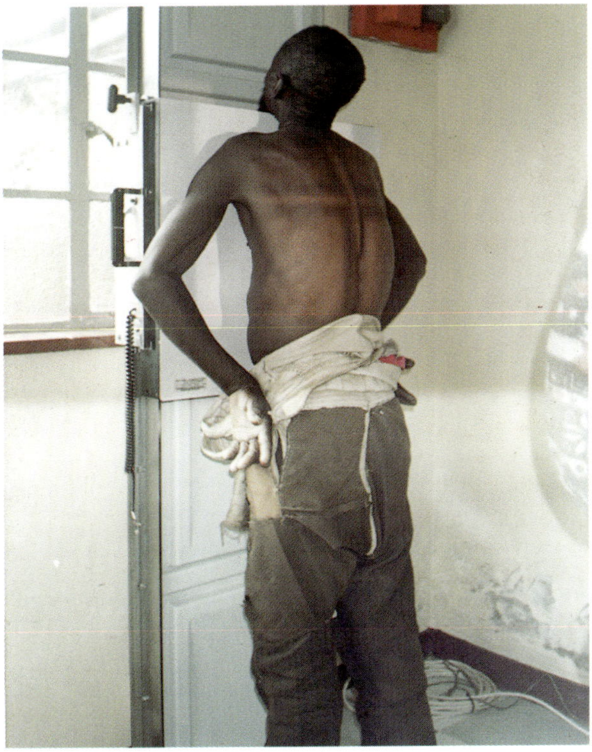

Plate 6 An ex-miner undergoing a chest radiograph for a medical benefits examination.

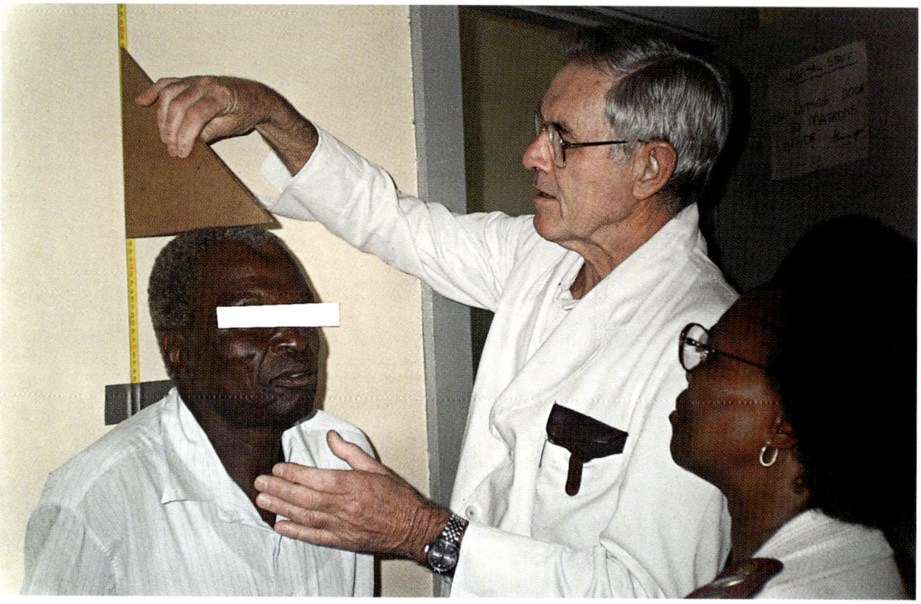

Plate 7 An ex-miner being evaluated by the nurse and physician during a medical benefits examination.

Plate 8 A mine worker performing spirometry during a routine yearly health evaluation.

Plate 9 These men are applying a mixture of emery and sand (silica) onto a wheel for grinding. Molasses is often used as an adhesive to bond the materials to each other and the wheel.

Plate 10 These men are grinding bicycle parts in Ludhiana, India. After the parts have been cast and removed from the molds, any irregularities and occasional sharp edges need to be ground. This is performed using the emery-silica wheel. Workers in these grinding jobs, as well as stone cutters and slate pencil workers, normally cover their nose and mouth with a simple cloth face mask, which is typically the end of any head gear worn. Women may cover their mouth and nose with a 'dupatta' (shoulder cloth) in order to lessen the dust exposure. This practice is of marginal help largely because the cover is unable to capture very small particles and may only partially cover the breathing zone. It may also be removed while eating, drinking, talking and resting at the work site. In these instances despite ceasing work the dust remains suspended and can be inhaled.

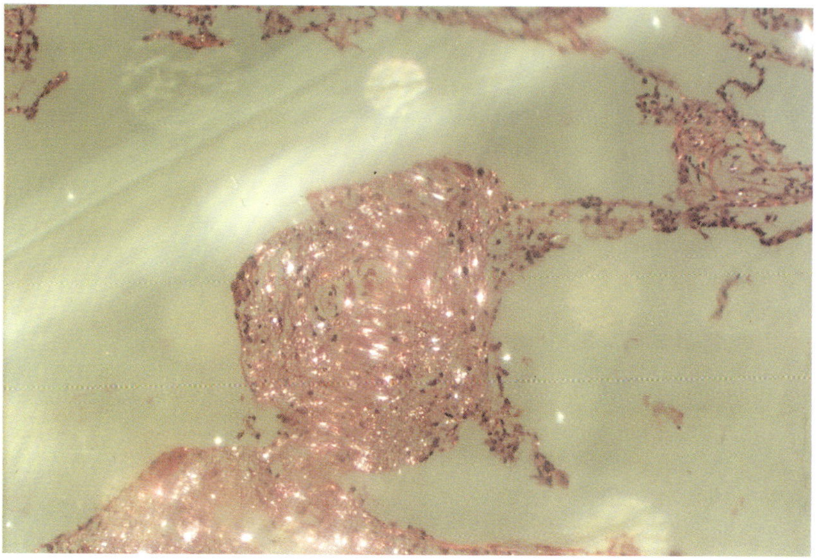

Plate 11 Birefringent crystals in a silica nodule seen by polarized light microscopy in a histologic section from a transbronchial lung biopsy. (Original magnification × 240.)

Plate 12 A patient in an exposure chamber undergoing challenge testing with soldering fumes.

Plate 13 The Union Carbide (India) plant in Bhopal, 1994.

Plate 14 Top-side coke oven. Top-side emissions are rich in polycyclic aromatic hydrocarbons and other related lung carcinogens. (Photograph courtesy of M.T. Wu and D.C. Christiani.)

Plate 15 Hard rock mining. Hard-rock miners are frequently exposed to lung carcinogens such as arsenic, radon, progeny and silica. (Photograph courtesy of T.J. Smith.)

Plate 16 Granite quarry. Granite quarrying results in silica exposure. Silica has been designated recently by IARC as a Class I carcinogen. (Photograph courtesy of D.C. Christiani.)

The physical characteristics of the individual isocyanates affect the ambient environmental levels and the chance for adverse respiratory effects. The ability of the different isocyanates to be dispersed into the ambient environment is determined by their vapor pressure. As a rule, the lesser-molecular weight isocyanates [TDI, hexamethylene diphenyl diisocyanate (HDI)] have similar vapor pressures and are readily volatilized into workplace air. Therefore, unless TDI and HDI are handled with great care, irritating (and potentially sensitizing) concentrations of vapor are the rule.

The ultimate use of the isocyanate also plays a role in determining the level of the agent in the environment. For example, the molecular weight of HDI, typically in the biuret or trimeric form (i.e. three molecules of HDI bound together) when used in HDI-based paints, is greater and the result is a diminution of vapor pressure. A lower concentration of isocyanate is likely to be produced by drying of HDI-based than by curing of TDI-based polyurethane foam.

While TDI exists as a liquid, pure methylene diphenyl diisocyanate (MDI) is solid at room temperature (although MDI is more often stored as a thick liquid combination of polymerized MDI and acetone). Higher-molecular weight compounds such as MDI exert a fraction of the vapor pressure of TDI or HDI. MDI must be pressurized or heated for vapors to be released, although droplets of MDI persist in the atmosphere following aerosolization. This realization has driven industry to adapt technologic processes that substitute MDI for TDI, the implication being that the lesser vapor pressure of MDI will result in lesser environmental exposures and fewer workers becoming sensitized. An example of this change is in the isocyanate injection molding process used to insulate doors and refrigerators. These changes, as well as the recognition of more uses for rigid foam, are reflected in the slope of increased production of rigid – compared with flexible – foam over the past years. Yet, in spraying and in certain foaming operations where isocyanates exist as an aerosol, substitution of higher-molecular weight agents for TDI may not effectively reduce exposure. In addition, in poorly ventilated areas these higher-molecular weight isocyanates may generate vapor concentrations sufficient to be both irritating and sensitizing.

Despite the knowledge that these agents can provoke asthma and sensitize individuals not previously atopic or asthmatic, only about 5% of workers exposed to isocyanates become sensitized (Tanser *et al.*, 1973). In a 1976 study, five of 112 workers employed in a newly opened TDI production facility developed asthma. A 1977 report of the same population showed 11 of 130 to be sensitized (Butcher *et al.*, 1977). Perhaps the explanation for this relatively high rate was the fact that the permissible exposure level was relatively excessive (National Institute for Occupational Safety and Health, 1978). This has resulted in a reduction of the permissible exposure levels in the US over the past several years.

Frequent exposure to large spills of isocyanates and upper respiratory tract infections are apparent predisposing factors to isocyanate sensitization. Among 332 TDI-exposed workers interviewed, 65% were exposed to spills (Brooks, 1982). The prevalence of airways obstruction and questionnaire-defined bronchitis and asthma increased in those exposed to 20 or more isocyanate spills. Yet, in another population of 63 workers undergoing laboratory-based isocyanate challenge tests to identify the cause of their asthma, the number of workers reporting exposure to spills were no different when comparing isocyanate-sensitive to isocyanate-insensitive workers (Banks *et al.*, 1989). Although the report of spills implies an excessive exposure, no one knows what level of exposure induces sensitization, or whether excessive exposures are necessary for sensitization. Chester *et al.* (1979) showed both isocyanate specific and non-specific bronchial hyperresponsiveness to be increased following upper respiratory tract viral illnesses, implying that the level of isocyanate exposure needed to induce sensitization is likely to be less in exposed workers when a viral infection is present.

In any given worker with a diagnosis of isocyanate–induced asthma, it is often impossible to identify the event(s) which have led to sensitization. How individuals become sensitized continues to remain a complex issue. Bernstein *et al.* (1993) reported the results of a cross-sectional survey of 243 employees of a polyurethane plant which used MDI exclusively for the 3 years the plant had been in operation. MDI levels were monitored 24 hours per day in multiple sites and no short-term exposure exceeded 0.005 ppm, although the authors frankly stated that they could not rule out excessive short-term exposures occurring outside of the monitored areas. Nine workers (4%) had respiratory complaints consistent with occupational asthma. Three of these nine had peak flow rate variability exceeding 15% over the work shift. If these three are considered to have objective evidence of isocyanate-induced asthma, then the prevalence in this population approximated 1%. Although this is less than the ‘expected’ rate of 5% of occupational asthma in the population, even this low level of exposure was insufficient to protect the worker.

Tarlo *et al.* (1997a) reviewed the ambient levels of MDI and TDI in Ontario workplaces and compared isocyanate exposures in workplaces where workers had been compensated for this illness to workplaces where cases were not recognized. Overall, approximately 1% of the isocyanate-exposed work force have isocyanate asthma compensated by the Workers' Compensation Board. There were 10 workplaces (and 17 cases of isocyanate-induced asthma among the 805 employees in these 10 workplaces) where cases of asthma were compensated as being workplace-induced. In these workplaces, there were no isocyanate samples which exceeded the permissible limit.

Finally, neither convincing epidemiologic evidence defining the effect of different isocyanates on the rate of lung function decline of a population in the workplace, nor an understanding of differences in the rates of sensitization associated with the use of different isocyanates within the workplace has been presented. Importantly, Tarlo *et al.* (1997b) have reported that the outcome of isocyanate asthma is not dependent on the sensitizing isocyanate.

In summary, beginning with the recognition of adverse respiratory reactions in the early 1950s, there has been a tremendous attempt to minimize isocyanate exposures (see Figure 1.8). Efforts have included alterations in technology so that a different isocyanate might be used. In this way, vapor pressures and ambient exposures might be lessened. In addition, extensive expenditures are often necessary to optimize engineering controls. Yet, despite the infusion of a tremendous amount of resources to make workplaces where isocyanate exposures occur safe, cases of isocyanate-induced asthma continue to be recognized even in workplaces where isocyanate levels have been measurably below permissible limits. These data imply that there is no safe level of isocyanate exposure.

SUMMARY

In conclusion, in the above brief comments I have focused on four issues which are relevant to the understanding of the respiratory health of many workers world-wide. The first is the recognition that there has been a traditionally acceptable health cost to workers which is accepted by society, as in Agricola's time. Although readers who disagree and can point to important public health achievements which have made work safer are certainly correct, the existence of a workers' compensation system allows society to define what is an acceptable health risk in a job. If the risks become 'excessive', then the employer is forced to pay increasing amounts into the compensation fund. The fund thereby controls the degree of risk as a great penalty (associated with a most hazardous job) may put the employer out of business. Second, although prevention can be implemented, it typically has required a serious outbreak of occupationally related illness for workers and managers to pay attention. Examples of this have been the continuing series of epidemics of silicosis in the US reported since the start of this century. Third, there are many industries where we can dramatically lessen the risk of respiratory injury, but we may never be able to protect the lung health of all employees. Workers who are exposed to isocyanates may be a very important example of this.

Finally, respiratory health risks associated with work situations where respiratory exposures are present appear to be more dangerous than they are 'officially' presented. The bulletin, the *Work-related lung diseases surveillance report*, reports an overall decline in lung disease attributed to exposures in the American workplace. Yet, reasonable concerns regarding the adequacy of record-keeping and the real usefulness of death certificate data remain. In some countries, physician-reporting of these kinds of illnesses provides a comprehensive and accurate surveillance system. Unless such a commitment is provided, an understanding of the true extent of the problem and the resources needed for an appropriate measure of response will remain elusive.

Acknowledgements

This work was supported in part by National Institute for Occupational Safety & Health Grants U60/CCU306149–01–1 and U60/CCU313230.

REFERENCES

Agricola, G. (1555) *De re metallica*. Translated by H. Hoover and L.H. Hoover, Dover Publications, Inc., New York, 1950 edition, p 6.

Attfield, M.D. and Althouse, R.D. (1992) Surveillance data on U.S. coal miners' pneumoconiosis, 1970–1986. *Am. J. Public Health*, **82**, 971–8.

Attfield, M.D. and Castellan, R.M. (1992) Epidemiologic data on U.S. coal miners' pneumoconiosis, 1960 to 1986. *Am. J. Public Health*, **82**, 964–70.

Banks, D.E., Morring, K., Boehlecke, B.A. *et al.* (1981) Silicosis in silica flour workers. *Am. Rev. Respir. Dis.*, **124**, 445–50.

Banks, D.E., Bauer, M.A., Castellan, R.N. *et al.* (1983) Silicosis in surface coalmine drillers. *Thorax*, **38**, 275–8.

Banks, D.E., Sastre, J., Butcher, B.T. *et al.* (1989) Role of inhalation challenge testing in the diagnosis of isocyanate-induced asthma. *Chest*, **95**, 414–23.

Bernstein, D.I., Korbee, L., Stauder, T. *et al.* (1993) The low prevalence of occupational asthma and antibody-dependent sensitization to diphenylmethane diisocyanate in a plant engineered for minimal exposure to isocyanates. *J. Allergy Clin. Immunol.*, **92**, 387–96.

Betts, W.W. (1990) Chalicosis pulmonum or chronic interstitial pneumonia induced by stone dust. *JAMA*, **34**, 70–4.

Brooks, S.M. (1982) The evaluation of occupational airways disease in the workplace. *J. Allergy Clin. Immunol.*, **70**, 56–66.

Butcher, B.T. and Banks, D.E. (1992) Immunologic and clinical features of occupational asthma induced by small molecular weight agents. *Immunol. Allergy Clinics*, **12**, 329–48.

Butcher, B.T., Jones, R.N., O'Neil, C.E. *et al.* (1977) Longitudinal study of workers employed in the manufacture of toluene diisocyanate. *Am. Rev. Respir. Dis.*, **116**, 411–22.

Cavarini, F., diPietro, A., Miceli, M. *et al.* (1995) Incidence of silicosis among ceramic workers in central Italy. *Scand. J. Work Environ. Health*, **21** (suppl. 2), 58–62.

Central statistical survey (1992) South African Labour statistics. Pretoria, Government printer.

Chester, E.H., Martinez-Catinchi, M., Schwartz, H.J. *et al.* (1979) Patterns of airway reactivity to asthma produced by exposure to toluene di-isocyanate. *Chest*, **75**, S229–31.

Corn, J.K. (1980) Historical aspects of industrial hygiene – silicosis. *Am. Ind. Hyg. Assoc. J.*, **41**, 125–32.

Cowie, R.L. (1994) The epidemiology of tuberculosis in goldminers with silicosis. *Am. J. Respir. Crit. Care Med.*, **150**, 1460–2.

Ellison, L. and Yeung, K.S. (1995) Policy development for compensating workers exposed to crystalline silica in Ontario, Canada. *Scand. J. Work Environ. Health*, **21** (suppl. 2), 111–14.

Fleming, D., Maynard, D., McKinney, B. *et al.* (1990) Silicosis: clusters in sandblasters – Texas, and occupational surveillance for silicosis. *Morbidity & Mortality Weekly Report*, **39**, 433–7.

Finkelstein, M.M. (1995) Silicosis surveillance in Ontario from 1979 to 1992. *Scand. J. Work Environ. Health*, **21** (suppl. 2), 55–7.

Freeman, C.S. and Grossman, E.A. (1995) Silica exposures in the United States between 1980 and 1992. *Scand. J. Work Environ. Health*, **21** (suppl. 2), 47–9.

Fuchs, S. and Valade, P. (1951) Etude clinique et experimentale sur quelques cad d'intoxication par les Desmodur T (diisocyanate de tolylene 1–2–4 et 1–2–6). *Archives des Maladies Professionnelles de Medicine du Travail et de Securite Sociale*, **12**, 191–6.

Glindmeyer, H.W., III and Hammad, Y. (1988) Contributing factors to sandblaster's silicosis: inadequate respiratory protection and standards. *J. Occup. Med.*, **30**, 917–21.

Graham, W.G.B., Ashikaga, T., Hemenway, D. *et al.* (1991) Radiographic abnormalities in Vermont granite workers exposed to low levels of granite dust. *Chest*, **100**, 1507–15.

Holandra, M.A., Holandra, M.A., Martins, M.P.S. *et al.* (1995) Silicosis in Brazilian pit diggers: relationship between dust exposure and radiologic findings. *Am. J. Ind. Med.*, **27**, 367–78.

Hosey, A.D., Trasko, V.M. and Ashe, H.B. (1957) *Control of silicosis in the Vermont granite industry: progress report*. Publication 557, U.S. Department of Health, Education, and Welfare. Washington, DC.

Liss, G.M., Tarlo, S.M. and Banks, D.E. (1997) Evidence for occupational asthma among compensation claimants at an isocyanate-using facility in Ontario. *Can. J. Public Health*, **87**, 401–3.

Malo, J.L. (1990) Compensation for occupational asthma in Quebec, in Workshop on Environmental and occupational asthma, (ed. J.A. Merchant). *Chest*, **98S**, 236S–239S.

McKay, G.R., Stettler, L.E., Kommineni, C. *et al.* (1980) Fibrogenic potential of slags used as substitutes for sand in abrasive blasting operations. *Am. Ind. Hyg. Assoc. J.*, **41**, 836–42.

Meijers, J.M., Swaen, G.M. and Slangen, J.J. (1996) Mortality and lung cancer in ceramic workers in The Netherlands: preliminary results. *Am. J. Ind. Med.*, **30**, 26–30.

Meiklejohn, A. (1951) History of lung disease of coalminers in Great Britain, part I, 1800–1875. *Br. J. Ind. Med.*, **8**, 127–37.

Meiklejohn, A. (1952a) History of lung diseases of coal miners in Great Britain, Part II, 1875–1920. *Br. J. Ind. Med.*, **9**, 93–8.

Meiklejohn, A. (1952b) History of lung diseases of coal miners in Great Britain. Part III, 1920–1952. *Br. J. Ind. Med.*, **9**, 208–20.

Meredith, S. (1993) Reported incidence of occupational asthma in the United Kingdom, 1989–1990. *J. Epidemiol. Community Health*, **47**, 459–63.

Murray, J., Kielkowski, D. and Reid, P. (1996) Occupational disease trends in black South African gold miners. *Am. J. Respir. Crit. Care Med.*, **153**, 706–10.

National Institute for Occupational Safety and Health (1974) Criteria for a recommended standard. Occupational exposures to crystalline silica. Washington, D.C. National Institute for Occupational Safety and Health. U.S. Public Health Service, Centers for Disease Control, pp. 54, 55.

National Institute for Occupational Safety and Health. (1978) Occupational exposure to Diisocyanates (NIOSH) criteria document for a recommended standard). Washington, D.C. U.S. Department of Health, Education and Welfare; DHEW publication 78–215.

Neukirch, F., Coorman, J., Korobaeff, M. *et al.* (1994) Silica exposure and chronic airflow limitation in pottery workers. *Arch. Environ. Health*, **49**, 459–63.

Partanen, T., Jaakkola, J. and Tossavainen, A. (1995) Silica, silicosis, and cancer in Finland. *Scand. J. Work Environ. Health*, **21** (suppl. 2), 84–6.

Provencher, S., Labreche, F.P. and De Guire, L. (1997) Physician-based surveillance system for occupational respiratory diseases: the experience of PROPULSE, Quebec, Canada. *Occup. Environ. Med.*, **54**, 272–6.

Rosenman, K.D., Reilly, M.J., Rice, C. *et al.* (1996) Silicosis among foundry workers. Implications for the need to revise the OSHA standard. *Am. J. Epidemiol.*, **144**, 890–900.

Rosenman, K.D., Reilly, M.J., Kalinowski, D.J. *et al.* (1997) Silicosis in the 1990's. *Chest*, **111**, 779–86.

Russell, A.E. (1941) *The health of workers in dusty trades – IV. Restudy of a group of granite workers.* Public Health Bull. No. 269, Federal Security Agency, U.S. Public Health Service, Washington, DC.

Russell, A.E., Britten, E.H., Thompson, L.R. *et al.* (1929) *The health of workers in dusty trades – II. Exposure to siliceous dusts (granite industry).* Public Health Bull. No. 187, U.S. Treasury Dept., Public Health Service.

Saiyed, H.N., Ghodasara, N.B., Sathwara, N.G. *et al.* (1995) Dustiness, silicosis, and tuberculosis in small scale pottery workers. *Indian J. Med. Res.*, **102**, 138–42.

Seaton, A., Legge, J.S., Henderson, J. *et al.* (1991) Accelerated silicosis in Scottish stonemasons. *Lancet*, **337**, 341–4.

Sherson, D. and Lander, F. (1990) Morbidity of pulmonary tuberculosis among silicotics and non-silicotic foundry workers in Denmark. *J. Occup. Med.*, **32**, 110–13.

Stanbury, M., Joyce, P. and Kipen, H. (1995) Silicosis and workers' compensation in New Jersey. *J. Occup. Environ. Med.*, **37**, 1342–7.

Starzynski, Z., Marek, K., Kujawska, A. *et al.* (1996) Mortality among coal miners with pneumoconiosis in Poland. *Int. J. Occup. Med. Environ. Health*, **9**(4), 279–89.

Steen, T.W., Gyi, K.M., White, N.W. *et al.* (1997) Prevalence of occupational lung disease among Botswana men formerly employed in the South African mining industry. *Occup. Environ. Med.*, **54**, 19–26.

Stettler, L.E., Proctor, J.E., Platek, S.F. *et al.* (1988) Fibrogenicity and carcinogenic potential of smelter slags used as blasting substitutes. *J. Toxicol. Environ. Health*, **25**, 35–6.

Suratt, P.M., Winn, W.C., Brody, A.R. *et al.* (1977) Acute silicosis in tombstone sandblasters. *Am. Rev. Respir. Dis.*, **115**, 521–9.

Tanser, A.R., Bourke, M.P. and Blandford, A.G. (1973) Isocyanate asthma: respiratory symptoms caused by diphenyl-methane diisocyanate. *Thorax*, **28**, 596–600.

Tarlo, S.M. and Broder, I. (1991) Outcome of assessment of occupational asthma. *Chest*, **100**, 329–35.

Tarlo, S.M., Liss, G., Corey, P. and Broder, I. (1995) A Workers' Compensation claim population for occupational asthma: comparison of subgroups. *Chest*, **107**, 634–41.

Tarlo, S.M., Liss, G.M., Diaz, C. and Banks, D.E. (1997a) Assessment of the relationship between isocyanate exposure and occupational asthma. *Am. J. Ind. Med.*, **32**, 517–21.

Tarlo, S.M., Banks, D., Liss, G. and Broder, I. (1997b) Outcome determinants for isocyanate-induced occupational asthma among compensation claimants. *Occup. Environ. Med.*, **54**, 756–61.

Work-related lung diseases surveillance report (1996) U.S. Dept. of Health and Human Services. October, 1996. Cincinnati, OH. DHHS (NIOSH) Publication No. 96–134.

2 Pulmonary function and disability

N. LeRoy Lapp

The lungs maintain an adequate exchange of oxygen and carbon dioxide between the external environment and the arterial blood which supplies the body tissues. Fresh air is brought into contact with a large alveolar–capillary bed where oxygen is exchanged for carbon dioxide. This is accomplished by three basic mechanisms. The first, ventilation, provides fresh air to the alveolar–capillary membranes. Next, gas transfer, or diffusion, is the process whereby molecules of oxygen and carbon dioxide are passively transferred through the alveolar–capillary membranes and fluids into the pulmonary circulation. Finally, pulmonary perfusion, transports oxygenated blood to the heart and systemic circulation to other metabolizing tissues. Each aspect of the process is important and can be affected by disease or dysfunction.

FUNCTIONAL ANATOMY OF THE RESPIRATORY SYSTEM

The respiratory system may be viewed as series of semi-rigid and compressible elastic tubes or dead space whose function is to heat, humidify and deliver fresh air to a gas exchanging unit extending from the alveolar duct, which is the first branch of the airways containing alveoli, to the alveolar sac. It includes all of the alveolar tissue, the capillaries and the supporting stroma (Figure 2.1). This gas-exchanging system is connected to the ambient atmosphere by a series of elastic tubes which revitalize the air in the gas-exchanging unit by delivering oxygen and expiring carbon dioxide.

The adequacy of ventilation in the lungs depends upon lung size, the elastic properties of the lungs and chest wall, and resistance to airflow within the airways. These factors can all be evaluated by appropriate tests.

LUNG VOLUMES

The first ventilatory test used a water displacement volume spirometer to measure lung function (Hutchinson, 1844). Hutchinson also described the relationships between vital capacity and body size, sex and disease states (Hutchinson, 1846). Interestingly, Hutchinson's spirometer was used in 1869 by Gould, a military surgeon, to measure the vital capacity of Negro soldiers in the Civil War, thus first noting the effect of ethnic origin or race on ventilatory capacity (Damon, 1966). There was little further clinical application of spirometry until the 1950s, at which time a nomenclature for the physiologic subdivisions of the lung was agreed upon by a group of physiologists (Pappenheimer *et al.*, 1950). In this schema, a volume cannot be further subdivided, whereas a capacity may include two or more lung volumes (Figure 2.2). Total lung capacity (TLC) is the amount of gas contained within the lung when the subject has inhaled maximally. Residual volume (RV) is the volume of gas remaining in the lung at the end of a maximal exhalation. Vital capacity (VC) is the volume of gas that can be expelled from the lung by forceful effort from maximal inspiration to maximal exhalation. Functional residual capacity (FRC) is the volume of gas remaining in the lungs at the resting expiratory

Occupational Lung Disease. Edited by Daniel E. Banks and John E. Parker. Published in 1998 by Chapman & Hall. ISBN 0 412 73630 6

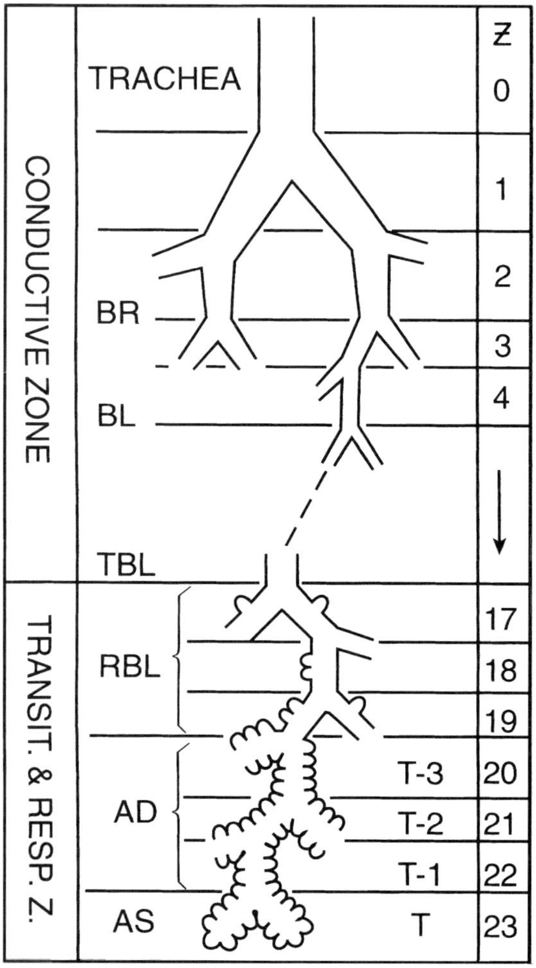

Figure 2.1 Schematic diagram of the lower respiratory system showing the conductive zone and the alveolar gas-exchanging zone. (Reproduced with permission from Weibel, E.R. (1963) *Morphometry of the Human Lung*, Fig. 82, p.111, Springer-Verlag.)

level. Tidal volume (TV) is the volume of gas inspired or expired during quiet breathing. Inspiratory reserve volume (IRV) is the maximal amount of gas that can be inspired from the resting end-inspiratory position. Expiratory reserve volume (ERV) is the maximal volume of gas that can be expired from the end-expiratory level. Together, TV, IRV and ERV make up the vital capacity (VC).

Closed-circuit method

Lung size is best quantified by measurements of the static lung volumes rather than from spirometry. Lung volumes may be determined from inert gas equilibration, such as helium, washout of resident nitrogen, or by body plethysmography. Helium equilibration utilizes Henry's law. If one knows the initial concentration and volume of gas in a spirometer, and can determine the final concentration at equilibration between the lungs and this spirometer, one can solve for the unknown volume in the lungs (Meneely and Kaltreider, 1949). Since the subject is usually turned into the system at the end of a normal tidal breath, the unknown volume will be FRC. By performing a VC maneuver, one can derive TLC and its subdivisions.

Open-circuit method

Similarly, by collecting the volume of gas exhaled from the lungs during a 7-minute washout with the subject breathing pure oxygen, and analyzing the quantity of endogenous nitrogen removed, it is possible to calculate the unknown lung volume at the beginning of the test, usually FRC (Darling *et al.*, 1940). One can determine TLC and all of its subdivisions by performing a VC maneuver.

Plethysmographic method

In 1956, Dubois and colleagues developed the whole body plethysmograph for measuring both lung volumes and airway resistance. In this method, the subject sits in a chamber. Pressures at the mouth and in the chamber relating to movement of the chest are recorded. The subject pants rapidly against a closed mouthpiece shutter to record lung volume and with no obstruction at the mouthpiece to record airflow resistance. Utilizing principles of Boyle's law, the thoracic gas volume (TGV) can be calculated. TGV is not synonymous with FRC, since subjects usually pant at a higher lung volume than end-expiration. By inhaling to maximal inspiration following the TGV panting maneuver, it is possible to derive TLC and all of its subdivisions. The plethysmograph is the most accurate method for determining TLC. This method measures all of the gas contained in the thorax, whether or not the gas communicates with the airways. In general, TLC by plethysmographic technique is slightly larger than TLC determined by inert gas equilibration or washout in normal subjects. In subjects with bullae or other conditions in which the gas within the thorax is not in free communication with airways, the differences between plethysmographic and inert gas measurements may be considerable.

Radiographic method

Several methods for measuring TLC from posteroanterior and lateral chest radiographs have been

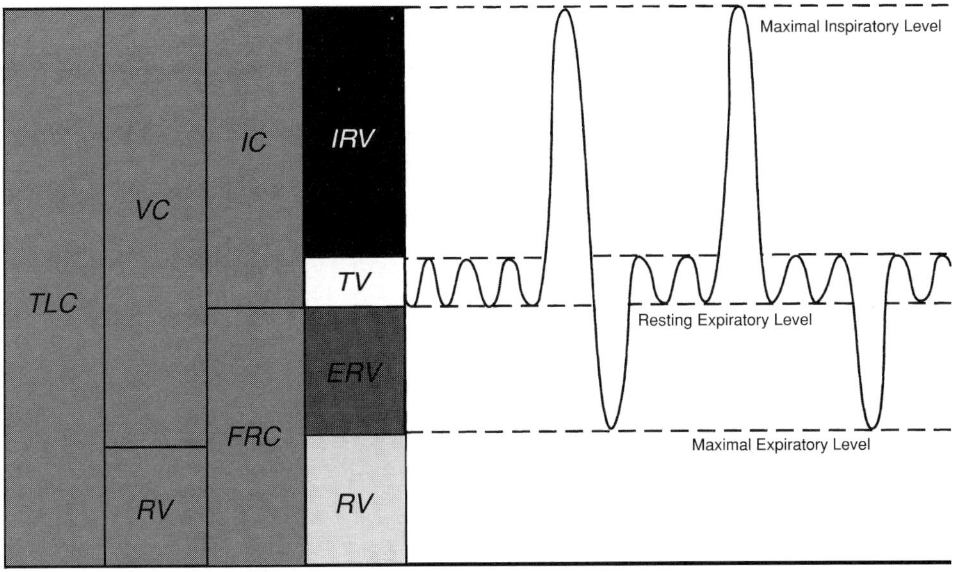

Figure 2.2 Diagram illustrating the various lung volumes and capacities that can be measured by pulmonary function tests. (Reproduced with permission from Comroe, J.H. *et al.* (1962) *The Lung, Clinical Physiology and Pulmonary Function Tests*, 2nd edn, Mosby Year Book, Inc., Chicago)

devised (Barnhard *et al.*, 1960; Harris *et al.*, 1971). The methods depend upon treating the lungs as elliptical cylindroids, calculating the volume of each, subtracting volumes attributable to the domes of the diaphragm, the heart, a volume for lung parenchyma (determined at autopsy to approximate 130 ml) and a blood volume related to the subject's height. The method is reasonably accurate provided the films are made in full inspiration. This method has been successfully applied in an epidemiological study of coal miners (Morgan *et al.*, 1971) and simplified by Reger and colleagues (Reger *et al.*, 1972). This method for determining TLC can be applied retrospectively to films made for other purposes (Ries, 1989).

Clinical significance of lung volume changes

Total lung capacity may be decreased owing to abnormalities of the lungs such as resection, pneumonia, atelectasis and diffuse infiltrative diseases. It may also be decreased due to extrapulmonary factors such as thoracic deformities, pregnancy, ascites, or by muscular weakness or diffuse neuromuscular disease (Forster *et al.*, 1986). An abnormal increase in plethysmographically determined TLC is reported in cases of severe air trapping such as emphysema, and transiently in cases of severe status asthmaticus (Meisner and Hugh-Jones, 1968).

FRC is often described as the lung volume at which the elastic recoil forces tending to collapse the lung are equally balanced in an opposite direction by the tendency of the chest wall to spring outward. Any change in the conditions on either side of this balance point will shift the FRC either upward or downward. Thus, loss of lung recoil due to emphysema will result in an increased FRC. Similarly, an undue stiffening of the chest wall in the presence of normal lung elastic recoil will cause an increased FRC. Loss of muscle tone of the chest wall in the presence of normal lung elastic recoil, such as occurs in diffuse neuromuscular diseases, will shift the FRC downward. Similarly, restriction of normal chest wall expansion due to massive obesity will, in the presence of normal elastic recoil, reduce the FRC. Thus, changes in FRC may be best thought of as reflecting the balance of mechanical forces either within the lungs or from the chest wall rather than solely hyperinflation of the lungs.

The RV is probably a better indicator of air trapping or hyperinflation even after a maximal expiratory effort, than is FRC or TLC. While in airway obstructive diseases the RV and FRC tend to increase together, early on one may see only an increase in RV. RV may be decreased in diffuse pulmonary restrictive diseases and in some diseases in which atelectasis is present. It also is decreased in massive obesity.

AIRWAY RESISTANCE

As noted earlier, the body plethysmograph is the most accurate means for measuring lung volumes. It is also useful for measuring airway resistance. A subject, seated in the plethysmograph, gently but rapidly pants through a mouthpiece into a pneumotachograph. By recording flow at the mouth during panting with an open shutter and relating it to pressure at the mouth during occlusion of the shutter (alveolar pressure) it is possible to measure airway resistance. This airway resistance predominantly reflects the behavior of the larger, non-compressed central airways. Macklem and Mead (1967) showed that while resistance in central airways made up approximately 85% of the total resistance, the remaining 15% was accounted for by the small, intrapulmonary airways. Thus, a subject might have diffuse disease of the small airways and yet have normal airway resistance.

In general, an increase in airway resistance and spirometric indices of reduced flows rates go together. However, some important exceptions have been encountered. This should not be surprising (Lapp and Hyatt, 1970) because resistance measurements reflect predominantly the pressure–flow behavior of the large intrathoracic and extrathoracic airways in their resting, non-compressed state whereas the maximal expiratory flows developed during the FVC maneuver clearly reflect a complex interplay among airway geometry, airway compliance or compressibility, and lung elastic recoil (Fry, 1968).

The ability rapidly to make repeated measurements of airway resistance using the plethysmograph renders it a useful tool for determining changes in airway caliber in response to bronchial challenge or administration of bronchodilator medication in normal as well as obstructed or restricted subjects. The body plethysmograph has been found most useful in the pulmonary function laboratory rather than in epidemiological or field studies.

SPIROMETRY

Timed lung volumes

The forced expiratory vital capacity maneuver is the basis upon which all of the of the timed lung volumes and most flow rates are based. The maneuver requires the subject to inhale maximally to TLC, then, with maximal effort and without interruption, exhale maximally to RV. Early studies of timed volumes based upon the FVC maneuver by Tiffeneau *et al.* (1949) and Gaensler (1951) noted that the volume of gas exhaled during the first second [now referred to as the Forced Expiratory Volume in the first second of the FVC maneuver ($FEV_{1.0}$)] was of value in discriminating between healthy subjects and those with airways obstruction.

Expiratory flow rates

Subsequently, other indices derived from the FVC maneuver were popularized. These included the FEV_1/VC × 100% (Miller *et al.*, 1959), the Maximal Expiratory Flow Rate (MEFR), now known as Forced Expiratory Flow between 200 and 1200 ml exhaled ($FEF_{200-1200}$) (Kory *et al.*, 1961) and the Maximal mid-Expiratory Flow Rate or Forced Expiratory Flow between 25% and 75% of the FVC maneuver ($FEF_{25-75\%}$) (Leuallen and Fowler, 1955) (Figure 2.3). The discriminatory value of the $FEV_{1.0}$ and $FEF_{25-75\%}$) in detecting differences between subjects with and without 'bronchitis' was demonstrated by Fairbairn and colleagues (Fairbairn *et al.*, 1962). Other flow rates derived from spirometric tracings have not been as helpful in discriminating between subjects with obstructive impairment and normal subjects.

The $FEF_{25-75\%}$ is more sensitive than $FEV_{1.0}$ to airflow obstruction, particularly in the small, intrapulmonary airways, owing to its inclusion of the lower portion of the FVC curve (Dawson, 1966). On the other hand, $FEF_{200-1200}$ includes only the early portion of the FVC curve that reflects the behavior of the larger airways which are supported by cartilage rather than the small, compressible, intrapulmonary airways (Lapp, 1973). Most guidelines recommend that the full FVC curve be recorded and that measurements of the $FEV_{1.0}$ and $FEF_{25-75\%}$ suffice for determination of impairment of function. A variety of measurements derived from different portions of the FVC maneuver have been proposed but have not received widespread acceptance.

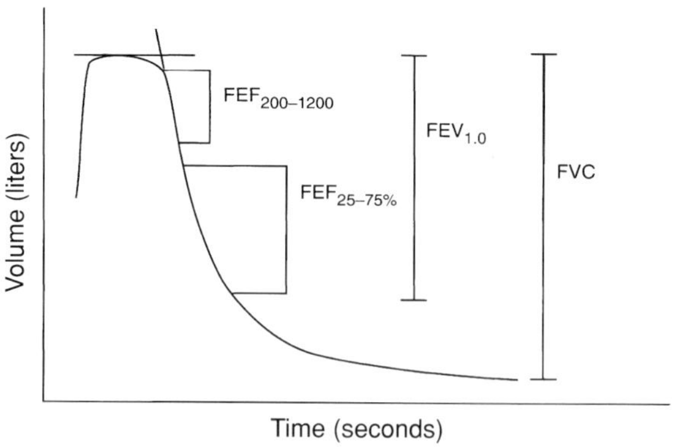

Figure 2.3 Volume versus time spirometric tracing of a forced vital capacity (FVC) maneuver illustrating the portions of the curve which are encompassed by the designated timed lung volumes and expiratory flows.

Types and standardization of measuring devices

There are two types of devices for measuring volume and airflow obstruction, the volume spirometer and the flow spirometer. Scanlon and Hankinson (1996) pointed out the advantages and disadvantages of each (Table 2.1).

At the request of the National Institute for Occupational Safety and Health, a consensus group was convened in 1979 at Snowbird, Utah, to recommend standardization for spirometry testing (Gardner, 1979). These recommended standards for equipment and performance of spirometry were updated in 1987 (American Thoracic Society, 1987a) and most recently in 1995 (American Thoracic Society, 1995a). In addition to proposing standards for equipment for lung function testing, these groups have also emphasized the need for a thorough quality control program which includes not only the equipment, but regular mechanical and biological calibration of equipment, training of technicians, careful instruction of subjects, and performance of multiple efforts until acceptable, reproducible maneuvers are obtained (American Thoracic Society, 1995b). Recently, the American Thoracic Society also made recommendations on the interpretation of spirometry values in the clinical setting (Table 2.2) (American Thoracic Society, 1991). In 1993, the European Respiratory Society published a statement on standardization of lung function tests (Quanjer *et al.*, 1993) which differs only slightly from the American Thoracic Society (1995b) statements.

Table 2.1 Features of volume and flow measuring devices

Positive features of volume spirometers
- Provide a mechanical linkage between the air collection reservoir and a recording device.
- Provide a greater degree of certainty of the volume collected, including a hard copy of the measured volumes, a necessary feature when the laboratory is performing disability evaluations.
- Relatively simple in design and relatively easy to maintain.

Negative features of volume spirometers
- Large size and weight of volume spirometers limit their portability.

Positive features of flow spirometers
- Generally portable, inexpensive, utilize inexpensive, disposable flow elements.
- Sensitive to rapid flow transients.
- Easy to clean.

Negative features of flow spirometers
- Errors in volume measurements if flows exceed calibrated linear flow ranges.
- Errors in vital capacity measurement if zero flow is not correctly sensed.
- Software errors in calculation or maneuver error detection may be difficult to detect.
- Loss of data may occur in the event of a computer malfunction or inadvertent erasure by the technician.

Source: Scanlon, P.D. and Hankinson, J.L. (1996) Physiologic methods, in *Occupational and Environmental Respiratory Disease* (eds P. Harber, M.B. Schenker and J.R. Balmes), Mosby, St. Louis.

A peak flow (PF) meter was developed by Wright and McKerrow (1959) in an attempt to provide a relatively portable instrument for use in epidemiologic studies. PF meter readings compare favorably with MEFR measured from the FVC curve. However, the PF meter suffers from the same deficiency as the MEFR, namely that it reflects the behavior of the larger, cartilage-supported airways and is not sensitive to changes in the small, compressible intrapulmonary airways.

Sources of variation of spirometry

The major sources of variation in pulmonary function identified to date are age, height, sex and race. Although body surface area (which is related to weight) was originally included in some equations for predicting normal function, it was felt to play a minor role and therefore excluded during most of the past several decades. Current work suggests that weight change (gain or loss) is an important factor in longitudinal or follow-up studies of lung function (Wang *et al.*, 1996).

GAS TRANSFER

Diffusing capacity

The rate of oxygen and carbon dioxide exchange between the blood and the alveolus depends on the relative pressure gradients on either side of the alveolar–capillary membranes. The process is one of passive diffusion of the molecules through the gaseous and liquid phases present in the lungs. Owing to the fact that carbon dioxide is about 24 times more soluble than oxygen in the liquid phase of the lungs, it diffuses more rapidly than oxygen. Thus, there are few, if any, disease states compatible with life while breathing room air, in which a lung diffusion limitation for carbon dioxide is the major problem. The oxygen diffusion limitation in lungs is most simply evaluated by the diffusion capacity (transfer factor) test.

The diffusing capacity for oxygen is seldom directly measured owing to the need to know the Po_2 of the mixed venous blood, and the Po_2 as it traverses the capillary. These can be derived through a series of

Table 2.2 Interpretative strategies for assessing the severity of abnormalities

A. Normal: the test is interpreted as 'within normal limits' if both the VC and the $FEV_{1.0}$/VC ratio are in the normal range.

B. Obstructive abnormality: this is interpreted when the $FEV_{1.0}$/VC ratio is below the normal range. The severity of the abnormality might be graded as follows:

'May be a physiological variant'	% Pred $FEV_{1.0}$ ≥100
'Mild'	% Pred $FEV_{1.0}$ <100 and ≥70
'Moderate'	% Pred $FEV_{1.0}$ <70 and ≥60
'Moderately severe'	% Pred $FEV_{1.0}$ <60 and ≥50
'Severe'	% Pred $FEV_{1.0}$ <50 and ≥34
'Very severe'	% Pred $FEV_{1.0}$ <34

C. Restrictive abnormality: this is most reliably interpreted on the basis of TLC. If this is not available, one may interpret a reduction in the VC without a reduction of the $FEV_{1.0}$/VC ratio as a 'restriction of the volume expansion of the lung'. The severity of the abnormality may be graded as follows:

Based on the TLC	
'Mild'	% Pred TLC <LLN but ≥70
'Moderate'	% Pred TLC <70 and ≥60
'Moderately severe'	% Pred TLC <60

Based on spirometry	
'Mild'	% Pred VC <LLN but ≥70
'Moderate'	% Pred VC <70 and ≥60
'Moderately severe'	% Pred VC <60 and ≥50
'Severe'	% Pred VC <50 and ≥34
'Very severe'	% Pred VC <34

LLN, lower limit of normal.

Reproduced with permission from the American Thoracic Society (1991) Lung function testing: selection of reference values and interpretive strategies. *Am. Rev. Respir. Dis.*, **144**, 1202–18.

complex mathematical calculations, and assumptions. Instead, carbon monoxide, which has a negligible P_{CO} in mixed venous blood except in heavy smokers, and has 210 times the affinity for hemoglobin, is generally used to measure the $D_{L}CO$. The diffusing capacity (transfer factor) is often expressed as a series of conductances (reciprocals of resistance) which represent its various components:

$$1/D_L = 1/D_M + 1/\theta V_c$$

where D_L is the diffusing capacity, D_M is the diffusing capacity of the membranes, θ is the reaction rate of hemoglobin with carbon monoxide, and V_c is the capillary blood volume.

(a) Breath-holding method

The technique for measuring diffusing capacity by the breath-hold method has been standardized in North America in an official statement by the American Thoracic Society (1987b). In this test, the subject inhales an inspiratory vital capacity of a test gas containing 0.3% CO, 10% helium, and 21% oxygen with the balance nitrogen. The gas is held in the lungs for approximately 10 s, followed by exhalation sufficient to wash out the mechanical and anatomic dead space. An alveolar sample is collected at this point. Helium dilution is used to calculate alveolar volume (V_A) and the initial alveolar concentration of CO. The driving pressure for CO uptake is estimated to be the initial alveolar pressure for CO. $D_{L}CO$ is calculated from the alveolar volume, breath-hold time, and the initial and final alveolar concentrations of CO (Van Kessel, 1982). The American Thoracic Society guides recommend that smokers be instructed to refrain from smoking for 24 hours before the test or measuring and correcting the diffusing capacity for the CO back-pressure (American Thoracic Society, 1995b). Adjustment of the diffusing capacity for COHb and CO back-pressure is recommended by the ATS statement by one of several methods. COHb can be measured directly in venous blood or CO back-pressure can be measured in expired air before performance of the $D_{L}CO$ maneuver (American Thoracic Society, 1995a).

The single-breath technique has several advantages. It is relatively simple to perform, it has been standardized and computerized, it is relatively insensitive to the back pressure of CO in capillary blood, and it is widely available. Disadvantages include the cost of the equipment, a requirement of an inspired volume of at least 1.3 liters, and the breath-hold time which may be difficult for some patients with significant impairment and concerns regarding the lack of standardization of measurements at altitude (Crapo and Forster, 1989). The breath-hold may be difficult, even at rest, for some patients. The single-breath technique is difficult to perform at more than low levels of exercise by healthy subjects.

(b) Steady-state method

In this method the subject breathes a low concentration of CO (0.1–0.2%) for several minutes until a steady state of CO exchange is achieved. A mixed expired sample is collected over several more minutes and the rate of CO uptake is calculated as the difference between the concentrations of inspired and expired CO per unit of time. In this method, back-pressure of CO can be directly measured before and following the test in the expired gas. Pco may be measured from end-tidal samples, or calculated by sampling arterial $Paco_2$ and using the Bohr equation to calculate Pco in the alveolar air (Van Kessel, 1982). Advantages of this method over the single-breath technique are that it requires less equipment, fewer calculations, less patient cooperation, and it can be performed during exercise. In patients with uneven distribution of inspired gas in relation to pulmonary capillary blood flow slight errors in the analysis of CO_2 and CO can lead to large errors in the calculation of the diffusing capacity (Crapo and Forster, 1989).

(c) Re-breathing method

This method requires the subject to re-breathe the test gas from a balloon, the volume of which approximately equals the subject's $FEV_{1.0}$. To begin, the subject exhales to either FRC or RV, a tap is switched, and the subject re-breathes from the balloon, which should be emptied with each inspiration. The re-breathing period is 30–45 s at a controlled rate of 30 breaths per minute. Gas concentrations are monitored continuously during the re-breathing period. CO uptake is determined from the system volume and the measured rate of CO disappearance from the system. Advantages of the re-breathing system are that it can be performed by subjects with a vital capacity less than 1.3 liters, and it can be adapted to exercise testing or in the intensive care setting (Crapo and Forster, 1989). A disadvantage is that some subjects have difficulty re-breathing at the fixed rate. Some authors feel the re-breathing method is the least useful means of determining the diffusing capacity (Van Kessel, 1982).

A number of confounding factors must be considered in interpreting measurements of D_Lco. Adjustment can be made for variations in age and body size, hemoglobin, carboxyhemoglobin and altitude. D_Lco increases with exercise, as one moves from a standing to sitting position (10–15%), and from a sitting to a supine position (15–20%) (Crapo and Forster, 1989). The American Thoracic Society (1995b) recommends adjustment of the Pio_2 to account for differences in altitude.

(d) Clinical use of diffusing capacity

D_Lco may be **increased** in a variety of conditions including obesity, asthma, polycythemia, fresh pulmonary hemorrhage (Lipscomb, 1978) and conditions associated with increased pulmonary capillary blood volume. D_Lco is typically **decreased** in disorders due to interstitial disease, in emphysema (in contrast to asthma where D_Lco may be either normal or increased), pulmonary thromboembolism, pulmonary hypertension, fat embolism and pulmonary edema (Scanlon and Hankinson, 1996).

Arterial blood gases

A complete review of oxygen and carbon dioxide transport and acid–base status is beyond the scope of this chapter. Basically, measurement of the pH will define the presence and magnitude of systemic acid–base disorders. The partial pressure of carbon dioxide ($Paco_2$) denotes whether the acid–base disorder is of metabolic or respiratory origin and provides a very specific measure of the adequacy of ventilation. A reduced $Paco_2$ indicates hyperventilation and an elevated $Paco_2$ signifies inadequate ventilation. The partial pressure of oxygen (Pao_2) serves as an indication of the adequacy of pulmonary oxygen exchange but is also influenced by non-respiratory factors. Oxygen saturation of hemoglobin (Sao_2) can be measured by a spectrophotometer as the amount of oxyhemoglobin relative to the amount of deoxyhemoglobin and reflects the amount of oxygen carried by the hemoglobin pigment in the blood. Development of a four-wavelength spectrophotometer which measures deoxyhemoglobin, oxyhemoglobin, carboxyhemoglobin and methemoglobin has provided a non-invasive means of monitoring these hemoglobins (Mohler *et al.*, 1982).

Details regarding drawing of the arterial blood, sample storage, calibration and maintenance of the analyzers are critical to the accuracy of the determinations and may seriously impact upon decisions regarding the results, as discussed by Mohler (Mohler *et al.*, 1982).

Pulse oximetry

Pulse oximeters measure pulsatile change in light transmission through living tissue to calculate arterial saturation, with the understanding that such change in light transmission is solely due to change in the intervening blood volume. The absorption of light by venous blood, skin pigments, tissue and bone is automatically eliminated from consideration (Wiedemann and McCarthy, 1989). The use of very small light-emitting diodes (LEDs) and photo detectors has permitted manufacturers to market lightweight, durable portable instruments. Empiric calibration of pulse oximeters is necessary because *in vivo* conditions do not exactly match the requirements of Beer's law. (Prologue, 1987) Pulse oximeters may produce spurious results in the presence of other hemoglobins (COHb, MetHb). Thus, COHb is viewed as though it were mostly HbO_2, accordingly the resulting reading is approximately the sum of COHb and HbO_2. Increasing amounts of MetHb may cause the pulse oximeter to underestimate when the true saturation is greater than and overestimate when the saturation is less than 85% (Wiedemann and McCarthy, 1989). Clinical studies have established the accuracy of pulse oximeters at ±3–5% at saturations above 70% (Ries, 1987; Severinghaus, 1987).

In spite of the limitations indicated above, pulse oximeters have found an important place in the clinical management of patients. A few examples include their use for real-time monitoring of oxygenation in unstable patients in the hospital setting, in pulmonary function laboratories to supplement measurement of diffusing capacity, in screening for gas exchange abnormalities, and in the operating room during anesthesia. Pulse oximeters are useful for monitoring changes in oxygenation during exercise tests (Scanlon, 1996). Pulse oximeter readings have not been universally accepted by certifying agencies as a basis upon which to make decisions regarding disability.

BRONCHIAL PROVOCATION TESTING

Bronchial hyper-responsiveness is a characteristic of subjects with asthma. In atopic individuals this is likely a consequence of genetic predisposition. However, the onset of asthma in later life in individuals without an atopic background suggests that bronchial hyper-responsiveness may be acquired. Thus, demonstration of bronchial hyperreactivity, particularly when subjects are without asthma symptoms and demonstrate normal spirometry, may be an important contribution to the diagnosis of asthma. In fact, a number of environmental stimuli have been shown both experimentally and in a natural setting to induce airway hyper-responsiveness.

Bronchial provocation tests may be performed with a variety of substances. Early studies used aero-allergens in an attempt to define hyper-responsiveness to specific allergens. More recently, there is a tendency to use agents such as histamine and methacholine which test for non-specific bronchial hyper-responsiveness.

The National Academy of Allergy and Immunology (Chai *et al.*, 1975) has recommended standards for bronchial challenge testing. These include the use of aqueous extracts of antigens or agents such as histamine or methacholine with the Johns Hopkins dosimeter and a DeVilbiss nebulizer with inhalation of the provocative agent for 0.6 s commenced from FRC to TLC. An equally popular method utilizes a Wright nebulizer with tidal breathing for 2 minutes. Both methods have been studied and produced comparable responses as measured by the dose at which a 20% decline in $FEV_{1.0}$ occurred (Ryan *et al.*, 1981). Both techniques resulted in the same dose to the lungs; however, the dosimeter produced more deposition of aerosol in the throat and central airways and less in the outer lung than the tidal breathing method. Table 2.3 summarizes most of the current methods for performing bronchial challenge tests.

Most clinicians consider a response to bronchial provocation positive when the $FEV_{1.0}$ declines by 20% or more from the baseline value. Other less frequently utilized positive responses are a 25% decrease in the $FEF_{25-75\%}$ or a 40% or more reduction in specific conductance, if the test is conducted in a body plethysmograph (Braman and Corrao, 1989). There does not seem to be uniform agreement by all investigators utilizing bronchial provocation tests as to what constitutes a positive response. Heir *et al.* (1995) used 'the geometric mean provocation concentration of methacholine causing a 10% fall in the forced expiratory volume in the first second (PC10)', whereas Annema *et al.*, 1995) regarded 'responsiveness to methacholine as a provocative dose producing a 20% fall in forced expiratory volume in one second ($PD20FEV_1$) of $\leq 8.6\,\mu mol$'.

Bronchial provocation is useful in confirming the diagnosis of asthma in subjects who complain of episodic wheezing, cough, chest tightness or dyspnea and who demonstrate normal spirometry at the time of examination. Bronchial challenge testing is considered an essential component in the diagnosis of occupational asthma (Bernstein, 1981). Workers with hyper-sensitive airways develop pulmonary function changes at doses that are 100 to 1000 times less than those that produce similar responses in normal subjects. Bronchial provocation may detect the presence of

Table 2.3 Comparison of methods for performing bronchial challenge tests

Authors	*Dosimeter*	*Power*	*Nebulizer*	*Dilutions*	*Cummulative dose Breath units*
Chai (1975)**	Yes	Air	De Vilbiss® # 646	10	225
Hendrick (1986)**	Yes	Air	De Vilbiss® # 646	13	640
Cockroft (1977)*	No	Air	Wright	9	
Chatham (1982)**	No	Air	De Vilbiss® # 646	2	150

*Histamine
**Methacholine
De Vilbiss® (De Vilbiss Co., Sommerset, PA.)
Table reproduced courtesy of Daniel Banks

hyperreactive airways in workers before or after exposure to harmful agents in the workplace.

In order to be most useful, bronchial challenge tests for asthma attributable to a specific workplace agent should be made using the agent or agents suspected of provoking the occupational asthma. This is difficult to accomplish in the laboratory, and very expensive. Additional problems with laboratory-based bronchial provocation testing are positive tests in those without occupational asthma when exposures are excessive, and negative in those with genuine occupational asthma when the exposure chamber does not reproduce the exposure at work (Gannon, 1997). Pepys developed a technique for exposing subjects with suspected occupational asthma in the laboratory by tipping powder containing the suspected agent plus inert vehicle from one pan to another, or painting liquid containing the suspected agent onto a board (Pepys and Hutchcraft, 1975). Others feel that the proper way to diagnose occupational asthma is to use the natural setting in the workplace as the exposure and to monitor the subject's response by recording peak flows before, during and following exposure. A peak flow meter is recommended for monitoring ventilatory function for 2 weeks, measuring PF before, during and after work on all days both at work and away from work for 2 weeks (Cartier *et al.*, 1984). Unfortunately, the quality control exerted in the pulmonary function laboratory is not present when peak flow is measured by the subject at the work site and problems with inaccurate and fabricated records may complicate the analysis and interpretation of the results of peak flow monitoring. Perhaps the use of a computerized peak flow meter will improve its reliability in monitoring ventilatory function in the work place in the future (Quirce *et al.*, 1995).

EXERCISE TESTING

A normal person at rest takes up approximately 250 ml of oxygen and produces approximately 200 ml of carbon dioxide per minute. In order to maintain metabolic homeostasis, approximately 8 liters of total ventilation are required per minute, approximately 6 liters for alveolar ventilation, and approximately 2 liters to wash out the dead space ('wasted ventilation'). When an individual begins to exercise, oxygen uptake and carbon dioxide production increase in relation to the level of increased activity. Similarly, the demand for alveolar ventilation increases in relation to the level of exercise in order to provide adequate gas exchange. At the point at which the body's energy needs are supplemented by anaerobic metabolism – the anaerobic threshold (AT) – ventilation increases out of proportion to the oxygen uptake required by the level of exercise, owing to accumulation of lactic acid and other metabolites (Zeballos *et al.*, 1985).

Adequate delivery of oxygen to the tissues during exercise depends on a balance of ventilation, diffusion and distribution of perfusion or blood flow in the lungs. Also necessary is a normally functioning cardiovascular system. Exercise testing protocols range from non-invasive monitoring of ventilation and heart rate on a stationary bicycle or treadmill at a controlled rate to the same exercise with collection of expired air and arterial blood gases. The choice of complexity in exercise testing depends upon the preferences of the operator and the precision needed to define the mechanism(s) involved in the impaired exercise response.

A number of authors, among them Jones and Campbell (Jones *et al.*, 1988) and Whipp (1994), have discussed the factors limiting aerobic work capacity. Basically, these factors involve one or more of the following: failure of the respiratory control center, inadequate ventilatory and gas exchange within the lungs, a failure of the cardiovascular system to deliver a sufficient amount of oxygen to the tissues, or defects in the electron transport chains necessary for aerobic energy transport in skeletal muscle.

Estimating cardiovascular limitation

In healthy subjects, maximal exercise is limited by the cardiac output and not by limitations of the respiratory reserve. Minute ventilation during maximal exercise does not usually exceed about 70% of the maximal voluntary ventilation (MVV). Measurement of maximal oxygen uptake ($\dot{V}O_{2max}$) is the most direct way to assess a subject's maximal exercise capacity. Often it is not feasible for a subject to perform maximal exercise, so indices are obtained from submaximal tests. Anaerobic supplementation to metabolism, the anaerobic threshold (AT), occurs at about 50–60% of predicted maximal oxygen uptake ($\dot{V}O_{2max}$) in sedentary individuals (Wasserman, 1988). The anaerobic threshold (AT) is one of the measurements that provides a non-invasive estimate of the adequacy of cardiovascular in response to an exercise stimulus. Other non-invasive indices of the cardiovascular response to exercise are the oxygen pulse (O_2 Pulse) and the heart rate–maximal oxygen uptake (HR–$\dot{V}O_{2max}$) relationship.

Estimating ventilatory limitation

Breathing reserve can be assessed by the relationship of the maximal minute ventilation during exercise ($\dot{V}_{Emax}$) to the Maximal Voluntary Ventilation (MVV) ($\dot{V}_{Emax}$/MVV%). A reduced or absent breathing reserve is one of the criteria often used for establishing ventilatory limitation to exercise. When normal subjects exercise the tidal volume (V_T) increases from its resting level to about 50–60% of the vital capacity (VC). Thereafter, increased demands for ventilation are met by increasing the frequency of breathing. The index V_T/VC reflects the relationship of tidal volume to vital capacity. In subjects with diffuse interstitial disease, V_T is reduced resulting in a low V_T/VC and associated with an exaggerated increase in breathing frequency.

Estimating pulmonary gas exchange limitations

Sampling of arterial blood permits estimation of the adequacy of pulmonary oxygen exchange. The usual indices considered are the arterial oxygen tension ($Pa{O_2}$), the alveolar–arterial oxygen difference ($P(A–a)O_2$) and the arterial oxygen saturation (SaO_2). These may increase, remain the same, or decrease during exercise in normal subjects (Zeballos, 1985). These arterial blood gas indices of adequacy of oxygenation are also dependent upon the PIO_2 (Weisman *et al.*, 1988), age (Wasserman *et al.*, 1987) and exercise type (Martin *et al.*, 1992).

In the restrictive/diffusion disorders, the PaO_2 and SaO_2 may be normal at rest but the $P(A–a)O_2$ is usually elevated. During exercise, the $P(A–a)O_2$ usually widens and both PaO_2 and SaO_2 fall with only modest work loads.

Subjects with severe obstructive impairment may demonstrate variable degrees and combinations of reduced PaO_2, SaO_2 and elevated $P(A–a)O_2$ with or without elevated $PaCO_2$. A variable response of these indices during exercise has been reported in both bronchitic and emphysematous patterns of obstructive airway diseases. Bates (1989) noted that no precise relationship can yet be adduced between morphologic change and the pattern of V/Q abnormality.

ASSESSING PULMONARY IMPAIRMENT

The cardinal symptom of disability due to respiratory disease is shortness of breath. Dyspnea may arise from a variety of sources including cardiovascular limitation, neuromuscular disorders and social or psychological factors. Since dyspnea is a subjective sensation it does not always correlate well with tests of pulmonary function. For this reason, physicians usually employ a battery of tests in an attempt to quantify factors that contribute to impairment or dysfunction of the respiratory system.

There is a fairly good relationship between obstructive impairment and tests that predominantly measure increased airflow resistance such as Raw or its reciprocal specific conductance (Sgaw), and spirometric tests such as the $FEV_{1.0}$. The MVV also has the additional advantage of testing for the effects of respiratory muscle blood flow, fatigue and air trapping, in addition to correlating well with the $FEV_{1.0}$ in detecting airflow obstruction (Gaensler and Wright, 1966). When airflow resistance is elevated there is an inability to move an adequate amount of air into or out of the lungs in relation to the perceived need during exercise. This is recognized as dyspnea. Many regard this situation as analogous to the concept of length tension inappropriateness proposed by Campbell and Howell (1961) in their studies of the origin of dyspnea.

In contrast to obstructive disorders, restrictive impairment seems to cause less dyspnea until the alteration in either lung or chest wall mechanical function is relatively advanced. This is illustrated by the observation that while some subjects may be visibly tachypneic at rest they do not complain of dyspnea. Some investigators felt that dyspnea is related to the increased work of breathing in restrictive disorders (Bader and Bader, 1955) whereas others tend to invoke the length tension inappropriateness model (Campbell and Howell, 1963). Comroe *et al.* (1962) cited several conditions that were inadequately explained by either theory and concluded that more widespread use of newer tests of pulmonary function might being us nearer to an understanding of dyspnea, the most common symptom in patients with cardiopulmonary disease.

Impairment versus disability

Gaensler and Wright (1966) defined impairment as 'a reduction of body or organ function and disability as a lack of ability for a certain level of specific type of performance'. Morgan and Seaton (1995) referred to impairment 'as a physiological abnormality of function which persists after treatment,' and disability as 'an inability to carry out a specific task or job, or the development of undue distress during the performance of the job or task'. Others have defined impairment and disability differently. The upshot is that most physicians recognize that there is a difference between impairment and disability. Many also recognize that the same degree of impairment does not translate into the same degree of disability. Thus, evaluating disability is a two-step process in which it is necessary to document reduction in respiratory function (impairment) which is sufficient to prevent the subject from engaging in gainful employment or activities of daily living (disability).

Many attempts have been made to correlate the objective and commonly used measurements of pulmonary function (spirometry, MVV, diffusing capacity and exercise tests), with dyspnea (Fletcher, 1952; Borg, 1970; Ferris, 1978; AMA guides, 1993). These have generally met with limited success (Epler *et al.*, 1980). Part of this failure may be related to ignorance of all the respiratory factors that are capable of producing dyspnea and part of other contributing factors such as cardiovascular dysfunction and to psychosocial factors which have no relationship to respiratory function.

Grading dyspnea

Fletcher (1952) was one of the first to attempt to quantify dyspnea by the use of a scale or grading system. Subsequent scales have ranged from the five grades of Fletcher to the relatively simple three-grade scale used by the American Medical Association (1993) (Table 2.4). Subjects tend to underestimate their level of performance (McGavin *et al.* 1978). Subjective rating of dyspnea tends to correlate better with physiological impairment if estimated following exercise (Borg, 1970). There is a generally good relationship between the symptom of dyspnea and the $FEV_{1.0}$ in chronic airflow obstruction. Epler and associates (1980) found the best correlation between dyspnea and measurements of the FVC and $D_{L}CO$ in their subjects with restrictive diseases.

Table 2.4 Dyspnea grading scales

*Fletcher's scale**	
Grade 1	Is the subject's breath as good as that of others his/her own age and build on the level, and on climbing hills or stairs?
Grade 2	Is the subject able to walk with normal persons of his/her own age and build on the level but unable to keep up on hills or stairs?
Grade 3	Is the subject unable to keep up with normal persons on the level, but able to walk about a mile or more at his/her own speed?
Grade 4	Is the subject unable to walk more than about 100 yards on the level without a rest?
Grade 5	Is the subject breathless on talking or undressing, or unable to leave his/her house because of breathlessness?
American Medical Association Scale†	
Mild	Do you have to walk more slowly on the level than people of your age because of breathlessness?
Moderate	Do you have to stop for breath when walking at your own pace.
Severe	Do you have to stop for breath after walking about 100 yards or for a few minutes on the level?
Very severe	Are you too breathless to leave the house, or breathless after washing or dressing?

* Data reproduced with permission from Fletcher, C.M. (1952) The clinical diagnosis of pulmonary emphysema – an experimental study. *Proc. R. Soc. Med.*, **45**, 577–8.

† Source: American Medical Association (1993) *Guides to the Evaluation of Permanent Impairment*, 4th edn, American Medical Association, Chicago, copyright 1996.

Malingering

Claimants seeking disability may exaggerate their symptoms for the purpose of secondary gain or other reasons. Thus, the relationship between the symptom of dyspnea and pulmonary function may become dissociated. Morgan (1979) demonstrated that coal miners seeking compensation had higher dyspnea scores than subjects being evaluated for chronic respiratory disease but not seeking compensation (Figure 2.4). A careful inspection of the tracings of the pulmonary function tests will usually reveal characteristics which indicate lack of cooperation or other faults which invalidate tests. On the other hand, one should be careful in attributing all discrepancies between the subject's symptoms of dyspnea and pulmonary physiology to malingering. Our lack of knowledge of all the factors that enter into causing the sensation of dyspnea should prevent us from becoming cavalier in assuming that a subject is malingering. Subjects seeking compensation or disability from respiratory causes are in the age range when cardiovascular disease has its onset. For this reason, subjects who complain of dyspnea out of proportion to either normal or near normal pulmonary function should undergo, at least, a non-invasive cardiopulmonary exercise test.

Optimal time for evaluation

The optimal time for evaluation of a subject's respiratory impairment is when the condition is stable. This requirement presents unique problems in the evaluation of subjects with asthma or bronchospasm and is discussed separately. Subjects with chronic airflow obstruction should be fully treated with bronchodilators and anti-inflammatory agents before attempting to determine their remaining respiratory capacity. In some

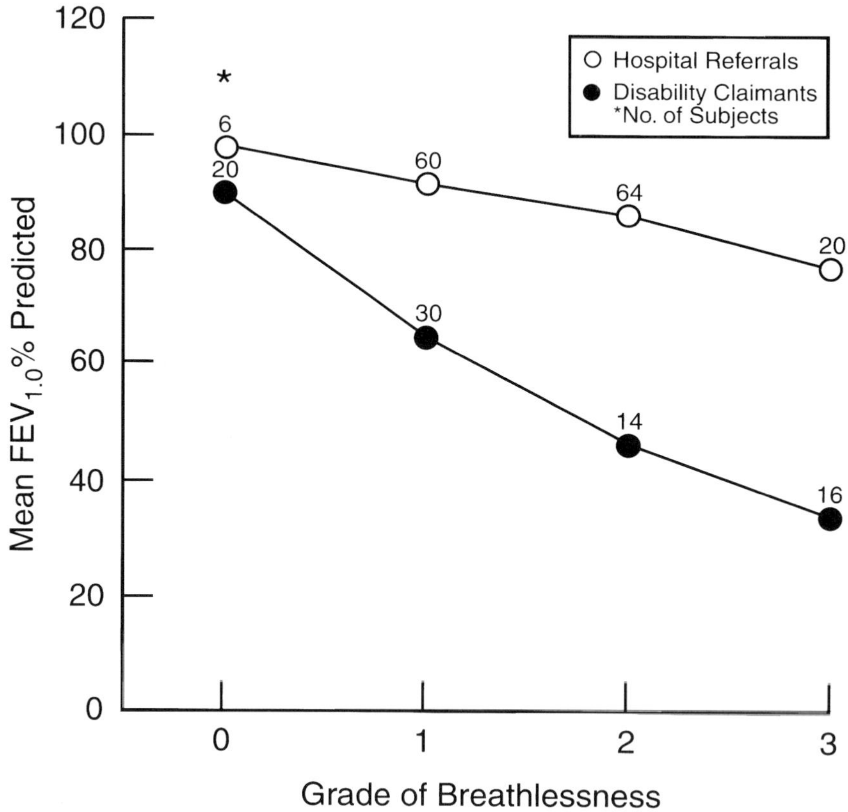

Figure 2.4 Diagram showing the $FEV_{1.0}$ percent predicted versus grade of breathlessness in hospital referrals and disability claimants. Numbers above each point indicate numbers of subjects. (Reproduced with permission from Morgan, W.K.C. (1979) Clinical significance of pulmonary function tests. Disability or disinclination? Impairment or importuning? *Chest*, **75**, 712–15.)

instances it may be necessary to postpone testing until a subject has been treated for a month or so with the above agents and then re-tested before arriving at an accurate estimate of the degree of respiratory impairment. Individuals should not be tested during episodes of upper or lower respiratory tract infections. Testing should be postponed at least 8 weeks after an episode of pneumonia or thoracic surgery.

The special case of asthma

Evaluation of impairment and disability in subjects with asthma represents a special case. Impairment in asthma differs from that of chronic, irreversible airflow obstruction. Asthma is characterized by variable airway obstruction which may be partially or completely reversible, and is associated with airway hyper-responsiveness to triggers such as dusts, fumes, gases or smoke. For these reasons, special guidelines are necessary in evaluating patients with asthma for respiratory disability (American Thoracic Society, 1993). The process includes an evaluation of symptoms consistent with a diagnosis of asthma. It includes measurement of spirometry to detect the presence of airflow obstruction. In the absence of demonstrable airflow obstruction, testing for airway hyper-responsiveness is performed. Once the diagnosis of asthma is substantiated, the subject's temporary impairment is determined and based upon post-bronchodilator response, reversibility of $FEV_{1.0}$, or degree of airway hyper-responsiveness and the minimum medication needed to maintain control or prevent exacerbations of asthma. Because asthma may improve or worsen, it is necessary to re-evaluate subjects at intervals to see whether their degree of impairment has changed.

Guidelines for disability evaluation by professional associations

There are a number of professional associations which have promoted schemas or outlines of systems for evaluating respiratory impairment. The American Thoracic Society guidelines (American Thoracic Society, 1987a, 1987b) recommend the use of ventilatory capacity tests, including FVC, $FEV_{1.0}$, $FEV_{1.0}$/FVC ratio, for evaluating obstructive impairment and single-breath D_Lco (transfer factor) for evaluating restrictive impairment. They specifically do not recommend the MVV ('markedly effort dependent') and miscellaneous tests of flow and volume ($FEF_{25-75\%}$, closing volume, closing capacity and volume of isoflow) 'because they may be abnormal when overall function as determined by $FEV_{1.0}$ is normal'. The rating of impairment is based on a non-linear reduction of ventilatory or diffusing capacity which is designated as mild commencing at 79% of predicted normal and is judged severe at 50% or less of ventilatory capacity and 40% or less of diffusing capacity (transfer factor).

The American Medical Association guidelines (1993) utilize values for FVC, $FEV_{1.0}$, $FEV_{1.0}$/FVC as the primary measures for obstructive impairment. The D_Lco is used as the primary index to evaluate restrictive disease. When the FVC, $FEV_{1.0}$ and $FEV_{1.0}$/FVC ratio are normal and the D_Lco is between 41% and 79%, then a $\dot{V}o_{2max}$ or measured exercise capacity is recommended. A non-linear scale of impairment based upon responses to the above tests is provided for estimating a percentage of impairment of the 'whole person' as a result of respiratory conditions. Mild impairment (10–25%) begins at approximately 79% of predicted ventilatory capacity and severe impairment (50–100%) when ventilatory capacity is reduced to approximately 40–50% and diffusing capacity is reduced to approximately 40% of predicted normal.

In 1990, a working group of the European Society for Clinical Respiratory Physiology reported on a rating procedure for respiratory disability (Cotes, 1990). These criteria utilize 1.64 SD below the reference value instead of 20% below the predicted for ventilatory and diffusion capacity (transfer factor) tests. Otherwise, the criteria for impairment/disability are similar to those proposed by the American Thoracic Society (1991). Slight impairment is designated when spirometry and diffusion capacity (transfer factor) is below the defined lower limit of normal. Severe impairment is defined by values below 50% for the FVC and less than 40% for the $FEV_{1.0}$ and D_Lco. The European guidelines direct performance of a symptom-limited exercise test except in those subjects who have gross impairment sufficient to preclude more than minimal activity. For the results of the exercise test to be used as evidence for respiratory disability, the exercise should be limited by breathlessness and not ischemic pain, changes in the electrocardiogram, leg muscle fatigue or other non-respiratory symptoms.

Entitlement programs

Many schema have been proposed over the years for the evaluation of respiratory impairment related to specific entitlement programs. The Canadian Medical Association's Task Force on Occupational Respiratory Disease (Ostiguy, 1979) recommended the use of FVC, $FEV_{1.0}$, D_Lco, $\dot{V}o_{2max}$ determined during exercise and arterial blood gases at rest for full evaluation of disability in workers exposed to asbestos, coal and silica. Class II (mild) impairment is diagnosed when the FVC and $FEV_{1.0}$ were less than 60% of predicted value, D_Lco less than 75% of predicted, and $\dot{V}o_{2max}$

less than 25 ml/min/kg. Class V (very severe) impairment is diagnosed when the FVC, $FEV_{1.0}$ and D_LCO are less than 40% predicted, or $\dot{V}O_{2max}$ is less than 7 min/kg or resting blood gases showed PaO_2 less than 50 mmHg and $PaCO_2$ greater than 45 mmHg.

The Black Lung Act in the United States set 60% of predicted normal ventilatory capacity as the level at which eligible coal miners could be declared totally and permanently disabled if they met the requisite number of years of employment in the nation's underground coal mines (Federal Register, 1980).

Each state or province has its own workers' compensation program with its individual requirements for determining disability. There may be separate programs for specifically identified groups such as railway workers, federal employees, and civilian employees of the US Navy (Barnhart, 1996). The physician participating in any of the systems for evaluating degree of impairment listed above should be able to provide the numerical assessment required for any of the specific entitlement programs.

APPROACH TO THE EVALUATION OF RESPIRATORY IMPAIRMENT

The approach to evaluating a subject for respiratory impairment is somewhat different from that for a patient seeking treatment for a medical condition. There is often a question of workers' compensation or alleged injury which needs to be explored in detail. In addition, medical records from this type of examination are bound to become public record at some point in the proceedings. Thus, the history of occupational and avocational activities, as well as personal habits, must be explored in considerable detail.

Medical history

A comprehensive medical history must be obtained, much in the same way as a new patient is evaluated. The cardinal respiratory symptoms of cough, phlegm, chest tightness and wheezing should be characterized in terms of onset, intensity, duration and progression. Dyspnea, or breathlessness, should be quantified in terms of one of the classification scales such as Fletcher's (1953) or the simpler version published by the American Medical Association (1993) (Table 2.4). The physician should inquire about symptoms related to the cardiovascular system, since dyspnea arising from disease in this system may limit exertion.

The occupational history should be recorded in chronological order, beginning with the first job on leaving school and progressing to the current job or retirement. It is desirable to attempt to get an idea of the worker's intensity as well as duration of exposure to substances for each job. Job titles may be unclear or misleading at times so it is wise to ask the subject to describe the activities and substances used in his/her work. One should inquire about the availability of and the worker's use of personal protective equipment.

A detailed smoking history must be obtained. This includes the duration and intensity of tobacco use. A preferable method of recording cigarette smoking is by identifying the number of cigarettes smoked per day and the number of years smoked. One can then calculate a measure of cumulative exposure which is usually expressed as pack-years. It is important to distinguish in pipe and cigar smokers whether or not they inhaled. Subjects may be exposed to harmful dusts, vapors or fumes, in pursuing their hobbies and these should be explored.

Physical examination

The physical examination is of limited usefulness in the assessment of respiratory impairment. Diffuse infiltrative diseases are often associated with tachypnea and a panting type of respiration at rest and yet the individual does not complain of dyspnea. Hyperresonance and inspiratory retractions of the suprasternal tissues are characteristic of subjects with severe obstruction. Decreased breath sound transmission with hyperresonance may be an indication of underlying emphysema. None of these findings may be present in early impairment of the respiratory system. The presence of continuous adventitious sounds such as wheezes and rhonchi (low-pitched wheezes) are consistent with disease in airways. Discontinuous sounds (crackles) are usually indicative of diffuse infiltrative diseases which alter the normal opening and closing of small, peripheral airways leading to areas of subsegmental collapse (Glaister, 1969).

Laboratory studies

Routine laboratory investigations are of little benefit in assessing respiratory impairment other than excluding anemia as a potential cause of a reduced D_LCO (transfer factor). Similarly, the chest radiograph is virtually useless in detecting disease within the airways. Signs of hyperinflation on a chest radiograph occur late in the course of the disease and correlate poorly with function. The chest radiograph in advanced cases of asbestosis may show a rough correlation with ventilatory function, but in most occupational diseases the relationship of radiographic change and function is poor. The chest radiograph is of value in helping to establish attribution in cases of occupational exposure. The standard resting electrocardiogram is not very sensitive in detecting early abnormality of the right ventricle due to pulmonary disease. It is valuable as a baseline for cardiopulmonary exercise testing.

Pulmonary function tests

Pulmonary function tests are of vital importance in diagnosing occupational respiratory disease and in objectively assessing the degree of impairment resulting from it. They are most useful in assessing impairment in chronic, irreversible conditions. The American Thoracic Society (1991) has proposed a useful strategy for interpreting the results of spirometry and lung volumes (see Table 2.2). The results of pulmonary function tests may be related to predicted values in order to determine how much capacity was lost (impairment), or conversely, how much capacity remains. This remaining capacity may be compared with the respiratory demands of specific jobs or activities to document disabling impairment.

Virtually all of the guidelines for evaluating respiratory impairment begin with measurements of spirometry and $D_{L}CO$. In order to obtain valid results, the laboratory obtaining the tests should use equipment meeting the specifications and techniques conforming to the guidelines set forth by the American Thoracic Society (1995a,b). Since spirometry and $D_{L}CO$ require a certain amount of subject cooperation, it is essential that test results meet the quality control criteria for reproducible, maximal efforts. The results should be reported as absolute values and expressed as percentages of predicted values appropriately corrected for age, height, gender and, in certain instances, race or ethnic origin. It is essential that the equipment provide hard copies of the actual tracings. In many instances, the tracings are required by the agency in order to validate claims for disability.

The American Thoracic Society guidelines (1993) indicate that the majority of subjects evaluated for respiratory impairment need not undergo exercise tests. In the event that the physician feels that exercise testing is needed, it should include, as a minimum, measurement of $\dot{V}O_{2max}$ as well as minute ventilation $\dot{V}_E$, tidal volume V_T and breathing frequency. The American Medical Association Guides (1993) indicate that exercise may be helpful in evaluating respiratory impairment in subjects whose dyspnea is more severe than spirometry or $D_{L}CO$ would indicate, those who state that they are unable to meet the demands of a specific job because of breathlessness, or who have not performed maximally or correctly in the spirometry or $D_{L}CO$ tests. Although it is not stated, the recommendations listing classes of respiratory impairment suggest that the exercise should be maximal ($\dot{V}O_{2max}$). One wonders whether if the subject did not perform maximally or correctly on spirometry and $D_{L}CO$ how likely he/she is to perform exercise sufficiently long enough to obtain valid data from an exercise test. The working group of the European Society for Clinical Respiratory Physiology (Cotes, 1990) accepted the World Health Organization definition of respiratory disability (WHO, 1980). Therefore, the group felt that a progressive exercise test to a symptom-limited maximum was the best way to assess respiratory disability. They also indicated that ventilation ($\dot{V}_E$), cardiac frequency, the electrocardiogram, expiratory gas concentrations and oxygen saturations should be measured throughout the test.

Most of the entitlement programs, i.e. Social Security, Railroad Employees, Federal Employees and Workers' Compensation, have promulgated their own requirements for what they want and/or accept as evidence of respiratory disability.

THE RESPIRATORY IMPAIRMENT EVALUATION REPORT

The examining physician's role in the disability determination process culminates in the preparation of a comprehensive report. This begins with a statement of the problem for which the examination was requested. Next comes a summary of the findings obtained from interrogation of the patient concerning cardiac and pulmonary symptoms, their onset, intensity, duration and progression. A detailed history of occupational and avocational exposures, and tobacco use follows. The presence or absence of physical findings pertinent to respiratory disease is noted. Laboratory findings including hemoglobin, chest radiograph findings, classified on the ILO classification of Pneumoconioses (International Labour Office, 1980), if appropriate, pulmonary function test results and an electrocardiogram, are the minimum elements to be included in the report. The diagnosis(es) arrived at should be stated using standard medical terminology. An assessment of degree of impairment of the respiratory system should be made. This assessment should identify the standards against which the subject's pulmonary function was compared and the overall level of impairment.

The report should state whether or not the subject has the remaining respiratory capacity to perform exercise (work) or activities of daily living, which have been determined to require this much ventilation and oxygen uptake. Finally, the report should discuss the relationship (if possible) to the workplace or other exposure. This may require reliance solely upon the history provided by the subject and one's knowledge of epidemiologic literature or studies of disease mechanisms. Several sources are available to assist the physician in assessing the potential exposures a worker may experience (Lapp, 1996) and the relative energy requirements for typical occupations (US Department of Labor, 1977). It may be necessary in certain cases either to consult with an industrial hygienist, toxicologist or chemist, or have an attorney provide reports from such experts, in order to arrive at an answer to the question of attribution (Richman, 1989). The ultimate opinion with regard to the question of attribution should be arrived at by standard methodology, derived by logical deduction from the data available and by comparison with the known literature on the subject. They should be stated as opinion(s) held with 'reasonable medical certainty' (Richman, 1989). It is generally not possible to precisely quantify in a given individual the relative contributions from multiple factors in the causation of occupational respiratory diseases.

REFERENCES

American Medical Association (1993) *Guides to the Evaluation of Permanent Impairment*, 4th edn, (Eds T.C. Dodge and T.P. Houston) The Respiratory System. American Medical Association, Chicago, pp. 154–67.

American Thoracic Society (1987a) Single breath carbon monoxide diffusing capacity (transfer factor): recommendations for a standard technique. *Am. Rev. Respir. Dis.*, **136**, 1299–307.

American Thoracic Society (1987b) Standardization of spirometry – 1987 update. *Am. Rev. Respir. Dis.*, **136**, 1285–98.

American Thoracic Society (1991) Lung function testing: selection of reference values and interpretive strategies. *Am. Rev. Respir. Dis.*, **144**, 1202–18.

American Thoracic Society (1993) Guidelines for the evaluation of impairment/disability in patients with asthma. *Am. Rev. Respir. Dis.*, **147**, 1056–61.

American Thoracic Society (1995a) Standardization of spirometry – 1994 update. *Am. J. Respir. Crit. Care Med.*, **152**, 1107–36.

American Thoracic Society (1995b) Single-breath carbon monoxide diffusing capacity (transfer factor). Recommendations for a standard technique – 1995 update. *Am. J. Respir. Crit. Care Med.*, **152**, 2185–98.

Annema, J.T., Sparrow, D., O'Connor, G.T. *et al.* (1995) Chronic respiratory symptoms and airway responsiveness to methacholine are associated with eosinophilia in older men: the Normative Aging Study. *Eur. Respir. J.*, **8**, 62–9.

Bader, M.E. and Bader, R.A. (1955) The work of breathing. *Am. J. Med.*, **18**, 851–4.

Barnhard, H.J., Pierce, J.A., Joyce, J.W. and Bates, J.H. (1960) Roentgenographic determination of total lung capacity. *Am. J. Med.*, **28**, 51–60.

Barnhart, S. and Balmes, J.R. (1996) Respiratory impairment and disability, in *Occupational and Environmental Respiratory Disease* (eds P. Harber, M.B. Schenker and J.R. Balmes), Mosby, St. Louis, pp. 868–82.

Bates, D.V. (1989) Severe Chronic Airflow Limitation in *Respiratory Function in Disease*, (Ed. D.V. Bates) 3rd edn., W.B. Saunders Company, Philadelphia, pp. 188–213.

Bernstein, I.L. (1981) Occupational asthma. *Clin. Chest Med.*, **10**, 255–72.

Borg, G. (1970) Perceived exertion as an indicator of somatic stress. *Scand. J. Rehab.*, **2–3**, 92–8.

Braman, S.S. and Corrao, W.M. (1989) Bronchoprovocation testing. *Clin. Chest Med.*, **10**, 165–76.

Campbell, E.J.M., Dinnick, O.P. and Howell, J.B.L. (1961) The immediate effects of elastic loads on the breathing of man. *J. Physiol.*, **156**, 260–73.

Cartier, A., Pineau, L. and Malo, J.L. (1984) Monitoring of maximum expiratory peak flow rates and histamine inhalation tests in the investigation of occupational asthma. *Clin. Allergy*, **14**, 193–6.

Chai, H., Farr, R.S., Froehlich, L.A. *et al.* (1975) Standardization of bronchial inhalation challenge procedures. *J. Allergy Clin. Immunol.*, **56**, 323–7.

Chatham, M., Bleecker, E.R., Norman, P. *et al.* (1982) A screening test for airways reactivity. An abbreviated methacholine inhalation challenge. *Chest*, **82**, 15–18.

Cockroft, D.W., Killian, D.N., Mellon, J.J.A., Hargrave, F.E. (1977) Bronchial reactivity to inhaled histamine: a method and clinical survey. *Clin. Allergy*, **7**, 235–43.

Comroe, J.H., Forster, R.E., Dubois, A.B. *et al.* (eds) (1962) Pulmonary Disability in *The Lung. Clinical Physiology and Pulmonary Function Tests*, Year Book Medical Publishers, Inc., Chicago, pp. 256–75.

Cotes, J.E. (1990) Rating respiratory disability: a report on behalf of a working group of the European society for clinical respiratory physiology. *Eur. Respir. J.*, **3**, 1074–7.

Crapo, R.D. and Forster, R.E. II. (1989) Carbon monoxide diffusing capacity. *Clin. Chest Med.*, **10**, 187–98.

Damon, A. (1966) Negro–White differences in pulmonary function (vital capacity, timed vital capacity, and expiratory flow rate). *Human Biol.*, **38**, 380–93.

Darling, R.C., Cournand, A. and Richards, D.W., Jr (1940) Studies on the intrapulmonary mixture of gasses. III. An open circuit method for measuring residual air. *J. Clin. Invest.*, **19**, 609–18.

Dawson A. (1966) Reproducibility of spirometric measurements in normal subjects. *Am. Rev. Respir. Dis.*, **93**, 264–8.

Dubois, A.B., Botelho, S.Y., Bedell, G.N., Marshall, R. and Comroe, J.H. (1956) A rapid plethysmographic method for measuring thoracic gas volume. *J. Clin. Invest.*, **35**, 322–6.

Epler, G., Saber, F. and Gaensler, E. (1980) Determination of severe impairment (disability) in interstitial lung disease. *Am. Rev. Respir. Dis.*, **121**, 647–59.

Fairbairn, A.S., Fletcher, C.M., Tinker, C.M. and Wood, C.H. (1962) A comparison of spirometric and peak flow measurements in men with and without chronic bronchitis. *Thorax*, **17**, 168–74.

Federal Register (1980) Appendix B Standards for administration and interpretation of ventilatory tests. *CFR*, **45**, 13694–712.

Ferris, B.G. (1978) Recommended standardized procedures for pulmonary function testing. *Am. Rev. Respir. Dis.*, **118**(Part 2), 55–88.

Fletcher, C.M. (1952) The clinical diagnosis of pulmonary emphysema – an experimental study. *Proc. R. Soc. Med.*, **45**, 577–84.

Forster, R.E., II, Dubois, A.B., Briscoe, W.A. and Fisher, A.B. (1986) The Lung Volumes in *The Lung. Physiologic Basis of Pulmonary Function Tests*, 3rd edn, Year Book Medical Publishers, Inc., Chicago, p. 8–24.

Fry, D.L. (1968) A preliminary lung model for simulating the aerodynamics of the bronchial tree. *Comp. Biomed. Res.*, **2**, 111–34.

Gaensler, E.A. (1951) Analysis of the ventilatory defect by timed capacity measurements. *Am. Rev. Tuberc.*, **64**, 256–78.

Gaensler, E.A. and Wright, G.W. (1966) Evaluation of respiratory impairment. *Arch. Environ. Health*, **12**, 146–89.

Gannon, P.F.G. and Burge, P.S. (1997) Serial peak flow measurement in the diagnosis of occupational asthma. *Eur. Respir. J.*, **10, Supp. 24**, 57s–63s.

Gardner, R.M. (1979) Chairman: ATS statement – Snowbird workshop on standardization of spirometry. *Am. Rev. Respir. Dis.*, **119**, 831–8.

Glaister, D.H. (1969) Lung collapse in aviation medicine. *Br. J. Hosp. Med.*, **2**, 635–42.

Harris, T.R., Pratt, P.C. and Kilburn, K.H. (1971) Total lung capacity measured by roentgenograms. *Am. J. Med.*, **50**, 756–63.

Heir, T., Aanestad, G., Carlsen, K.H. *et al.* (1995) Respiratory tract infection and bronchial responsiveness in elite athletes and sedentary control subjects. *Scand. J. Med. Sci. Sports*, **5**, 94–9.

Hendrick, D.J., Fabbri, L.M., Hughes, J.M. *et al.* (1986) Modification of the methatholine inhalation test and its epidemiologic use in polyerethane workers. *Am. Rev. Respir. Dis.*, **133**, 600–4.

Hutchinson, J. (1844) Lecture on vital statistics, embracing an account of a new instrument for detecting the presence of disease in the system. *Lancet*, **2**, 567–70 and 594–86.

Hutchinson, J. (1846) On the capacity of the lungs, and on the respiratory functions, with a view of establishing a precise and easy method of detecting disease by the spirometer. *Med. -Chir. Trans. (London)*, **29**, 137–252.

International Labour Office (1980) *Guidelines for the use of ILO international classification of radiographs of pneumoconiosis*, Geneva, International Labour Office, iii–48.

Jones, N.L., Campbell, E.J.M, Edwards, R.H.T. and Robertson, D.G. (1988) *Clinical Exercise Testing*, 3rd edn, W.B. Saunders, Philadelphia.

Kory, R.C., Callahan, R., Boren, H.G. and Syner, J.C. (1961) The veterans administration–army cooperative study of pulmonary function. I. Clinical spirometry in normal men. *Am. J. Med.*, **30**, 243–58.

Lapp, N.L. (1973) Physiological approaches to detection of small airway disease. *Environ. Res.*, **6**, 253–64.

Lapp, N.L. (1996) Mining, in *Occupational and Environmental Respiratory Disease* (eds P. Harber, M.B. Schenker and J.R. Balmes), Mosby, St. Louis, pp. 654–64.

Lapp, N.L. and Hyatt, R.E. (1970) Evaluation of lung mechanics in epidemiological studies. *Bull. Physiopath. Resp.*, **6**, 595–604.

Leuallen, E.C. and Fowler, W.S. (1955) Maximal midexpiratory flow. *Am. Rev. Tuberc.*, **72**, 783–800.

Lipscomb, D.J., Patel, K., Hughes, J.M. (1978) Interpretation of increases in the transfer coefficient for carbon monoxide (TLCO or KCO). *Thorax*, **6**, 728–33.

Macklem, P.T. and Mead, J. (1967) Resistance of central and peripheral airways measured by a retrograde catheter. *J. Appl. Physiol.*, **22**, 395–401.

Martin, T.W., Zeballos, R.J. and Weisman, I.M. (1992) Use of arm crank exercise in the detection of abnormal pulmonary gas exchange in patients at low altitude. *Chest*, **105**, 168–74.

McGavin, F.R., Artvinli, M., Naoe, H. *et al.* (1978) Dyspnea, disability, and distance walked: comparison of estimates of exercise performance in respiratory disease. *Br. Med. J.*, **2**, 241–3.

Meisner, P. and Hugh-Jones, P. (1968) Pulmonary function in bronchial asthma. *Br. Med. J.*, **1**, 470–5.

Meneely, G.R. and Kaltreider, N.L. (1949) The volume of the lung determined by helium dilution. Description of the method and comparison with other procedures. *J. Clin. Invest.*, **28**, 129–39.

Miller, W.F., Johnson, R.L., Jr and Wu, N. (1959) Relationships between fast vital capacity and various timed expiratory capacities. *J. Appl. Physiol.*, **14**, 157–63.

Mohler, J.D., Collier, C.R., Brand, W. *et al.* (1982) Blood Gases, in *Pulmonary Function Testing Guidelines and Controversies* (ed. J.L. Clausen), Academic Press, New York, pp. 223–57.

Morgan, W.K.C. (1979) Clinical significance of pulmonary function tests. Disability or disinclination? Impairment or Importuning? *Chest*, **75**, 712–15.

Morgan, W.K.C., Burgess, D.B., Lapp, N.L. and Seaton A. (1971) Hyperinflation of the lungs in coal miners. *Thorax*, **26**, 585–90.

Morgan, W.K.C. and Seaton, A. (1995) Pulmonary physiology – its application to the determination of respiratory impairment and disability in industrial lung disease in *Occupational Lung Diseases*, 3rd edn, (eds. W.K.C. Morgan and A. Seaton), W.B. Saunders, Philadelphia, pp. 35–81.

Ostiguy, G.L. (1979) Summary of task force report on occupational respiratory disease (pneumoconiosis). *Can. Med. Assoc. J.*, **121**, 414–21.

Pappenheimer, J.R., Comroe, J.H., Cournand, A. *et al.* (1950) Standardization of definitions and symbols in respiratory physiology. *Fed. Proc.*, **9**, 602–5.

Pepys, J. and Hutchcroft, B.J. (1975) Bronchial provocation tests in etiologic diagnosis and analysis of asthma. *Am. Rev. Respir. Dis.*, **112**, 829–59.

Prologue, J.A. (1987) Pulse oximetry: technical aspects of machine design. *Int. Anesthesiol. Clin.*, **25**, 137–53.

Quanjer, P.H., Tammeling, G.H., Cotes, J.E. *et al.* (1993) Lung volumes and forced ventilatory flows: report of working party, standardization of lung function tests, European Community for Steel and Coal – official statement of the European Respiratory Society. *Eur. Respir. J.*, **6** (suppl. 16), 5–40.

Quirce, S., Contreras, G., Dybuncio, A. *et al.* (1995) Peak expiratory flow monitoring is not a reliable method for establishing the diagnosis of occupational asthma. *Am. J. Respir. Crit. Care Med.*, **152**, 1100–2.

Reger, R.B., Young, A. and Morgan, W.K.C. (1972) An accurate and rapid radiographic method to determine total lung capacity. *Thorax*, **27**, 163–8.

Richman, S.I. (1989) Compensating victims of occupational lung disease: the physician's role in the system. *J. Occup. Med.*, **31**, 335–8.

Ries, A.L. (1987) Oximetry – know thy limits. *Chest*, **91**, 316.

Ries, A.L. (1989) Measurement of lung volumes. *Clin. Chest Med.*, **10**, 177–86.

Ryan, G., Dolovich, M.B., Roberts, R.S. *et al.* (1981) Standardization of inhalation provocation tests: two techniques of aerosol generation and inhalation compared. *Am. Rev. Respir. Dis.*, **123**, 195–9.

Scanlon, P.D. and Hankinson, J.L. (1996) Physiologic methods, in *Occupational and Environmental Respiratory Disease* (eds P. Harber, M.B. Schenker and J.R. Balmes), Mosby, St. Louis, pp. 39–54.

Severinghaus, J.W. (1987) History, status and future of pulse oximetry. *Adv. Exp. Med. Biol.*, **220**, 3–8.

Tiffeneau, R., Bousser, J. and Drutel, P. (1949) Capacité vitale et capacité pulmonaire utilisable à l'effort. Critères statique et dynamique de la ventilation pulmonaire. *Paris Medical*, **137**, 543–7.

U.S. Department of Labor (1977) *Dictionary of Occupational Titles*, Fourth Edition. U.S. Government Printing Office, Washington, DC.

Van Kessel, A.L. (1982) Pulmonary diffusing capacity for carbon monoxide, in *Pulmonary Function Testing Guidelines and Controversies*, (ed. J.L. Clausen), Academic Press, New York, pp. 165–85.

Wang, M., McCabe, L., Hankinson, J.L. *et al.* (1996) Longitudinal analysis of lung function in steel workers. *Am. J. Respir. Crit. Care Med.*, **153**, 1907–13.

Wasserman, K. (1988) New concepts in assessing cardiovascular function. *Circulation*, **78**, 1060–70.

Wasserman, K., Hansen, J.E., Sue, D.Y. *et al.* (eds) (1987) *Principles of Exercise Testing and Interpretation*, Lea & Febiger, Philadelphia, pp. 112–31.

Weisman, I.M., Zeballos, R.J. and Johnson, B.D. (1988) The effect of moderate inspiratory hypoxia on exercise performance in sickle cell trait. *Am. J. Med.*, **84**, 1033–41.

Whipp, B.J. (1994) The bioenergetics and gas exchange basis of exercise testing. *Clin. Chest Med.*, **15**, 173–92.

World Health Organization (1980) International Classification of Impairments, Disabilities and Handicaps, WHO, Geneva.

Wiedemann, H.P. and McCarthy, K. (1989) Noninvasive monitoring of oxygen and carbon dioxide. *Clin. Chest Med.*, **10**, 239–54.

Wright, B.M. and McKerrow, C.B. (1959) Maximum forced expiratory flow rate as a measure of ventilatory capacity. *Br. Med. J.*, **2**, 1041–6.

Yan, K., Salome, C., Woolcock, A.J. (1983) Rapid method for measurement of bronchial responsiveness. *Thorax*, **38**, 760–5.

Zeballos, R.J., Weisman, I.M., Johnson, B. *et al.* (1985) $P(A\text{–}a)o_2$ during exercise in healthy young blacks with sickle cell trait (SCT) and controls. *Fed. Proc.*, **44**, 1368.

3 Identification, monitoring and control of dust exposures

Frank J. Hearl

INTRODUCTION

The recognition, evaluation and control of dust exposures in occupational environments can be complex. An occupational hygienist must assess the nature of the hazard, the likely target of the dust, and the properties of that dust responsible for the biological response of interest. Dust measurement and dust hazard evaluation is complex because of the need to characterize properties beyond the intensity of exposure, i.e. the dust concentration. It may be necessary to describe the exposure in terms of the particle size distribution, and the often inhomogeneous chemical or morphological properties of the dust. For fibrous minerals such as asbestos, particle shape may have profound impact on the toxicity of the material. Several descriptors of the particles may be used to characterize the concentration including: the mass of the particles, the mass of one chemical species in the particles, the active surface area, the number of particles, or the crystalline properties of the particles.

Some studies show that freshly generated dust containing crystalline silica will exhibit increased toxicity in lung cells compared with aged dust, due to the recent creation and presence of surface free radicals (Castranova *et al.*, 1996). Therefore, specific knowledge about the process that generated the dust and the interactions of the aerosol with the environment provide important information. Particles of varying sizes deposit in different regions of the lungs based on their aerodynamic size. The largest of the particles have settling velocities greater than the air velocity at the nostril and drop harmlessly away. Other smaller particles may be captured by the hair and mucus-covered surfaces of the nose and be quickly expelled or ingested. Still smaller particles can travel to the furthest depths of the lungs and deposit in the alveolar regions where gas exchange takes place with the blood.

The large particles captured in the nose, mouth or upper airways are cleared to the gastrointestinal tract within minutes or hours. They may contain soluble toxic materials that can be absorbed producing systemic toxicity, or they may produce local effects such as nasal cancer. Particles depositing in the upper airways may be responsible for chronic obstructive pulmonary disease, asthmatic reactions or lung cancer. The smallest particles – those that deposit in the alveolar regions of the lungs – are often associated with hypersensitivity pneumonitis (extrinsic allergic alveolitis), pneumoconiosis, including coal workers pneumoconiosis, asbestosis and silicosis.

In the preceding discussion, we have touched on the complexities of describing dust concentrations and the importance of characterizing particle size distributions, chemical properties, morphologic features, and the crystallinity of the inhomogeneous mixture we call 'dust'. Developing a strategy for appropriate identification, evaluation and control of dust must necessarily involve some or all of these factors. Monitoring

Occupational Lung Disease. Edited by Daniel E. Banks and John E. Parker. Published in 1998 by Chapman & Hall. ISBN 0 412 73630 6

exposure concentrations usually has one of three purposes:

- finding a dose–response relationship or mechanism of action as part of an epidemiologic study;
- evaluating the effectiveness of an engineering exposure control system; or
- checking compliance with an accepted exposure limit.

Typically, monitoring strategies developed for epidemiologic purposes are more complex than those needed for the other purposes. Reasonable feed-back and control strategies for recognizing and characterizing significant dust exposures often make assumptions about the size and composition of the dust being monitored to allow for rapid assessment. Likewise, enforcement monitoring strategies are usually based on some standardized sampling systems; however, they may require specific information about the chemical composition, morphology or crystallinity to demonstrate compliance or non-compliance with specific standards. Generally, complete characterization of the aerosol is not feasible or necessary for these routine dust exposure monitoring scenarios. However, strategies for characterizing exposure for assessing health risk in epidemiologic studies, animal exposure experiments, cell-culture studies and health risk assessments are often intricate. To improve the chances of detecting a true dose–response relationship in these studies, the most appropriate exposure metric must be used. Since it is often difficult or impossible to predict which metric will be the best correlate to the biological response, full characterizations of the exposure are usually required. The following sections will discuss these exposure metrics and the circumstances under which under each should be or has been used.

EXPOSURE METRICS

To quantify aerosol exposures, a variety of exposure metrics has been used. Some common descriptors of the exposure include:

- the total *mass* of particles in a given volume of air;
- the mass of particles with a particular size or aerodynamic property in a given, volume of air, usually *respirable* dust;
- the total *number* of particles in a given volume of air;
- the total number of particles with a particular size and/or shape (e.g. length to width *aspect* ratio) in a given volume of air;
- the mass of specific *chemical species* contained in a collected mass of particles in a given volume of air;
- the *relative composition* of a specific species compared with another species or the total mass or count (percentage composition);
- the *crystalline structure* of the dust particles; and
- the *active* surface area or number of surface functional groups in a given volume of air.

Over the years, each of these measures has been used to describe workplace exposures to particulate. Most of these metrics are quantified in terms of mass or number per stated volume of air, and have customarily been reported as milligrams per cubic meter (mg/m^3), or million particles per cubic foot (mppcf), or fibers per cubic centimeter (f/cc).

Generally, the sampling technique involves collecting and concentrating the particles from a measured volume of air. Some early dust collection systems relied upon gravitational settling of the dust onto an open-faced collection surface. Later techniques included accelerating the dust-laden air stream through a glass tube and impinging the particles in the bottom of a liquid-filled glass bottle. Other devices used thermal precipitation or electrostatic precipitation to attract the airborne particles toward surfaces that could later be analyzed. Similar to impinging the particles into a liquid-filled device, particles may be collected by impaction onto a flat-plate substrate that is often coated with grease or other sticky and non-volatile substance. Impactors are *size-selective samplers*, with the size of particles collected determined by the velocity with which the air stream is accelerated toward the collection surface. Perhaps the most common method for collecting dust samples is to use an air-mover (pump) to draw a known volume of dust-laden air through a filter for further analysis. If the air passes through a size-selective device, such as an impaction surface or a cyclone, before collection by a filter the largest particles are removed and only the small particles are collected on the filter.

Particle count

Once the dust is removed from the air and is concentrated in either a liquid impinger solution, non-volatile impactor grease, glue, or on a filter, it must be quantified. Generally, dusts collected using the impinger method were quantified by particle counting, while dusts collected using impactors or filters were analyzed for particle mass. A notable exception is the method used for counting fibers. In this case, a special filter is used that has properties that allow it to be 'cleared', that is it becomes transparent to visible light so that the particles can be seen, sized and counted by optical microscopy (NIOSH, 1994a).

Dust particles collected in the impinger solution are counted using a counting cell and a method adapted from those used to count plankton in water (Ayer, 1969). Rules for counting have not been standardized. The main steps in particle counting begin with the removal of a known portion of the solution containing the collected dust particles. This aliquot is transferred to a counting cell, which is a glass microscope slide constructed with a 1 mm-deep well to contain the solution. Particles that settle to the floor of the cell are counted using light field microscopy with a 10-power objective lens (Hall *et al.*, 1965). Opaque particles as small as 0.5 μm can be detected by trained observers under these conditions. The total particle count can then be computed, noting the ratio between the withdrawn portion and the total volume of the sample. Concentrations reported from impinger samples were usually stated in million particles per cubic foot, mppcf, computed by:

$$C = \frac{\text{COUNT}}{Q \times t}$$

where C is the concentration, COUNT is the number of particles in the impinger solution, Q is the pump flow rate, and t is the sampling time.

One investigator contacted 45 industrial hygiene laboratories and documented significant differences in their standard procedures involving: collection medium (alcohol or water or mixtures thereof), type of illumination (dark field versus light field microscopy), length of time between collection and counting, method of counting (i.e. counting only those particles settled on the bottom of the slide or focussing through the entire 1 mm depth of field), and settling time of the sample before counting was begun (Halley, 1946). As understanding developed among industrial hygienists that it was the respirable fraction that was most closely associated with pneumoconiosis, many industrial hygienists rejected counts of particles with observed diameters larger than 10 μm since these were 'not respirable' (Lippmann, 1983). Impinger count data were perceived to be highly inaccurate and subject to training and experience bias among counters (Ayer, 1969). With few exceptions, most occupational exposure limits and standards for dusts enforced since 1970 have been based on mass sampling methods. The preceding section was presented primarily to put historical dust data in proper context.

Particle mass

Particle mass is measured by *gravimetric* analysis. The particle collection substrate, which is usually either a filter or a greased impactor substrate, is pre-weighed on a precision microbalance under controlled conditions. This *tare* weight is carefully recorded along with the filter number. After sampling, the substrate is reweighed, under the same controlled environmental conditions. The difference between the final weight and the tare weight is the mass of particles collected.

For gravimetric mass measurements, the airborne concentration is computed by dividing the collected mass by the known volume of air from which they were collected. The air volume is usually computed by using a flow control device in conjunction with the pump to regulate the air flow rate, and to sample for a measured period of time, computing the concentration, usually reported as mg/m^3, by:

$$C = \frac{W_f - W_i}{Q \times t}$$

where C is the concentration, W_f is the final weight of the substrate, W_i is the initial weight of the substrate, Q is the pump flow rate, and t is the sampling time.

PARTICLE SIZE CRITERIA

Total dust

The Greenburg–Smith impinger, developed in 1925 by the US Bureau of Mines, the American Society of Heating and Ventilating Engineers, and the US Public Health Service, along with the more convenient size midget impinger (Figure 3.1), was used before 1970 as the dust sampling method of choice in the United States. The impinger sampler collects dust particles with diameters larger than 0.7 μm aerodynamic diameter with efficiencies ranging from 68 to 95%, depending on the particle density and the specific solution used in the impinger (Jacobson *et al.*, 1970). Impingers collect *total dust* samples. A common total dust sampler currently in use is the 37-mm or 25-mm

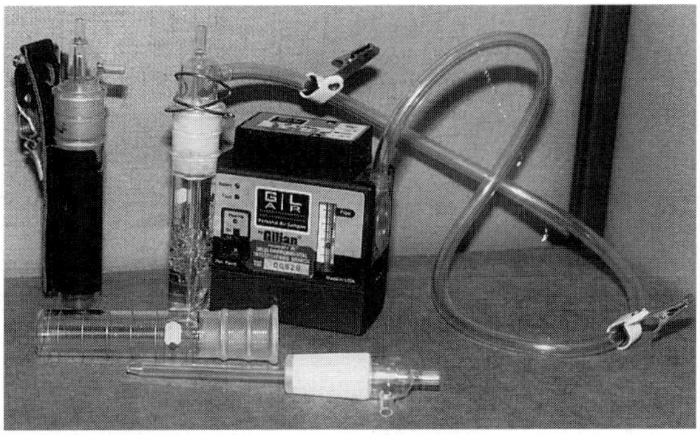

Figure 3.1 Total dust impinger sampler.

Figure 3.2 Total dust filter cassettes. Left, closed-faced; right, open-faced.

diameter filter housed in a plastic cassette (Figure 3.2), connected to an air sampling pump. The sample is collected by removing the protective end caps and sampling through the small 4-mm cassette opening for a known time, at a known air flow rate, usually from 1–2 l/min. Alternatively, the filter may be mounted in a three-piece cassette such that the entire end section of the cassette is removed for sampling giving a 37-mm or 25-mm open-faced filter for collection of the sample. Both sampling configurations have their respective problems. When sampling through the cassette's small-diameter opening, some large particles are not efficiently sampled. In the open-faced configuration, the sample is subject to contamination since some particles can be propelled onto the filter. Samples collected using mannequins in a wind tunnel showed that the amount of dust that enters the nose of the mannequin can be substantially greater than the amount of dust sampled by total dust samplers (Mark *et al.*, 1994). Clearly, the two configurations described are working as size-selective samplers and not collecting a true total dust samples. However, with the mannequin studies previously reported, it was shown that for large particles, up to 100 μm in diameter, the nose is only 50% efficient for collection of those particles (Vincent and Armbruster, 1981; Armbruster and Breuer, 1982; Vincent and Mark, 1982). Therefore, a more appropriate assessment of whole-body dust burden might be *inhalable dust*, where the sampler is designed to collect particles as efficiently as the nose. Such inhalable dust samplers have been designed and described elsewhere (Figure 3.3) (Mark and Vincent, 1986; Vincent, 1995).

Another problem with characterizing total dust concentrations occurs because the size distribution is constantly changing. Dust may be generated

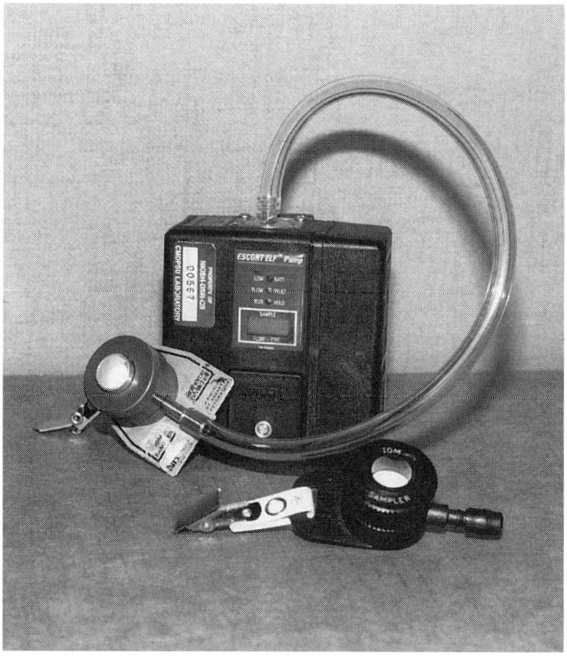

Figure 3.3 Inhalable dust sampler.

continuously or intermittently by the dust-generating process with some particular size-distribution. However, once airborne, the particles begin to settle due to gravitational forces according to the relationship:

$$c_s = \frac{\rho_p \, d_p^2 \, g}{18\,\mu}$$

where c_s is the particle settling velocity, ρ_p is the particle density, d_p is the particle diameter, g is the gravitational constant, and μ is the viscosity of the air (Friedlander, 1977). Since the settling velocity of particles is dependent on particle size, over time with no new dust generation, the airborne size distribution will appear depleted of large particles and enriched in small-sized particles, perhaps setting up a vertical spatial concentration gradient in the environment.

In spite of these inherent problems, measures of total dust have been successfully used for exposure control monitoring, even where the health risk is primarily associated with the respirable dust fraction. For example, in the Peoples Republic of China, a total dust device has been used for over 30 years to track and control dust exposures (Figure 3.4). In the US, for many years the prescribed sampling system for silica dust monitoring was the impinger combined with particle counting. By controlling the total dust concentration so that the levels were below recommended

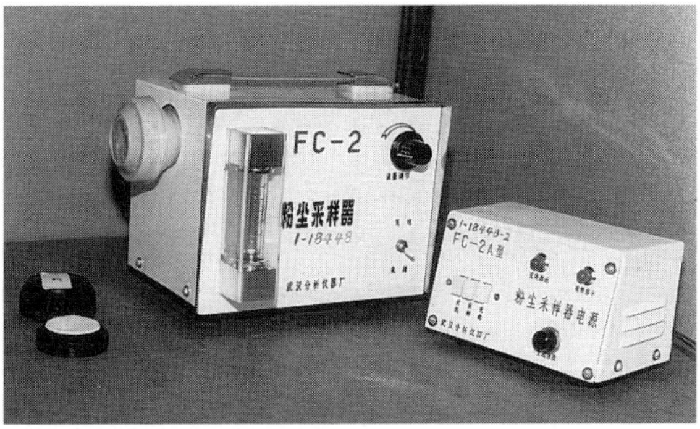

Figure 3.4 Total dust sampler – Chinese sampler.

limits, the incidence of silicosis was significantly reduced (Costello and Graham, 1988). This sampling strategy was successful because the concentration of total dust and the concentration of respirable dust and its chemical components are often sufficiently correlated for a particular dust-generating operation (Wu *et al.*, 1992). It is this principle that simplifies dust characterization tasks for occupational hygienists involved in controlling dust exposures and those involved in monitoring compliance and non-compliance with specific standards. Additionally, where systemic toxicity of soluble particles or local effects in the head or upper airways may be a concern, exposure control should be based on the total or inhalable dust concentration.

Respirable dust

In 1970, the American Conference of Governmental Industrial Hygienists, (ACGIH) proposed a specific definition for respirable dust based on human pulmonary deposition data. This definition specified the fraction of particles that could pass through a size-selective pre-separator and be collected on a filter for gravimetric analysis by specifying several points on the respirable size-selection curve (Lippmann, 1970):

Aerodynamic diameter (*μm*)	*% Passing selector*
≤ 2	90
2.5	75
3.5	50
5.0	25
10	0

Another early approach to the definition of respirable dust was to define it as that dust collected by a particular instrument. In the United States, the 10-mm

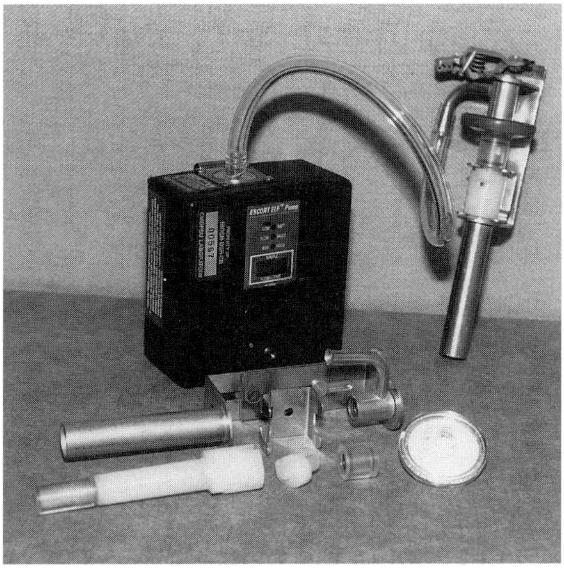

Figure 3.5 Respirable dust sampler – 10-mm nylon cyclone.

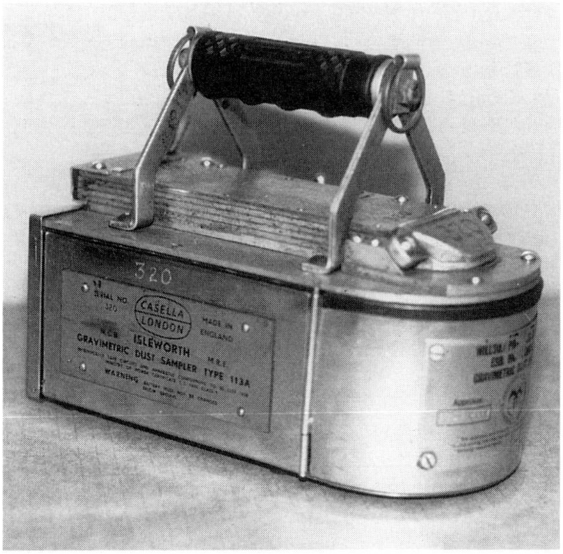

Figure 3.6 Respirable dust sampler – horizontal elutriator (MRE).

nylon cyclone (Figure 3.5) has been used as the size-selective pre-separator which defines respirable dust. The British Medical Research Council (BMRC) adopted the horizontal elutriator, also known as the Medical Research Establishment instrument or MRE (Figure 3.6), to provide the basis for the respirable dust definition in Britain. The MRE instrument is portable, battery operated, and has collection properties such that particles less than 1 μm aerodynamic diameter are collected with nearly 100% efficiency, particles with 5 μm aerodynamic diameters are collected with 50% efficiency, and particles larger than 7 μm aerodynamic diameter are excluded from collection on the filter.

Defining respirable dust by specifying a particular sampling system is not without its problems. For example, the 10-mm nylon cyclone's collection efficiency curve shifts with changes in operating flow rate. From the outset, there has been controversy about whether the flow rate should be 1.4 l/min, 1.7 l/min or 2.0 l/min (Bartley and Brauer, 1982; Caplan *et al.*, 1977a; Knuth, 1969; Ettinger *et al.*, 1970; Tomb and Raymond 1970; Tomb and Treaftis, 1976). After examining the flow rate bias curves, which measure the difference between the ACGIH respirable dust definition and the sampler's observed performance, the US National Institute for Occupational Safety and Health (NIOSH) specified 1.7 L/min as the flow rate for respirable dust sampling using the 10-mm nylon cyclone preseparator (NIOSH, 1994b). However, in selecting a sampler for underground coal mines, where the US coal mine dust standard was based on data obtained from Britain (using MRE-collected dust), it was observed that if a cyclone sampler was operated at 2.0 L/min, and a correction factor of 1.38 was applied, an acceptable correlation to MRE-collected samples was obtained for a wide range of coal mining settings (Tomb *et al.*, 1973; Caplan *et al.*, 1977b). This procedure, using the 10-mm nylon cyclone operated at 2.0 L/min and multiplying the concentration by a constant factor of 1.38, has been the legal definition of respirable coal mine dust in US coal mines since the early 1970s. The use of the '1.38 correction factor' has been disputed, with some investigators noting that these same data could be used to obtain different correction factors (NRC, 1980). The use of correction factors to adjust measurements made with samplers having varying particle collection characteristics can create some unusual biases in the data. For example, the 38% positive correction applied to cyclone-collected respirable dust provides adjustment for dust in the size range from 3.5 to 7 μm, where the MRE instrument has a higher collection efficiency than the cyclone. However, if the mine environment contains a surplus of particles in the smallest size ranges, below 2 μm, where both instruments collect essentially 100% of the particles, then the 1.38 factor introduces a positive 38% bias in the measurement (Bartley and Breuer, 1982). This scenario becomes particularly relevant with the introduction of diesel-powered equipment into coal mining environments. Diesel equipment produces a particle fume that is usually smaller than 1 μm and can represent a sizable fraction of the collected dust (McCawley and Cocalis, 1986; Marple *et al.*, 1991).

In summary, whenever two samplers, such as the MRE or the 10-mm nylon cyclone, are used to sample the same environment, the conversion between measured concentrations will be dependent on the particle size distribution.

International size-selective dust definition

Recognizing the limitations of the existing respirable dust definitions, the international community has attempted to standardize the absolute definition of respirable dust (ISO, 1981). These efforts also extend the definitions to other physiologically significant fractions, and provide a criterion for assessing the adequacy of instruments used to evaluate concentrations. The ACGIH Air Sampling Procedures Committee developed criteria for size-selective sampling (Phalen, 1984). They defined the rationale for size-selective sampling and the recommended definitions and sampler performance characteristics for sampling inhalable (formerly called *inspirable*), thoracic and respirable mass fractions (Hinds, 1984; John, 1984; Lioy *et al.*, 1984; Soderholm, 1984). An accuracy criterion that could be applied to specific instruments which meet the various size-selective criteria was provided (McCawley, 1984). The committee also established a logical framework for selecting the appropriate size fraction based on the dust-hazard being controlled (Stuart *et al.*, 1984). Others tested these criteria against a range of observed work place particle-size distributions and found that the acceptability criteria were too broad and found two samplers acceptable even though they may report dust concentrations as different as 0.71 mg/m^3 and 4.3 mg/m^3 (Bartley and Doemeny, 1986).

The International Standards Organization (ISO), Comité Européen de Normalisation (CEN) and ACGIH cooperated in the most recent efforts to establish specific definitions for size-selective dust measurements (Soderholm, 1989). In an attempt to harmonize the various definitions in common use, e.g. the ACGIH and the BMRC definitions of respirable dust, several compatible dust size definitions were developed and adopted by the various national and international organizations (Figure 3.7) (ISO, 1991; CEN, 1993; ACGIH, 1995a).

Inhalable (or inspirable) fraction

The *inhalable* or *inspirable* fraction describes those particles that enter through the nose or mouth during inhalation. The definition was developed based on data collected using mannequins of human heads and torsos in a wind tunnel (Vincent and Armbruster, 1981; Armbruster and Breuer, 1982; Vincent and Mark,

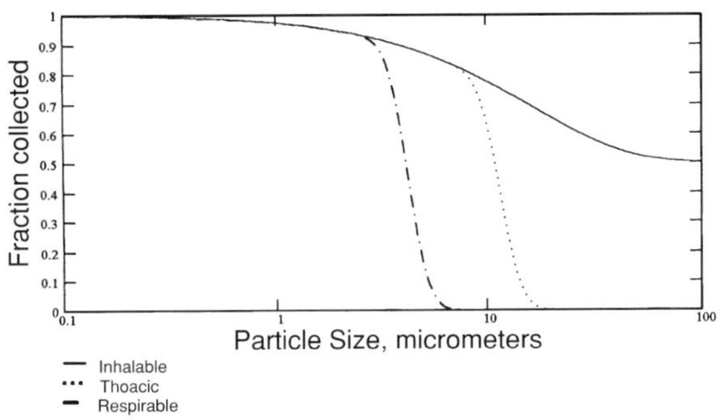

Figure 3.7 International standard definitions for inhalable, thoracic and respirable dust.

1982). The inhalable fraction is described by the equation:

$$SI(d) = 0.5\ (1 + e^{-0.06d})$$

where $SI(d)$ is the inhalable particle sampling efficiency for particles with aerodynamic size of d in μm. Using the inhalable size fraction for monitoring control of dust exposures is appropriate for substances that are hazardous or toxic when deposited anywhere in the respiratory system including nose and head, or when they are ingested. This creates some complications with regards to prior occupational exposure limits that may have been based on total dust sampling. The ratio of side-by-side concentrations measured by total dust sampling and inhalable dust sampling have been reported to range between factors of 0.7 to 3.4 (Vinzents *et al.*, 1995). As with the case of different respirable dust samplers, the relationship between total dust sampling and inhalable dust sampling is dependent on other factors such as the overall particle size distribution and wind speed and direction with respect to the sampling port (Lidén and Kenny, 1994).

Thoracic size fraction

The thoracic fraction was previously defined by a cumulative lognormal function with a mass median diameter of 10 μm and a geometric standard deviation, σ_g of 1.5 (Raabe, 1984). This definition includes those particles small enough to penetrate past the nasal–pharyngeal passages of the head, and includes particles that deposit in the trachea–bronchial tree and the alveolar region of the lung. The ISO, CEN and ACGIH definition of the thoracic fraction begins by considering only those particles first determined to be inhalable, and then applying the cumulative normal frequency distribution as:

$$ST(d) = SI(d\ [\ 1 - F(x)\ ]$$

where $ST(d)$ is the thoracic particle sampling efficiency for particles with an aerodynamic diameter of d, Γ = 11.64 μm (the geometric mean of the log-normal distribution), Σ = 1.5 (the geometric standard deviation of the log-normal distribution), $SI(d)$ is the inhalable dust sampling efficiency, and the function, $F(x)$ which is defined by:

$$F(x) = \int_0^x \frac{e^{-(x^2/2)}}{\sqrt{2\pi}}\, dx$$

where,

$$x = \frac{\ln(d / \Gamma)}{\ln(\Sigma)}$$

Respirable size fraction

The respirable dust definition that was adopted by the ISO, CEN and ACGIH is based on the cumulative probability density function, $F(x)$, of a standardized normal random variable, e.g. based on this function, the respirable fraction is defined by:

$$SR(d) = SI(d)\ [\ 1 - F(x)\ ]$$

where Γ = 4.25 μm (the geometric mean of the log-normal distribution), and Σ = 1.5 (the geometric standard deviation of the log-normal distribution), and $SI(d)$ is the inhalable dust sampling efficiency. The function $F(x)$, and x are as defined in the preceding section.

Dust deposition

Each of these definitions describes ‘penetration’ curves for particles, based on their aerodynamic diameters. Particles with aerodynamic diameters much smaller than the median cut-point are collected with almost 100% efficiency. Samplers designed to measure penetration do not measure ‘deposition’ in any particular region of the lung. Alveolar deposition is less than 100% for particles between 0.1 and 1.0 μm aerodynamic diameter because many particles in this size range are subsequently exhaled (Chan and Lippmann, 1980; Stahlhoffen, *et al.*, 1980). Respirable dust samplers will overestimate the fraction deposited in the alveolar region of the lung for sub-micrometer size particles. The fraction of dust which deposits in the lungs shows a marked reduction from 80–95% deposition of 5 μm particles, dropping to 20–40% deposition for 0.5 μm particles, then rising to between 50–70% deposition for particles smaller than 0.1 μm (Chan and Lippmann, 1980). The deposition curve shifts downward as settling and impaction become less important for depositing particles in the smaller size range, while the process of diffusive deposition becomes more important for the particles smaller than 0.1 μm. For mechanically generated dusts which exist primarily in the 2 to 15 μm size-range, this issue may be insignificant. However, for fumes generated through diesel combustion or by welding processes, there could be a significant difference between measured exposure and deposited dose (Soderholm and McCawley, 1990). Likewise, tracheobronchial deposition is not measured by the thoracic fraction samplers because those samplers include and collect particles that would be deposited in the alveolar region, i.e. the respirable fraction. Extrathoracic deposition (the fraction deposited in the head and pharynx, down to and including the larynx, i.e. the 'nasal–pharyngeal' fraction) is not measured with an inhalable fraction sampler because that fraction includes particles that would penetrate further and be deposited in the tracheobronchial and alveolar regions of the lung (Raabe, 1984).

Figure 3.8 PM_{10} sampler – environmental aerosols.

Other size-selective fractions

(a) Ambient sampling using the PM_{10} sampling system

The US Environmental Protection Agency (EPA) specified in its National primary and secondary air quality standards for particulate matter, that particulate matter shall be measured in the ambient air as PM_{10} particles with an aerodynamic diameter less than or equal to a nominal 10 μm (EPA, 1995). The requirements for a PM_{10} sampler (Figure 3.8) were based on the lung deposition curve proposed by Chan and Lippmann (1980), Kashdan *et al.* (1986) and Liu and Pui (1986).

(b) Vertical elutriator for cotton dust sampling

Another instrument-specified size fraction is collected by the Lumsden–Lynch vertical elutriator (Figure 3.9). This device was designed to collect airborne cotton dust with a mass-median diameter of 15 μm. The vertical elutriator is specified in US regulation as the standard method for measuring airborne cotton dust in textile mills and selected secondary cotton processing industries (OSHA, 1995). The size-selection criteria is similar to the thoracic fraction adopted by the ACGIH, CEN and the ISO.

Criteria for selection of an appropriate size

The selection of an appropriate sampling system should be based on the criteria that the measured quantities are related to the risk being controlled (Roach, 1953; Ayer, 1969). The selection of one or more of the size-selective criteria could be guided by the following classification scheme:

- Respirable dust sampling – suggested when the aerosol is insoluble and the disease is associated with the gas exchange regions of the lungs, as in fibrotic diseases like silicosis or coal workers' pneumoconiosis, or emphysema.
- Thoracic dust sampling – suggested when the aerosol is believed to produce asthma, to aggravate chronic bronchitis, or to be responsible for inducing bronchogenic tumors. Thoracic sampling is appropriate if the material which deposits in the head airways does not contribute to its toxicity.
- Inhalable or inspirable dust – suggested when the aerosol is likely to be absorbed by the body wherever it deposits, or if the dust is associated with nasal cancer as in wood workers (ISO, 1981; Stuart *et al.*, 1984).

As an example, for environmental particulate exposures, concern should be based on health effects including asthma, bronchitis, emphysema, fibrosis and pneumoconiosis. Based on these broad needs, the

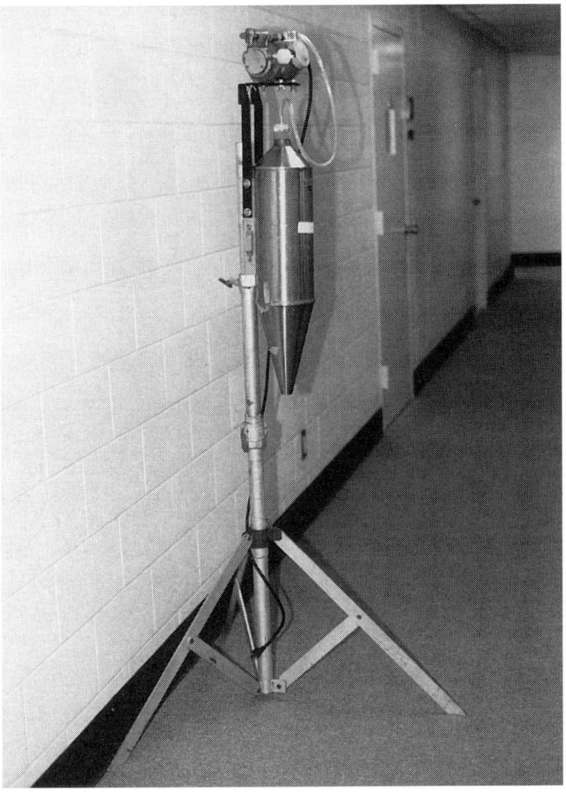

Figure 3.9 Lumsden–Lynch vertical elutriator.

selection of PM_{10} sampling for the National Ambient Air Quality Standard for particulate is appropriate. Alternatively, the most appropriate size-selective criterion for measuring silica concentrations, exclusive of other particulate exposures, would be the respirable dust criterion. If the sampled aerosol has a relatively stable size distribution, then thoracic sampling (PM_{10}) would provide an acceptable surrogate for respirable mass sampling when measuring ambient exposures.

THE COMPOSITION OF DUST

Crystalline silica in dust particles

Because the earth's crust is predominantly composed of crystalline silica, whenever earth is disturbed through mining, milling, or earth moving operations, there exists a potential for exposure to crystalline silica. Historically, the development of dust sampling programs and systems revolve around the need to sample quartz-containing dusts.

The ACGIH recognized that the health risk of exposure to dust was directly related to the total dust concentration and to the concentration of crystalline silica, also known as quartz, in the dust. In 1962, after reviewing available data on dust-exposed populations, the ACGIH adopted a Threshold Limit Value (TLV) for dust containing quartz (ACGIH, 1986):

$$\mathrm{TLV} = \frac{250}{\%\mathrm{quartz}+5}\ \mathrm{mppcf}$$

To apply this exposure limit, the concentration of crystalline silica (%quartz) was quantified through chemical analysis of collected airborne samples. The TLV for that exposure setting was then calculated using the formula above. The total dust level was then measured using the impinger sampler, and that result was compared with the calculated TLV. For example, if the %quartz was computed as 5%, the calculated TLV would be 25 mppcf. Any dust sample over this amount would exceed the exposure limit. In 1970, to make the TLV consistent with a nuisance dust standard of 30 mppcf, the formula was revised to (ACGIH, 1986):

$$\mathrm{TLV} = \frac{300}{\%\mathrm{quartz}+10}\ \mathrm{mppcf}$$

For both of these formulas, the measurement method was the impinger, coupled with particle count analysis (ACGIH, 1986). Based on growing body of literature in the US, summarized by Morrow (1964), and the findings of the Johannesburg Conference (Orenstein, 1960), in 1968 the ACGIH proposed a respirable dust TLV based on size-selective gravimetric sampling. The revised formula, based on the particle count formula, would be designed to provide approximately equal protection from dusts containing quartz, and to limit the exposure to total airborne dust. Based on findings of the US Public Health Service, a value of $0.1\ \mathrm{mg/m^3}$ of quartz was found equivalent to the 10 mppcf exposure concentrations observed from the Vermont Granite Shed study where the quartz was 25% (Ayer *et al.*, 1968; Ayer, 1995). Besides the quartz limit, it was decided to limit size-selected respirable dust to $5\ \mathrm{mg/m^3}$, so the total dust concentration would not exceed $15\ \mathrm{mg/m^3}$, based on an assumption that the respirable dust was approximately one-third of the total dust concentration. Combining these limits into an analogous formula led the ACGIH to adopt the following formula for control of respirable dust (ACGIH, 1986):

$$\mathrm{TLV} = \frac{10}{\%\mathrm{quartz}+2}\ \mathrm{mg/m^3}$$

The %quartz value in the formula was to be measured from airborne samples except where it was decided that other methods were acceptable (Ayer,

1969). In 1986, the ACGIH adopted a revised silica standard that requires the direct measurement of respirable crystalline silica, comparing the result against a 0.1 mg/m^3 TLV (ACGIH, 1986). The revised TLV was proposed without reference to the mixture formula, that was implicitly included in the former TLV formula (Hearl, 1996).

Particle size dependence in analysis of crystalline silica

There are three principal analytical methods used to analyze crystalline silica. The wet-chemistry method is based on the differential solubility of crystalline silica, various silicates, and amorphous silica when treated with hot phosphoric acid. Following the dissolution of the amorphous silica and silicates, the remaining crystalline 'free silica' was quantified either by weighing the residue (Talvitie, 1951) or, more precisely by colorimetric determination (Talvitie and Hyslop, 1958; Talvitie, 1964). In an attempt to find suitable alternatives to the Talvitie method, investigators examined other spectral methods such as infrared spectroscopy (IR) and X-ray diffraction (XRD). One study examined particle counting using a light microscope with dispersion staining (Edwards, 1965). In a comparative study of IR and XRD methods, it was found that both IR and XRD could quantify crystalline silica with approximately equal precision and ruggedness (NIOSH, 1983). However, both methods were subject to interference by other common minerals, and both exhibited a response dependent on the particle size of the sample. When a pure silica standard (Min-U-Sil®) was used to establish a calibration curve for the XRD method, the instrumental response for the 10 μm Min-U-Sil® was 1.6 times the response obtained for the 5 μm Min-U-Sil® (NIOSH, 1983). Similarly, it was found that the quartz absorbance peaks observed using the IR method varied by as much as 30% for particle diameters varying from 0.65 μm to 5.8 μm (Lorberau, 1989). Due to these response characteristics of the methods, the choice of a suitable reference standard was critical to making accurate measurements of silica concentrations (Verma *et al.* 1992). To improve the standards for XRD analysis of silica, the National Bureau of Standards produced a Standard Reference Material (NBS-SRM#1878) by cycloning Min-U-Sil® 5, which was an α-quartz sold by U.S. Silica, West Virginia. The particle size distribution of the SRM approximated that found on respirable dust samples, thus reducing systematic bias from particle-size effects on instrument response. However, where the particle size distribution has a low dispersion (i.e. a small geometric standard deviation of the distribution, approaching a monodisperse aerosol) the size-dependence of the analysis could become important. For example with ground silica (silica flour) samples, the particles tend to be very small and the X-ray diffraction response is weak. Comparison of such finely ground samples to the larger respirable dust standard reference material will result in an underreporting of the true concentration in the sample. Selection of an appropriately-sized standard reference is essential where crystalline silica analysis is required (Bhaskar *et al.*, 1994).

Coal dust sampling

The selection of an appropriate sampler for coal mine dusts is based on the basic criteria that the measured quantity should be related to the risk of developing occupational lung disease from that exposure (Roach, 1953; Ayer, 1969). A variety of devices has been used over the years to quantify airborne dust levels including: filters, impactors, impingers, elutriators, electrostatic precipitators, thermal precipitators and cyclones (Lippmann, 1983; Lioy *et al.*, 1984; Glenn and Kraft, 1986).

In British coal mines, thermal precipitators were used beginning in 1936, to monitor airborne dust levels (Fay and Rae, 1959; Roach, 1959; Fay and Ashford, 1964). The samples were analyzed microscopically. Dust concentrations were assessed by particle counting, reporting the number of particles in the size range 1–5 μm diameter per cubic foot of mine air. Following the recommendations from the Johannesburg Pneumoconiosis Conference that mine dust sampling be based on respirable size-selective gravimetric measurements (Orenstein, 1960; Chamberlain *et al.*, 1971), the British National Coal Board (NCB) chose to use the Medical Research Establishment MRE dust sampler, a horizontal elutriator described earlier. Since 1965, the MRE system has been used to measure dust exposures in British mines.

Based on the toxicity of coal mine dust reported in British studies (Jacobsen *et al.*, 1971) the US Federal Coal Mine Health and Safety Act of 1969 as amended in 1977 (US Code, 1977) specified that the US standard for coal mine dust would be 3.0 mg/m^3 on the effective date of the Act (1969), and the standard would be reduced to 2.0 mg/m^3 three years after the date of enactment of the Act. The Act did not explicitly define respirable dust, stating only that respirable dust means dust measured with a device approved by the Government. As noted earlier, the chosen instrument was a 10-mm nylon cyclone and a 37-mm filter in series with a pump operated at 2.0 l/min air flowrate.

The sampling procedure used to assess a coal miner's exposures to crystalline silica in US mines is indirect. A respirable dust sample is analyzed for crystalline silica. If the coal mine dust sample contains greater than 5% crystalline silica, a reduced respirable dust standard is calculated and compared with the respirable dust concentration measured by gravimetric methods. The MSHA Respirable Dust Standard (RDS) is reduced according to the formula (MSHA, 1995):

$$\text{RDS} = \frac{10\,\text{mg/m}^3}{\text{Percent Quartz}}$$

A RDS established for a particular work place may be enforced for extended periods of time (up to one year) without additional analysis of the percentile composition of silica. This approach assumes that the concentration of silica is relatively constant in the material being processed. Some have argued that this is a poor assumption for many mining situations (Corn *et al.*, 1985).

MIXED DUST EXPOSURES

Other toxic substances may be present in the environment, and may exist as airborne particulate. Particulate, liquid aerosol or vapors capable of adsorbing on the surface of airborne particles may work as carriers for the condensed vapors, transporting the substances and depositing them in the deepest regions of the lungs. Depending on environmental conditions, particle-adsorbed materials may, after collection on a filter, be desorbed back into the environment (Soderholm and Ferron, 1992). This would introduce a *negative* bias as an artifact of the sampling system. These types of problems have been observed even when sampling high-molecular weight materials such as some polynuclear aromatic hydrocarbons that are often associated with diesel emissions. As a result, sampling methods for such substances may require the use of a sorbent back-up tube placed in series behind the particulate filter in order to capture vapors desorbing from particle surfaces (NIOSH, 1994c; Soderholm, 1995).

Early control limits on dust, in the generic sense, were not based on toxicological data. Maximum allowable concentration limits were set for generic dust, termed 'nuisance dust' or 'inert or nuisance particulate' at 50 mppcf or 15 mg/m^3 in order to gain some control over excessively dusty operations. Part of the definition of 'nuisance dust' was that they did not produce significant disease or toxic effect (e.g. air spaces remain intact, no fibrosis or scarring, and tissue reactions are reversible) when inhaled in controlled amounts. Recent inhalation studies have shown that high exposures to dust, generically, can significantly retard the lungs' protective clearance mechanisms, leading to various additional dysfunctional and pathologic changes (Morrow *et al.*, 1991). Based on these findings there is justification for sampling and controlling respirable dust exposure *per se*, even when there are no specific observable toxic constituents in the dust.

Some specific effects caused by dust exposures with specific chemical compositions are cataloged in Tables 3.1 and 3.2 (Langer, 1986). From the description of the expected health response, the appropriate size-selective sampling criteria can be selected using the decision logic outlined on p. 37.

INTERNATIONAL COMPARABILITY OF DUST MEASUREMENTS

As described in the preceding sections, there are a number of choices available for exposure metrics, for instruments to use in quantifying those exposures, and for subsequent analysis to further describe the composition, crystallinity and morphology of the dust particles. However, there is an additional factor that confounds attempts to achieve international comparability for dust measurements and exposure standards. The sampling strategy and protocol itself can have a dramatic influence on the measured exposure. These elements of the sampling strategies that can influence measurements include:

- the duration for sample collection;
- the location for sample collection; and
- the frequency of sample collection and statistical treatment of the results.

Sample duration is sometimes dictated by the sampling system being used as well as the prevailing exposure concentration. For measurement of very low concentrations by gravimetric methods, large air sample volumes are required to concentrate a sufficient mass on the sampling substrate for quantification. In cases of high exposure concentrations, relatively short sampling periods are required to avoid overloading the sampling media, resulting in an unusable or unreliable quantitative result.

By regulation, US coal mine dust samples are required to be collected on a 'portal-to-portal' basis, i.e. sampling the entire work-shift from the time the worker enters the workplace to the time they leave (MSHA, 1995). Generally, industrial hygienists in the US will attempt to collect full-shift dust samples for comparison with 8-hour time-weighted average exposure limits. However, as is done in Chinese dusty trades, the sampling strategy may be limited to those periods of time when the dust-generating tasks are in progress (Wu *et al.*, 1992). Obviously for such task-based sampling schemes, higher levels will generally

Table 3.1 Natural materials associated with human disease: minerals

Rock type	*Components/uses*	*Comments*
Amosite Anthophyllite Crocidolite Chrysotile	Used as asbestos	Asbestosis, lung cancer; cancer of the gastrointestinal tract; pleural and peritoneal mesothelioma; possible increase in other malignancies.
Bauxite	Hydrated aluminum oxides	Some reports of lung scarring
Biotite Muscovite Phlogopite	Insulation and filler	Pulmonary fibrosis; silicosis; strong association with free silica, likely a major factor in producing fibrosis.
Coal	Anthracite Bituminous ± quartz	Coal workers' pneumoconiosis; 'miner's consumption'; focal emphysema; anthracosis, silicoanthracosis; may be related to rank (anthracite >bituminous); 'melanosis' or 'black lung'; greater incidence of tuberculosis; may progress to 'massive fibrosis'.
Diatomaceous earth	Opaline diatom fragments	Some lung scarring developed; much greater in processing when calcined (conversion of opal to cristobalite); silicosis with progressive massive fibrosis.
Feldspar	Ceramics	Silicosis; often attributed to the included crystalline silica.
Fluorite	Industrial flux	Fluorosis; silicosis.
Fullers earth	Butonites and related morillonites ± quartz	Pneumoconiosis without massive fibrosis and nodules (diffuse); mottled X-ray appearance; some related silicosis.
Granite Quartzite Sandstone Slate	All with large amounts of quartz	Silicosis; silico-tuberculosis; nodular silicosis; fibrosis; enlarged and hardened lymph glands. Silicotic nodules in spleen
Graphite	Crucibles; high temperature facings; electrodes; paints	Graphite pneumoconiosis; tuberculosis; resembles silicosis.
Gypsum Anhydrite	Gypsum ± evaporites Anhydrite ± sulfur	Some bronchitis; when calcined for industrial purposes, toxic effect to skin increases some irritation to eyes, nose and pharynx.
Kaolin	Ceramics; filler	Some lung scarring observed, but only in areas associated with crystalline silica.
Kyanite	Ceramics	Some lung scarring observed.
Limestone Marble Dolomite	Calcite ± quartz Dolomite ± quartz	Some bronchitis; some emphysema, some scarring reported; when calcined for industrial purposes, toxicity increases; caustic burns; dermatitis; ulceration of skin; injury to conjunctiva and cornea.
Nepheline	Industrial uses	Nephelosis; some lung scarring.
Olivine	Industrial uses	Silicosis; some lung scarring.
Pumice	Volcanic glass, some devitrification to quartz	Resembles silicosis; some linear scarring.
Silica Quartz Cristobalite Tridymite Chalcedony Flint	Abrasive; industrial uses	Silicosis; 'potters' asthma'; 'potters' consumption'; silicotic nodules in spleen; silico-tuberculosis; progressive pulmonary fibrosis; progressive pulmonary fibrosis. Cristobalite and tridymite are more fibrogenic than quartz.
Talc	Industrial uses; filler	Talcosis; talc pneumoconiosis.

From Langer (1986).

Table 3.2 Natural materials associated with human disease: elements

Metal	*Mineral*	*Comments*
Aluminum	Bauxite ore Corundum	Aluminosis; some lung scarring; pneumoconiosis; emphysema; crystalline silica associated with corundum deposits.
Arsenic	Cobaltite Enargite Realgar Orpiment Arsenopyrite Smaltite	Local skin and mucous irritant; carcinogen; anemia (hemolytic agent); hemoglobinuria; mixed dust pneumoconioses.
Beryllium	Beryl Chrysoberyl Bertrandite	Berylliosis; pulmonary lesions; acute poisoning; granuloma; pneumonitis; ore minerals are associated with crystalline silica.
Cadmium	Greenockite	Renal and pleural involvement.
Chromium	Chromite	Some reports of lung cancer; ore may be associated with chrysotile asbestos.
Cobalt	Smaltite Linnaeite Cobaltite Erythrite	Some reports of excess deaths due to lung cancer and cancer of the main bronchus; hard metal disease.
Iron	Hematite Magnetite Taconite	Siderosis; some reports of increased lung cancer among taconite ore miners.
Lead	Galena Cerussite Anglesite	Pneumoconiosis; diseases of the central nervous system; nephritis; 'plumbism'; anemia.
Manganese	Pyrolusite Braunite Manganite Hausmanite Rhodochrosite	Pneumoconiosis; affects the central nervous system (Parkinsonism syndrome); usually high rate of pneumonia among ore processors.
Mercury	Cinnabar	A range of systemic diseases; nephrosis (renal lesions); salivation; vertigo; paralysis; 'hatters shakes'; erethism; stomatitis; 'mercury poisoning'; perforation of nasal septum.
Nickel	Pentlandite Niccolite Millerite Garnierite	Pneumoconiosis; some observations indicate a higher than normal lung cancer rate; higher incidence of lung and nasopharynx cancer.
Phosphorus	Apatite	Minor scarring of lungs; some neurological disorders; phosphine and pesticide forms toxic-to-lethal.
Platinum	Platinum Sperrylite	Some lung scarring observed which resembles asbestosis.
Selenium	Tiemannite Guanajuatite Clausthalite Naumannite Eucairite Chalcomenite	Pneumoconiosis; severe irritation of nose and eyes; gastrointestinal disorders; dental caries.
Silver	Silver Acanthite	Argyria-discoloration of the skin to a pale gray or blue-gray.

continued

Table 3.2 continued

Metal	*Mineral*	*Comments*
Tellurium	Montanite Emmonsite Durdenite Tetradymite	Pneumoconioses; some lung scarring; gastrointestinal disorders; renal disorders.
Tin	Cassiterite	Severe X-ray changes in miners; tin pneumoconiosis.
Titanium	Rutile Sphene Ilmenite	Severe lung scarring, titaniosis; lung scarring;
Tungsten	Wolframite Tungstite	Lung scarring; tungsten pneumoconiosis.
Uranium	Uraninite Carnotite Pitchblende Thorium ores Vanadium ores	Excess lung cancers.
Vanadium	Vanadinite Carnotite	Lung scarring; respiratory irritant; irritant to eyes; increased susceptibility to pneumonia.
Zinc	Sphalerite	Shortness of breath; some minor lung changes; pneumoconiosis.

be observed compared with systems that sample and average exposures over a full work shift. Where discretion is allowed, some occupational hygienists use their professional judgement to remove the samplers or otherwise stop the sampling and discount certain non-operational portions of a work shift, such as during the lunch hour.

Sampling location, which has a significant impact on the measured dust level, varies in different countries. In German mines, respirable dust is measured at the location of a working area in which the maximum dust concentration is expected. In British coal mines, dust samples are collected at a fixed point, located in the return airway approximately 70 m behind the face. In the former Soviet Union, dust with particle-sizes up to 74 μm are collected at a fixed point, determined by the type of mining, generally located near the cutting machine. In US coal mines, respirable dust is collected using personal samplers, attached to the 'high risk' miners on the working section. Since these measurement locations vary as noted, exposure limits and reported dust levels cannot be directly compared between these countries (Prinz and Stolz, 1990).

The sampling frequency and the statistical treatment of measured concentrations can also affect the reported dust values. For example, when partial-period sequential samples are collected, there are two accepted practices for dealing with the unsampled portions of the shifts. Compliance officers in the US may be required to assume the unsampled periods have 'zero' exposure when computing the 8-hour time-weighted average. However, if the data are reported from sources not related to regulatory compliance sampling, the occupational hygienist might well have decided to assume that the average concentration measured during the partial-shift period, extended during the balance of the shift. This is a reasonable assumption when the work operation continued uninterrupted during these unsampled time intervals. In some instances, data handling differences may not be noted with the reported data, making direct comparisons difficult or impossible.

DUST CONTROL

Control of dust exposure follows some common themes regardless of industry. Exposure control can be achieved by:

- substitution of less toxic materials in the process;
- suppression of dust generation;
- capture of the particles at their source;
- separating the dust generation source from the worker;
- diluting the concentration with supplied fresh air; and
- using personal protective equipment.

These various techniques for exposure control have been well described elsewhere (Talty, 1988; Burgess, 1994).

Water has often been used to suppress dust generation. Wetting down roadways, using water sprays on cutting machines, industrial drills and other devices where material is mechanically broken or ground are common dust control measures. Where material is drilled or cut, using sharpened bits can often reduce the amount of dust generated in the process. Good maintenance of cutting tools provides an easy means of dust control, preventing dust from being generated.

Perhaps the most common techniques for dust control involve ventilation systems (ACGIH, 1995b). By providing exhaust air movement away from the workers at sufficient velocity, small particles entrained in the air can be exhausted safely where no one is exposed. Barriers containing dusty operations, such as sandblasting booths, glove boxes or brattice curtains and permanent stoppings in mines, serve to separate the workers from the dust generation sources. Generally less effective are systems that supply copious amounts of fresh air to dilute the concentration of dust. Due to the high costs of air conditioning or heating the large quantities of fresh air that is needed for adequate dilution ventilation, this method is not frequently used.

When the use of engineering controls is not possible, or during times when the control systems are being installed, maintained or repaired, respiratory protective equipment may be used. This type of personal protective equipment may include: supplied air respirators, self-contained breathing apparatus, powered-air filter respirators, cartridge filter particulate respirators and disposable (paper mask) type respirators. These each have operational limits that should be researched before recommending an appropriate unit for a particular application.

REFERENCES

ACGIH (1986) *Documentation of the Threshold Limit Values and Biological Exposure Indices*, Fifth edition, American Conference of Governmental Industrial Hygienists (ACGIH), Cincinnati, Ohio, pp. 523–5.

ACGIH (1995a) *1995–1996 Threshold Limit Values (TLVs®) for Chemical Substances and Physical Agents and Biological Exposure Indices (BEIs®)*, American Conference of Governmental Industrial Hygienists (ACGIH), Cincinnati, Ohio.

ACGIH (1995b) *Industrial Ventilation: A Manual of Recommended Practice*, 22nd edition. American Conference of Governmental Industrial Hygienists (ACGIH), Cincinnati, Ohio.

Armbruster, L. and Breuer, H. (1982) Investigations into defining inhalable dust. *Ann. Occup. Hyg.* **26**(1–4), 21–32.

Ayer, H.E. (1969) The proposed ACGIH mass limits for quartz: review and evaluation. *Am. Ind. Hyg. Assoc. J.*, **30**, 117–25.

Ayer, H.E. (1995) Origin of the U.S. respirable mass silica standard. *Appl. Occup. Environ. Hyg.*, **10**(12), 1027–30.

Ayer, H.E., Sutton, G.W. and Davis, I.H. (1968) Size-selective gravimetric sampling in dusty industries. *Am. Ind. Hyg. Assoc. J.*, **29**, 336–41.

Bartley, D.L. and Breuer, G.M. (1982) Analysis and optimization of the performance of the 10 mm cyclone. *Am. Ind. Hyg. Assoc. J.*, **43**, 520–8.

Bartley, D.L. and Doemeny, L.J. (1986) Critique of 1985 ACGIH report on particle size-selective sampling in the workplace. *Am. Ind. Hyg. Assoc. J.*, **47**(8), 443–7.

Bhaskar, R., Li, J. and Xu, L. (1994) A comparative study of particle size dependency of IR and XRD methods for quartz analysis. *Am. Ind. Hyg. Assoc. J.*, **55**(7), 605–9.

Burgess, W.A. (1994) Philosphy and management of engineering control, in *Patty's Industrial Hygiene and Toxicology*, 3rd edn, Volume 3, Part A, (eds R.L. Harris, L.J. Cralley and L.V. Cralley), John Wiley & Sons, Inc., New York, pp. 129–71.

Caplan, K.J., Doemeny, L.J. and Sorenson, S.D. (1977a) Performance characteristics of the 10 mm cyclone respirable mass sampler: Part I – monodisperse studies. *Am. Ind. Hyg. Assoc. J.*, **38**, 83–95.

Caplan, K.J., Doemeny, L.J. and Sorenson, S.D. (1977b) Performance characteristics of the 10 mm cyclone respirable mass sampler: Part II – coal dust studies. *Am. Ind. Hyg. Assoc. J.*, **38**, 162–73.

Castranova, V., Dalal, N.S. and Vallyathan, V. (1996) Role of surface free radicals in the pathogenicity of silicosis, in *Silica and Silica-Induced Lung Diseases* (eds V. Castranova, V. Vallyathan and W.E. Wallace). CRC Press, Inc., New York, pp. 91–105.

CEN (1993) Comité Européen de Normalisation. Standard on workplace atmospheres. Size fraction definitions for the measurement of airborne particles in the workplace, EN 481. Brussels, Belgium.

Chamberlain, E.A.C., Makower, A.D. and Walton, W.H. (1971) New gravimetric dust standards and sampling procedures for British coal mines, in *Inhaled Particles III*, (ed. W.H. Walton), Unwin Brothers Ltd., Vol. 2, Old Woking, UK, pp. 1015–30.

Chan, T.L. and Lippmann, M. (1980) Experimental measurements and empirical modelling of the regional deposition of inhaled particles in humans. *Am. Ind. Hyg. Assoc. J.*, **41**, 399–409.

Corn, M., Breysse, P., Hall, T., Chen, G., Risby, T. and Swift, D.L. (1985) A critique of MSHA procedures for determination of permissible respirable coal mine dust containing free silica. *Am. Ind. Hyg. Assoc. J.*, **46**, 4–8.

Costello, J. and Graham, W.G.B. (1988) Vermont Granite workers' mortality study. *Am. J. Ind. Med.*, **13**, 483–97.

Edwards, G.H. (1965) Comparison of X-ray diffraction, chemical (phosphoric acid) and dispersion staining methods of determination of quartz in dust. *Am. Ind. Hyg. Assoc. J.*, **26**, 532–6.

EPA (1995) Environmental Protection Agency. Code of Federal Regulations: Protection of Environment, Title 40: Parts 50 and 53, U.S. Government Printing Office, Washington, DC.

Ettinger, H.J., Partridge, J.E. and Royer, G.W. (1970) Calibration of two-stage air samplers. *Am. Ind. Hyg. Assoc. J.*, **31**, 537–45.

Fay, J.W.J. and Ashford, J.R. (1964) Environmental exposures of British coal miners. *Arch. Environ, Health*, **9**, 144–59.

Fay, J.W.J. and Rae, S. (1959) The pneumoconiosis field research of the National Coal Board. *Assoc. Occ. Hyg.*, **1**, 149–61.

Friedlander, S.K. (1977) *Smoke, Dust and Haze*, John Wiley and Sons, New York.

Glenn, R.E. and Kraft, B. (1986) Air sampling of particulates, In *Occupational Respiratory Diseases* (eds J.A. Merchant, B.A. Boehlecke, G. Taylor and M. Pickett-Harner. DHHS (NIOSH) Publication No. 86–102, Superintendent of Documents, Washington, DC, pp. 69–82.

Hall, F.E., Kupel, R.E. and Harris, R.L., Jr (1965) Particle settling times in ethyl alcohol–water mixtures as affected by variables in impinger sampling. *Am. Ind. Hyg. Assoc. J.*, **26**, 537–43.

Halley, P.D. (1946) Are dust counts made by different laboratories comparable? *Ind. Hyg. Quarterly*, **7** (2), 15–22.

Hearl, F.J. (1996) Guidelines and limits for occupational exposure to crystalline silica, in *Silica and Silica-Induced Lung Diseases* (eds V. Castranova, V. Vallyathan, and W.E. Wallace), CRC Press, Inc., New York, pp. 15–22.

Hinds, W.C. (1984) Sampler efficiencies: inspirable mass fraction. *Ann. Am. Conf. Ind. Hyg.*, **11**, 67–74.

ISO (1981) International Organization for Standardization: Size Definitions for Particle Sampling: Recommendations of Ad Hoc Working Group Appointed by Committee TC 146 of the International Standards Organization. *Am. Ind. Hyg. Assoc. J.*, **42**, A64–8.

ISO (1991) International Organization for Standardization: Air Quality – Particle Size Fraction Definitions for Health-Related Sampling, CD 7708, Geneva.

Jacobsen, M., Rae, S., Walton W.H. and Rogan, J.M. (1971) The relation between pneumoconiosis and dust-exposure in British coal mines, in *Inhaled Particles III*, Vol. 2, (ed. W.H. Walton), Unwin Bros., Old Woking, UK.

Jacobson, M., Terry, S.L. and Ambrosia, A. (1970) Evaluation of some parameters affecting the collection and analysis of midget impinger samples. *Am. Ind. Hyg. Assoc. J.*, **31**, 442–5.

John, W. (1984) Sampler efficiencies: thoracic mass fraction. *Ann. Am. Conf. Ind. Hyg.*, **11**, 75–9.

Kashdan, E.R., Madhav B.A.R., Purdue, L.J. and Rehme, K.A. (1986) Interlaboratory evaluation of two inlets of sampling particles less than 10 μm. *Environ. Sci. Technol.*, **20**(9), 911–16.

Knuth, R.H. (1969) Recalibration of size-selective samplers. *Am. Ind. Hyg. Assoc. J.*, **30**, 379–85.

Langer, A. (1986) Characterization and measurement of the industrial environment: Mineralogy, in *Occupational Respiratory Disease*, (eds J.A. Merchand, B.A. Boehlecek, G. Taylor and M. Pickett-Harner), DHHS (NIOSH) Publication No. 86–102, U.S. Government Printing Office, Washington, DC.

Lidén, G. and Kenny, L.C. (1994) Errors in inhalable dust sampling for particles exceeding 100 μm. *Ann. Occup. Hyg.*, **38**(4), 373–84.

Lioy, P.J., Lippmann, M. and Phalen, R.E. (1984) Rationale for particle size-selective air sampling. *Ann. Am. Confl. Ind. Hyg.*, **11**, 27–34.

Lippmann, M. (1970) 'Respirable' dust sampling. *Am. Ind. Hyg. Assoc. J.*, **31**, 138–159.

Lippmann, M. (1983) Size-selective health hazard sampling, in *Air sampling Instruments*, (eds P.J. Lioy, and M.J.Y. Lioy), 6th edn, ACGIH, Cincinnnati, Ohio, pp. H1–H22.

Liu, B.Y.H. and Pui, D.Y.H. (1986) Aerosol sampling and sampling inlets, in *Aerosols – Research, Risk Assessment and Control Strategies*, (eds S.D. Lee, T. Schneider, L.D. Grant and P.J. Verkerk), Lewis Publishers, Inc., Chelsea, Michigan, pp. 175–83.

Lorberau, C. (1989) *Identification of the Determination of Respirable Quartz on Filter Media Using Fourier Transform Infrared Spectrophotometry*, National Technical Information Service, Springfiedl, Virgine, PB90–130105.

Mark, D. and Vincent, J.H. (1986) A new personal sampler for airborne total dust in workplaces. *Ann. Occup. Hyg.*, **30**(1), 89–102.

Mark, D., Lyons, C.P., Upton, S.L. and Kenny, L.C.(1994) Wind tunnel testing of the sampling efficiency of personal inhalable aerosol samplers. *J. Aerosol Sci.*, **25**(Suppl.1), S339–40.

Marple, V.A., Robow, K.L. and Tao, Y. (1991) A personal diesel exhaust sampler, in *Third Symposium on Respirable Dust in the Mineral Industries*, (eds R.L. Frantz and R.V. Romani) Society for Mining Metallurgy and Exploration, Inc., Littleton, Colorado, pp. 73–91.

McCawley, M.A. (1984) Performance considerations for size-selective samplers. *Ann. Am. Conf. Ind. Hyg.*, **11**, 97–100.

McCawley, M.A. and Cocalis, J. (1986) Diesel particulate measurement techniques for use with ventilation control strategies in underground coal mines. *Ann. Am. Conf. Ind. Hyg.* **14**, 271–81.

Morrow, P.E. (1964) Evaluation of inhalation hazards based upon the respirable dust concept and the philosophy and application of selective sampling. *Am. Ind. Hyg. Assoc. J.*, **25**, 213–36.

Morrow, P.E., Muhle, H. and Mermelstein, R. (1991) Chronic inhalation study findings as a basis for proposing a new occupational dust exposure limit. *J. Am. Coll. Tox.*, **10**(2), 279–89.

MSHA (1995) Mine Safety and Health Administrations. Code of Federal Regulations: 'Mineral Resources', Title 30: Parts 56, 57, 70, 71 and 75, US Government Printing Office, Washington, DC.

NIOSH (1983) Collaborative Tests of Two Methods for Determining Free Silica in Airborne Dust. DHHS (NIOSH) Publication No. 83–124 (NIOSH Publications Dissemination, 4676 Columbia Parkway, Cincinnati, OH 45226).

NIOSH (1994a) *Manual of Analytical Methods, 4th edition. Method 7400 – Asbestos and Other Fibers by PCM*, Revised 08/15/94. DHHS (NIOSH) Publication No. 94–113 (NIOSH Publications Dissemination, 4676 Columbia Parkway, Cincinnati, OH 45226).

NIOSH (1994b) *Manual of Analytical Methods, 4th edition. Method 0600 – Particulates Not Otherwise Regulated, Respirable*, Revised 08/15/94. DHHS (NIOSH) Publication No. 94–113 (NIOSH Publications Dissemination, 4676 Columbia Parkway, Cincinnati, OH 45226).

NIOSH (1994c) *Manual of Analytical Methods, 3rd edition. Method 5506 – Polynuclear Aromatic Hydrocarbons by HPLC*, Revised 08/15/94. DHHS (NIOSH) Publication No. 94–113 (NIOSH Publications Dissemination, 4676 Columbia Parkway, Cincinnati, OH 45226).

NRC (1980) Measurement and Control of Respirable Dust in Mines. Report of the Committee on Measurement and Control of Respirable Dust, National Materials Advisory Board, Commission on Socieotechnical Systems, National Research Council. NMAB-363, National Academy of Sciences, Washington, DC.

Orenstein, A.J. (1960) *Proceedings of the Pneumoconiosis Conference, Johannesburg, 1959*, Little, Brown & Co., Boston.

OSHA (1995) *Occupational Safety and Health Administration. Code of Federal Regulations: 'Labor'*, Title 29: Parts 1000.1043, US Government Printing Office, Washington, DC.

Phalen, R.F. (1984) Particle size-selective sampling in the workplace – Report of the ACGIH Technical Committee on air sampling procedures – Introduction and Recommendations. *Ann. Am. Conf. Ind. Hyg.*, **11**, 23–6.

Prinz, B. and Stolz, R. (1990) Effect of the measuring strategy on the determination of the respirable dust concentration in the breathable air at underground workplaces, in *Proceedings, VIIth International Pneumoconiosis Conference*, Pittsburgh, Pennsylvania. DHHS(NIOSH)-Publication No. 90–108, Part II, pp. 1148–1151.

Raabe, O.G. (1984) Size-selective sampling criteria for thoracic and respirable mass fractions. *Ann. Am. Conf. Ind. Hyg.*, **11**, 53–65.

Roach, S.A. (1953) Relating the incidence of pneumoconiosis to airborne dust exposure. *Br. J. Ind. Med.*, **10**(4), 220–6.

Roach, S.A. (1959) Measuring dust exposure with thermal precipitator in collieries and foundries. *Br. J. Ind. Med.*, **16**(2), 104–22.

Soderholm, S.C. (1984) Size-selective sampling criteria for inspirable mass fraction. *Ann. Am. Conf. Ind. Hyg.*, **11**, 47–52.

Soderholm, S.C. (1989) Proposed international conventions for particle size-selective sampling. *Ann. Occup. Hyg.*, **33**(3), 301–20.

Soderholm, S.C. (1995) Particle and gas phase interactions in air sampling, in *Air Sampling Instruments*, American Conference of Governmental Industrial Hygienists, Inc., Cincinnati, Ohio, Chapter 4, pp 67–80.

Soderholm, S.C. and Ferron, G.A. (1992) Estimating effects of evaporation and condensation on volatile aerosols during inhalation exposures. *J. Aerosol Sci.*, **23**(3), 257–77.

Soderholm, S.C. and McCawley, M.A. (1990) Should dust samplers mimic human lung deposition? *Appl. Occup. Environ. Hyg.*, **5**(12), 829–35.

Stahlhoffen, W., Gebhart, J. and Heyder, J. (1980) Experimental determination of the regional deposition of aerosol particles in the human respiratory tract. *Am. Ind. Hyg. Assoc. J.*, **41**, 385–98.

Stuart, B.O., Lioy, P.J. and Phalen, R.F. (1984) Use of size-selection in establishing TLVs. *Ann. Am. Conf. Ind. Hyg.*, **11**, 85–96.

Talty, J.T. (1988) *Industrial Hygiene Engineering: Recognition, Measurement, Evaluation and Control*, 2nd edn. Noyes Data Corporation, Park Ridge, New Jersey.

Talvitie, N.A. (1951) Determination of quartz in presence of silicates using phosphoric acid. *Anal. Chem.*, **23**(4), 623–6.

Talvitie, N.A. (1964) Determination of free silica: gravimetric and spectrophotometric procedures applicable to air-borne and settled dust. *Am. Ind. Hyg. Assoc. J.*, **25**, 169–78.

Talvitie, N.A. and Hyslop, F. (1958) Colorimetric determination of siliceous atmospheric contaminants. *Ind. Hyg. J.*, **19**, 54–8.

Tomb, T.F. and Raymond, L.D. (1970) Evaluation of the penetration characteristics of a horizontal plate elutriator and of a 10-mm nylon cyclone elutriator. US Bureau of Mines Report of Investigations No. USBM-RI-7367.

Tomb, T.F. and Treaftis, H.N. (1976) Review of published experimental calibrations performed on the 10 millimeter nylon cyclone. Mining Enforcement and Safety Administration, Informational Report 1040.

Tomb, T.F., Treaftis, H.N., Mundell, R.L. and Parobeck, P.S. (1973) Comparison of respirable dust concentration measured with MRE and modified personal gravimetric sampling equipment. US Bureau of Mines Report of Investigation, USBM-RI-7772.

US Code (1977) Federal Mine Safety and Health Act of 1977: 30 U.S.C. §§ 801–961 (1971 and Supp. 1978); Public Law 91–173 as amended by Public Law 95–164. (Federal Coal Mine Health Act of 1969 [PL 91–173] amended by the Federal Mine Safety and Health Amendments Act of 1977 [PL 95–164], Superintendent of Documents, Government Printing Office, Stock 024–019–00023–3, Washington, DC.

Verma, D.K., Sebestyen, A., Julian, J.A. and Muir, D.C.F. (1992) A comparison of two methods of sampling and analysis for crystalline silica α-quartz. *Appl. Occup. Environ. Hyg.*, **7**(11), 758–63.

Vincent, J.H. (1995) Progress towards implementation of new aerosol industrial hygiene standards, with special reference to the aluminium industry. *Sci. Total Environ.*, **163**, 3–9.

Vincent, J.H. and Armbruster, L. (1981) Short communication on the quatitative definition of the inhalablity of airborne dust. *Ann. Occup. Hyg.*, **24**(2), 245–8.

Vincent, J.H. and Mark, D. (1982) Applications of blunt sampler theory to the definition and measurement of inhalable dust. *Ann. Occup. Hyg.*, **26**(1–4), 3–19

Vinzents, P.S., Thomassen, Y. and Hetland, S. (1995) A method for establishing tentative occupational exposure limits for inhalable dust. *Ann. Occup. Hyg.*, **39**(6), 795–800.

Wu, Z., Hearl, F.J., Peng, K.L. *et al.* (1992) Current occupational exposures in Chinese iron and copper mines. *Appl. Occup. Environ. Hyg.*, **7**(11), 735–43.

4 Defining workplace exposures to vapors and gases

Roy J. Rando

INTRODUCTION

Respiratory toxicants may be divided into two general categories which are defined by physical state: aerosols and gases/vapors. Aerosols are airborne dispersions of solids or liquid particulates, whereas vapors are airborne dispersions of molecules of a substance that is liquid or solid in its normal state. Certain toxicants have sufficiently low normal vapor pressures that they may exist either as a mixture of vapor and aerosol or, if the vapor pressure is sufficiently small, as an aerosol only. Particulate pollutants are normally considered to be dispersions of solids or liquids that have essentially no vapor phase present, for example, coal dust, crystalline silica particles and asbestos fibers. However, some of the most important respiratory toxicants are present only in the gas phase. For example, the war gases chlorine and phosgene have vapor pressures above normal atmospheric pressure at room temperature and thus exist only in the gas phase in normal workplace environments.

Assessing the exposure of workers to gas/vapor phase pulmonary toxicants is one of the primary roles of the industrial hygienist. An industrial hygienist may be broadly defined as a scientist or engineer who is trained in the anticipation, recognition, evaluation and control of workplace hazards resulting from chemical, biological and physical agents. Defining occupational exposures is an integral part of the process of evaluating a workplace hazard and is one of the first steps in ultimately eliminating the hazard or controlling it. Furthermore, exposure assessment is a significant component of several important activities of the industrial hygienist: demonstrating compliance with appropriate exposure guidelines, such as the permissible exposure limits (PEL) of the Occupational Safety and Health Administration (OSHA) and the threshold limit values (TLV) of the American Conference of Governmental Industrial Hygienists (ACGIH) or International Occupational Exposure limits (OEL); investigating outbreaks of workplace disease or illness; performing epidemiologic research in occupational diseases of the respiratory and other bodily systems; and evaluating the effectiveness of exposure control procedures.

EXPOSURE DISTRIBUTIONS

The workplace is a dynamic environment. As such, airborne levels of respiratory toxicants are likely to vary considerably in both time and space. The resulting exposure to workers shows a similar variability. If a sufficiently large number of samples are taken to define the overall exposure pattern for a given worker, a skewed sample distribution usually results (Figure 4.1). If the same data are first transformed by taking the logarithm of the individual sample values, the distribution becomes symmetrical. Sample distributions that exhibit this behavior are said to be log-normally distributed.

Occupational Lung Disease. Edited by Daniel E. Banks and John E. Parker. Published in 1998 by Chapman & Hall. ISBN 0 412 73630 6

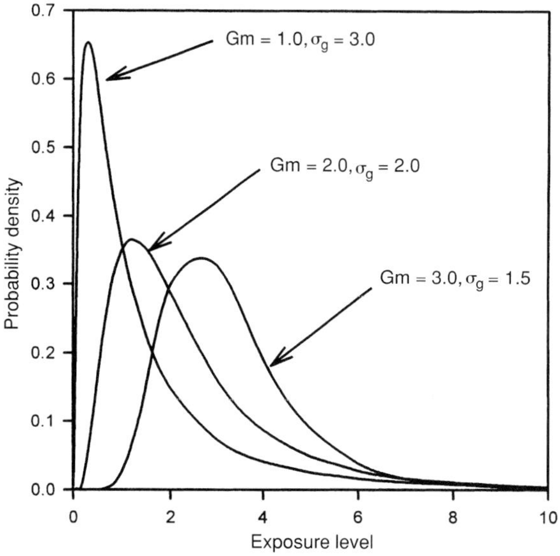

Figure 4.1 Examples of log-normal distributions. GM, geometric mean; σ_g, geometric standard deviation. Occupational exposures typically exhibit log-normality.

The probability density function for the log-normal distribution is given as follows:

$$f(x) = \frac{1}{x \ln \sigma_g \sqrt{2\pi}} \exp\left\{\frac{-\frac{1}{2}(\ln x - \ln GM)^2}{\ln^2 \sigma_g}\right\} \quad (4.1)$$

where x is the magnitude of the individual exposures, $f(x)$ is the probability density (frequency of occurrence) for the value x, and GM and σ_g are the geometric mean and geometric standard deviation, respectively. The geometric mean of the samples is calculated as follows:

$$GM = \exp\left\{\frac{\sum_{i=1}^{N} \ln x_i}{N}\right\} \quad (4.2)$$

where N is the sample size. The geometric standard deviation is calculated as:

$$\sigma_g = \exp\left\{\left(\frac{\sum_{i=1}^{N} (\ln GM - \ln x_i)^2}{N-1}\right)^{1/2}\right\} \quad (4.3)$$

In addition to occupational exposure values, many other types of concentration measurements for environmental constituents are best represented by the log-normal distribution. Four conditions are frequently associated with the occurrence of log-normality in environmental concentration measurements (Esmen and Hammad, 1977):

- The concentrations cover a wide range of values, often several orders of magnitude.
- The concentrations lie close to a physical limit (zero concentration).
- The variability of the measured concentration is of the order of magnitude of the size of the measured concentrations.
- There is a finite probability of very large values occurring.

For occupational exposures, the inherent log-normality is commonly a result of there being specific, well-defined sources of the pollutant in the workplace. The source often is not continuous in emission rate, so that there are periods of relatively high emission, intermingled with periods of low or no emission. In effect, when this situation occurs, the overall exposure distribution may actually be a combination of several individual exposure distributions associated with the respective operating conditions of the source. Also, when the source emission rate decreases or increases, the corresponding atmospheric concentration profile shows an exponential decay or rise as a function of time. This is a direct consequence of the physics of the dilution mechanism of pollutant in the workplace atmosphere (Ishizu, 1980; Wadden and Scheff, 1987; and Eq. 4.6).

The geometric mean is a measure of central tendency and one of the primary parameters of interest for any log-normal distribution. The geometric mean and geometric standard deviation are sufficient to describe completely any log-normal distribution. Furthermore, the corresponding logarithms of the geometric mean and geometric standard deviation should be used in performing the usual statistical tests and procedures on a log-normally distributed dataset, for example: calculation of confidence intervals, comparing two sets of log-normally distributed data for differences, estimating the percentage of samples below some threshold value or OEL, and so on.

While the geometric mean is the most statistically appropriate measure of central tendency of a log-normally distributed exposure distribution, the arithmetic mean is often of equal importance when examining exposure–response relationships, or performing epidemiologic research. For a given worker's occupational exposure profile, which exhibits random variation in exposure level as a function of time, the resulting exposure dose, D (exposure concentration – exposure time product) is usually defined as:

$$D = \int_{o}^{T} C dt = C_{avg} \cdot T \quad (4.4)$$

Thus, the exposure dose is equivalent to the product of the *arithmetic* mean concentration, C_{avg} and the total exposure time, T.

The decision as to whether to use the arithmetic or geometric mean when analyzing exposure response relationships is governed by the study design and objectives. Seixas *et al.* (1988) have examined the relative merits of using the geometric and arithmetic means in occupational epidemiology. For a linear effects model, in which response is directly proportional to exposure, it is suggested that use of the arithmetic mean results in Berkson-type grouping (Cochran, 1968) of individual exposure measurements, thus reducing the overall bias originating from errors in measurements of the exposure variable. Conversely, if the exposure effects model is non-linear, e.g. log-linear or logistic (dichotomous outcome), use of the arithmetic mean as the estimator of the overall exposure will introduce bias into the analysis.

The arithmetic mean of a log-normal distribution may be calculated in the usual manner for any set of data. Alternatively, in those circumstances where there is a priori knowledge of the geometric mean and geometric standard deviation, it is readily calculated using the following conversion:

$$AM = GM \cdot \exp\left\{\frac{(\ln \sigma_g)^2}{2}\right\} \tag{4.5}$$

For the example of log-normal distributions illustrated in Figure 4.1, the corresponding arithmetic means are listed in Table 4.1. It is seen that as the geometric standard deviation decreases, the geometric and arithmetic means tend to converge. In the trivial case where the geometric standard deviation is unity, the geometric mean and arithmetic mean are identical. When σ_g is equal to one, there is no dispersion in the distribution and all samples have the same value.

Table 4.1 Comparison of geometric mean, geometric standard deviation and arithmetic mean for example log-normal distributions

Geometric mean	*Geometric standard deviation*	*Arithmetic mean*
1.0	1.5	1.09
1.0	2.0	1.27
1.0*	3.0	1.83
2.0	1.5	2.17
2.0*	2.0	2.54
2.0	3.0	3.66
3.0*	1.5	3.26
3.0	2.0	3.81
3.0	3.0	5.49

* Illustrated in Figure 4.1.

FACTORS AFFECTING OCCUPATIONAL EXPOSURES

Occupational exposures are typically log-normally distributed as a reflection of variation in their magnitude, duration and frequency. The nature of occupational exposures is dependent on a number of factors which may be broadly divided into two categories: intrinsic and extrinsic. Intrinsic factors are those related directly to the toxicant itself, while extrinsic factors are determined by the nature of the workplace and also by the worker. This is illustrated schematically in Figure 4.2.

Intrinsic physical properties

The physical and chemical properties of a pulmonary toxicant are important determinants of occupational exposures. These properties may have profound effects on the propensity of a toxicant to be emitted into the workplace atmosphere, the physical state and behavior of the toxicant once it has entered the atmosphere, and the uptake and deposition of the toxicant once it is inhaled by the worker. Among the more important physical/chemical properties of the toxicant are its vapor pressure, solubility, and reactivity.

(a) Vapor pressure

The vapor pressure of a toxicant will affect both the propensity of a material to become airborne and its physical state once it is emitted into the workplace atmosphere. For a volatile material, the rate at which it

Intrinsic

- vapor pressure
- surface affinity
- solubility
- chemical reactivity

Extrinsic

Worker	Process	Workplace
• personal protective equipment • personal hygiene • work rate • breathing rate	• exhaust ventilation • temperature • mechanical energy	• temperature • humidity • dilution ventilation

Figure 4.2 Factors affecting the nature and magnitude of occupational exposures may be classified as intrinsic or extrinsic.

evaporates from the liquid phase into the atmosphere and thus its ultimate airborne concentration will be a function of its vapor pressure. For a semi-volatile material, the vapor pressure will determine its saturated vapor concentration in air or the maximum concentration of vapor phase material in the atmosphere at a given temperature. At concentrations above this value, the material will partition into the vapor and condensed (or aerosol) phases in the atmosphere.

For example, Table 4.2 lists the vapor pressures and resulting saturated vapor concentrations of a variety of isocyanate compounds. The isocyanates are important commercial chemicals used in the production of polyurethane paints, foams and plastics, as well as in the manufacture of certain carbamate pesticides. As a class, the isocyanates are widely recognized pulmonary irritants and sensitizers.

Methylisocyanate (MIC), the isocyanate involved in the notorious Bhopal incident, is the smallest and most volatile member of this family of chemicals and has a vapor pressure of 348 mmHg at normal ambient temperature. This results in a saturated vapor concentration of almost 50% in air. Its saturated vapor concentration is nearly 23 million times as large as some published OEL (0.02 ppm). Clearly there is no possibility of MIC existing in aerosol form in the workplace atmosphere. In contrast, the vapor pressure of 4,4′-methylene-bis-(phenylisocyanate) [MDI] is equivalent to a saturated vapor concentration of only 0.008 ppm (Brochhagen and Schall, 1986). This is less than the OSHA PEL of 0.02 ppm. In the case of MDI then, exposures at or near the PEL are likely to be to a mixture of condensed and vapor phase material (Dharmarajan and Well, 1978; Dharmarajan, 1979; Rando and Poovey, 1994). The other isocyanates listed in Table 4.2 have vapor pressures intermediate between MIC and MDI, so that the possibility of partitioning between vapor and aerosol phases will be a function of their individual concentrations and the ambient temperature. In any case, isocyanate material mechanically introduced into the atmosphere by nebulization processes such as spray painting may have a finite lifetime in the condensed phase as the aerosols absorb heat from the environment and volatilize. Thus, the volatility of a material may have a profound effect on its behavior in the atmosphere and the resulting patterns of deposition and uptake in the respiratory tract upon inhalation.

(b) *Vapor/particle interactions*

Partitioning of lung toxicants between the vapor phase and the condensed phase may occur to some extent depending on the volatility of the particular compound, the ambient conditions in which the material is emitted, and the amount and character of any existing particulates in the atmosphere. For example, Pankow *et al.* (1994) examined the gas/particle partitioning of two homologous series of n-alkanes (C_{16} to C_{22}) and polynuclear aromatic hydrocarbons (fluorene to chrysene) in environmental tobacco smoke (ETS). The partitioning between the vapor and condensed phases was shown to be a function of the vapor pressures of the individual compounds within the two compound classes. Moreover, the overall relationship of the partition constant to vapor pressure was quite different

Table 4.2 Overview of characteristics of some commercial isocyanates

	Vapor pressure (mmHg, 25°C)	*Saturated vapor concentration (ppm)*	*PEL [*TLV] (ppm)*	*Ratio: SVC/PEL or SVC/TLV*
Methylisocyanate (MIC)	348	457 895	0.02	2.29×10^7
1,6-Hexamethylene-diisocyanate (HDI)	0.05	66	*0.005	1.32×10^4
Toluene diisocyanate (2,4-TDI; 2,6 TDI)	0.0167	22	0.02	1.1×10^3
4,4′-Dicyclohexylmethane-diisocyanate (HMDI)	0.001	1.3	*0.005	260
Isophorone diisocyanate (IPDI)	0.0003 (20°C)	0.4	*0.005	80
4,4′-Diphenylmethane-diisocyanate (MDI)	6.3×10^{-6}	0.008	0.02	0.4

*TLV: threshold limit value

for these two compound classes, with the polynuclear aromatics being more strongly adsorbed to the smoke particles than the alkanes.

The partitioning of nicotine between vapor and condensed phases is also of particular interest since it is one of the more commonly used markers of tobacco smoke. Indeed, nicotine has been suggested as a suitable surrogate material for an occupational standard for environmental tobacco smoke exposure (Repace and Lowrey, 1993). It is widely recognized that greater than 90% of nicotine in both ETS and sidestream smoke exists in the vapor phase (Eatough *et al.*, 1989; Koutrakis *et al.*, 1989). In contrast, measurements in mainstream smoke indicate significantly more of the nicotine to be in the condensed phase. This phenomenon is a result of the slightly acidic environment of mainstream smoke which results in protonation of the nicotine, thus lowering its volatility. In contrast, the alkaline nature of sidestream smoke maintains nicotine as the free base which is volatile so that it partitions between vapor and particle-bound phases. This difference in physical state of nicotine in mainstream smoke and sidestream smoke/ETS could have a profound influence on the pattern and rate of uptake of nicotine in active smokers vis à vis those passively exposed to tobacco smoke in the workplace or other environments.

The adsorption of very volatile pulmonary irritants onto particulates can increase their toxicity and facilitate injury to the deeper parts of the respiratory tract, since the scrubbing action of the upper airways may be effectively by-passed. Reports in the literature have demonstrated this effect for hydrochloric acid vapor adsorbed onto smoke particles from burning polyvinyl chloride, and for sulfur dioxide adsorbed onto atmospheric aerosols. For both of these materials, the lower respiratory tract irritancy is enhanced in comparison to vapor-only mixtures. In these cases, the particles act as a vehicle to carry the irritant vapor into the lower respiratory tract and prevent the normal scrubbing-out of these very soluble vapors by the moist surfaces of the upper respiratory system.

Vapor/particle interactions are also likely to play a role in the etiology of certain types of hypersensitivity pneumonitis of occupational origin. The isocyanates are occasionally associated with this disease although occupational asthma occurs more frequently with exposure. A scan of the medical literature over the past 10-year period will yield on the order of 20 case reports detailing hypersensitivity pneumonitis associated with exposure to isocyanates in the workplace. Essentially all of these cases are linked either to exposure to MDI or to TDI, with the former being predominant. As we have seen, MDI has exceedingly low volatility and tends to partition into the aerosol phase at atmospheric concentrations below the occupational exposure standard. Furthermore, in all the case reports linking TDI exposure to hypersensitivity pneumonitis, there was evidence of exposure to a mixture of TDI and particulate matter, usually from spray painting of polyurethane paint containing TDI, but also from other sources such as dust from the manufacture of cork products containing TDI-based binders. It is apparent that the existence of isocyanate in the condensed phase, either as aerosols of the pure material or perhaps adsorbed onto the surface of otherwise inert particulate matter, facilitates penetration of these semi-volatile toxicants into the alveolar compartment where they may be deposited in a concentrated and localized manner.

(c) Solubility

For vapor-phase toxicants, the aqueous solubility is a primary determinant of the uptake and resulting pulmonary exposure. The upper airways act as fairly efficient gas scrubbers to remove soluble gases from the air as it is inhaled and passes into the lower respiratory tract. For example, Speizer and Frank (1966) showed that virtually all of the sulfur dioxide was absorbed in the noses of volunteers inhaling 16.1 ppm, with only 0.3 to 0.4 ppm reaching the pharynx. As a result, exposure to highly water-soluble toxicants is less likely to result in damage to the alveoli and bronchioles than insoluble materials for equivalent exposure levels and times. This concept is further illustrated in Figure 4.3 which presents the correlation

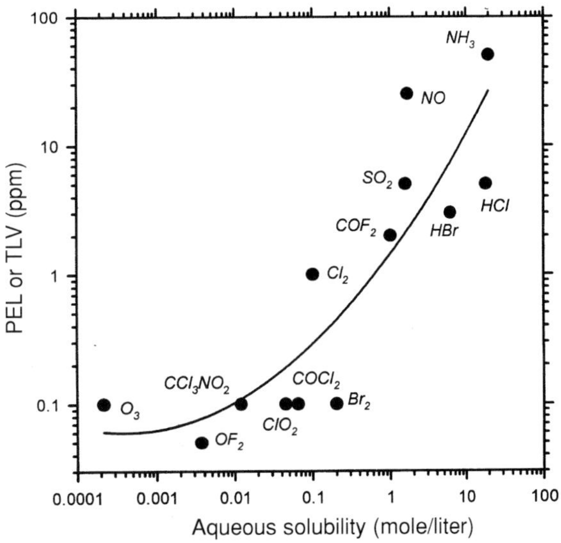

Figure 4.3 The relationship between solubility and allowable exposure limits for the primary inorganic irritants.

of aqueous solubility with the current United States occupational exposure guideline [either the OSHA permissible exposure limit (PEL), or the ACGIH threshold limit value (TLV)] for the primary inorganic irritant gases. Overall, there is a clear trend of increasing PEL/TLV with increasing aqueous solubility among these toxicants. In addition to the solubility, the inherent toxicity of the individual compounds also plays a role in defining what is an acceptable exposure level. In light of this, the correlation of solubility and allowable exposure for compounds which fall within a common chemical class is even more illustrative: for example, hydrogen chloride and hydrogen bromide; phosgene ($COCl_2$) and fluorophosgene (COF_2).

(d) Reactivity

In the case of some respiratory toxicants, the chemical reactivity, or lack thereof, has a significant effect on resulting occupational exposures. Reaction of a toxicant with other chemical species in the environment may serve to generate other toxic species or may result in inert by-products, in which case the reaction serves as a removal mechanism or chemical sink for the toxicant. The reactivity may also exert an effect on the identity of the toxicant which is ultimately emitted into the workplace atmosphere.

An illustrative example of this is provided by toluene diisocyanate (TDI), a diisocyanate monomer commonly used in the production of flexible polyurethane foams. It is a potent respiratory sensitizer and is a common cause of occupational asthma (Banks *et al.*, 1989). Commercial grades of TDI are usually a mixture of two positional isomers: 2,4-toluene diisocyanate and 2,6-toluene diisocyanate:

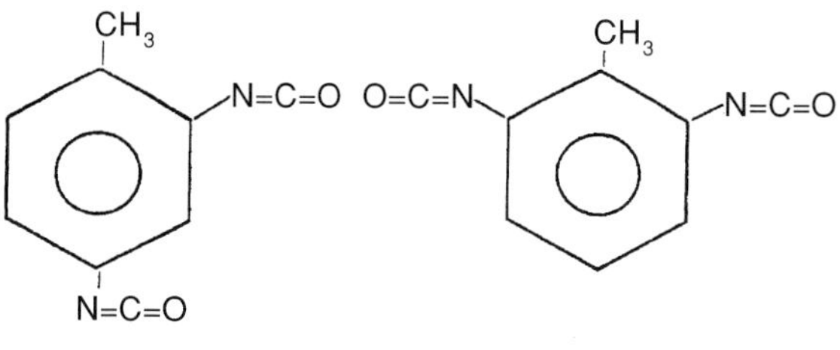

The most common mixture contains 80% of the 2,4-isomer and 20% of the 2,6 isomer. 2,4-TDI is much more reactive than 2,6-TDI as a consequence of steric hindrance of the isocyanate moiety in either the '2' or '6' position from the methyl group in the '1' position. During the production of polyurethane foam, TDI is reacting with the polyols and forming urethane cross-links between its isocyanate groups and the alcohol groups of the polyol. Because of the difference in reactivity of the isomers of TDI, 2,4-TDI is being consumed at a more rapid rate than is the 2,6-isomer. As a result, as the foam is being formed, there is a shift in the relative amounts of 2,4- and 2,6-TDI, with the 2,6-isomer becoming enriched in the unreacted polyurethane foam mixture. This in turn results in enhanced emissions of 2,6-TDI in comparison with 2,4-TDI.

Figure 4.4 (Rando *et al.*, 1984) illustrates this phenomenon for TDI exposures in a flexible polyurethane foam slab-stock plant. The starting materials used in this process consisted of the usual 80% 2,4-/20% 2,6-TDI mixture (ratio of 2,6- to 2,4-TDI = 0.25). Area monitoring was conducted on the mixing end of the foam lines as well as on the finishing end. On the mixing end, the TDI, polyols and other ingredients are initially mixed and laid out onto a moving sheet of Kraft paper. The foam bat begins to form and rise and then proceeds into a curing oven, and finally exits at the finishing end where the foam is cut into appropriately sized loaves. In this survey, less than 10% of the samples showed a ratio of 2,4- to 2,6-TDI equivalent to the starting materials on the mixing end of the process. On the finishing end, all of the samples exhibited a ratio larger than the value of 0.25 for the starting materials. The shift in isomer composition of airborne TDI was much more pronounced in the area of the finishing end of the process with there being about six times as much 2,6-TDI, on average. On the finishing end, TDI is only emitted from the hot foam

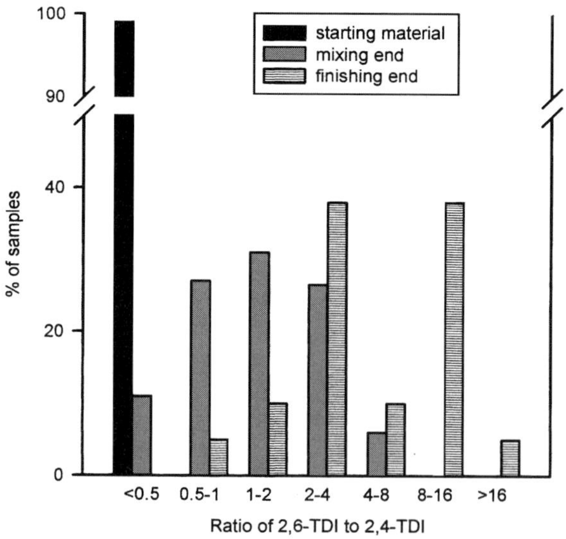

Figure 4.4 The isomeric composition of airborne TDI in the atmosphere of flexible polyurethane foam plants. 2,6-TDI is emitted in larger quantities than the more reactive 2,4-TDI. (Source: Rando *et al.* 1984.)

bats emerging from the curing oven. At this point, 2,4-TDI has been consumed in forming the polyurethane to a much larger extent than the 2,6-TDI as result of their relative reactivities. Correspondingly, most of the TDI offgassing is 2,6-TDI. In contrast, on the mixing end, TDI is likely to be emitted from spills and leaks from the chemical handling equipment as well as offgassing from the freshly poured foam mixture, thus airborne TDI should exhibit an isomer composition close to that of the starting materials.

The importance of this difference in occupational exposures to the isomers of TDI was demonstrated in subsequent studies by Banks *et al.* (1989) which suggested a possible difference in the pattern and intensity of asthmatic response to the isomers. Among a group of TDI-sensitized polyurethane foam workers, challenge exposure to 2,6-TDI appeared to elicit increased airway reactivity in comparison to the 2,4-isomer. It was suggested that the difference was a result of prevailing exposure conditions in the subjects' workplaces (TDI-using industry) and that sensitized workers in the TDI-manufacturing industry, where exposure is likely to be predominately to the 2,4-isomer, would possibly show the opposite response to the isomers of TDI upon exposure.

Extrinsic factors affecting occupational exposures

The concentration of a pollutant in the workplace atmosphere, and the resulting exposure to workers is influenced by many extrinsic factors which are related to the nature of the source and of the workplace itself. If the workplace is considered to be a space that approaches the conditions of a completely mixed reactor, then the concentration–time profile of a pollutant may be modeled as a function of its rate of generation, the rate of dilution with workplace air, and losses due to reaction and filtration as follows (Wadden and Scheff, 1987):

$$C_i = \frac{k[Q_0(1-F_0) + Q_2]C_0 + S - R}{k(Q_0 + Q_1F_1 + Q_2)} \times \left[1 - \exp\left\{\frac{k}{V}(Q_0 + Q_1F_1 + Q_2)t\right\}\right] + C_s \exp\left\{\frac{k}{V}(Q_0 + Q_1F_1 + Q_2)t\right\} \qquad (4.6)$$

where C_i is the concentration in the workplace, C_0 is the concentration outdoors (which in many cases will be essentially equal to zero), C_s is the concentration in the workplace at the beginning of some modeling period when time (t) is zero, Q_0 is the volumetric flow rate of makeup or dilution air in the workplace, Q_1 is the volumetric flow rate of recirculated air, Q_2 is the volumetric flow rate of infiltration, F_0 is the filtration efficiency for the makeup air, F_1 is the filtration efficiency for recirculated air, V is the workplace air volume, S is the source pollutant emission rate, R is the workplace sink removal rate (likely to be zero for 'unreactive' or 'inert' pollutants), and k is an empirical mixing factor which accounts for the fact that in most real-world situations, incoming air is not completely nor instantaneously mixed with the air already present in the workplace. The value of the mixing factor, k, is equal to unity when mixing inside the workplace is perfect.

(a) Source characteristics

The source emission rate, S, is directly dependent on the intrinsic properties of the pollutant as well as on the characteristics of the source itself. In addition to any temporal variability in the timing and frequency of the source output, the source characteristics which affect emission rate include the energy of the source, its geometry (for example the surface area of a degreasing tank), and the degree to which the source is enclosed or isolated from the surrounding workspace. The energy of the source is one of the principal governing factors in defining the quantity and rate of pollutant release. Volatile pollutants require heat input for evaporation so that hot processes logically generate larger volumes of volatile pollutants than would similar sources at a lower temperature. Likewise, for sources that mechanically generate pollutant, such as paint spray operations, the more energetic a source (mechanical energy as work), the greater will be the value of the source emission factor, S. For those pollutants of low volatility, mechanical energy supplied through spray operations or other aerosolization processes, may be essential in getting them airborne.

An illustration of the effects on exposure from both the intrinsic properties of the toxicant and the source character is provided by an industrial hygiene survey of 47 different industrial spray and non-spray operations conducted over a period of 8 years (Myer *et al.*, 1993). The survey included facilities manufacturing polyurethane paints based on 1,6-hexamethylene diisocyanate (HDI), as well as facilities using such paints. The operations which were surveyed included the manufacture and refinishing of transportation vehicles, painting of large military and civilian equipment, industrial finishing operations, maintenance and construction operations, and wood and furniture manufacture. Exposures to both HDI (saturated vapor concentration approximately 66 ppm or 453 mg/m^3 at room temperature) and HDI-biuret (a trimer of HDI with a saturated vapor concentration less than 100 ppb or 2 mg/m^3 at room temperature) were measured by a total of 466 area and personal monitoring samples.

For the spray and non-spray operations, the geometric mean concentrations of HDI measured in the personal samples were approximately 5.2 ppb and 0.8 ppb, respectively. This represents an increase of about a factor of 5 for the spraying operations. In contrast, 93% of the personal samples collected in the non-spray operations showed no detectable amounts of HDI-biuret, with the overall geometric mean being less than 0.03 mg/m^3. However, for the spray applications, HDI-biuret was undetected in only 8% of the samples and the overall geometric mean exposure was 0.94 mg/m^3 – an increase of over 31 times compared with the non-spray operations.

For the area samples, there was little difference in measured HDI levels for the spray and non-spray operations (geometric means of 1.2 ppb and 0.8 ppb, respectively). But again for HDI-biuret, there was approximately a 5-fold increase in area concentrations for the two types of operations – a geometric mean of approximately 0.16 mg/m^3 for spraying operations, in contrast to less than 0.03 mg/m^3 for the non-spray operations. These data illustrate the importance of the source characteristics – in this case, spray versus non-spray applications – in defining exposures, particularly for toxicants exhibiting a wide range in volatility.

(b) Use of personal protective equipment

The worker can also have a great effect on his or her individual exposure profile. Personal factors that influence exposure dose received from a contaminated work environment include work rate, attention to personal hygiene, breathing rate and use of personal protective equipment. The use of personal protective equipment is a common confounding factor in defining occupational exposures to respiratory toxicants. Correction of personal environmental exposure for respirator use must account for both frequency of use and the workplace respirator protection factor. These elements are often quite different between groups of homogeneously exposed workers and, in addition, may also be unique for the individuals within the groups and may vary from day to day.

Determination of the personal exposure of a worker who uses respiratory protective equipment may involve simultaneous sampling inside and outside of the respirator to determine the workplace respirator protection factor (RPF), along with documentation of the frequency and duration of respirator use to determine an effective protection factor (EPF). These terms are defined as follows:

$$\mathrm{RPF} = \frac{C_{\mathrm{out}}}{C_{\mathrm{in}}} \tag{4.7}$$

where C_{out} is the exposure concentration outside the respirator (the environmental exposure) and C_{in} is the exposure concentration inside the respirator (the personal exposure). The effective protection factor (EPF) has the same form as the RPF, but is based on average concentrations which include those periods of time when the respirator is not worn. During such periods, the personal and environmental exposure concentrations are the same ($C_{\mathrm{out}} = C_{\mathrm{in}}$).

Determination of the RPF can be technically difficult and must be done without perturbing the fit characteristics of the respirator which could cause leakage and infiltration of workplace air into the mask. Frequency of use is usually difficult to ascertain and requires constant surveillance of the worker during the monitoring period. Several studies examining the variability of these factors in actual workplaces have appeared in the literature and are illustrative of the difficulties and problems involved.

Harris *et al.* (1974) evaluated the protection afforded by half-mask air-purifying respirators in the bituminous coal mining industry. A median EPF of 3.2 for coal dust was noted. In addition, observed average frequency of respirator use ranged from 42 to 60% of the work shift for various jobs. Respirator protection factors were also examined in a study of cadmium workers using half-mask air-purifying respirators (Smith *et al.*, 1980). A geometric mean EPF of 5.6 (σ_g = 2.62) was found for multiple tests on nine subjects. One subject had consistently higher protection factors (geometric mean of 103) than the rest of the population. Without this individual, the geometric mean EPF was 3.9. The authors concluded that a large part of the variability in the measured EPFs was due to intermittent respirator usage.

Moore and Smith (1976) assessed the respiratory protection for SO_2 afforded by three brands of air-purifying respirators used by workers in a copper smelter. The median RPFs ranged from 9.6 to 15. However, a subsequent study determined that the workers generally used the respirators less than 1 hour per shift and 30% used them less than 10 minutes per shift (Smith *et al.*, 1977). Thus, it is likely that the EPFs for this population of users would have been significantly lower than the measured RPFs if the low frequency of use had been factored into the analysis.

In contrast to the relatively low protection factors noted above, a study of lead smelter workers (Lenart and Campbell, 1984) resulted in geometric mean EPFs of 180 and 380 for a negative-pressure air-purifying and a powered air-purifying respirator, respectively. However these results were significantly influenced by several factors: the tested population consisted of workers who were able to achieve a quantitative laboratory fit test RPF of at least 250, and the workers were required to wear the respirators during the entire work shift.

In a recent study examining potential chronic respiratory impairment in a group of chlorine production workers, exposure to chlorine gas was examined with reference to the effect of respiratory protective devices (Rando and Lefante, 1992). A total of 149 workers were enrolled in this study. Full-shift continuous monitoring of chlorine exposure was performed intermittently over a period of 4 years. Chlorine monitoring was accomplished with an electrochemical sensor attached to a datalogger. A respirator mounting switch flagged the data anytime the worker used the respirator during the monitoring period. For 456 full-shift monitoring periods, the geometric mean time weighted average concentration of chlorine in the breathing zone of the workers was 40.6 ppb with a geometric standard deviation of 4.2. For the same samples, assuming complete removal of chlorine by the respirator filters, the reduced exposures had a geometric mean and standard deviation of 31.6 ppb and 4.2, respectively. This is equivalent to an overall effective protection factor of 1.28.

Overall, 66% of the monitoring periods exhibited no respirator use. This was a reflection of the overall low exposure levels seen in this survey. However, there was a clear correlation of increasing respirator use with increasing exposure level. Figure 4.5 illustrates this observation. For exposure levels in the range of 0.1 ppm, or one-tenth the allowable exposure level, the average reduction in exposure dose through respirator use was on the order of 10%. In contrast, for exposures near the OSHA permissible exposure limit of 1 ppm, there was a 30 to 50% reduction in exposure dose by use of respiratory protective equipment.

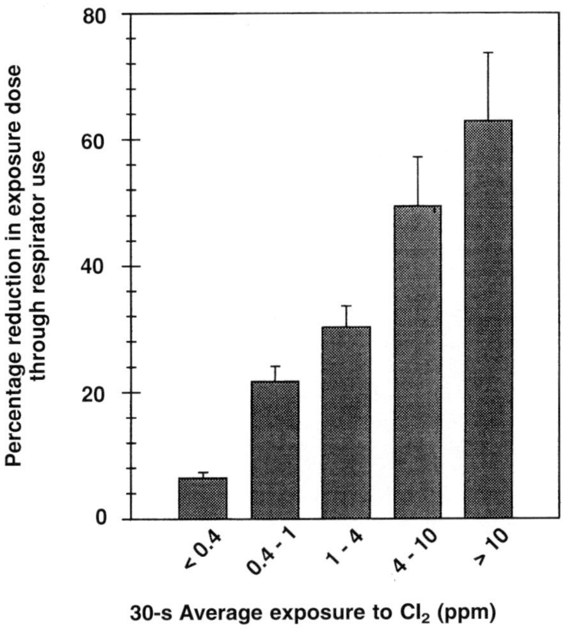

Figure 4.5 Respirator use among workers in a brine-electrolysis plant. The frequency and duration of respirator use increases as the exposure to chlorine increases.

The studies described above illustrate several important points concerning the use of respiratory protective equipment in the workplace and its effect on estimating worker exposure: patterns of respirator usage can be highly variable among workers, even those working in the same areas and performing the same tasks; the frequency and extent of respirator use tends to increase as the level of exposure increases, particularly for those toxicants with good warning properties, such as odor or irritation; and effective protection factors in the actual work environment are significantly lower than traditionally assigned protection factors which are solely based on respirator fit testing in the laboratory or clinical setting, or are extrapolated from the nominal filtration efficiency of the respirator cartridges.

EXPOSURE ASSESSMENT STRATEGIES

Design of an appropriate exposure assessment/sampling strategy is one of the more difficult and important tasks for the industrial hygienist. High degrees of experience, skill and competence are needed in developing a satisfactory exposure assessment scheme that answers the posed question in a satisfactory manner. There are several different scenarios which might serve as the initiating event in developing a sampling strategy. Some of the more important are accident/incident investigations, demonstration of compliance with exposure guidelines, and epidemiologic investigations. Each of these situations requires a somewhat unique approach in strategy.

In accident/incident investigations, the question is often what were the identity and quantity of the worker's exposure? This question is frequently asked in the context of a legal proceeding. As such, the object of the sampling strategy is usually to prove the presence (or absence) of some suspected toxicant and to measure its concentration reliably in the work environment. Initial investigations concerning the nature of the exposure and related information are invaluable. Preliminary questions to be posed include: What were the materials used in the workplace? Where was the worker stationed? Did the worker have direct contact with the suspect material? How long did the potential exposure last? What were the operating conditions in the workplace (for example, production rates, chemical raw material consumption rates, ventilation characteristics and quantities)? The information retrieved from this initial investigation is integrated into a preliminary assessment of the conditions and nature of the exposure. The sampling strategy then entails conducting exposure monitoring on the worker (or other workers who perform the same tasks and would therefore have the same exposure pattern) under conditions as identical as possible to those established in the preliminary investigation. During the actual sampling activity, replicate measurements are commonly done to provide some measure of statistical reliability to the results.

(a) Homogeneous exposure groups

In developing a sampling strategy for both demonstration of compliance and epidemiologic research, breaking the workforce down into homogeneous exposure groups (HEGs) is useful. A HEG is a grouping of workers who, by similarity of tasks, proximity to exposure sources, job title and work area, etc., are expected to have similar if not identical exposure patterns and magnitudes. An example of the development of a HEG scheme is shown in Table 4.3 in which the workforce of two facilities producing flexible polyurethane foam has been divided up into four HEGs consisting of foam line workers, foam finishing workers, maintenance workers, and R&D personnel (Rando *et al.*, 1987; Jones *et al.*, 1992). Because of the relatively crude grouping and the large numbers of unique job titles in each of the groups, there was a wide range in exposures within some of these particular HEGs. For the purposes of compliance sampling, this may be sufficient in conjunction with further scrutiny of the workforce, as discussed below. For epidemiologic purposes, further refinement of the HEG scheme, including subdividing the existing HEGs into smaller groupings with more similar exposures, would be in order. This process is aided by concomitant exposure monitoring of targeted workers within the HEGs. Initial quantitative assessments of exposure developed during these surveys is invaluable in refining the composition of the HEGs.

(b) Sample size requirements

The development of required sample sizes is undoubtedly one of the more important components of an exposure assessment strategy. The number of samples that are collected and analyzed impacts directly on the statistical reliability of the quantitative exposure values ultimately assigned to members of a HEG.

Table 4.3 Example breakdown of workers into homogeneous exposure groups (HEGs): flexible polyurethane foam production

Group	*Job titles/tasks*	*TDI exposure potential*
Foam production line	Chemical handler Cutter/trimmer Foam machine operator Pourman	Continuous low-level; frequent high-level Mean = 2.9–4.5 ppb, 1–4% samples > 20 ppb
Foam finishing	Flame laminator Foam cutter and trimmer Foam reticulator Mattress production Vinyl coater	Occasional low-level; infrequent high-level Mean = 1.4–1.5 ppb, < 0.2% samples > 20 ppb
Maintenance	Electrician Machinist Mechanic Millwright Oiler	Frequent low-level; occasional high-level Mean = 1.7–2.1 ppb, < 0.3% samples > 20 ppb
Reasearch & Development	Chemist Engineer Laboratory technician Pilot plant operator	Widely variable in both frequency and magnitude Mean = 4.1 ppb, 4.7% samples > 20 ppb

Source: Jones *et al.* (1992) and Rando *et al.* (1987).

Furthermore, the issue of sample size is also one of economics, in that the collection and subsequent laboratory analysis of samples is often a significant component of the overall cost of an investigation, in terms of labor, capital equipment and supplies. A careful assessment of the costs versus the benefits of projected sample size requirements is therefore a valuable exercise in developing the exposure assessment strategy.

Sample size requirements in compliance monitoring are dependent on the ability of the industrial hygienist to appropriately target those individuals for sampling who are likely to have the highest potential for exposure. If this targeting is feasible, then individual exposure monitoring samples collected on the targeted individuals may be analyzed to determine whether the results are within the allowable exposure limit with a certain degree of statistical confidence (Leidel *et al.*, 1977). If the industrial hygienist is unable subjectively to define the workers at highest risk of exposure, then probability theory must be used to guide the determination of required sample size. The objective is to define a sample size large enough to contain with a high degree of probability at least one worker with 'high' exposure if one exists. From the theory of random sampling from a group without replacement, a sample of size, n, randomly selected from a group of size, N, may be determined that fulfills the above criteria at some statistical confidence level, $1-\alpha$, with τ being the proportion or percentile of the exposures within the group that are defined as 'high'.

Table 4.4 shows a representative listing of required sample sizes for various group sizes, exposure proportion percentiles and confidence levels. For example, for a homogeneous exposure group consisting of 50 individuals, it would take a random sample of 21 of these to ensure that at least one of the 21 samples is among the highest 10% in terms of exposure, at a statistical confidence level of 95%. The rather large sample sizes, particularly in proportion to the smaller homogeneous group sizes, illustrates the value of making an informed assessment of the exposure potentials among the members of a HEG, rather than resorting to a random sampling strategy.

Table 4.4 Required random sample size (n) for obtaining samples with highest exposure potential from a HEG

N	τ, *1–α*			
	10%, 0.9	*10%, 0.95*	*20%, 0.9*	*20%, 0.95*
10	9	10	7	7
20	13	15	8	10
50	18	21	9	12
∞	22	29	11	14

Source: Leidel *et al.* (1977).

In contrast to compliance sampling where the objective is to define the highest potential exposures among a HEG, the objective in sampling for epidemiologic or other investigative purposes, is typically to define the appropriate measure of central tendency of exposure with some degree of statistical reliability. As we have seen, this is complicated by the common need to use the arithmetic mean of the log-normally distributed exposures of the HEG, as the measure of central tendency. In this case, a sample size should be chosen to meet the objective of estimating the arithmetic mean with a certain level of precision, β, at some statistical confidence level, $1-\alpha$. An approximation of the required sample number based on confidence interval width and large-sample theory through the Central Limit Theorem has been developed (Perez, 1995). The sample size, n, is approximated as follows:

$$n = \left(\frac{z_{\alpha/2}}{\beta)}\right)^2 \left(\sigma_g{}^{(\ln\sigma_g)} - 1\right) \tag{4.8}$$

In this case, $z_{\alpha/2}$ is the z-value corresponding to the desired confidence level, usually taken as 1.96 for the 95% confidence level. For example, if one wanted to estimate the arithmetic mean of a log-normally distributed exposure variable to within ±30% of the true value at the 95% confidence level and the geometric standard deviation of the distribution is 3, the required sample size would be 100.

A deficiency in the approximation procedure shown above is that one has to have previous knowledge of the true geometric standard deviation of the distribution. In reality, this is never known for certain but is only approximated from sampling from the distribution. Hewett (1995) recently addressed this problem. In Hewett's procedure, the 't' statistic is used rather than the z-value and the geometric standard deviation is estimated from data gathered during a small pilot study involving a few samples collected solely for this purpose. With these preliminary data in hand, the calculation of the required overall sample size using Hewett's procedure is still computationally complex. However, Hewett produced a summary table illustrating representative sample sizes for various combinations of σ_g, β and pilot sample size, n_{pilot}. This is reproduced in abbreviated form in Table 4.5. For the previously cited example, if a pilot study consisting of 10 samples were first conducted and yielded an estimated geometric standard deviation of 3, the required sample size for a precision level of 30% at the

Table 4.5 Sample size requirements for estimating the arithmetic mean of a log-normally distributed variable

n_{pilot}	σ_g	β = 20%	β = 30%	β = 50%
5	1.5	34	15	6
	2	119	53	19
	3	452	201	72
	4	1124	500	180
10	1.5	23	10	4
	2	79	35	13
	3	300	133	48
	4	746	332	119
20	1.5	20	9	3
	2	79	35	13
	3	257	114	41
	4	639	284	102
50	1.5	18	8	3
	2	62	28	10
	3	234	105	38
	4	589	261	94
>50	1.5	17	8	3
	2	59	26	9
	3	225	100	36
	4	560	249	90

Source: Hewett (1995).

95% confidence level would be 133. This compares favorably with the result of 100 obtained using equation 4.8 (Perez and Lefante, 1997).

SAMPLING FOR VAPORS AND GASES

Once an appropriate exposure monitoring strategy which addresses the questions of where, when and who to sample and how many samples to take has been developed, the choice of monitoring technique must be made. Sample collection and analysis methods for various respiratory toxicants abound in the industrial and environmental hygiene literature. Any technique which meets target criteria for accuracy, precision, sensitivity and specificity may be suitable for use in compliance, investigative or epidemiologic monitoring. Table 4.6 lists various factors that can influence the performance of a monitoring technique or result in random or systematic errors. These factors must be addressed in defining the characteristics of a satisfactory air sampling and analysis method such as that outlined by NIOSH (Hull, 1984):

- The mean concentration measurement of the samples must be within 10% of the true concentration (in reality, the concentration determined by some independent and validated method). In addition, any uncorrected analytical bias, in conjunction with the sampling and measurement precision, must meet the ±25% criterion discussed below.
- The pooled relative standard deviation for unbiased methods must be less than 0.105 for the total analytical error to be less than 25% in at least 95% of the samples.
- Analytical recovery of collected toxicant for exposures equal to 0.5, 1 and 2 times the appropriate exposure standard or target level should be greater than or equal to at least 75% but preferably greater than 90%.
- Stability of samples collected at 80% relative humidity and at an exposure level equal to the standard or target level should be such that there is less than a 10% difference in recovery of analyte between those samples analyzed immediately and those stored for 1 week at room temperature or those stored for 2 weeks under refrigeration.
- For OSHA compliance sampling, the analytical range of the method should allow analysis of samples ranging from at least one-half to two times the exposure standard of the toxicant of interest.

Table 4.6 Factors affecting the precision and accuracy of air monitoring methods

Type of error	*Potential sources*
Random	• quality of equipment • skill of laboratory and field personnel • natural exposure variability
Bias	• chemical interferences • contaminated collection media • calibration of sample flow rate • pollutant collection efficiency • recovery of collected pollutant • loss of pollutant during sample storage • standardization of analytical response

'Standard' methods for certain toxicants have been developed and promulgated by various North American agencies including NIOSH, OSHA and the EPA. A listing and general description of OSHA and NIOSH methods for some important respiratory toxicants is shown in Table 4.7. As illustrated in the table, a wide range of analytical techniques have been applied to the measurement of collected respiratory toxicants, but in particular gas chromatography, liquid chromatography (HPLC and ion chromatography) and spectrophotometry. The collection methods for these materials may simply involve adsorption or absorption, such as collection on charcoal or in a water or solvent-filled

Table 4.7 Examples of standard methods employed in the United States for measurement of occupational exposure to some important respiratory toxicants

Respiratory toxicant	*NIOSH Method**	*OSHA Method†*
Ammonia	6701: passive collection in 0.01 N H_2SO_4; determination as ammonium cation by ion chromatography and conductivity detection	–
Chlorine dioxide	–	ID 202: active collection in carbonate-buffered KI solution; determination as chlorite ion by chromatography with conductivity detection
Formaldehyde	2541: active collection on XAD2 coated with hydroxylethyl-piperidine; desorption with toluene; determination as oxazolidine derivative by gas chromatography with flame ionization detection	ID 205: passive collection on bisulfite-treated paper, desorption with dilute sulfuric acid and reaction with chromotropic acid; determination by spectrophotometry at 580 nm.
Methylisocyanate	–	ORG 54: active collection on XAD7 treated with 1-(2-pyridyl)-piperazine; desorption with acetonitrile; determination as urea derivative by HPLC with UV or fluorescence detection.
Nitrogen dioxide	6014: active collection on molecular sieve coated with triethanolamine; desorption and reaction with solution of naphthylethylene diamine and sulfanilamide; determination by spectrophotometry at 540 nm.	–
Sulfur dioxide	6004: active collection on cellulose filter treated with Na_2CO_3; desorption with carbonate–bicarbonate buffer, determination as sulfite by ion chromatography with conductivity detection.	ID 104: active collection in hydrogen peroxide solution; determination as total sulfate by ion chromatography with conductivity detection.
Toluene diisocyanate (TDI)	2535: active collection on glass wool coated with *N*-4-nitrobenzyl-*N*-*n*-propylamine; desorption with methanol; determination as urea derivative by HPLC with UV detection.	ORG 18: active collection in toluene solution of *N*-4-nitrobenzyl-*N*-*n*-propylamine; determination as urea derivative by HPLC with UV detection.

* NIOSH (1994).
† OSHA Analytical Laboratory (1985).

bubbler, or they may involve simultaneous collection and reaction with a suitable reagent which converts the toxicant to a more chemically stable derivative. The latter approach is far more prevalent for respiratory toxicants which by nature tend to be chemically reactive and often require immediate stabilization upon collection (e.g. ozone, isocyanates, chlorine).

(a) Active and passive sampling

Active sampling methods employ a suitable sampling pump in conjunction with a collection device for the toxicant of interest. The sampling pump draws workplace air into or through the collection device at a known rate. The collection device either integrates the sample air into a homogeneous sample, as in sampling into an impermeable gas sampling bag, or selectively strips the toxicant from the moving air stream, concentrating it on the collection medium, as in charcoal tube sampling. In general, exposure monitoring should be accomplished by personal sampling in which the pump and collection device are placed on the worker with the inlet of the collection device positioned in the breathing zone.

While active sampling has long formed the basis of many classical and standardized methods for assessing workplace exposure, passive sampling devices have become increasingly popular for monitoring of exposure to vapors and gases in the workplace, as well as in the indoor and other environments. This is a result of their simplicity of design and ease of use. In contrast to active samplers, a passive sampler uses no external means of air movement for collection of pollutant. Rather, pollutant is collected through the mass transfer processes of either diffusion or permeation. Thus, no sampling pump or source of electricity is required. In consequence, passive samplers are small, lightweight and make no noise. In effect, a passive sampler is the chemical equivalent of a radiation badge.

The performance of a generic passive sampler (shown schematically in Figure 4.6) is governed by Fick's law of diffusion:

$$m = \frac{DA}{l} \cdot C \cdot t \tag{4.9}$$

where m is the mass of pollutant collected over time, t at average concentration, C. The passive sampler's

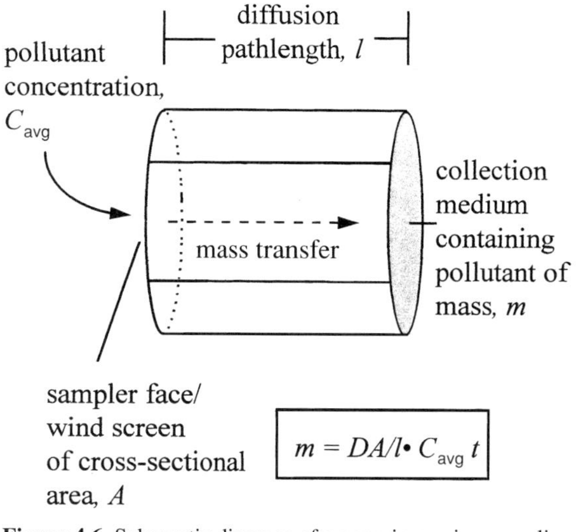

Figure 4.6 Schematic diagram of a generic passive sampling device. The passive sampler collects pollutant by the 'passive' mass transfer process of diffusion.

cross-sectional area is A, and the distance between the collecting face of the sampler and the collection medium constitutes the diffusion pathlength, l. D is the diffusion coefficient of the pollutant in air under the ambient conditions of atmospheric temperature and pressure.

The first passive samplers were described in the early 1970s (Palmes and Gunnison, 1973; Reisner and West, 1973). Since then, the application of passive sampling to the solution of environmental and occupational exposure problems has grown exponentially. There is a host of well-designed and characterized passive samplers now available from various commercial sources. There are even a limited number of 'standard' methods for respiratory toxicants which utilize passive sampling devices, as illustrated in Table 4.7.

While passive samplers have become widely accepted and utilized, there are certain unique limitations to the devices which should be mentioned:

- *Reduced sensitivity.* This limitation derives from the relatively low equivalent sampling rate of most passive sampling devices. The equivalent sampling rate is defined by the term DA/l and represents an imaginary flow of sample air through the device which would result in an equivalent amount of pollutant being collected. The equivalent sampling rates of passive samplers are in the range of 0.01 L/min. In contrast, active sampling devices usually are operated in the range of about 0.1 to 2 L/min. As a result, passive samplers tend to collect less than 10% of pollutant collected by an active sampling device for the same sampling time and exposure concentration. This can sometimes be a problem when the analytical finish used for measuring the collected pollutant lacks sufficient sensitivity. This limitation also makes passive samplers less applicable to short term monitoring, for example sampling for documentation of compliance with 10- or 15-minute short-term excursion limit (STEL) exposure standards. In general, passive samplers are more appropriate for extended monitoring periods, such as full shift, in order to allow for collection of sufficient toxicant for analytical measurement.
- *Air stagnation.* Passive samplers should never be used in situations where there is no external air movement, such as in enclosed spaces, in the corners of rooms, etc. This problem derives from the formation of a boundary layer at the passive sampler's face as air moves over it (Palmes and Lindeboom, 1979; Williams *et al.*, 1995). The boundary layer effectively increases the magnitude of the diffusion pathlength, l, so that the sampling rate is proportionately decreased. The boundary

layer depth is inversely related to the external wind speed, so that at high wind speeds, it is negligible and as the wind velocity decreases, the boundary layer depth grows. To counteract this effect, passive samplers generally employ wind screens and long internal diffusion pathlengths, so that the boundary layer depth must become very large (at very low wind speeds) before it has a significant effect on sampler performance. This approach results in a trade-off with sensitivity since the longer internal diffusion pathlength results in a concomitant decrease in sampling rate. Most commercial passive sampling devices are designed to operate properly at wind speeds of at least 25 feet/minute.

(b) Direct-reading devices

Direct-reading devices are useful adjuncts to the standard active or passive sampling methods used for measuring workplace exposure. These devices may range from relatively simple colorimetric detector tubes to sophisticated electronic devices that automatically collect a sample, analyze it, and store the data for later inspection, all on a continuous basis. Direct-reading devices are particularly valuable in investigative monitoring, when an assessment of exposure might be needed on an almost immediate basis, in initial exposure surveys for compliance monitoring where the object is to demonstrate no measurable exposure, and in those instances where an essentially continuous profile of exposure during the course of a workday is needed in order to identify peak exposure levels and perhaps correlate these with a specific operation, job task or location.

A thorough discussion of the myriad of commercially available direct reading monitors is beyond the scope of this work and the reader is directed to supplemental materials presented below under Further reading (p.68). It is worthwhile to note however that there has been much activity directed towards the development of performance criteria for direct reading devices (e.g. ANSI, 1990). This is an important endeavor in that there is a natural tendency towards falling into the 'Black Box Syndrome' when using these devices. In this, the device is considered to be a 'black box' whose output is considered to be infallible and absolute. In reality however, no direct-reading device is absolute. Rather, a signal (electrochemical, colorimetric, chemiluminescent, etc.) is generated in response to exposure to the targeted agent. Ideally, the magnitude of the signal will be linearly related to exposure concentration. In any case, the actual magnitude of the response/exposure relationship is arbitrary and must be periodically calibrated and standardized, either at the point of manufacture for those direct reading devices used once (colorimetric detector tubes) or by the end user for other types including various electronic devices. The re-calibration/standardization of these devices is the most important component of a quality control/quality assurance program in using any direct reading device.

REFERENCES

ANSI (1990) *American National Standard for Gas Detector Tube Units – Short Term Type for Toxic Gases and Vapors in Working Environments*, American National Standards Institute, New York.

Banks, D.E., Sastre J., Butcher, B.T. *et al.* (1989) Role of inhalation challenge testing in the diagnosis of isocyanate-induced asthma. *Chest*, **95**, 414–23.

Brochhagen, F.K. and Schal, H.P. (1986) Diphenylmethane diisocyanate: the concentration of its saturated vapor. *Am. Ind. Hyg. Assoc. J.*, **47**, 225–8.

Cochran, W.G. (1968) Errors of measurement in statistics. *Technometrics*, **10**, 637–66.

Dharmarajan, V. (1979) Occupational exposures to methylene bis-phenylisocyanate: gaseous or aerosol? *J. Environ. Pathol. Toxicol.*, **2**, 1–8.

Dharmarajan, V. and Weil, H. (1978) Physical state of airborne 4,4′-diphenylmethane diisocyanate and its measurement. *Am. Ind. Hyg. Assoc. J.*, **39**, 737–44.

Eatough, D.J., Benner, C.L., Bayona, J.M. *et al.* (1989) The chemical composition of environmental tobacco smoke. I. Gas phase acids and bases. *Environ. Sci. Technol.*, **23**, 679–87.

Esmen, N.A. and Hammad, Y.Y. (1977) Log-normality of environmental sampling data. *J. Environ. Sci. Health*, **A12**, 29–41.

Harris, H.E., DeSieghardt, W.C., Burgess, W.A. and Reist, P.C. (1974) Respirator usage and effectiveness in bituminous coal mining operations. *Am. Ind. Hyg. Assoc. J.*, **35**, 159–64.

Hewett, P. (1995) Sample size formulae for estimating the true arithmetic or geometric mean of lognormal exposure distributions. *Am. Ind. Hyg. Assoc. J.*, **56**, 219–25.

Hull, R.D. (1984) Development and evaluation of methods, in *NIOSH Manual of Analytical Methods*, 3rd edn, National Institute for Occupational Safety and Health, Cincinnati, pp. 29–34.

Ishizu, Y. (1980) General equation for the estimation of indoor pollution. *Environ. Sci. Technol.*, **14**, 1254–7.

Jones, R.N., Rando, R.J., Glindmeyer, H.W. *et al.* (1992) Abnormal lung function in polyurethane foam producers: weak relationship to toluene diisocyanate exposures. *Am. Rev. Respir. Dis.*, **146**, 871–7.

Koutrakis, P., Fasano, A.M., Slater, J.L., Spengler, J.D., McCarthy, J.F. and Leaderer, B.P. (1989) Design of a personal annular denuder sampler to measure atmospheric aerosols and gases. *Atmos. Environ.*, **23**, 2767–73.

Leidel, N.A., Busch, K.A. and Lynch, J.R. (1977) *Occupational Exposure Sampling Strategy Manual*, National Institute for Occupational Safety and health, Cincinnati.

Lenart, S.W. and Campbell, D.L. (1984) Assigned protection factors for two respirator types based upon workplace performance testing. *Ann. Occup. Hyg.*, **28**, 173–82.

Myer, H.E., O'Block, S.T. and Dharmarajan, V. (1993) A survey of airborne HDI, HDI-based polyisocyanate and solvent concentrations in the manufacture and application of polyurethane coatings. *Am. Ind. Hyg. Assoc. J.*, **54**, 663–70.

Moore, D.E. and Smith, T.J. (1976) Measurement of protection factors of chemical cartridge half-mask respirators under working conditions in a copper smelter. *Am. Ind. Hyg. Assoc. J.*, **37**, 453–8.

NIOSH (1994) *NIOSH Manual of Analytical Methods*, 4th edn, National Institute for Occupational Safety and Health, Cincinnati.

OSHA Analytical Laboratory (1985) *OSHA Analytical Methods Manual, Vol. I, II, III,* American Conference of Governmental Industrial Hygienists, Cincinnati.

Palmes, E.D. and Gunnison, A.F. (1973) Personal monitoring device for gaseous contaminants. *Am. Ind. Hyg. Assoc. J.*, **33**, 78–81.

Palmes, E.D. and Lindeboom, R.H. (1979) Ohm's Law, Fick's Law, and diffusion samplers for gases. *Anal. Chem.*, **51**, 2400–1.

Pankow, J.F., Isabelle, L.M., Buchholz, D.A., Luo, W. and Reeves, B.D. (1994) Gas/particle partitioning of polycyclic aromatic hydrocarbons and alkanes to environmental tobacco smoke. *Environ. Sci. Technol.*, **28**, 363–5.

Perez, A. and Lefante, J.J. (1997) Sample size determination and the effect of censoring when estimating the arithmetic mean of a log-normal distribution. *Communications in Statistics, Theory and Methods*, **26**, 2779–801.

Rando, R.J. and Lefante J.J. (1992) Assessing respirator use in exposure surveys: chlorine as a paradigm for irritant gas exposure. Presented at The American Industrial Hygiene Conference & Exposition, Boston, June 1–5.

Rando, R.J. and Poovey, H.G. (1994) Dichotomous sampling of vapor and aerosol of methylene-bis-(phenylisocyanate) [MDI] with an annular diffusional denuder. *Am. Ind. Hyg. Assoc. J.*, **55**, 716–21.

Rando, R.J., Abdel-Kader H. and Hammad, Y.Y. (1984) Isomeric composition of airborne TDI in the polyurethane foam industry. *Am. Ind. Hyg. Assoc. J.*, **45**, 199–203.

Rando, R.J., Abdel-Kader, H., Hughes, J. and Hammad, Y.Y. (1987) Toluene diisocyanate exposures in the flexible polyurethane foam industry. *Am. Ind. Hyg. Assoc. J.*, **48**, 580–5.

Reisner, K.D. and West, P.W. (1973) Collection and determination of sulfur dioxide incorporating permeation and West-Gaeke procedure. *Environ. Sci. Technol.*, **7**, 526–32.

Repace, J.L. and Lowrey, A.H. (1993) An enforceable indoor air quality standard for environmental tobacco smoke in the workplace. *Risk Analysis*, **13**, 463–75.

Seixas, N.S., Robins, T.G. and Moulton, L.H. (1988) The use of geometric and arithmetic mean exposures in occupational epidemiology. *Am. J. Ind. Med.*, **14**, 465–77.

Speizer, F.E. and Frank, N.R. (1966) The uptake and release of SO_2 by the human nose. *Arch. Environ. Health*, **12**, 725–8.

Smith, T.J., Peters, J.M., Reading, J.C. and Castle, C.H. (1977) Pulmonary impairment from chronic exposure to sulfur dioxide in a smelter. *Am. Rev. Respir. Dis.*, **116**, 31–9.

Smith, T.J., Ferrell, W.C., Varner, M.O. and Putnam, R.D. (1980) Inhalation exposure of cadmium workers: effects of respirator usage. *Am. Ind. Hyg. Assoc. J.*, **41**, 624–9.

Wadden, R.A. and Scheff, P.A. (1987) *Engineering Design for the Control of Workplace Hazards*, McGraw-Hill, New York.

Williams, C.E., Pintauro, P.N. and Rando, R.J. (1995) A transient model of mass transfer and kinetics in a passive vapor sampler. *Am. Ind. Hyg. Assoc. J.*, **56**, 1074–82.

FURTHER READING

Burgess, W.A. (1995) *Recognition of Health Hazards in Industry*, 2nd edn, John Wiley & Sons, Inc., New York.

Cohen, B.S. and Hering, S.V. (1995) *Air Sampling Instruments*, 8th edn, American Conference of Governmental Industrial Hygienists, Cincinnati.

Cohen, B.S., McCammon, C.S. and Vincent, J.H. (eds) (1993) Proceedings of the International Symposium on Air Sampling Instrument Performance, *Appl. Occup. Environ. Hyg.*, **8**, 225–411.

Hawkins, N.C., Norwood, S.K. and Rock, J.C. (1991) *A Strategy for Occupational Exposure Assessment*, American Industrial Hygiene Association, Akron.

5 Airways obstruction and occupational inorganic dust exposure

Mei-lin Wang and Daniel E. Banks

INTRODUCTION

Occupational exposure to silica, coal, asbestos and/or mixed industrial dusts has long been recognized as a cause of pneumoconiosis. The association between chronic dust exposure and loss of lung function, even in the absence of simple pneumoconiosis, has found acceptance in the 1980s. Therefore, dust exposures have the potential adversely to alter lung function by inducing industrial bronchitis, or to add to existing bronchitis which may be attributable to smoking (Morgan, 1978). However, disagreement exists on the degree of lung function decline attributed to dust exposure, and whether occupational dust exposure alone can cause 'clinically important' airways obstruction and disabling impairment.

Airways obstruction discovered on a routine examination or in a population survey is commonly described as airflow limitation. Chronic obstructive pulmonary disease (COPD) is a response by the bronchi and the lungs which depends on the nature and intensity of the adverse environmental exposure, as well as the host's susceptibility. COPD is characterized by a decrease in the expiratory flow rates and is defined as a reduction in the forced expiratory volume in one second (FEV_1) occurring without a decrease in lung volume. Asthma, chronic bronchitis, emphysema and overlaps of these entities are generally accepted under this heading and although these are distinct entities with different implications for prognosis, they often occur together (Burrows, 1981, 1990). Although the risk of death from COPD is strongly related to the degree of airways obstruction, the presence or absence of mucus hypersecretion does not affect mortality (Peto *et al.*, 1983). Since the term COPD describes a physiologic parameter rather than an etiologically defined disease, the cause explaining these abnormal pulmonary parameters in any single individual cannot be made without a clinical evaluation.

For the purposes of this chapter, a clinically important degree of impairment is defined as an FEV_1 <65% of predicted. Others have suggested that this level of FEV_1 reflects 'shortness of breath when walking with other people of one's own age on level ground' (Soutar *et al.*, 1993). This can be reasonably acknowledged as an important degree of impairment. As a rule, the extent of reduction in FEV_1 relates to symptom severity, the frequency of complications, and the mortality attributable to COPD. The FEV_1 is the most important objective clinical tool for assessing the severity of airways obstruction. It should be noted that an FEV_1 decline may also be attributable to restriction of lung volumes. Restriction may

Occupational Lung Disease. Edited by Daniel E. Banks and John E. Parker. Published in 1998 by Chapman & Hall. ISBN 0 412 73630 6

be attributable to numerous explanations, notably pulmonary fibrosis, muscle or nerve weakness, lung resection or obesity.

In the general population, cigarette smoking accounts for the overwhelming proportion of patients with severe airways obstruction. However, since most smokers show little excess loss of lung function [smokers average a 10–20 ml/year excess rate of FEV_1 decline which returns to normal when the individual ceases smoking (Stenton and Hendrick, 1992)], with just 10–15% of the smoking population severely affected (Fletcher *et al.*, 1976), it is clear that there is a very wide range of individual susceptibility. The explanation for this individual susceptibility remains elusive.

Numerous epidemiologic studies have addressed the issue of whether occupational dust exposure causes COPD, and the relative importance of dust versus cigarette smoke exposure, but have arrived at different conclusions. Reports have estimated the effect of dust exposure on airways obstruction to be one-third, equivalent, no less, or more severe than the effect of cigarette smoking (Morgan, 1986).

EFFECTS OF DUST ON AIRWAYS OBSTRUCTION

Silica dust

The evidence for airways obstruction, as it relates to silica, is found primarily in the epidemiologic studies of South African gold miners. For the most part, investigators of these populations have found an association between silica dose and airways obstruction in workers exposed to silica dust (Becklake *et al.*, 1958; Sluis-Cremer *et al.*, 1967a,b; Cowie and Mabena, 1991; Hnizdo *et al.*, 1991; Hnizdo, 1992).

In a cross-sectional sample of 1973 middle-aged white gold miners, chronic bronchitis was equally prevalent in men with radiologically diagnosed silicosis matched with those without silicosis, and the mean forced vital capacity (FVC) did not differ between the two groups. Men with silicosis had the same amount or excess airways obstruction compared with the men without silicosis (when adjusted for total dust exposure), leading to the conclusion that the association between dust exposure and airways obstruction was a dust, rather than a disease (silicosis), effect (Irwig and Rocks, 1978).

A 5-year follow-up of lung function from 1249 white South African gold miners indicated that the contribution of tobacco smoking to COPD was substantially higher than the contribution of dust (Hnizdo, 1992). Mortality attributable to airways obstruction occurred only in dust-exposed smokers (Hnizdo, *et al.*, 1990; Reid and Sluis-Cremer, 1996; Hnizdo, 1992).

A cross-sectional study of 1197 working black South African gold miners showed that simple silicosis was associated with significant loss of lung function. A greater percentage of the reduction in FEV_1 and FEV_1/FVC % was attributed to the dust exposure than the amount of tobacco smoked [after 25 years in mining, workers averaged an FEV_1 loss of 172 ml due to smoking (25 pack-years) and 200 ml due to dust (Cowie and Mabena, 1991)]. Commenting on the above paper, Hnizdo *et al.* (1991) suspected that the effect of smoking may have been underestimated by the regression model applied, since smoking was represented by cigarette pack-years only. The fact that the effect of cigarette smoking in current smokers is likely greater than the effect on ex-smokers was not a part of this model.

Graham *et al.* (1994) provided epidemiologic information regarding the rate of lung function decline in Vermont granite workers studied serially from 1979 to 1987. Exposure and lung function data from these workers has formed the backbone for the silica standard in the US. Silica dust exposures were closely monitored and mean values were below the current Occupational Safety and Health Administration (OSHA) standard of 100 $\mu g/m^3$, generally in compliance with the current silica standard. The results obtained by analysis of covariance showed that lung function decline over an 8-year period attributable to dust exposure was not significantly different from those without dust exposures.

Since 1952, COPD has been a compensable disease for South African miners. Gold mine dust contains roughly 30% free silica, and inhaled silica particles could be involved in a pathogenic processes leading to COPD. First, as with other dusts, chronic bronchitis can develop and be associated with a decline in flow rates. Second, inhaled small silica particles impact on the bifurcations of the terminal conducting airways, leading to small airway wall (peribronchiolar) fibrosis and thickening, adversely affecting small airway flow rates (Churg *et al.*, 1985). Third, silica particles can be engulfed by alveolar macrophages, which release proteolytic enzymes and toxic free radical species. These have the potential to destroy alveolar walls and cause histologic change which may eventually lead to emphysema (Churg *et al.*, 1989), although a description of emphysema in silicosis is conspicuously absent in other reports (Gibbs and Wagner, 1988).

Becklake *et al.* (1987) reported the demographic determinants of autopsy-proven emphysema in 44 South African gold miners with emphysema and compared with an age-matched group of 42 miners without emphysema. Multiple logistic regression analysis indicated that age, smoking and dust exposure were three strong and independent predictors of emphysema. The presence of silicosis did not increase the risk for emphysema and bronchitis. When compared with a non-smoker, a smoker of 20 cigarettes a day has an approximately 30 times greater chance of having grade 2 or more emphysema; and compared with a miner who has never worked in high dust, one who has worked in dust for 20 years has an approximately 13 times greater chance of developing grade 2 or more emphysema. Yet, in a much larger autopsy series of 242 South African gold miners who never smoked but were exposed to silica dust for many years, explanations for accelerated lung function declines were heart disease, silicosis and obesity (Hnizdo *et al.*, 1994). In these miners, there was an insufficient degree of emphysema to result in a statistically significant impairment in lung function.

Computed tomography (CT) scans were performed on silicotic and asbestotic workers (ILO category 0 or 1) evaluated for pneumoconiosis (Begin *et al.*, 1995). All six of the silicotics who were non-smokers had CT scan evidence of emphysema. Of the 60 current and ex-smokers with silicosis, 32 (53%) had emphysema. For all silicotics, the FEV_1 was 96 ± 3% predicted and FVC was 110 ± 2% predicted. Although the emphysema scores of those with silicosis and those without silicosis were not different, those without silicosis had a significantly greater emphysema score when compared with those without asbestosis.

Coal mine dust

For the most part, cross-sectional and longitudinal studies have shown a loss in lung function in miners using parameters of coal mine dust exposure such as years of employment, cumulative dust exposure and job category (Enterline and Lainhart, 1967; Higgins *et al.*, 1968; Kibelstis *et al.*, 1973; Rogan *et al.*, 1973; Morgan *et al.*, 1974). Perhaps the most common adverse effect of dust exposure is the development of industrial bronchitis. Miners are at considerable risk for the development of this process through their dust exposures in the mine environment (Rogan *et al.*, 1973; Morgan, 1978; Attfield and Hodgson, 1986). Categories of simple pneumoconiosis were not consistently associated with a decline in FEV_1 unless another process (typically bronchitis) was present, while complicated pneumoconiosis was generally associated with a significant decline in FEV_1 (Morgan *et al.*, 1974). Not all investigators have identified an accelerated decline related to dust exposure. The FEV_1 in Nigerian coal miners assessed cross-sectionally was no different than in the general population (Jain and Patrick, 1981). In the report by Seixas *et al.* (1993) describing a coal mining population studied over a 13-year period (between Rounds 2 and 4 of the US National Study for Coal Workers' Pneumoconiosis) there was no accelerated decline in FEV_1 attributable to dust. The loss of lung function detected from the Round 4 survey was thought to have occurred before the start of the longitudinal assessment of this cohort.

The coal macule, the basic pathologic feature of coal workers' pneumoconiosis, is associated with the development of centriacinar emphysema. Ryder *et al.* (1971) showed a significant excess of emphysema in coal miners with pneumoconiosis compared with non-miners, but did not comment on the prevalence of emphysema in miners without pneumoconiosis. Coal miners in South Wales showed more centrilobular emphysema at autopsy compared with non-miners; however, much of this difference was explained by current smokers over the age of 60 years, a group where PMF was common (Cockcroft *et al.*, 1982). Lyons *et al.* (1981) reported similar findings.

Centrilobular emphysema was by far the commonest type of emphysema encountered in both smokers and non-smokers; with less decline in FEV_1 in the non-smokers than smokers. Ruckley *et al.* (1984), in a large autopsy of miners from 24 British coal mines, showed a very strong relationship between smoking and emphysema. As expected, centrilobular emphysema increased with smoking and age. In the absence of pneumoconiosis (n = 72 miners), there was no detectable effect of dust leading to emphysema. Even though the groups with simple pneumoconiosis (n = 149) and PMF (n = 121) (groups with an increased exposure to coal dust) had an increased chance of centriacinar emphysema, no clear relationship could be demonstrated between the amount of dust in the lung and the extent of emphysema.

Leigh *et al.* (1994) noted that use of the pathologic assessment of the lung to determine the relationship between emphysema and coal dust might induce bias because the coal macules might draw attention to emphysematous spaces, resulting in the perspective that emphysema is more common in those with pneumoconiosis. They measured the actual coal content of the lungs and found that in smokers, the emphysema score was driven by the coal content in the lungs, cigarette smoking and age. In non-smokers, the coal content and age were most important. Questionnaire data on smoking were reported and correlations between the emphysema score and the questionnaire and lung function performed during life were attempted. In 28 non-smokers, the relationship between the decline in FEV_1 % and the increase in emphysema score approached significance (P = 0.09).

There has been little work done in miners which addresses the response of lung function after ceasing dust exposure. Bates *et al.* (1985) showed no change in the annual rate of lung function decline in miners who retired from mining compared with the rate at which they declined while mining. Those who were rapid decliners continued at that pace. Bates attributed this to dust loading of the lung. Although perhaps not relevant to the mining population, Kauffman *et al.* (1982) noted that the rate of lung function decline lessened or returned to normal when dust-exposed Paris factory workers moved to non-dusty jobs.

Asbestos dust

Restriction in lung volume and impairment in diffusion capacity ($D_{L}CO$) are the major physiologic features resulting from long-term asbestos exposure (Kleinfeld *et al.*, 1966; Becklake, 1976; Selikoff and Lee, 1978). In numerous studies, a high incidence of airflow limitation, primarily in the small airway flow rates, was also found (Muldoon and Tumer-Warwick, 1972; Weill *et al.*, 1975 Harless *et al.*, 1978, Rodriguez-Roisin *et al.*, 1980). However, when the effects of smoking were taken into account, the evidence of airway obstruction in asbestos-exposed workers was considerably weakened (Murphy *et al.*, 1972; Zitting *et al.*, 1978).

Several studies have sought to demonstrate a link between asbestos exposure and obstructive lung disease. Some have emphasized a causal role of cigarette smoking, whereas others suggest that airways disease caused by asbestos is undetectable in its early stages by usual testing methods. In surveys of asbestos workers where the effects of cigarette smoking were addressed, most airway obstruction could be explained by smoking. For example, Samet *et al.* (1979) used multivariate analysis techniques to separate the effects of asbestos exposure from those of cigarette smoking. They found additive and independent effects of both factors on FVC and $D_{L}CO$, but did not report the effects on FEV_1 or other flow rates. No evidence of synergism was identified. Pearle (1981) reported the results of lung function testing in 88 asbestos workers. In 21 non-smokers and 56 smokers (> 10 pack-years), there were significantly more smokers with an FEV_1 <80% predicted. Only one non-smoker (5%) had a % FEV_1/FVC less than 70% as compared with 17 (30%) of the smokers.

Peripheral airways obstruction in asbestos workers has been investigated. In a small number of men exposed to asbestos without chest radiographic abnormalities, Jodoin *et al.* (1971) used an intraesophageal balloon to show that flow rates were less at a given transpulmonary pressure, suggesting that small airway air flow resistance was increased. Begin *et al.* (1983) also suggested that narrowing of the small airways may limit air flow by finding a higher upstream resistance in 17 life-long non-smoking asbestos workers. They concluded that small airway obstruction is associated with asbestos exposure and is independent of smoking.

Lung function was measured in 296 boilermakers with 15 or more years' shipyard exposure to asbestos and showed that airway obstruction occurs after 15 years of exposure and before the slightest radiographic evidence of asbestosis. The small airway flow rates lessen as the profusion of irregular opacities increase (Kilburn and Warshaw, 1990). Sue *et al.* (1985) also assessed the contribution of smoking to lung function abnormalities of asbestos-exposed shipyard workers applying for disability. Non-smokers had significantly higher VC, FEV_1, FEV_1/VC, and $D_{L}CO$ values than smokers. Only 3/73 (4%) of the non-smokers, but 23/73 (32%) of the smokers, had an FEV_1/VC below the 95% confidence limit of predicted value.

Wang and Lu (1985) studied lung function of 155 asbestos textile workers and 130 controls. Lung function impairment in asbestos workers was predominantly restrictive, appearing even before radiographic changes. There was no significant difference in small airway function when comparing exposed workers and controls.

Wright and Churg (1984) compared autopsy findings in the small airways of asbestos workers matched with cigarette smokers without asbestos exposure. The authors concluded that a characteristic increase in fibrous tissue often accompanied by pigment in the respiratory bronchioles and alveolar ducts was found in some asbestos workers. These are different from the small airway lesion attributable to cigarette smoking and the authors suggested that this be termed 'asbestos airways disease'.

QUANTITATIVE ESTIMATES OF DUST EFFECTS ON AIRWAYS OBSTRUCTION

The approach of multiple regression analysis in cross-sectional or longitudinal data assessment provides quantitative estimates of lung function decline attributable to dust, smoking exposures and to other risk factors.

In Figure 5.1, the predicted FEV_1 was plotted versus years of dust exposure (at specific dust concentrations) using regression equations derived from several cross-sectional data analyses (Hnizdo, 1992; Attfield and Hodous, 1992; Wang *et al.*, 1996a). The predicted

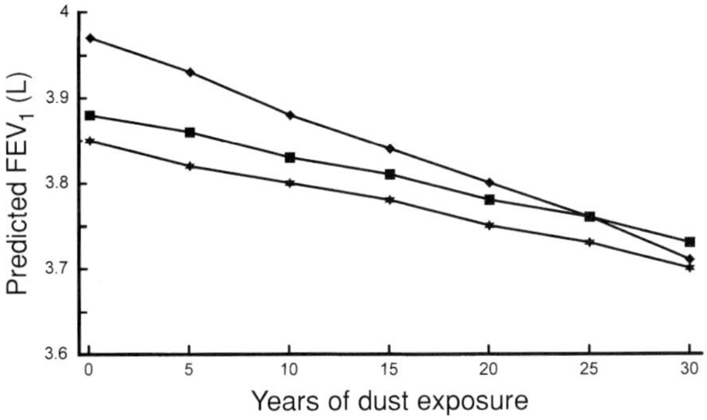

Figure 5.1 Lung function which would be expected to be present in a 50-year-old white man, non-smoker, 71 inches (1.8 m) tall, with different years of dust exposure in a gold mine in South Africa (silica), a coal mine in the United States (coal), or in a United States steel mill (Hnizdo, 1992; Attfield and Hodous, 1992; Wang *et al.*, 1996a) ◆, lung function values from silica workers exposed to 0.3 mg/m^3; ■, lung function in coal workers exposed to 4.0 mg/m^3; *, lung function in steel mill workers who worked in a dusty job. For the silica and coal mine dust exposures, dust exposures in units of mg/m^3-year were converted to gh/m^3, using a factor of 1740 (hours of employment per year)/1000 (mg per gram). The average silica dose used in this plot is three times the current US standard and the average coal dust exposure used is twice the current US standard. Exposure measurements were not available for the steel mill. In this workplace, exposure was expressed as years worked in a dusty job.

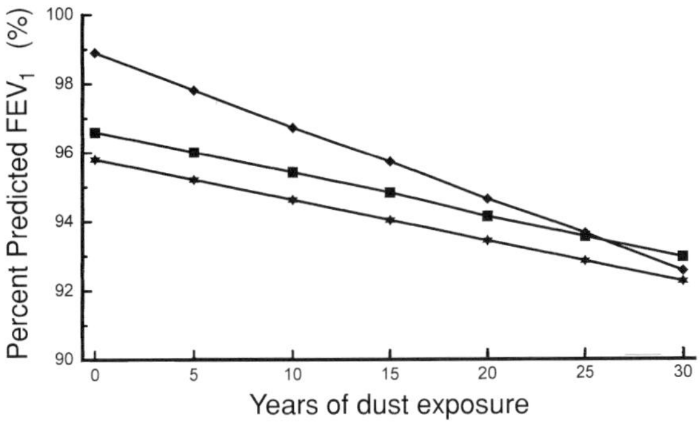

Figure 5.2 Adjusted lung function for populations described in Figure 5.1. The FEV_1 on the figure is reported as the percent of predicted FEV_1. It is the result of dividing the FEV_1 derived from regression equations used in Figure 5.1 by the value from Knudson *et al.* (1983). Symbols are as for Figure 5.1.

FEV_1 was calculated from these equations for non-smoker white men standardized to age 50, 71 inches (1.8 m) tall, with dust exposures ranging from none to 30 years. Compared with coal dust, silica dust has a greater adverse effect on lung function. The percent predicted FEV_1 (the FEV_1 calculated by the above equations divided by the value calculated by Knudson's equation, 1983) was also plotted versus years of dust exposure (Figure 5.2). For non-smokers, after 30 years' dust exposure the percent predicted FEV_1 exceeded 90%. Silica dust exposure at a mean concentration of 0.3 mg/m^3 for 30 years resulted a 6.4% reduction in FEV_1, while exposure to coal dust at a mean concentration of 4 mg/m^3 resulted in a 3.6% decline. By comparison, employment in a dusty job in a steel mill is estimated to result in a 3.7% decline in FEV_1. Further information is available in the legends for Figures 5.1 and 5.2.

Table 5.1 reviews longitudinal epidemiologic studies which reported regression models and showed the regression coefficients. The estimates of annual decline of FEV_1 attributable to smoking and dust exposure, and their relative importance have been summarized. Smoking was constantly a strong predictor for annual decline of FEV_1. In studies where both cross-sectional and longitudinal data analysis was used (Hnizdo, 1992; Seixas *et al.*, 1993; Wang *et al.*, 1996a), the relationship between dust exposure and lung function was less evident when longitudinal compared with cross-sectional approaches were applied in the same population. Four of the six longitudinal studies listed in Table 5.1 were unable to detect a statistically significant dust effect on lung function. Only one of four models in one report (Attfield, 1985) showed a relationship between intersurvey at face work and annual decline of FEV_1 ($P = 0.01$). In the reports cited in Table 5.1, the estimated annual loss of FEV_1 attributed to smoking ranged from 8.5 to 15.5 ml/year, and to dust exposure varied from 3.8 to 20.0 ml/year, depending upon the level.

In addition, large unexplained variations in the rate of FEV_1 decline in longitudinal studies have been found by other investigators. Love and Miller (1982) examined the lung function of 1677 miners who took part in three surveys in British coal mines. Regression analysis showed that age, height and smoking each made a strong contribution to explaining the observed loss of FEV_1. After adjustment for these factors, the rate of FEV_1 declined about 3.8 ml/year with the average observed dust exposure (117 g·h/m^3). The author indicated 'inclusion of all the factors including age, height, smoking status, mine and previous dust, accounted for just over 6% of the total variance (R^2) in loss of FEV_1, the remainder being unexplained. This was typical of all other similar regression analyses carried out; none of them explained more than 7% of the total variance'. The variance explained by the models listed in Table 5.1 ranged from 6% to 12%.

Table 5.1 Longitudinal studies estimating annual decline of FEV_1 attributable to smoking and dust exposure and their relative importance in multiple linear regression models*

Study	*Study details*	*Change FEV_1 (ml/yr) attributed to smoking*	*Change FEV_1 (ml/yr) attributed to dust*	*Relative importance (smoking vs. dust)*
Love and Miller (1982)	1,677 British coal miners studied for approximately 11 yrs., beginning in 1957	Current smoker versus Non-smoker −11.1 ml/yr (P<0.001)	Previous dust at average exposure of 117 gh/m^3 −3.8 ml/yr (P <0.05)	Ratio = 2.92
			Previous dust at likely maximal level for a working life of 245 gh/m^3 −7.9 ml/yr (P <0.05)	Ratio = 1.41
			Previous maximal observed dust of 628 gh/m^3 −20.0 ml/yr (P<0.05)	Ratio = 0.56
			Intersurvey dust exposure	Dust effect not significant
Attfield (1985)	1072 US coal miners, between 1969–1971 to 1977–1981, adjusted to 11 years	Model I: Current smoker versus Non-smoker −8.6 ml/yr (P <0.001)	Intersurvey underground work −3.4 ml/yr (P = 0.29)	Dust effect not significant
		Model II: Current smoker versus Non-smoker −8.7 ml/yr (P <0.001)	Intersurvey at face work −7.3 ml/yr (P = 0.01)	Ratio = 1.19
		Model III: Current smoker versus Non-smoker −8.5 ml/yr (P <0.001)	Intersurvey dust concentration −2.5 ml/mg/m^3/yr (P = 0.12)	Dust effect not significant
		Model IV: Current smoker versus Non-smoker −8.7 ml/yr (P <0.001)	Underground work before initial survey −2.4 ml/yr (P = 0.15)	Dust effect not significant
Seixas *et al.* (1993)	977 US coal miners, beginning work in or after 1970, between 1972–1975 (Round 2) to 1985–1988 (Round 4), approximately 13 yrs	Current smoker versus Non-smoker −8.7 ml/yr (P <0.001)	Cumulative exposure before initial survey +0.7 ml/yr per mg/m^3 (P = 0.06)	A small increase in FEV_1, with higher pre-survey cumulative exposure
			Intersurvey mean exposure +2.8 ml/mg/m^3/yr (P >0.2)	Dust effect not significant
Hnizdo (1992)	White South African gold miners studied between 1968–71 and 1975, approximately 5 yrs	Current smoking status and cigarette equivalent pack-years significantly related to longitudinal loss of lung function	Cumulative dust exposure not related to longitudinal loss of lung function	The corresponding regression coefficients were not reported. Results reported in accord with figure 1, page 476 of the original publication

Table 5.1 Continued

Study	*Study details*	*Change FEV_1 (ml/yr) attributed to smoking*	*Change FEV_1 (ml/yr) attributed to dust*	*Relative importance (smoking vs. dust)*
Graham *et al.* (1994)	711 US granite workers studied from 1979 to 1987	Current smoker versus Non-smoker –15.5 ml/yr (P <0.001)	Shed vs. Office –7.1 ml/yr (P = 0.197)	Dust effect not significant
			Quarry vs. Office –9.8 ml/yr (P = 0.163)	Dust effect not significant
Wang *et al.* (1996a)	475 US steel workers studied from 1982 to 1991	Current smoker versus Non-smoker –14.5 ml/yr (P <0.001)	Intersurvey years of dusty job < –1.0 ml/yr (P >0.2)	Dust effect not significant. Weight gain over time was an important predictor.

* R^2 for the studies reported in this table range from 6–12%.

'CLINICALLY IMPORTANT' AIRWAYS OBSTRUCTION IN DUST-EXPOSED WORKERS

Many of the above studies have suggested a statistically significant association between dust exposure and the loss of lung function; however, these 'average effects' provide little indication of the importance to individual miners. Attempts have been made to identify miners who suffered severe impairment of lung function. Although the effect of dust in smokers is more difficult to quantitate, one study reports 31 non-smokers out of 611 miners applying for 'Black Lung' compensation. Of six with an FEV_1 below 60% predicted or an FEV_1/FVC ratio below 60%, five had medical conditions to explain this impairment. One of 31 (3%) had unexplained severe obstruction (Lapp *et al.*, 1994).

Soutar and Hurley (1986) studied current and ex-miners in Britain. Cross-sectional analysis of the relationship between lung function and individual cumulative exposure to respirable dust was estimated in 1867 miners and in 2192 ex-miners. A questionnaire was administered which allowed classification of the worker as bronchitic, or not, and reported the smoking history. If those with progressive massive fibrosis (PMF) are excluded, the average loss of FEV_1 due to moderately high dust exposures (a cumulative exposure of 300 g·h/m^3) was 228 ml. In a select group of 199 ex-miners with chronic bronchitis, the estimated effect of this exposure was greater and in 35 ex-smokers (1.6% of the ex-miners), the mean loss was 940 ml of FEV_1 (leaving the remaining lung function to approximate 65% of the predicted FEV_1). The authors observed, 'In general, the effect of dust on lung function was not affected by smoking habit, and the reason for the excessive effect of dust in some men must be sought elsewhere. A third factor (or more), whether intrinsic susceptibility to the effects of dust or an environmental factor not so far identified, must be sought to explain why these men suffered such severe lung damage in response to their exposure to respirable dust'.

Marine *et al.* (1988) analyzed the results of lung function from 3380 British miners (2837 smokers and 543 non-smokers) using a linear logistic model with an average miner's age of 47 years. Assuming a 'high' or 'zero' exposure, respectively, the estimated prevalence of miners who smoked and had an FEV_1 less than 65% predicted was 14.2% and 5.0%, and in non-smoking miners was 7.7% and 3.2%. The 'high' exposure (348 g·h/m^3) implied average dust concentrations in the range 6.1 to 7.2 mg/m^3. Some 9% of the miners had accumulated exposures exceeding 348 g·h/m^3 when they were surveyed in the 1960s. Overall, the observed prevalence of FEV_1 less than 65% predicted in non-smokers and smokers was 5.7% and 10.3%, respectively.

Similarly, Soutar *et al.* (1993) showed that in a British colliery where lifetime dust exposures could be shown to approximate 400 g·h/m^3 (35 years of employment at 7 mg/m^3 for 1631 hours per year), approximately 20% of the non-smokers and 45% of the smokers had a 'clinically important' deficit in lung function. The authors suggest that these values are greater than those in other reports because miners with PMF were included in this study. In two other collieries, where the exposures were less, such clinically important declines were not shown. This was thought due to less dust and possible confounding between dust and age.

Oxman *et al.* (1993) reviewed studies between 1966–1991 where dust measurements were available and where the relationship between dust exposure and outcome was determined while controlling for age and smoking. Although 13 studies were available, these encompassed only four different cohorts (three of the cohorts were coal miners and one was gold miners). All showed a relationship between lung function and dust exposure. In two papers (Marine *et al.*, 1988; Hnizdo *et al.*, 1990), data were re-analyzed to show the number of workers by smoking category with clinically significant declines in FEV_1. Approximately 1.2% of non-smoking coal miners ($P = 0.053$) and 2.3% of smoking coal miners ($P < 0.001$) with a cumulative dust exposure of 122.5 g·h/m^3 (35 years of work with a mean dust exposure of 2 mg/m^3) could be expected to have an percent predicted FEV_1 less than 65%. Although the maximal cumulative dust concentration in the South African gold mines was only one-fifth of 122.5 g·h/m^3 (the average dust level in the coal mines), the risk for a clinically important loss in FEV_1 in non-smoking gold miners was estimated to be three times as great.

Three of 203 (1.5%) and 17 of the remaining 200 (8.5%) never-smoking steel workers undergoing yearly spirometry twice or more over the decade of the 1980s had final FEV_1 values < 65% or between 65% and 80% predicted, respectively (Banks *et al.*, 1996). Of the three < 65% predicted, one was morbidly obese (body mass index of 39.6 kg/m^2) (FEV_1/FVC of 70%), the second had a history of trauma with fractured ribs, arm and pelvis, and a diaphragmatic hernia following a severe crush injury (FEV_1/FVC of 88%), and the third had an FEV_1 of 64% and FVC of 59% (FEV_1/FVC of 89%). In this third case, the explanation for this apparent restriction is unclear. In the 17 non-smokers with FEV_1 values between 65% and 80% predicted, one chart was unavailable for review, one had chest trauma, one was obese with asthma, seven were obese, three had asthma and two had hay fever. Medical conditions which might explain this decline in the remaining two workers were not found (Banks *et al.*, 1996).

Becklake (1985) summarized the relevant epidemiologic studies during 1973–1985 and observed that 'airflow limitation, almost certainly multifactorial in etiology, occurs in those engaged in dusty occupations as well as in those who are not; among the environmental factors concerned, cigarette smoking is clearly one of the most important, but occupational exposures are increasingly implicated; it is also clear that not all with similar exposures are affected, pointing to the importance of host or personal factors'. In this connection, it is natural to consider the variation in responses to dust, and why some workers respond to dust exposures by developing severe functional changes (Hurley and Soutar, 1986).

OTHER POTENTIAL RISK FACTORS AFFECTING FEV_1 IN DUST-EXPOSED POPULATIONS

Our inability to define the reasons for the various rates of lung function decline within a population points out the need of further research to identify both environmental and constitutional risk factors for lung function loss in an occupationally dust-exposed population. The commonly identified risk factors for the rate of lung function decline (aging, smoking and the usual measurements of dust exposure, such as the years of dusty job, cumulative dust exposure and job categories) do not explain most of the decline in any given worker. Other possible explanations are noted below.

(a) Host susceptibility factors

Individual susceptibility to develop airways obstruction is associated with an 'asthmatic predisposition' and this host factor may be important in determining individuals who will develop COPD. An asthmatic predisposition includes bronchial hyperreactivity and eosinophilia as well as classic features of atopy such as skin test reactivity (Burrows, 1981). α_1-Anti-trypsin deficiency (α_1-anti-trypsin levels <50 mg/dl) is associated with early-onset emphysema (Eriksson, 1965; Kueppers and Black, 1974; Larsson, 1978). Evidence continues to mount that the protease–antiprotease hypothesis is relevant to the pathogenesis of panacinar emphysema (that due to α_1-anti-trypsin deficiency) but also to the more common centriacinar emphysema. For example, evidence for the role of cigarette smoke as an oxidator and inactivator of α_1-anti-trypsin is incontrovertible (Carp *et al.*, 1982).

The tendency to form IgE is, in part, genetically determined. In turn, the amount of serum IgE is related to the likelihood that an individual will show evidence of 'atopy', that is, that he or she will develop positive allergy skin tests, allergic rhinitis or allergic reactions in the airway (Abboud and Morton, 1975).

(b) Familial history of atopy

A familial history of asthma, allergic rhinitis, hay fever or infantile eczema may affect lung function. These possible effects of atopy on airways obstruction have been well recognized (Kueppers *et al.*, 1977).

(c) Childhood illnesses
Childhood illnesses such as asthma or wheezy bronchitis-like illnesses lead to an increased susceptibility to COPD in the adult. Acute childhood illnesses are associated with mild functional abnormalities later in life. Severe early childhood illnesses are associated with an increased frequency of cough in early adult life (Kieran *et al.*, 1976). Perhaps children with asthma or wheezy bronchitis-like illnesses enter adult life with slightly impaired lung function but then show excessive rates of decline in function during adult life. These declines may be greatly magnified if the individual smokes (Burrows *et al.*, 1977). Tobacco smoke heightens bronchial reactivity in children. Passive smoking may increase the frequency of bronchial responsiveness and atopy, thereby increasing the risk of asthma in childhood, particularly in boys (Martinez *et al.*, 1988). Social factors, such as the quality of housing and nutrition, significantly determine respiratory status through their relationship with respiratory infections in childhood and later life (Myers, 1984).

(d) Obesity and excessive weight gain
Lifelong non-smoking, white South African gold miners with silicosis were studied at necropsy (Hnizdo *et al.*, 1994). The pulmonary pathologic features identified included the extent and type of emphysema, the presence of airway disease, and the degree of silicosis affecting the lung parenchyma and pleura. These features were related to lung function tests done a few years before death, to the extent of impairment, and to the cumulative silica dust exposure. In miners with a moderate or greater degree of limitation of airflow, the main explanation for these findings were silicosis, heart disease and obesity. In some miners the FEV_1/FVC ratio was <70%. In these individuals, the authors considered obesity and a general lack of fitness could have contributed to airflow limitation.

In a sample of 611 US 'Black Lung' claimants, 80 were obese (Lapp *et al.*, 1994). In 1087 Polish steel mill workers, there was a strong relationship between the FVC, the FEV_1, and the degree of obesity (Kolarzyk *et al.*, 1985a). In a second report on this population, 642 of these steel workers without respiratory or circulatory disease were classified according to relative body mass. In workers with more than 124% of optimal body mass, the VC was adversely affected. The FEV_1 was not adversely affected until the body mass reached 135% of optimal. Significantly lower values of VC and FEV_1 were found when comparing obese to non-obese counterparts (Kolarzyk *et al.*, 1985b). Cotes *et al.* (1966) showed that coal miners with pneumoconiosis, when submitted to a 3-month period of inactivity, had an average increase of weight of 4.1 kg, together with a decrease in both the FVC and of $FEV_{0.75}$ of 90 ml.

Excessive weight gain, in the absence of obesity, may alter lung function rate of decline. Reports (Schoenberg *et al.*, 1978; Bande *et al.*, 1980) show the relationship between weight gain and loss of lung function over time. Most recently, a matched-pair comparison of data from the US National Coal Study was conducted to identify the potential risk factors in a group of miners with clinically important decline of FEV_1 (Wang *et al.*, 1996b). Subjects were chosen from this longitudinally studied cohort. All miners with at least 6 years' follow up ($N = 5900$) were grouped into strata based on age, height, smoking status, gender and initial FEV_1. Within these strata, pairs of miners were selected whose annual rate of FEV_1 decline differed by more than 60 ml/year. A total 344 pairs were entered into the analysis. Excessive weight gain was found to be one of the risk factors for the rapid decline of lung function. Rapid decliners gained 8.2 pounds (3.7 kg) more weight than their matched partners over an average of 11 years' follow up ($P < 0.001$). A similar study design was applied to a sample of 541 steelworkers. The matching process resulted in 66 pairs. The slope of weight gain for the rapid decliners was 1.83 lb/year (0.8 kg) and 0.62 lb/year (0.3 kg) for their partners. Rapid decliners gained 1.21 lb/year (0.5 kg) more than their partners ($P < 0.0058$) (Banks *et al.*, 1995). Multiple linear regression models obtained from longitudinal and cross-sectional analysis of lung function in 475 steel workers suggested that weight gain was an important determinant for longitudinal lung function decline and which explained approximately 6% of the total variation (Wang, 1996a).

(e) Intercurrent respiratory infection
Subjects reporting frequent infections tend to develop an excess of respiratory disorders (Lebowitz and Burrows, 1977).

(f) Mine effect
The term 'mine effect' reflects the significant differences in the prevalence of pneumoconioses and the decrement of lung function between the geographic regions. Mine effect has been well described in the large epidemiologic studies in both US and UK; however, the explanation for this effect has never been fully identified.

(g) Environmental exposures
Environmental exposures outside of the workplace may occur to a variety of bronchial irritants. Numerous agents, such as indoor air pollution, as well as environmental pollutants have been implicated as a

cause for adverse respiratory effects. For example, chronic exposure to low levels of oxides of nitrogen, such as those found in homes with gas stoves and central heating may lead to respiratory impairment, especially in children (Melia *et al.*, 1977; Speizer *et al.*, 1980).

(h) Socioeconomic status

An individual's socioeconomic status can be mediated by factors such as crowding, the frequency of respiratory illness, nutrition, the availability of medical care, etc. Using Registrar General's statistics, bronchitis was, and continues to be, a far more frequently certified cause of death in lower socioeconomic groups. This phenomenon was particularly evident in coal miners who had worked at the coal face. This was also associated with a dramatically higher prevalence of bronchitis in miners' wives than in wives of non-miners (Morgan, 1978). Enterline and Lainhart (1967) showed that coal miners had an excess of wheezing and dyspnea, and a more frequent history of pneumonia and pleurisy when compared with other manual workers. Their wives also had an excess of dyspnea and previous pneumonia when compared with wives of other manual workers, raising doubt as to the etiology of some of the excess respiratory disease symptoms among coal miners. Educational level was strongly related to symptoms of respiratory disease among both the miners and their wives.

CONCLUSIONS

Several approaches have been used to address the rates of lung function decline in a population. These include multiple regression analyses, assessment of mean lung function declines in specific pulmonary parameters over time, and the attempt to determine the number of workers who have developed significant disease or changes in a specific parameter over time.

Multiple regression analysis has been used to explore risk factors which influence lung function declines in the population. This modern statistical technique provides parameter estimation for quantitative assessment of the relationship between response variable and various regressors. It is also used as an instrument to detect the degree of importance of each variable in explaining the variance in response. This approach has supplanted analysis of workplace groups by smoking category or job description because of the ability to better understand the contribution of different variables to the lung function decline.

Numerous investigators have used the approach of multiple regression analysis to study longitudinal changes in lung function in populations. Becklake (1989) commented on the strengths of multiple regression analysis in longitudinal assessments of lung function. She considered this the most powerful approach to understanding the reasons which contribute to the rate of a population's lung function decline. Others (Smidt *et al.*, 1977; Bates *et al.*, 1985) have expressed concern that the value of the multiple regression in determining the explanation for lung function declines in longitudinal studies are not yet clear because of the inherent variability of the data.

Large unexplained differences in the rates of decline commonly exist among individual members in any given population. When multiple linear regression models are developed to address the rate of lung function decline over time in a population, parameters such as gender, aging, smoking and dust exposure explain only about 10% of the differences in the rates of decline among the members of the group (variation or R^2). Thus, the reasons why some workers have an exaggerated rate of decline in lung function remain largely unexplained.

Nevertheless, we did not expect that the relationship between dust exposure and the decrements in pulmonary function was weaker in longitudinal analyses compared with cross-sectional analyses. Survey biases and large variability in estimates of dust exposure might be responsible, in part, for the inability to detect the dust effect and the lower model R^2 in longitudinal data analysis. When dust exposures are measured they are typically only measured infrequently, and often for just part of a work shift. How accurate these measures are in describing a given worker's exposure are not clear. It is also possible that the rate of lung function decline may not be linear in any given population and may sometimes show an accelerated decline early, and then plateau; thus, much of the dust-related decline occurs soon after a worker is hired or perhaps before the start of the study (Hodous and Hankinson, 1990; Seixas *et al.*, 1993).

In summary:

1. There is a statistically significant relationship between mean FEV_1 decline and dust exposure. Within these populations of dust-exposed workers, a small proportion of workers develop a clinically significant decline in FEV_1. In the population of those who decline, there can be evidence for both apparent restrictive pulmonary disease and airways obstruction. Illnesses such as asthma, bronchospasm, bronchiectasis, chronic bronchitis, airway responsiveness, emphysema and pneumoconiosis may play a role.

We wish to express several cautions. First, attributing this effect to dust alone in any individual worker may not be reasonable unless specific information regarding the overall health of each worker is available. Briefly cited clinical reviews of workers with decreased FEV_1 values in this report reveal the numerous medical reasons responsible for an accelerated decline. In any individual case, the explanation for these significant declines may not be apparent. An assessment of the individual is necessary to understand the relationship between dust exposure, lung function decline, and other medical problems. Second, since chronic bronchitis is frequent in dust-exposed workers, it is likely to be frequently responsible for FEV_1 declines. Further studies of the lung function of miners who have chronic bronchitis and leave mining would help us to better understand the role of chronic bronchitis in the decline of FEV_1 of this population. Finally, although reports of the statistically expected number of workers with 'clinically significant' declines in FEV_1 exist in populations with dust exposures, the number of workers with such declines attributable to dust have not been frequently reported. Nor have the number with such declines been reported among 'blue-collar' workers in whose working environment dust is absent.

2. Emphysema can be found in dust-exposed nonsmokers without pneumoconiosis. Dust may somehow trigger air space enlargement, but the relationship between the histologic features of emphysema, the clinical features of dust-induced chronic bronchitis, and the rate of lung function decline of the population is not clear. The relationship between dust exposure and emphysema is stronger if pneumoconiosis is present.

3. The longitudinal assessment of lung function using multiple regression analyses is expected to be more powerful than cross-sectional analyses. We would anticipate that the most convincing evidence relating dust to FEV_1 decline should come from longitudinal studies. When the regression equations have been made available, this has not consistently been the case (Table 5.1). Further comparisons between cross-sectional and longitudinal surveys are in order. In the studies cited, smoking remains a much more powerful risk for FEV_1 decline than dust exposure.

4 Only a few populations of dust-exposed workers have been studied world-wide. Numerous other cohorts of dust-exposed workers exist and are available for study. We could develop a better understanding of this topic if different populations were evaluated.

5. There is a need to improve statistical methodology and to develop more appropriate modeling methods in order to better estimate the lung's reaction to environmental exposures, mainly smoking and dust. The studies cited in Table 5.1 only explain 6–12% of the variability of FEV_1 decline within any given population, even considering the effect of dust and cigarette smoke. Other factors which are not yet accounted for apparently play an important role.

REFERENCES

Abboud, R.T. and Morton, J.W. (1975) Comparison of maximal mid-expiratory flow, flow volume curves, and nitrogen closing volumes in patients with mild airway obstruction. *Am. Rev. Respir. Dis.*, **111**, 405–17.

Attfield, M.D. (1985) Longitudinal decline in FEV_1 in United States coalminers. *Thorax*, **40**, 132–7.

Attfield, M.D. and Hodgson, M.J. (1986) Respiratory symptoms and chest illness in underground coal miners. *Ann. Am. Conf. Gov. Ind. Hyg* **14**, 91–100.

Attfield, M.D. and Hodous, T.K. (1992) Pulmonary function of U.S. coal miners related to dust exposure estimates. *Am. Rev. Respir. Dis.*, **14**, 605–9.

Bande, J., Clement, J. and Van De Woestijne, K.P. (1980) The influence of smoking habits and body weight on vital capacity and FEV_1 in male air force personnel: a longitudinal and cross-sectional analysis. *Am. Rev. Respir. Dis*, **122**, 781–90.

Banks, D.E., Wang, M.L., McCabe, L., Shammsain, M.H. and Hankinson, J.L. (1995) Weight gain and the longitudinal measurement of lung function in steel workers. *Chest*, **108**, 137S.

Banks, D.E., Wang, M.L. and McCabe, L. (1996) Clinically important declines in FEV_1 in non-smoking steel workers. *Chest*, **108**, 189S.

Bates, D.V., Pham, Q.T., Chau, N., Pivoteau, C., Dechoux J. and Sadoul, P.A. (1985) A longitudinal study of pulmonary function in coal miners in Lorraine, France. *Am. J. Ind. Med.* **8**, 21–32.

Becklake, M.R., duPreez, L. and Lutz, W. (1958) Lung function in silicosis of the Witwatersrand gold miners. *Am. Rev. Tuberculosis*, **77**, 400–12.

Becklake, M.R. (1976) Asbestos-related diseases of the lung and other organs: their epidemiology and implications for clinical practice. *Am. Rev. Respir. Dis.*, **114**(1),187–227.

Becklake, M.R. (1985) Chronic airflow limitation: Its relationship to work in dusty occupations. *Chest*, **88**(4),608–17.

Becklake, M.R. (1989) Occupational exposures: evidence for a causal association with chronic obstructive pulmonary disease. *Am. Rev. Respir. Dis.*, **140**(3),S85–91.

Begin, R., Cantin, A., Berthiaume, Y., Boileau, R., Peloquin, S. and Masse, S. (1983) Airway function in lifetime-nonsmoking older asbestos workers. *Am. J. Med.*, **75**, 631–7.

Begin, R., Filion, R. and Ostiguy, G. (1995) Emphysema in silica- and asbestos-exposed workers seeking compensation – A CT scan study. *Chest*, **108**, 647–55.

Burrows, B. (1981) An overview of obstructive lung diseases. *Med. Clin. North Am.*, **65**, 455–71.

Burrows, B. (1990) Airways obstructive diseases: pathogenetic mechanisms and natural histories of the disorders. *Med. Clin. North Am.*, **74**(3), 547–60.

Burrows, B., Knudson, R.J. and Lebowitz, M.D. (1977) The relationship of childhood respiratory illness to adult obstructive airway disease. *Am. Rev. Respir. Dis.*, **115**, 751–60.

Carp, H., Miller, F., Hoidal, J.R. and Janoff, A. (1982) Potential mechanism of emphysema: alpha-1-proteinase inhibitor recovered from lungs of cigarette smokers contains oxidized methionine and has decreased elastase inhibitory capacity. *Proc. Natl. Acad. Sci. USA*, **79**, 2041–5.

Churg, A., Wright, J.L., Wiggs, B., Pare, P.D. and Lazar, N. (1985) Small airways disease and mineral dust exposure. Prevalence, structure, and function. *Am. Rev. Respir. Dis.*, **131**, 139–43.

Churg, A., Hobson, J. and Wright, J. (1989) Functional and morphological comparison of silica-and elastase-induced airflow obstruction. *Exp. Lung Res.*, **15**, 813–22.

Cockcroft, A., Wagner, J.C., Ryder, R., Seal, R.M.E., Lyons, J.P. and Andersson, N. (1982) Post-mortem study of emphysema in coalworkers and non-coalworkers. *Lancet*, September 11, pp. 600–3.

Cotes, J.E., Gilson, J.C. and John, C. (1966) Effects of inactivity and a gain in weight upon lung function and indices of submaximal exercise. *J. Physiol.*, **188**, 36P–38P.

Cowie, R.L. and Mabena, S.K. (1991) Silicosis, chronic airflow limitation, and chronic bronchitis in South African gold miners. *Am. Rev. Respir. Dis.*, **143**, 80–4.

Enterline, P.E. and Lainhart, W.S. (1967) The relationship between coal mining and chronic nonspecific respiratory disease. *Am. J. Public Health*, **57**, 484–95.

Eriksson, S. (1965) Studies in alpha-antitrypsin deficiency. *Acta Med. Scand.*, **432**, 5–85.

Fletcher, C.M., Peto, R., Tinker, C. and Speizer, F. (1976) *The Natural History of Chronic Bronchitis and Emphysema*, Oxford, Oxford University Press.

Gibbs, A.R. and Wagner, J.C. (1988) *Pathology of Occupational Lung Disease*, 1st edn (eds A. Churg and F.H.Y. Green), Igaku – Shoin, Publ., New York, Tokyo, pp. 155–77.

Graham, W.G.B., Weaver, S., Ashikaga, T. and O'Grady, R.V. (1994) Longitudinal pulmonary function losses in Vermont granite workers – A reevaluation. *Chest*, **106**, 125–30.

Harless, K.W., Watanabe, S. and Renzetti, A.D., Jr (1978) The acute effects of chrysotile asbestos exposure on lung function. *Environ. Res.*, **16**(1–3), 360–72.

Higgins, I.T.T., Gilson, J.C., Ferris, B.G., Waters, M.E., Campbell, H. and Higgins, M.W. (1968) Chronic respiratory disease in an industrial town: A nine-year follow-up study. Preliminary report. *Am. J. Public Health*, **58**, 1667–76.

Hnizdo, E. (1992) Loss of lung function associated with exposure to silica dust and with smoking and its relation to disability and mortality in South African gold miners. *Br. J. Ind. Med.*, **49**, 472–9.

Hnizdo, E., Baskind, E. and Sluis-Cremer, G.K. (1990) Combined effect of silica dust exposure and tobacco smoking on the prevalence of respiratory impairments among gold miners. *Scand. J. Work Environ. Health*, **16**, 411–22.

Hnizdo, E., Hessel, P.A. and Sluis-Cremer, G.K. (1991) Chronic airflow limitation in South African gold miners (letter). *Am. Rev. Respir. Dis.*, **144**, 1423–4.

Hnizdo, E., Sluis-Cremer, G.K., Baskind, E. and Murray, J. (1994) Emphysema and airways obstruction in non-smoking gold miners with long exposure to silica dust. *Occup. Environ. Med.*, **51**, 557–63.

Hodous, T.K. and Hankinson, J.L. (1990) Prospective spirometric study of new coal miners, in *Proceedings of the International Symposium on Pneumoconioses, 1988, Shenyang*, PRC Institute of Occupational Medicine, Chinese Academy of Preventive Medicine, Beijing, PRC, pp. 206–11.

Hurley, J.F. and Soutar, C.A. (1986) Can exposure to coalmine dust cause a severe impairment of lung function? *Br. J. Ind. Med.*, **43**, 150–7.

Irwig, L.M. and Rocks, P. (1978) Lung function and respiratory symptoms in silicotic and nonsilicotic gold miners. *Am. Rev. Respir. Dis.*, **117**, 429–35.

Jain, B.L. and Patrick, J.M. (1981) Ventilatory function in Nigerian coal workers. *Br. J. Ind. Med.*, **38**, 275–80.

Jodoin, G., Gibbs, G.W., Macklem, P.T., McDonald, J.C. and Becklake, M.R. (1971) Early effects of asbestos effect on lung function. *Am. Rev. Respir. Dis.*, **104**, 525–35.

Kauffmann, F., Drouet, D., Lellouch, J. and Brille, D. (1982) Occupational exposures and 12-year spirometric changes among Paris area workers. *Br. J. Ind. Med.*, **39**, 221–32.

Kibelstis, J.S., Morgan E.J., Reger, R., Lapp, N.L., Seaton, A. and Morgan, W.K.C. (1973) Prevalence of bronchitis and airways obstruction in American bituminous coal miners. *Am. Rev. Respir. Dis.*, **108**, 886–93.

Kieran, K.E., Colley, J.R.T., Kouglas, J.W.B. *et al.* (1976) Chronic cough in young adults in relation to smoking habits, childhood environment and chest illness. *Respiration*, **33**, 236–44.

Kilburn, K.H. and Warshaw, R.H. (1990) Airway obstruction in asbestos-exposed shipyard workers: with and without irregular opacities. *Respir. Med.*, **84**, 449–55.

Kleinfeld, M., Messite, J., Kooyman, O. *et al.* (1966) Effect of asbestos dust inhalation on lung function. *Arch. Environ. Health*, **12**, 741–6.

Knudson, R.J., Lebowitz, M.D., Holberg, C.J. and Burrows, B. (1983) Changes in the normal maximal expiratory flow–volume curve with growth and aging. *Am. Rev. Respir. Dis.*, **127**, 725–34.

Kolarzyk, E., Kiec E. and Wiater, M. (1985a) Effect of obesity on the ventilatory capacity of the respiratory system. I. Relation between basic spirometric indicators: Vital capacity (VC) and forced expiratory volume (FEV_1) and obesity. *Med. Pr.*, **36**, 87–95.

Kolarzyk, E., Kiec, E., Galuszka, Z. and Wiater, M. (1985b) Effect of obesity on the ventilatory capacity of the respiratory system. II. Reference spirometry values of VC and FEV_1 for overweight males. *Med. Pr.*, **36**, 184–90.

Kueppers, F. and Black, L.F. (1974) Alpha1-antitrypsin and its deficiency. *Am. Rev. Respir. Dis.*, **110**, 176–94.

Kueppers, F., Miller, R.D., Gordon, G. *et al.* (1977) Familial prevalence of chronic obstructive pulmonary disease in a matched pair study. *Am. J. Med.*, **63**, 336–42.

Lapp N.L., Morgan W.K.C. and Zaldivar G. (1994) Airways obstruction, coal mining, and disability. *Occup. Environ Med.*, **51**, 234–8.

Larsson, C. (1978) Natural history and life expectancy in severe alpha-1-antitrypsin deficiency. *Piz. Acta. Med. Scand.*, **204**, 345–51.

Lebowitz, M.D. and Burrows, B. (1977) The relationship of acute respiratory illness history to the prevalence and incidence of obstructive lung disorders. *Am. J. Epidemiol.*, **105**, 544–54.

Leigh, J., Driscoll, T.R., Cole, B.D., Beck, R.W., Hull, B.P. and Yang, J. (1994) Quantitative measurement between emphysema and lung mineral content in coalworkers. *Occup. Environ. Med.*, **51**, 400–7.

Love, R.G. and Miller, B.G. (1982) Longitudinal study of lung function in coal-miners. *Thorax*, **37**, 193–7.

Lyons, J.P., Ryder, R.C., Seal, E.M.E. and Wagner, J.C. (1981) Emphysema in smoking and non-smoking coalworkers with pneumoconiosis. *Bull. Eur. Physiopath. Resp.*, **17**, 75–85.

Marine, W.M., Gurr, D. and Jacobsen, M. (1988) Clinically important respiratory effects of dust exposure and smoking in British coal miners. *Am. Rev. Respir. Dis.*, **137**, 106–12.

Martinez, F.D., Antognoni, G., Macri, F. *et al.* (1988) Parental smoking enhances bronchial responsiveness in nine-year-old children. *Am. Rev. Respir. Dis.*, **138**, 518–23.

Melia, R.J.W., Florey, C., du Valtman, D.G. and Swan, A.V. (1977) Association between gas cooking and respiratory disease in children. *Br. Med. J.*, **11**, 149–52.

Morgan, W.K.C. (1978) Industrial bronchitis. *Br. J. Ind. Med.*, **35**, 285–91.

Morgan W.K.C. (1986) On dust, disability, and death. *Am. Rev. Respir. Dis.*, **134**, 639–41.

Morgan, W.K.C., Handelsman, L., Kibelstis, J. *et al.* (1974) Ventilatory capacity and lung volumes of U.S. coal miners. *Arch. Environ. Health*, **282**, 182–9.

Muldoon, B.C. and Turner-Warwick, M. (1972) Lung function studies in asbestos studies. *Br. J. Dis. Chest*, **66**, 121–32.

Murphy, R.L., Jr, Gaensler, E.A., Redding, R.A. *et al.* (1972) Low exposure to asbestos. Gas exchange in ship pipe coverers and controls. *Arch. Environ. Health*, **25**, 253–64.

Myers, J.E. (1984) Differential ethnic standards for lung functions, or one standard for all? *S. Afr. Med. J.*, **65**, 768–72.

Oxman A.D., Muir, D.C.F., Shannon, H.S., Stocvk, S.R., Shizdo, E. and Lange, H.J. (1993). Occupational dust exposure and chronic obstructive pulmonary disease. *Am. Rev. Respir. Dis.*, **148**, 38–48.

Pearle, J. (1981) Exercise performance and functional impairment in asbestos-exposed workers. *Chest*, **80**(6),701–5.

Peto, R., Speizer, F.E., Cochrane, A.L. *et al.* (1983) The relevance in adults of air-flow obstruction, but not of mucus hypersecretion, to mortality from chronic lung disease – Results from 20 years of prospective observation. *Am. Rev. Respir. Dis.*, **128**, 491–500.

Reid, P.J. and Sluis-Cremer, G.K. (1996) Mortality of South African gold miners. *Occup. Environ. Med.*, **53**, 11–16.

Rodriguez-Roisin, R., Merchant, J.E., Cochrane, G.M., Hickey, B.P., Turner-Warwick, M. and Clark, T.J. (1980) Maximal expiratory flow volume curves in workers exposed to asbestos. *Respiration*, **39**(3),158–65.

Rogan, J.M., Attfield, M.D., Jacobsen, M. *et al.* (1973) Role of dust in the working environment in development of chronic bronchitis in British coal miners. *Br. J. Ind. Med.*, **30**, 217–26.

Ruckley, V.A., Gould, S.J., Chapman, J.S. *et al.* (1984) Emphysema and dust exposure in a group of coalworkers. *Am. Rev. Respir. Dis.*, **129**, 528–32.

Ryder, R.C., Dunnill, M.S. and Anderson, J.A. (1971) A qualitative study of bronchial mucus gland volume, emphysema, and smoking in a necropsy population. *J. Pathol.*, **104**, 59–71.

Samet, J.M., Epler, G.R., Gaensler, E.A. and Rosner, B. (1979) Absence of synergism between exposure to asbestos and cigarette smoking in asbestosis. *Am. Rev. Respir. Dis.*, **120**, 75–82.

Schoenberg, J.B., Beck, G.J. and Bouhuys, A. (1978) Growth and decay of pulmonary function in healthy Blacks and Whites. *Resp. Physiol.*, **33**, 367–93.

Seixas, N.S., Robins, T.G., Attfield, M.D. and Moulton, L.H. (1993) Longitudinal and cross sectional analyses of exposure to coal mine dust and pulmonary function in new miners. *Br. J. Ind. Med.* **50**, 929–37.

Selikoff, I.J. and Lee, D.H.K. (1978) *Asbestos and Disease*, Academic Press, New York, NY, pp. 152–64.

Sluis-Cremer, G.K., Walters, L.G. and Sichel, H.S. (1967a) Chronic bronchitis in miners and nonminers: an epidemiological survey of a community in the gold-mining area in the Transvaal. *Br. J. Ind. Med*, **24**, 1–12.

Sluis-Cremer, G.K., Walters, L.G. and Sichel, H.S. (1967b) Ventilatory function in relation to mining experience and smoking in a random sample of miners and nonminers in a Witwatersrand town. *Br. J. Ind. Med.*, **24**, 13–25.

Smidt, V., Worth, G. and Bielert, D. (1977) Lung function and clinical findings in cross sectional and longitudinal studies in coal workers from the Ruhr area. *Int. Arch. Occup. Environ. Health*, **40**, 45–70.

Soutar, C.A. and Hurley, J.F. (1986) Relation between dust exposure and lung function in miners and ex-miners. *Br. J. Ind. Med.*, **43**, 307–20.

Soutar, C., Campbell, S., Gurr, D., Lloyd, M., Love, R., Cowie, H., Cowie, A. and Seaton, A. (1993) Important deficits of lung function in three modern colliery populations. Relations with dust exposure. *Am. Rev. Respir. Dis.*, **147**, 797–803.

Speizer, F.E., Ferris, B., Bishop, Y.M.M. and Spengler, J. (1980) Health effects of indoor NO_2 exposure, preliminary results, in (ed. S.D. Lee). *Nitrogen Oxides and their Effects on health* Ann Arbor, Michigan, pp. 343–61.

Stenton, S.C. and Hendrick, D.J. (1992) Airflow obstruction and mining. State of the Art Review. *Occup. Med.*, **8**, 155–70.

Sue, D.Y., Oren, A., Hansen, J.E. and Wasserman, K. (1985) Lung function and exercise performance in smoking and nonsmoking asbestos-exposed workers. *Am. Rev. Respir. Dis.*, **132**(3), 612–18.

Wang, M.L. and Lu, P.L. (1985) Lung function studies of asbestos workers. *Scand. J. Work Environ. Health*, **11**(suppl. 4), 34–42.

Wang, M.L., McCabe, L, Hankinson, J.L., Shammsain, M.H., Gunel, E., Lapp, N.L. and Banks, D.E. (1996a) Longitudinal and cross sectional analyses of lung function in steel workers. *Am. J. Respir. Crit. Care Med.*, **153**, 119–29.

Wang, M.L., Petsonk, E.L., Attfield, M.D., Short, S.R., Beeckman, L.F., Bonnett, B. and Hankinson, J.L. (1996b) Miners with clinically important declines in FEV1: Analysis of data from the US national coal study. *Appl. Occup. Environ. Hyg.*, **11**(7),989–95.

Weill, H., Ziskind, M.M., Wagenspack, C. and Rossiter, C.E. (1975) Lung function consequences of dust exposure in asbestos cement manufacturing plants. *Arch. Environ. Health*, **30**, 88–97.

Wright, J.L. and Churg, A. (1984) Morphology of small-airway lesions in patients with asbestos exposure. *Hum. Pathol.*, **15**(1),68–74.

Zitting, A., Huuskonen, M.S., Alanko, K. and Mattsson, T. (1978) Radiographic and physiological findings in patients with asbestosis. *Scand. J. Work Environ. Health*, **4**(4),275–83.

6 Workers' compensation for occupational lung disease: German–US parallels

Claus Piekarski, Elizabeth A. Jennison and John E. Parker

THE BEGINNINGS OF WORKERS' COMPENSATION IN GERMANY

In this chapter, we address the development of a 'workers' compensation' system in Germany, perhaps the first nation to develop such a plan, and relate these developments to the formation of such a system in the United States. The foundation of the social welfare system for the worker was developed by workers in the mining industry. These initiatives have led governmental bodies throughout the world to recognize the importance of providing care to injured workers and to those workers who develop chronic disease in association with their employment.

The social security system for German workers dates to the middle ages. In 1225, the Iglau Mining Rules (Iglauer Bergrecht) provided the first codified regulations protecting a miner's health. For the first time, work of children underground, as well as their night and shift work, was forbidden. In addition, these regulations also addressed the health care of the miners, mandating free care or low-cost care to miners and their families by designated physicians. Later, Paracelsus and Georgius Agricola, noted authors of occupational medicine treatises, worked as mining physicians and described health care and preventive medicine for miners and smelters of the 16th century.

The implementation of these regulations was funded by mineral royalties granted by German Emperors to the individual sovereigns of the different states within the German Reich. These regulations of Saxony were considered to be the most progressive, and were later freely copied in other German lands. With time, these plans became the bases of further mine regulations in Prussia and the foundation for mining legislation of today's Germany.

In professions such as the traditional crafts, apprentices and workers belonged to their master's household and the master was expected to care for his workers' health and social welfare when illness or disability developed. Rules for this system were established and supervised by the local Guilds, but they differed widely due to local conditions. Overall, these regulations were comparably weak, since there was no legal basis or general law to enforce this social care for the individual. The most infirm or elderly craftsmen often turned to welfare systems of the townships or the church for help.

With the beginnings of industrialization at the end of the 18th century, a great number of people left rural areas to work in relatively urban areas, where a social security system did not exist, leaving them and their families to poverty in case of illness, disability or death. It was again the mining industry which took the

Occupational Lung Disease. Edited by Daniel E. Banks and John E. Parker. Published in 1998 by Chapman & Hall. ISBN 0 412 73630 6

lead to improve the situation. The passage of the General Mining Act in 1865 prohibited underground work for women and initiated a miners' social security system ('Knappschaftskasse'), requiring the mine proprietor to provide financial support to the miner not only during illness, but also after accidents or injuries resulting in chronic impairment. The intentions were good but the support provided by this insurance plan was insufficient. The German Liability Insurance Act of 1871 promised additional security for the whole industrial work force. Here, the employer or the mine owner paid the costs of work accidents whenever he was found guilty in a legal action. Yet, because the burden of bringing a court case generally fell to the employee (who had to pay for all judicial expenses, unless the employer was found guilty), the ability of the employee to pursue compensation for health damages remained relatively weak. Because of this Liability Act of 1871, a great number of plant or mine owners joined so-called 'insurance fraternities' to share their potential compensation risks.

Still, this did not provide a satisfactory solution to compensating a worker for injuries or illnesses occurring in the workplace. Subsequent court cases and disputes resulting from accidents troubled the industrial peace and aggravated social tension. Pensions and workers' compensation went unpaid if employers went bankrupt or out of business. Recognition of these problems led to several initiatives in the kingdom of Prussia in an attempt to improve the social situation. Notable social reformers, including the Princess Royal Victoria, daughter of Queen Victoria of England, played a role in facilitating these changes. On November 12, 1881, Chancellor Bismarck presented his recommendations for social change in his 'Imperial Message' (Kaiserliche Botschaft), delivered before the Parliament of the German Reichstag; this provided a major stimulus for the Industrial Accidents Act of 1884. This act laid the basis for a network of social security systems in Germany which included the Systems of Health Insurance Act in 1883 and led to the Invalidity and Old Age Insurance Act (pension insurance) of 1889.

The Industrial Accidents Act was based on a cooperative concept. This placed the insurance system under the federal government, with costs borne only by the employers. This act also established equal representation by employers and employees in the administration of this program. This compulsory insurance system covered occupational accidents in the industry irrespective of whether the injury occurred due to negligence of the employer or of the employee. The financial contributions of the employers were based on insurance risk. In this way, accident prevention became a fundamental aim of this system. Based on the perceived risk of injury or disease associated with a specific industry, different industries were grouped within the insurance system, forming the individual compensation authorities ('Berufsgenossenschaften'). Each of the different Berufsgenossenschaften established the regulations necessary for accident prevention in their industry and structured a program giving inspectors the authority to enforce these rules.

The establishment of the Miners' Compensation Authority ('Knappschafts–Berufsgenossenschaft') in 1885 created an insurance system for miners. Compensation for mine accidents, and later for industrial diseases in miners was provided by the Authority. Representatives of the Compensation Authority had roles which paralleled those of government inspectors. A dual system of government and compensation mine inspectors was thus created. Both groups were concerned with the safety and health of the miners and had goals of reducing the hazards of underground mining. This program still exists today.

The 'Knappschafts–Berufsgenossenschaft' also pioneered skilled medical treatment and rehabilitation for miners with workplace-related injuries and illnesses. Between 1890 and 1910, the miners' hospital 'Bergmannsheil' (miners' care) in Bochum, and the miners clinics 'Bergmannstrost' (miners' conciliation) in Halle and in Leipzig Schkeudiz were established. Later, in 1939, the miners' clinic 'Bergmannssegen' (miners' blessing) in Hindenburg/Silesia was founded.

MODERN GERMAN COMPENSATION APPROACHES

Beginning in 1925, accidents occurring in travel to and from work were compensated in the same manner as accidents at the workplace. In addition to accidents, specific occupational illnesses were included under the protection of the Industrial Accidents Act. Another administrative action, the General Insurance Act of the German Reich (Reichsversicherungsordnung, RVO), authorized the German government to issue an official list of occupational diseases to be compensated under the rules of the Industrial Accidents Act. In 1963, other illnesses related to the workplace which were not a part of the original official list but, according to general scientific opinion met the standards of an occupational disease, were also eligible to be compensated. The General Insurance Act (RVO) was replaced in 1997 by the Social Statutes Code (Sozialgesetzbuch, SGB) which incorporates the regulations concerned in volume seven (SGB VII).

The official list includes nearly 60 occupational diseases and details the recommended evaluation of the workplace and the workers' health in an individual case. The Berufsgenossenschaft is required 'to take all suitable means to reduce the danger of an occupational disease'. Employees who are unable to continue work due to an occupational disease are to be compensated for the amount of former income lost due to their disability for up to 5 years.

German law mandates that when an occupational disease is suspected, the diagnosing physician, as well as the employer, must inform the representatives of Berufsgenossenschaft responsible for this worker. Representatives of the medical services of the work inspectorates (Staatlicher Gewerbearzt, Staatl. Gewerbeaufsicht, Landesamt fur Arbeitsschutz) evaluate the worker's employment background, as well as his or her degree of impairment, in order to determine the amount of compensation and the potential for rehabilitation, with a perspective that retraining for another profession may be needed. Costs for worker evaluation, as well as financial compensation and necessary retraining, are covered by the Berufsgenossenschaften.

In general, the degree of impairment attributable to an accident or an occupational illness defines the amount of compensation. If the worker is 100% disabled, the employee receives two-thirds of his annual earnings. Using social security legislation, the degree of impairment (Grad der Behinderung = GdB) is primarily used to describe the health of a worker. In the system of Accident Insurance, the reduction in earning power (Minderung der Erwerbsfahigkeit = MdE) is preferred. With these two systems, different reduction percentages can be used to calculate the degree of the reduction in earning power necessary to determine the amount of compensation. For example: disease A (40%) + impairment B (30%) = Compensation 70% MdE. A reduction in earning power of 20% is the minimal percentage to be compensated. In the case of a fatal accident or the fatal outcome of an occupational disease, the widow receives 60% of her husband's compensation.

In addition to this system, there are over 100 similar institutions in agriculture and the public sector, all under the administration of Statutory Accident Insurance.

The structures of the different Berufsgenossenschaften are generally similar, and are mainly characterized by a system of equal representation of employers and employees on the one side and a system of self administration on the other. The main elements of this structure are formed by the representatives' assembly (which serves as the Parliament), by the Board (comparable with the Government), and by the Directorium (formed by the Managing Director and his Deputy).

The unification of Germany has required common compensation legislation for occupational diseases for all workers. For the first time, the consequences of secret uranium mining were made public to facilitate lawful compensation. Between 1954 and 1989, a work force of 400 000 to 600 000 miners worked in the East German uranium mines. Approximately 11 000 of these workers with bronchial carcinoma have been recognized, with a projected annual incidence of 250 and 300 cases for the next 30 years.

The Central Federation of Industrial Professional Associations (Hauptverband der gewerblichen Berufsgenossenschaften) serves as the central compensation authority for occupational diseases. It has established a central registry ('Zentrale Erfassungs – und Betreuungsstelle Wismut, Zeß Wis') for the medical information of uranium miners and uranium processing workers. This registry offers medical examinations at a maximum of 5-year intervals to all surviving miners that are a part of this registry. When an occupational disease is suspected, or at the discretion of the examining physician, these intervals can be shortened. Workers with lung carcinoma and a radiation burden of at least 200 Working Level Months (WLM) are recognized as having radiation-induced carcinoma. Those meeting this definition are provided with expedited administrative proceedings. Other workers with lung cancer but with radiation burdens between 50 and 200 WLM, are considered on a case-by-case basis for compensation. Compensation for other cancerous diseases which may occur in association with the radon burden of underground mining in the Wismut uranium mines is also considered.

Silicosis, silico-tuberculosis and coal workers' pneumoconiosis have been the main respiratory health hazards in underground mining for centuries. Strict dust control, combined with frequent medical examinations, have decreased the adverse health effects. Since the end of the Second World War, a dramatic reduction of silicosis has been quite evident. There have been no new cases of coal workers' pneumoconiosis in the past 15 years or so in the West German coal industry. Yet, even as these disorders have become increasingly controlled, other respiratory diseases in miners require careful vigilence. The compensation system in Germany allows for identification of new disorders. Of note, there are a growing number of compensation cases for asthma due to exposures to diisocyanates used for rock consolidation in the mining industry. Finally, chronic bronchitis and emphysema in coal miners has been introduced into the official list of occupational diseases after a long scientific and political dispute.

PARALLELS IN THE US SYSTEM

Even though a modern workers' compensation program was adopted in Germany as early as 1884, it would be 65 years before Mississippi became the last American state to adopt such a plan. Initially limited to coverage for industrial accidents (injuries), workers' compensation programs have expanded in scope in terms of the type and number of persons and occupations covered, as well as adding coverage of occupational diseases (Auerbach *et al.*, 1961).

The German plan had three components and was administered by representatives of employers and employees under government supervision. The first component was a sickness fund to which workers contributed two-thirds and employers contributed one-third; it provided benefits for the first 13 weeks of sickness or disability due to an accident. Disability after the first 13 weeks was handled by the accident fund, which was funded entirely by employer contributions. The third component, disability insurance, was funded equally by employers and employees and provided for payments due to old age or other causes not specifically covered elsewhere.

In the United States, the Occupational Safety and Health Act of 1970 prompted the appointment of a task force to study the various state Workers' Compensation Acts. One of the 19 essential recommendations of that task force was that all occupational diseases in every state be given the same benefits as accidental injuries. Although all states now have some type of occupational disease coverage, considerable variation persists. For example, there are instances where benefits may be limited or the statute of limitations may be so restrictive as to effectively preclude recovery of claims (Stephenson, 1981).

Compensation for occupational respiratory diseases is further complicated by the diagnostic uncertainties associated with many of these entities. Most states initially covered occupational diseases with schedules or special language statutes outlining covered conditions. Schedules of covered diseases provided clear legislative direction to courts and compensation boards regarding the extent of available compensation. They also eliminated some of the uncertainty about whether a disease is to be considered occupational or an ordinary disease of life; this tends to reduce litigation and increase the number of worthy claims that are compensated. Disease schedules need to be updated frequently as previously unrecognized links are discovered between work and disease, as work environments are modified by new processes or substances, and as new industries locate within the covered jurisdiction. Overall, disease schedules tend to be restrictive and inflexible; their use has been abandoned in the United States but continues in many international settings (Barth and Hunt, 1980).

Several authors have noted that medical criteria seem to be applied arbitrarily and inconsistently by workers' compensation boards (Hughes, 1981; White *et al.*, 1992; Nevitt, *et al.*, 1994). Ideally, standards could be developed which would include criteria for determining whether an illness is work-related and the degree of impairment resulting from the disease. Decisions could then be based on objective medical tests that are standardized and widely available, and not in and of themselves dangerous to the worker (Brown, 1981). At present, there are barriers to establishing clear definitions of the work-relatedness of diseases. These include long latencies between exposure and disease and uncertainties about exposure as a cause of illness. This is further complicated by the multifactorial nature of many occupational diseases and the similarities of clinical and pathologic expression of occupational and non-occupational disease. Some analysts estimate that only 5–10% of justifiable claims for disabling occupational diseases due to all causes are remunerated through workers' compensation (Ramsey and Rosenstock, 1994).

WORKERS' COMPENSATION AND OCCUPATIONAL ASTHMA

Occupational asthma presents a number of difficulties from the standpoint of workers' compensation. First are the inherent problems with defining occupational asthma, including the difficulty of diagnosis, the aggravation of pre-existing asthma, the multicausal nature of asthma, and the persistence of occupational asthma after removal from the workplace (Seligman and Bernstein, 1993). Sensitization without asthma raises medicolegal issues. Circumstances may arise in which workers are sensitized to substances found at work and are removed from workplace contact with the agent before developing symptoms. Should sensitization, and thus, the potential for impairment, be regarded as an occupational disease? Should there be compensation if an individual is sensitized through workplace exposures but develops impairment only after subsequent non-occupational exposures (Barth and Hunt, 1980; Stephenson, 1981)? If there has been sensitization and development of occupational asthma, the problem becomes only slightly easier. Some authors have suggested that, once a diagnosis of occupational asthma is confirmed, the worker should be considered as having a 100% permanent impairment in terms of the job that caused the illness, as well as for other jobs entailing exposure to the same causative agent (Banks *et al.*, 1990; DeWitte *et al.*, 1994).

Diagnosis remains a problem. For example, more than half of the patients in one study population had terminated their employment by the time of their medical work-up, limiting their physician's ability to establish a connection between work and symptoms (Tarlo *et al.*, 1995). Frequently the medical history provides the only information on which the diagnosis of occupational asthma is based (Stenton *et al.*, 1995). Despite this difficulty, a review of Ontario Workers' Compensation claims for the years 1984–88 revealed an 87% concordance in decisions between the workers' compensation board and the assessing physician for individuals with occupational asthma and no prior history of asthma (Stenton *et al.*, 1995; Tarlo *et al.*, 1995).

Compensation for occupational asthma is further complicated by the fact that asthma is viewed as a common disease with multiple causalities. Physicians may be unaware of an association between the development or aggravation of asthma and a patient's work despite estimates that as much as 15% of adult-onset asthma is of occupational origin (Chan-Yeung and Malo, 1995). In a British review of workers applying for compensation for occupational asthma, just over 10% had been advised by their medical practitioner to submit a claim. The majority had decided to apply on the suggestion of a union representative, coworker, or on their own accord (Stenton *et al.*, 1995). This indicates that there is a need to educate physicians about both occupational asthma and its compensability.

Determining disability due to occupational asthma can be a lengthy process. Some have suggested that assessment should be made about 2 years after the last exposure to the offending agent. This is a point at which improvement in spirometry and bronchial responsiveness seems to plateau, at least for snow-crab asthma, a variety of occupational asthma attributable to exposure to a high-molecular weight allergen. The same authors recommend that subjects be assessed when their asthma is relatively stable, as defined by the absence of nocturnal awakenings due to asthma, the use of inhaled beta-adrenergic agents less than twice a day, and a daily variability in peak expiratory flow rate of less than 20%. In Quebec, the permanent disability rating is based on the degree of bronchial obstruction, the baseline level of bronchial hyperresponsiveness, and the need for medication (Gannon *et al.*, 1993; DeWitte *et al.*, 1994).

Since many workers with occupational asthma remain symptomatic even when removed from exposure to the inciting agent, a significant number of them may be limited in their ability to find or perform other employment, and will need workers' compensation. In the Ontario study, 54% of the claimants who could be reached for follow-up were unemployed (Stenton *et al.*, 1995; Tarlo *et al.*, 1995). Follow-up of 112 occupational asthma patients from a British pulmonary clinic revealed that 35% were no longer employed. Another 35% had changed jobs since their diagnosis, most as a consequence of their occupational asthma. Some 74% of those who had changed jobs lost income, with the mean loss estimated to be 54% of their annual income. Even the 30% of workers who remained in the same job with the same company, presumably with the same exposures, were at risk for losing income; 44% reported losing a mean of 35% of their annual income. For individuals who had received workers' compensation, the mean annual settlements fell well below their estimated loss of earnings. Thus, it appears likely that workers may choose to remain at work with continuing exposures for financial reasons (Gannon *et al.*, 1993). This may have further adverse effects on their health since continuous exposure in affected workers is associated with a worsening of asthma, while early removal from exposure increases the likelihood of recovery (Chan-Yeung and Malo, 1995).

WORKERS' COMPENSATION AND BYSSINOSIS

The history of workers' compensation coverage for byssinosis in Great Britain provides a good example of how coverage for an occupational disease has expanded in scope as medical knowledge has increased. The first Byssinosis Workmen's Compensation Scheme became effective on May 1, 1941, and was very restrictive. It applied only to men employed on or after that date in cotton chambers, blowing rooms, or cardrooms in which raw cotton was prepared for spinning. Claimants had to have been employed in the trade for at least 20 years and be totally and permanently disabled. Subsequent revisions extended covered areas of mills, allowed women to be covered, removed the requirements for minimum duration of employment, and no longer required the worker to show any permanent loss of faculty (Rooke, 1981).

At present, compensability in Great Britain is determined by a medical board which reviews the case and makes an assessment within 1–2 months after the claim is made. This still requires subjective judgment as there is no legally established definition of byssinosis (Rooke, 1981). North Carolina also uses an expert panel to decide compensation cases involving byssinosis; in this case, the reviewing entity is the Textile Occupational Disease Panel of the North Carolina Industrial Commission. This group has been criticized for its seemingly arbitrary and inconsistent application of medical criteria in the awarding of compensation. Not only is the burden of proof placed on each individual claimant to prove the work-relatedness of their condition, but the strong epidemiologic association between byssinosis and cotton dust may be considered irrelevant in evaluating a given case (Hughes, 1981).

Case definition of byssinosis is also a problem in South Africa where the Workmen's Compensation Act recognizes occupational diseases but does not define them. When a panel of the National Center for Occupational Health used set criteria to independently review 32 decisions of the Workmen's Compensation Commission, it was felt that 17 of the claims had been adversely affected, either by denial or reduced award and/or by inconsistent or arguable decisions on the part of the commission. In South Africa, filing a workers' compensation claim for byssinosis is a lengthy and expensive process. The medical costs involved in initiating a claim average more than three times the usual weekly wage of a textile worker and take an average of 29.8 months from initiation to finalization of appeals (White *et al.*, 1992).

WORKERS' COMPENSATION AND THE PNEUMOCONIOSES

Although there is strong epidemiologic linkage between exposure to pneumoconiotic dusts and the radiologic picture of pneumoconiosis, there is still little standardization of compensation for these diseases. Many workers have had exposure to pneumoconiotic dusts in multiple jobs, often involving more than one workers' compensation jurisdiction. There is a long latency period between first exposure and the development of disease, and concomitant non-occupational health conditions that impair respiratory function are common (Nevitt *et al.*, 1994). In addition, some state statutes contain special provisions which limit workers' compensation claims, especially for silicosis and asbestosis. Minimum exposure rules may mandate a minimum length of employment with a particular employer in a job with dust exposure. Rules requiring recent exposure bar claims for disease caused by hazards encountered on the job more than a specified number of years before the claim is made. There also may be minimum time requirements for exposure within the state in which the claim is being made (Barth and Hunt, 1980; Stephenson, 1981).

As with most occupational diseases, it appears that workers' compensation claims are filed by only a small percentage of employees with pneumoconiosis. Between 1980 and 1985, New York state had an average of 599 hospital discharges per year with a diagnosis of pneumoconiosis. By contrast, workers' compensation benefits for pneumoconiosis were initiated for an average of 52 individuals per year between 1979 and 1982. Although an individual may have more than one hospitalization in a year, only the most severe cases are likely to require hospitalization. Thus, the 599 hospital discharges per year is probably a significant underestimation of the burden of pneumoconiosis in New York (Markowitz *et al.*, 1989).

SILICOSIS

In the 1940s, silicosis became the first chronic disease to be compensable under workers' compensation in the United States. The recognition, and subsequent acceptance, of silicosis as an occupational disease occurred in response to the large number of silicosis deaths associated with construction of a tunnel at Gauley Bridge, West Virginia, in the 1930s, an episode which has been termed, 'America's Worst Industrial Disaster' (Cherniack, 1986).

A review of silicosis cases in New Jersey from 1979–1992 revealed that only 31% of patients who were able to be contacted had filed a workers' compensation claim. While 84% of the settled claims were awarded, there was no statistically significant association between radiologic findings and the awarding of compensation payment. Pulmonary function data were not available for this population. The authors speculated as to why so many of these individuals, all of whom had received a diagnosis of silicosis from their treating physician, did not file workers' compensation claims (Stanbury *et al.*, 1995). Possible explanations include the inability to locate the responsible employer and presence of coexisting medical problems. Other possibilities, such as a failure to associate disease with employment or the lack of physician support in filing a claim reflect the need for more physician training in the field of occupational medicine. Finally, individuals may have been receiving adequate medical care paid for by either private insurance or Medicare. Shifting the burden of medical costs from the workers' compensation system has been called a 'hidden tax' which is paid by other health insurers (Brown, 1981; Ramsey and Rosenstock, 1994; APHA, 1984). Companies that are not compensating silicotics for medical care lack the economic incentive to implement good health and safety practices in their workplaces (Stanbury *et al.*, 1995).

COAL WORKERS' PNEUMOCONIOSIS

Several authors (Smith, 1981; Kippen, 1995) have described the political and social background leading to the acceptance of lung disease in miners as being of occupational origin. In the United States, acceptance of such disease, and compensation for it, has waxed and waned with the changes in power between mine operators, miners, and the involvement of the federal government. During the 1860s and 1870s, a few physicians in the anthracite region of eastern Pennsylvania published articles about miners' respiratory difficulties and speculated about their relation to the inhalation of dusts and gases in the workplace. At the same time, miners were developing mutual benefit associations that provided financial assistance in cases of sickness, disability and death. Subsequent actions by mine operators resulted in the dissolution of these benefit associations. They were replaced by a system of company physicians who subscribed to the medical ideology that the respiratory symptoms experienced by miners were natural, inevitable and non-disabling consequences of work in the mines. Not until a protracted strike and federalization of the mines in 1946 did miners win a health care plan that was union controlled and industry financed. The Welfare and Retirement Fund, financed by a royalty assessed on each ton of coal, provided pensions, hospitalization and medical care for miners and their families. About this time, a few physicians began to recognize and write about coal workers' pneumoconiosis (CWP); they remained a minority within the medical establishment, however, and coal miners in most states continued to be denied workers' compensation for occupational lung disease (Smith, 1981).

By late 1968, there was pressure on the West Virginia legislature to pass a bill recognizing 'black lung' as a compensable disease under its state Workers' Compensation statutes. The law passed in March 1969, not only recognized 'black lung' as an occupational disease, but also made any miner with any respiratory impairment and who had 10 or more years of mining experience eligible for workers' compensation (Richman and Smith 1993). The Federal Coal Mine Health and Safety Act of 1969, including a section entitled 'Title IV-Black Lung Benefits,' grew out of this West Virginia legislation, and had the initial purpose of providing benefits for miners who were 'totally disabled by pneumoconiosis'. *Pneumoconiosis* has been defined by statute as 'a chronic dust disease of the lung and its sequelae, including respiratory and pulmonary impairments arising out of coal mine employment'.

The poor correlation between respiratory symptoms and radiographic findings, especially for simple CWP, has caused confusion in the compensation arena. The 1969 Act legislatively declared any miner, even one still working, to be 'totally disabled' if he had radiographic evidence of complicated CWP or was unable to 'engage in any substantial activity' due to pneumoconiosis. The 1972 amendments revised this definition to include miners who could no longer perform their regular coal mine jobs but who might be able to work in other capacities. For miners with 15 years or more of underground employment, the presence of totally disabling pulmonary impairment, even coupled with chest radiographs that did not show complicated CWP, could provide the basis for a claim, although this was a rebuttable presumption. The 1977 Act further redefined 'totally disabled' to include individuals who were still mining but who had a 'reduced ability' to perform their usual work. The most recent amendment in 1981 permits 'totally disabled' miners to be employed in other occupations while receiving benefits, and also allows a finding of 'total disability' when regular coal mine work is being performed if the miner's work capability is reduced. There is no requirement that a miner have lost income to be eligible for compensation (Richman and Smith 1993).

WORKERS' COMPENSATION AND ASBESTOSIS

In the United States, the history of workers' compensation for asbestosis has involved complex interactions between workers, labor organizations, government officials and members of the medical and legal professions. As early as 1930, employees began suing the Johns-Manville Corporation, the largest asbestos producer in the United States, alleging disability from asbestos-induced lung damage. At that time, asbestosis, like most occupational diseases, was not covered by the majority of state workers' compensation statutes. In 1934, the Supreme Court of North Carolina determined that the occupational disease asbestosis, contracted unexpectedly and gradually over a long time, if due to the employer's failure to provide a means of prevention, was an injury 'by accident' and was compensable under the state workmen's compensation act. The next year an Illinois court threw out several cases against Johns-Manville, ruling that workers had no right to either bring such suits under common law or to file workers' compensation claims for asbestos-related disease. In response, the Illinois legislature expanded the state compensation law to include occupational diseases. Other states gradually recognized asbestosis as a compensable occupational disease; however, as late as 1978, two states had not yet awarded anyone compensation for asbestosis (Castleman, 1984).

As with other occupational diseases, there are no firm guidelines for establishing diagnosis or causation of asbestosis. It has been suggested that a national panel of experts develop and periodically modify guidelines and criteria for the adjudication of claims for asbestos-associated diseases. Case-by-case determinations would be handled by regional panels which would include health professionals knowledgeable in occupational pulmonary diseases (Weill, 1983). At present, physiologic and radiographic findings do not always correlate well with compensation awards.

Among a group of patients from the University of Washington Occupational Medicine Clinic who filed workers' compensation claims for non-malignant asbestos disease in 1982–1986, the likelihood of claim acceptance was unrelated to the severity of asbestos disease as judged on chest X-ray. There also was no significant association between claims outcome and either concurrent obstructive pulmonary disease or past or current tobacco use (Nevitt *et al.*, 1994).

In Great Britain, asbestosis was formally included under workers' compensation in 1931, but with several significant restrictions. A successful claim could only be made if the employee had worked in an industry or process specifically covered by the compensation scheme. While asbestos-processing industries were covered, the scheme did not explicitly include users of asbestos products. Workers who had left the industry three or more years before filing a claim were ineligible (Castleman, 1984). Currently, compensation for asbestos-related disease in the United Kingdom is for one of four conditions: pneumoconiosis, mesothelioma, bilateral diffuse pleural thickening, and lung cancer in the presence of either asbestosis or diffuse pleural thickening. Compensability is determined by medical boards which disagree with the diagnosing physician in about one-third of cases (Seaton, 1990).

WORKERS' COMPENSATION AND PULMONARY MALIGNANCIES

It has been estimated that less than 10% of occupational cancers occurring in Canada are compensated by Workers' Compensation Boards (Teschke and Barroetavena, 1992). Patterns of compensation differ significantly for pleural malignancies versus other lung tumors. In a study of fatal disease among asbestos insulation workers in Ontario (1967–1985), 88% of claims filed for mesothelioma were awarded compared with 40% for other lung cancers (Finkelstein, 1989). The ability to associate disease with workplace exposure impacts both the likelihood that a claim will be filed and the probability that compensation will be awarded. Physicians and employees are both likely to associate mesothelioma with occupational exposure; the association between mesothelioma and asbestos exposure has been well publicized and few other causal associations have been identified. It is estimated that 24–36% of mesotheliomas are compensated in Canada while the population-attributable risk (PAR) is estimated at 47–76%. By comparison, only an estimated 1% of incident lung cancer cases received compensation while the PAR for lung cancer due to occupational exposure is estimated at 3–17%. Given the strong epidemiologic link between cigarette use and lung cancer, physicians may not question cancer etiology further in patients who are smokers (Teschke and Barroetavena, 1992).

As with other occupational diseases, issues arise regarding minimum duration of exposure and latency between exposure and disease. The 1984 Ontario guidelines for adjudication of asbestos-related claims require a minimum of 10 years' occupational exposure for both mesothelioma and other lung cancers. Minimum latency is 10 years for non-pleural cancers and 15 years for mesothelioma (Finkelstein, 1989). Asbestos was listed as the causative agent for 36% of occupational lung cancers in Canada; other agents listed included silica, 'dust' and arsenic (Teschke and Barroetavena, 1992).

WORKERS' COMPENSATION, RESPIRATORY DISEASE AND RADIATION EXPOSURE

In the late 1940s, the US Public Health Service began a survey of uranium miners in the Colorado Plateau region. This study demonstrated an excess rate of lung cancer in underground miners exposed to radon, both in uranium mines and in other radon-contaminated mines. On the basis of this and other epidemiologic data, radon was confirmed as an occupational carcinogen and regulations to limit exposure were implemented in many countries. In addition to the malignant effects of radon exposure, non-malignant respiratory diseases such as pulmonary fibrosis and emphysema were plausibly linked to radon exposure and may be compensable (Samet, 1992).

In October 1990, the US Congress passed the Radiation Exposure Compensation Act, Public Law 101–426. In addition to compensating specified classes of individuals affected by fallout from above-ground nuclear weapons testing, the Act provides for compensation of individuals who had been employed in underground uranium mines that provided ore for the nuclear weapons program. Benefits were to be paid from a federally established Radiation Exposure Compensation Trust Fund which received an initial $100 million appropriation. Eligible individuals must have worked in an underground uranium mine in Colorado, Utah, New Mexico, Arizona or Wyoming at any time between January 1, 1947, and December 31, 1971. Exposure criteria are based on the concept of a working level month (WLM) which means radiation exposure equal to one working level every work day for a month. A working level corresponds to a concentration of short half-life radon daughters that will release 1.3×10^5 million electron volts of alpha energy per liter of air. Eligibility is based on an individual's exposure history, disease process and smoking status, but all accepted claims are paid equally at $100 000.

To receive compensation for lung cancer, a non-smoker must have 200 (or greater) WLM of exposure. For smokers, the exposure requirement is 300 WLM for those developing cancer before age 45, while those with exposures of 500 WLM and above can be compensated for lung cancer at any age. Compensation for non-malignant respiratory disease, defined as fibrosis of the lung and cor pulmonale related to fibrosis of the lung, requires identical exposure levels and written medical documentation of disease. Specific medical criteria are defined in the Act. In addition, moderate or severe silicosis or other pneumoconiosis is compensable provided the claimant worked in a uranium mine on or within an Indian reservation for sufficient time to meet the WLM criteria.

SUMMARY

The story of worker compensation in Germany reflects nearly 800 years of development of the relationship among workers and employers. This relationship stabilized when the third party, the federal government, became a mediator and administrator of the programs. A system is in effect which mandates that the physician diagnosing the disease, as well as the employer, must report the illness to the federal government.

In the United States, only a percentage of individuals with occupational respiratory disease receive workers' compensation. A number of factors contribute to this finding. For example, there has been considerable debate about what are appropriate diagnostic criteria for occupational asthma; diagnosis is confounded by the fact that many workers have ceased exposure before receiving medical evaluation. Even with the pneumoconioses, considerable disagreement frequently exists among physicians who are interpreting the chest radiograph. Furthermore, the poor correlation between pulmonary impairment and X-ray findings may add to the diagnostic difficulties. When medical criteria are defined, workers' compensation boards have been criticized for applying them arbitrarily and inconsistently. Workers and their physicians may fail to link disease with occupational exposure, especially for common diseases, diseases with long latency, or diseases with multiple causalities. If the association is made between disease and work, workers and physicians may not be aware that the condition is compensable. Difficulties with reimbursement from the compensation system may provide disincentives for physicians to assist patients in filing claims. Finally, compensation, when received, may be inadequate to provide for the financial needs of the affected individual.

Solutions have been proposed for several of these problems. The APHA (1984) has proposed that Congress establish procedures for the promulgation of definitions of work-related diseases, their coverage, protective methods and presumptive standards for compensation on a national level. They have also recommended that a compensation fund be established, financed by manufacturers, insurers, importers and users of hazardous materials, to compensate claimants in cases where the responsible party(s) cannot be established or held accountable.

The Institute of Medicine (IOM) has identified inadequate training at all levels (undergraduate, graduate and post-graduate) of medical education as a major factor limiting the role of physicians in the diagnosis, treatment and prevention of occupational and environmental disease. At a minimum, 'All primary care physicians should be able to identify possible occupationally or environmentally induced conditions and make the appropriate referrals for follow-up'. In response, proposals were made to increase the awareness of practicing physicians of occupational and environmental conditions. In addition to increasing the number of trained specialists in occupational medicine, it was also proposed that the number of primary care physicians with special expertise in occupational and environmental medicine be increased (Rosenstock, 1990).

There also have been proposals to remove what some consider 'disincentives' in motivating physicians to participate in the claims filing process. Financial disincentives are well recognized. Physician payment is often withheld until a workers' compensation dispute is resolved; payment delays of up to 3 years may be encountered for chronic disease claims that are eventually accepted. Proponents of a plan initially to reimburse occupational disease and injury claims through third-party health insurance believe that this will both cut workers' compensation costs as well as making physicians more willing to be involved in the claims process (Brown, 1981; APHA, 1984; Ramsey and Rosenstock, 1994).

While workers' compensation programs are needed to compensate workers who are injured or become ill as a result of their employment, the appropriate means of dealing with occupational diseases is to prevent them. As the APHA policy statement 'Compensation for and Prevention of Occupational Disease' points out, 'A failure to construct a compensation system which significantly contributes to the prevention of disease poses the danger of institutionalizing occupational disease' (APHA, 1984).

The ultimate prevention of work-related lung disease in all nations occurs in a specific or unique milieu which includes social, political, economic, and legal

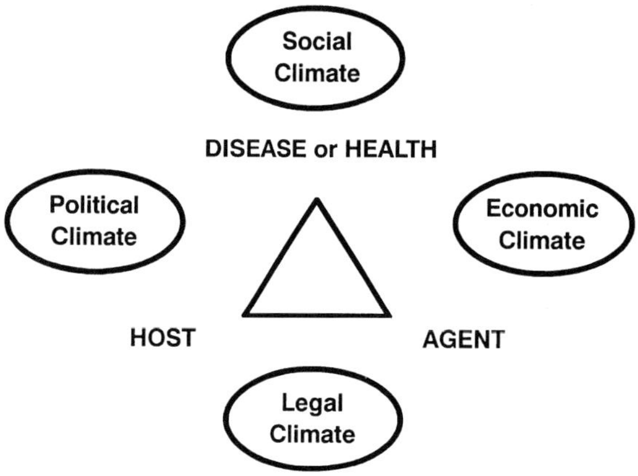

Figure 6.1 The context of prevention.

influences (Figure 6.1). Consequently, the prevention of occupational lung disease requires careful attention not only to host, agent and disease interaction, but also requires the implementation of strategies that reflect these non-medical influences.

A model of prevention used in the United States (Figure 6.2) shows the important interactions between workers and employers, regulatory agencies [Occupational Safety and Health Administration (OSHA) and Mine Safety and Health Administration (MSHA)], and research organizations [National Institute for Occupational Safety and Health (NIOSH) and academic institutions]. Each of these groups is an important player in driving the prevention of occupational lung diseases, and each of these groups also must strive to reduce primary hazardous exposures to avoid 'the danger of institutionalizing occupational lung diseases' (APHA, 1984)

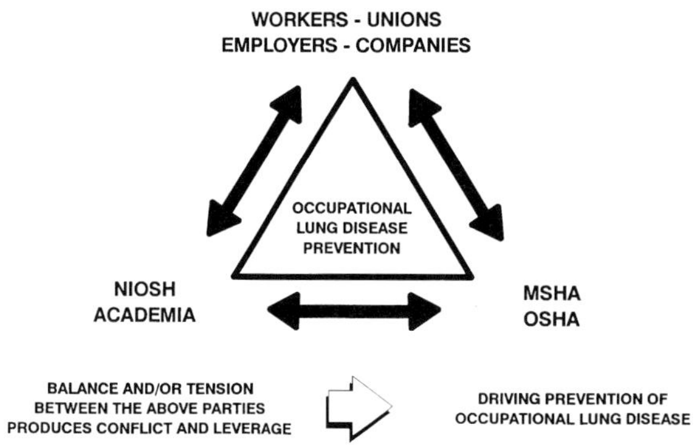

Figure 6.2 Occupational lung disease. The triangle of prevention: United States model.

REFERENCES

APHA (1984) Policy Statement 8329 (PP): Compensation for and prevention of occupational disease. *Am. J. Public Health*, **74**,292–5.

Auerbach, C.A., Garrison, L.K., Hurst, W. and Mermin, W. (1961) *The Legal Process: An Introduction to Decision-Making by Judicial, Legislative, executive, and Administrative Agencies*, Chandler Publishing Co., San Francisco.

Banks, D.E., Rando, R.J. and Barkman, H.W., Jr (1990) Persistance of toluene diisocyanate-induced asthma despite negligible workplace exposures. *Chest*, **97**,121–125.

Barth, P.S. and Hunt, H.A. (1980) *Workers' Compensation and Work-Related Illnesses and Diseases*, MIT Press, Cambridge.

Brown, T.C. (1981) Evaluating work relatedness of disease. *Chest*, **79**,S127–9.

Castleman, B.I. (1984) *Asbestos: Medical and Legal Aspects*, Harcourt Brace Jovanovich, New York.

Chan-Yeung, M. and Malo, J.-L. (1995) Current concepts: occupational asthma. *N. Eng. J. Med.*, **333**,107–12.

Cherniack, M. (1986) *The Hawk's Nest Incident*. Yale University Press, New Haven.

DeWitte, J.-D., Chan-Yeung, M. and Malo, J.-L. (1994) Medicolegal and compensation aspects of occupational asthma. *Eur. Respir. J.*, **7**,969–80.

Finkelstein, M.M. (1989) Analysis of mortality patterns and workers' compensation awards among asbestos insulation workers in Ontario. *Am. J. Ind. Med.*, **16**,523–8.

Gannon, P.F.G., Weir, D.C., Robertson, A.S. and Burge, P.S. (1993) Health, employment, and financial outcomes in workers with occupational asthma. *Br. J. Ind. Med.* **50**,491–6.

Hughes, J.T. (1981) Diagnosing byssinosis: the medical controversy. *Chest*, **79**,S129–30.

Kippen, S. (1995) The social and political meaning of the silent epidemic of Miners' Phthisis, Bendigo 1860–1960. *Soc. Sci. Med.*, **41**,491–9.

Markowitz, S.B., Fischer, E., Fahs, M.C., Shapiro, J. and Landrigan, P.J. (1989) Occupational disease in New York state. *Am. J. Ind. Med.*, **16**,417–35.

Nevitt, C., Daniell, W. and Rosenstock, L. (1994) Workers' compensation for nonmalignant asbestos-related lung disease. *Am. J. Ind. Med.*, **26**,821–30.

Ramsey, S. and Rosenstock, L. (1994) Proposal to reimburse occupational medicine disease and injury claims through third party health insurance. *Am. J. Ind. Med.*, **26**,147–54.

Richman, S.I. and Smith, C.J. (1993) Legal aspects of impairment and disability in pneumoconiosis. *Occupational Medicine: State of the Art Reviews*, **8**,71–92.

Rooke, G.B. (1981) Compensation for byssinosis in Great Britain. *Chest*, **79**,S124–7.

Rosenstock, L. (1990) Role of the physician in environmental and occupational asthma. *Chest*, **98**,S162–4.

Samet, J.M. (1992) Diseases of uranium miners and other underground miners exposed to radon, in *Environmental and Occupational Medicine* (ed. W.N. Rom), Little, Brown & Company, Boston pp. 1085–91.

Seaton, A. (1990) Asbestos diseases and compensation. *Br. Med. J.*, **301**,453–4.

Seligman, P.J. and Bernstein, I.L. (1993) 'Medicolegal and compensation aspects,' in *Asthma in the Workplace*, (eds I.L. Bernstein, M. Chan-Yeung, J.-L. Malo and D.I. Bernstein), Marcel Dekker Inc., New York, pp. 323–40.

Smith, B.E. (1981) Black Lung: the social production of disease. *Int. J. Health Services*, **11**,343–59.

Stanbury, M., Joyce, P. and Kipen, H. (1995) Silicosis and workers' compensation in New Jersey. *J. Occup. Env. Med.*, **37**,1342–7.

Stenton, S.C., Sandhu, P.S. and Hendrick, D.J. (1995) Industrial injury benefit for occupational asthma in north east of England. *Br. Med. J.*, **318**,1299–300.

Stephenson, W.H. (1981) Administrative law problems with byssinosis. *Chest*, **79**,S132–3.

Tarlo, S.M., Liss, G., Corey, P. and Broder, I. (1995) A workers' compensation claim population for occupational asthma. Comparison of subgroups. *Chest*, **107**,634–41.

Teschke, K. and Barroetavena, M.C. (1992) Occupational cancer in Canada: what do we know? *Can. Med. Assoc. J.*, **147**,1501–7.

Weill, H. (1983) Asbestos-associated diseases: science, public policy, and litigation. *Chest*, **84**,601–8.

White, N.W., Cheadle, H. and Dyer, R.B. (1992) Workmens' compensation and byssinosis in South Africa: a review of 32 cases. *Am. J. Ind. Med.*, **21**,295–309.

FURTHER READING

Bulow, W. (1905) Das Knappschaftswesen im Ruhrkohlenbezirk bis zum allgemeinen preussischen Berggesetz vom 24. Juni, 1865. Borna, Leipzig.

Boyer, J. (1995) Unfallversicherung und Unternehmer im Bergbau. Die Knappschafts-Berufsgenossenschaft 1885–1945, Beck, Munchen.

Piekarski, C. (1991) Aspekte der arbeitsmedizinischen Betreuung des Bergmannes. *Zbl. Arbeitsmed.*, **41**,373–8.

Bundesberggesetz vom 13 August, 1980, *Bundesgesetzblatt Teil*, **1**,1330.

Morfeld, P., Vautrin, H.-J., Kosters, A. *et al.* (1997) Limit value assessment for respirable coal mine dust in Germany. *Appl. Occup. Environ. Hyg.*, **12**, 840–8.

Lehmann, F. and Renner, H.J. (1994) Belastung durch ionisierende Strahlung im Uranerzbergbau der ehemaligen DDR. *Kompass J.*, **312**,649–55.

Renz, K. (1993) Das Lungenkrebsrisiko ehemaliger Beschaftigter der SDAG Wismut. *Die BG*, **1**,46–8.

Schulz, H. (1994) ZeBWis und die Archive der Wismut. *Die BG*, **6**,398–401.

Schuttmann, W. and Arndt, D. (1994) Die Begutachtung der Schneeberger Lungenkrankheit in der ehemaligen DDR. *Die BG*, **6**,403–5.

The financing of the Berufsgenossenschaften. Hauptverband der gewerblichen Berufsgenossenschaften, St. Augustin (1994).

7 Occupational health, occupational illness: tuberculosis, silicosis and HIV on the South African mines

Brian Williams, Catherine Campbell, Nokuzola Mgogi and Immo Kleinschmidt

INTRODUCTION

South Africa and gold have long been synonymous, and mining plays a key role in the economy. Gold contributes about 4% of the gross domestic product directly, twice as much if indirect multipliers are included, and provides almost 30% of foreign exchange earnings (Baxter, 1996). The industry employs about 350 000 workers, about 90% of whom are black. The 10% who are white occupy mainly supervisory positions (Hnizdo, 1991; COM, 1994). With some gold mines reaching depths of more than 3 km, in seams that may be only 1–2 m deep, gold mining remains labor intensive and working conditions are dangerous and stressful. In 1993, out of every 100 000 gold miners, 113 died in accidents, 2000 suffered a reportable injury (COM, 1994), 1100 developed active tuberculosis and, of these, 25 died (Goldfields, 1993); in 1990 about 500 were certified as having silicosis (MBOD, 1990).

The majority of mineworkers in South Africa are migrant laborers, either from within South Africa or from neighboring states. Migration occurs in the context of a labor recruiting system which has been in place for more than a century (Crush *et al.*, 1991). Each contract of employment lasts for one year, but may be renewed many times and 90% of mineworkers spend more than 10 years in the mines (Leon *et al.*, 1996). A large proportion of black mineworkers are still housed in single men's hostels living and working under conditions that put them at high risk of disease, trauma and death. An increasing duration of service in the mines (Murray *et al.*, 1996) leads to an increased lifetime exposure to dust, radiation, noise and other

Occupational Lung Disease. Edited by Daniel E. Banks and John E. Parker. Published in 1998 by Chapman & Hall.
ISBN 0 412 73630 6

contaminants. The long delay between exposure and the onset of many occupational diseases means that illnesses may develop long after men have left the mines and returned to their rural homes, where even the most rudimentary medical services are often lacking.

Tertiary health care in the mines is, however, excellent (Plate 5). The industry operates 41 hospitals offering 7000 beds with approximately 350 full-time doctors and 3500 health care staff (Fourie, 1996). While these facilities provide a valuable national resource, we argue below that existing biomedical interventions are inadequate for containing diseases such as silicosis, tuberculosis and HIV in the particular social context of the mines, and that greater attention needs to be paid to early case detection, evaluation of the delivery of health care services, and the development of innovative approaches to primary health care, as well as to the social and living conditions in which such diseases flourish.

Unfortunately, there are little publicly available data on the incidence, prevalence or mortality due to work-related and other diseases on the mines. Gold Fields of South Africa publishes an annual report which contains basic health statistics; the other five major mining houses have discontinued the publication of annual medical reports (Leon *et al.*, 1996). The health department of the Chamber of Mines of South Africa collects morbidity and mortality data but does not make these publicly available (Leger, 1992). The Medical Bureau for Occupational Diseases, under the National Department of Health, publishes data on claims for compensable disease. From these and some other sources researchers have produced proxy rates of work-related respiratory diseases in the mining industry (Leger, 1992). It must be stressed that not only are the available numerator data poor, but good denominator data are also lacking and most rates are aggregated over workers in different mines, in different jobs, and with different levels of exposure and risk.

THE HISTORY OF SILICOSIS AND TUBERCULOSIS IN THE MINES

In gold mines where the concentration of silica in the ore is high, silicosis has long been responsible for high levels of mortality and morbidity among miners. This disease continues to take a heavy toll. Very little is known about the history or prevalence of silicosis among black mine workers beyond the fact that, as Katz (1994) notes, during the first decades of this century, silicosis among black mine workers was largely ignored, with doctors arguing, for example, that black workers who contracted simple silicosis were restored to health by rest at their rural homes. However, more is known about silicosis in white mine workers. Early in this century the annual mortality from 'phthisis', including silicosis, among white miners was estimated to be 1100 per 100 000 underground workers (Katz, 1994), leading a mining journal to note that 'phthisis has killed such a large number of young, strong laborers, that many miners in Europe and America prefer to stay in their own countries, earn lower wages, and live, rather than come to the Rand, earn a higher wage, and die' (Katz, 1994).

Before the 19th century there seems to have been little tuberculosis among residents of southern Africa (Anderson, 1990). During that century, however, many Europeans suffering from tuberculosis came to South Africa to take advantage of the climate. The arrival of large numbers of infectious people, combined with the social changes associated with the discovery of minerals and especially the need to recruit migrant labor on a large scale, provided ideal conditions for the spread and increase of tuberculosis (Anderson, 1990; Packard, 1990). The increased risk of tuberculosis among those with silicosis (Cowie, 1994) meant that mine workers were particularly vulnerable.

In the 1920s, a confidential report by the mine inspectorate noted that the mines were returning an average of 1000 workers with newly diagnosed tuberculosis to the rural areas each year (Anderson, 1990). The tuberculosis rates among gold miners were extremely high, particularly among men from Botswana, Lesotho and southern Mozambique. Annual incidences were 2350, 1340 and 1490 per 100 000, respectively (DOPH, 1990). The impact on rural communities was highlighted by a study of 112 patients who had returned to the Transkei and the Ciskei after being diagnosed as having tuberculosis on the mines. Within five years, 65 had died. Of the 47 still alive, nearly half were incapable of working again (Allan, 1924).

In the early decades of this century, crowded and unsanitary living conditions, combined with stressful working conditions, led to very high rates not only of silicosis and tuberculosis, but also of pneumonia among miners. This, combined with the high levels of migrancy, probably contributed to the high rates of these diseases in rural areas of southern Africa (Gorgas, 1914). Living conditions continue to be problematic today. In 1995, members of a governmental commission of inquiry investigated the regulation of occupational health and safety in the mining industry. They visited three hostels on two mines (Leon *et al.*, 1996). Each room was occupied by between 12 and 20 men, giving an average of just over 5 m^2 per person. The assessors were 'shocked by the conditions in which food was prepared' and by ablution facilities 'so squalid as to shock the most hardened'.

TUBERCULOSIS

Tuberculosis has long been a major cause of morbidity and mortality in the general population of South Africa. In 1980, this accounted for an estimated 7% of all deaths among blacks between 15 and 64 years of age (Yach, 1987). In the most recent review of tuberculosis in South Africa, the authors concluded that in 1994 the overall incidence was 311 per 100 000 per year (DOH, 1996b). After trauma, tuberculosis claims the largest number of lives among working miners, with a mortality rate of 25 per 100 000 in 1993. The industry-wide incidence of tuberculosis in the same year was 1000 per 100 000 (COM, 1994). There is evidence that the incidence of tuberculosis among miners is increasing (Murray *et al.*, 1996). HIV enhances the spread of tuberculosis but other factors, including the high prevalence of silicosis, longer periods of service in mining and the associated increase in the age of the work force, the physical stress of long shifts under conditions of high humidity and heat, as well as the difficult living conditions, increase the risk of tuberculosis in miners.

As a result of the high incidence of tuberculosis in the mining industry, the Mines Phthisis Act of 1916 made provisions for the compensation of tuberculous miners who were then repatriated to their place of origin (Katz, 1980). However, in 1977 a study to evaluate the regulations that prevented men with tuberculosis from continuing to work as gold miners was initiated by Cowie (1987), who subsequently argued that tuberculous mine workers receiving suitable short-course chemotherapy could safely return to work underground. In 1984, the regulations were changed and men who were under treatment for pulmonary tuberculosis with an approved regimen were allowed to return to work (Cowie, 1987). In 1994, the Occupational Disease in Mines and Works Act was revised to reduce the entitlement to miners with tuberculosis. Since then, miners with tuberculosis are only compensated for shifts lost when they are hospitalized, when they are found to have second-degree tuberculosis, or when they show permanent lung damage due to tuberculosis (ODMWA, 1994).

The mining industry has excellent medical facilities which use modern drugs and competent physicians who employ treatment regimes recommended by the World Health Organization. Mine medical clinics are close to the compounds or hostels where miners live. In one typical group of mines, workers have an annual fitness examination (Plates 6–8); miniature chest radiographs are used to screen miners for tuberculosis. All suspected cases of tuberculosis are confirmed bacteriologically. Nevertheless, a number of questions arise concerning tuberculosis case detection. Sluis-Cremer (1980) observed that it was not unusual to see men with advanced pulmonary tuberculosis less than 9 months after their chest radiographs appeared entirely normal, and it may be that the policy of screening miners once a year is inadequate. The hearts and lungs of miners who die while in employment are examined at the National Centre for Occupational Health for evidence of occupational disease including tuberculosis. A post-mortem study of black mineworkers showed that of those who were shown to have tuberculosis at autopsy, 40% were not detected in life (Lowe and Murray, 1994).

There is an urgent need to improve tuberculosis control on the mines. Beginning in 1996, the mining industry has followed the new National Tuberculosis Control Programme guidelines (DOH, 1996a), adopted from the World Health Organization Global Tuberculosis Programme. The emphasis is on treating smear-positive cases under Directly Observed Therapy, Short Course Treatment (DOTS). While DOTS is now accepted and followed, a recent study on one group of mines (Mgogi *et al.*, 1996) showed that while the apparent rate of compliance assessed from clinic records is 99.8%, the actual rate of compliance – when assessed by random urine tests – is only 85% among the 40% of patients who returned to give urine samples. In this study, 21 ± 6% of patients whose treatment regime did not include isoniazid had positive urine tests for isoniazid and 35 ± 7% not on rifampicin had positive urine tests for rifampicin.

The epidemic of HIV complicates the problem of tuberculosis control. In a study in Kinshasa, for example, extending tuberculosis treatment among HIV-positive patients from 6 to 12 months reduced the relapse rate at 2 years from 9% to 2% (Perriens *et al.*, 1995). In one group of South African mines, acquired isoniazid and multi-drug resistance have both increased since 1985 and now occur in 7% and 0.8% of tuberculosis patients, respectively (Churchyard, 1996).

SILICOSIS

The relationship between exposure to dust containing silica and the occurrence of silicosis has been thoroughly investigated in the South African mining industry (Beadle, 1971). The Leon Commission (Leon *et al.*, 1996) concluded that a person working for 20 years or more in high-dust areas, such as drilling or shaft sinking, has a 20–30% chance of developing simple silicosis. Silicosis compounds the problem of tuberculosis. In one autopsy study, even a slight degree of silicosis not detected radiologically during life was significantly associated with an increased prevalence of pulmonary tuberculosis compared with gold miners without silicosis (Sluis-Cremer, 1980). In a cohort study of South African gold miners the annual incidence of tuberculosis was 1% among those without silicosis but increased to 2.2%, 2.9% and 6.3% for those with categories 1, 2 and 3 silicosis, respectively (Cowie, 1994). The association with silicosis increases the need to ensure early case detection and high levels of compliance for tuberculosis.

A retrospective cohort study of over 2000 white South African gold miners with an average of 24 years of service (Hnizdo and Sluis-Cremer, 1993) showed that the risk of silicosis increased exponentially with cumulative dust dose. In these white miners, those with an exposure of 15 mg/m^3-years (the highest exposure group), which represents approximately 37 years of gold mining at an average respirable dust concentration of 0.4 mg/m^3, the cumulative risk for developing silicosis reached 77%. In a large proportion of the miners, radiologic signs of silicosis developed only after they had left the mines.

An autopsy study of black gold miners found a prevalence of silicosis of nearly 12% (Sluis-Cremer, 1980), although only 3% of these were 'moderate' or 'marked' in severity. A more recent study confirmed the prevalence of silicosis diagnosed at autopsy and reported an increasing trend over the past 15 years (Murray *et al.*, 1996). For comparison, the prevalence of silicosis diagnosed during life in a cohort of black gold miners was 1.4% (Cowie, 1987). Using International Labour Organization category 1/1 or greater to define silicosis radiographically, a study of silicosis comparing radiological findings in life with pathologic findings at autopsy showed that only one-third of the pathologically confirmed cases were diagnosed on the chest radiograph, taken on average 2.7 years earlier. The sensitivity of the radiologic findings improved with increasing degree of silicosis at autopsy but, even among those with a moderate or marked degree of silicosis, a large proportion were not diagnosed radiologically (Hnizdo *et al.*, 1993).

Simple, chronic silicosis does not appear to impair lung function (Irwig and Rocks, 1978), but more advanced silicosis does, even after controlling for the effects of smoking (Cowie and Mabena, 1991; Cowie *et al.*, 1993). Studies on white South African miners have shown that smoking acts synergistically with silica dust to cause respiratory impairment (Hnizdo *et al.*, 1990). There are no studies on the effect of silicosis on life expectancy of black South African mine workers, but it is likely to be substantial, particularly in view of the resulting increase in the lifetime risk of tuberculosis.

Jobs performed by black workers generally involve higher levels of dust exposure than those done by white workers. This is reflected in the fact that the average length of service before certification of silicosis is 15 years for black workers, and 26 years for white workers (Leger, 1992). Studies of former black mine workers have shown that silicosis has been a serious problem. In Botswanan men who had worked in South Africa mines, more than one-quarter had pneumoconiosis and 6% had progressive massive fibrosis (Steen *et al.*, 1997). According to preliminary results of a recent study in the Libode District of the Transkei, among a random sample of retired miners recruited between 1969 and 1980, 32% were diagnosed as having pneumoconiosis, 13% as having tuberculosis, and 23% as having both (Trapido *et al.*, 1996).

HIV

In a country with already very high levels of tuberculosis, there is now a rising epidemic of HIV infection. The only available statistics on the nationwide burden of HIV in South Africa are those kept by antenatal clinics. Between 1990 and 1994, the nationwide prevalence of HIV infection among women attending antenatal clinics rose from less than 1% to 7.4% with a mean doubling time of 15 months, although with wide provincial variation (Swanevelder, 1996).

The most important factors associated with severe epidemics of HIV/AIDS are well known and include high rates of sexually transmitted diseases, migrancy, social instability, disruption of social support mechanisms and family structures, and opportunistic infections, especially tuberculosis (Campbell and Williams, 1996). The extent of migrancy and the lack of social support among miners, as well as the high levels of tuberculosis, have been described above. The high rates of sexually transmitted disease (STD) are indicated by data from a clinic serving one group of mines where the annual incidence of new STD cases is 16 100 per 100 000 miners (Ballard, 1996). As in the rest of Africa, the main mode of HIV transmission is heterosexual. A recent pilot study suggested that in a culture which promotes high levels of sexuality among male migrant workers, most of whom live some distance from their wives and families, multiple sexual encounters with a number of partners are common. Furthermore, levels of condom use appears to be low. Within such a context, it is likely that the impact of HIV/AIDS on the mining population will be substantial.

The heterosexual epidemic of HIV/AIDS in South Africa began to take off in the 1990s. Unfortunately, there are no published data on the current overall prevalence of HIV in the mines in South Africa. In 1987, HIV testing was done on workers in the South African gold mines and correlation made between HIV status and country of origin. The prevalence among mineworkers from Malawi was 4%, while the prevalence in those from other countries, including South Africa, was about 0.03% (Brink and Clausen, 1987). As a result, the Chamber of Mines stopped recruiting novices from Malawi (James, 1992) and the number of Malawians employed in the South African mines declined from 13 090 in 1988 to 2212 in 1989 (Crush *et al.*, 1991). Since then no further prevalence studies have been published. Using the 1987 prevalence of 0.03% among non-Malawian workers, and assuming a doubling time of 15 months, as in the antenatal clinic surveys, the prevalence of HIV among mineworkers could now be as high as 15% and informal estimates by mine medical officers place the current prevalence between 15 and 20%. In 1994, 19% of the tuberculosis patients in one group of mines were HIV-positive (Churchyard, 1996).

The potential impact of HIV infection when prevalence reaches these levels is shown by a study of a mature epidemic in Uganda (Mulder *et al.*, 1994) where the prevalence of infection among adults was 8%. In this rural population adult mortality was increased 15 times by HIV infection. Against this background, the lack of an appropriate surveillance system in the mines is particularly worrying given that such data are essential for monitoring the course of the epidemic and planning adequate health care services in the future.

The association between tuberculosis and HIV is well established. Studies in other African countries have shown that the relative risk of developing active tuberculosis may be increased 8 to 12 times among those with HIV infection. Approximately one-third of all cases of tuberculosis may be related to HIV infection (Van den Broek *et al.*, 1993; Richards *et al.*, 1995). In many countries where tuberculosis was thought to be under control it is now becoming epidemic. Increased levels of HIV infection in the mines are likely to reduce immunity and resistance to tuberculosis, making it more difficult to treat those who are diagnosed with this infection (Packard and Coetzee, 1995). Between 1990 and 1992, the annual incidence of tuberculosis in one group of mines was steady at about 600 per 100 000 but had increased to about 1000 per 100 000 in 1994 (Churchyard, 1996). At least part of this increase is likely to be attributable to HIV infection.

THE SOCIAL CONTEXT OF DISEASE

Against the background of the lack of success in stemming the tide of STDs (including HIV) and tuberculosis in southern Africa using conventional biomedical interventions, there is a growing argument that unless miners understand the nature of disease transmission and are actively involved in programs to manage both HIV and tuberculosis, there is little hope that the levels of transmission will be significantly reduced (Campbell, 1996). In a recent study of mine workers' perceptions of health (Campbell, 1997) very few mine workers had any first-hand experience of AIDS. This made it difficult for them to regard the disease as a threat to their own lives. Living in a dangerous and stressful environment with few opportunities for intimacy and little social support reinforced a 'macho' notion of sexuality in which regular sex, often with commercial sex workers, and involving 'flesh-to-flesh contact' is regarded as essential. This continued to place them at high risk of acquiring HIV. While all of the mine workers were aware of tuberculosis and many had experienced the diseases themselves, few of them knew that tuberculosis is an infectious disease. Indeed, the association with silicosis led them to conclude that tuberculosis was caused by the dust.

Evidence suggests that miner's understandings of illness such as tuberculosis and HIV may often be at variance with those of the medical profession and that such variances may play a role in the failure of excellent biomedical facilities to make an impact on the growing incidence of these disease. Furthermore, miners have conceptualized notions of health and illness in a variety of healing systems that embraced both traditional and modern views of healing. It is not surprising then that the vast majority of them consult both biomedical practitioners and traditional healers when they are ill. Attempts to deal with STDs, HIV and tuberculosis in the mines need to involve a far more active dialogue between mineworkers, traditional healers and those who are part of the formal medical structures (Abdool Karim *et al.*, 1994).

OTHER DISEASES

There are a number of other occupational lung diseases, some of which are associated with mines other than gold mines, and this section provides a brief overview of these.

Asbestos mining in South Africa has created an epidemic of asbestos-related disease (Felix *et al.*, 1994; White, 1994) among asbestos workers as well as in the broader community due to large-scale environmental contamination. Although very few people are still employed in the asbestos industry, tens of thousands of people have been occupationally or environmentally exposed and a proportion will develop diseases such as mesothelioma. Between 1980 and 1990, about 400 cases of mesothelioma were certified annually and it is expected that these adverse health effects will continue well into the next century. Since the asbestos industry has all but disappeared it is not clear who will pay for the cost of rehabilitation and compensation (White, 1994).

While asbestos mining has declined greatly, coal mining remains important in the economy. Certifications for coal workers' pneumoconiosis have declined from approximately 600 per 100 000 workers in 1980, to around 400 per 100 000 in 1989, numbers similar to the silicosis rate for gold miners (White, 1994). Unfortunately, separate data are not available for silicosis and coal workers' pneumoconiosis after 1989. Routine post-mortem examinations of black coal miners (White, 1994) yielded a prevalence of coal workers' pneumoconiosis of approximately 7%.

Apart from the diseases discussed above, there were 60 certifications for chronic obstructive pulmonary disease per 100 000 workers in the mining industry (MBOD, 1993). Smaller numbers of workers were also affected by platinum salt sensitivity, scleroderma associated with dust exposures, and hard-metal pneumoconiosis (Leon *et al.*, 1996). Much larger numbers of workers were affected by non-occupational, but arguably work-related, diseases such as pneumonia (817 incident cases and 20 deaths per 100 000 workers) and neoplasia (200 incident cases and 21 deaths per 100 000 workers). The Chamber of Mines also reports 62 deaths per 100 000 workers as due to 'ill defined diseases and disorders' (Goldfields, 1993).

CONCLUSIONS

Probably the most important diseases currently facing miners in South Africa are tuberculosis, especially through its association with silicosis, and the rising epidemic of HIV/AIDS. However, tuberculosis is treatable, silicosis is preventable, and there are interventions that may reduce the incidence of HIV/AIDS (Grosskurth *et al.*, 1995).

There is an urgent need to set up proper surveillance systems for these diseases (Leon *et al.*, 1996). It is important to know considerably more about the distribution and determinants of disease. Improving the compilation and reporting of existing health records would be an important first step, but epidemiologic surveys and monitoring should be carried out on a regular basis in order to identify risk situations and factors for the most important diseases. Such surveys will require agreement between labor and employers to establish an effective and ethical approach. If these disease are to be addressed effectively, it is important to monitor and assess the effectiveness of any interventions so that those involved in disease control in the mines will know whether they are succeeding and/or whether they should attempt alternative strategies.

The mining industry has good medical facilities, well-trained medical personnel and access to the most modern drugs. The tertiary interventions that are in place tend to address the clinical rather than epidemiologic causes of disease. Good primary health care interventions are needed that focus on prevention and the early detection of problems. This will not only be more cost effective but will also create links between the medical aspects of disease and the social context of disease transmission in the mines. This will also provide a basis for more effective participation by labor in health interventions in the mines.

The prevention of silicosis requires the monitoring and reduction of dust levels (Murray *et al.*, 1996). This is primarily a technical problem, but again it is important that the control of disease be addressed jointly by the mine management, the mine medical services, and the trade unions. The control of tuberculosis is closely linked to the containment of HIV. HIV is as much a social as a biomedical disease and the management of HIV depends critically on bringing about changes in an individual's sexual behavior. There is no evidence that traditional information-based HIV education programs, favored by many of the mines, have succeeded in promoting large-scale behavior change (Crisp, 1996). There are a number of initiatives planned to develop more innovative and participatory 'peer education' programs in the context of mining. These are health education programs which aim to take account of the fact that health behavior is governed not only by factual information and individual choice, but by the cultural construction of health and illness, group norms and the routine organization of everyday life (Dube and Wilson, 1996).

The World Health Organization contends that if 85% of people who contract tuberculosis comply fully with their treatment, the incidence of tuberculosis will decline. Unfortunately, it is often the case that non-compliance is blamed on the patients. The study by Mgogi *et al.* (1997) shows that the primary failure may be the medical program. For a disease in which compliance and early case detection are the most important determinants of successful control, active participation of the workers in addressing the control of disease is essential.

The Leon Commission of Enquiry into the Health and Safety of Mineworkers (Leon *et al.*, 1996) felt justified in concluding that 'there is no evidence to indicate a decline in the prevalence or severity of any occupational disease in the mining industry during the past twenty years . . . radical steps are required to deal with the serious occupational health problems described in the evidence presented to the Commission'. The challenge is not only to manage the current problems and to deal with the consequences of past neglect but also to ensure that in future, the levels of disease are substantially reduced.

REFERENCES

Abdool Karim, S., Ziqubu-Page, T. and Arendse, R. (1994) *Bridging the Gap: Potential for a Health Care Partnership between African Traditional Healers and Biomedical Personnel in South Africa*, Medical Association of South Africa, Pinelands, South Africa.

Allan, P. (1924) *Report of Tuberculosis Survey of the Union of South Africa*, Government Printer, Cape Town, UG 18–24 (cited in Anderson, 1990).

Anderson, N. (1990) Tuberculosis and social stratification in South Africa. *Int. J. Health Services*, **20**, 141–65.

Ballard, R. (1996) STDs on the mines: the nature of the problem, in *HIV/AIDS Management in South Africa: Priorities for the Mining Industry* (eds B.G. Williams, and C.M. Campbell), ERU, Johannesburg, pp. 111–14.

Baxter, R. (1996) The economics of the South African mines, in *HIV/AIDS Management in South Africa: Priorities for the Mining Industry* (eds B.G. Williams and C.M. Campbell), ERU, Johannesburg, pp. 42–50.

Beadle, D.G. (1971) The relationship between the amount of dust breathed and the development of radiological signs of silicosis: an epidemiological study of South African gold miners, in *Inhaled Particles III* (ed. W.H. Walton), Pergamon Press, Oxford, pp. 953–66.

Brink, B. and Clausen, L. (1987) The acquired immune deficiency syndrome. *J. Mine Medical Officers Assoc.*, **63**, 10–17.

Campbell, C.M. (1996) The psychosocial context of HIV transmission on the gold mines: implications for HIV education programmes, in *HIV/AIDS Management in South Africa: Priorities for the Mining Industry* (eds B.G. Williams and C.M. Campbell), ERU, Johannesburg, pp. 77–91.

Campbell, C.M. (1997) Migrancy, masculine identities and AIDS: the psychosocial context of HIV transmission on the South African gold mines. *Soc. Sci. Med.*, **45**, 273–81.

Campbell, C.M. and Williams, B.G. (1996) Academic research and HIV/AIDS in South Africa. *S. Afr. Med. J.*, **86**, 55–60.

Churchyard, G. (1996) Of soil and seed: HIV related TB on the mines, in *HIV/AIDS Management in South Africa. Priorities for the Mining Industry* (eds B.G. Williams and C.M. Campbell), ERU, Johannesburg, pp. 117–19.

COM (1994) *Chamber of Mines Statistical Tables, 1993*, Chamber of Mines, Johannesburg.

Cowie, R.L. (1987) *Pulmonary Tuberculosis in South African Gold Miners. Determinants of Relapse after treatment*, MSc Thesis, McGill University, Montreal.

Cowie, R.L. (1994) The epidemiology of tuberculosis in gold miners with silicosis. *Am. J. Resp. Crit. Care Med.*, **150**, 1460–2.

Cowie, R.L. and Mabena, S.K. (1991) Silicosis, chronic airflow limitation, and chronic bronchitis in South African gold miners. *Am. Rev. Respir. Dis.*, **143**, 80–4.

Cowie, R.L., Hay, M. and Thomas, R.G. (1993) Association of silicosis, lung dysfunction, and emphysema in gold miners. *Thorax*, **48**, 746–9.

Crisp, J. (1996) AIDS programmes in the mining industry: an overview, in *HIV/AIDS Management in South Africa. Priorities for the Mining Industry*, (eds B.G. Williams and C.M. Campbell), ERU, Johannesburg, pp. 91–3.

Crush, J., Jeeves, A. and Yudelman, D. (1991) *South Africa's Labour Empire: a History of Black Migrancy to the Gold Mines*, Westview Press, Boulder.

DOH (1996a) *The South African Tuberculosis Control Programme. Practical Guidelines*, Department of Health, Pretoria.

DOH (1996b) *Draft Report of the Review of the Tuberculosis Control Programme of South Africa.* Unpublished.

DOPH (1990) *Department of Public Health: Annual Report* (cited in Anderson, 1990).

Dube, N. and Wilson, D. (1996) Peer education programmes among HIV-vulnerable communities in southern Africa, in *HIV/AIDS Management in South Africa: Priorities for the Mining Industry* (eds B.G. Williams and C.M. Campbell), ERU, Johannesburg, pp. 30–40.

Felix, M.A., Leger, J. and Ehrlich, R. (1994) *The Identification and Control of Environmental and Occupational Diseases*, Princeton Scientific, New Jersey.

Fourie, I. (1996) Health care in the mining industry, in *HIV/AIDS Management in South Africa. Priorities for the Mining Industry* (eds B.G. Williams and C.M. Campbell), ERU, Johannesburg, pp. 53–8.

Goldfields (1993) *1993 Annual Report of the Medical Division of Gold Fields of South Africa Limited*, Goldfields, Johannesburg.

Gorgas, W.C. (1914) Recommendation as to sanitation concerning employee of the mines on the Rand made to the Transvaal Chamber of Mines. *Report to the Transvaal Chamber of Mines.*

Grosskurth, H., Mosha, F., Todd, J. *et al.* (1995) *Lancet*, **346**, 530–6.

Hnizdo, E. (1991) *Combined effects of dust exposure and tobacco smoking on the occurrence of lung diseases in South African gold miners*, PhD Thesis, University of the Witwatersrand, South Africa.

Hnizdo, E. and Sluis-Cremer, G.K. (1993) Risk of silicosis in a cohort of white South African gold miners. *Am. J. Ind. Med.*, **24**, 447–57.

Hnizdo, E., Baskind, E. and Sluis-Cremer, G.K. (1990) Combined effect of silica dust exposure and tobacco smoking on the prevalence of respiratory impairments among gold miners. *Scand. J. Work, Environ. Health*, **16**, 411–22.

Hnizdo, E., Murray, J., Sluis-Cremer, G.K. and Thomas, R.G. (1993) Correlation between radiological and pathological diagnosis of silicosis: an autopsy population based study. *Am. J. Ind. Med.*, **24**, 427–45.

Irwig, L.M. and Rocks, P. (1978) Lung function and respiratory symptoms in silicotic and non-silicotic gold miners. *Am. Rev. Respir Dis.*, **117**, 429.

James, W.G. (1992) *Our Precious Metal: African Labour in South Africa's Gold Industry*, 1970–1990, David Philip, Cape Town.

Katz, E. (1980) Silicosis on the South African gold mines, in *Economics of Health in South Africa: Hunger, Work and Health* (eds F. Wilson and G. Westcott), Raven Press, Johannesburg, pp. 187–244.

Katz, E. (1994) *The White Death: Silicosis on the Witwatersrand Gold Mines 1886–1910*, Witwatersrand University Press, Cape Town.

Leger, J.P. (1992) Occupational diseases in South African mines – A neglected epidemic? *S. Afr. Med. J.*, **81**, 197–201.

Leon, R., Davies A., Salomon, M. and Davies J. (1996) *Leon Commission of Enquiry into Safety and Health in the Mining Industry*, Government Printers, Pretoria.

Lowe, R. and Murray, J. (1994) *Tuberculosis at autopsy in Black Mine Workers.* National Centre for Occupational Health, NCOH Report No. 6/94, Johannesburg.

MBOD (1990) Report of the Medical Bureau for Occupational Disease for the period 1 April 1989 to 31 March 1990, Government Printers, Pretoria.

MBOD (1993) Report of the Medical Bureau for Occupational Disease for the year 1992, Government Printers, Pretoria.

Mgogi, N., Churchyard, G., Kleinschmidt, I. and Williams, B.G. (1997) Attendance versus compliance with tuberculosis treatment in an occupational setting: a pilot study. *S. Afr. Med. J.*, **87**, 1517–21.

Mulder, D.W., Nunn, A.J., Kamali, A., Nakiyingi, J., Wagner, H.U. and Kengeyakayondo, J.F. (1994) Two-year HIV-1-associated mortality in a Ugandan rural population. *Lancet*, **343**, 1021–3.

Murray, J., Kielkowski, D. and Reed, P. (1996) Occupational disease trends in black South African gold miners. *Am. J. Respir. Crit. Care Med.*, **153**, 706–10.

ODMWA (1994) *Occupational Disease in Mines and Works Act, 1973 (Act No. 78 of 1973) and Regulations*, Issue No. 1, Aquilia Publications CC.

Packard, R.M. (1990) *White Plague, Black Labour: Tuberculosis and the Political Economy of Health and Disease in South Africa*, James Currey, London.

Packard, R.M. and Coetzee, D. (1995) White plague, black labour revisited: TB and the mining industry, in *Crossing Boundaries: Mine Migrancy in a Democratic South Africa* (eds J. Crush and W. James), Creda Press, Cape Town, pp. 101–15.

Perriens, J.H., St. Louis, M.E., Mukadi, Y.B. *et al.* (1995) Pulmonary tuberculosis in HIV-infected patients in Zaire – a controlled trial of treatment for either 6 or 12 months. *N. Engl. J. Med.*, **332**, 779–84.

Richards, S.B., St. Louis, M.E., Nieburg, P. *et al.* (1995) Impact of the HIV epidemic on trends in tuberculosis in Abidjan, Côte d'Ivoire. *Tubercle Lung Dis.*, **76**, 11–16.

Sluis-Cremer, G.K. (1980) Active pulmonary tuberculosis discovered at post-mortem examination of the lungs of black miners. *Br. J. Dis. Chest*, **74**, 374–8.

Steen, T., Gyi, K. White, N., Gabosianelwe, T. *et al.* (1997) Prevalence of occupational lung disease amongst Botswana men formerly employed in the South African mining industry. *Occup. Environ. Med.*, **54**, 19–26.

Swanevelder, R. (1996) The national HIV antenatal surveillance programme: spatial and temporal trends in South Africa, in *HIV/AIDS Management in South Africa. Priorities for the Mining Industry*, (eds B.G. Williams, and C.M. Campbell), ERU, Johannesburg, pp. 61–6.

Trapido, A.S.M., Mqoqi, N.P., Macheke, C.M., Williams, B.G., Davies, J.C.A. and Panter, C. (1996) Occupational lung disease in ex-mineworkers – sound a further alarm! *S. Afr. Med. J.*, **86**, 559.

Van den Broek, J., Borgdorff, M.W., Pakker, N.G., Chum, H.J., Klokke, A.H., Senkoro, K.P. and Newell, J.N. (1993) HIV-infection as a risk factor for the development of tuberculosis – a case-control study in Tanzania. *Int. J. Epidemiol.*, **22**, 1159–65.

White, N. (1994) *Health Hazards in the mining industry: an overview.* Report submitted to the Leon Commission of Enquiry into the Health Hazards in the Mining Industry; June 1994.

Yach, D. (1987) Tuberculosis deaths in South Africa, 1980. *S. Afr. Med. J.*, **72**, 149–51.

8 Occupational lung diseases in Brazil

Eduardo Algranti

INTRODUCTION

Over the past decades Brazil's fertile land, abundant minerals and sufficient water have attracted investments. This has resulted in an evolutionary change from an agrarian to an industrial society. Large-scale industrialization began in the 1950s and peaked in the 1970s. Regrettably, social development has not paralleled economic development. After decades of industrialization, social inequalities have arisen, reflected by poor housing, poor education and poor general health for an important percentage of the population. Within this context, occupational diseases have emerged as a public health problem. Even though Brazilians have established occupational research and regulatory programs, there are limited means to implement proper surveillance and prevention programs because of the lack of a public health infrastructure.

The 1991 census estimated the Brazilian population to be 147 million (IBGE, 1993). Almost 60% of the population is concentrated in the south-eastern and southern regions, the most developed areas. Most Brazilians (75.4%) live in urban areas; Figure 8.1 shows the population distribution in the Brazil's five regions (IBGE, 1993). The northern and north-eastern regions have a huge contingent of rural people, mainly because of geographic and economic characteristics. Figure 8.2 shows employment by industry in 1990 (IBGE, 1993).

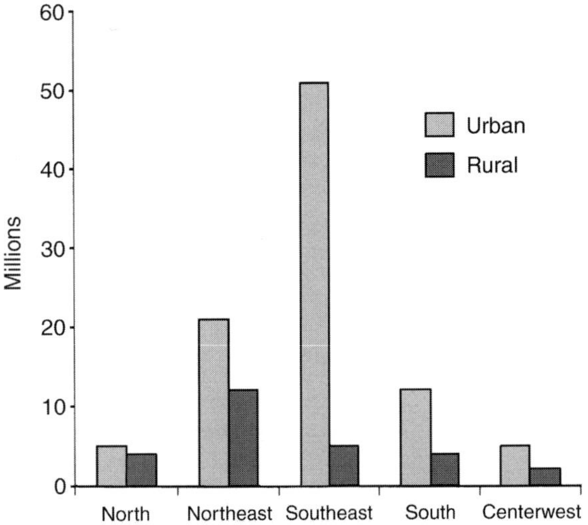

Figure 8.1 Estimated urban and rural population in Brazil, 1990. In this census, the population of Brazil numbered 147 053 940 (IBGE, 1993).

Occupational Lung Disease. Edited by Daniel E. Banks and John E. Parker. Published in 1998 by Chapman & Hall. ISBN 0 412 73630 6

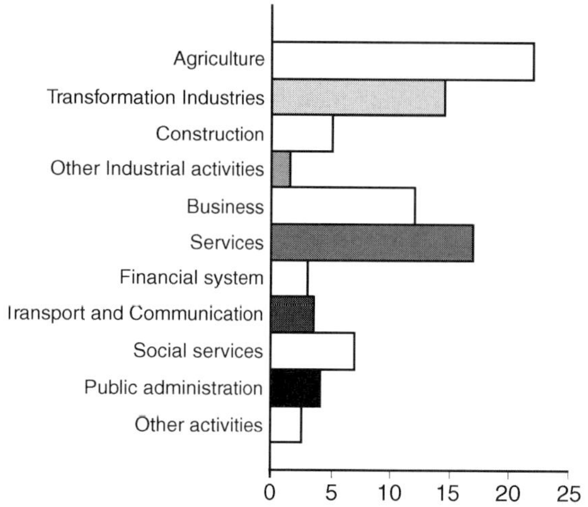

Figure 8.2 Percent of workers employed in different segments of the economy, 1993 (IBGE, 1993).

Mining and prospecting

Mining and prospecting involves a modest number of Brazilians, relative to the dimension of metal and non-metal mineral reserves. There are huge deposits of iron, bauxite, zinc, manganese, gold, calcium, potassium, phosphate, asbestos, granite, quartz, clay, sand and other minerals. Mining represented 2.72% of the official Gross Domestic Product in 1990 (including oil and gas) (DNPM, 1991). This probably underestimates the true figures, because of underegistration of production.

Figure 8.3 shows the distribution of population and employment in mining and prospecting, by region. The 1991 census estimated 100 000 registered miners, though this does not take into account those unregistered and employed under third-party contracts. There are approximately 300 000 working prospectors (DNPM, 1993), but it is likely that many more than 500 000 Brazilians are involved in mining and prospecting.

Although the law mandates that registered miners have access to medical screening, prospectors often do not have such access because they are self-employed, they frequently move from area to area, and, in general, they work far from urban centers.

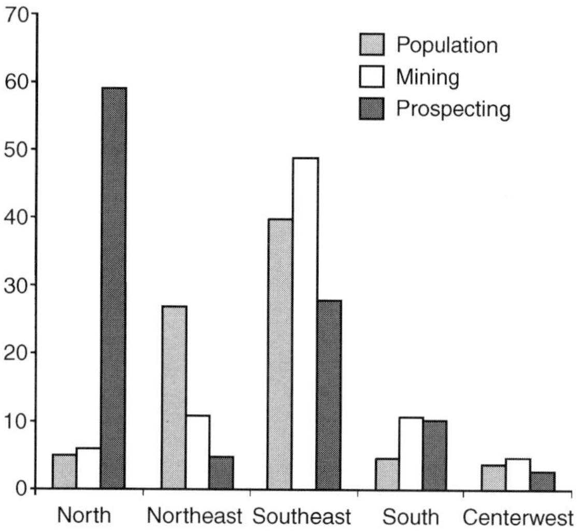

Figure 8.3 Percent distribution of the Brazilian population by region, and relative percent of employment in mining and prospecting (DNPM, 1993; IBGE, 1993).

Manufacturing industry: a workforce in transition

Workers employed in industrial manufacturing industries in Brazil number some 9.5 million, with another 3.8 million directly involved in construction. Together, these groups represent 21.3% of the Brazilian working population (IBGE, 1993) (Figure 8.2). In the transformation industry, about 17% of workers are employed in small enterprises (less than six workers), compared with 53% of construction workers. Such workers in relatively small businesses have had difficulties in gaining access to health care services because of their limited economic power and lack of unionization. This has led to changes in the code of regulatory practices in occupational health (NR 18, 1995).

Nearly half of the workers in the transformation industry may be exposed (at least potentially) to mineral dusts, organic dusts and chemicals, non-metal minerals, metals, toxic agents in paper mills, rubber, chemical, plastics and textiles. Among those workers, those in the agro-industry represent the single largest group (13%).

Workers in agriculture

Workers in agriculture represent around 23% of the working population (IBGE, 1993), though this percentage is underestimated because of exclusion of the Northern region. About three-quarters of agricultural workers work on, or own, properties with less than six employees and also have limited access to health care.

Legislation regarding exposure to mineral dusts

Regulations for acceptable workplace exposures are issued by the Ministry of Labour. Enforcement is the responsibility of representatives (physicians, engineers, lawyers and economists) of the Ministry of Labour. In addition, employees of the Secretaries of Labour of the States may assist with the inspections. It is also possible, under the Brazilian Sanitary Code, that health surveillance teams from the health secretaries of the States and Municipalities inspect worksites, but they have no legal power to fine employers or close plants.

Regulations (NR 7, 1994) provide for medical screening for workers at risk for occupational hazards. This has led to major developments regarding occupational lung diseases:

- The concept of fibrogenic and non-fibrogenic dusts has been introduced and rules regarding the frequency of radiological screening made.
- Chest radiographs from workers exposed to hazardous dusts or fumes are interpreted according to the ILO International Classification of Radiographs of Pneumoconiosis.
- Spirometry is mandated at the beginning of employment and every 2 years for workers exposed to dusts and other respiratory hazards.

Since this revision, the Ministry of Labour, Ministry of Health, and the Brazilian Society of Pneumology and Tisiology have joined to improve compliance with permissible exposure limits and to improve the quality of screening. This has not been entirely successful because of regional differences in the availability of screening tests and the relatively paucity of Ministry of Labour inspectors to address workplace hazards.

Permissible dust levels of worksite asbestos, manganese and silica are regulated (NR 15, 1978; 1991). Use of amphiboles is forbidden unless 'substitution is not possible'. Permission for amphibole use may be granted by the competent authority after consultation with representatives of the employers and employees. Asbestos spraying is forbidden. The threshold Limit Values (TLV) for respirable chrysotile is 2.0 fibers/cm^3.

The TLV for respirable silica is calculated by the following equation:

$$\text{TLV (mg/m}^3) = \frac{8}{\%\text{quartz} + 2}$$

The above formula is widely employed for worksites with dusts with variable silica content.

Underground work is permitted at a minimum age of 21 years. Special retirement is granted for underground miners, who have had higher exposures in their jobs after 15 years of work. Consequently, in some mining areas, the population comprises a considerable proportion of fairly young, retired men.

OCCUPATIONAL LUNG DISEASES IN BRAZIL

Coalworkers' pneumoconiosis (CWP)

Coal mining is concentrated in the southern region of Brazil. Brazilian coal is mainly bituminous and rich in sulphur, and is typically destined for transformation to coke for use in steel plants and as fuel in small and medium-sized thermoelectric plants.

Production and employment in the Brazilian coal industry reached a peak in the 1980s, in accordance with a policy to increase the national production of energy – an answer to the oil crisis of the 1970s. In 1985 there were 14 000 coal workers, 70% of whom worked underground (Ministerio de Minas e Energia, 1986). This number has steadily decreased as many mines were closed under the pressures of economic change and privatization, and there are now approximately 4000 coal workers. The approach to mining and working conditions in the mines has not changed substantially over the past 10 years, mostly because of unclear economic perspectives.

CWP was first described in the Brazil in a coal miner showing the radiographic features of this illness and who presented with bilateral spontaneous pneumothoraces (Giglio, 1941). However, it was not until 1981 that a detailed clinical description of 536 cases of CWP was presented (Souza Filho *et al.*, 1981). In 1981, CWP was estimated to affect between 5–8% of miners, a figure likely to be low as it was based only on referred cases. In 1984, the Ministry of Labour sponsored a comprehensive study of the health of coal miners. Approximately 10% of the underground work force (956 miners) were randomly selected from six mining companies and stratified by job. Data on respiratory symptoms, lung function, chest radiology and dust exposure were collected. The prevalence of CWP among active miners was 5.6%, with a mean exposure of 8.4 years (Algranti and Souza Filho, 1990). More than half (52.2%) of the mine samples exceeded 2.0 mg/m^3 of respirable dust, and 40.5% of the samples contained more than 10% quartz (Valenti, 1985).

In 1989, a follow-up study was conducted, though due to fewer mining jobs and limited participation by ex-miners, only 42.5% of the original sample was evaluated. Most of these were current miners. Over these 5 years, the incidence of new cases of CWP was 11.4 per 1000 miners. Miners with 15 years of exposure at the coal face had an approximately 20% chance of developing CWP (Algranti, 1991; Algranti *et al.*, 1993a). Figure 8.4 shows the probability of developing CWP based on a semi-cumulative exposure index, derived from 1984 dust measurements. This index attributes different weights to underground jobs. A value of 1 was given to jobs with low exposure, 2 to jobs with moderate exposure, and 4 to jobs with high

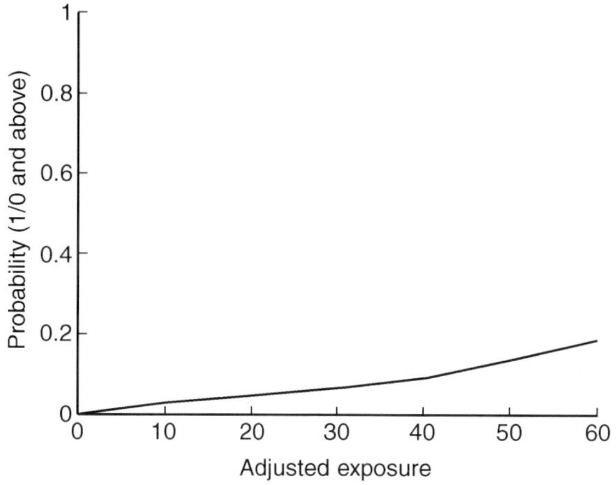

Figure 8.4 Probability of a Brazilian underground coal miner developing profusion category 1/0 or greater pneumoconiosis by exposure. The exposure has been adjusted to reflect the sum of the number of years underground times the value assigned for the dust exposure associated with each job (low, intermediate, or high).

dust exposure. The adjusted exposure is the sum of the number of years in each job category times the corresponding weight of each job's exposure.

Both studies (Algranti and Souza Filho, 1990; Algranti, 1991) showed that although prevalence of CWP was slightly greater than that reported in the UK (Muir *et al.*, 1977) and US (Attfield *et al.*, 1984), the mean time for developing the disease was between one-half and one-third of the US experience (Merchant, 1986). This likely underestimated the prevalence of CWP, because of the failure to evaluate ex- and retired miners. The interpretation of Figure 8.4 is disconcerting. Although the adopted exposure index differs from other studies, the risk of developing CWP far exceeds the risk outlined in reports from the UK (Hurley *et al.*, 1982) and Germany (Vautrim *et al.*, 1990). The high levels of quartz and high respirable dust levels in Brazilian coal mines are likely responsible for these differences.

Silica dust exposure and silicosis

Silicosis is by far the most common pneumoconiosis in Brazil. The first description of cases dates back to 1939, when Mignone (cited in Mendes, 1979) commented on autopsies performed on workers with silicosis. Many situations where dust exposure with variable silica content exist, ranging from classical industrial exposures to exposures unique to Brazilian society. Data presented below include cases of mixed dust pneumoconiosis and other pneumoconiosis that radiologically resemble silicosis. The diagnosis of silicosis in Brazil, whether for compensation claims, clinical evaluation or epidemiology, is based on occupational history and radiologic appearance.

The estimated number of workers at risk for silicosis can be derived from official census figures. Mining and prospecting involves at least 500 000 workers, while those employed in the non-metal and metal, chemical and rubber industries number approximately 2 300 000. By including 3 800 000 construction workers, a total of approximately 6 600 000 workers (4.5% of the entire population) have potential exposure to silica-containing dusts.

There are scanty data available on the prevalence of silicosis. The first attempt to estimate the number of cases (Mendes, 1978) was the result of interviews with 3440 patients hospitalized for tuberculosis therapy. Of these, 327 had a history of occupational exposure to silica, and 119 were considered to have silicotuberculosis (3.5% of the cases). Minervino *et al.* (1964) reported a 3.3% prevalence of silicotuberculosis in a chest radiograph screening program of workers exposed to silica. Using these two estimates, Mendes estimated that there were 30 000 cases of silicosis in the country. This number seems modest, because the measured incidence of tuberculosis in silicotics is larger than the 3.3% utilized in these calculations. Although a considerable amount of information remains unpublished, data on silicosis in Brazil continues to accumulate (Table 8.1).

The largest source of exposure leading to silicosis comes from gold mining. At present, there are more than 4500 diagnosed cases of silicosis in the state of Minas Gerais alone (A. Silveira, personal communication).

Information on dust exposure is available for masonry work and pit digging and shown in Table 8.1. In the first report, only 16/29 samples had <1 mg/m^3 of dust with concentrations of quartz ranging from 5 to 10.3% in the respirable fraction. In pit digging – the hand-digging of wells in regions of drought in the north-eastern part of the country (see Chapter 1) – diggers encounter a layer of rock (97% quartz) before reaching the water. This leads to the generation of dusts in a confined space (Holanda *et al.*, 1995). Most with silicosis in pit digging have radiologic appearances compatible with accelerated form of silicosis.

Silicosis in tire repair workers merits special comment. After an incidental finding of silicosis in a tire repair worker (Souza Filho and Alice, 1992), 85 men employed in tire repair shops were evaluated. The men had a mean of 15 years of employment. Eleven (13.7%) had radiologic evidence of silicosis. The source of silica exposure was in the handling and repairing of

Table 8.1 Selected Brazilian studies on the epidemiology of silicosis

Industry	*Reference*	*Study design*	*No. of workers affected*	*Rates (numbers)*	*Description*
Urban industry	Minervino *et al.* (1964)	Descriptive	278		Number of cases
				3.3%	Prevalence of silicotuberculosis
Masonry	Franco (1974)	Cross-sectional	200	3.0%	Prevalence
General	Mendes (1978)	Prevalence	30 000	(–)	Estimated number of silicotics in the country
Ceramics	Oliveira (1988)	Cross-sectional	4000	3.9%	Prevalence
Tire repair	Souza Filho and Alice (1992)	Cross-sectional	85	13.7%	Prevalence
Pit Digging	Holanda *et al.* (1995)	Cross-sectional	687	27.0%	Prevalence
General (mainly ceramics)	Bagatin *et al.* (1995)	Descriptive	818	(–)	Number of cases
Naval industry (sandblasting)	Comissão Tecnica, Estadual, R.J. (1995)	Cross-sectional	728	23.6%	Prevalence

tires. Excessive respirable exposures were attributable to the widespread use of compressed air to clean the tires and the shop floor.

Asbestos exposure and asbestos-related diseases

The risk of lung disease from asbestos exposure embodies the best example of the concerns regarding occupational lung disease in developing countries. Perhaps due to the long latency period, pulmonary fibrosis or malignancy are rarely recognized as occupational illnesses. In addition, there are inadequate public (and private) health measures employed to control these exposures. These difficulties, in association with the double standards employed by multinational enterprises which export risks to countries with lax industrial and environmental controls, raise strong feelings in Brazil, as well as in other developing countries.

Brazil is one of the top five producers of asbestos (ABRA, 1995). Apart from Canada and South Africa, the other major asbestos producers are developing countries. Figure 8.5 shows trends of asbestos fiber production for the major producers (the former Soviet Union, the leading producer, is excluded to amplify the other curves).

There are five asbestos deposits in the country, but commercial exploitation comes from the state of Goias. Asbestos production ranges from 200 000 to 250 000 tons/year. Brazilian asbestos is nearly pure chrysotile without amphibole contamination (ABRA, 1995). Production comes from one mine, which is one of the largest open-cast asbestos mines in the world. Approximately one-third of the production is exported, mainly to Japan, Thailand, India and other South American countries (ABRA, 1995). The great majority of the remaining asbestos (85–90%) is destined for asbestos cement products, mainly roofing tiles, boards and water reservoirs. Some 10% goes to friction materials, and a smaller quantity to textiles, paper and paperboards, paints and others. There are some asbestos imports from Canada and South Africa, but information regarding the amount and type is unavailable. Asbestos consumption in Brazil is estimated at 1400 g per capita per year (Giannasi and Thebaud-Mony, 1994). Paradoxically, Canada, one of the world's leading producers, and the champion of a controlled use policy, consumes around 1000 g per capita per year.

The asbestos cement industry began in the 1940s. Large-scale asbestos fiber production started in the 1970s, but it was not until the 1980s that efforts to control dust exposure were initiated. Before the 1970s, asbestos imports were responsible for an important percentage of Brazilian consumption and included amphiboles used in the asbestos cement and friction material plants. Twelve companies with 33 plants manufacture asbestos cement products, with

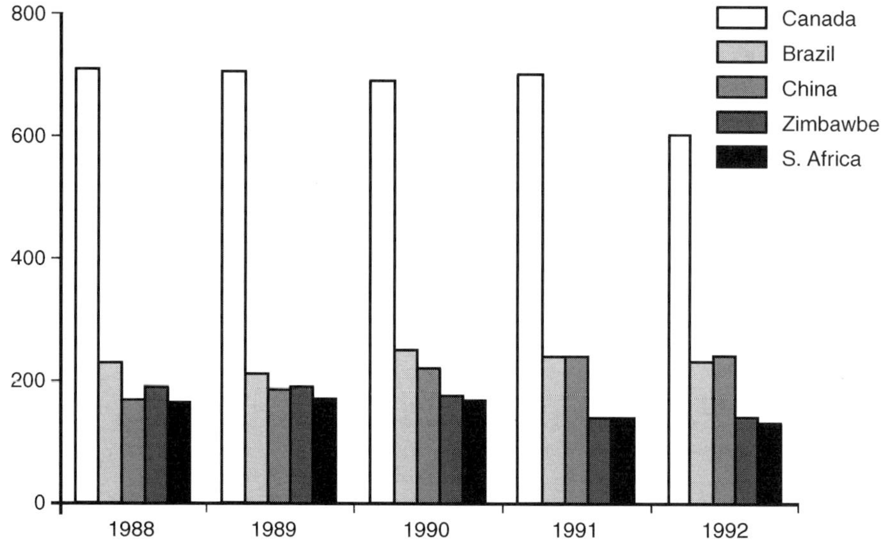

Figure 8.5 Trends in asbestos fiber production for selected countries for 1988–1992 (ABRA, 1995).

production concentrated in the state of São Paulo. Plants producing friction materials are more numerous and disperse, but smaller in size and in economic power, resulting in less of an investment in working conditions (Bonciani, 1994).

An accurate estimate of the number of workers exposed to asbestos is difficult. There are about 20 000–24 000 workers directly engaged in mining, transport, handling and manufacture of asbestos-containing products. There are no rules which regulate handling of finished asbestos products and consequently this has become a very extensive problem. About half of Brazilian households and industrial plants have asbestos roofing (ABRA, 1995). In addition, about 300 000 workers are engaged in brake and clutch repair shops.

Medical literature on asbestos-related diseases is scanty in Brazil. The first case of asbestosis was published by Nogueira *et al.* (1975). Since then, three other studies focusing on asbestosis have been published (Table 8.2). Costa (1983) called attention to a problem in asbestos cement plants by identifying 14 cases of asbestosis in a local registry. In 1987, representatives of employers, unions and government determined that a high prevalence of asbestosis occurred in workers with more than 10 years of exposure in medium-sized asbestos cement plants in the state of São Paulo (Amâncio *et al.*, 1988). From this, recommendations regarding medical screening for asbestos exposed workers were published (Comite de Estudos do Amianto, 1989) and a recommendation to lower the limit to 1 fiber/ml was adopted for the asbestos cement industry.

Lung cancer and asbestos exposure was first reported by Algranti *et al.* (1989) and mesothelioma was reported very recently (De Capitani *et al.*, 1997).

Table 8.2 Selected studies on asbestosis in Brazil

Industry	*Reference*	*Study design*	*No. of subjects*	*Rates*
Asbestos-cement	Costa (1983)	Case descriptive	86	14/86
Asbestos-cement	Amâncio *et al.* (1988)	Cross-sectional	386	5.8%* 18.7%†
Friction materials	Noronha *et al.* (1992)	Cross-sectional	150	16%

* For the whole group.
† Workers with more than 10 years of exposure.

Asbestos-related diseases will continue to be an important issue at the end of this century, and for decades of the next century in Brazil and in other developing countries. Further discussions relating the health effects of asbestos to politics, sociological aspects, governmental actions, and an industry point of view are available (Berman, 1986; Bonciani, 1994; Giannasi and Thebaud-Mony, 1994; ABRA, 1995).

Other pneumoconiosis due to mineral dusts

Other forms of pneumoconiosis have been reported since the 1980s (Table 8.3). As noted, it is likely that some cases previously classified as silicosis are, in fact, pneumoconiosis attributable to other causes. As an example, of four cases of siderosis (Algranti *et al.*, 1985), two had been compensated as silicosis, until the industrial hygienist of the plant questioned the diagnosis on grounds that the industrial process was silica-free.

Seventy-three phosphate rock miners were investigated for lung disease. Of these, 20 (27.4%) had chest radiographs compatible with pneumoconiosis and two underwent lung biopsies which showed extensive pigmentation and dust accumulation, but little fibrosis (De Capitani, 1989). Results from investigations in an aluminum corundum plant showed 13/106 (12.3%) workers to have chest radiologic changes (De Capitani *et al.*, 1993a). The etiology of these changes are unclear, but may be attributable to exposure to cristobalite and trydimite generated in the fusion of bauxite.

Respiratory exposure to talc remains a problem. Chibante *et al.* (1990) reported the histologic findings of a talc miner and showed asbestos-free talcosis. Recently, a restorer, an individual who re-shaped and polished a plastic mass, presented with advanced pleural fibrosis. Light microscopy analysis of respirable samples showed fibers. Samples of the plastic contained amphiboles, possibly attributable to the use of contamined industrial talc in the production of plastic. Shortly after the recognition of this case, a large talc deposit in the state of Minas Gerais was recognized to be contaminated with amphiboles.

Dental technicians are at risk of respiratory diseases due to dusts and fumes generated during their work. The fibrogenic nature of the agents, the working practices (generally in small shops with limited ventilation), and the lack of knowledge of occupational risks place these workers in jeopardy. An advanced case of apparent simple silicosis was recognized in a worker at a university dental laboratory who polished molds with an abrasive thought to be 'volcanic rock'. X-ray diffractometry of the abrasive disclosed this to be nearly pure quartz (Algranti and Chibante 1993).

Table 8.3 Reports of pneumoconiosis due to other dusts

Industry or exposure	*Pneumoconiosis*	*Study design (no. of subjects involved)*	*Histology (no. of subjects with histology)*	*Reference*
Iron oxide plant	Siderosis	Case descriptive (4)	Yes (4)	Algranti *et al.* (1985)
Phosphate rock storehouse	Phosphate rock pneumoconiosis	Cross-sectional (73)	Yes (2)	De Capitani, 1989
Talc grinding	Talcosis	Case descriptive (7)	Yes (7)	Chibante *et al.* (1990)
Silicon carbide plant and abrasive wheels*	Silicon carbide pneumoconiosis	Case descriptive (22)	No	Cukier *et al.* (1991)
Tool grinding	Hard-metal pneumoconiosis	Case descriptive (1)	Yes	De Capitani *et al.* (1993b)
Aluminum corundum plant	Under investigation	Cross-sectional (106)	No	De Capitani *et al.* (1993a)
Dental laboratories	Mixed dust pneumoconiosis	Case descriptive (2)	Yes (1)	Algranti *et al.* (1993a)
Domestic welding and abrasives	Berylliosis	Case descriptive (2)	Yes (2)	Morrone *et al.* (1982) De Capitani *et al.* (1995)

* Some subjects had also been exposed to aluminum corundum.

There are two cases of berylliosis in the Brazilian literature. The first report described a worker with exposure to beryllium fumes from a welding alloy (Morrone *et al.*, 1982). The second described a man who made abrasive wheels (De Capitani *et al.*, 1995). In a recent investigation of a worker with berylliosis employed in the rubber industry (unpublished), high levels of beryllium were found in the dust generated from grinding wheels. Exposure to dust generated from grinding wheels is common in numerous Brazilian industries.

Exposure to organic dusts

Brazil is a leading grain producer, the most important crops being soy, sugar cane and coffee. Although there are many workers in agriculture, few studies have addressed the health effects of exposure to organic dusts in these workers. One reason may be the limited access of rural residents to medical care, particularly to occupational health care. In addition, most of the rural workers are engaged in small enterprises which are not covered by labour legislation. The respiratory effects of exposure to organic dusts are summarized in Table 8.4.

There are no published reports of exposure to bagasse, although it is extensively employed as fuel in the agro-industry. Nor is there a single well-documented case of extrinsic allergic alveolitis due to organic dust exposure. Since a major stimulus for studies of occupational lung diseases has been sentinel cases, it is unlikely that organic dust disease in rural workers will be an area of future studies due to the limited access of these workers to medical care.

Occupational airways disease

Although silicosis is the best known occupational lung disease in Brazil, occupational airways disease are more prevalent in the urban industrialized areas. Mendonça *et al.* (1994) reported the 10-year experience of the FUNDACENTRO outpatient clinic. In 778 workers, 248 cases had occupational asthma (OA) and 13 non-smokers or negligible smokers were diagnosed as having occupational bronchitis (OB). Workers in the chemical and plastic industry accounted for 39% of the cases of OA. Epidemiologic assessments of airways disease in occupational groups were reported in coal mining and a flax mill (Table 8.5).

Both chronic bronchitis and OA were assessed by a translated Medical Research Council (UK) questionnaire with additional questions to explore OA (Algranti, 1991; Algranti *et al.*, 1993b). In coal mining, statistical models showed that smoking was significantly related to bronchitis, but not dust exposure, whereas the reverse occurred with grade III shortness of breath (Algranti, 1991).

STRATEGIES TO CONTROL OCCUPATIONAL LUNG DISEASES IN BRAZIL

Physicians interested in worker health in Brazil, as an example of a developing country, are in a peculiar situation. They have the knowledge of occupational lung diseases, but only modest means of implementing programs of prevention, screening and surveillance. We can disseminate this knowledge, but we lack epidemiologic support to define the most pressing needs for intervention.

Strategies for federal and state governments, groups of employers and workers can be developed. The governmental agencies could work on several fronts. Not only is it important to develop and implement an adequate system of worksite inspections and identify workplace populations at risk, but programs that deal with prevention of occupational illnesses and compensation of injured workers also need to be developed and implemented. Academic investigators and governmental representatives need to work together to

Table 8.4 Selected epidemiologic studies on respiratory effects of exposure to organic dusts

Industry	*Disease*	*Reference*	*Study design*	*No. of subjects*	*Rates*
Cotton mills	Byssinosis	Nogueira *et al.* (1973)	Cross-sectional	99 exposed 135 controls	18%*
Grain storage	Respiratory symptoms	Tietboehl *et al.* (1994)	Cross-sectional	235 exposed 58 controls	OR†
Flax mills	Byssinosis	Mendonça *et al.* (1995)	Cross-sectional	63 exposed	30%*

* Percentage with grade 1 byssinosis (Schilling criteria).

† Adjusted odds-ratio for cough, phlegm, chronic bronchitis and conjunctivitis was significantly higher in the exposed group.

Table 8.5 Epidemiologic studies of airways disease in occupational groups

Industry	*Disease*	*Reference*	*Study design*	*No. of subjects*	*Prevalence (%)*
Coal mining	Chronic bronchitis	Algranti *et al.* (1993b)	Cross-sectional	280	28.5
Coal mining	Occupational asthma	Algranti *et al.* (1993b)	Cross-sectional	280	12.1
Flax mill	Chronic bronchitis	Mendonça *et al.* (1995)	Cross-sectional	63	15
Flax mill	Occupational asthma	Mendonça *et al.* (1995)	Cross-sectional	63	11

generate epidemiologic information. Employers need to update factories and industrial sites where old equipment and inadequate ventilation systems exist, provide adequate training and information for workers, and increasingly concern themselves with occupational health issues. Workers will need to be educated and trained about proper work processes and to be aware of their righ to a safe workplace.

In summary, Brazilian legislatures have passed the proper exposure and screening regulations for workers at risk, but, the regulatory agencies are underfunded and compliance is poor. This situation has implications in topics ranging from the safe handling of hazardous substances to respirable dust exposures. Although there is a commitment by the government to take measures to control respiratory risks, the successful implementation of these measures for all Brazilian workers remains our goal. Even if we were to achieve that goal today, occupational respiratory diseases will remain a major public health problem for the coming years.

Acknowledgments

We are indebted to Dr Bernardo Bedrikow and Dr Elizabete Medina Coeli de Mendonça for the careful reading and suggestions to the text.

REFERENCES

ABRA, Associação Brasileira do Amianto (1995) O amianto no Brasil. Ebart Editorial, São Paulo.

Algranti, E. (1991) *Doenças respiratórias associadas à mineração de carvão-estudo de coorte de 5 anos*, PhD Thesis, Faculdade de Saúde Pública, Universidade de São Paulo.

Algranti, E. and Chibante, A.M. (1993) Pneumoconiose em trabalhadores de laboratório odontológico. *Revista da Associação Paulista de Cirurgiões Dentistas*, **47**(1), 969–72.

Algranti, E. and Souza Filho, A.J. (1990) Correlations between radiology, respiratory symptoms and spirometry in active underground coal miners in Brazil in, *Proceedings, VII International Pneumoconioses Conference*, US Department of Health and Human Services (NIOSH Publication 90–108, Part I), pp. 105–11.

Algranti, E., Lima, C.Q.B. and Vieira, A.V. (1989) Asbesto e carcinoma broncogêrico: pesquisa de fibras em tecido pulmonar de três pacientes com carcinoma broncogênico. *Rev. Paul. Med.* **107**(3), 133–8.

Algranti, E., Morrone, L.C., Morrone, N. *et al.* (1985) Siderose pulmonar por óxido de ferro em trabalhadores: uma poeira inerte? *Rev. Paul. Med.*, **103**(5), 259–64.

Algranti, E., Mendonça, E.M.C., Silva, R.C.C. *et al.* (1993a) Incidence, radiological progression and probability of coalworkers' pneumoconiosis, in *Proceedings, Eighth International Conference on Occupational Lung Diseases*, Vol. II (eds J. Hurych, M. Lesage and A. David), ILO, Prague, pp. 580–7.

Algranti, E., Mendonça, E.M.C. and Silva, R.C.C. (1993b) Bronchial reactivity, asthma and FEV_1 decline in underground coalworkers, in *Proceedings, Eighth International Conference on Occupational Lung Diseases*, Vol. III (eds J. Hurych, M. Lesage and A. David), ILO, Prague, pp. 919–28.

Amâncio, J.B., Bonciani, M. and Urquiza, S.D. (1988) Avaliação radiológica de trabalhadores da indústria de fibrocimento do estado de São Paulo. *Rev. Bras. Saúde Ocupacional*, **16**(63), 51–5.

Attfield, M., Reger, R. and Glenn, R. (1984) The incidence and progression of pneumoconiosis over nine years in US coal miners: I. Principal findings. *Am. J. Ind. Med.*, **6**(6), 407–15.

Bagatin, E., Jardim, J.R.B., Nery, L.E. *et al.* (1995) Ocorrência de silicose pulmonar na região de Campinas-SP. *Jornal de Pneumologia*, **21**(1), 17–26.

Berman, D.M. (1986) Asbestos and health in the third world: the case of Brazil. *Int. J. Health Serv.*, **16**(2), 253–63.

Bonciani, M. (1994) O asbesto no Brasil, in *Isto é trabalho de gente?* (eds J.T.P., Buschinelli, L.E.

Rocha and R.M. Rigotto), Vozes, Petrópolis, pp. 550–67.

Chibante, A.M., Padilha, C.P., Bethlem, E.P. *et al.* (1990) Pneumoconiose dos moedores de talco-estudo de sete casos. *Jornal de Pneumologia*, **16**(2), 57–61.

Comissão Técnica Estadual de Pneumopatias Ocupacionais do Estado do Rio de Janeiro (1995) A silicose na indústria naval do Rio de Janeiro: análise parcial. *Jornal de Pneumologia*, **21**(1), 13–16.

Comite de Estudos do Amianto (1989) *Avaliação médica dos trabalhadores expostos ao amianto*, SSST/FUNDACENTRO, São Paulo.

Costa, J.L.R. (1983) *Estudo da asbestose no municipio de Leme (SP)*, MSc Thesis, Faculdade de Ciências Médicas, UNICAMP.

Cukier, A., Algranti, E., Terra Filho, M. *et al.* (1991) Pneumoconiose em trabalhadores de indústria de abrasivos. *Rev. Hosp. Clin. Fac. Med. Sao. Paulo.* **46**(4), 180–3.

De Capitani, E.M. (1989) Prevalência de pneumoconiose em trabalhadores expostos a rocha fosfática. *Rev. Saúde Pública*, **23**(2), 98–106.

De Capitani, E.M., Amancio, J.B. and Bagatin, E. (1993a) Pneumoconiosis in an aluminium corundum plant in Brazil, in *Proceedings, Eighth International Conference on Occupational Lung Diseases*, Vol. II (eds J. Hurych, M. Lesage and A. David), ILO, Prague, pp. 766–71.

De Capitani, E.M., Saldiva, P.H. and Pereira, M. (1993b) Hard metal subacute pneumopathy: a case report, in *Proceedings, Eighth International Conference on Occupational Lung Diseases*, Vol. III (eds J. Hurych, M. Lesage and A. David), ILO, Prague, pp. 1026–31.

De Capitani, E.M., Altemani, A.M.A., Kavakama, J.I. *et al.* (1995) Beriliose pulmonar: revisão de literatura e relato de caso. *Jornal de Pneumologia*, **21**(3), 135–42.

De Capitani, E.M., Metze, K., Frazzato, C., Jr *et al.* (1997) Mesotelioma maligno de pleura com associação etiológica a asbesto: a propósito de 3 casos clinicos. *Rev. Assoc. Méd. Bras.*, **43**(3), 265–72.

DNPM, Departamento Nacional de Produção Mineral (1991) *Anuário mineral brasileiro*, DNPM, Brasília.

DNPM, Departamento Nacional de Produção Mineral (1993) *Levantamento nacional dos garimpeiros*, DNPM, Brasília.

Franco, A.R. (1974) *Silicose pulmonar em trabalhadores de pedreiras na cidade de Ribeirão Preto-estado de S.P*, MSc Thesis, Faculdade de Medicina de Ribeirão Preto, Universidade de São Paulo.

Giannasi, F. and Thebaud-Mony, A. (1994) Asbestos in Brazil (sociological aspects) in, *Sourcebook on Asbestos Diseases* (eds G.A. Peters and B.J. Peters) Butterworths, Salem, New Hampshire, pp. 315–38.

Giglio, J.S. (1941) Pneumotórax bilateral espontâneo por pneumoconiose. *Rev. Méd. Paraná*, **10**, 208–40.

Holanda, M.A., Holanda, M.A., Martins, M.P.S. *et al.* (1995) Silicosis in Brazilian pit diggers: relationship between dust exposure and radiological findings. *Am. J. Ind. Med.*, **27**, 367–78.

Hurley, J.F., Burns, J., Copland, L. *et al.* (1982) Coalworkers' simple pneumoconiosis and exposure to dust at 10 British coalmines. *Brit. J. Ind. Med.*, **39**, 120–7.

IBGE, Instituto Brasileiro de Geografia e Estatistica (1993) *Anuário estatistico do Brasil*, IBGE, Brasília.

Mendes, R. (1978) *Epidemiologia da silicose na região sudeste do Brasil*, PhD Thesis, Faculdade de Saúde Pública, Universidade de São Paulo.

Mendes, R. (1979) Pneumoconioses-bibliografia brasileira comentada, 1886–1976. *Rev. Associ. Méd. Bras.*, **25**(11), 406–10.

Mendonça, E.M.C., Algranti, E., Silva, R.C.C. *et al.* (1994) Ambulatório de pneumopatias ocupacionais da FUNDACENTRO-resultados após 10 anos. *Rev. Bras. Saúde Ocupacional*, **22**(84), 7–14.

Mendonça, E.M.C., Algranti, E., Silva, R.C.C. *et al.* (1995) Bissinose e asma ocupacional em trabalhadores de uma fiação de linho em Sorocaba-São Paulo. *Jornal de Pneumologia*, **21**(1), 1–8.

Merchant, J.A. (1986) *Occupational respiratory diseases*, US Department of Health and Human Services (NIOSH Publication 86–102), Washington, DC.

Minervino, D.M., Garrafa, N.M. and De Stefano, I.J. (1964) A silicose pulmonar nas indústrias de São Paulo, in *Anais do Congresso Americano de Medicina do Trabalho*, São Paulo, pp. 268–80.

Ministério de Minas e Energia (1986) *Informativo anual da indústria carbonífera 1985*, DNPM, Brasília.

Morrone, N., Campos Neto, J.S., Shibata, N.P. *et al.* (1982) Importância da cintilografia com gálio (GA 67) em pneumologia. *Rev. Paul. Med.*, **100**(3), 30–3.

Muir, D.C.F., Burns, J., Jacobsen, M. *et al.* (1977) Pneumoconiosis and chronic bronchitis. *Br. Med. J.*, **2**, 424–7.

Nogueira, D.P., Goelzer, B.F., Cox, J.W. *et al.* (1973) Bissinose no município da capital do estado de São Paulo, Brasil. *Rev. Saúde Pública*, **7**(3), 251–72.

Nogueira, D.P., Certain, D.A. Uesugui, S.J. *et al.* (1975) Asbestose no Brasil: um risco ignorado. *Rev. Saúde Pública*, **9**(3), 427–32.

NR 15, Norma Regulamentadora-15 (1978) Atividades e operações insalubres. Portaria 3214, Diário Oficial da União de 08/06/78 and Portaria 1, Diário Oficial da União de 28/05/91.

NR 7, Norma Regulamentadora-7 (1994) Programa de contole médico de saúde ocupacional. Portaria 24, Diário Oficial da União de 30/12/94.

NR 18, Norma Regulamentadora-18 (1995) Condições e meio ambiente do trabalho na indústria da construção. Portaria 4, Dirio Oficial da União de 07/07/95.

Noronha, A., Jansen, J.M. and Tavares, J.L. (1992) Levantamento de casos de asbestose numa indústria de artefatos de amianto (abstract). *Jornal de Pneumologia*, **18**(suppl. 2), 102–3.

Oliveira, J.I. (1988) Prevalence of silicosis among ceramic industry workers in the city of Pedreira, Brazil (Abstract) in, *Abstracts of Communications, VII International Pneumoconioses Conference*, p.114.

Souza Filho, A.J. and Alice S.H. (1992) Pneumoconiose em borracheiros (abstract). *Jornal de Pneumologia*, **18**(suppl. 2), 101–2.

Souza Filho, A.J., Alice, S.H. and De Luca, V. (1981) Pneumoconiose dos trabalhadores das minas de carvão. *Jornal de Pneumologia*, **7**(2), 57–66.

Tietboehl, C.N., Moreira, J.S. and Edelweiss, M.K. (1994) Os efeitos respiratórios da exposição à poeiras de grãos de cereais: uma revisão sucinta da literatura e um estudo epidemiológico em trabalhadores de silos do Rio Grande do Sul. *Jornal de Pneumologia*, **20**(4), 193–203.

Valenti, I.F. (1985) *Levantamento das condições de segurança e higiene das indústrias carboniferas (SC)*, Technical Report, FUNDACENTRO, Porto Alegre.

Vautrim, H.J., Morfeld, P. and Kampmann, B. (1990) Prevalence, incidence densities and cummulative incidences of pneumoconiotic changes for two groups of miners of a mine in Western Germany coal mining, in *Proceedings, VII International Pneumoconioses Conference*, US Department of Health and Human Services (NIOSH Publication 90–108, Part 1) pp. 132–5.

Part Two Pneumoconioses

9 The diagnosis of pneumoconiosis and novel therapies

Jeffrey L. Wilt, John E. Parker and Daniel E. Banks

DEFINITION OF PNEUMOCONIOSIS

Pneumoconioses are lung diseases which result from mineral dust deposition in the lung and the subsequent host response. This term is derived from the Greek language (*pneumo*, the Greek root for lung and *konis*, the root for dust) and has been applied in clinical medicine since the early 19th century (Meiklejohn, 1951). These are illnesses rooted in antiquity, which have had their highest human toll during the industrial age, yet it remains a tragedy of our time that these preventable diseases have not disappeared.

A number of respiratory manifestations associated with the inhalation, retention and tissue reaction to mineral dusts have been described and reflect the different effects of these dusts on the lung (Balaan *et al.*, 1993). Although a wide variety of mineral dusts are implicated in disease, all share common diagnostic criteria. The focus of this report is the description of diagnostic criteria for the most frequently occurring disorders. These are illnesses attributable to coal dust inhalation [coal workers' pneumoconiosis (CWP)], inhalation of dusts containing crystalline silica (silicosis), and inhalation exposure to asbestos fibers (asbestosis and other asbestos-related pulmonary disorders). Novel, but unproven, therapeutic approaches which may alter the natural history of pneumoconioses will also be considered.

There are three major criteria for the diagnosis of pneumoconiosis. The first is a sufficient exposure to a mineral dust known to cause pneumoconiosis with an appropriate latency period. Exposures to silica, coal or asbestos occur most commonly in an occupational setting. The second criteria is the recognition of a characteristic chest radiograph which meets published standards for the diagnosis of pneumoconiosis. Although respiratory symptoms and impairment in lung function commonly occur in workers diagnosed to have pneumoconiosis, neither is requisite for the diagnosis. The third is the absence of an illness which might mimic pneumoconiosis. Disseminated mycobacterial or fungal disease might mimic CWP or silicosis, as may sarcoidosis. Asbestosis is not infrequently confused with idiopathic pulmonary fibrosis, bronchiolitis obliterans with organizing pneumonia, or interstitial lung disease associated with collagen vascular disease (Banks *et al.*, 1993; Wilt and Banks, 1995). The diagnosis of pneumoconiosis, therefore, is a clinical one and only in the very unusual situation is a lung biopsy indicated to make the diagnosis.

A HISTORY OF EXPOSURE TO MINERAL DUST

The important issues when considering the diagnosis of pneumoconiosis are listed in Table 9.1 and relate to the suspected occupational exposures, the use of

Occupational Lung Disease. Edited by Daniel E. Banks and John E. Parker. Published in 1998 by Chapman & Hall. ISBN 0 412 73630 6

Table 9.1 Important concerns in the history of a worker with suspected pneumoconiosis

1. Are the suspected and recognized occupational exposures, as well as the respiratory protective devices, noted?
2. Is this an epidemic?
3. Are all of the jobs listed?
4. Are there important non-occupational exposures?

respiratory protective devices, and the consideration of whether dust exposure was sufficient to induce respiratory disease. Because the diagnosis of pneumoconiosis is a clinical diagnosis, answers to these concerns serve to strengthen or weaken the attribution of disease to work place exposure for a patient and determine whether additional invasive investigations are needed.

In the context of reporting a worker's exposure, the physician may be obliged to include a specific job description of each place of employment, as well as the nature and duration of exposure to the suspected offending agents. Perhaps the most effective approach is to begin with the individual's first job and proceed on a year-by-year, job-by-job basis. It is not enough, for example, to understand that the worker served in the Navy, when such a term of duty could range from land-based clerk, to shipbuilder with clinically important asbestos exposure, to sandblaster with sufficient silica exposure to induce pneumoconiosis.

The first criterion which must be addressed by the physician, determining whether the worker's occupational exposure is sufficient to cause pneumoconiosis, requires knowledge of the work environment and the worker's exposures. The most important data which aid in determining whether the exposures were sufficient to cause disease include knowledge regarding the daily job activities, a description of the nature of the exposure associated with that activity, the years when employed and the length of employment, exposure measurements (if available), and whether the worker was provided effective respiratory protection, and if it was used or not.

As an example, a reasonable starting point would be the understanding that silicosis occurs in association with industrial processes where the silica particle is made respirable size. Without manipulation of the silica particle, the particle generally remains large and is trapped in the upper airway defenses. Making the particle respirable size is the common thread among occupations where silicosis is a clearly recognized risk (Table 9.2). Furthermore, underground coal mining jobs often require drilling. Roof bolters, in particular, are at increased for silicosis, because the drilling needed to erect supports to hold up the roof is done beyond the coal seam and exposure is not just to coal dust, but to hard rock or sandstone which borders the seam. In this instance, it may be silicosis, rather than CWP which develops in the underground miner.

Why some workers develop pneumoconiosis (or greater or lesser extents of this illness) while employed in the same general work area may be related to the adequacy of respiratory protection. Pneumoconiosis can certainly occur in workers even though personal respiratory protection is used. To begin with, since each respirator has a limit beyond which the dust level exceeds the protective capacity (expressed as a multiple of the permissible exposure limit), the worker may be using a respiratory protective device which is inadequate for the amount of dust present (Glindmeyer and Hammad, 1988). Second, the worker may be using the wrong type of respirator. Respiratory protective devices are designed to protect for specific exposures. If a respirator designed to block solvent fumes from entering the lung were to be used for dust protection, the effectiveness would be compromised. Third, the respirator may be ill-fitting. Finally, and probably most importantly, protective devices increase the work of breathing. They are hot, they make it impossible for the worker to speak to co-workers, and they must be removed to smoke cigarettes. These features are relevant to the worker's decision as to whether or not they will wear these protective devices. Overall, protection by respirator is much less likely to be successful as compared with properly constructed engineering controls.

Information regarding dust levels in the work place can also provide important information. However, this approach cannot be guaranteed to represent the complete picture. For example, acute silicosis in surface coal mine drillers has been reported where the measured dust levels were within normal limits (Banks, *et al.*, 1983). The measurements were not representative of the overwhelming dust exposures which induced very aggressive silicosis, nor were there measurements of the percentage of free crystalline silica in the dust. Although fraud has been reported in coal mine sampling in US underground mines (Hearl and Hewitt 1993; Weeks, 1993), perhaps the US surface coal mines are the most notorious in their approach to collecting dust samples. Not only were dust exposures during active drilling infrequently measured, but when they were measured, they were not always representative of work place exposures. On the days of sampling, the worker might be away from the drill when personal sampling was performed, alter the amount of dust generated by changing the drill parameters, or use a wet drilling process. Because of these potential abuses and inconsistencies, laws mandating sampling at surface mines have been changed and now, if the drills are generating plumes of dust, the mine is considered to be out of compliance and a citation issued (Code of Federal Regulations, 1994).

Therefore, although dust measurements can be helpful, understanding the conditions under which the samples were collected and having confidence that these measurements accurately represent the work place environment is essential. Sometimes this can only be understood by visiting the work place to see how work is undertaken.

COAL WORKERS' PNEUMOCONIOSIS

In order to diagnose CWP, a history of exposure to coal mine dust must be obtained. Underground coal mine employment is an important risk factor for the development of CWP. Miners who work near the face of the coal seam underground are generally exposed to the highest dust levels and are at greatest risk for developing the disease. Coal may also be found close to the surface and can be mined after removing the soil and hard rock overburden. Respirable dust levels at surface mines are generally lower than in underground mines, although silica dust levels can be excessive (Amandus and Piacitelli, 1987; Lapp and Parker, 1992).

The time from initial exposure to coal mine dust and the subsequent development of CWP is variable and depends in part on the intensity of exposure. The recognition of CWP is uncommon in the first 10–15 years after initial dust exposures.

SILICOSIS

Millions of workers world-wide are exposed to silica dust and are believed to be at risk for silicosis (Wilt and Banks, 1995). Again, an occupational history is of paramount importance in diagnosing this form of pneumoconiosis. Occupations when silicosis occurs include mining (metal, coal, and non-metallic minerals) as well as working with clay, stone and glass. Hazardous exposures also occur in other workplaces such as in sandblasting, pottery making, stone masonry and foundry work (Table 9.2). Virtually every ore or mineral is embedded in silica-containing rock and thus the mining and quarrying industry, in general, have had a long association with the disease. Both acute and chronic silicosis has been reported in surface coal miners (Banks *et al.*, 1983, 1993) (Figures 9.1a, b). Workers exposed to finely ground silica seem to be at especially high risk (Banks *et al.*, 1981). The rate of

Table 9.2 Major industries with silica exposure

Occupation	*Exposure*
Sandblasting	Shipbuilding, oil rig maintenance, preparing steel for painting
Mining	Surface coal mining, roof bolting, shot firing, underground mining
Milling	Silica flour workers
Glassmaking	Polishing with sand and enameling
Pottery	Crushing flint and fettling
Foundry worker	Mold making and mold cleaning
Quarrying and stone work	Cutting of slate, sandstone and granite
Abrasive workers	Inhaling fine particles during grinding

progression of dyspnea and the decline in lung function appears to depend on the amount of dust exposure (Hughes *et al.*, 1982). The clinical characterization of silicosis is often related to the duration of exposure in relation to clinical presentation (Chapter 13).

ASBESTOSIS

Asbestos has been used in many work sites with numerous applications throughout the world. Unlike silicosis, it is a disease which was first recognized in this century (Murray, 1907; Pancoast *et al.*, 1917; Oliver, 1927). Although the use of asbestos has decreased in many of the developed countries of the world, cases are still reported. The long latency period from the inhalation exposure of asbestos fibers to the development of adverse health effects is well known and a search for a history of exposure in individuals with interstitial lung disease is an important part of their clinical investigation. Because of its widespread use, it has been estimated that over 25 million workers in North America alone are at risk for developing asbestos-related disease (Murphy *et al.*, 1986).

Asbestos is still commonly used in the developing world, and many more workers remain at risk to develop asbestosis and asbestos-related disease. Asbestos has been used in automotive brakes, in boiler rooms and furnaces, in shipbuilding, in spray paints, in the construction of houses and buildings, and in asbestos cement products.

Asbestos fiber inhalation induces distinct clinical scenarios, some of which have never been adequately

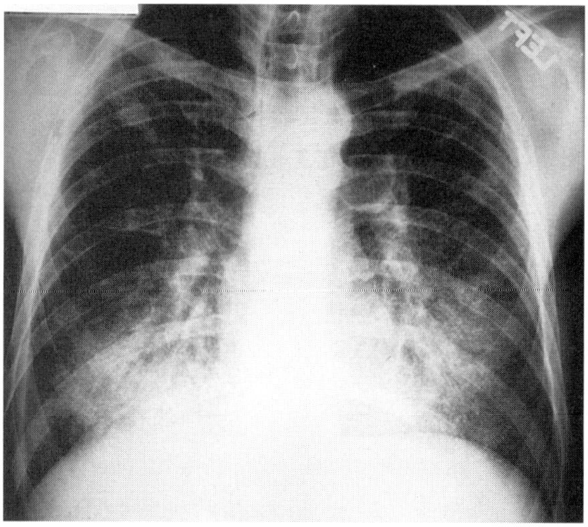

(a)

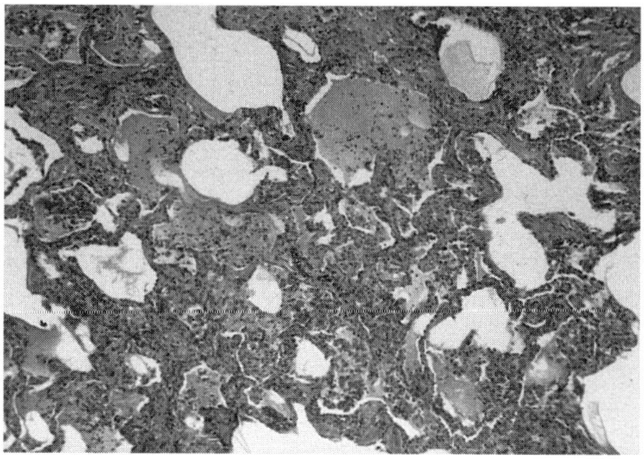

(b)

Figure 9.1 A 34-year-old man was hospitalized with a 6-month history of progressive shortness of breath, a dry cough, weight loss of 30 pounds (13 kg), and Raynaud's phenomenon. He had worked as a surface coal mine driller for the previous 5 years but 2 years previously had had a normal physical examination and chest radiograph. Physical examination revealed a cachectic man who was tachypneic at rest. There were coarse crackles at the bases, stellate scars on the fingertips, and sclerodactyly. Laboratory information showed microscopic hematuria and proteinuria with mildly impaired renal function, a decreased serum albumin, a normal level of complement, and circulating immune complexes. Lung function showed moderate restriction and a diffusing capacity only 18% of predicted. The radiograph (a) showed an alveolar filling pattern in both lower lung zones, a presentation consistent with acute silicosis. An open-lung biopsy was performed, from which the lung tissue showed features described as acute silicosis (b). The alveolar spaces are filled with a relatively acellular proteinaceous material, there are inflammatory cells in the interstitium with thickening of the interstitium, but negligible development of silicotic nodules within the interstitium. Because of elevated serum levels of circulating immune complexes, the worker received daily oral cyclophosphamide at doses of 2 mg/kg. Despite 8 months of therapy, he became more dyspneic and his FVC and FEV_1 declined to approximately half of the initially measured values. He was unable to perform a diffusing capacity maneuver due to dyspnea. A chest radiograph taken 14 months after the initial film showed conglomerate mass lesions in both lungs and loss of volume in the right upper lobe with tracheal distortion, blunting of both costophrenic angles and bilateral pleural thickening. The patient died of respiratory failure 26 months after his initial evaluation. (Reproduced with permission from Banks, D.E., Bauer, M.A., Castellan, R.M. and Lapp, N.L. (1983) Silicosis in surface coal mine drillers, *Thorax*, **38**, 275–78.)

explained. The first is a pleural effusion that is usually exudative and benign. This has the shortest latency, usually 10–15 years. Subsequent thickening or calcification of the pleura is probably the most commonly encountered asbestos-related clinical syndrome and is the result of inflammation and fibrosis as a sequelae of these effusions. Asbestos-related primary lung cancer and mesothelioma generally typically have latencies of greater than 20 years. Asbestosis occurs predominantly in the lower zones (Figure 9.2). This is frequently described by an exaggerated profusion of lower zone irregular opacities. Historically, caution has been recommended if 1/1 International Labour Office (ILO) profusion category, or greater, is not present. The American Thoracic Society has developed criteria for a non-invasive (i.e. without pathologic specimen examination) diagnosis (Murphy *et al.*, 1986). It is important to note that asbestos-related pleural disease may occur independently from the parenchymal scarring as well as to point out that the latency period for asbestosis, as well as the manifestations of other illness, is variable, but generally develops 10 or more years after initial exposure (Figure 9.3). In those with clinical or radiographic features suggestive of asbestosis, but with no obvious exposure history, it is important to ask questions regarding any unsuspected exposure. For example, spouses of asbestos workers have been known to present with asbestos-related pulmonary disease.

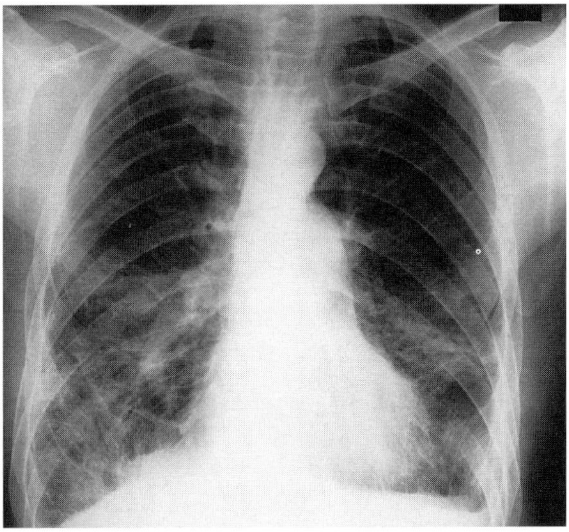

Figure 9.2 This chest radiograph from an asbestos-exposed worker shows extensive bilateral interstitial abnormalities in lower and mid-zones, classified as ILO profusion category 3/2 and opacity size and shape s/t. Blunting of the left costophrenic angle consistent with diffuse pleural thickening and a likely pleural plaque on the medical aspect of the right hemidiaphragm is apparent. Underlying emphysema is also noted.

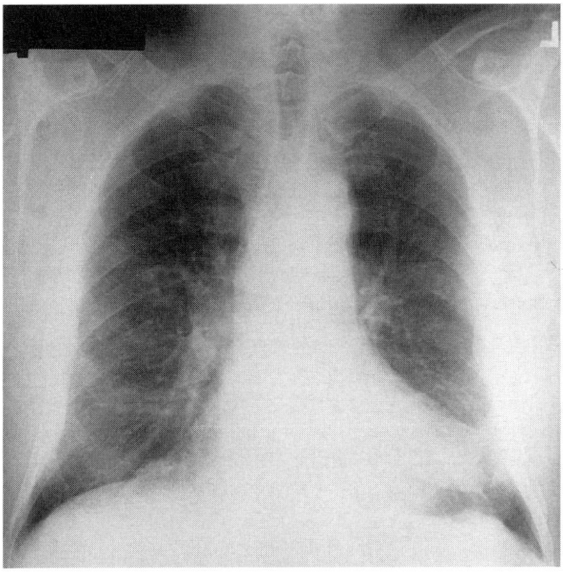

Figure 9.3 This radiograph demonstrates bilateral chest wall pleural plaques with sparing of the costophrenic angles. This is referred to as 'circumscribed pleural thickening'. If the costophrenic angles were blunted and these changes contiguous with chest wall pleural thickening, a finding such as this would be described as 'diffuse pleural thickening'. In addition, there is a discrete plaque on the left hemidiaphragm and obliteration of the right cardiophrenic angle by pleural thickening. The lateral aspect of the left heart border is obscured by a face-on pleural plaque. There is also parenchymal fibrosis manifest by an increase in irregular interstitial markings in the mid- and lower-lung zones bilaterally. The right hilum, in particular, is prominent, raising concerns regarding pulmonary hypertension and neoplasm. When extensive pleural disease including face-on plaques are present, the precision of profusion classification of parenchymal fibrosis is difficult and can be improved by the use of other imaging techniques such as high-resolution chest tomography.

CHEST RADIOGRAPHIC EVALUATION OF PNEUMOCONIOSIS

Radiography showing evidence of dust-induced disease is paramount in making the diagnosis of pneumoconiosis. The current standard method was last updated in 1980 by the ILO (International Labour Office, 1980). This system describes radiographic changes seen on the chest radiograph of exposed individuals. The classification calls for recognizing and grading of both parenchymal and pleural abnormalities, and allows for the detection and recording of abnormalities that are possibly unrelated to occupational exposure. In addition, the technical quality of the radiograph is also scored. The standard films published by the ILO are required for comparison with films being read or classified and this side-by-side method of observation is integral for determining small opacity profusion.

The parenchymal abnormalities are classified by the presence (or absence) of opacities, the size and shape of these abnormalities, as well as their anatomic location. In addition, the profusion level is characterized. Parenchymal opacities less than 1 cm in diameter are defined as small lesions and are subdivided into two distinct categories – rounded or irregular. On the basis of opacity diameter, they are labeled *p* (<1.5 mm), *q* (1.5–3 mm) and *r* (3–10 mm). The irregular-shaped opacities are also defined by their width and are labeled *s* (fine linear opacities up to 1.5 mm width), *t* (fine linear opacities from 1.5–3 mm width) and *u* (fine linear opacities from 3–10 mm width). Both round and irregular opacities may be seen and described on the same film.

The profusion of opacities denotes the relative number of abnormalities (i.e. either rounded or irregular), per unit area of lung. The four major profusion categories (0, 1, 2 or 3) are determined by comparing the radiograph to known ILO standards. These major categories can be subdivided into 12 minor categories, utilizing two numbers of the major categories, much like when grading the size of opacities. The first number is the category felt by the reader to be most representative of the standard ILO film. The second number is the major category which was considered to be similar to the film being read, but not eventually chosen. For example, a film will be read as 1/1 if it most resembles a standard ILO film of category 1. A film will be read as 3/2 if it mostly resembles standard film category 3, but a similarity to a standard film of category 2 was seriously considered.

A large opacity is a parenchymal abnormality greater than 1 cm diameter in size. These are labeled as A, B or C, once again denoting size of the abnormality. Large opacities considered to be A size are each greater than 1 cm in diameter, with the sum of their diameters not exceeding 5 cm. Large opacities considered to be B size are greater than 5 cm in diameter, but their combined area is less than the volume encompassed by the right upper lobe. Large opacities considered to be C size represent lesions whose combined area exceeds the volume encompassed by the right upper lobe (Figure 9.4).

Pleural abnormalities are similarly described by the type of lesion, the anatomic location and the approximate size of the lesion. Pleural disease encompassed by the ILO classification system includes a pleural effusion, pleural calcification, diffuse pleural thickening and pleural plaques or circumscribed pleural thickening (Figure 9.5). A scheme for recording the types of pleural abnormality (i.e. in profile and en-face), as well as its site, width, extent and presence or absence of calcification, is part of the ILO classification.

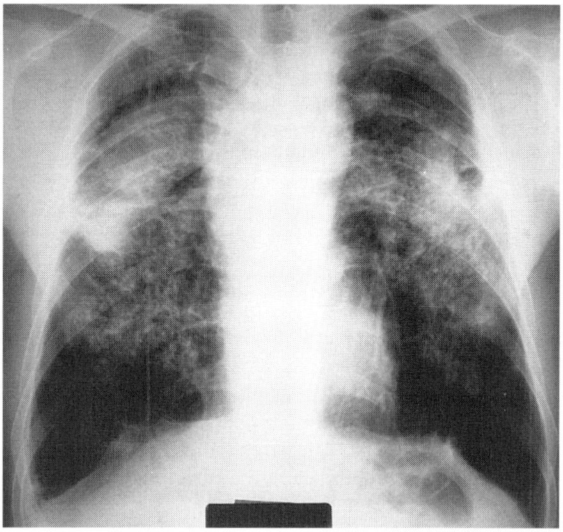

Figure 9.4 This radiograph, taken from a 44-year-old surface coal mine driller with accelerated silicosis, demonstrates upper and mid-zone large opacities on a background of small rounded opacities of ILO category 2/3. Note also the dramatic compensatory basilar emphysema, retraction of the hilum, and distortion of the mediastinal structures. He died of respiratory failure approximately 2 years after this radiograph was made.

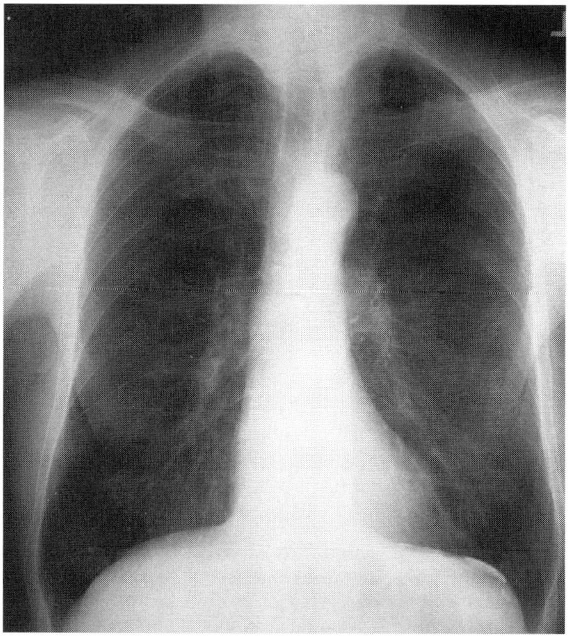

Figure 9.5 This radiograph demonstrates a calcified pleural plaque on the dome of the left hemidiaphragm. Underlying emphysema is present. The pleural change is a not infrequent finding of asbestos exposed workers who have a considerable latency period. This pleural change serves as a useful marker of asbestos exposure.

The ILO classification scheme has proved to be a very useful method to describe and to classify chest radiographs. It has improved and standardized epidemiologic research, fostered the identification of health hazards, and served as an important aid in establishing exposure–response relationships in the work place. It has frequently been useful for screening and surveillance programs in industry, as it helps to help identify early radiographic abnormalities as well as to monitor health trends.

HIGH-RESOLUTION CHEST TOMOGRAPHY AND OTHER IMAGING TECHNIQUES (with special emphasis on asbestos-related diseases)

The non-invasive assessment of the lungs of dust-exposed workers was the near exclusive purview of the traditional chest radiograph for over 75 years. More recently, advanced imaging techniques have provided new opportunities for enhanced visualization of lung and chest wall structures. These techniques, conventional computed chest tomography (CT), high-resolution CT (HRCT) and magnetic resonance imaging (MRI), have expanded the array of options for clinicians and researchers to evaluate chest diseases, including those diseases resulting from silica, coal and asbestos dust exposures (Staples *et al.*, 1989). Most importantly, chest CT and HRCT are recognized to be sensitive and specific tools in separating the pleural and parenchymal abnormalities resulting from asbestos exposure from normal lung structures such as extrapleural fat or competing parenchymal shadows (Begin *et al.*, 1989).

Although there is a relatively good correlation between lung pathology findings and the chest radiographic interpretation for dust-exposed workers (particularly at higher profusion levels), traditional radiographs are insensitive to early changes resulting from dust exposure (Epler *et al.*, 1978). For example, nearly 20% of asbestos-exposed workers with fibrotic changes on pathologic examination have no detectable radiographic abnormalities (Kipen *et al.*, 1987). The radiograph may be normal in some cases of asbestosis confirmed by functional abnormality and histopathology. Importantly, imaging techniques such as CT or HRCT provide better views of the pleural surface and more detail of the pulmonary parenchyma than traditional chest films, though reports do exist of even normal HRCT scans with histologic evidence of fibrosis (Gamsu *et al.*, 1995). Both HRCT and ILO classification of plain films are excellent predictors of clinical abestosis (Harkin *et al.*, 1996).

CT of the chest plays an important role in detecting pleural plaques, as well as defining the extent of malignant disease. HRCT images detail the pulmonary parenchyma and are useful in supplementing, but not replacing, the traditional chest radiograph when detection of the earliest changes of fibrosis is desirable. For example, asbestos-related pulmonary parenchymal fibrosis is considered to be present when the previously described radiographic findings are apparent, in the context of a consistent exposure history. A chest CT along with a few HRCT cuts may confirm the findings on plain film, as well as identify pleural or parenchymal abnormalities which might not have been appreciated on the plain film. When seen together, pleural and parenchymal abnormalities strengthen the accuracy of the diagnosis of asbestos-related disease.

MRI currently plays little role in studying asbestos-related diseases of the chest, although it is a particularly helpful technique in helping to answer neurologic and orthopedic concerns. Gallium scanning can detect inflammation, infection and tumors, but is not currently an important modality in the evaluation of asbestos-related chest diseases. The same is true of other nuclear medicine studies such as ventilation or profusion scans. The role of position emission tomography (PET) scans also requires more study and evaluation.

Chest CT has improved sensitivity for detecting diffuse lung disease, such as asbestosis, by eliminating superimposition of structures and enhancing attenuation discrimination. HRCT provides even greater detailed images of pulmonary architecture by thin collimation (1 to 1.5 mm-thick sections) (Naidich, 1989; Remy-Jardin *et al.*, 1992). Detection of basilar parenchymal abnormalities in the prone position is of particular value (Aberle *et al.*, 1988). Additional clinical and research utilization of HRCT has led to the delineation of several important findings (Akira *et al.*, 1991; Webb *et al.*, 1993; Gamsu *et al.*, 1995):

- Centrilobular thickening, or radiating fine lines within 1–2 cm of the pleural surface, likely from peribronchiolar fibrosis.
- Interlobular septal thickening, or septal lines; these parallel short lines or small nodules are thickened single or branching lines less than 2 cm long, radiating from the subpleural parenchyma extending into the lung, and likely represent thickened interlobular septa.
- Curvilinear subpleural lines running from 1 to 10 cm parallel to the pleura, especially in conjunction with septal lines.
- Parenchymal bands (long scars), non-tapering densities 2–5 cm in length that contact a pleural surface; these represent long fibrotic scars.
- Honeycombing, small cyst-like spaces less than 1 cm in diameter with discrete thickened walls.
- A ground-glass pattern, which persists when altering position, and may correspond to histologically minimal thickening of the alveolar walls and filling of the interlobular septa, possibly reflecting alveolitis.

Unlike the standardized approach with the ILO system for classifying the plain film, HRCT is awaiting a widely accepted and standardized method for classification. Several methods using expanded, subjective semi-quantitative scoring, grading and coding schemes and descriptors, as well as grid systems for classifying HRCT studies, have been proposed (Oksa *et al.*, 1994; Gamsu *et al.*, 1995; Harkin *et al.*, 1996; Kraus *et al.*, 1996). Such schemes need further development and study.

The current role of CT and HRCT should be reserved for the further clinical evaluation of difficult individual cases, and should not be applied to routine screening and surveillance of asbestos-exposed populations. There is no established role for HRCT in screening for lung cancer, although the CT remains a valuable tool for staging and evaluation of suspected lung cancer or in determining the extent of previously recognized malignancy.

The chest CT and HRCT are helpful in a number of specific clinical settings. Examples include evaluation of the parenchyma for fibrosis when lung function is abnormal and the chest radiograph shows normal or nearly normal parenchymal findings (Oksa *et al.*, 1994); evaluation of the pleura when no pleural abnormalities are seen on plain film but radiographic fibrosis is suspected; evaluation of the parenchyma when a good view of the lower lung fields on radiograph is limited by extensive plaques, soft tissue shadows from muscular development or obesity, cardiac enlargement or dysfunction, or other competing shadows; and excluding the presence of emphysema or other diseases. Other clinical indications can also be envisioned. In general, these more sophisticated radiographic techniques are acceptable to workers undergoing evaluation. Nevertheless, the routine use of these scans for screening or surveillance may be limited due to high cost, and lack of availability, as well as by lack of standardization in interpretation and reporting of abnormalities. However, for diagnostic purposes, the high level of sensitivity to abnormality and the promise of specificity in the recognition of certain dust effects, including pleural disease and parenchymal fibrosis, makes these important for both clinical and research activities.

To summarize, in an individual worker with suspected asbestos-related lung disease, the traditional chest radiograph, supplemented by the chest CT and HRCT, is useful in establishing the presence or absence of pleural abnormalities and provide sufficient detail of the parenchyma to determine whether asbestosis is present, in most cases (Klass, 1993; Harkin *et al.*, 1996). Yet, since the interstitial changes on these imaging techniques only reflects the disturbance of normal pulmonary anatomy by dust-related pathology, it is not surprising that the sensitivity and specificity of tissue analysis exceeds radiologic studies. The histologic diagnosis of asbestosis requires the presence of diffuse interstitial fibrosis deep to the subpleural zone of the parenchyma *plus* the presence of asbestos bodies, and in early asbestosis, the histologic fibrosis is concentrated in the respiratory bronchioles and then spreads to the alveolar septa in a patchy distribution (Craighead *et al.*, 1982). Since, the findings on imaging studies are the result of these pathologic changes, they have similar patterns and descriptions. Finally, with all due respect to the clinicians and the radiologists, tissue examination is sometimes necessary and remains the 'gold standard' for establishing asbestos-related fibrosis.

EXPERIMENTAL APPROACHES TO TREATMENT OF PNEUMOCONIOSES

There is no proven therapy which alters the natural history of pneumoconiosis. 'Compassionate use' medications are not available for even the most aggressive cases. Cases continue despite aggressive enforcement policies by regulatory agencies, dissemination of very readable educational materials aimed at workers and management (Centers for Disease Control, 1992a, b), respiratory dust exposure standards which are in place and appear, for the most part, protective (Graham *et al.*, 1991) [although there are instances of individuals leaving the industry with normal chest radiographs and developing disease in retirement (Kreiss and Zhen, 1996)] and the public health perspective that pneumoconioses can be prevented by engineering controls (Socie *et al.*, 1995). Indeed, we are unaware of any society or culture which has abolished pneumoconiosis.

In European and North American cultures, both the physician and the affected worker are helpless when a diagnosis of pneumoconiosis is made. Dust-induced fibrosis is regarded as an irreversible process. Yet, because pneumoconiosis is a world-wide occupational health problem, other cultures – in particular the Chinese – have considered these diseases in a different light. Silicosis is one of the most serious occupational lung diseases in China, and as such, there have been important efforts to develop therapies for this disease (Li, 1995).

The persistence of dust, the stimulated macrophage with release of many mediators, and the role of other effector cells and their products in the alveolar and interstitial space are the keys to this process. Attempts to alter the natural history should focus on these processes of fibrosis. This may be done by preventing the development of associated diseases (such as mycobacterial infection), by pharmaceutically controlling and theoretically reversing or halting the progression of fibrosis, or by removing the fibrogenic stimuli of dust, activated macrophages and other stimulated effector cells, and cytokines which perpetuate the fibrotic process (Table 9.3). While the effectiveness of experimental treatments have not been proven, these studies have yielded insight as to how therapies for these illnesses might be approached.

It is our intent to discuss experimental therapies which may alter the natural history of silicosis and coal workers' pneumoconiosis. Although those with asbestosis with active alveolitis have been considered responsive to steroid therapy (Rosenstock, 1994), to our knowledge, there have been no structured attempts to alter the natural history of asbestos-related non-malignant diseases.

Table 9.3 Potential means to alter the natural history of silicosis and coal workers' pneumoconiosis

	Mechanism of action	*Therapeutic agent or procedure*
1.	Altering the surface properties of the silica particle	Aluminum citrate complex, PVNO (poly-2-vinylpyridine-*N*-oxide)
2.	Control of the immune and inflammatory response	Immunosuppressive and corticosteroid therapy
3.	Control of fibroproliferative activity	Tetrandrine, xinin
4.	Prevention of associated diseases	Anti-mycobacterial therapy, pneumococcal vaccine
5.	Dust, macrophage and cytokines removal	Whole-lung lavage
6.	Monoclonal antibody therapy	Anti-cytokine therapy
7.	Lung transplantation	Lung transplantation

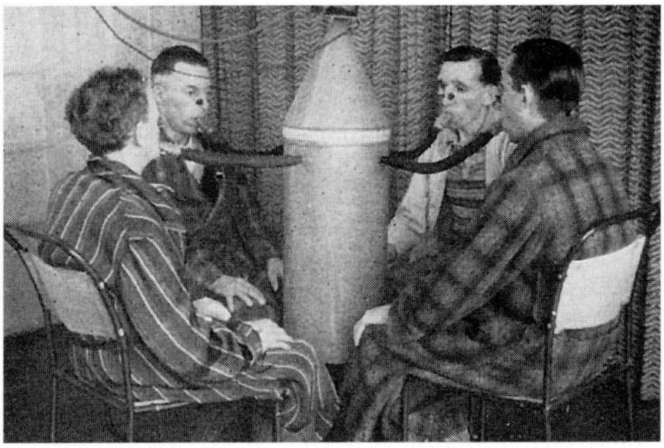

Figure 9.6 Inhalation of aluminum dust by miners with silicosis or CWP. The rate of dust generation was adjusted by trial and error until the desired amount of dust was collected over 15 minutes. The air compressor discharged into the chamber at a rate sufficient to keep the dust in suspension. A more detailed explanation is available in Kennedy, M.C.S. (1956) Aluminum powder inhalations in the treatment of silicosis of pottery workers and pneumoconiosis of coal-miners. *Br. J. Ind. Med.*, **13**, 85–101 (reproduced by permission).

Aluminum inhalation

Inhalation of aluminum was thought to alter the surface properties of silica in the lung and in that manner protect the macrophage. This notion was recognized early this century. Not only did small quantities of metallic aluminum inhibit the solubility of siliceous material in a test tube, but when rabbits inhaled quartz particles coated with aluminum they failed to produce silicosis, whereas uncoated particles caused frank fibrosis (Denny *et al.*, 1937). Using this same model, inhalation of 1% aluminum citrate independent of the time when silica was inhaled, resulted in a marked decrease in fibrosis. It was hypothesized that prevention of silicosis was the result of a silica–aluminum reaction resulting in a silica particle coated with a protective gelatinous aluminum hydroxide capsule (Denny *et al.*, 1939). Further support of this hypothesis was gleaned from additional animal studies (King *et al.*, 1950). Serial assessment of workers with silica exposure who inhaled aluminum showed few, if any, adverse effects, with some improvement in respiratory symptoms in selected individuals (Crombie *et al.*, 1944; Berry, 1948).

Prospective human trials followed. In 1956, an important prospective report addressed the effectiveness of aluminum inhalation treatment in 120 pottery workers followed for 4 years (Kennedy, 1956) (Figure 9.6). One group of workers inhaled aluminum at what was considered to be an 'effective dose' while the comparison group inhaled an 'ineffective dose'. Although most of the workers reported less dyspnea and symptomatic improvement, there was no significant improvement in pulmonary function or radiograph with inhalation of the 'effective dose'. These data, combined with other studies reporting adverse effects associated with the long-term inhalation of aluminum powder (Shaver, 1948; Jordan, 1961), have been an important reason why these compounds are not used as therapeutic agents.

Since then, other therapeutic trials of inhaled aluminum have been undertaken. In 1973, an aluminum citrate complex developed in China showed a clear protective effect against the development of silicosis in rats. However, significant toxicity existed, including a significant number of nodular deposits of aluminum in hepatic Kuppfer cells. In humans with pneumoconiosis, however, administration of this compound for two 3-month periods, separated by a 3-month period off drug, showed variable symptomatic improvement, and otherwise inconclusive results (Liu and Zou, 1985).

Yet there remains interest in inhaled aluminum as a means to alter the surface properties of silica. In a sheep model of silica inhalation, aluminum inhalation was associated with a reduction in the disease process and accelerated particle clearance from the lung (Dubois *et al.*, 1988). Moreover, in this same model, inhalation of aluminum 1 month after quartz exposure significantly suppressed alveolitis and lessened fibrosis

(Begin *et al.*, 1993). Although aluminum may be considered an effective prophylactic agent in the animal model, the proven adverse effects in humans make it unacceptable for use.

Corticosteroids

Lymphocytic inflammation, in association with an increased number of stimulated macrophages, plays an important role in the development of the silicotic nodule. The use of broncho-alveolar lavage (BAL) to sample the alveolar lining fluid in asymptomatic silica-exposed workers reflects inflammation with elevated numbers of lymphocytes and an increased level of immunoglobulins. BAL findings in workers with chronic silicosis show an increased number of lymphocytes and activated macrophages (Begin *et al.*, 1987; Rom *et al.*, 1987). Based on clinical approaches recognized in other granulomatous diseases, the presence of activated macrophages and an elevated number of lymphocytes (in this instance attributable to the presence of silica dust) are associated with the formation of granuloma in the lung and have the potential to be altered by corticosteroid therapy (Kotloff and Rossman, 1992).

Although theoretically attractive, the evidence for corticosteroid therapy as an effective long-term agent in altering the outcome of silicosis is essentially untested and primarily anecdotal. A 33-year-old surface miner of 12 years' work experience presented with an abnormal chest radiograph showing an alveolar filling pattern, clinical features of inflammation and progressive dyspnea (Goodman *et al.*, 1992) (Figure 9.7a–c). No infectious pathogens were found and an open-lung biopsy showed acute silicosis. Postoperatively, he received high-dose prednisone and empiric anti-mycobacterial therapy. He responded to this therapeutic approach with relatively rapid defervescence, chest radiograph and lung function improvement, and weight gain of 25 pounds (11 kg). Unfortunately, the worker could not tolerate prolonged high-dose corticosteroid therapy and the inevitable decrease in steroid therapy resulted in rapid progression of silicosis. He ultimately succumbed to the disease.

To our knowledge, the effect of steroid therapy on chronic silicosis has been evaluated in only one study. Thirty-four Indian workers with silicosis (ILO category 1 or greater) and VC, TLV or $D_{L}co$ <80% predicted initially received prednisolone (30 mg/day) for 6 weeks, which was then tapered to 0.25 mg/kg/day for a total of 6 months of therapy (Sharma *et al.*, 1991). Each worker served as his own control. Respiratory complaints were determined and BAL was performed before entry into the study.

Results were independent of cigarette smoking history or the presence of simple or complicated silicosis, although those with a longer duration of dust exposure were less likely to respond. BAL results showed less inflammation post-therapy. Lung volume measurements post-steroid therapy showed a statistically significant improvement in mean FVC, RV/TLC, FEV_1, PEFR, $D_{L}co$ and Po_2 values, but no differences in compliance, K (D_L/V_A) or TLC. Moreover, 24/34 showed a clinical response, defined as improvement in cough and exertional dyspnea, and lung function (FVC and/or $D_{L}co$) of $\geq$ 15%. Yet, several patients described weight gain, striae and moon facies – recognized complications of steroid therapy. The authors point out that while the results were promising, the potential long-term adverse effects of steroid therapy may not outweigh the benefits.

In summary, corticosteroids appear to lessen inflammation in the short-term in silicotics. Whether these drugs should be indicated for long-term usage in workers with chronic silicosis would require further assessment and raises reasonable concerns regarding the risk for corticosteroid complications. This risk versus benefit ratio should be assessed in a case-by-case basis if such therapy is considered. If steroid therapy is considered, it is reasonable to monitor the worker closely and to also consider empiric mycobacterial prophylaxis.

PVNO (poly-2 [or 4]-vinylpyridine-*N*-oxide); P204

PVNO is a member of a class of large-molecular weight polyvinylpyridine-*N*-oxides first reported by Schlipkoter and Brockhaus, (1960, 1961) to exhibit an inhibitory effect on the development of silicosis in an animal model. The mechanism of drug action has been attributed to alteration of surface properties of the silica particle; specifically, the adsorption of PVNO onto silica particles apparently lessens the level of macrophage activation after the particles are phagocytized (Beck *et al.*, 1963; Fetzer, 1967). Microscopic inspection reveals that the alveolar macrophages concentrate the PVNO–silica particles within lipid inclusions within the cell (Schiller, 1970).

In 1967, the first large-scale clinical initiative of the treatment of pneumoconiosis was begun in China. The 394 workers who participated were categorized according to the ILO profusion category on the chest radiograph and received P204 (molecular weight = 200 000 Da, known in China as 'Kexiping', meaning an antisilicotic) either intramuscularly or by inhalation for four 3-month treatment periods, each separated by a 1-month non-treatment period (Chen and Lu, 1970). Therapy was associated with fewer complaints of chest pain, cough and lesser dyspnea on exertion. During this

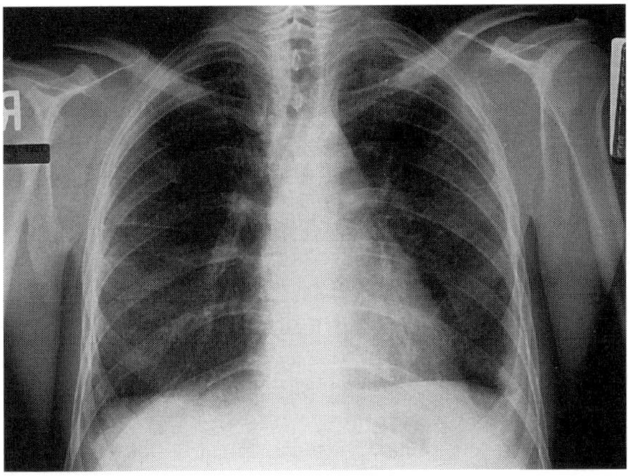

(a)

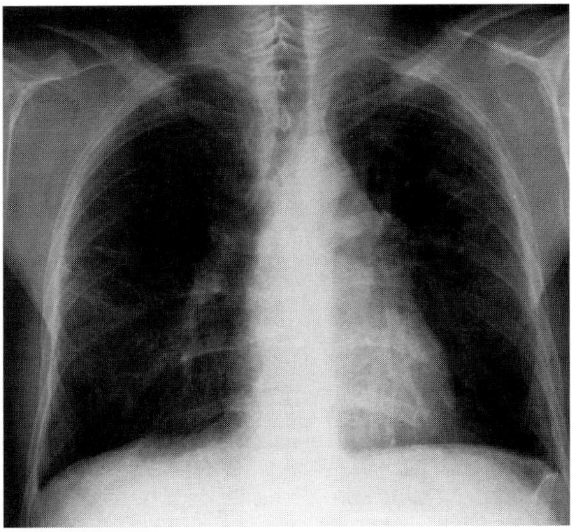

(b)

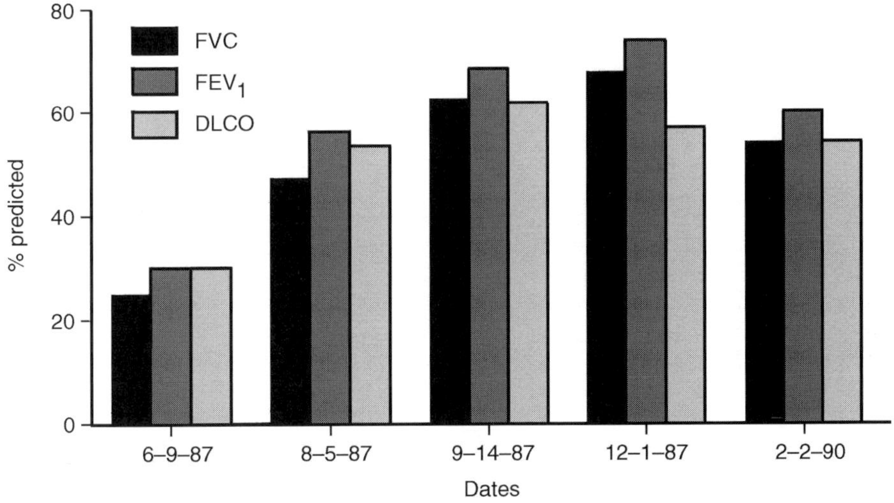

(c)

Figure 9.7 A 33-year-old man employed as a surface coal mine driller presented with a 50-pound (22 kg) weight loss, malaise, dyspnea, arthralgias, fever and night sweats. All drilling had been done with a wet process up until 4 years before this evaluation. He was a thin, febrile man who looked chronically ill. Chest examination showed poor expansion and crackles. A chest radiograph (a) showed an infiltrate at the left base and bilateral small rounded opacities. The right major fissure is markedly thickened. An open-lung biopsy showed interstitial inflammation and fibrosis with early silicotic granulomata. Electron microscopic assessment showed needle-shaped and polyhedral intracytoplasmic fragments consistent with silica particles. An infectious etiology was not recognized. Postoperatively, he received high-dose prednisone daily with two anti-mycobacterial medications for 3 months. He defervesced and experienced subjective improvement with an improved exercise tolerance and lung function tests. A chest radiograph taken 3 months after open biopsy (b) showed the profusion score and opacity size to have markedly lessened. His steroid therapy was gradually tapered, and because of intolerance to side effects, was discontinued after 8 months. (c) Lung function tests. These serial tests reflect significant improvement in association with steroid therapy. After 8 months, steroid therapy was stopped and testing several years later shows worsening of these parameters. Despite this apparent improvement over the short term, the radiograph inevitably worsened and progressive massive fibrosis developed and he died 4 years after his initial evaluation for silicosis. (Reproduced with permission from Goodman, G.B., Kaplan, P.D., Stachura, I., Castranova, V., Pailes, W.H. and Lapp, N.L. (1992). Acute silicosis responding to corticosteroid therapy. *Chest*, **101**, 366–70.)

relatively short time, approximately 70% of the radiographs had stabilized and were without progression.

PVNO, however, has been shown to cause severe chronic toxicities in laboratory animals. Chronic subcutaneous injection of PVNO in rats resulted in a significant hepatic and renal accumulation (Cheng *et al.*, 1970; Hasciri and Holt, 1977). Liver toxicity included impaired liver function, formation of foamy cells, abnormal morphologic structures of the hepatic lobules, and necrosis. Cirrhosis was detected in animals as late as 2 years after stopping the drug. Chronic injection of PVNO in rats was also associated with tumor formation at the injection site (Schmadl, 1969). Although intramuscularly delivered ^{14}C-PVNO reached plasma peak concentration 60 hours post-injection (Li and Yao, 1970), much of the toxicity may be due to the long half-life of the drug (only one-third of the dose is excreted in the urine and stool in 30 days) apparently attributable to the large-molecular weight (80 000 to 180 000 Da) PVNO molecules.

Attempts to minimize this toxicity were recently reported in a clinical trial where the outcome of 53 silicotics who received 144 ml of a 1% solution of PVNO instilled intratracheally once or twice and 200 mg of tetrandrine daily over a 2 year period were compared with 27 untreated silicotics. Although symptoms were less in the treated group, changes in the treated group's lung function were no different. Chest radiographs progressed less in the treated group. Therapy was most effective in those with silicosis and a history of exposure to silica alone. The main side effects were pigmentation (72%) and itching (8%) (Zhou *et al.*, 1996).

Tetrandrine

Tetrandrine, a bisbenzylisoquinoline alkaloid isolated from the root of the plant *Stephania tetrandra*, has been in use in Chinese herbal medicine for maladies including hypertension and angina. In these instances its effectiveness may be due to its calcium channel blocking effect (King *et al.*, 1988). It is also the first drug reported to reverse silica-induced fibrosis as determined by a reduction of the number of silicotic nodules in the lung, the total lung weight, and the lung collagen content in rat lung (Li *et al.*, 1981; Yu *et al.*, 1983).

Tetrandrine has a marked inhibitory effect on many of the important features of inflammation and fibrosis. These include inhibitory actions on lymphoproliferation and immunoglobulin synthesis (Seow *et al.*, 1988a), chemotaxis of neutrophils, monocytes, and alveolar macrophages into the alveolar space (Seow *et al.*, 1988b), and particle-induced respiratory burst and production of fibrogenic factors. In addition, the inhibitory action of tetrandrine on the dust-induced respiratory burst and production of interleukin-1 from alveolar macrophages and on fibroblast proliferation mediated by macrophage-derived growth factors (Castranova *et al.*, 1991; Ma *et al.*, 1991, 1992; Kang *et al.*, 1992), and synthesis and secretion of collagen, glycosaminoglycans and phospholipids (Chen *et al.*, 1982; Liu *et al.*, 1983) has been recognized. This drug, as well as PVNO, has the ability to block the increase of type I and type III collagen mRNA, a very early stage in the fibrotic process, caused by inhalation of silica dust in the rat model (Liu *et al.*, 1994). Altering collagen metabolism is a critical approach because collagen is a major constituent of the silicotic nodule.

The diverse action of tetrandrine may be due to its structural specific interactions with the plasma membrane of various cell types via its calcium channel blocking effect (King *et al.*, 1988). Tetrandrine binds strongly to the microtubular cytoskeletal system of alveolar macrophages, an effect which can be inhibited by the combined use of agents known to alter the microfilament and microtubular structure (Chen *et al.*, 1988; Liu *et al.*, 1988; Ma *et al.*, 1991, 1992). This binding action prevents cells from activation by either non-specific stimuli or receptor-mediating factors (Kang *et al.*, 1992).

Tetrandrine is excreted from urine and feces mainly as intact drug (Yin *et al.*, 1982). Toxicity appears to be species dependent. In rats given tetrandrine in 120–150 mg/kg doses once weekly for 1–3 months, there was no liver or kidney toxicity, yet, when these same animals received doses of 120–150 mg/kg/week over 5–9 months, histologic evidence of hepatocyte fatty degeneration and necrosis was observed, without impairment of liver or kidney function (Yu *et al.*, 1983). In dogs receiving the same therapy, there was an increased erythrocyte sedimentation rate and elevated levels of serum alanine transferase (ALT), aspartate aminotransferase (AST) and alkaline phosphatase enzymes. Histologic assessment of the chronic toxicity showed fatty degeneration and necrosis of liver cells and necrosis of the renal epithelial cells of the convoluted tubule (Li *et al.*, 1983).

In a 1980 open clinical trial with tetrandrine, 240 Chinese workers with silicosis (Lu and Li, 1983) were divided into three groups and received an oral tetrandrine dose of 100, 200 or 300 mg/day six times weekly. The regimen included a 3-month treatment period alternating with a 2-month resting period for a total of 1 to 3 years. Fifty-three of the 188 workers receiving the 200 or 300 mg/day doses exhibited radiographic improvement, as both small and large opacities were described to appear thinner and smaller and the bronchial shadows were described as sharper and more clearly defined. Workers receiving therapy, in general, had fewer complaints of chest tightness, dyspnea, cough, sputum production and upper respiratory infection. After the drug had been discontinued for 3 months, no further improvement was observed and, in some instances, there was radiographic progression. The results also showed that tetrandrine may prevent the progression of silicosis. In patients treated with 200 or 300 mg/day doses, progression of silicotic lesions was observed in only 13% and 4%, compared with a 53% progression rate in those treated with the 100 mg/day dosage. Although improvement of the chest radiograph was reported in only a relatively small percentage of the total patients (22% of all who received any therapy), this study has received attention because of the apparent ability of this drug to alter the natural history of silicosis.

There were, however, several concerns regarding this impressive work. First, the symptomatic improvement, which may indicate the anti-inflammatory action of the drug, did not appear to be dose-dependent. More workers receiving the 200 mg dose improved compared with those who received the 300 mg dose. There were suggestions that some of the improvement may have resulted from better patient care. Second, there was no untreated group of silicotics, so we cannot know what the placebo effect might have been. Third, it is not clear whether the drug effect has a relationship with the pretreatment disease state of the patients (i.e. it would seem intuitive that those with the least amount of fibrosis might be most sensitive to the effects of tetrandrine). Furthermore, since this study was carried out in numerous clinics with diverse facilities and personnel, it is difficult to judge whether the chest radiograph qualities and their interpretations were comparable and internally reproducible.

Nevertheless, this is exciting work, and has formed the starting point for continuing research into the use of this drug as a therapy of silicosis. However, despite its apparently strong antifibrogenic activity *in vitro*, the therapeutic value of tetrandrine is not yet defined. This may be attributed, in part, to the persistence of dust in the lungs. Perhaps recent advances, like packaging tetrandrine in a water-in-oil-in-water emulsion, will allow equally effective therapy with lesser amounts of tetrandrine with less toxicity (Chao *et al.*, 1996).

Xinin

Xinin, a Chinese word which carries the meaning of 'rendering silica harmless', is another example of the exploratory efforts by Chinese scientists to develop an effective anti-fibrotic agent. This drug is a bisbasic ethoxyfluorene derivative similar in structure and activity to tilorone but with reduced systemic toxicity (Cheng *et al.*, 1989). In experimental animals, the anti-silicotic activity of xinin is comparable with that of tetrandrine (Han *et al.*, 1989; Cheng *et al.*, 1990), yet *in vivo* studies have shown it to have less potential for chronic toxicity than tetrandrine (Han *et al.*, 1987; Cheng *et al.*, 1992).

Xinin is distributed mainly in the liver, adrenal glands, kidney and lungs, and eliminated as an intact molecule through urinary exertion. Administration of oral doses (240 mg/kg/week) of xinin to rats for a 6-month period did not adversely alter blood, liver or kidney function. Histologic examination of the liver showed mild formation of foamy cells, an effect which reversed after the termination of drug administration (Han *et al.*, 1987). Xinin's mechanism of action against silicosis has been attributed to its ability to prevent silica-induced alveolar macrophage death and to inhibit biosynthesis of phospholipids and collagen. The cellular action of xinin is similar to that or tetrandrine. Although to a lesser extent than tetrandrine, xinin interacts with alveolar macrophages and has an inhibitory effect on particle-stimulated macrophage respiratory burst, macrophage productive production of interleukin-1 in response to stimulation by liposaccharides, and release of growth factor-mediated thymocyte proliferation (Kang *et al.*, 1992).

Whole-lung lavage (WLL)

The cited reports have focussed on altering the natural progression of pneumoconiosis by intervening at one or more of several points in the fibrotic process. While the outcomes of therapeutic trials of pneumoconiosis have remained largely inconclusive (often due to inadequate testing of the hypotheses), these studies have added insight into potentially effective strategies for the treatment of pneumoconiosis. Yet, while these various agents may retard a certain step in the fibrotic process occurring after dust inhalation, the therapeutic effect(s), when recognized, are transient and only temporize an inevitable clinical effect. This is likely due to the persistence of dust which perpetuates the inflammatory and fibrotic processes. This is the rationale for considering WLL as a method of treatment for pneumoconiosis. Removing dust, activated cells and inflammatory mediators is a logical step and may serve as a therapy in itself, or in association with drug therapy, potentiate the effect(s) of a therapeutic drug.

In addition to removing dust, WLL removes activated cells and cytokines, agents which amplify and perpetuate the stimulus for fibrosis. In an animal model exposed to tagged aluminosilicate particles, most of the particles removed by large volume were located in macrophages (Ellender *et al.*, 1992). If an appreciable amount of the particle load in the human lung is harbored within the macrophages, then removal of a considerable number of macrophages may mean that slowing the rate of dust-induced progression is a possibility.

WLL is the removal of alveolar material by repeated large volume saline lavage (up to a total of 15 liters or more per lung) of each lung. It is performed under general anesthesia. As early as 1963, Ramirez *et al.* (1963) reported the therapeutic efficacy of WLL in patients with pulmonary alveolar proteinosis (PAP). Since then, this technique has also been proposed for the removal of alveolar filling material associated with alveolar microlithiasis (Palombini *et al.*, 1981), acute silicosis (Lavandier *et al.*, 1989), accidental inhalation of radioactive particles (Nolibe *et al.*, 1989), and even obstructive lung diseases (Kylstra *et al.*, 1971). The value of WLL in the exploration and management of interstitial lung diseases has recently been updated in reports of the European Task group on BAL (Klech and Hutter, 1990; Klech *et al.*, 1992).

At this time, WLL is only an accepted therapy of PAP, a rare idiopathic disorder with a pattern of lung injury which closely resembling acute silicosis. In PAP, WLL has been found to be safe, with no adverse effects attributable to this procedure in patients followed for as long as 25 years (Green and Griner, 1987). Complications, when they infrequently happen, occur during or soon after the procedure, and include hydropneumothorax, bronchospasm, pneumonia and lung abscess (Claypool *et al.*, 1984; Watters, 1988; Nolibe *et al.*, 1989).

Mason *et al.* (1982) first reported the use of WLL in the treatment of dust-induced lung disease in a foundry worker. Although the worker had symptomatic improvement after the procedure, there was no associated lung function change. Information regarding long-term changes in lung function or respiratory symptoms was not reported.

Chinese studies involving a larger number of CWP patients have been reported (Liang *et al.*, 1992). Under general anesthesia the authors inserted a double lumen endotracheal tube into the airways (ventilating one lung while lavaging the other), infused saline at a rate determined by gravity at a level 30 cm above the patient's chest, and then drained the fluid by gravity. Each lung was lavaged an average of 10 times with 1000–2000 ml saline each time. Close auscultatory monitoring of the chest is essential to ensure that fluid from the lavaged lung does not overflow into the ventilated lung. Liang and colleagues also showed that applying positive pressure (administered as 100% oxygen) during WLL via positive-pressure bag ventilation to the lavaged lung improved oxygenation, perhaps by lessening a shunt-like effect, as well as significantly increasing the amount of dust and cells removed.

In those with minimal disease (ILO category 1) and no or negligible impairment, one lung was lavaged after the other during the same period of general anesthesia. In this situation, the recovery period between lavages (typically lasting 1–2 hours) was sufficient to allow the compliance and peak ventilator pressures in the lavaged lung to approximate pre-lavage values; thereafter, the second lung was lavaged. When both lungs are lavaged during one period of general anesthesia, the procedure typically lasted 5 or 6 hours. When pneumoconiosis is more severe or is associated with significant functional impairment, each lung was lavaged on a separate occasion. Overall, the average recovery of the total amount of lavage fluid exceeded 95% of that instilled. When the fluid was in the lung, added pressure, delivered as manual compressions of 100% oxygen by 'ventilation-bag', resulted in a two- to three-fold increase in dust removal and about 500-fold increase in the removal of alveolar macrophages from the lungs. We modified the technique of WLL to include a pressure manometer attached to the positive-pressure bag ventilation of the lavaged lung to minimize the possibility of barotrauma (Wilt *et al.*, 1996) (Figure 9.8).

As in Mason's work, the Chinese study reported improvement in respiratory symptoms in nearly all workers. Six-month follow-up showed the procedure to be without residual adverse effects; only one patient developed a respiratory tract infection. The average pulmonary function and arterial tension measurements from 16 of the 25 workers who had undergone whole lung lavage was obtained one week or 3–6 months after the treatment and were not significantly different when compared with pre-treatment values. There was no long-term follow-up and the results were not controlled, therefore, conclusions regarding whether this procedure altered the natural history of pneumoconiosis are not possible. In addition, studies which addressed whether the cellularity, inflammatory mediators and dust content of the alveolar lining fluid either before, or following WLL, was changed were not performed. Similarly, work by Borm *et al.* (1988) which showed differences in blood cytokine levels in miners with abnormal radiographs compared to those with no abnormality, could be applied to attempt to measure alterations in the lung which could be measured systemically before and after the procedure.

Several comments may be made regarding Liang's work (1992). First, most workers undergoing WLL had minimal disease which may never have progressed or become disabling. Second, it is difficult to assess whether the use of average values of these physiologic

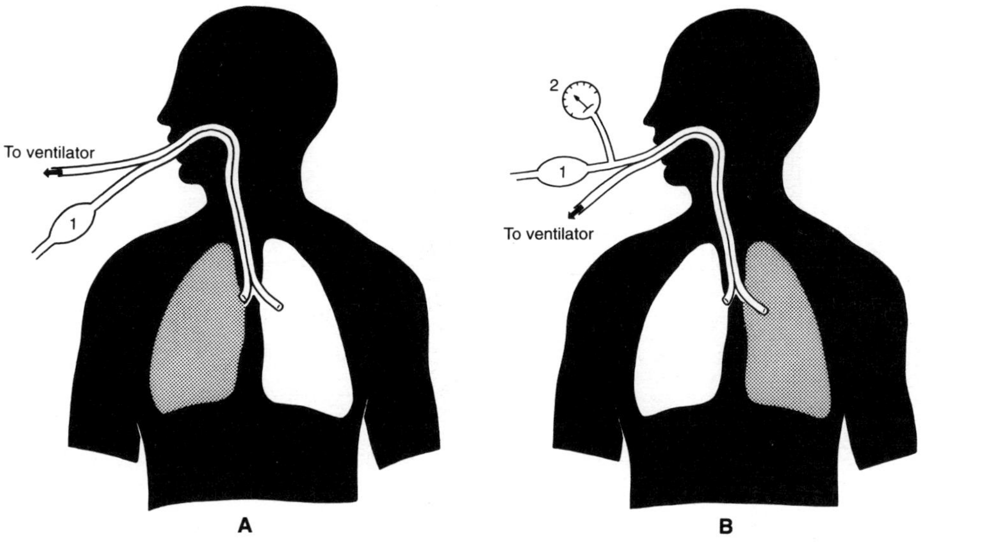

Figure 9.8 Diagram of the WLL aparatus used to limit peak airway pressure. Panel A represents the set up for WLL of the right lung without a pressure manometer to minimize the risk of barotrauma. Panel B represents a modification of this set up with an in-line manometer. This limits the peak airway pressure of the lung being lavaged to 40 cm H_2O. 1, anesthesia bag; 2, manometer. (Reproduced by permission from Wilt, J.L., Banks, D.E., Weissman, D.N. *et al.* (1996). Reduction of lung dust burden in pneumoconiosis by whole-lung lavage. *J. Occup. Environ. Med.*, **38**, 619–24.)

measurements truly represent the effect of this procedure in all workers. Improvement in an individual miner might have been missed by this type of data analysis. Third, the results were reported at the end of 6 months, likely too short a time to show a clinically important effect. Nevertheless, the manuscript breaks important ground in therapeutic approaches.

In two patients (four lungs), dust removal and cellularity has been evaluated (Wilt *et al.*, 1996). Although discrepant amounts of dust, cells and cytokines were recovered in the two reported cases (as well as between lungs within the same individual), all lungs showed on-going inflammation. Nearly 2 g of dust were removed from a lung of a coal miner with ILO category 2 pneumoconiosis by WLL (Figure 9.9). This approximates the mean mineral dust burden in the lung of coal miners with a similar pneumoconiosis profusion score as assessed by ashing the lung at autopsy (Rivers *et al.*, 1960). A critical issue is whether dust in the lung is in a 'dynamic state', i.e. is a significant amount of dust transported intracellularly into and out of the interstitium? Studies in animals have shown that dust particles in the interstitial spaces can be carried into the alveolar space by migrating neutrophils (Adamson and Bowden, 1992). Furthermore, repeated lavages in animals continue to remove dust which may originate in the interstitial spaces and progress into the alveolar spaces with recurrent washings (Brain, 1980). Overall, the dynamics of dust transport within the lung is not well understood (Sorokin and Brain 1975; Brody and Overby, 1989; Scheule and Holian, 1991).

Since WLL has not been widely tested in independent laboratories, critical review of its clinical significance is not yet possible. Whether the risks outweigh the benefits is not yet clear; however, this appears to be a relatively safe clinical procedure which is capable of removing a significant amount of dust and activated cells. WLL, in association with effective antifibrotic drug therapy, may be an approach which may be beneficial.

WITH AN EYE TO THE FUTURE

Fibrosis represents the end-stage of lung inflammation. However, targeting inflammatory areas around parenchymal fibrosis may be important. Treatment strategies will likely be aimed at removing the dust-induced stimulus of inflammation and interrupting the pathogenetic mechanisms. *In vivo* studies could employ animal models of fibrosis in conjunction with *in situ* probes which better determine sites of cellular activation associated with dust exposures (Scheule and Holian, 1991). For example, collagen mRNA probes might determine which fibroblasts are actively depositing matrix. Studies designed to target specific responses

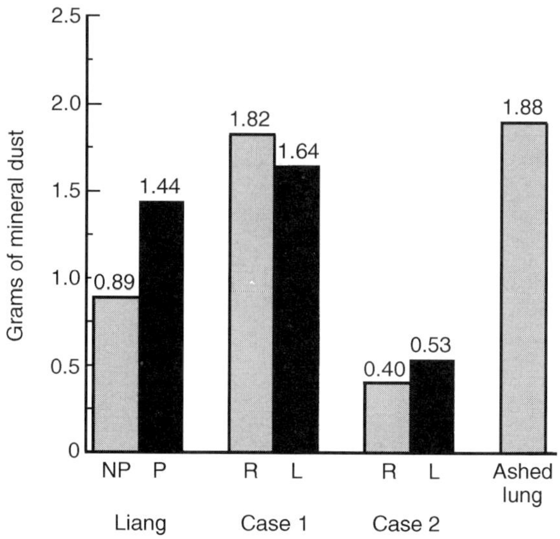

Figure 9.9 Amounts of dust removed using the technique of Liang *et al.* (1992), which first compared the amount of dust removed by infusing and then draining saline instilled into the lung (NP or no pressure) with 'agitating the fluid' within the lung by gentle compression of a ventilation bag (P or pressure) from different lungs within the same individual. These values are then compared with the amount of dust removed in two cases who underwent WLL with gentle ventilation bag bagging [described as the right (R) and left (L) lung from cases 1 and 2] (Wilt *et al.*, 1996), and then to mean data from an autopsy study by Rivers *et al.* (1960) who ashed lungs of coal miners with a comparable degree of pneumoconiosis and measured the amount of mineral dust removed. It appears that a significant amount of dust can be removed by the using WLL, although the effect of removing this dust on the natural history of pneumoconiosis is unknown.

might delineate mechanisms which can be affected by therapeutic interventions. Yet, even this process has limitations because *in vitro* techniques often lack adequate the complex cellular milieu present in the organism. Other evaluations could target mediators derived from additional cells involved in the pathogenesis of fibrosis.

It is likely that there will need to be more than one therapeutic strategy. One phase could be the removal of dust, using the procedure of WLL. A second phase could be the use of multiple drugs which might modulate the inflammatory response associated with silica exposure. For example, in silicotic rats receiving tetrandrine and PVNO, greater amounts of products of collagen breakdown were measurable in the serum compared with the levels in silicotic animals which did not receive these drugs (Yu *et al.*, 1995). This implies that the use of multiple drugs may accelerate collagen breakdown. Yet, before reaching decision points regarding what drugs might be indicated in workers who develop silicosis, well-designed controlled studies will need to be performed to determine the effects of these interventions on the natural history of these illnesses.

Acknowledgments

This work was supported in part by National Institute for Occupational Safety & Health Grants U60/CCU306149–01–1 and U60/CCU313230.

REFERENCES

Aberle, D.R., Gamsu, G., Ray, C.S. and Feuerstein, I.M. (1988) Asbestosis-related pleural and parenchymal fibrosis: detection with high-resolution CT. *Radiology*, **166**, 729–34.

Adamson, I.Y.R. and Bowden, D.H. (1992) Clearance of interstitial silica reduces pulmonary fibrosis in mice. *FASEB J.*, **6**,A1910.

Akira, M., Yokoyama, K., Yamamoto, S. *et al.* (1991) Early asbestosis: evaluation with high-resolution CT. *Radiology*, **178**, 409–16.

Amandus, H.E. and Piacitelli, G. (1987) Dust exposures at U.S. surface coal mines in 1982–1983. *Arch Environ. Health*, **42**, 374–81.

Balaan, M.R. Weber, S.L. and Banks, D.E. (1993) Clinical aspects of coal worker's pneumoconiosis and silicosis. *Occupational Medicine, State of the Art Review*, **8**, 19–34.

Banks, D.E., Morring, K., Boehlecke, B.A., Althouse, R. and Merchant, J.A. (1981) Silicosis in silica flour workers. *Am. Rev. Respir. Dis.* **124**, 445–50.

Banks, D.E., Bauer, M.A., Castellan, R.M. and Lapp, N.L. (1983) Silicosis in surface coalmine drillers. *Thorax*, **38**, 275–8.

Banks, D.E., Cheng, Y.H., Weber, S.L. and Ma, J.K.H. (1993) Strategies for the treatment of pneumoconiosis. *Occupational Medicine, State of the Art Review*, **8**, 204–32.

Beck, E.G., Bruch, J. and Brockhaus, A. (1963) Die Beeinflussung der cytotoxischen Quarzwirkung au mausefibroblasten (strain L) durch polyvinylpyridin-*N*-oxyde (P204). *Z. Zellforsch*, **59**, 568–76.

Begin, R.O., Cantin, A.M., Boileau, R.D. and Bisson, G.Y. (1987) Spectrum of alveolitis in quartz-exposed human subjects. *Chest*, **92**, 1061–7.

Begin, R., Cantin, A. and Masse, S. (1989) Recent advances in the pathogenesis and clinical assessment of mineral dust pneumoconioses: asbestosis, silicosis and coal pneumoconiosis. *Eur. Respir. J.*, **2**, 988–1001.

Begin, R., Cantin, A., Masse, S., Dufresne, A., Perreault, G. and Sebastian, P. (1993) Aluminum inhalation in sheep silicosis. *Int. J. Exp. Pathol.*, **74**, 299–307.

Berry, J.W. (1948) Aluminum therapy in advanced silicosis. *Am. Rev. Tuberc.*, **57**, 557–73.

Borm, P.J.A., Palmen, N., Engelen, J.J.M. and Burman, W.A. (1988) Spontaneous and stimulated release of TNF-alpha from blood monocytes of miners with coal workers' pneumoconiosis. *Am. Rev. Respir. Dis.*, **138**, 1589–94.

Brain, J.D. (1980) Macrophage damage in relation to the pathogenesis of lung disease. *Environ. Health Perspect.*, **35**, 21–30.

Brody, A.R. and Overby, L.H. (1989) Incorporation of tritiated thymidine by epithelial and interstitial cells in bronchiolar-alveolar regions of asbestos-exposed rats. *Am. J. Path.*, **134**, 133–40.

Castranova, V., Kang, J.H., Ma, J.K.H., Malanga, C.J., Moore, M.D., Schwegler-Berry, D. and Ma, J.Y.C. (1991) Effect of bisbenzylisoquinoline alkaloids on alveolar macrophages: correlation between binding affinity, inhibitory potency and antifibrotic potential. *Tox. Appl. Pharmacol.*, **108**, 242–52.

Centers for Disease Control Alert (1992a) Request for assistance in preventing silicosis and deaths from sandblasting. Public Health Service. Centers for Disease Control. Department of Health and Human Services (National Institutes for Occupational Safety and Health) Publication No. 92–102.

Centers for Disease Control Alert (1992b) Request for assistance in preventing silicosis and deaths in rock drillers. Public Health Service. Centers for Disease Control. Department of Health and Human Services (National Institute for Occupational Safety and Health) Publication No. 92–107.

Chao, D.-H., Ma, J.Y.C., Malanga, C.J., Banks, D.E., Hubbs, A.F., Rojanasakul, Y., Castranova, V. and Ma, J.K.H. (1996) Multiple emulsion-mediated enhancement of the therapeutic effect of tetrandrine against silicosis. *Appl. Occup. Environ. Hyg.*, **11**, 1008–18.

Chen, N.M., Liu, N.F. and Li, Y.R. (1982) Observations of the inhibitory effect of tetrandrine on silica-induced fibrogenesis by culture system *in vitro*. *J. Hyg. Res.*, **11**(4), 79–84.

Chen, N., Liu, L., Gai, G. *et al.* (1988) Studies on the effect of tetrandrine on microtubules. II. Observations with immunofluorescence techniques. *Ecotoxicol. Environ Safety*, **15**, 149–52.

Chen, S.Y. and Lu, X.R. (1970) Clinical studies of the therapeutic effect of kexiping on silicosis, in *Proceedings of the Therapeutic Effect of Kexiping on Silicosis*, Institute of Occupational Medicine, CAPM Press, Beijing, pp. 162–5.

Cheng, Y.H., Han, S.I. and Zhang, Z.L. (1970) Studies of the therapeutic effect of Kexiping on silicosis, in *Proceedings of the Therapeutic Effect of Kexiping on Silicosis*, Institute of Occupational Medicine, CAPM Press, Beijing, pp. 40–5.

Cheng, Y.H., He, Q.N., Han, S.L., Zhang, Z.L., Zhao, S.W., Du, C.C., Yao, P.P. and Chin, X.Y. (1989) Studies on the curative effects of tilorone analogs on experimental silicosis in rats. *Chinese J. Ind. Hyg. Occup. Med.*, **7**, 288–91.

Cheng, Y.H., Han, S.L. and Zhang, Z.L. (1990) Studies on the curative effect of xinin on silicosis in rats, in *Proceedings of the China–Japan Joint Symposium on Occupational Health*, Tanshen, pp. 46–7.

Cheng, Y.H., Qin, L.H. and Huan, R.H. (1992) Pharmacokinetic studies of xinin. *J. Hyg. Res.*, **21**, 1–4.

Claypool, W., Rogers, R. and Matuschak, G. (1984) Update on the clinical diagnosis management and pathogenesis of pulmonary alveolar proteinosis. *Chest*, **85**, 550–8.

Code of Federal Regulations (1994) Title 30, part 56. Air quality health standards for abrasive standards and drill dust control, February 18.

Craighead, J.E., Abraham, J.L., Churg, A. *et al.* (1982) The pathology of asbestos-associated diseases of the lungs and pleural cavities: diagnostic criteria and proposed grading schema. *Arch. Pathol. Lab. Med.*, **106**, 544–96.

Crombie, D.W., Blaisdell, J.L. and MacPherson, G. (1944) The treatment of silicosis by aluminum powder. *Can. Med. Assoc. J.*, **50**, 318–28.

Denny, J.J., Robson, W.D. and Irwin, D.A. (1937) The prevention of silicosis by metallic aluminum. I. *Can. Med. Assoc. J.*, **37**, 1–11.

Denny, J.J., Robson, W.D. and Irwin, D.A. (1939) The prevention of silicosis by metallic aluminum. II. *Can. Med. Assoc. J.*, **40**, 213–28.

Dubois, F., Begin, R., Cantin, A. *et al.* (1988) Aluminum inhalation reduces silicosis in a sheep model. *Am. Rev. Respir. Dis.*, **137**, 1172–9.

Ellender, M., Hodgson, A., Wood, K.L. and Moody, J.C. (1992) Effect of bronchopulmonary lavage on lung retention and clearance of particulate material in hamsters. *Environ. Health Perspect.*, **97**, 209–13.

Epler, G.R., McLoud, T.C., Gaensler, E.A., Mikus, J.P. and Carrington, C.B. (1978) Normal chest roentgenograms in chronic diffuse infiltrate lung disease. *N. Engl. J. Med.*, **298**, 934–9.

Fetzer, J. (1967) Accumulation of polyvinylpyridine-*N*-oxide in lysosomes. *Naturwissenschaften*, **54**, 518–24.

Gamsu, G., Salmon, C.J., Warnock, M.L. and Blanc, P.D. (1995) CT Quantification of interstitial fibrosis in patients with asbestosis: a comparison of two methods. *Am. J. Roentgenol.*, **164**, 63–8.

Glindmeyer, H.W. and Hammad, Y. (1988) Contributing factors to sandblaster's silicosis: inadequate respiratory protection equipment and standards. *J. Occup. Med.*, **30**, 917–21.

Goodman, G.B., Kaplan, P.D., Stachura, I., Castranova, V., Pailes, W.H. and Lapp, N.L. (1992) Acute silicosis responding to corticosteroid therapy. *Chest*, **101**, 366–70.

Graham, W.G.B., Ashikaga, T., Hemenway, D., Weaver, S. and O'Grady, R.V. (1991) Radiographic abnormalities in Vermont granite workers exposed to low levels of granite dust. *Chest*, **100**, 1507–14.

Green, D. and Griner, G. (1987) Twenty five years follow-up of patient treated with lung lavage for pulmonary alveolar proteinosis. *N. Engl. J. Med.*, **317**, 839–40.

Han, S.L., Zhang, Z.L. and Cheng, Y.H. (1987) Xinin is non-mutagenic by Salmonella test. *J. Hyg. Res.*, **6**(3), 8–10.

Han, S.L., Zhang, Z.L. and Cheng, Y.H. (1989) The curative effect of xinin on experimental silicosis. *Chinese J. Ind. Hyg. Occup. Med.*, **7**, 331–4.

Harkin, T.J., McGuinness, G., Goldring, R., Cohen, H., Parker, J.E., Crane, M., Naidich, D.P. and Rom, W.N. (1996) Differentiation of the ILO boundary chest roentgenograph (0/1 to 1/0) in asbestosis by high-resolution computed tomography scan, alveolitis, and respiratory impairment. *J. Occup. Environ. Med.*, **38**, 46–52.

Hasciri, V.N. and Holt, P.F. (1977) Poly-2-vinylpyridine-1-oxide in silicosis therapy: effect of molecular weight. *Int. Arch. Occup. Environ. Health*, **38**, 177–88.

Hearl, F.J. and Hewitt, P. (1993) Problems in monitoring dust levels within mines. *Occupational Medicine State of the Art Reviews*, **8**, 93–108.

Hughes, J.M., Jones, R.N., Gilson, J.C. *et al.* (1982) Determinants of progression of silicosis in sandblasters silicosis. *Ann. Occup. Hyg.*, **26**, 701–12.

International Labour Office (1980) *Guidelines for the use of ILO international classification of radiographs of pneumoconioses*, ILO, Geneva.

Jordan, J.W. (1961) Pulmonary fibrosis in workers using an aluminum powder. *Br. J. Ind. Med.*, **18**, 21–3.

Kang, J.H., Lewis, D.M., Castranova, V., Rojanasakul, Y., Banks, D.E., Ma, J.Y.C. and Ma, J.K.H. (1992) Inhibitory action of tetrandrine on macrophage production of interleukin-1 (IL-1) like activity and thymocyte proliferation. *Exp. Lung Res.*, **18**, 715–29.

Kennedy, M.C.S. (1956) Aluminum powder in the treatment of silicosis of pottery workers and pneumoconiosis of coal-miners. *Br. J. Ind. Med.*, **13**, 85–101.

King, E.J., Wright, B.M., Ray, S.C. and Harrison, C.V. (1950) Effect of aluminum on the silicosis-producing action of inhaled quartz. *Br. J. Ind. Med.*, **7**, 27–36.

King, V.F., Garcia, M.L., Himmel, D. *et al.* (1988) Interaction of tetrandrine with slow inactivation calcium channels: characterization of calcium channel modulation by an alkaloid of Chinese medicinal herb origin. *J. Biol. Chem.*, **263**, 2238–44.

Kipen, H.M., Lilis, R., Suzuki, Y., Valciukdas, J.A. and Selikoff, I.J. (1987) Pulmonary fibrosis in asbestos insulation workers with lung cancers: a radiological and histopathological evaluation. *Br. J. Ind. Med.*, **44**, 96–100.

Klass, V. (1993) A diagnostic approach to asbestosis utilizing clinical criteria, high resolution computed tomography, and gallium scanning. *Am. J. Ind. Med.*, **23**, 801–9.

Klech, H. and Hutter, C. (1990) Clinical guidelines and indications for bronchoalveolar lavage (BAL). Report of the European Society of Pneumology Task group on BAL. *Eur. Respir. J.*, **3**, 937–69.

Klech, H., Hutter, C. and Costabel, U. (1992) The clinical role of BAL. *Eur. Respir. Rev.*, **2**, 47–127.

Kotloff, R.M. and Rossman, M.D. (1992) Sarcoidosis. *Immunol. Allergy Clin. North. Am.*, **12**, 421–49.

Kraus, T., Raithel, H.J. and Hering, K.G. (1996) Evaluation and classification of high-resolution computed tomographic findings in patients with pneumoconiosis. *Int. Arch. Occup. Environ. Health*, **68**, 249–54.

Kreiss, K. and Zhen, B. (1996) Risk of silicosis in a Colorado mining community. *Am. J. Ind. Med.*, **30**, 529–39.

Kylstra, J., Bauch, D., Hall, K. and Spock, A. (1971) Volume-controlled lung lavage in the treatment of asthma, bronchiectasis and mucoviscidosis. *Am. Rev. Respir. Dis.*, **103**, 651–65.

Lapp, N.L. and Parker, J.E. (1992) Coal Workers' Pneumoconiosis. *Clin. Chest Med.*, **13**, 243–52.

Lavandier, M., Belleau, R., Desmeules, M. and Benazera, G.A. (1989) Echec des lavages pulmonaires massifs dans les silicoproteinases alveolaires. *Rev. Pneumol. Clin.*, **45**, 42–3.

Li, Q., Xu, Y., Zhou, Z., Chewn, X., Huang, X., Chen, S. and Zhun, C. (1981) The therapeutic effect of tetrandrine on silicosis. *Chin. J. Tuberc. Respir. Dis.*, **4**, 321–8.

Li, T.L., Hu, T.Y. and Zou, C.Q. (1983) Studies on the chronic toxicity of tetrandrine in dogs. *Ecotoxicol. Environ. Safety*, **6**, 528–34.

Li, Y.R. (1995) Prevention and treatment of pneumoconiosis in China. *Biomed. Environ. Science*, **8**, 211–17.

Li, Y.R. and Yao, P.P. (1970) Studies of the absorption, distribution and excretion of PVNO-C, in *Proceedings of the Therapeutic Effect of Kexiping on Silicosis*, Institute of Occupational Medicine, CAPM Press, Beijing, pp. 143–56.

Liang, Y.P., Sun, Y., Chen, C.Y. *et al.* (1992) Clinical evaluation of massive whole lung lavage for treatment of coal workers' pneumoconiosis. *He Bei Liao Yang (J. Hebei Convalescence)*, **1**, 1–9.

Liu, B., Zou, C.Q. and Li, Y.R. (1983) Studies on the contents of glycosaminoglycans from lungs of silicotic rats and tetrandrine-treated silicotic rats. *Ecotoxicol. Environ. Safety*, **7**, 323–9.

Liu, B.-C., He, Y.-X., Miao, Q., Wang, H.-H. and You, B.-R. (1994) The effects of tetrandrine (TT) and polyvinylpyridine-*N*-oxide (PVNO) on gene expression of type I and type III collagens during experimental silicosis. *Biomed. Environ. Sciences*, **7**, 199–204.

Liu, L., Chen, N., Gai, G., Li, Z., Yang, J. and Li, Y.R. (1988) Studies on the effects of tetrandrine on microtubules. 1: Biochemical observations and electron microscopy. *Ecotoxicol. Environ. Safety*, **15**, 142–8.

Liu, S.Q. and Zou, S.G. (1985) Studies on the prophylactic and therapeutical effect of aluminum citrate on silicosis and coal worker's pneumoconiosis, in *Proceedings of Therapeutical Effect of Aluminum Citrate on Silicosis and Coal Worker's Pneumoconiosis*, Beijing University Press, Beijing, pp. 74–9.

Lu, X.R. and Li, Q.R. (1983) Clinical studies of the therapeutic effect of tetrandrine on silicosis. *Chin. J. Ind. Hyg. Occup. Med.*, **2**, 106–10.

Ma, J.K.H., Mo, C.G., Malanga, C.J., Ma, J.Y.C. and Castranova, V. (1991) Binding of bisbenzylquinoline alkaloids to phosphatidylcholine vesicles and alveolar macrophages: relationship between binding affinity and antifibrogenic potential. *Exp. Lung Res.*, **17**, 1061–77.

Ma, J.Y.C., Barger, M.W., Ma, J.K.H. and Castranova, V. (1992) Inhibition of respiratory burst activity in alveolar macrophages by bisbenzylquinoline alkaloids: characterization of drug–cell interaction. *Exp. Lung Res.*, **18**, 829–43.

Mason, G.R., Abraham, J.L., Hoffman, L., Cole, S., Lippman, M. and Wasserman, K. (1982) Treatment of mixed-dust pneumoconiosis with whole lung lavage. *Am. Rev. Respir. Dis.*, **126**, 1102–7.

Meiklejohn, A. (1951) History of lung diseases of coal miners in Great Britain, part II. *Br. J. Med.*, **9**, 93–8.

Murray, M. (1907) Departmental committee for compensation for industrial diseases, Cd 3495 and 3496, His Majesty's Stationery Office, London, p. 127.

Murphy, R.L., Becklake, M.R., Brooks, S.M. *et al.* (1986) The diagnosis of nonmalignant diseases related to asbestos: official statement of the American Thoracic Society. *Am. Rev. Respir. Dis.*, **134**, 363–8.

Naidich, D.P. (1989) Pulmonary parenchymal high-resolution CT: to be or not to be? *Radiology*, **71**, 22–4.

Nolibe, D., Metivier, H., Masse, R. and Chretien, J. (1989) Benefits and risks of bronchopulmonary lavage: a review. *Radiat. Prot. Dosimetry*, **26**, 337–43.

Oksa, P., Suoranta, H., Koskinen, H., Zitting, A. and Nordman, H. (1994) HRCT in the early detection of asbestosis. *Int. Arch. Occup. Health*, **65**, 299–304.

Oliver, T. (1927) Clinical aspects of pulmonary asbestosis. *Br. Med. J.*, **2**, 1026.

Palombini, B., Da Silva Porto, N., Wallan, C.U. and Camargo, J.J. (1981) Bronchopulmonary lavage in alveolar microlithiasis. *Chest*, **80**, 242–3.

Pancoast, H.K., Miller, T.G. and Landris, H.R.M. (1917) A roentgenologic study of the effects of dust inhalation on the lung. *Trans. Soc. Am. Physicians*, **32**, 97.

Ramirez, R.J., Schultz, R.B. and Dutton, R.E. (1963) Pulmonary alveolar proteinosis. A new technique and rationale for treatment. *Arch. Intern. Med.*, **112**, 419–31.

Remy-Jardin, M., Remy, J., Farre, I. and Marquette, C.H. (1992) Computed tomographic evaluation of silicosis and coal worker's pneumoconiosis. *Radiol. Clin. North. Am.*, **30**, 1155–76.

Rivers, D., Wise, M.E., King, E.J. and Nagelschmidt, G. (1960) Dust content, radiology, and pathology in simple pneumoconiosis of coalworkers. *Br. J. Indust. Med.*, **17**, 87–108.

Rom, W.N., Bitterman, P.B., Rennard, S.I., Cantin, A. and Crystal, R.G. (1987) Characterization of the lower respiratory tract inflammation of nonsmoking individuals with interstitial lung disease associated with chronic inhalation of inorganic dusts. *Am. Rev. Respir. Dis.*, **136**, 1429–34.

Rosenstock, L. (1994) Asbestosis and asbestos related diseases, in *Textbook of Clinical Occupational and Environmental Medicine* (eds L. Rosenstock and M. Cullen), W.B. Saunders, Co., Philadelphia, pp. 254–63.

Scheule, R.K. and Holian, A. (1991) Immunologic aspects of pneumoconiosis. *Exp. Lung. Res.*, **17**, 661–85.

Schiller, E. (1970) Experimental pathology and pharmacology of dust exposure: an electronic microscopic study, In *Inhaled Particles III* (ed. C.N. Davis), Perganon Press, pp. 403–14.

Schlipkoter, H.W. and Brockhaus, A. (1960) The action of polyvinylpyridine on experimental silicosis. *Dtsch. Med. Wochenschr.*, **85**, 920–3.

Schlipkoter, H.W. and Brockhaus, A. (1961) Inhibition of silicosis by subcutaneous injection of polyvinylpyridine-*N*-oxide. *Klin. Wochenschr.*, **22**, 1182–9.

Schmadl, D. (1969) Prufung von polyvinylpyridine-N-oxide auf die carcinogene wirkung bei ratten und Mausen. *Arzneim. Forsch.*, **19**, 1313–14.

Seow, W.K., Ferrante, A., Goh, D.B.H., Chalmers, A.H., Li, S.Y. and Thong, Y.H. (1988a) *In vitro* immunosuppressive properties of the plant alkaloid tetrandrine. *Arch. Allergy Appl. Immunol.*, **85**, 410–15.

Seow, W.K., Ferrante, A., Li, S.Y. and Thong, Y.H. (1988b) Antiphagocytic and antioxidant properties of plant alkaloid tetrandrine. *Arch. Allergy Appl. Immunol.*, **85**, 404–9.

Sharma, S.K., Pande, J.N. and Verma, K. (1991) Effect of prednisolone treatment in chronic silicosis. *Am. Rev. Respir. Dis.*, **143**, 814–21.

Shaver, C.G. (1948) Pulmonary changes encountered in employees engaged in the manufacture of alumina abrasives. *Occup. Med.*, **5**, 718–28.

Socie, E., Migliozzi, A., Wagner, S. and Halpin, T. (1995) Occupational Silicosis–Ohio, 1989–1994. *MMWR*, **44**, 61–4.

Sorokin, S.P. and Brain, J.D. (1975) Pathways of clearance in mouse lungs exposed to iron oxide aerosols. *Anat. Rec.*, **181**, 581–626.

Staples, C.A., Gamsu, G., Ray, C.S. *et al.* (1989) High resolution computed tomography and lung function in asbestos-exposed workers with normal chest radiographs. *Am. Rev. Respir. Dis.*, **139**, 1502–8.

Watters, L.C. (1988) Chronic alveolar filling disease, in *Interstitial Lung Disease* (eds M.I. Schwartz and T.E. King), Decker, Toronto, pp. 239–91.

Webb, W.R., Muller, N.L. and Naidich, D.P. (1993) Standardized terms for high-resolution computed tomography of the lung: a proposed glossary. *J. Thorac. Imaging*, **8**, 167–75.

Weeks, J.L. (1993) From explosions to black lung: a history of efforts to control coal mine dust. *Occupational Medicine*: *State of the Art Reviews*, **8**, 1–17.

Wilt, J.L. and Banks, D.E. (1995) Reactions of inorganic dusts, in *Immunology and Management of Interstitial* Lung Disease (eds E.H. Walters and R.M. duBois), Chapman & Hall, London, pp. 199–222.

Wilt, J.L., Banks, D.E., Weissman, D.N. *et al.* (1996) Reduction of lung dust burden in pneumoconiosis by whole lung lavage. *J. Occup. Environ. Med.*, **38**, 619–24.

Yin, R.G., Zou, C.Q. and Hu, T.X. (1982) Observation of the therapeutic effect and toxicity of tetrandrine in monkeys. *J. Hyg. Res.*, **11**(4), 55–65.

Yu, L., Zou, C.-Q., Li, Y.-R., Qu, L., Xing, K.-J. and Du, Q.-C. (1995) A biochemical study of combined treatment of experimental silicosis with tetrandrine-PVNO and tetrandrine-QOP. *Biomed. Environ. Sciences*, **8**, 265–8.

Yu, X.F., Zou, C.Q. and Lim, M.B. (1983) Observations of the effect of tetrandrine on experimental silicosis of rats. *Ecotoxicol. Environ. Safety*, **7**, 306–12.

Zhou, Z., Wang, F.Y., Hu, Y.B., Sun, J.L., Sun, W.J., Huang, X.Q., Wen, J. and Hu, T.X. (1996) Clinical investigation on therapeutic effects of combined use of polyvinylpyridine-*N*-oxide and tetrandrine. *J. Labour Medicine*, **13**, 1–7.

10 Mechanisms of pneumoconiosis

Ken Donaldson

INTRODUCTION

In this article a sequence of mechanistic cellular events are suggested for the pathogenesis of pneumoconiosis based on data obtained from studies with the main pneumoconiotic particles – asbestos, silica and coal mine dust. There is without doubt some particle specificity in the responses of the lung, but the pathogenesis given below is intended to represent the response to an idealized 'pneumoconiotic particle' although, where specific examples are given, the nature of the particle will be given. A general sequence of postulated events leading to pneumoconiosis can be seen in Figure 10.1.

THE SITE OF DEPOSITION AND TARGET CELLS

The site of deposition of particles is important in defining the target cells that will be exposed. The ground-breaking studies of Brody and his co-workers have confirmed that the key site of deposition of both compact particles (Brody and Roe, 1983) and fibres (Brody *et al.*, 1981) is the terminal bronchiolar region and the proximal alveolar region, including the ridges of the alveolar duct bifurcations. In this region the terminal bronchiolar epithelium has a much less effective mucociliary clearance system than is found in the upper airways. Here, and on the alveolar ducts and proximal alveoli, the macrophages form the principal defensive cells and clearance is much slower than in the upper airways. In the rat, following inhalation, the deposited particles can be seen littering the surfaces of the airspaces and being phagocytosed by alveolar macrophages (Figure 10.2). It is in this area, highly focussed in the terminal bronchioles and alveolar ducts of rats (Chang *et al.*, 1988) and respiratory bronchioles and proximal alveoli of humans (Gibbs *et al.*, 1984), that the early lesions of pneumoconiosis arise. Using electron microscopy to assess morphometry of first alveolar duct bifurcations, Brody and co-workers described a progressive increase in interstitial volume, myofibroblasts and extracellular matrix 1 month after a single inhalation exposure to chrysotile asbestos (Chang *et al.*, 1988). The target cells that particles can interact with in this area are shown in Table 10.1.

Of the different potential target cells, research has focussed on the alveolar macrophage, the epithelial cell and to a lesser extent on the fibroblast. Little is known in experimental terms of the interstitial macrophage although, as discussed below, its response is probably central to pathogenic events and interstitial macrophages are found in significantly increased numbers in Brody's evolving alveolar duct lesions described above.

Table 10.1 Potential target cell types in the pathogenesis of pneumoconiosis

Terminal bronchioles	*Alveoli*
Alveolar macrophage	Alveolar macrophage
Ciliated cell	Type 1 Epithelial cell
Mucus-secreting cell	Type 2 Epithelial cell
Clara cell	Interstitial macrophage
Interstitial macrophage	Fibroblast
Fibroblast	

Occupational Lung Disease. Edited by Daniel E. Banks and John E. Parker. Published in 1998 by Chapman & Hall. ISBN 0 412 73630 6

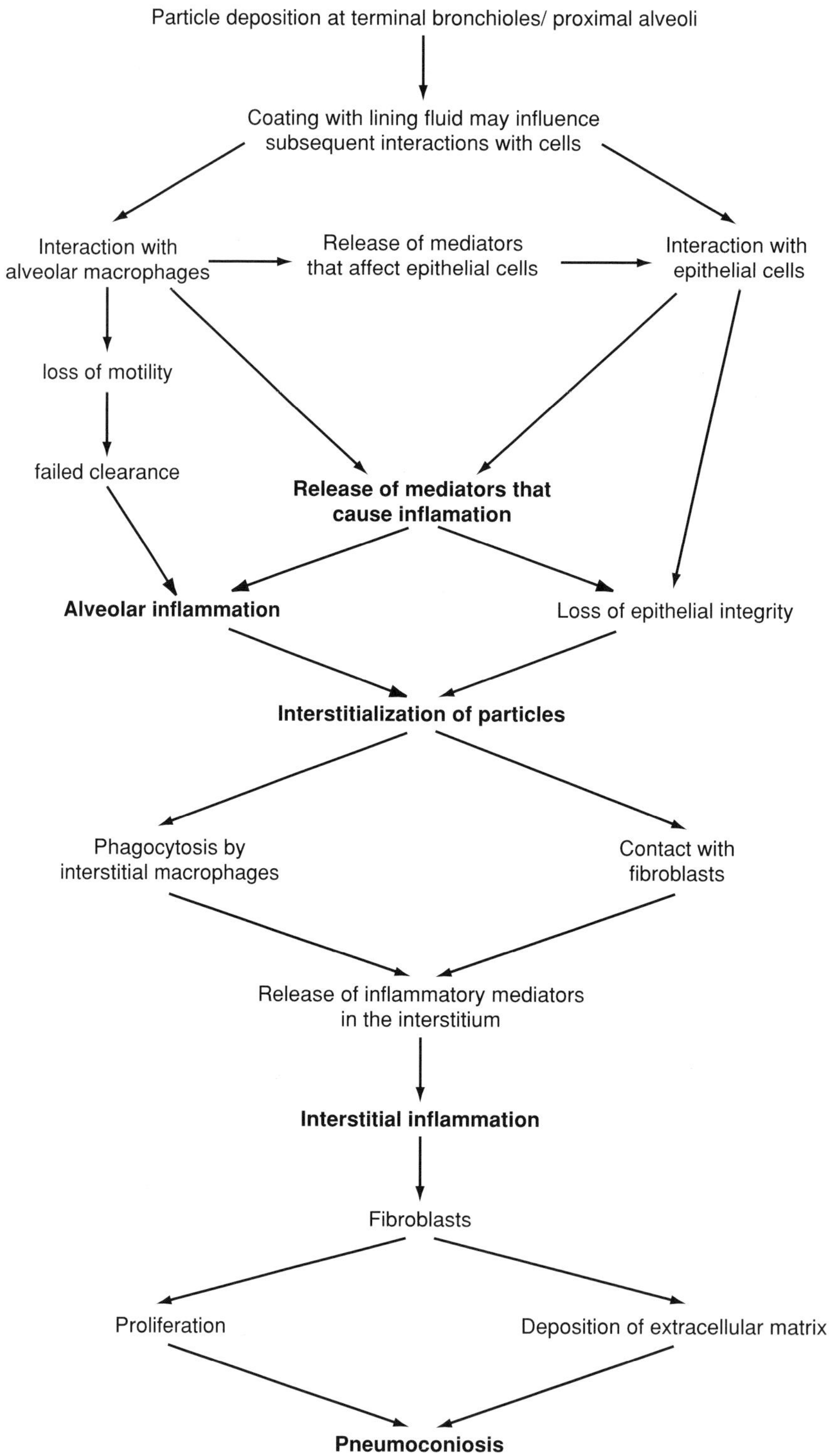

Figure 10.1 Postulated sequence of cellular events that lead to pneumoconiosis following the deposition of particles.

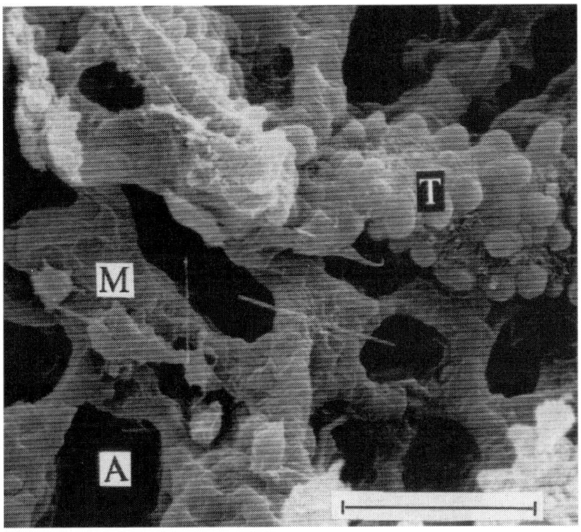

Figure 10.2 Scanning electron microscope image of the transition region between the terminal airway (T) and the alveolar duct (A) in a rat which has been exposed to amosite asbestos for 7 hours. Fibers can be seen free on the alveolar surface and phagocytosed by macrophages (M). Original magnification, ×700; scale bar = 40 μm.

ACTIVATION OF TARGET CELLS – A CENTRAL HYPOTHETICAL MECHANISM IN PNEUMOCONIOSIS

Early work with pneumoconiotic particles emphasized the ability of particles to kill cells and evoked the likely role of this cell death in the subsequent fibrogenic response. Developments in understanding pathogenic processes in general, and interestitial lung disease in particular, have highlighted the role of secretory products such as cytokines, oxidants and proteases. In the pneumoconioses therefore, it is now recognized that, at the relatively low ratios of particles to cells that arise in the occupational situation, activation of cells dominate the response to particles. There are a number of potential pathways that could result in cell activation, involving direct contact with particles, activation by mediators and a combination of these and the activated phenotype may differ depending on which route of activation is involved.

These types of activational responses are likely to vary in ways that are somewhat cell-specific but include classical intracellular signaling pathways that lead to Ca ion mobilization, generation of lipid mediators, activation of latent enzyme systems, exocytosis of preformed mediators, kinase activation, gene expression for inflammatory mediators and proliferation. Examples of the the cell-specific consequences of activation by particles are shown in Table 10.2.

An example of an important enzyme system that is likely to be activated by particles is the NADPH oxidase in macrophages which is responsible for initiating the respiratory burst (Bastian and Hibbs, 1994), as well as kinases (Jabbour *et al.*, 1996). Multiple intracellular signaling events have been reported in cells treated with asbestos (Sesko *et al.*, 1990; Faux *et al.*, 1995) including PKC activation (Donaldson *et al.*, 1992; Scheule and Holian, 1991). Activation of proto-oncogenes (Quinlan *et al.*, 1994) and subsequent gene expression have also been reported (Savici *et al.*, 1994; Rosenthal *et al.*, 1994) in response to various pneumoconiotic particles. Mechanisms whereby these event occur are discussed below. A dominant hypothesis for the mechanism of pneumoconiosis is therefore that the pneumoconiotic particles cause activation of target cell types.

Table 10.2 The effect of activation of the various target cell types

Cell type	*Consequence of cell activation in particle-exposed lung*	*Role in pneumoconiosis*
Macrophage	Assembly of NADPH oxidase and production of superoxide anion; expression of cytokine genes; eicosanoid generation	Oxidative stress; autocrine stimulation and activation of target cells; chemotaxis and other proinflammatory events
Epithelial cell	Expression of chemokine genes; proliferation	Inflammatory cell recruitment; hyperplasia
Fibroblasts	Expression of genes for ECM components; proliferation	Fibroplasia and accumulation of connective tissue

ECM, extracellular matrix.

POSSIBLE INVOLVEMENT OF THE CLASSIC IMMUNE SYSTEM IN THE MECHANISM OF PNEUMOCONIOSIS

Evidence of immune system activation has been reported in workers exposed to asbestos and quartz (reviewed in Scheule and Holian, 1991). These could arise from non-specific adjuvant effects on alveolar macrophage cytokine release and the overlapping roles of these mediators in immunity and inflammation (Figure 10.3). However, it is also possible that host proteins that adsorb onto particle surfaces are 'denatured' in some way and become antigenic, resulting in autoimmunity. However, there is no convincing evidence that this is the case, and more research is needed in this area.

INTERACTION OF PARTICLES WITH ELEMENTS OF THE LUNG LINING FLUID

Before making contact with cells, deposited particles interact with lung lining fluid. In the key site of deposition, the terminal bronchioles and proximal alveoli, this fluid largely comprises the secretions of the Type 2 alveolar epithelial cells and transudate from plasma. The interaction of pneumoconiotic particles with lung lining fluid has not received the attention that it deserves and the science of biomaterials has much to offer in aiding an understanding of this key area (see for instance Kasema and Lausmaa, 1988). Biomaterials scientists have long considered, and indeed defined, the factors dictating the interaction between a non-biological surface and its interface with tissue (Kasemo and Lausmaa, 1988).

This lining fluid in the centri-acinar region comprises phospholipids and proteins of various types (Haagsman and van Golde, 1985). Immunoglobulin G (IgG) is a normal component that is increased during inflammation (discussed in Donaldson *et al.*, 1992). Coating of particles with IgG has been shown to enhance the ability of macrophages to release cytokine and oxidant in response to asbestos (Donaldson *et al.*, 1992; Hill *et al.*, 1995) and to generate an oxidative burst in response to quartz (Nyberg and Klockars, 1990). This is no doubt due to a classical opsonizing effect in coating the fibers with a molecule, the Fc portion of the IgG molecule, for which macrophages have a receptor. The role that other elements of the lung lining fluid might have on the response to particles is in need of investigation and such effects might include enhanced stimulation of cells, as described above, or protection, e.g. by masking of free radical sites or diminution of the availability of iron (see below) (Wallace *et al.*, 1988) There could be competition between different molecules for binding sites on the particle surface and the net opsonizing outcome might be a complex one that depends on the state of the lung lining fluid and the nature of the particle. Consequently, the degree of inflammation at the time of deposition of a particle may have a profound effect on the biological activity of the particle. More research is required in this key area.

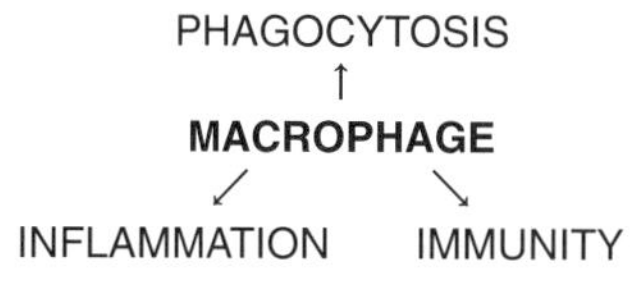

Figure 10.3 Multiple overlapping roles of the alveolar macrophage.

RECEPTORS INVOLVED IN THE INTERACTION BETWEEN PNEUMOCONIOTIC PARTICLES AND CELLS

Relatively little is known about the specific molecules on the surface of target cells with which pneumoconiotic particles interact. Where particles are coated with well-known opsonins, then it is apparent which receptors are involved, e.g. coating particles with IgG undoubtedly enhances the activation of macrophages via the Fc receptor (Hill *et al.*, 1995). In addition, the macrophage scavenger receptor has been shown to be capable of binding to crocidolite asbestos (Resnick *et al.*, 1993) and so may be important in mediating its internalization. Since the response of macrophages differs greatly depending on which type of receptor is engaged, then it is important to determine the receptors involved in phagocytosis of pneumoconiotic particles.

Virtually nothing is known of the specific receptors that are involved in the binding of pneumoconiotic particles to epithelial cells or fibroblasts.

THE KEY ROLE OF THE MACROPHAGE IN THE RESPONSE OF THE LUNG TO PNEUMOCONIOTIC PARTICLES

Because of the central role of the alveolar macrophage in the defense against particles in the lung parenchyma, and its other roles in defense (Figure 10.3), a great deal of research has been directed towards the responses of macrophages following phagocytosis of pneumoconiotic particles.

The macrophage has these multiple effects because of its specialized function and multiple secretory products. In evolutionary terms, the macrophage has been 'designed' to kill bacteria or viruses, digest them to antigenic fragments, and then 'present' the antigen to the immune system. In the lung, however, the normal day-to-day killing of the small numbers of bacteria that deposit is carried out by macrophages without stimulating inflammation or immunity, and indeed normal alveolar macrophages are immunosuppressive (Kusaka *et al.*, 1990). The normal alveolar macrophage is suppressive since stimulation of inflammation or immunity under these 'normal' circumstances would be hazardous to the delicate gas exchange alveolar/capillary membrane because of the accompanying fluid and cellular exudate. It is only in the case of overwhelming infection, e.g. pneumonia, exacerbations of COPD, that inflammation/immunity does occur in the lung parenchyma, and both conditions are life-threatening.

In the case of long-term exposure to airborne particles, however, as occurs in pneumoconiosis, the constant phagocytic load results in the chronic activation of the macrophages to release mediators at low levels, resulting in chronic inflammation.

Attraction of macrophages to the site of particle deposition

Studies in the lungs of rats inhaling chrysotile asbestos demonstrated that deposition of the fibres at the alveolar duct bifurcations was associated with activation of complement in the lung lining fluid, with generation of C5a, and that this was responsible for attracting macrophages (Warheit *et al.*, 1986). This does not, however, seem to be a universal mechanism for attracting macrophages to deposited particles to allow efficient phagocytosis, since complement does not appear to be involved in the leukocyte response to silica *in vivo* (Warheit *et al.*, 1991).

(a) Alveolar macrophage-derived inflammatory mediators

Because of the large secretory repertoire of the alveolar macrophage, including many proinflammatory mediators, which are generally secreted during phagocytosis, a role for these mediators in pneumoconiosis has frequently been suggested. The most likely candidates among the macrophage-derived mediators that could lead to inflammation and pathologic change in pneumoconiosis are oxidants, proteases, arachidonic acid metabolites and cytokines. While these are an absolute requirement for normal defense and lung tissue balance, their release in exaggerated amounts by macrophages chronically stimulated by pneumoconiotic particles can be seen to be potentially harmful (Table 10.3).

Table 10.3 Role of macrophage-derived mediators in the normal defense functions of macrophages and how these functions could become subverted to a pathogenic role in pneumoconiosis

Mediator	*Role in normal defense*	*Role in pneumoconiosis*
Oxidants	Killing of microbes	Tissue damage, oxidant stress
Proteases	Killing and digestion of microbes	Tissue damage and remodeling
Arachidonic acid metabolites	Mobilizing the inflammatory response	Chemotaxis and activation of inflammatory leukocytes
Cytokines	Modulating the normal function of lung cells and leukocytes to maintain tissue balance	Promoting inflammation and stimulating the growth of epithelial cells and fibroblasts

(b) Transfer of particles to the interstitium and the role of the interstitial macrophages

The alveolar macrophage is clearly the pivotal cell in the preliminary response of the lung to particles deposited on the terminal bronchiolar/proximal alveolar surface of the lung. However, interstitialization of particles can be seen as a pivotal event in determining the outcome of such a deposition. If the particle is phagocytosed by macrophages and is then successfully transported to the mucociliary escalator for removal from the lung, then this can be seen as a successful outcome in terms of lung defense. Particles may also remain in macrophages which accumulate in the airspaces. Macrophage activation and secretion that accompany phagocytosis by alveolar macrophages in the alveolar space are most likely to influence the adjacent macrophages in the alveolar spaces and also the lining epithelial cells. It seems unlikely that release of mediators in the alveolar space could have profound effects in the interstitium, given the tightness of the epithelial barrier. However, this presupposes that the integrity of the epithelial barrier is not compromised and there is evidence, e.g. from the accumulation of protein in the alveolar space following particle deposition (Donaldson and Brown, 1993), that this is not the case and that loss of epithelial integrity is an early event following the deposition of particles in the lung. It is possible, therefore that alveolar macrophage secretions could have a direct effect in the interstitium if the epithelial integrity is breached.

If the particles do cross the epithelium and enter the lung interstitium, then this can be seen as a highly adverse event in the sense that the particles will not be cleared from the lung, becoming part of the retained dose (unless there is dissolution of the particles) and their fate is now likely to be phagocytosis by interstitial macrophages (IMs). This will result in either transport to the mediastinal lymph nodes, inside macrophages, or chronic release of inflammatory mediators in the interstitium. The close proximity of the IMs to the fibroblasts would suggest intuitively that the presence of particles in IMs in long-term residence in the interstitium, is likely to be the most adverse scenario leading to pneumoconiosis. In instillation studies, Bowden *et al.* (1989) demonstrated that, in mice whose alveolar space defense was compromised by whole-body irradiation, there was increased interstitialization of quartz with an accompanying increase in fibrosis. Following inhalation exposure, Velan *et al.* (1993) reported dose responsive differences in the quality and quantity of the response to silica. In particular they noted that while low dose exposure to fine silica caused no interstitialization or fibrosis, high-dose exposure was accompanied by interstitialization of the silica particles and severe, progressive fibrosis.

THE ROLE OF NEUTROPHILS

There is overwhelming evidence in occupationally exposed humans and in experimental animals that the inhalation of pneumoconiotic dusts causes neutrophilic inflammation as assessed by bronchoalveolar lavage (Begin, 1986; Donaldson *et al.*, 1988b, 1990a). Therefore neutrophils are present in the bronchoalveolar spaces during the development of pneumoconiosis. The role that these cells play in the mechanism of pneumoconiosis is not well established although in rodent models, at least, there has been a suggestion that neutrophils contribute to the acute toxicity and oxidant stress accompanying an instillation of quartz (Henderson *et al.*, 1991). However, in another study (Gavett *et al.*, 1992), the biochemical indicators of lung injury measured in lavage following high-dose inhalation exposure to quartz, were not affected by neutrophil depletion. In one well-documented model of pneumoconiosis following short-term, high-dose inhalation of asbestos (Chang *et al.*, 1988) which is characterized by the early appearance of interstitial hyperproliferative lesions, and activated alveolar macrophages showing increased cytokine transcription, there is no recruitment of neutrophils detectable in the bronchoalveolar lavage. In the study by Piguet *et al.* (1993), antibody against the CD11 integrins inhibited quartz-mediated fibrosis in a mouse model, as measured by lung hydroxyproline, but did not ameliorate the recruitment of neutrophils to the lung. It would appear, therefore that neutrophils may be attracted to pneumoconiotic lung but that the principal events, *viz.* macrophage, epithelial and fibroblast activation with accumulation of connective tissue cells and their products, occurs independently of neutrophils.

LEUKOCYTE-DERIVED OXIDANTS

Leukocyte-derived oxidants are likely to be important in the pathogenesis of pneumoconiosis but are not the only source of oxidants. The particles themselves are an important source of oxidants via free radicals generated at the surface. This source of oxidative stress is discussed in more detail later.

The respiratory burst of the phagocyte is a complex process involving the active assembly of the membrane-associated NADPH oxidase which carries out a one-electron reduction of molecular oxygen to form superoxide anion. In a series of subsequent reactions, H_2O_2 and hydroxyl radical are formed (Figure 10.4). These reactive oxygen species (ROS) are all oxidants and hydroxyl radical is the most common pathogenic free radical formed in biological systems and is highly toxic. The production of all of these oxidants by macrophages exposed to pneumoconiotic particles has been described (Donaldson *et al.*, 1988a; Castranova *et al.*, 1996). BAL cells from patients with coal workers' pneumoconiosis (CWP) release more superoxide anion *in vitro* than control BAL (Wallaert *et al.*, 1987). Recently, the generation of the multi-functional free radical NO· has been reported in quartz (Blackford *et al.*, 1994) and asbestos (Thomas *et al.*, 1994) exposed macrophages by the reaction shown in Figure 10.4 involving NO synthase. NO can interact with superoxide anion to produce the peroxynitrite radical, which is also highly reactive (Gutteridge and Halliwell, 1994).

Mossman *et al.* (1990) produced compelling evidence that oxidative stress was important in asbestos fibrosis by intervening with the anti-oxidant enzyme catalase and ameliorating the fibrogenicity and inflammogenicity of inhaled asbestos. In these studies rats were fitted subcutaneously with mini-pumps delivering a 'long-life' version of catalase and superoxide dismutase (SOD) made by conjugating the enzymes with

$$O_2 \xrightarrow{\textbf{Membrane NADPH oxidase}} O_2^{-\bullet} \text{ (superoxide anion)}$$

$$2O_2^{-\bullet} + 2H^+ \longrightarrow H_2O_2 \text{ (hydrogen peroxide)} + O_2$$

$$\text{L-arginine} \xrightarrow{\textbf{NO synthase}} \text{L-citrulline} + NO^\bullet \text{ (nitric oxide)}$$

$$O_2^{-\bullet} + NO^\bullet \longrightarrow ONOO^- \text{ (peroxynitrite)}$$

$$ONOO^- + H^+ \longrightarrow OH^\bullet \text{ (hydroxyl radical)} + NO_2^\bullet$$

Figure 10.4 Phagocytic generation of oxidants.

Table 10.4 Inflammatory parameters and fibrosis (hydroxyproline) in bronchoalveolar lavage and lung tissue of asbestos-exposed rats concomitantly treated with osmotic pump-delivered PBS (control) or polyethylene glycol-conjugated catalase

	Asbestos inhalation + PBS	*Asbestos inhalation + PEG-catalase*
Total cells ($\times 10^6$)	9.1 (0.4)	7.3 (1.2)**
PMN (%)	18.8 (1.2)	9.3 (0.3)**
Protein (mg/ml)	418 (86)	338 (25)
Hydroxyproline (μg/right lung)	1275 (129)	999 (78)*

Significant difference between treatments, * $P<0.005$;** $P<0.001$.
PBS, phosphate–buffered saline; PEG, polyethylene glycol; PMN, polymorphonucleocytes.

Reproduced from Mossman, B.T. *et al.* (1990) *Am. Rev. Resp. Dis.*, **141**, 1266–71, with the permission of the publishers.

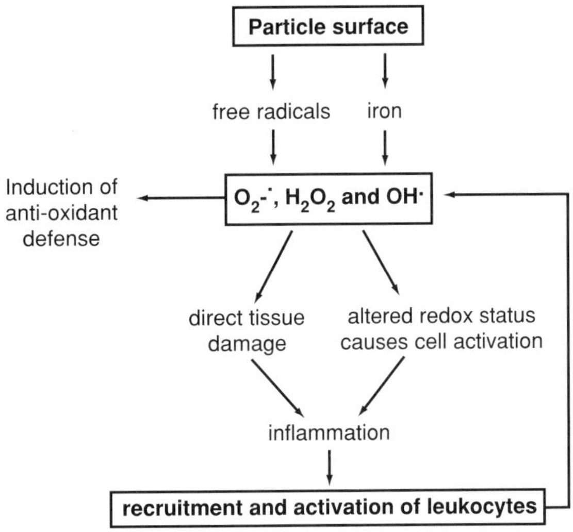

Figure 10.5 Sumary of the likely roles of free radicals in the mechanism of pneumoconiosis.

polyethylene glycol (PEG): PEG-catalase and PEG-SOD. PEG-SOD had no effect. The highest dose of PEG-catalase (8000 IU/day) however, inhibited the increase in lung hydroxyproline, a measure of collagen, seen in the asbestos-exposed group and also lowered total number of macrophages and neutrophils in lavage fluid (Table 10.4).

These data strongly support a role for hydrogen peroxide in mediating and enhancing pneumoconiosis in a rat model. Because of the potential that H_2O_2 has for generating more free radicals by interacting with particle-derived iron (see below) these data do not necessarily mean that H_2O_2 itself is necessarily the most important oxidant.

The effects of oxidants are likely to be at least two-fold, direct cellular injury or activation of cells and these could be mediated by both particle surface-mediated and leukocyte-derived oxidants, as shown in Figure 10.5.

- Oxidative damage: these toxic and proinflammatory effects are described in Table 10.5.
- Activation of cells: oxidative stress may result in the activation of gene transcription via oxidative stress-sensitive transcription factors as described in detail later.

(a) Anti-oxidant defenses

Evidence of oxidative stress in the lungs of rats inhaling pneumoconiotic particles has been shown in the study of Janssen *et al.*, (1994a). Increased expression of the gene for the antioxidant defense enzyme Mn-SOD was demonstrated in whole lung after inhalation of silica using both Northern analysis of mRNA and Western analysis of the enzyme protein itself. Increased mRNA for glutathione peroxidase has also been detected in particle-exposed rat lungs (Janssen *et al.*, 1992), further suggesting a need for the particle-exposed lung to deal with increased peroxides.

The important role of non-enzymatic anti-oxidant defense has been demonstrated in the study of Lombard-Gillooly and Hubbard (1993). In this study mice were treated with buthionine sulfoximine, an inhibitor of GSH synthesis, and the inflammatory response to instilled silica was compared with that in control mice. When the GSH levels were depleted, the inflammatory responses to silica, as detected by the BAL cell profile, were markedly enhanced. This strongly suggests a role for GSH in the normal defense of the lung against silica.

Table 10.5 The effects of ROS (Reactive Oxygen Species) on biological molecules

Biological molecule	*Effect of oxidant*
Lipid	Lipid peroxidation
Protein	Cross-linking by disulfide formation; protein strand scission, enzyme inactivation; receptor modification
DNA	DNA adduct formation

Taken together, these and other studies provide evidence that particles have the potential to cause stimulation of phagocytes to release oxidants and cause oxidative stress following deposition of pneumoconiotic dust. However, it is not yet clear whether the important oxidative stress arises from the particles themselves, from the inflammatory response to the particles or, as seems quite likely, from some combination of the two (see Figure 10.5). Additionally, the relative importance of the two different effects of oxidative stress, direct oxidative damage leading to inflammation and oxidative stress-stimulated cell activation, in the causation of pneumoconiosis are unknown. Intuitively it might be anticipated that, at the high doses most often seen in experimental pneumoconiosis, the direct particle-mediated effects might predominate, but at low levels of exposure the cell activation by particles may be most important.

OXIDATIVE STRESS AND THE EPITHELIUM

The maintenance of an epithelial barrier is central to normal lung function, and loss of epithelial integrity could be an important factor in precipitating the early events that lead to pneumoconiosis. Work by the author and collaborators has demonstrated the importance of oxidative stress in leading to loss of epithelial integrity. Depletion of intracellular GSH causes loss of epithelial integrity *in vitro* and *in vivo* (Li *et al.*, 1994, 1995). Therefore, oxidative stress in the lungs following deposition of particles, either as a direct effect of free radical generation at the particle surface or indirectly from inflammatory leukocytes, could lead to loss of epithelial integrity. In addition, macrophages phagocytosing particles on the epithelial surface and releasing cytokines (see below) could add to this oxidative stress, since cytokines also deplete GSH in epithelial cells (Rahman *et al.*, 1995) and directly cause epithelial permeability (Li *et al.*, 1995).

THE IMPORTANCE OF PARTICLE SIZE, SHAPE AND CHEMISTRY

(a) Size

One standard approach to the study of pneumoconiosis has been to compare the biological effects of particles that cause pneumoconiosis with particles that do not. A particle that is commonly used in such studies is titanium dioxide since it is not associated with pneumoconiosis. It is notable however, that TiO_2 that is in the form of 20-nm diameter particles is highly pathogenic (Ferin *et al.*, 1992) compared with the normal TiO_2 particle which is 0.2 μm in diameter. This has given rise to the proposition that very fine particles are toxic, even for materials that in the form of larger particles are not toxic to the lung. The mechanism that underlies the toxicity of ultrafine particles remains obscure, but may be related to the ability of very fine particles to gain access to the interstitium (Ferin *et al.*, 1991), where interstitial inflammation may result. This in turn may be a result of overwhelming of the normal phagocytic defenses at the alveolar surface by the huge number of particles and their accompanying large surface area. Failure to effectively remove this particle burden, by phagocytosis, could allow enhanced interaction between particles and the epithelium. Such a prolonged interaction between the particles and the epithelial cells may be permissive for interstitialization. The likely importance of interstitial inflammation for the development of fibrosis is discussed elsewhere in this review.

Donaldson *et al.* (1995) also demonstrated that the ultrafine TiO_2 and ultrafine carbon (not published) has greater surface free radical activity than the larger-sized TiO_2 or carbon, and oxidant stress may be a factor in leading to loss of epithelial integrity and interstitial access of ultra-fine particles. While pneumoconiotic particles are not generally considered to be ultrafine, the evidence showing that ultrafine particles have high biological activity means that size should be an important factor to consider in dissecting the toxicity of any particle sample. In addition, the ultrafine particle experience may help to dissect the mechanism of pneumoconiosis, particularly in highlighting the importance of the epithelial barrier and interstitialization.

(b) Shape

For asbestosis and also the fibrosis caused by man-made fibers in experimental animals, the fibrous shape is likely to be very important because of the impact that incomplete phagocytosis could have on the macrophage. The presence, within the macrophage, of a fiber that is larger than the macrophage (Figure 10.6) will result in incomplete phagocytosis and very likely result in impairment of the ability to migrate to clear the fiber. In support of this contention, there is abundant evidence that long fibers can cause more inflammation (Donaldson *et al.*, 1989) and release of macrophage oxidants (Hill *et al.*, 1995) and cytokines (Donaldson *et al.*, 1992) than short fibers. However, many short-fiber samples are prepared by milling a long-sample fiber and this may result in a change in the surface activity of the fibers (Hill *et al.*, 1995) and hence the ability to generate free radicals at the fiber surface (Gilmour *et al.*, 1995). Therefore it is not always simple to separate the effects of fiber length from those of surface chemistry.

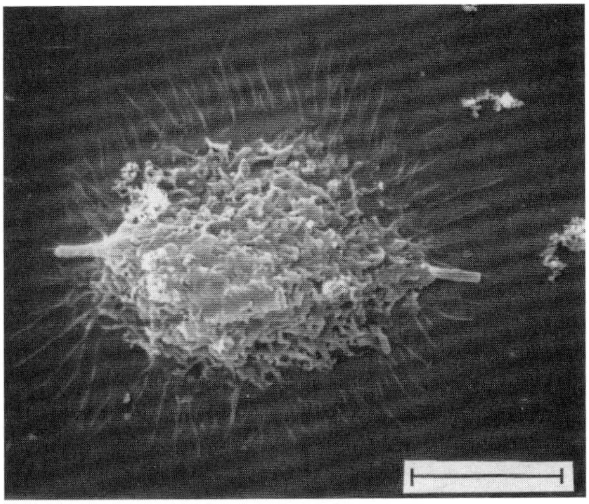

Figure 10.6 Alveolar macrophage from a rat inhaling amosite asbestos for 7 hours. The macrophage has phagocytosed a fiber that is protruding from each end of the cell. Original magnification, ×4000; scale bar = 7.5 μm.

(c) Particle surface chemistry

Compelling evidence for the role of the particle surface in an animal model of silicosis was provided by the demonstration of Brown *et al.* (1990) that incubating quartz briefly in aluminium lactate, a soluble aluminium salt, dramatically attenuated the ability of 1 mg of instilled quartz to cause sustained inflammation in the rat lung (Figure 10.7). This simple demonstration of the importance of the quartz surface in its pathogenicity raises questions as to how the surface chemistry of pneumoconiotic particles cause cell activation or gene expression, in line with the hypotheses raised above.

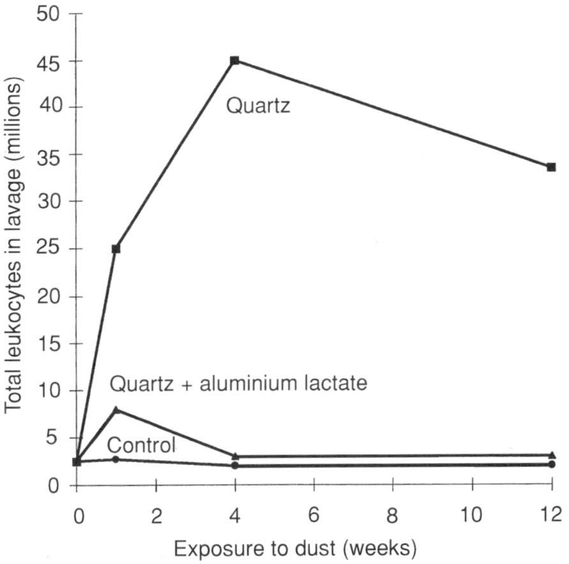

Figure 10.7 Inflammatory response in the bronchoalveolar lavage of rats instilled with 1 mg of quartz or 1 mg of quartz that has been pre-treated with aluminium lactate. (Reproduced with permission from Dr Geraldine Brown.)

FREE RADICAL ACTIVITY AT THE SURFACE OF PNEUMOCONIOTIC PARTICLES: A UNIFYING MECHANISM IN CAUSING CELL ACTIVATION?

In trying to find a common factor for the ability of particles to cause pneumoconiosis, based on the surface characteristics of the particles, the free radical activity is one that immediately presents itself. There is a large body of evidence demonstrating free radicals at the surface of different pneumoconiotic particles (Table 10.6).

We have recently provided data on: (i) long and short amosite asbestos; and (ii) normal and ultrafine TiO_2. These pairs of particles are important because long amosite causes fibrosis and short amosite does not (Davis *et al.*, 1986), while ultrafine TiO_2 is fibrogenic and normal TiO_2 is non-fibrogenic (Ferin *et al.*, 1992). When these pairs of particles were compared, there was a good correlation between ability to cause hydroxyl radical-mediated damage to naked plasmid DNA and fibrogenic activity (Donaldson *et al.*, 1995; Gilmour *et al.*, 1995).

The effects of surface free radicals could be severalfold. If the levels of free radicals are high enough then direct oxidation of cellular macromolecules, lipids, proteins and nucleic acids could result (Table 10.5).

In addition to causing oxidative damage to cells, free radicals can also cause more subtle effects on cells that can culminate in gene expression. The genes that have been studied in this regard are those that are proinflammatory (e.g. cytokine genes) or proliferative (proto-oncogenes). Gene expression is mediated by transcription factors, proteins that bind to DNA in the

Table 10.6 Reports describing free radical activity on the surface of particles that can cause fibrosing lung disease

Particle	*Reference*
Quartz	Vallyathan *et al.* (1988)
Asbestos	Gilmour *et al.* (1995)
Coal mine dust	Dalal *et al.* (1989)
Ultrafine TiO_2	Donaldson *et al.* (1995)
Man-made fibers	Gilmour *et al.* (1995)

promoter region of a gene and initiate its transcription by providing the correct conditions for RNA polymerase to function. The activation of some transcription factors, particularly those that are involved in immune responses and in proliferation, can be activated in reponse to oxidative stress. For example the transcription factor NF-κB is important in the activation of transcription of many cytokine genes, as well as other proinflammatory genes such as those for adhesion molecules (Lee, 1994). NF-κB is present in an inactive form in the cytoplasm, bound to its inhibitor IκB. When the cell comes under oxidative stress, and particularly when there is oxidation of GSH to GSSG (Droge *et al.*, 1994), NF-κB separates from IκB (Brown *et al.*, 1995) and translocates to the nucleus to bind to the NF-κB motif in the promoter region. Binding of the transcription factor then leads to transcription. The c-fos, c-jun dimerized transcription factor AP-1, which plays a role in the transcription of genes involved in cell division and other key functions (Angel and Karin 1991) is also under redox control (Meyer *et al.*, 1994). Direct activation of these transcription factors may be anticipated to occur in the case of particles with free radical activity, by the GSH-GSSG pathway shown in Figure 10.8. Driscoll has described the activation of NF-κB by a range of pneumoconiotic particles (Driscoll *et al.*, 1995) and Janssen *et al.* (1995) have demonstrated that addition of thiols can inhibit the activation of c-fos and c-jun by asbestos in cultured mesothelial cells, presumably by preserving the GSF/GSSG balance.

In addition, pneumoconiotic particles such as silica and asbestos have been found to activate multiple signaling pathways in cells leading to hydrolysis of inositol phospholipids (Sesko *et al.*, 1990) increase in intracellular Ca^{2+} (Faux *et al.*, 1994; Jabbour *et al.*, 1996), activation of protein kinase C (Perderiset *et al.*, 1991; Donaldson *et al.*, 1992), activation of MAP kinase (Jabbour *et al.*, 1996), and c-fos/c-jun (Janssen *et al.*, 1994b) (Figure 10.8). Parts of the mechanism

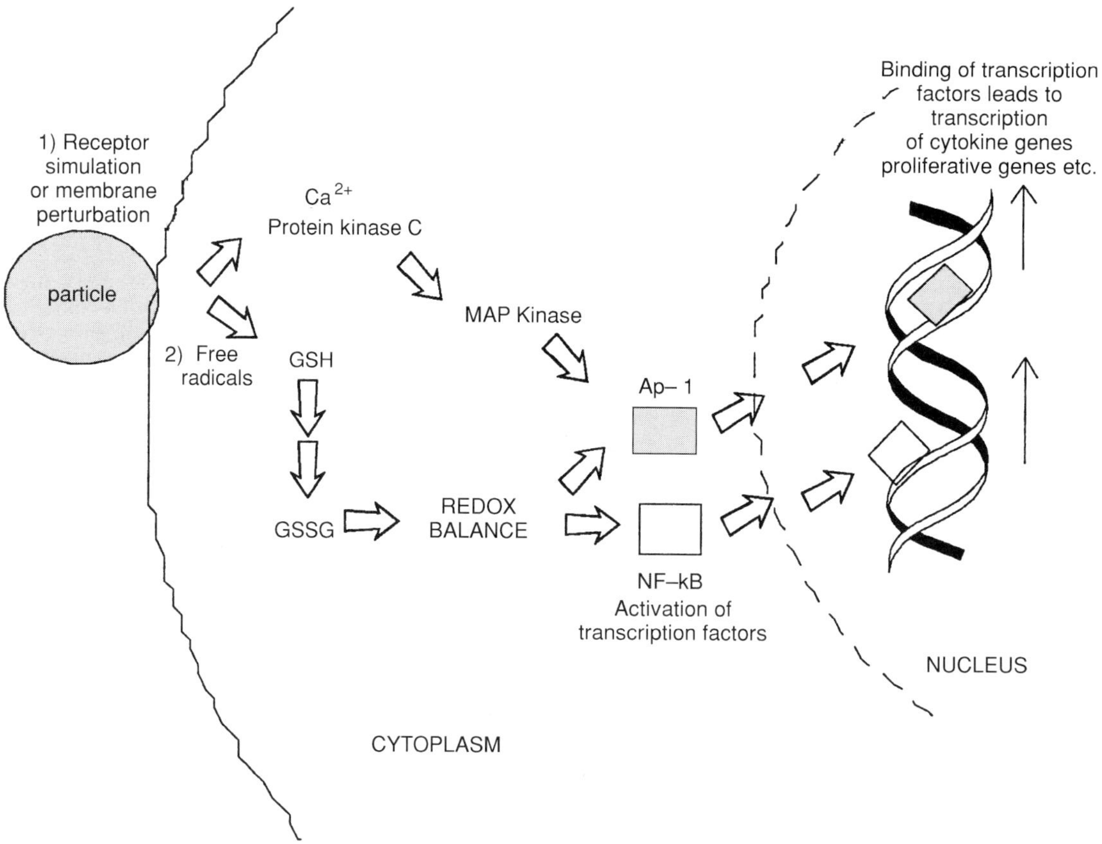

Figure 10.8 Potential pathways of intracellular signaling in cells making contact with pneumoconiotic particles. Pathway 1 is the result of membrane perturbation while pathway 2 is a result of particle surface-derived free radicals. Both pathways could be activated by a particle, with the added involvement of local inflammatory leukocyte-derived ROS.

explaining how contact between pneumoconiotic particles and cells can lead to activation and gene transcription is now known. However, neither the specific receptors nor the exact pathways stimulated by specific particles leading to the transcription of specific genes is clearly established.

The importance of particle surface iron chemistry was demonstrated in the study of Simeonova and Luster (1995) when chelation of iron at the surface of asbestos completely abolished the ablity of the asbestos fibers to stimulate TNF release from rat alveolar macrophages. These workers also demonstrated that an oxidative stress could substitute for asbestos in causing the stimulation of TNF release. This dramatic demonstration of how the chemistry, and not the shape, of asbestos dictates a key biologic activity, points the way to future research on the role of the particle surface in pneumoconiosis.

THE ROLE OF PARTICLE-DERIVED IRON

The relatively high iron content of some pneumoconiotic dusts such as asbestos and coal mine dust and trace levels in quartz (Castranova *et al.*, 1996) raises the possibility that iron may be important in particle bioactivity. Although this has most often been discussed in relation to asbestos (e.g. Kamp *et al.*, 1992) the same arguments can hold true for other pneumoconiotic dusts. Iron, as seen for other transition metals, has the ability to generate free radicals via Fenton chemistry that is well characterized. The state of the iron is all important, in particular the amount of Fe(II) is central since this is the directly harmful species. Consequently, total iron is not necessarily informative as to the biologically active iron. The iron must redox-cycle to be capable of causing major injury to macromolecules and this is accomplished by a reductant in the region of the fiber, e.g. GSH, ascorbate, NADH or or even superoxide anion (Figure 10.9). This means that the presence of antioxidants in the inflammatory milieu is a double-edged sword (see above).

By the sequence of events shown in Figure 10.9 the highly toxic hydroxyl radical can be formed. The hydroxyl radical may be involved in diffusion-limited reactions which lead to the formation of various carbon-centered radicals, peroxyl, alkoxyl and thiyl radicals, all of which have harmful consequences for cells. Because different reductants could have different potencies in causing reduction of Fe(III), and because of the known role of chelating agents, the microenvironment of the lung where the fiber is present could be all-important in determining how much reactive iron is present at the fiber surface. In addition, the particle may accumulate biologic iron which can also have free radical-generating activity (Lund *et al.*, 1994).

LEUKOCYTE-DERIVED PROTEASES

The dramatic remodeling of the lung parenchyma that characterizes pneumoconiosis is likely to be accompanied by proteolysis of structural proteins in the parenchyma. In addition, the focal emphysema

Fe(III) (from particles) + reductant species (e.g. GSH, ascorbate, superoxide anion)

↓

Fe(II) + oxidized species

FE(II) (from particles or from the above reaction) + O_2

↓

Fe(III) + $O_2^{-\bullet}$ (superoxide anion)

$2O_2^{-\bullet}$ (from the above reaction or from macrophages) + $2H^+$

↓

O_2 + H_2O_2 (hydrogen peroxide)

Fe(II) + H_2O_2 (from macrophages or from the above reaction)

↓

Fe(III) + OH^- + $OH^\bullet$ (hydroxyl radical)

Figure 10.9 The generation of harmful free radicals by reactions between particle-derived iron and elements of the lung milieu. Note the redox cycling of iron.

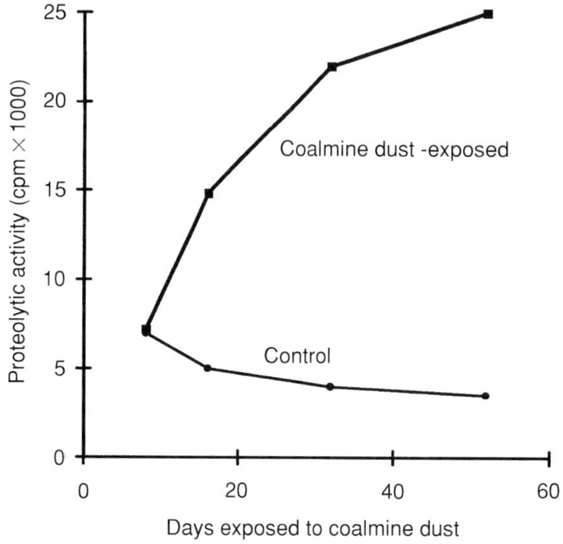

Figure 10.10 Proteolytic activity of the BAL cells from the lungs of rats inhaling coal mine dust (10 mg/m^3). (Reproduced with permission from Dr Geraldine Brown.)

that accompanies CWP can be considered to be caused in part at least by increased proteolytic activity of the inflammatory leukocytes recruited to the sites of dust deposition. There has been relatively little research into this area but Brown and Donaldson (1988) used a rat model to investigate the ability of bronchoalveolar leukocytes from control and quartz-inflamed lung to cause degradation of connective tissue components including fibronectin, laminin and collagen. The control cells (macrophages alone) were able to cause some degradation of connective tissue components but this was greatly increased by inflammatory cells (macrophages and neutrophils) obtained from the bronchoalveolar spaces after quartz deposition. Further studies showed the sustained ability of quartz to cause increased proteolytic activity in BAL cells after a single injection (Brown *et al.*, 1991) and during prolonged inhalation of coal mine dust as shown in Figure 10.10 (Brown and Donaldson, 1990). Rom (1990) has demonstrated increased proteolytic activity in inflammatory BAL cells from patients with CWP.

CYTOKINES

The cytokines are a group of pleiotropic effector proteins which have proinflammatory, mitogenic and immunostimulatory effects. Cytokines are released by macrophages, epithelial cells, lymphocytes, fibroblasts

Table 10.7 Cytokines implicated in pneumoconiosis

Particle	*Cytokine*	*Reference*
Quartz	TNFα, IL-1	Driscoll *et al.* (1990), Piguet *et al.* (1990), Schmidt *et al.* (1984)
	MIP-1α, MIP 2	Driscoll (1994)
Asbestos	TNFα, PDGF	Donaldson *et al.* (1992), Baumann *et al.* (1990)
Coal mine dust	TNFα	Borm *et al.* (1988)

TNF, tumor necrosis factor; IL-1, interleukin-1; MIP, macrophage inflammatory protein; IFN, interferon; PDGF, platelet-derived growth factor.

and other cells. A number of cytokines have been implicated in the pneumoconioses and in fibrosing lung disease caused by inhaled agents as shown in Table 10.7.

The implied role of cytokines in pneumoconiosis relates to:

- the demonstration that cytokine release from alveolar macrophages can be stimulated by particles; and
- the ability of cytokines such as TNFα to cause up-regulation of adhesion receptors on endothelial cells and leukocytes, stimulate chemokine and cytokine expression in target cells, activate and prime leukocytes for arachidonic acid metabolism and the respiratory burst and cause proliferation of epithelial cells, fibroblasts and endothelial cells (Driscoll, 1996).

Cytokine gene expression

In unstimulated cells, cytokine gene transcription is at a low level and it is only following cellular activation that proinflammatory levels of these proteins are produced following stimulation of gene transcription and translation. Because of the pleiotropic effects of cytokines in mediating proinflammatory and mesenchymal-activating events (Driscoll, 1996) the ability of particles to cause cytokine gene expression in appropriate target cells can be seen as a defining event in the pneumoconiotic process. In particular TNFα has been shown to have many effects that could contribute to pneumoconiosis (discussed in Driscoll *et al.*, 1996) and so has been the focus of attention in many studies of pneumoconiosis (see Table 10.7).

Early evidence for the role of TNFα in an animal model of silicosis came from the study of Piguet *et al.* (1990) who reported increased mRNA for TNFα, using Northern analysis, 3 and 15 days after instillation of silica into mouse lung. Interestingly, in this model neither TGFβ nor IL-1 mRNA were increased, suggesting that there could be a narrow spectrum of important cytokine responses involved in pneumoconiotic processes. In a companion experiment, these workers inserted osmotic mini-pumps in the mice and delivered either normal immunoglobulin or rabbit anti-TNFα to groups of mice that were instilled with quartz.

As shown in Figure 10.11, the anti-TNFα prevented the quartz-mediated fibrosis as shown by the fact that hydroxyproline levels were the same in the quartz + anti-TNFα group as in the control group. As can be seen in panel (b) of Figure 10.11, simply instilling recombinant TNFα (rTNFα) into normal mice did not cause lung fibrosis, but giving quartz + rTNFα was a super-stimulus to fibrosis (panel (c) of Figure 10.11). The latter finding suggests that the evolving quartz lesion provides an environment where infused TNFα can cause stimulatory effects, presumably on populations of primed macrophages, epithelial cells and fibroblasts. Demonstration of the activation of the TNF promoter region by silica in a myelomonocytic cell-line was provided by Savici *et al.* (1994) who transfected the THP1 cells with a plasmid construct containing a fragment of the TNFα 5′-untranslated promoter region upstream of a chloramphenicol transferase (CAT) gene. In such a system, the transcription of the host TNFα gene promoter would be accompanied by the transcription of the construct and eventual translation of the CAT gene, which can be detected as CAT-mediated conversion of CM to CM-3-AC. As shown in Figure 10.12 (lower panel) there is activation of the TNFα promoter by silica, in a dose-responsive manner, compared with latex beads as the control. This was confirmed by Northern blot analysis of TNFα mRNA (Figure 10.12, upper panel). A system such as this allows dissection of the transcriptional control of cytokine genes and the effects of particles and hopefully there will be future studies along similar lines.

While macrophages can be seen as a major source of cytokines in dust-exposed lung, the interaction between particles and epithelial cells may also be important in leading to proinflammatory events. Rosenthal *et al.* (1994) reported that interaction of asbestos fibers with

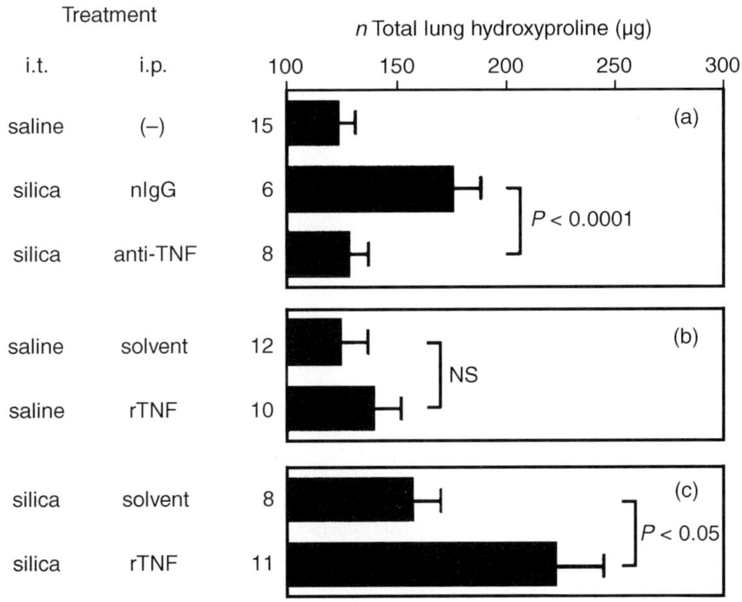

Figure 10.11 (a) Influence of anti-TNF antibody on lung fibrosis caused by silica instillation in mice. Instillation of silica plus treatment with normal IgG caused significant increase in lung hydroxyproline compared with saline instillation while silica instillation + anti-TNF caused a significant amelioration of the fibrogenic effect of silica. (b) Recombinant TNF alone (rTNF) did not cause fibrosis. (c) rTNF plus silica instillation was a super-stimulus to fibrosis, compared with silica alone. Groups of 6–15 mice instilled with 2 mg quartz 12–15 days previously. (Reproduced with permission from Piguet, P.F. *et al.* (1990) *Nature*, **344**, 245–7, MacMillan Magazines.)

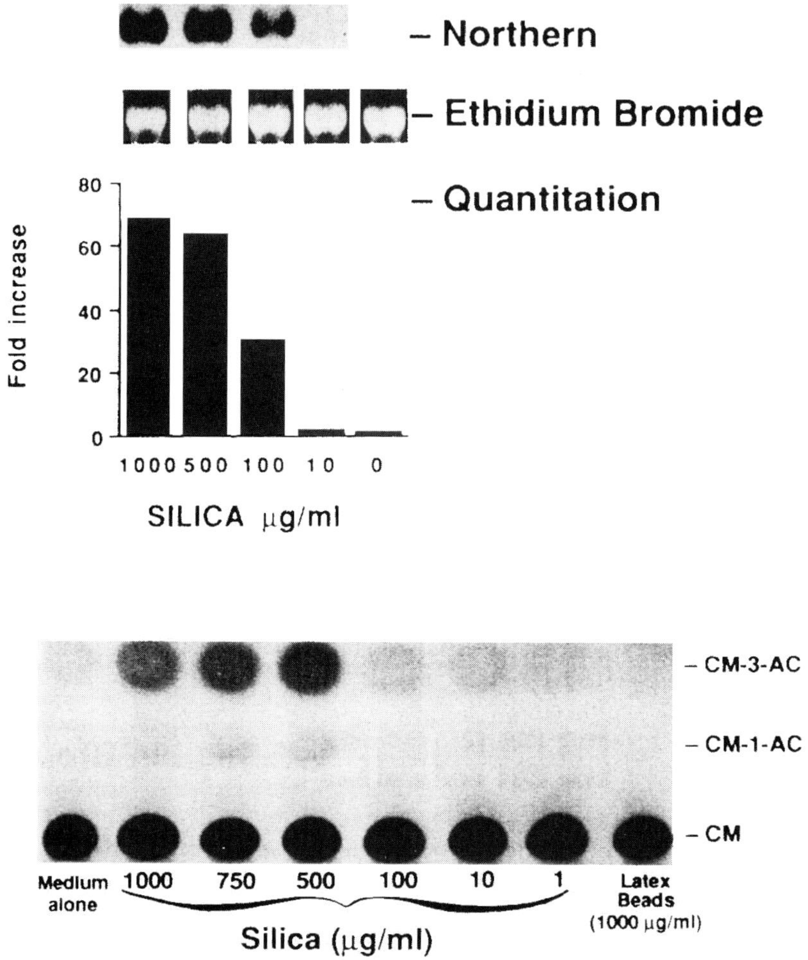

Figure 10.12 Upper panel: Effect of silica on the accumulation of TNF mRNA in THP-1 myelomonocytic cells. Top: Northern blot analysis of TNF mRNA 3 hours after stimulation of THP-1 cells with various amounts of silica. Middle: ethidium bromide staining to check equal loading with mRNA. Bar graph shows densitometric analysis of the TNF mRNA blots, confirming silica dose-dependent stimulation of mRNA accumulation. Lower panel: Activation of the TNF promoter with increasing dose of silica in THP-1 cells, as shown by TNF-CAT reporter system. Activation of the TNF promoter is accompanied by increased CAT activity, evident as accumulation of the product CM-3-AC (top row). Note failure of latex beads to stimulate the TNF promoter. (Reproduced from Savici, D. *et al.* (1994). *Exp. Lung Res.*, **20**, 613–26, with permission.)

cells from the human Type 2 alveolar epithelial cell line A549, caused transcription of the gene for the chemokine IL-8, a very potent chemotactic factor for neutrophils. As shown in Figure 10.13, both chrysotile and crocidolite asbestos at 10 μg/cm^2 caused transcription of increased amounts of mRNA for IL-8 as detected by PCR. It was important to note that this was accompanied by release of IL-8 protein and that two non-pathogenic dusts, wollastonite and titanium dioxide failed to stimulate the production of IL-8, confirming the specificity of this response to pneumoconiotic dust.

Further confirmation of the likely importance of the chemokine family of cytokines comes from the work of Driscoll and co-workers who have demonstrated increased activity of the genes for MIP-1α and MIP-2 in whole lung following inhalation and instillation of silica and asbestos (Driscoll *et al.*, 1993). Figure 10.14 shows the activity of the MIP-1α and MIP-2 genes mRNA 6 hours following instillation of quartz. It is clear that the activity of MIP-1α is up-regulated in both the BAL cells, due very likely to macrophage activation, while MIP-2 is up-regulated in both the BAL cells

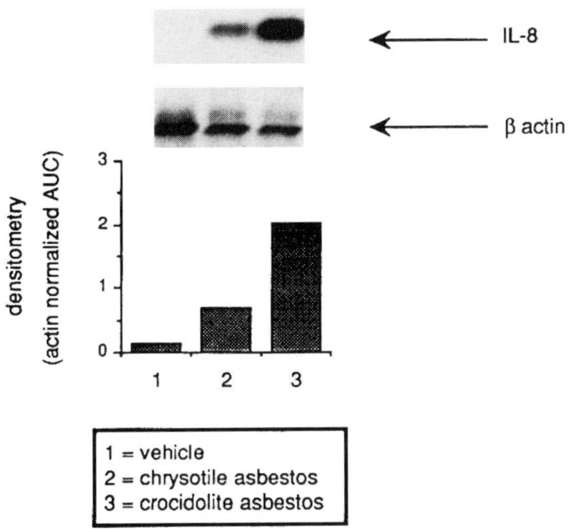

Figure 10.13 Accumulation of IL-8 mRNA in A549 epithelial cells exposed to 10 μg/cm^2 of chrysotile or crocidolite asbestos. Top row shows RT–PCR product of IL-8 mRNA. Middle row shows β-actin product to check for equal loading. Bar graph shows densitometric analysis of the top row, normalized to the β-actin bands. (Reproduced with permission from Rosenthal, G.J. *et al.* (1994) *J. Immunol.*, **153**, 3237–44. © 1994 The American Association of Immunologists.)

and the lung tissue itself. Furthermore, confirmation of a likely cytokine network was evident in the demonstration that TNFα can up-regulate the chemokine genes. Given the demonstration of the role of TNFα in the response to pneumoconiotic dust described above, it seems likely that activation of the TNFα gene is a key early event in the interaction of particles with macrophages. This, in turn leads on to the activation of other cytokine genes, as does direct interaction of the particle with epithelial cells and orchestrates a full-blown inflammatory response with the involvement of epithelial cells. A possible network of cytokines in pneumoconiosis is shown in Figure 10.15.

The final activation of the mesenchymal cells can be seen as being stimulated by various cytokines but PDGF and TGFβ are two cytokines with very powerful stimulatory activity to fibroblasts (Brody, 1991). Up-regulation of growth factor cytokine in macrophages and other cells in asbestos-exposed lung have been reported (Perdue and Brody, 1994) and this could be an important contribution to the increase in mesenchymal tissue characteristic of pneumoconiosis.

LEUKOCYTE-DERIVED ARACHIDONIC ACID DERIVATIVES

Perturbation of the macrophage membrane can also cause activation of arachidonic acid metabolism pathways that culminate in activation of phospholipase A2. This is the first step in an important series of events involving increases in intracellular Ca^{2+} and release of arachidonic acid from the membrane leading to the cyclo-oxygenase and lipoxygenase cascades and the generation of multiple proinflammatory mediators (Van Dyke *et al.*, 1996). This pathway has been demonstrated after cells have been activated *in vitro* by quartz (Englen *et al.*, 1989; Van Dyke *et al.*,

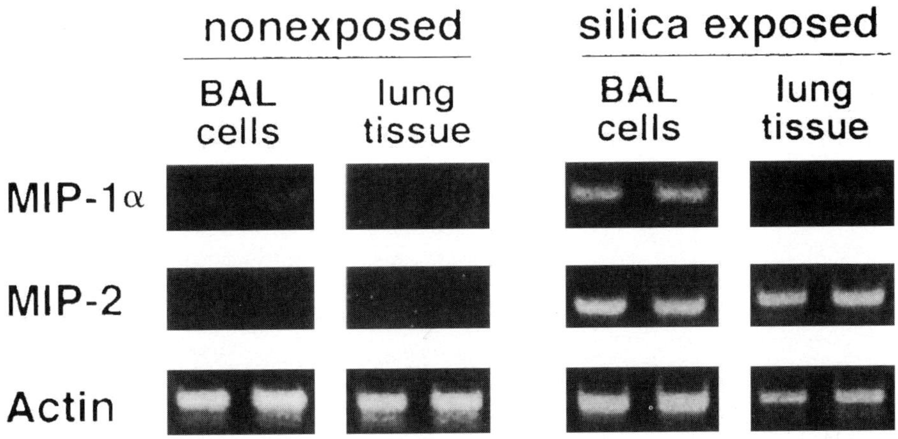

Figure 10.14 Accumulation of mRNA for the chemokines MIP-1α and MIP-2 in BAL cells or lavaged lung-tissue of groups of two rats instilled with 2–4 mg of quartz 6 hours previously. On the left, in the non-exposed group there is very little signal for transcription of either cytokine. On the right, the silica-exposed group shows increased transcription of the MIP-1α gene in the BAL cells whereas the MIP-2 gene transcription is increased in both the BAL cells and the lung tissue. (Reproduced from Driscoll, K.E. *et al.* (1993) *Am. J. Resp. Cell Mol. Biol.*, **8**, 311–18 with permission.)

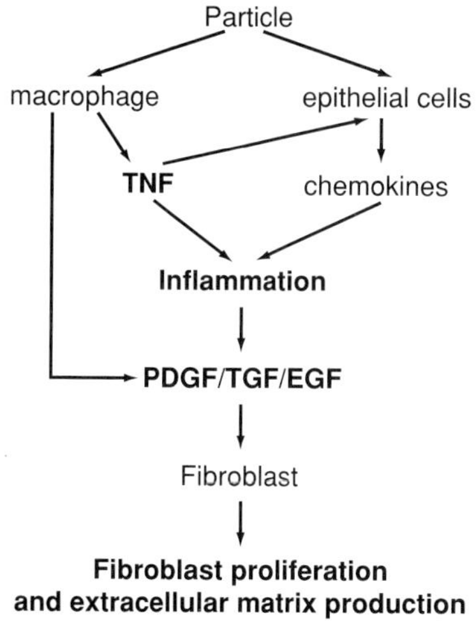

Figure 10.15 Potential cytokine network in pneumoconiotic lung.

1996) asbestos (Kouzan *et al.*, 1985) and coal mine dust (Kuhn *et al.*, 1990).

As shown in Table 10.8, the products of the cyclooxygenase and lipoxygenase cascades are proinflammatory and so could be important in the mechanism of pneumoconiosis.

THE MACROPHAGE AND CLEARANCE OF PNEUMOCONIOTIC PARTICLES

Clearance of deposited particles from the terminal brochioles and proximal alveoli is via phagocytosis by alveolar macrophages. Normally, clearance from this compartment depends on effective phagocytosis, followed by movement of the alveolar macrophages up to the start of the mucociliary escalator for removal. This in turn depends on the ability of the macrophages to detect a chemotactic gradient via receptors and on an intact locomotory apparatus/cytoskeleton. Donaldson *et al.* (1990b) demonstrated that bronchoalveolar leukocytes from rats exposed by inhalation to several different pneumoconiotic particles, coal mine dust, quartz and asbestos, had impaired ability to migrate towards a standard chemotaxin. There was no impairment of motility in bronchoalveolar cells from rats with protracted exposure to the non-pathogenic particle TiO_2.

A further failure of normal clearance of particles arises when the particles are transported to the lymph nodes. In this case the macrophage-mediated transport of the particles to the lymph nodes can be seen as a classic subversion of a defensive response. It is beneficial for the normal daily killing of low numbers of microbes that deposit in the lung to be executed without the involvement of the immune or inflammatory responses, and indeed for the normal brocholaveolar leukocytes to be immunosuppressive (Kusaka *et al.*, 1990). However, if there is an infection, it is necessary for the bronchoalveolar macrophages to change from being immunosuppressive to being stimulatory and to take the 'antigen' to the lung lymph nodes so that an immuno-inflammatory response can be mounted. When a particulate stimulus is sufficient to cause inflammation, either because of the high dose of particles or because of its inherent toxicity, then this is met by the evolved response and macrophages become stimulatory and carry the particles to the lymph nodes. There is evidence that this proposition is true from the demonstration that:

- the immunosuppressive effects of normal bronchoalveolar leukocytes are reversed, so becoming immunostimulatory, after quartz exposure (Kusaka *et al.*, 1990); and
- transport of particles to the lymph nodes coincides with the onset of inflammation in rat inhalation studies with pneumoconiotic dusts (Vincent and Donaldson, 1990).

Thus pneumoconiosis can be viewed as the result of the lung having a limited spectrum of response that has evolved to deal with microbial insult and

Table 10.8 Arachidonic acid metabolites and their proinflammatory effects

Mediator	*Platelet aggregation*	*Vasodilation*	*Chemotaxis*	*Increased permeability*
Prostacyclin	✓	✓		✓
Prostaglandins		✓		✓
Thromboxanes	✓			
Leukotrienes			✓	✓
HPETE, HETE			✓	

that evocation of this response to particles is inappropriate. The normal degradative functions in the lymph node would clear the transported material if it were antigen, whereas with a durable pneumoconiotic dust there is sequestration of dust in the lymph nodes. The consequences of this lymph dust sequestration for for normal lung functioning can only be speculative, but a localized adjuvant effect might be anticipated and the immunomodulating effects of iron (Weiss *et al.*, 1995) in the lymph node may also be of consequence.

PROLIFERATIVE RESPONSES IN PNEUMOCONIOSIS

Pneumoconiotic lung is characterized by increased numbers of macrophages, Type 2 alveolar epithelial cells and fibroblasts. Macrophages are considered to be derived principally by recruitment from the monocyte pool but Type 2 alveolar epithelial cells and fibroblasts undergo proliferation.

Epithelial cell proliferation

Following quartz instillation in the rat, dramatic changes in the phenotype of the Type 2 epithelial cells are found, with increased size and phospholipid metabolism (Miller and Hook, 1988) and increased DNA synthesis that is continued into culture when the Type 2 cells are isolated from the rat lung (Panos *et al.*, 1990). This change in the activity of the Type 2 cells can be seen to be a response to localized accumulation of macrophages in close proximity to the epithelial cells (Lee, 1985) with high focal concentrations of cytokines, mitogenic factors or toxic factors, leading to proliferation. Direct effects of particles on the epithelium is also a strong possibility and asbestos fibers have been shown to directly activate proliferative pathways in epithelial cells, including proto-oncogene activation, as described above.

Fibroblast proliferation

Proliferation of mesenchymal cells is clearly a key response in pneumoconiosis and the macrophage-derived cytokines that are potent in causing fibroblast proliferation are very likely to be involved in the increase in fibroblast number and the increased production of extracellular matrix components (Table 10.9) (Brody, 1991). Alveolar macrophages release fibroblast growth factors in response to asbestos fibers *in vitro* (Schapira *et al.*, 1990) and *in vivo* (Perdue and Brody, 1994). However, as discussed above the importance of interstitialization of particles and interactions with the interstitial macrophages leading to release of cytokines, in close proximity to the target fibroblasts, should be emphasized (Brody *et al.*, 1992).

Table 10.9 The properties of selected macrophage-derived cytokines regarding ability to attract fibroblasts, mediate their proliferation or activate them for production of extracellular matrix components

Cytokine	*Proliferation*	*ECM production*
Platelet-derived growth factor	✓	✓
Transforming growth factor α	✓	✓
Tumor necrosis factor α	✓	✓
Epidermal growth factor	✓	
Interleukin -1	✓	✓

Modified from Scheule and Holian (1991).

Recent studies showing that asbestos can directly up-regulate the PDGF receptor on the surface of fibroblasts (Bonner *et al.*, 1993) and stimulate the fibroblasts to synthesize and release PDGF (Lasky *et al.*, 1995) raises the possibility of an autocrine loop in fibroblast activation. Whether this is a general phenomenon in the pneumoconioses presents a fascinating area for future research.

FUTURE RESEARCH ON MECHANISMS IN PNEUMOCONIOSIS

There is never a right time to write a review in an area of science that is advancing fast, and the mechanistic scheme described here is written at a time when enormous advances are being made in understanding the molecular biology of disease. The study of pneumoconiosis stands on the edge of this advance and is poised to undergo the same revolution in understanding as other diseases. For instance, there is comparatively little mention here of the role of cellular oncogenes and kinases in the present review, despite growing awareness that they will almost certainly be central to understanding a proliferative disorder like pneumoconiosis. There is no mention of apoptosis, cyclins or checkpoints since there are no published data on their role in penumoconiosis at the time of writing, but these are also areas of great potential growth. Laboratory approaches to studying disease mechanisms at the cell and molecular level has seen exciting developments in the past 10 years and will undoubtedly make even more advances in the decade ahead. The application of these modern approaches to a further understanding of pneumoconiosis at the molecular and genetic level should hold out the best hope yet of a full understanding and hopefully, treatment, of this most ancient of occupational lung diseases.

Acknowledgments

The author would like to acknowledge the contribution of past and present colleagues in the Institute of Occupational Medicine, Edinburgh, The Unit of Respiratory Medicine, University of Edinburgh and in Napier University, Edinburgh in studying the mechanisms of particle-related lung disease. I would like to thank Dr William MacNee, Dr Harry Beswick and Dr Irfan Rahman for reading the manuscript and for making helpful corrections and suggestions.

REFERENCES

Angel P. and Karin M. (1991) The role of Jun, Fos and the AP-1 complex in cell-proliferation and transformation. *Biochim. Biophys. Acta*, **1072**, 129–57.

Bastian, N.R. and Hibbs, J.B. (1994) Assembly and regulation of NADPH oxidase and NO synthase. *Curr. Opin. Biol.* **6**, 131–3.

Baumann, M.D., Jetten, A.M., Bonner, J.C., Kumar, R.K., Bennett, R.A. and Brody, A.R. (1990) Secretion of a platelet growth factor homologue by rat alveolar macrophages exposed to particulates *in vitro*. *Eur. J. Cell Biol.*, **51**, 327–34.

Begin, R. (1986) Assessment of disease activity by Gallium 67 scan and lung lavage in the pneumoconioses. *Semin. Resp. Dis.*, **7**, 275–80.

Blackford, J.A., Antonini, J.M., Castranova, V. and Dey, R.D. (1994) Intratracheal instillation of silica up-regulates inducible nitric oxide synthase gene expression and increases nitric oxide production in alveolar macrophages and neutrophils. *Am. J. Resp. Cell Mol. Biol.*, **11**, 426–31.

Bonner, J.C., Goodell, A.L., Coin, P.G. and Brody, A.R. (1993) Chrysotile asbestos upregulates gene expression and production of alpha receptors for platelet-derived growth factor (PDGF-AA) on rat lung fibroblasts. *J. Clin. Invest.*, **92**, 425–30.

Borm, P.J.A., Palen, N., Engelen, J.J.M. and Buurman, W.A. (1988) Spontaneous and stimulated release of tumor necrosis factor alpha (TNF) from blood monocytes of miners with coalworkers pneumoconiosis. *Am. Rev. Respir. Dis.*, **138**, 1589–94.

Bowden, D.H., Hedgecock, C. and Adamson, I.Y.R. (1989) Silica-induced pulmonary fibrosis involves the reaction of particles with interstitial rather than alveolar macrophages. *J. Pathol.*, **158**, 73–80.

Brody, A.R. (1991) Production of cytokines by particle-exposed lung macrophages, in *Cellular and Molecular Aspects of Fiber Carcinogenesis*, Cold Spring Harbor Laboratory Press, NY, pp. 83–102.

Brody, A.R. and Roe, M.W. (1983) Deposition pattern of inorganic particles at the alveolar level in the lungs of rats and mice. *Am. Rev. Respir. Dis.*, **128**, 724–9.

Brody, A.R., Hill, L.H., Adkins, B. and O'Connor, R.W. (1981) Chrysotile asbestos inhalation in rats: deposition pattern and reaction of alveolar epithelium and pulmonary macrophages. *Am. Rev. Respir. Dis.*, **123**, 670–9.

Brody, A.R., Bonner, J.C., Overby, L.H., Badgett, A., Kalter V., Kumar, R. and Bennett, R.A. (1992) Interstitial pulmonary macrophages produce platelet-derived growth factor that stimulates rat lung fibroblast proliferation *in vitro*. *J. Leukocyte Biol.*, **51**, 640–8.

Brown, G.M. and Donaldson, K. (1988) Degradation of connective tissue components by lung-derived leukocytes *in vitro*. the role of proteases and oxidants. *Thorax*, **43**, 132–9.

Brown, G.M. and Donaldson, K. (1990) Inflammatory responses in lungs of rats inhaling coalmine dust: enhanced proteolysis of fibronectin by bronchoalveolar leukocytes. *Br. J. Ind. Med.*, **46**, 866–72.

Brown, G.M., Donaldson, K. and Brown, D. (1990) Broncholaveolar leukocyte response in experimental silicosis: modulation by a soluble aluminium compound. *Toxicol. Appl. Pharmacol.*, **101**, 95–105.

Brown, G.M., Brown, D.M., Slight, J. and Donaldson, K. (1991) Persistent biological reactivity of quartz in the lung: raised protease burden compared with a non-pathogenic mineral dust and microbial particles. *Br. J. Ind. Med.*, **48**, 61–9.

Brown, K., Gerstberger, S., Carlson, L., Franzoso, G. and Siebenlist, U. (1995), Control of IκB-alpha proteolysis by site-specific, signal induced phosphorylation. *Science*, **267**, 1485–9.

Castranova, V., Antonini, J.M., Reasor, M.J., Wu, L. and VanDyke, K. (1996) Oxidant release from pulmonary phagocytes, in *Silica and silica-induced Disease*, (eds V. Castranova, V. Vallyathan, and W.E. Wallace), CRC Press, Boca Raton, pp. 185–95.

Chang, L.-Y., Overby, L.H., Brody, A.R. and Crapo, J.D. (1988) Progressive lung cell reactions and extracellular matrix production after a brief exposure to asbestos. *Am. J. Pathol.*, **131**, 156–69.

Dalal, N.S., Suryan, M.M., Vallyathan, V., Green, F.H.Y., Jafari, B. and Wheeler, R. (1989) Detection of reactive free radicals in fresh coal mine dust and their implications for pulmonary injury. *Ann. Occup. Hyg.*, **33**, 79–84.

Davis, J.M.G., Addison, J., Botton, R.E., Donaldson, K., Jones, A.D. and Smith, T. The Pathogenicity of long versus short fibre samples of amosite asbestos administered to rats by inhalation and intraperitoneal injection. *Br. J. Exp. Path.*, **67**, 415–30.

Donaldson, K. and Brown, G.M. (1993) Bronchoalveolar lavage in the assessment of the cellular response to fiber exposure, in *Fiber Toxicology* (ed. D.B. Warheit), Academic Press, pp. 117–38.

Donaldson, K., Bolton, R.E., Jones, A.D. *et al.* (1988a) Kinetics of the bronchoalveolar leukocyte response in rats during exposure to equal airborne mass concentration of quartz, chrysotile asbestos or titanium dioxide. *Thorax*, **43**, 159–62.

Donaldson, K., Slight, J. and Bolton, R.E. (1988b) Oxidant production by control and inflammatory bronchoalveolar leukocyte populations treated with mineral dust *in vitro*. *Inflammation*, **12**, 231–43.

Donaldson, K., Brown, G.M., Brown, D.M., Bolton, R.E. and Davis, J.M.G. (1989) The inflammation generating potential of long and short fibre amosite asbestos samples. *Br. J. Ind. Med.*, **46**, 271–6.

Donaldson, K., Brown, G.M., Brown, D.M., Robertson, M.D., Slight, J., Cowie, H., Jones, A.D., Bolton, R.E. and Davis, J.M.G. (1990a) Contrasting bronchoalveolar leukocyte responses in rats inhaling coalmine dust, quartz or titanium dioxide: effects of coalrank, airborne mass concentration, and cessation of exposure. *Environ. Res.*, **52**, 62–76.

Donaldson, K., Brown, G.M., Brown, D.M., Slight, J., Robertson, M.D. and Davis, J.M.G. (1990b) Impaired chemotactic responses of bronchoalveolar leukocytes in experimental pneumoconiosis. *J. Pathol.*, **160**, 63–9.

Donaldson, K., Li, X.Y., Dogra, S., Miller, B.G. and Brown, G.M. (1992) Asbestos-stimulated tumour necrosis factor release from alveolar macrophages depends on fibre length and opsonisation. *J. Pathol.*, **168**, 243–8.

Donaldson, K., Gilmour, P.S., MacNee, W. and Oberdorster, G. (1995) Free radical activity associated with ultrafine titanium dioxide particles and PM10 material. *Am. J. Resp. Crit. Care Med.*, **151**, A64.

Driscoll, K.E. (1994) Macrophage inflammatory proteins: biology and role in pulmonary inflammation. *Exp. Lung Res.*, **20**, 473–90.

Driscoll, K.E. (1996) The role of interleukin-1 and Tumor Necrosis Factor alpha in the lung's response to silica, in *Silica and Silica-induced Disease*, (eds V. Castranova, V. Vallyathan and W.E. Wallace), CRC Press, Boca Raton, pp. 163–84.

Driscoll, K.E., Higgins, J.M. Leytart, M.J. and Crosby, L.L. (1990) Differential effects of mineral dusts on the in vitro activation of alveolar macrophage eicosanoid and cytokine release. *Toxic. in Vitro*, **4**, 284–8.

Driscoll, K.E., Howard, B.W., Hassenbein, D.G., Janssen, Y.M.W. and Mossman, B.T. (1995). Oxidative stress and silica induced increases in nuclear, NF-κB binding activity in alveolar epithelial cells. *Am. J. Resp. Crit. Care Med.*, **151**, A712.

Driscoll, K.E., Hassenbein, D.G., Carter, J., Poynter, J., Asquith, T.N., Grant, R.A., Whitten, J., Purdon, M.P. and Takigiku, R. (1993) Macrophage inflammatory proteins 1 and 2: expression by rat alveolar macrophages, fibroblasts and epithelial cells and in the rat lung after mineral dust exposure. *Am. J. Respir. Cell Mol. Biol.*, **8**, 311–18.

Driscoll, K.E. Howard, B.W., Carter, L.M. and Hassenbein, D.G. (in press) Inflammatory cytokine production by rat lung epithelial cells *in vitro*: particles, NF-κB and oxidative stress. To be published in the proceedings of the 5th International Inhalation Symposium, Hannover Medical School.

Droge, W., Schultze-Ostoff, K., Mihm, S., Glater, D., Shenk, H., Eck, H.P., Roth, S. and Gmunder, H. (1994) Functions of glutathione and glutathione disulphide in immunology and immunopathology. *FASEB J.*, **8**, 1131–8.

Englen, M.D., Taylkor, S.W., Laegreid, W.W., Liggit, H.D., Silflow, R.M., Breeze, R.C. and Leid, W. (1989) Stimulation of arachidonic acid metabolism in silica-exposed alveolar macrophages. *Exp. Lung Res.*, **15**, 511–16.

Faux, S.P., Michelangeli, F., Levy, L.S. (1994) Calcium chelator Quin–2 prevents crocidolite–induced DNA strand breakage in human white blood cells. *Mutation Res.*, **311**, 209–15.

Ferin, J., Oberdorster, G., Soderholm, C. and Gelein, R. (1991) Pulmonary tissue access of ultrafine particles. *J. Aerosol Med.*, **4**, 57.

Ferin, J., Oberdorster, G. and Penney, D.P. (1992) Pulmonary retention of ultrafine and fine particles in rats. *Am. J. Respir. Cell Mol. Biol.*, **6**, 535–42.

Gavett, S.H., Carakostas, M.C., Belcher, L.A. and Warheit, D.B. (1992) Effect of circulating neutrophil depletion on lung injury induced by inhaled silica particles. *J. Leukocyte Biol.*, **51**, 455–61.

Gibbs, A.R., Seal, R.M.E. and Wagner, J.C. (1984) Pathological reactions of the lung to dust, in *Occupational Lung Diseases* (eds W.K.C. Morgan and A. Seaton), 2nd edn W.B. Saunders, Philadelphia, pp. 129–62.

Gilmour, P., Beswick, P.H. and Donaldson, K. (1995) Detection of surface free radical activity of respirable industrial fibres using supercoiled $\phi \times 174$ plasmid DNA. *Carcinogenesis*, **16**, 2973–9.

Gutteridge, J.M.C. and Halliwell, B. (1994) *Antioxidants in Nutrition, Health and Disease*, Oxford University Press.

Haagsman, H.P. and van Golde, L.M.G. (1985) Lung surfactant and pulmonary toxicity. *Lung*, **163**, 275–303.

Henderson, R.F., Harkema, J.R., Hotchkiss, J.A. and Boehme, D.S. (1991) Effect of blood leukocyte depletion on the inflammatory response of the lung to quartz. *Toxicol. Appl. Pharmacol.*, **109**, 127–36.

Hill, I., Beswick, P.H. and Donaldson, K. (1995) Differential release of superoxide anion by macrophages treated with long and short fibre amosite asbestos is a consequence of differential affinity for opsonin. *Occup. Environ. Med.*, **52**, 92–6.

Jabbour, A.J., Iyer, R., Scheule, R.K. and Holian, A. (1996) Potential intracellular messengers involved in silica stimulation of alveolar macrophages, in *Silica and Silica-Induced Disease* (eds V. Castranova, V. Vallyathan, and W.E. Wallace), CRC Press, Boca Raton, pp. 137–49.

Janssen, Y.M.W., Marsh, J.P., Absher, M.P., Hemenway, D., Vacek, P.M., Lelie, K.O., Borm, P.J.A. and Mossman, B.T. (1992) Expression of antioxidant enzymes in rat lung after inhalation of asbestos or silica. *J. Biol. Chem.*, **267**, 10625–30.

Janssen, Y.M., Marsh, J.P., Driscoll, K.E., Borm, P.J. Oberdorster, G. and Mossman, B.T. (1994a) Increased expression of manganese-containing superoxide dismutase in rat lungs after inhalation of inflammatory and fibrogenic minerals. *Free Radical Biol. Med.*, **16**, 315–22.

Janssen, Y.M.W. Heintz, N.H., Marsh, J.P., Borm, P.J.A. and Mossman, B.T. (1994b) Induction of c-fos and c-jun proto-oncogenes in target cells of the lung and pleura by carcinogenic fibres. *Am. J. Respir. Cell Mol. Biol.*, **11**, 522–30.

Janssen, Y.M.W., Heintz, N.H. and Mossman, B.T. (1995) Induction of c-fos and c-jun proto-oncogene expression by asbestos is ameliorated by *N*-acetyl-L-cysteine in mesothelial cells. *Cancer Res.*, **55**, 2085–9.

Kamp, D.W., Gracefa, P., Pryor, W.A. and Weitzman, S.A. (1992) The role of free radicals in asbestos-induced diseases. *Free Radical Biol.*, **12**, 293–315.

Kasemo, B. and Lausmaa, J. (1988) Biomaterials from a surface science perspective, in *Surface Characterisation of Biomaterials* (ed. B.D. Rattner), Elsevier, pp. 1–12.

Kouzan, S., Nettesheim, P., Eling, T. and Brody, A.R. (1985) Production of arachidonic acid metabolites by alveolar macrophages treated with asbestos *In vitro effects of mineral dusts* (eds E.G. Beck and J. Bignon), Springer-Verlag, Berlin, pp. 291–5.

Kuhn, D.C., Stanley, C.F., El-Ayouby, N. and Demers, L.M. (1990) Effect of *in vivo* coal dust exposure on arachidonic acid metabolism in the rat alveolar macrophage. *J. Toxicol. Environ. Health*, **29**, 157–68.

Kusaka, Y., Brown, G.M. and Donaldson, K. (1990) Alveolitis caused by exposure to coalmine dusts: production of interleukin-1 and immunomodulation by bronchoalveolar leukocytes. *Environ. Res.*, **53**, 76–89.

Lasky, J.A., Coin, P.G., Lindroos, P.M., Ostrowski, L.E., Brody, A.R. and Bonner, J.C. (1995) Chrysotile asbestos stimulates platelet-derived growth factors – AA production by rat lung fibroblasts *in vitro*: evidence for an autocrine loop. *Am. J. Respir. Cell Mol. Biol.*, **12**, 162–70.

Lee, J.C. (1994) Transcription factor NF-κB: an emerging regulator of inflammation. *Ann. Reports Medicinal Chem.*, **29**, 235–44.

Lee, K.P. (1985) Lung responses to particulates with emphasis on asbestos and other fibrous dusts. *CRC Crit. Rev. Toxicol.*, **14**, 33–86.

Li, X.Y., Donaldson, K., Rahman, I. and MacNee, W. (1994) An investigation of the role of glutathione in the increased epithelial permeability induced by cigarette smoke *in vitro* and *in vivo*. *Am. J. Respir. Crit. Care Med.*, **149**, 1518–25.

Li, X.Y., Donaldson, K., Brown, D. and MacNee, W. (1995) The role of neutrophils, tumour necrosis factor and glutathione in increased airspace epithelial permeability in acute lung inflammation. *Am. J. Respir. Cell Mol. Biol.*, **13**, 185–95.

Lombard-Gillooly, K. Hubbard, A.K. (1993) Modulation of silica-induced lung injury by reducing lung non-protein sulfhydryls with buthionine sulphoximine. *Toxicol. Lett* **66**, 305–15.

Lund, L.G., Williams, M.G., Dodson, R.F. and Aust, A.E. (1994) Iron associated with asbestos bodies is responsible for the formation of single strand breaks in X174RF1 DNA. *Occup. Environ. Med.*, **51**, 200–4.

Meyer, M., Pahl, H.K. and Bauerle, P.A. (1994) Regulation of the transcription factors NF-κB and AP-1 by redox changes. *Chem. Biol. Interact.*, **91**, 91–100.

Miller, B.E. and Hook, G.E.R. (1988) Isolation and characterisation of hypertrophic Type II cells from the lungs of silica-treated rats. *Lab. Invest.*, **58**, 565–75.

Mossman, B.T., Marsh, J.P., Sesko, A. *et al.* (1990) Inhibition of lung injury, inflammation and interstitial pulmonary fibrosis by polyethylene glyco-conjugated catalase in a rapid inhalation model of asbestosis. *Am. Rev. Respir. Dis.*, **141**, 1266–71.

Nyberg, P. and Klockars, M. (1990) Effect of immunoglobulins on mineral dust-induced production of reactive oxygen metabolities by human macrophages. *Inflammation*, **14**, 621–9.

Panos, R.J., Suwabe, A., Leslie, C.C. and Mason, R.J. (1990) Hypertrophic alveolar Type 2 cell from silica-treated rats are committed to DNA synthesis *in vitro. Am. J. Respir. Cell Mol. Biol.*, **3**, 51–9.

Perderiset, M., Marsh, J.P. and Mossman, B.T. (1991) Activation of protein kinase C by crocidolite asbestos in hamster tracheal epithelial cells. *Carcinogenesis*, **12**, 1499–502.

Perdue, T.D. and Brody, A.R. (1994) Distribution of transforming growth factor β1, fibronectin and smooth muscle actin in asbestos-induced pulmonary fibrosis in rats. *J. Histochem. Cytochem.*, **42**, 1061–70.

Piguet, P.F., Collart, M.A., Grau, G.E., Sappino, A.P. and Vassalli, P. (1990) Requirement of tumour necrosis factor for the development of silia-induced pulmonary fibrosis. *Nature*, **344**, 245–7.

Piguet, P.F., Rosen, H., Vesin, C. and Grau, G.E. (1993) Effective treatment of the pulmonary fibrosis elicited in mice by bleomycin with anti-CD-11 antibodies. *Am. Rev. Respir. Dis.*, **147**, 435–41.

Quinlan, T.R., Marsh, J.P., Janssen, M.W., Leslie, K.O., Hemenway, D., Vacek, P. and Mossman, B.T. (1994) Dose-responsive increases in pulmonary fibrosis after inhalation of asbestos. *Am. J. Respir. Crit. Care Med.*, **150**, 200–6.

Rahman, I., Li, X.Y., Donaldson, K. and MacNee, W. (1995) Cigarette smoke, glutathione metabolism and epithelial permeability in rat lungs. *Biochem. Soc. Trans.*, **23**, 235S.

Resnick, D., Freedman, N.J. and Krieger, M. (1993) Secreted extracellular domains of macrophage scavenger receptors form elongated trimers which specifically bind crocidolite asbestos. *J. Biol. Chem.*, **268**, 3538–45.

Rom, W. (1990) Basic mechanisms leading to focal emphysema in coal worker's pneumoconiosis. *Environ. Res.*, **53**, 16–28.

Rosenthal, G.J., Germolec, D.R., Blazka, M.E. *et al.* (1994) Asbestos stimulates IL-8 production from human lung epithelial cells. *J. Immunol.*, **153**, 3237–44.

Savici, D., He, B., Geist, L.J., Monick, M.M. and Hunninghake, G.W. (1994) Silica increases tumor necrosis factor (TNF) production in part by upregulating the TNF promoter. *Exp. Lung Res.*, **20**, 613–25.

Schapira, R.M., Osornio, A.R., Bonner, J.C. and Brody, A.R. (1990) Platelet-derived growth factor secretion by alveolar macrophages is stimulated by carbonyl iron and chrysotile asbestos *in vitro*, in *The Physiological and Pathological Effects of Cytokines*, Wiley-Liss Inc., pp. 189–94.

Scheule, R.K. and Holian, A. (1991) Immunologic aspects of pneumoconiosis. *Exp. Lung Res.*, **17**, 661–85.

Schmidt, J., Pliver, C.N., Lepe-Zuniga, J.L., Green, I. and Grey, I. (1984) Silica-stimulated monocytes release fibroblast proliferation factors identical to IL-1. A potential for IL-1 in the pathogenesis of silicosis. *J. Clin. Invest.*, **73**, 1462–72.

Sesko, A., Cabot, M. and Mossman B. (1990) Hydrolysis of inositol phospholipids precedes cellular proliferation in asbestos-stimulated tracheobronchial epithelial cells. *Proc. Natl. Acad. Sci. USA*, **87**, 7385–9.

Simeonova, P.P. and Luster, M.I. (1995) Iron and reactive oxygen species in the asbestos-induced tumor necrosis factor-alpha response from alveolar macrophages. *Am. J. Respir. Cell Mol. Biol.*, **12**, 676–83.

Thomas, G. Ando, T., Verma, K. and Kagan, E. (1994) Asbestos fibres and IFN-γ up-regulate nitric oxide production in rat alveolar macrophages. *Am. J. Respir. Cell Mol. Biol.*, **11**, 707–715.

Vallyathan, V., Shi, X., Dalal, N.S., Irr., W. and Castranova, V. (1988) Generation of free radicals from freshly fractured silica dust: potential role in acute silica-induced lung injury. *Am. Rev. Respir. Dis.*, **138**, 1213–19.

Van Dyke, K., Gutiererez, J., VanDyke, C. and Wu, L. (1996) Early role of inflammatory mediators in silica toxicity: possible methods to detoxify silica, in *Silica and Silica-Induced Disease*, (eds V. Castranova, V. Vallyathan and W.E. Wallace), CRC Press, Boca Raton, pp. 209–28.

Velan, G.M., Kumar, R. and Cohen, D.D. (1993) Pulmonary inflammation and fibrosis following subacute inhalational exposure to silica: determinants of progression. *Pathology*, **25**, 282–90.

Vincent, J.H. and Donaldson, K. (1990) A dosimetric approach for relating the biological response of the lung to the accumulation of inhaled mineral dust. *Br. J. Ind. Med.*, **47**, 302–7.

Wallace, W.E., Keane, M.J., Vallyathan, V., Hathaway, P., Regad, E.D., Castranova, V. and Green, F.H.Y. (1988) Suppression of inhaled particle cytotoxicity by pulmonary surfactant and re-toxification by phospholipase: distinguishing properties of quartz and kaolin. Inhaled particles Conference VI (eds J. Dodgson, R.I. McCallum, M.R. Bailey and D.R. Fisher). *Ann. Occup. Hyg.*, **32**, 291–8.

Wallaert, B., Bart, F., Aerts, C., Ouaissi, M.A., Neyrink, J.L. and Voisin, C. (1987) Spontaneous superoxide anion, neutrophil chemotactic activity and fibronectin secretion by AM in pneumoconiotic patients, in *Silicosis and Mixed Dust Pneumoconiosis*, Colloque INSERM (ed. L. Lebouffant), Vol. 155, pp. 43–60.

Warheit, D.B., Hill, L.H., George, G. and Brody, A.R. (1986) Time course of chemotactic factor generation and the corresponding macrophage response to asbestos inhalation. *Am. Rev. Respir. Dis.*, **134**, 128–33.

Warheit, D.B., Carakostas, M.C., Bamberger, J.R. and Hartsky, M.A. (1991) Complement facilitates macrophage phagocytosis of inhaled iron particles but has little effect in mediating silica-induced lung inflammatory and clearance responses. *Environ. Res.*, **56**, 186–203.

Weiss, G., Wachter, H. and Fuchs, D. (1995) Linkage of cell-mediated immunity to iron metabolism. *Immunol. Today*, **16**, 495–500.

11 Lung diseases in coal workers

John E. Parker and Daniel E. Banks

INTRODUCTION

Coal is not a pure mineral. It is formed by the accumulation of vegetable matter covered by sedimentary rock (thereby sealing it from air) and subjected to pressure and temperature over the ages. This causes the physical and chemical properties of the matter to change. The matter dries, becomes warmer, and loses oxygen content, all the while increasing the relative carbon content (Stahl, 1989).

The first step in this conversion of vegetable matter to coal is the formation of peat, a moist spongy material. This transformation of organic deposit can occur in a stagnant water bed relatively quickly (at a rate of approximately 1 foot (30 cm) per 100 years). An approximately 100-ft (30-m) accumulation of peat compresses to form a 1 ft-wide coal seam. In the most simple terms, coal is comprised of moisture (which lessens with time), pure coal (carbon) and mineral matter (Thomas, 1992).

The process of conversion (coalification) from organic matter follows a transformation of wood → peat → lignite → bituminous coal → anthracite coal. The terms brown coal and black coal refer to coals of low and high rank, respectively. *Rank* describes the extent of change from vegetation to mineral-free coal. Rank is directly related to the percent of carbon in coal, the completeness of the transformation from vegetation to coal, and the geologic age of the deposit, with new deposits forming above the old. Indeed, Hilt (1873) observed that 'in a vertical sequence, at any one locality in a coal field, the rank of the coal seam rises with increasing depth'. The *type* of coal relates to the plant materials which form the coal. The *grade* refers to the purity of the coal – or the amount of inorganic material (including ash and sulfur) released in the burning of coal. Sulfur is a frequent contaminant of coal and is derived from sulfur in the plants which have formed the coal or as a result of infusion of water containing sulfur during the formation of coal. In today's world, sulfur released as a by-product of coal burning is an important air pollutant and boiler corrosive, making coal with a high sulfur content less desirable.

Other important descriptive terms include a quantitative measurement of coal *moisture* (the percentage of water that is contained in coal and released with moderate heat) and the amount of *volatile matter* (that percentage of substances, mainly gases and coal tar, lost when a sample is well heated). The residue of combusted material remaining after a coal sample is completely burned is coal *ash*. The ash content has two sources. The first is the inherent minerals in the decaying wood and the second is material which was blown or entered the swamp while the peat was being formed (Haught, 1955). Coal is classified primarily by rank (Table 11.1).

The composition of coal mine dust varies with the coal seam. Most of the dust is composed of carbon, although in some seams it may only approximate 60% of the dust (Coal data, 1995), with more than 50 different elements and their oxides (Coates, 1981). Dusts of high rank typically have more silica than dusts of lesser rank because the anthracite seams often have roofs and floors of quartz which contaminate the coal during mining (Mutmansky, 1984)

During this century coal has become less important as an energy source, with petroleum and natural gas

Occupational Lung Disease. Edited by Daniel E. Banks and John E. Parker. Published in 1998 by Chapman & Hall. ISBN 0 412 73630 6

Table 11.1 Coal characteristics

Coal rank	*Volatile matter (%)*	*Carbon (%)*	*Ash (%)*	*Moisture (%)*	*BTU*/pound*
Peat (dried)	70	25	5	10	8000–12 000
Lignite	41	36	9	14	5500–7000
Bituminous	32	55	10	3	12 000–14 000
Anthracite	5	87	5	3	> 13 000

* One BTU (British thermal unit) is equivalent to the amount of heat necessary to raise the temperature of 1 lb water from approximately 4° to 5° F (1 BTU/lb°F= 4.1868 kJ/kg°C).

Source: adapted from Haught (1955).

used increasingly as a substitute. In 1900, 94% of the world's energy requirements were met by coal. By 1940, coal accounted for 75% of the world's energy consumption. In the 1950s and 1960s, cheap oil became available and this further curtailed coal use. By 1967, coal utilization had fallen to 40%, and by 1978, coal was only responsible for 31% of fuel use (Stahl, 1989).

In the US, electric power generation accounts for the utilization of more than 80% of the coal produced each year. Coal can also be processed and made into liquid and gaseous fuels. These products have the potential to compete with natural gas and petroleum-derived fuels depending on the cost. Nowadays, natural gas and petroleum-derived fuels are cheaper, and the use of liquid and gaseous coal fuels is not cost effective. Coke, necessary for the production of steel, is also an important use of coal, as is coal tar, a hydrocarbon by-product released by the formation of coke. For about 75 years, until the end of the Second World War, a great percentage of the organic chemical industry used coal tar as the substrate to form additional agents. More recently, the great majority of organic compounds have been derived from petroleum-based products, although some specialty chemicals are still made from coal tar (National Research Council, 1995).

A BRIEF HISTORY OF THE STUDY OF LUNG DISEASE IN COAL MINERS

As one might expect, the earliest coal mines were outcroppings of coal along the banks of rivers. As such, they could be described as surface mines. The drive for coal production on a large scale was associated with the manufacture of iron and derived from the development of the steam engine in the early part of the 18th century. It was not until 1762 that a steam engine was used to work a colliery in Scotland (Meiklejohn, 1951). Yet, even underground mining grew quickly as proper methods for ventilating fresh air and toxic gases from the mines, as well as effectively pumping water from the mines, were put into place. As of 1866, in Scotland, there were 472 coal mines, 41 000 miners and 12 million tons of coal produced yearly.

Meiklejohn (1951, 1952a,b) has provided a wonderful summary of the history of the health of United Kingdom coal miners. Lung disease in miners has been referred to as miners' asthma, phthisis, anthracosis, and in Scotland, miners' black lung. In 1831, the first report of 'black lungs' attributed to employment in coal mines was published (Gregory, 1831). He described a 59-year-old man who had been employed as a miner for 10 or 12 years and was hospitalized for generalized anasarca. He soon died from progressive heart failure. The lungs were examined, and a picture consistent with progressive massive fibrosis (PMF) with cavitation of these large lesions on a background of simple coal workers' pneumoconiosis (CWP) was described.

> *'At necropsy, the lungs were universally adherent to the chest wall; the pleura was thickened and in places ossified. When cut into, both lungs presented one uniform black carbonaceous color, pervading every part of their substance . . . The left lung did not appear to contain any cavities, but was condensed and loaded with black serum. Some minute hard points could be felt in various parts of both lungs'.*

Opinions regarding the explanation for this black pigment included gunpowder, inhalation of lamp-black, or soot from the oil lamps of the miners. Marshall (1833–1834), just several years later, concluded that: 'The true explanation of the origin of this disease in colliers seems to be, that it is in consequence of the inhalation of fine coal dust, and its deposition in the substance of the lung. That coal may float through the air in particles sufficiently fine to be inhaled without immediate irritation and that it is thus inhaled is a matter of common observation'.

Yet, for the rest of the 19th century, there was considerable controversy and disagreement regarding the impact on coal dust inhalation on the survival of miners. Overall, data showed the mortality of coal miners only minimally exceeded the mortality of the most healthy men, such as farmers or agricultural laborers. This lack of coal dust effect on survival was thought to be attributable to better mine ventilation and lesser exposures to coal dust. Yet, even then, the relatively less fibrogenicity of coal dust compared with silica was well recognized (Ogle, 1885; Meiklejohn, 1952a).

Early in the 20th century, the understanding of lung disease in miners again changed. Collis (1915) opined that coal dust did not produce pneumoconiosis (and even served to protect against tuberculosis) but was a marker which showed the path by which dust enters and travels in the lung. He attributed dust disease in miners to silica inhalation (associated with working of thinner seams with likely silica contamination) and reported the disease that developed to be silicosis. Yet, this view failed to take into account increasing mechanization in the mines and deepening of the mines, with the resulting increase in coal dust generation. The adverse effects of these factors more than compensated for the positive effect of increased mine ventilation (Meiklejohn, 1952b) (Figure 11.1).

It was not until washed coal, free of silica, was recognized to produce dust disease in the lungs in stevedores, those employed loading and leveling coal in the holds of ships, that CWP was widely accepted as being pathologically distinct from silicosis (Collis and Gilchrist, 1928; Gough, 1940; Hepplestone, 1947). Later, King *et al.* (1956) showed that the radiologic and histologic severity of pneumoconiosis in coal miners was related to the total amount of dust and not the silica content of the coal. Although it was recognized that mining was associated with lung disease, only with the development of specialized techniques such as chest radiography, the discovery of the tubercle bacillus, and the application of the histologic examination of lung tissue that the specific causation of respiratory diseases affecting miners was identified.

With the development of medical tools necessary to make an accurate diagnosis of pneumoconiosis, the emphasis from the 1930s to the present time has been on epidemiologic studies of the prevalence of pneumoconiosis and attempts to understand the basic mechanism of both CWP and silicosis.

HOW COAL IS MINED TODAY

Safe mining requires close attention to numerous aspects of coal production. Although we are primarily interested in the respiratory consequences of coal mine dust inhalation, mining is inherently dangerous with recognized risks of underground fires and explosions, toxic gas excess and rock falls. Therefore, safety measures must emphasize adequate ventilation so that methane is removed and the risk of explosion and fire

Figure 11.1 The approach to mining in the earlier part of this century where coal was removed with a pick and shovel and hand loaded into a shuttle car.

minimized; protection from roof falls; and the control of dust. Each of the different methods of mining presents their own challenge in controlling the workplace environment.

Coal may be found in outcroppings and in seams that are sometimes just a few feet below the surface. In these deposits coal can be readily obtained by simply scraping away the surface or overburden and extracting the coal with earth-moving equipment. This is called surface or strip mining. Dust levels in the air at surface mines are generally considerably lower than in underground mines with few notable exceptions. In

Figure 11.2 (a) Typical rock drilling for overburden removal at an above ground mine in West Virginia. Highly respirable dust is generated, often with a high silica content. The operator is required to stand near the drill to control the drilling operation. (b) A damaged rubber skirting which surrounds the drill at the level of the ground. This is used for local dust control. In this instance, the skirt is damaged and dust generated by drilling is not captured. Such exposures may be hazardous to the drill operators and the driller helpers. (c) The operator standing in an intense dust cloud while operating the drill. (d) A driller helper, equipped with a hard hat and hearing protection, but without respiratory protection, standing near an on-going drilling process. (Figure 11.2 (a–d) are courtesy of Joseph C. Cocalis, P.E., CIH.)

surface miners and coal plant cleaning workers in the anthracite mining area of the US, the mean FVC, FEV_1 and peak flow rates was not related to the number of years worked in coal-cleaning plants (Amandus *et al.*, 1989). Yet, there is a group of these workers at special risk for pneumoconiosis. Workers who operate the large drills (drillers and driller helpers) to make holes in which explosives are placed at surface coal mines so that the overburden might be removed are exposed to silica and are at risk for the development of silicosis rather than CWP (Banks *et al.*, 1983) (Figure 11.2 (a–d)).

To a certain extent, workers in some exclusively above-ground operations may be exposed to coal dust. These are the workers at the tipples where crushing, sizing, washing and blending of coal is done; at surface coal mine sites away for the drilling operations; and at locations where coal is loaded into ships, railroad cars or river barges. Although each case needs to be evaluated on an individual basis, the great majority of these exposures are relatively small and insufficient to cause disease.

On hillsides where the coal seam may be many feet below the surface of the hill but some outcropping occurs from the side of the hill, a feasible method of mining coal is to bore into the outcropping with an auger. Depending upon how far into the hillside the seam extends, this method of mining is a combination of surface and underground mining (Figure 11.3).

Where coal seams are buried deep, typically greater than 200 feet underground, it is not economically feasible to strip away the overburden. The only practical way of mining the coal is to sink shafts from the surface to the coal seam and then follow the seam

Figure 11.3 An auger fracturing overburden at a hillside mine.

with a series of more or less horizontal tunnels. This has been the most prevalent method of mining coal in this century (Figure 11.4 (a–d)).

Modern underground coal mining methods are categorized as room and pillar mines or long-wall mines. A room and pillar mine is a series of parallel rooms where mining has occurred separated by pillars of unmined coal. As mining goes on this looks increasingly like a honeycomb. Two types of mining are performed in room and pillar mining. The first is *conventional mining*, and is the least mechanized, accounting for less than 10% of US coal mine production (Coleman, 1992). Here, incisions are cut in the sides and bases of the seam and charges inserted through holes drilled in these areas. The roof is first supported, then the charges are exploded, and the coal loaded and removed from the face. The cycle is typically performed sequentially in adjacent areas to maximize coal production (Taylor and Thakur, 1993).

Alternatively, *continuous mining* is a mechanized process which accounts for approximately 60% of US coal production. This approach uses a machine with a rotating head to break coal from the face. The operator advances this machine through the seam, moving the cutting head of the machine up and down to maximize coal breakaway from the face. A conveyor, incorporated into the continuous mining machine, moves the broken coal from the face to waiting underground shuttle cars. The continuous miner may include roof bolter units located just behind the rotating cutter head or have a roof bolting machine nearby. Workers remain here while mining is conducted.

In this method of mining, between one-and two-thirds of the coal remains within the pillars after the rooms are mined. To recover the coal within the pillars, each pillar is mined using timber for temporary roof support. The timber is then removed, thereby allowing the roof in that area to collapse. This approach is called retreat mining because after mining of the pillars begins, no attempt is made to go back into the block. The roof caves in and the mine is abandoned (Stout, 1980).

Longwall mining is the most mechanized and the most productive means of underground mining. The number of underground mines employing longwall mining techniques continues to increase in the US (and now is used in approximately 100 mines). This accounts for approximately 30% of US coal production. Longwall mining produces up to 40 tons of coal per minute, and in some instances, nearly 4000 tons per work shift. In this process, the continuous miner cuts out a longwall of coal which may be 600–800 feet (180–250 m) wide and 5000–12 000 feet (1500–3600 m) long. When the wall is formed, the hydraulically powered roof supports, the cutting

(a) (b) (c) (d)

Figure 11.4 (a) A coal cutter. This is an integral part of room and pillar mining. The 'saw' may be 15–20 feet (4–6 m) in length. The operator sits behind the saw in the motorized open cab. (b) A cutting 'ball' operating at the edge of a coal seam. Frequently, a conveyor is placed at the seam to capture and remove the cut coal. (c) Multiple cutting heads on a coal cutting machine. The operator sits behind the machine in close proximity to the freshly fractured coal. (d) Roof supports with jacks and preparation for additional roof bolting. This underground job is associated with silica exposure and a potential for the development of silicosis. The risk for silicosis in an underground coal miner is primarily related to the extent of roof bolting, and underground 'drilling' in general.

machine or shearer (often with built-in water sprays to minimize dust generation), and the conveyors are put into place. The shearer moves across the coal face in a bi-directional manner with a depth of cut of about 2 ft (0.6 m). A conveyor belt, which runs the length of the longwall, captures the newly excavated coal. As each pass is made by the shearer (which requires one or two workers to operate and travels up to 60 ft/min), the roof supports and the conveyor are advanced. As these supports are advanced, the roof behind these supports caves in, another example of retreat mining (Euler, 1981).

Finally, it should be noted that longwall mining has several disadvantages. First, if the cutting machine or any machine necessary for the process fails, production of coal from the entire system stops. Second, a coal face of 600–800 feet (180–250 m) may have numerous different characteristics. Attempting to mine this extensive block may lead to mining difficulties. Third, developing and setting up a longwall requires considerable time and worker effort. Finally, the large number of roof supports for a longwall are very costly. Using fewer supports would cut costs. This has led operators to combine the roof support system of the longwall mine with the flexibility of the continuous mining machine. This has been named *shortwall mining*. With this approach, the face is shortened to about one-third of the length of the longwall, production is increased relative to room and pillar mining, and the miner works under the hydraulic roof support the entire time, thereby lessening the hazards of roof falls. Coal production has increased for several decades in the US, while the number of working miners has decreased (NIOSH, 1995) (Figure 11.5 (a–c)).

CLINICAL AND EPIDEMIOLOGIC FEATURES OF COAL-INDUCED LUNG DISEASE

Coal inhalation can result in industrial bronchitis and CWP, presenting as either simple pneumoconiosis or complicated CWP (PMF). In addition, there has been considerable discussion regarding the relationship between coal dust exposure and clinically significant airways obstruction in the recent literature. This relationship between inorganic dust exposure, including inhalation of coal dust and airways obstruction is discussed in detail in Chapter 5.

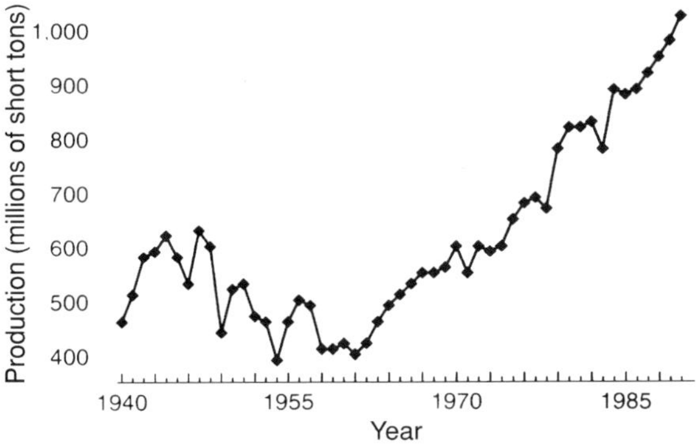

(a)

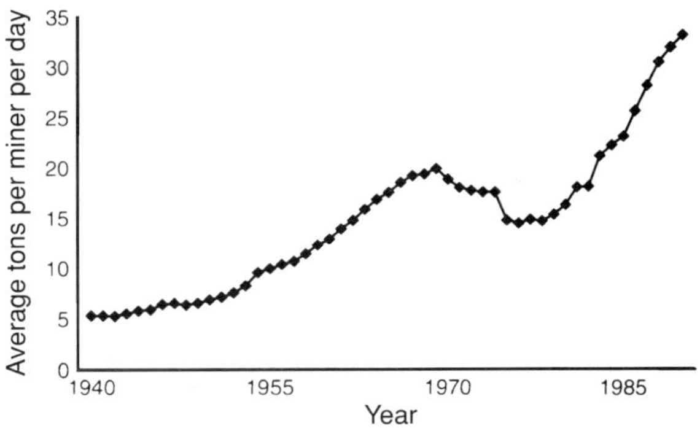

(b)

(c)

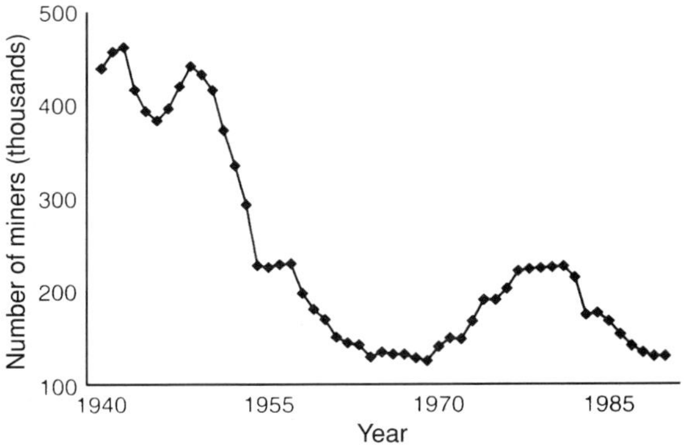

Figure 11.5 The effect of industrialization on coal output in the US over the past 50 years. There has been a dramatic rise in coal production (a,b) in association with a large decrease in the number of miners (c) (NIOSH, 1995).

Industrial bronchitis is a common diagnosis among workers exposed to dusts, including coal dust (Morgan, 1978). It manifests as a productive cough which persists for at least 3 months per year for at least 2 years (chronic bronchitis) associated with workplace dust exposure. The relationship between coal mining and bronchial mucous gland dimensions was presented in an autopsy study by Douglas *et al.* (1982). Coal dust exposure resulted in an increase in the maximal gland : wall ratio independent of smoking, but no relationship between mucous gland size and lung dust or pneumoconiosis was found. This suggested that pneumoconiosis is related to respirable dust exposure, and mucous gland enlargement is related to the inhalation of larger (non-respirable) dust particles which present a chronic burden to the mucociliary escalator and act as irritants to the airway.

Early reports showed that the prevalence of bronchitis in coal miners varied by smoking history, age and job (Kibelstis *et al.*, 1973). In all reports, smoking miners had bronchitis more frequently than non-smokers, and when studied, showed bronchitis to increase with age (a surrogate for years of dust exposure in the workplace). Marine *et al.* (1988) identified 543 lifetime non-smokers among 14 888 British coal miners (3.6%) studied between 1953 and 1967 who were less than 65 years of age, worked at the coal face in most instances, did not have PMF, and who had participated in three surveys over 10 years. Of these, 17% had bronchitis at the third survey. In another report of coal miners, 16% had bronchitis (Seixas *et al.*, 1992). There is little information on the natural history of chronic bronchitis in non-smoking miners once they are removed from dust, but it is thought to resolve when dust exposure ends. In those who develop chronic bronchitis due to dust, there is often a measurable, but not clinically significant, decline in FEV_1.

CWP results from the inhalation and deposition of coal mine dust, and the lung's reaction to its presence. Three criteria are necessary for this diagnosis (Balaan *et al.*, 1993). They include:

1. A chest radiograph consistent with the features of CWP.
2. A work history (typically that of underground coal mining) which is sufficient in exposure and latency to result in pneumoconiosis.
3. The absence of other illnesses which may mimic CWP.

Therefore, the diagnosis of CWP is a radiologic diagnosis which can be made with confidence without histologic confirmation, by an adequate history of coal dust exposure and a characteristic chest radiograph. Clinical features such as dyspnea, cough and sputum production are important in addressing the degree of a miner's overall respiratory impairment, but are not a part of the diagnostic criteria. The radiographic appearance of CWP is similar to that of silicosis, and miners may develop disease from exposure to both dusts during their employment as a coal miner.

Although numerous radiographic classifications have been used, the first attempt at an international consensus regarding classification of the radiographs of pneumoconiosis was put forth by Cochrane and colleagues in 1951. Since then, a series of approaches have been taken, and although this is recognized to be an imperfect system (Wagner *et al.*, 1993), the International Labour Office 1980 classification is the most widely accepted (ILO, 1980). This is discussed in detail in Chapter 9.

Typically, CWP is described as simple pneumoconiosis or PMF. The radiograph in simple pneumoconiosis shows small opacities, ranging in size from pinhead to 1 cm in diameter. Rounded nodules predominate and tend to appear first in the upper zones and then the middle and lower zones as the number of opacities increase. With prolonged excessive exposure, these small opacities may coalesce and form larger opacities, recognized as PMF lesions and characterized by one or more large opacities greater than 1 cm in diameter. As the disease progresses, the upper zone nodules begin to coalesce and distort the lung architecture, a feature which gradually becomes more prominent. This typically presents as deviation of the trachea and major airways to the side of the most prominent area of coalescence, loss of upper zone lung volume, elevation of the hila, and basilar emphysema (typically of a panacinar type), attributable to traction placed on the lower lung zones by the elevation of the hila.

There are no symptoms or physical signs associated with simple CWP. The not infrequent presence of a chronic cough and sputum production, even in the presence of CWP, is attributable to 'industrial bronchitis'. Alternatively, these same clinical features in a smoking miner may be partially attributable to bronchitis caused by the inflammatory stimulus of cigarette smoke. Finger clubbing is not a feature of CWP and, if noted, should prompt further investigations.

When PMF is recognized on the chest radiograph, the worker frequently describes dyspnea, cough and sputum production, although it is well recognized that the degree of impairment or the presence or absence of symptoms does not always correlate well with the extent of chest radiographic abnormalities.

The effect of pneumoconiosis on the heart has been described by several investigators. Lapp *et al.* (1971) performed cardiac catheterization during rest and exercise in 43 miners with CWP (23 with airways obstruction and 24 without). In seven of 12 miners, either chronic obstructive lung disease or PMF appeared to explain the increase in pulmonary artery pressures. The other five with increased pressures had the smaller type of pinpoint (p) opacity on their radiograph. It is likely pertinent that the pinpoint type of opacity is associated with diminished diffusion and a diminution of the vascular bed (Lyons *et al.*, 1967). A later autopsy study of 215 British coal miners (100 with simple CWP or no pneumoconiosis and 115 with PMF) was designed to address the relationship between coal dust exposure and right ventricular hypertrophy (RVH). In this population, the prevalence of RVH was the same in those with simple CWP as in those without pneumoconiosis (approximately 15%) and was related to the extent of airways obstruction. Overall, RV enlargement did not occur unless the coal miner was a smoker with severe airflow obstruction or had developed PMF (Fernie *et al.*, 1983).

CWP alone does not appear significantly to increase the risk for development of coexisting mycobacterial infection. Miners, however, often have exposure to silica in drilling, a job typically associated with silica, rather than coal dust exposure. Miners with silica exposure in the workplace may also be at risk for mycobacterial infection. The appearance of a cavity in a PMF lesion or an aggressive and unexplained rate of radiographic progression should prompt examination of the sputum for mycobacteria.

The chest radiograph in simple CWP correlates with the amount of dust in the lung at autopsy (Rivers *et al.*, 1962). This is not true of the complicated form of the disease, suggesting that inadequately defined host factors play a role in the development of this lesion. Hypotheses which might explain the differences in tissue response to coal dust in those who remain with simple CWP and those who progress to PMF include:

- inhalation of significant amounts of silica in addition to coal in mine dust;
- infection with *M. tuberculosis* or an atypical mycobacterial organism; and
- development of specific immunologic factors which are responsible for progression.

Yet, none of these explanations appears to explain satisfactorily the development of PMF. First, PMF has been reported in carbon electrode workers. This suggests that silica exposure is not necessary for PMF to occur (Collis and Gilchrist, 1928). Treatment of miners with simple CWP in Wales with anti-tuberculous drugs did not prevent the occurrence of PMF (Watson *et al.*, 1959). Autoantibodies reactive to lung tissue antigens have been demonstrated in the sera of miners, suggesting a role in the tissue response. There are factors driving the development of PMF beyond the amount of dust in the lung.

A special case of nodular lung reaction occurs in dust-exposed individuals who either have rheumatoid arthritis or who develop rheumatoid arthritis within the subsequent 5 to 10 years (Caplan, 1953; Caplan *et al.*, 1962). This leads to discussion regarding the potential role of immunologic factors in mineral dust pneumoconioses. The nodules of Caplan's syndrome vary in diameter from 0.5 to 5 cm, and are usually multiple, bilateral and situated peripherally. Grossly, the lesions resemble a giant silicotic nodule. Microscopically, the amount of dust in the lesion is small, there is a necrotic area in the center, and there is a surrounding cellular zone infiltrated with lymphocytes and plasma cells. In many nodules there is a peripheral zone of active inflammation with neutrophils and a few macrophages. This observation was further enhanced when similar radiographic appearances in miners without arthritis, but in whom circulating rheumatoid factor was demonstrated, was described. An increased incidence of circulating positive rheumatoid factor among miners with PMF has also been reported. In a group of 60 consecutive cases of scleroderma, Rodnan *et al.* (1967) described 26 with a history of employment as a coal miner or employment in an occupation where silica exposure was encountered. Although the course of scleroderma did not vary in the two groups, it appears that chronic fibrogenic dust exposure may be a risk factor for the development of scleroderma.

The clinician may be presented with the diagnostic dilemma of attempting to distinguish a primary or metastatic neoplasm from an unusual presentation of PMF. When large opacities of PMF occur bilaterally on a background of simple CWP, one can be reasonably confident that the lesions are less likely to represent neoplastic disease. When there is a sparse background of simple CWP or none, or there are crops of nodules (as in Caplan's syndrome), the differentiation from neoplasm may indeed be difficult.

EPIDEMIOLOGY

A series of studies in the 1960s showed that the prevalence of CWP varied by region of the country, the type (rank) of coal mined (Lainhart *et al.*, 1969), and the duration and intensity of exposure (McBride *et al.*, 1963, 1966). Among underground miners, those working at the face and exposed to higher concentrations of coal mine dust had a higher prevalence of CWP than surface workers or those whose jobs caused them to enter the face area intermittently (Jacobsen, 1971). For example, miners in Pennsylvania anthracite seams had a considerably greater prevalence of CWP than those miners in the western plateau of Colorado and Utah who mined a lower rank of coal. The institution and enforcement of dust control measures has reduced the attack rate of CWP and may have lessened progression among those with radiographic opacities who continued to work in mines under the US dust standard which was begun in 1969 at 3 mg/m^3 and reduced to 2 mg/m^3 in 1972 (Coal Mine Health & Safety Act, 1969).

A series of prevalence studies of miners with CWP have been conducted by the US Public Health Service. The first, round one of the US coal workers' surveillance program, was performed on data collected from 1970 to 1973. The most recent, round five of this same program, was undertaken from 1987 to 1991. Data from these studies have been presented in summary form in an impressive series of reports (Attfield and Castellan, 1992; Attfield and Althouse, 1992; Attfield, 1992). The mines were chosen to represent different geographic areas, coal seams and coal mining methods. Briefly, although the data are complicated by chest X-ray reader variation and concerns regarding worker self-selection for examination, there appears to be a reduction in the prevalence of CWP in the miners who have participated in this study over time. When comparing studies from the mid-1970s with the mid-1980s, the number of cases with both simple CWP and PMF declined in all groups with at least 10 years of mining. In one group, miners with a mining tenure of between 15 and 24 years, the prevalence decreased to less than half. Similarly, in those with a lifetime of coal mining, 11% had category 2 simple CWP in round one compared with 2% in round 4.

The most important concern about these data, however, is whether the recent work is representative of the miners' respiratory health and whether this decline in pneumoconiosis reflects a true decrease or self-selection inducing bias. Most importantly, in the years 1970–1973, approximately three-fourths of the nation's 100 000 miners participated in the program. Participation decreased to less than 30% of the mining work force in the study performed from 1987–1991.

With this diminished level of participation, death certificate data may be the most accurate measure of the number of miners with CWP. Over the next decade, a decline in the number of death certificates showing CWP may confirm the conclusions suggested by epidemiologic studies (NIOSH, 1994). In this report, the number of US death certificates mentioning CWP appears to have initially increased from the numbers reported in 1968, to have changed little over 20 years, and to have perhaps begun a minimal downward trend. This leads one to be concerned whether the current coal dust standard is sufficient to block the development of CWP in miners with coal dust exposure remains a concern (Figure 11.6).

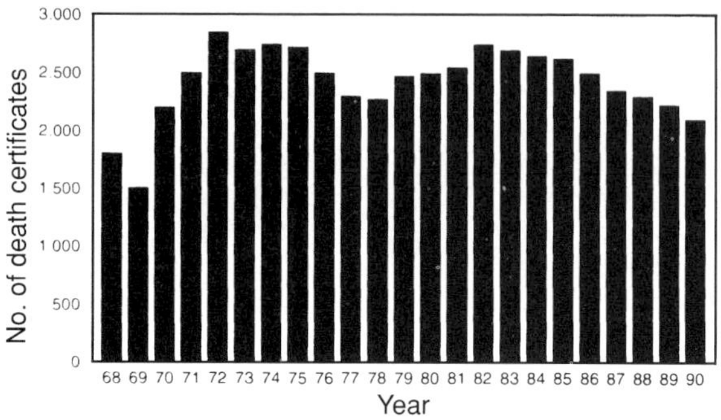

Figure 11.6 Numbers of death certificates in the US with mention of coal workers' pneumoconiosis by year. (Adapted from the NIOSH Work Related Lung Disease Surveillance Report, 1994.)

Data from British studies have also clearly shown that the attack rate (incidence of new cases) and the probability of progressing to a higher category of simple CWP were related to the mass of respirable dust to which the miner was exposed during his lifetime (McLintock *et al.*, 1971). Yet, once the miner with simple CWP leaves dust exposure, the chest radiograph is not likely to progress. The same cannot be said for the PMF. The rate of progression to PMF appears to be influenced chiefly by the age at which the miner begins to show radiographic changes of CWP. Progression may also be influenced by the presence of a rheumatoid diathesis (Davies and Mann, 1949).

Does the chronic inhalation of coal dust increase the risk for a premature death? A number of studies of mortality in coal miners CWP have been primarily conducted in Europe and the US. Perhaps the best to summarize the rates of mortality in the first part of this century was Enterline (1972), who noted that it was not until 1950 that relatively accurate death certificate data which described an individual's usual occupation were available. In US Public Health Service data collected from death certificates for the year 1950 (Enterline, 1964), there was a gross excess in mortality for all ages for coal miners compared with the general population. For example, in miners aged 45–64 years, the Standard Mortality Ratio (SMR, the ratio between the actual number of deaths and the observed number of deaths) approximated 2.0.

To emphasize further the mortality risk of being a coal miner between the years 1949 and 1963, the Society of Actuaries (1967) published SMRs on 44 US occupations using life insurance policy data. Of all of these occupations, coal mining had the highest SMR. The SMR for coal miners was 1.5. The SMR for death attributable to respiratory disease exceeded 40 using death certificate data and 11 using life insurance policy information. Overall, most SMRs for both the death certificate data and the life insurance data for all of the causes of death were increased in miners, with the only clear consistency being the discrepancy between the death certificate study showing an SMR of 1.7 for deaths due to cancer compared with only 0.8 in the insurance policy study. Both the SMR from death certificates which measured mortality attributable to accidents at work and the SMR from insurance company data due to accidents and homicides exceeded 6.0. Yet, even when accidents and respiratory disease were eliminated as causes of death, the SMR for coal miners ranged between 1.4 and 1.7 times the death rate for all working men. Even as late as 1979, a coal miner was recognized to have a two-fold risk of being killed at work compared with the risk of the average worker (Editorial, 1979).

The mortality experience of more than 6000 miners and ex-miners residing in the Rhondda Fach (Cochrane *et al.*, 1978; Atuhaire *et al.*, 1985), a mining community in South Wales, also showed an excess mortality. In the absence of pneumoconiosis, or when simple pneumoconiosis, or even PMF of category A was present, the SMR approximated 1.2. The number of deaths did not vary by pneumoconiosis category unless category B or C large opacities were present. When this was the case, the SMR was almost double that of the comparison group. Chronic bronchitis was identified twice as often as a cause of death among the miners. Additionally, carcinoma of the stomach increased with pneumoconiosis category, a feature frequently reported in other studies. Ortmeyer *et al.* (1973) reviewed mortality for compensated Pennsylvania miners by category of opacity on the chest radiograph and recorded similar results.

Stocks (1962) was one of the first to present the relationship between mining and gastric cancer, but his report made it clear that it was necessary to compare the mining risks for illness with local controls. He showed that the rate of gastric cancer among miners in the United Kingdom varied, with it being most frequent in South Wales, intermediate in northern England, and least in the Midlands. The relationship between this malignancy and coal dust exposure was also suggested in a study of gastric carcinoma in Utah, where gastric malignancy occurred nearly three times as often in residents of two Utah coal mining counties, leading the authors to speculate that swallowed coal particles (containing polyaromatic hydrocarbons, including benzpyrene) increases the carcinogenic risk (Matolo *et al.*, 1972).

Further understanding of the mortality experience of US coal miners on the rolls of the United Mine Workers Health and Retirement Fund from 1959 to 1971 was presented by Rockette (1977). Of the 23 232 men enrolled, 7741 died. The overall SMR for this population was 1.02, a dramatic improvement compared with the above-cited death certificate study (Enterline, 1964). Yet, the SMRs for several specific categories of illness remained elevated. Specifically, the SMR for non-malignant respiratory disease was 1.6 and that for accidents was 1.8. Although mortality due to all cancers was not increased, the SMRs for stomach and lung cancer were excessive (1.4 and 1.1, respectively).

In a mortality report of Appalachian coal miners and ex-miners followed from 1963 to 1971, the mortality of current miners was 7% less than expected, although ex-miner mortality was 24% greater than expected (Ortmeyer *et al.*, 1974). Simple pneumoconiosis did not affect life expectancy, while the presence of PMF did.

Meijers *et al.* (1991) postulated that coal workers have a greater risk of lung cancer due to their exposure to coal mine dust, a material containing various potentially carcinogenic organic (i.e. hydrocarbons) and inorganic (i.e. trace elements such as cadmium and chromium) compounds. In support of this hypothesis, crystalline silica – a component of coal dust – had been declared a probable carcinogen (IARC, 1987). The cause of death of 334 miners from a cohort of 5400 miners with CWP employed in 11 Dutch coal mines from 1956 to 1960 was evaluated in 1983. Data showed a higher than expected mortality, primarily attributed to stomach and large and small intestine malignancy, as well as more frequent non-malignant respiratory diseases.

Exposure data were available in 26 363 coal miners from 20 collieries in England and Wales who first attended medical screening between 1953 and 1958 (Miller and Jacobsen, 1985). A chest radiograph, respiratory questionnaire and coal mining history was recorded and estimates of exposure made. An assessment of the causes of 8489 deaths was begun in 1980. First, the general mortality was 13% less on average than in English and Welsh men in the same region. The survival of miners with category A large opacities was less than that of men without pneumoconiosis, yet the mortality rate did not increase with increasing category of simple pneumoconiosis. Mortality due to all causes, as well as that attributable to pneumoconiosis, bronchitis and emphysema, increased with cumulative dust exposure. Furthermore, in those with greater dust exposures at the start of the evaluation, particularly those of an older age group, a shorter survival was reported. There was no association between lung cancer and coal mining, but the risk for cancer of the digestive system approached significance. Similarly, there was no relationship between gastric mortality and survival in Australian coal miners (Armstrong *et al.*, 1979).

The mortality experience of US coal miners with exposure estimates was reported by Kuempel *et al.* (1995). In this population, the mortality experience of

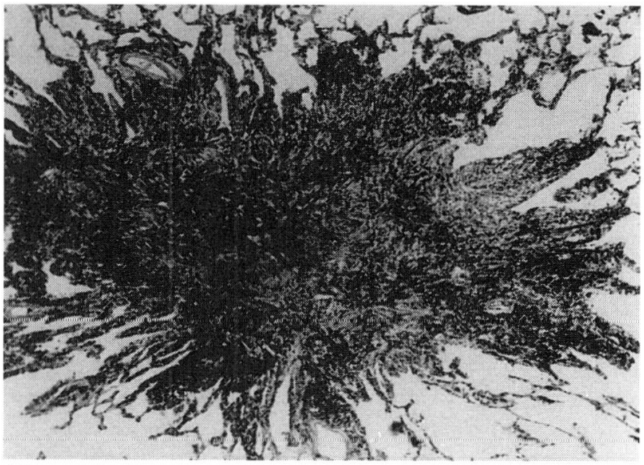

(a)

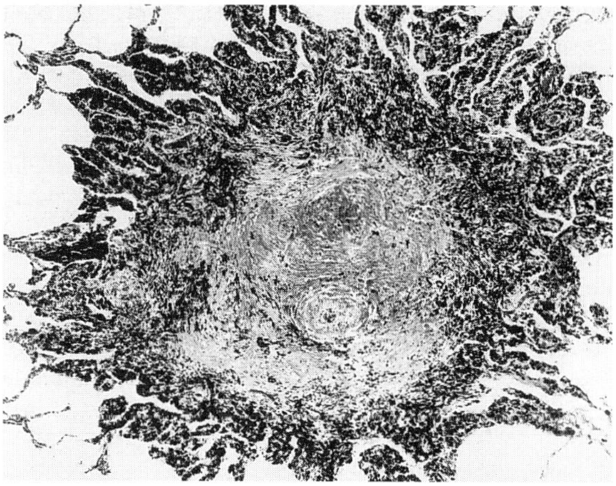

(b)

Figure 11.7 (a) Photomicrograph showing a coal macule. There is coal dust, reticulin fibers, and coal dust-laden macrophages, with minimal fibrosis. The periphery of the nodule shows air space enlargement. (b) Photomicrograph of a lung biopsy from a roof bolter with histologic evidence of both a coal macule and a silicotic nodule. This reflects the coal dust and silica dust exposures which have occurred over his working lifetime. On the periphery, there is coal dust and coal-laden macrophages. Centrally, there are whorls of fibrosis characteristic of a silicotic nodule. The cellularity is most prominent in the periphery of the lesion. (Figure courtesy of V. Vallyathan, PhD.)

793 men among a population of 8878 miners enrolled between 1969 and 1971 and followed until 1979 was evaluated. Overall, the SMR for all causes was 0.85, the SMRs for bronchitis, emphysema, lung cancer and stomach cancer were not increased, but the SMR for pneumoconiosis was 3.7. No statistical relationship between deaths due to lung cancer or stomach cancer was reported.

In summary, a series of mortality reports have not convincingly shown that simple CWP is associated with premature mortality, but that PMF adversely affects survival, most apparent when category B and C large opacities are present. Although there appears to be no clear difference in the SMR for lung cancer between miners and non-miners (Vallyathan *et al.*, 1985), a number of studies show an increase in the SMR for digestive cancer in general, and stomach cancer in particular.

PATHOLOGY OF COAL WORKERS' PNEUMOCONIOSIS

As the normal dust clearance mechanisms of the lung are overwhelmed, dust deposition increases. With the initiation and progression of fibrosis, the lung lesions increase in size and number. A focal collection of coal dust in pigment-laden macrophages around dilated respiratory bronchioles which tapers off towards the alveolar duct is initially apparent (Hepplestone, 1954). This is the coal macule, the characteristic lesion of CWP (Figure 11.7). A fine network of reticulin within this collection of cells may be visible early on. Focal emphysema is a specific entity that is an integral part of the simple lesion of simple CWP (Figure 11.8). It is characterized by enlargement of

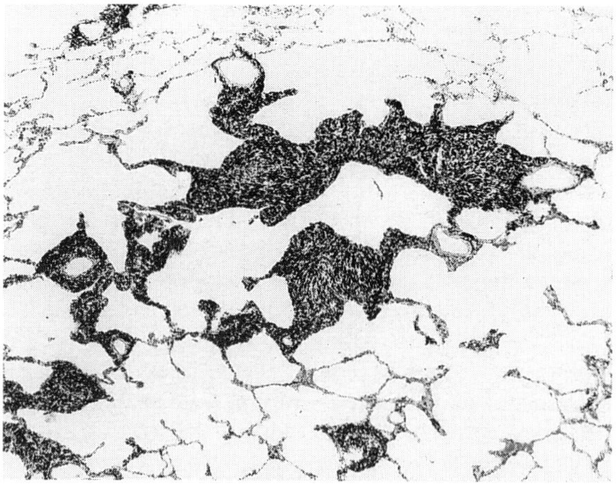

Figure 11.8 A coal macule surrounding a respiratory bronchiole. As the macule enlarges, the smooth muscle atrophies, and the bronchiole enlarges. Of interest in this photomicrograph are the enlarged air spaces surrounding the macule, a feature described as focal emphysema. (Figure courtesy of V. Vallyathan, PhD.)

the air spaces immediately adjacent to the dust macule (Kleinerman *et al.*, 1979).

What impact this has on lung function has been the subject of considerable discussion. In non-smoking, non-obstructed miners, Morgan *et al.* (1971) showed an increase in the residual volume (RV) in miners. In miners without CWP, the RV was 105% of a group of non-miners. This increased to 108% and 114% of controls in miners with category 1 or category 2 and 3 simple CWP. Although Morgan considered that this may be attributable to focal emphysema, he thought that unlikely because pathologic disruption of the bronchioles was absent. Rather, he considered this hyperinflation to be the result of coal macules narrowing the airways with alteration of the peripheral flow rates by and increasing resistance.

On gross examination of the lung, larger collections of dust are described as coal nodules, classified as micronodular if they are 7 mm in diameter or less, and macronodular if they are larger than this. These are palpable whereas coal macules are not.

PMF is diagnosed when one or more nodules attain a size of 2 cm or greater in diameter, typically on a background of simple CWP (Shennan *et al.*, 1981; Hodous, 1990) (Figures 11.9 and 11.10). The 2 cm diameter is an arbitrary choice of a minimal diameter that has allowed better correlation with clinical and radiographic measurements. Gross examination of the lung in PMF reveals a solid, heavily pigmented lung which is rubbery-to-hard in texture. These features are most common in the apical posterior portions of the upper lobes or the superior segments of the lower

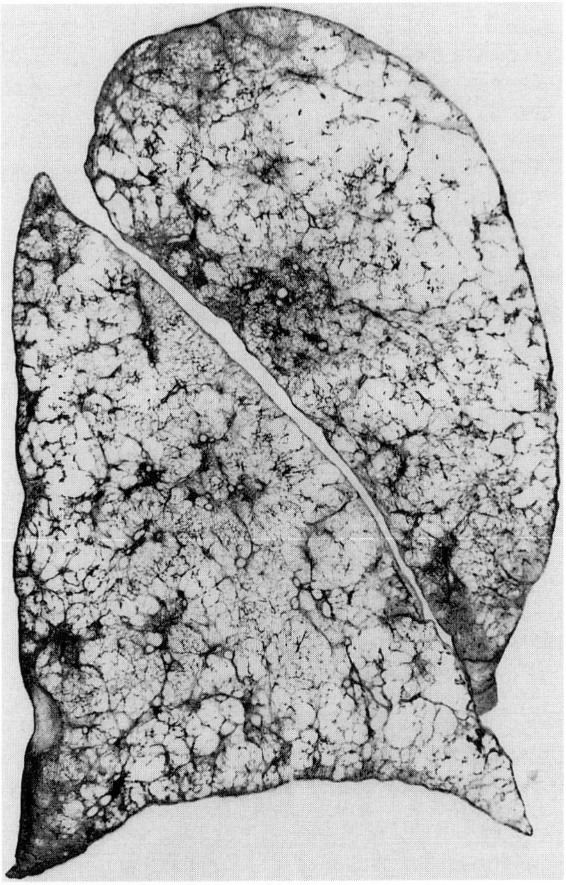

Figure 11.9 Simple Coal Workers' Pneumoconiosis (category 1) on a background of diffuse emphysema. The black deposits are the coal macules. The degree of emphysema is widespread and does not appear associated with the coal macules. (Figure courtesy of V. Vallyathan, PhD.)

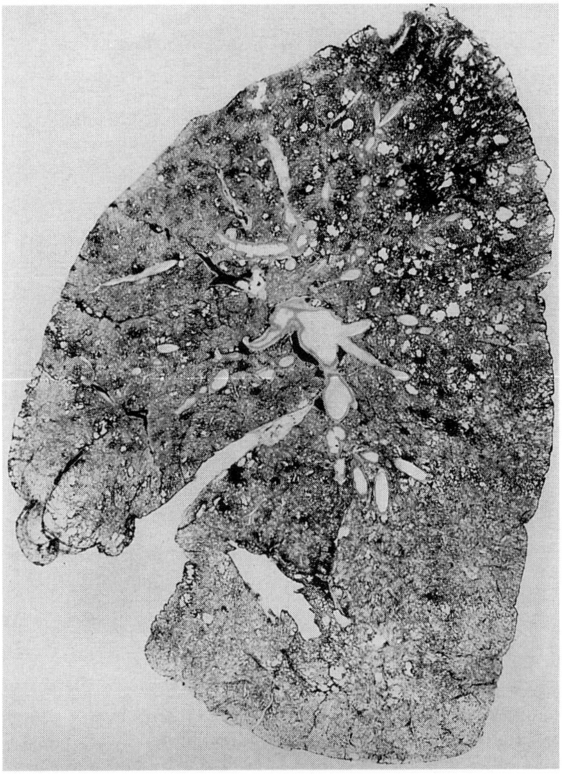

Figure 11.10 Progressive Massive Fibrosis in the upper zone on a background of coal macules, a few of which show focal emphysema. (Figure courtesy of V. Vallyathan, PhD.)

lobes. These lesions tend to occur asymmetrically, occasionally showing first in one lung and then the other, leading to a suspicion of malignancy. When these lesions are ashed, they appear to be composed of varying amounts of coal, silica, calcium and other substances. About one-fourth of the proteinaceous material in the center of these lesions is collagen (Wagner, 1972). These lesions may also cavitate and the worker may expectorate an ink-like fluid (a clinical sign described as melanoptysis), or when these cavitary lesions are cut, they may drain ink-like fluid. Airways and vessels adjacent to the lesions are distorted and destroyed within the lesions.

Silica exposures, in surface coal mine drillers, can also induce massive fibrotic lesions and result in a miner's death because of cor pulmonae and respiratory failure (Figure 11.11).

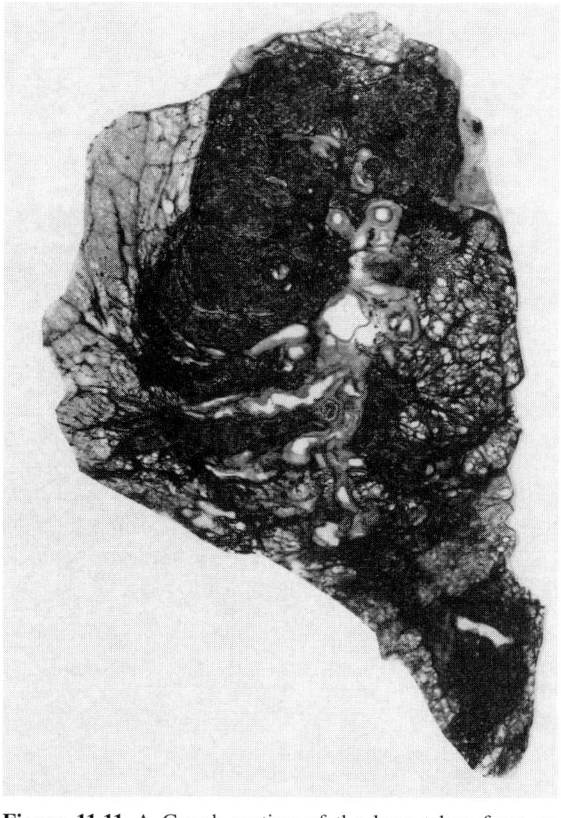

Figure 11.11 A Gough section of the lung taken from an autopsy of a 46-year-old West Virginian surface coal miner. His primary job was hard rock drilling to remove overburden. He died of silicosis with cor pulmonale and respiratory failure. The section demonstrates massive scarring with secondary emphysema and parenchymal distortion attributable to the pulmonary response to silica inhalation.

HOW CWP DEVELOPS: CELLULAR AND IMMUNOLOGIC FACTORS

CWP is the result of coal dust-induced cell damage with activation of the fibrotic process. Lapp and Castranova (1993) provided a well-outlined approach to addressing how coal dust causes lung damage. The potential mechanisms include:

- direct cytotoxicity of coal dust;
- release of oxidants, enzymes and cell membrane constituents from alveolar macrophages in association with cell death due to coal dust exposure; and
- stimulation of cytokine release from alveolar macrophages to recruit effector cells (other macrophages or neutrophils) and stimulate fibroblast proliferation and collagen synthesis in the area of coal dust deposition.

Coal dust is much less fibrogenic than silica. As an example, a mixture of 10% silica and 90% coal is far more cytotoxic to macrophages than pure coal dust (Adamis and Timlar, 1978). However, both dusts, when cleaved, show surface radicals by electron spin resonance spectroscopy. The free radicals generated by crushing anthracite coal are more numerous than those generated from crushing bituminous coal, leading to speculation regarding free radical release from different coal ranks and the suggestion that exposure to anthracite coal increases the risk for the development and progression of disease (Dalal *et al.*, 1989).

Long-term coal dust exposure in animals increases the number of alveolar cells recovered by lavage (Castranova *et al.*, 1985). In addition, an elevation in the number of blood monocytes and an increased rate of mitosis has been recognized in cells from animals undergoing chronic coal dust inhalation (Adamson and Bowden, 1978). This suggests that there was recruitment of cells into the lung from the alveolar capillaries and the interstitium of the lung. Recruited 'young' macrophages appear to be more phagocytically active than older macrophages. This may imply a more effective clearance of particles (Castranova *et al.*, 1985).

Exposure of alveolar macrophages to particles can result in the release of proteolytic enzymes, reactive oxygen species [i.e. hydrogen peroxide (H_2O_2), the hydroxyl radical ($OH^{\bullet}$) and superoxide anion (O_2–)], and leukotrienes via breakdown of arachadonic acid in the cell membrane. Again, silica is a much stronger stimulus to the release of these agents than coal dust (Vallyathan *et al.*, 1988). Excessive release of these reactive oxygen species have the potential to overwhelm the naturally protective antioxidant system within the lung and begin the process of inflammation and fibrosis.

Activated macrophages secrete a wide variety of mediators which can attract neutrophils into the affected area and then stimulate them to release reactive oxygen species and enzymes (Borm and Henderson, 1990). Similarly, the macrophage-derived inflammatory factors can act as chemotactic agents for neutrophils (e.g. tumor necrosis factor, interleukin-8 and leukotrienes), and increase neutrophil adherence and reactive enzyme release (platelet-derived growth factor and platelet activating factor). The presence of these factors escalates the inflammatory process. Finally, secretion of many of the above factors by the activated macrophages also enhance fibroblast growth and/or stimulate the production of collagen.

Although alveolar macrophages in animals are activated after chronic exposure to coal dust, there is little lung damage. Protein and lysosomal levels in the acellular lavage fluid of these animals are not increased (Castranova *et al.*, 1985). However, when as little as 2% silica is added, the fibrotic process begins (Ray *et al.*, 1951).

Development of bronchoalveolar lavage (BAL) as a means for sampling lung cells and fluid has provided an opportunity to study the lung's response to inhalation of various dusts. Studies of the mechanisms of inflammation and fibrosis using BAL-derived material have opened the possibility for studying at least some of the mechanisms in the pulmonary reactions in CWP.

Lapp *et al.* (1991) compared 12 asymptomatic, lifelong non-smoking coal miners without pneumoconiosis with a mean of 17 years of underground mining exposure using BAL with the results from 18 controls. There was no difference in BAL mean total or individual cell counts, IgA or IgG, or spontaneous or PMA stimulated chemiluminescence. The BAL protein levels and particle-stimulated chemiluminescence was less in the miners. Although scanning electron microscopy showed a marked increase in alveolar macrophage cell surface ruffling, and transmission electron microscopy showed particles in macrophages consistent with coal, alveolitis was not present. These data are best explained by recognizing that the workplace dust exposures were insufficient to overwhelm the underlying clearance mechanisms.

Rom *et al.* (1987) studied 15 symptomatic, nonsmoking coal miners with simple CWP by BAL. They found no significant difference between miners with CWP and controls in the number of cells recovered, the percentage distribution of cells, and in the release of superoxide anion or hydrogen peroxide. This contrasted with the findings in subjects with asbestosis and silicosis, whose values for spontaneous release of oxidant superoxide and hydrogen peroxide were significantly greater than those of controls. BAL levels of fibronectin and alveolar macrophage-derived growth factor (AMDGF) were elevated but not different from the values obtained in subjects with asbestosis and silicosis. While fibrosis has not been a major part of the pathology of simple CWP, there is evidence of the presence of fibronectin in pneumoconiotic lesions (Wagner *et al.*, 1982). Wallaert *et al.* (1990) demonstrated a significantly increased number of total cells in the BAL recovered from miners with simple and complicated CWP, as well as increased percentages of alveolar macrophages, lymphocytes and neutrophils. Alveolar cells from miners with simple CWP and PMF spontaneously released significantly more superoxide, in contrast with the levels generated in the comparison group. However, when these alveolar cells were stimulated with phorbol myristic acetate (PMA), only the cells recovered from the miners with PMF showed significantly increased amounts of superoxide release.

There have been few studies of the peripheral lymphocytes in coal miners. Dauber *et al.* (1976) examined lymphocyte function of 15 miners with anthracosilicosis. They found decreased numbers of both T and B lymphocytes in the peripheral blood in the miners compared with controls. Lymphocyte functional capacity, as determined by response to stimulation by concavalin A, was less in miners with PMF compared with the response in miners with simple CWP or controls.

Overall, coal dust is much less fibrogenic than inhalation of silica dust, yet coal dust is a sufficient stimulus for the secretion and release of macrophage products. These activated macrophages can release enzymes, reactive oxygen species, cytokines and factors which cause fibroblast proliferation. All are important factors in inflammation and the development of CWP.

A less clear-cut relationship occurs when one addresses the role of the immunologic system in CWP. Soutar *et al.* (1974) reported circulating ANA and RF levels among 109 miners with radiographic evidence of pneumoconiosis. The authors did not identify the region of the country from which the miners came, or in which mines they worked. They found positive ANA in 17% and RF in 10% of the miners. The expected prevalence in the male general population of a positive ANA approximates 2–3%. The prevalence of ANA was least (9%) in simple CWP and 27% in those with category C (PMF). A similar but less striking trend was seen with RF, ranging from 6% in simple CWP to 18% in category C. Combining both ANA and RF resulted in prevalence of positive results in 13% of the miners with simple and 45% of those with category C PMF.

In 1973, Lippmann *et al.* (1973) reported a prevalence study of circulating antinuclear factor (ANA) and rheumatoid factor (RF) among coal miners in the

United States. Sera from 207 coal miners were examined. Of the 196 miners with pneumoconiosis, nine were positive for RF and 34% had positive ANA. There were regional variations in ANA that seemed to parallel the prevalence of radiographic changes; namely, a higher prevalence in the anthracite miners and a lesser prevalence in bituminous miners. These authors did not find the increased prevalence of RF in miners with PMF compared with simple CWP that Wagner and McCormick (1967) reported in Welsh miners.

Benedek *et al.* (1976) were unable to confirm the high prevalence of serum immunoglobulins and rheumatoid factor among US miners reported by European investigators. These authors matched 55 pairs of coal miners with rheumatoid arthritis with non-miners with rheumatoid arthritis. The miners and controls did not differ in respect to serum concentrations of IgG or IgA. Elevated levels of IgM were more frequent among the miners but did not reach significance. Latex agglutination test reactions were positive in 82% of miners and only 64% of non-miners ($P < 0.05$), but there was no correlation with category of pneumoconiosis.

Studies of serum immunoglobulins were also conducted by Hahon *et al.* (1980) among 155 US coal miners with chest radiographs demonstrating simple pneumoconiosis, Caplan's Syndrome and PMF. They found significantly higher serum concentrations of C3, alpha-1-anti-trypsin, and IgA and IgG in anthracite miners than in bituminous miners with PMF. There were few differences in these serum proteins among the miners with only simple CWP. They did find that compared with normal controls, the miners' C3, alpha-1-anti-trypsin and IgG values were elevated. The authors did not find an association between the elevated immunoglobulins and FEV_1.

There has been some attention paid to the role of tissue autoantibodies in CWP. Burrell (1972) used an anti-globulin consumption test to identify autoantibodies in the sera of coal miners directed at lung collagen and in the reticulin in the parenchyma and basement membranes of the lung. The lung autoantibodies tend to reside in the serum IgA. It is not clear at this time whether these autoantibodies participate in the CWP reaction in the lungs or simply represent epiphenomena. Animals injected with purified lung autoantibodies and then challenged with *M. tuberculosis* demonstrate much more extensive lesions of tuberculosis than animals receiving the *M. tuberculosis* without the prior injections of lung autoantibodies.

It has been suggested that constitutional differences may help to explain the variation in response to inhaled dusts, especially the attack rates and progression of CWP. One study looked at the prevalence of histocompatibility antigens in coal miners (Heise *et al.*, 1979). So far it has not been possible to confirm any definite association between these factors and CWP.

MANAGEMENT OF CWP

There is no proven therapy for CWP. The primary prevention of lung disease in miners must include continuing efforts at reducing coal mine dust exposure. Management is best directed at prevention, early recognition and treatment of complications. The major challenges to the physician are the recognition and management of airflow obstruction, respiratory infection, hypoxemia, respiratory failure, cor pulmonale, arrhythmias and pneumothorax.

Improved mining methods and lower dust levels appear to be reducing exposures and new cases of both simple and complicated pneumoconiosis in the US. Medical surveillance programs, using chest radiographs, allow the early recognition of workers with simple pneumoconiosis. Workers with simple pneumoconiosis should be encouraged to transfer from jobs with high dust exposure to jobs with low dust exposure. Any worker with the unexpected finding of PMF should be carefully advised about the hazards of further dust exposures.

Workers presenting with respiratory symptoms should have careful evaluation. The initial history and examination should be supplemented by chest radiography, spirometry with bronchodilators, diffusing capacity, electrocardiogram and resting arterial blood gas measurement as indicated. A thorough initial database allows accurate assessment of the worker's respiratory health and serves a starting point for observing the response to therapy or progression of disease.

Smoking cessation is important regardless of symptoms of respiratory disease, chest radiograph abnormalities or pulmonary function status. Physician encouragement to stop smoking should be supplemented by support from smoking cessation groups, use of nicotine substitutes, and behavior modification techniques.

Symptomatic reversible airflow obstruction should be treated with inhaled bronchodilator therapy as indicated. Patients with severe obstruction and inadequate improvement from the usual measures should be considered for a trial of glucorticoids.

Hypoxemia can be a serious complication in those with PMF. It is typically present first during exercise, but can occur with rest and during sleep. Chronic hypoxemia can lead to the complications of polycythemia, pulmonary hypertension, cor pulmonale and cerebral dysfunction. Therapy with low flow oxygen is indicated when the arterial oxygen tension is less than 55 mmHg or when clinical evidence of cor pulmonale is present.

Workers with significant airflow obstruction or PMF should receive appropriate immunization with influenza and pneumococcal vaccines. Bacterial and viral episodes of bronchitis or pneumonia should be promptly recognized and appropriately treated. Similarly, miners with concomitant exposure to silica dust (most often roof bolters, drillers and motormen) deserve special attention with regard to mycobacterial infection. Symptoms of weight loss, fever, sweats, a change in sputum production, or malaise should be promptly investigated with a chest radiograph and examination of the sputum through stain and culture for AFB. Occasionally, the sputum may be negative and the infecting mycobacterial organisms can only be obtained by fiberoptic bronchoscopy with brushings and washings. Active tuberculosis in this population can be usually successfully treated with the usual drug regimens, provided that rifampin is one of the drugs used (Ball *et al.*, 1969; Dubois *et al.*, 1977). In coal miners with a significant history of concurrent silica exposure, the treatment for tuberculosis may need to be more aggressive, and long-term follow-up is indicated in view of reports of recurrent pulmonary tuberculosis in patients with PMF after completion of apparently adequate therapy (Morgan, 1979).

Pneumothorax can be particularly troublesome events in miners with pneumoconiosis. Those with bullous disease in the presence of advanced complicated pneumoconiosis appear to be at the greatest risk. Typically, once the lung collapses, it is difficult to expand, a feature attributable to the decreased compliance associated with interstitial lung disease. Therapy with one or several chest tubes is often therapeutic; however, recurrent pneumothorax which cannot be expanded may require an open procedure and pleurodesis.

Respiratory failure may complicate advanced PMF as it does in other chronic respiratory diseases. Ventilatory support measures are indicated when the failure is precipitated by a treatable complication. The application of ventilatory support measures should be discussed with the patient before the need arises. In general, miners with advanced pneumoconiosis are poor candidates for long-term mechanical ventilatory therapy.

PREVENTION STRATEGIES FOR COAL WORKERS' LUNG DISEASES

The control of coal workers' lung diseases in both the developed and developing world requires comprehensive prevention strategies, including exposure control, medical surveillance, research and education. Example approaches include:

- Major efforts must be directed to improving work methods or work practices, including engineering controls, to progressively reduce dust exposures to acceptable levels. These efforts are labelled *primary prevention*. Personal respiratory protection can be utilized as a secondary measure, to reduce exposures during short-term operations or unusual/emergency conditions, or while exposure control measures are being modified or improved. The use of respirators will only be effective when it is part of a comprehensive respiratory protection program, and should never be relied upon outside of such a program.
- Primary prevention should involve ongoing environmental dust monitoring, and assure mechanisms for feed-back to improve and modify working conditions if exposures above acceptable levels are observed.
- *Secondary prevention* through medical screening and surveillance should be designed to benefit the individual worker and future workers. Illness identified through medical screening represents a failure of primary prevention, and thus should trigger feedback to those involved in environmental monitoring and work practice evaluations.
- Education about the respiratory health hazards from uncontrolled exposures to silica and coal mine dust must be available to workers, employers, managers and health care providers.
- Information on the cumulative burden of disease should be monitored over time for CWP control.
- Research into mining-related lung diseases should be encouraged to improve recognition, monitoring, exposure reduction and therapy, and to increase understanding of pathogenesis. Research efforts should supplement, not displace, attention to dust control.
- Clinicains who recognize coal-related diseases in their patients should attempt to determine whether ongoing workplace exposures present a continuing risk to current workers, while maintaining the confidentiality of the patient–physician relationship. Assistance in this can often be obtained through local or regional health departments, occupational medicine groups or central governmental agencies. Reporting of occupational diseases is required in some jurisdictions.

SUMMARY

CWP is a disease distinct and separate from silicosis. Although the coal particle is not nearly as fibrogenic as the silica particle, excessive exposures over a period of time can overwhelm effective clearance mechanisms and initiate chronic interstitial lung disease. Simple CWP is clearly related to the amount of dust deposited within the lungs. PMF most often occurs on a background of simple CWP and is the result of dust deposition plus other inadequately defined host factors. Immunologic and local cellular factors may contribute to the development of this form of the disease.

PMF is clearly associated with alterations in ventilatory, mechanical and vascular function of the lungs. These abnormalities in PMF contribute to the premature morbidity and mortality of this disease. Similarly, there is no specific therapy for CWP. Prevention remains the cornerstone of eliminating this occupational lung disease. The education of workers and employers regarding the hazards of coal dust exposure and measurement and effective control of dust exposure remains the sole means of eliminating this disease.

REFERENCES

Adamis, Z. and Timlar, T. (1978) Studies on the effect of quartz, bentonite, and coal dust mixtures on macrophages *in vitro*. *Br. J. Exp. Pathol.*, **59**, 411–19.

Adamson, I.Y.R. and Bowden, D.H. (1978) Adaptive responses of the pulmonary macrophagic system to carbon: II. Morphologic studies. *Lab. Invest.*, **38**, 430–8.

Amandus, H.E., Petersen, M.R. and Richards, T.B. (1989) Health status of anthracite surface coal miners. *Arch. Environ. Health*, **44**, 75–81.

Attfield, M.D. (1992) British data on coal miners' pneumoconiosis and relevance to US conditions. *Am. J. Public Health*, **82**, 978–83.

Attfield, M.D. and Althouse, R.B. (1992) Surveillance data on US coalminers' pneumoconiosis, 1970–1986. *Am. J. Public Health*, **82**, 971–77.

Attfield, M.D. and Castellan, R.M. (1992) Epidemiological data on US coal miners' pneumoconiosis, 1960–1968. *Am. J. Public Health*, **82**, 964–70.

Armstrong, B.K., McNulty, J.C., Levitt, L.J., Williams, K.A. and Hobbs, M.S.T. (1979) Mortality in gold and coal miners in Western Australia with special reference to lung cancer. *Br. J. Ind. Med.*, **36**, 199–205.

Atuhaire, L.K., Campbell, M.J., Cochrane, A.L., Jones, M. and Moore, F. (1985) Mortality of men in the Rhondda Fach 1950–1980. *Br. J. Ind. Med.*, **42**, 741–5.

Balaan, M.R., Weber, S.L. and Banks, D.E. (1993) Clinical aspects of coal workers' pneumoconiosis and silicosis. *Occup. Med.*, **8**, 19–34.

Ball, J.D., Berry, G., Clarke, W.G., Gilson, J.C. and Thomas, J. (1969) A controlled trial of anti-tuberculosis chemotherapy in early complicated pneumoconiosis of coal workers. *Thorax*, **24**, 399–406.

Banks, D.E., Bauer, M.D., Castellan, R.M. and Lapp, N.L. (1983) Silicosis in surface coal mine drillers. *Thorax*, **38**, 275–8.

Benedek, T.G., Zwadzki, Z.A. and Medsger, T.A., Jr (1976) Serum immunoglobulins, rheumatoid factor and pneumoconiosis in coal miners. *Arth. Rheum.*, **19**, 731–6.

Borm, P.J.A. and Henderson, P.T. (1990) Symposium on the health effects of occupational exposures to inorganic dusts. *Exp. Lung Res.*, **16**, 1–3.

Burrell, R. (1972) Immunological aspects of coal workers' pneumoconiosis. *Ann. N.Y. Acad. Sci.*, **200**, 94–105.

Caplan, A. (1953) Certain unusual radiological appearances in the chest of coal miners suffering from rheumatoid arthritis. *Thorax*, **8**, 29–37.

Caplan, A., Payne, R.B. and Withey, J.L. (1962) A broader concept of Caplan's syndrome related to rheumatoid factors. *Thorax*, **17**, 205–12.

Castranova, V., Bowman, L., Reasor, M. *et al.* (1985) The response of rat alveolar macrophages to chronic inhalation of coal dust and/or diesel inhalation. *Environ. Res.*, **36**, 405–19.

Coal data: a reference (1995) Washington, DC. Energy Information Administration. Office of coal, nuclear, electric, and alternate fuels. US Dept of Energy, Washington, DC, GPO stock no. 061–003–00901–8.

Coal Mine Health and Safety Act (1969) U.S. Pub. L No. 91–173, Stat 2917.

Coates, D.R. (1981) Energy and fossil fuels, in *Environmental Geology*, John Wiley and Sons, New York.

Cochrane, A.L., Davies, I. and Fletcher, C.M. (1951) 'Entente Radiologique'. A step towards agreement on the classification of radiographs in pneumoconiosis. *Br. J. Ind. Med.*, **8**, 244–54.

Cochrane, A.L., Haley, T.J.L., Moore, F. and Hole, D. (1978) The mortality of men in the Rhondda Foch, 1950–1970. *Br. J. Ind. Med.*, **36**, 15–22.

Coleman, L.L. (1992) *Coal Data*, 1992 edn, The National Coal Association, Washington, DC.

Collis, E.L. (1919) Industrial pneumoconiosis (Milroy Lectures, 1915) His Majesty's Stationary Office, 1919. (Reprinted from *Public Health, London*, **28**, 252,292, and **29**, 11, 37.)

Collis, E.L. and Gilchrist, J.C. (1928) Effects of dust upon coal trimmers. *J. Ind. Hyg. Toxicol.*, **10**, 101–9.

Dalal, N.S., Suryan, M.M., Vallayathan, V. *et al.* (1989) Detection of reactive free radicals in fresh coal mine dusts and their implication for pulmonary injury. *Ann Occup Hyg.*, **33**, 79–84.

Dauber, J.H., Finn, D.R. and Daniele, R.P. (1976) Immunologic abnormalities in anthracosilicosis. *Am. Rev. Respir. Dis.*, **113**, 94.

Dubois, P., Gyselen, A. and Prignot, J. (1977) Rifampicin-combined chemotherapy in coal-workers' pneumoconio-tuberculosis. *Am. Rev. Respir. Dis.*, **115**, 221–8.

Davies, I. and Mann, K.J. (1949) *Proc. 9th Inter. Congr. Industr. Med.*, London, p. 768.

Douglas, A.N., Lamb, D. and Ruckley, V.A. (1982) Bronchial gland dimensions in coalminers: influence of smoking and dust exposure. *Thorax*, **37**, 760–4.

Editorial (1979) Coal mining and mortality. *Br. Med. J.*, **2**, 1168–9.

Enterline, P.E. (1964) Mortality rates among coal miners. *Am. J. Public Health*, **54**, 758–68.

Enterline, P.E. (1972) A review of mortality data for American coal miners. *Ann. N.Y. Acad. Sci.*, **200**, 260–72.

Euler, W.J. (1981) in *Coal Handbook* (ed. R.A. Myers), Marcel Dekker, Inc., New York, pp. 75–114.

Fernie, J.M., Douglas, A.N., Lamb, D. and Ruckley, V.A. (1983) Right ventricular hypertrophy in a group of coalworkers. *Thorax*, **38**, 436–42.

Gough, J. (1940) Pneumoconiosis of coal trimmers. *J. Pathol. Bacteriol.*, **51**, 277–85.

Gregory, J.C. (1831) Case of peculiar black infiltration of the lungs resembling melanosis. *Edinburgh Med. Surg. J.*, **36**, 389–92.

Hahon, N., Morgan, W.K.C. and Petersen, M. (1980) Serum immunoglobulin levels in coal workers' pneumoconiosis. *Ann. Occup. Hyg.*, **23**, 165–74.

Haught, O.L. (1955) *Coal and coal mining in West Virginia*, West Virginia Geological and Economic Survey, Morgantown, WV, pp 1–32.

Heise, E.R., Mentnech, M.S., Olenchock, S.A., Kutz, S.A., Morgan, W.K.C., Merchant, J.A. and Major, P.C. (1979) HLA-A1 and coalworkers' pneumoconiosis. *Am. Rev. Respir. Dis.*, **119**, 903–8.

Hepplestone, A.G. (1947) The essential lesion of pneumoconiosis in Welsh coal workers. *J. Pathol. Bacteriol.*, **59**, 453–60.

Hepplestone, A.G. (1954) The pathogenesis of simple pneumoconiosis in coal workers. *J. Pathol. Bacteriol.*, **67**, 51–63.

Hilt, C. (1873) Die Beiziehungen zwischen der Zusammensetzung und der technischen Eigenschaften der Steinkohle. *Bezirksvereinigung Ver. Deutsch. Ingenieure Zeitschrift*, **17**(4),157–69.

Hodous, T.K. and Attfield, M.D. (1990) Progressive Massive Fibrosis developing on a background of minimal simple pneumoconiosis, in *Proceedings of the VIIth International Pneumoconiosis Conference*, U.S. Department of Health and Human Services, Public Health Service, Center for Disease Control, National Institute for Occupational Safety and Health, DHHS (NIOSH) publication No. 90–108, part 1, pp. 122–6.

International Agency for Research on Cancer (IARC) (1987) IARC monographs on the evaluation of carcinogenic risks to humans. Supplement no. 7: Overall evaluations of carcinogenity: An updating of IARC monographs, Lyons: IARC, Vol. 1–42.

International Labour Office. (1980) Guidelines for the use of ILO international classification of radiographs of pneumoconiosis, revised ed. International Labour Office, Geneva.

Jacobsen, M., Rae, S., Walton, W.H. and Rogan, J.M. (1971) The relation between pneumoconiosis and dust-exposure in British coal mines, in *Inhaled Particles III*, (ed. W.H. Walton), Unwin Brothers, Woking, pp. 903–17.

Kibelstis, J.A., Morgan, E.J., Reger, R., Lapp, N.L., Seaton, A. and Morgan, W.K.C. (1973) Prevalence of bronchitis and airway obstruction in American bituminous coal miners. *Am. Rev. Respir. Dis.*, **108**, 886–93.

King, E.J., Maguire, B.A. and Nagelschmidt, G. (1956) Further studies of the dust in the lungs of coal-miners. *Br. J. Ind. Med.*, **13**, 9–13.

Kleinerman, J., Green, F., Lacquer, W., *et al.* (1979) Pathology standards for coal workers' pneumoconiosis. *Arch. Path. Lab. Med.*, **103**, 374–432.

Kuempel, E.D., Stayner, L.T., Attfield, M.D. and Buncher, C.R. (1995) Exposure-response analysis of mortality among coal miners in the United States. *Am. J. Ind. Med.*, **28**, 167–84.

Lainhart, W.S., Doyle, H.M., Enterline, P.E., Henschel, A. and Kendrick, M.A. (1969) *Pneumoconiosis in Appalachian bituminous coal miners*, US Department of Health, Education and Welfare, US Government Printing Office, Washington, D.C.

Lapp, N.L. and Castranova, V. (1993) How silicosis and coal workers' pneumoconiosis develops – a cellular assessment. *Occupational Medicine: State of the Art Reviews*, **8**, 35–65.

Lapp, N.L., Seaton, A., Kaplan, K.C., Hunsaker, M.R. and Morgan, W.K.C. (1971) Pulmonary haemodynamics in coal workers' pneumoconiosis, in *Inhaled Particles III*, (ed. W.H. Walton), Unwin Brothers, Woking, pp. 645–657.

Lapp, N.L., Lewis, D., Schwegler-Berry, D. *et al.* (1991) Bronchoalveolar lavage in asymptomatic underground coal miners, in *Proceedings of Respiratory Dusts in the Mineral Industry*, (ed. H. Ramani) Society of Mining Engineering, pp. 159–69.

Lippmann, M., Eckert, H.L., Hahon, N. and Morgan, W.K.C. (1973) Circulating antinuclear and rheumatoid factors in United States coal miners. *Ann. Intern. Med.*, **79**, 807–11.

Lyons, J.P., Clarke, W.G., Hall, A.M. and Cotes, J.E. (1967) Transfer factor (diffusing capacity) for the lung in simple pneumoconiosis of coal workers *Br. Med. J.*, **4**, 772–4.

Marine, W.M., Gurr, D. and Jacobsen, M. (1988) Clinically important respiratory effects of dust exposure and smoking in British coal miners. *Am. Rev. Respir. Dis.*, **137**, 106–12.

Marshall, W. (1833–1834) Cases of spurious melanosis of the lungs of or phthisis melanotica. *Lancet*, **2**, 271–4.

Matolo, N.A., Klauber, M.R., Gorishek, W.M. and Dixon, J.A. (1972) High incidence of gastric carcinoma in a coal mining region. *Cancer*, **29**, 733–7.

McBride, W.W., Pendergrass, E. and Lieben, J. (1963) Pneumoconiosis study of western Pennsylvania bituminous-coal miners. *J. Occup. Med.*, **5**, 376–88.

McBride, W.W., Pendergrass, E. and Lieben, J. (1966) Pneumoconiosis study of Pennsylvania anthracite miners. *J. Occup. Med.*, **8**, 365–76.

McLintock, J., Rae, S. and Jacobsen, M. (1971) The attack rate of progressive massive fibrosis in British coal miners, in *Inhaled Particles III*, (ed. W.H. Walton), Unwin Brothers, Woking, pp. 933–50.

Meijers, J.M.M., Swaen, G.M.H., Slangen, J.J.M., van Vliet, K. and Sturmans, F. (1991) Long-term mortality in miners with coal workers' pneumoconiosis in the Netherlands: a pilot study. *Am. J. Ind. Med.*, **19**, 43–50.

Meiklejohn, A. (1951) History of lung disease of coalminers in Great Britain, part I, 1800–1875. *Br. J. Ind. Med.*, **8**, 127–37.

Meiklejohn, A. (1952a) History of lung diseases of coal miners in Great Britain, Part II, 1875–1920. *Br. J. Ind. Med.*, **9**, 93–8.

Meiklejohn, A. (1952b) History of lung diseases of coal miners in Great Britain. Part III, 1920–1952. *Br. J. Ind. Med.*, **9**, 208–20.

Miller, B.G. and Jacobsen, M. (1985) Dust exposure, pneumoconiosis, and mortality of coalminers. *Br. J. Ind. Med.*, **42**, 723–33.

Morgan, W.K.C. (1978) Industrial bronchitis. *Br. J. Ind. Med.*, **35**, 285–91.

Morgan, E.J. (1979) Silicosis and tuberculosis. *Chest*, **75**, 202–3.

Morgan, W.K.C., Burgess, D.B., Lapp, N.L., Seaton, A. and Reger, R.B. (1971) Hyperinflation of the lungs in coal miners. *Thorax*, **26**, 585–90.

Mutmansky, J.M. and Lee, C. (1984) An analysis of coal and geologic variables related to coal workers' pneumoconiosis, in *Proceedings of the Lakeview Conference*, October 8, 1984, Morgantown, W.V. Center for Generic Respirable Disease Studies.

National Research Council (U.S.) (1995) Committee on the strategic assessment of the US Department of Energy's coal program. Coal: Energy for the future. National Academy Press, pp. 38–63.

NIOSH (1994) Work-related lung disease surveillance report. US Department of Health and Human Services. Public Health Service. Centers for Disease Control. National Institute for Occupational Safety & Health, DHHS number 94–120, pp. 28–46.

NIOSH (1995) Criteria for a recommended standard: occupational exposure to respirable coal mine dust. US Department of Health and Human Services. Public Health Service. DHHS (NIOSH) publication No. 95–106, Cincinnati, Ohio.

Ogle, W. (1885) Supplement of the 45th annual report of the Registrar General for England and Wales, Code 4564, Her Majesty's Stationery Office, London.

Ortmeyer, C.E., Baier, E.J. and Crawford, G.M., Jr (1973) Life expectancy of Pennsylvania coal miners compensated for disability. *Arch. Environ. Health*, **27**, 227–30.

Ortmeyer, C.E., Costello, J., Morgan, W.K.C., Swecker, S. and Peterson, M. (1974) The mortality of Appalachian coal miners, 1963–1971. *Arch. Environ. Health*, **29**, 67–72.

Ray, S.C., King, E.J. and Harrison, C.V. (1951) The action of small amounts of quartz and large amounts of coal and graphite on the lungs of rats. *Br. J. Ind. Med.*, **8**, 68–73.

Rivers, D., Wise, M.E., King, E.J. and Nagelschmidt, G. (1962) Dust content, radiology, and pathology in simple pneumoconiosis of coalworkers. *Br. J. Ind. Med.*, **152**, 298–306.

Rockette, H.E. (1977) Cause specific mortality of coal miners. *J. Occup. Med.*, **19**, 795–801.

Rodnan, G.P., Benedek, R.G., Medsger, T.A. and Cammarata, R.J. (1967) The association of progressive systemic sclerosis (scleroderma) with coal miners' pneumoconiosis and other forms of silicosis. *Ann. Intern. Med.*, **66**, 323–34.

Rom, W.N., Bitterman, P.B., Rennard, S.I., Cantin, A. and Crystal, R.G. (1987) Characterization of the lower respiratory tract inflammation of nonsmoking individuals with interstitial lung disease associated with chronic inhalation of inorganic dust. *Am. Rev. Respir. Dis.*, **136**, 1429–34.

Seixas, N.S., Robins, T.G., Attfield, M.D. and Moulton, L.H. (1992) Exposure–response relationships for coal mine dust and obstructive lung disease following enactment of the Federal Coal Mine Health and Safety Act of 1969. *Am. J. Ind. Med.*, **21**, 715–34.

Shennan, D.H., Washington, J.S., Thomas, D.J., Dick, J.A., Kaplan, Y.S. and Bennett, J.G. (1981) Factors predisposing to the development of Progressive

Massive Fibrosis in British coal miners. *Br. J. Ind. Med.*, **38**, 321–6.

Society of Actuaries (1967) *Occupational Study*, Society of Actuaries, Chicago, Illinois.

Soutar, C.A., Turner-Warwick, M. and Parkes, W.R. (1974) Circulating antinuclear antibody and rheumatoid factor in coal pneumoconiosis. *Br. Med. J.*,**3**, 145–7.

Stahl, R. (1989) Coal and derivatives, in *Encyclopedia of Occupational Health and Safety*, Third (revised) edition, International Labour Organisation, Geneva.

Stocks, P. (1962) On the death rates from cancer of the stomach and respiratory diseases in 1949–1953 among coal miners and other male residents in counties of England and Wales. *Brit. J. Cancer*,**16**, 592–8.

Stout, K.S. (1980) *Mining Methods and Equipment*, McGraw-Hill, Inc., New York.

Taylor, L.D. and Thakur, P.C. (1993) Recent developments in coal mining technology and their impact on miners' health. *Occupational Medicine: State of the Art Reviews*, **8**, 109–26.

Thomas, L. (1992) *Handbook of Practical Geology*, John Wiley & sons, Chichester.

Vallyathan, V., Green, F.H.Y., Rodman, N.F., Boyd, C.B. and Althouse, R. (1985) Lung carcinoma by histologic type in coal miners. *Arch. Pathol. Lab. Med.*, **109**, 419–23.

Vallyathan, V., Schwegler, D., Reasor, M. *et al.* (1988) Comparative *in vitro* toxicity and relative pathogenicity of mineral dusts. *Ann. Occup. Hyg.*, **32**, 279–89.

Wagner, G.R., Attfield, M.D. and Parker, J.E. (1993) Chest radiography in dust-exposed miners: promise and problems, potential and imperfections. *Occupational Medicine: State of the Art Reviews*, **8**, 127–41

Wagner, J.C. (1972) Etiologic factors in complicated coal workers' pneumoconiosis. *Ann. N.Y. Acad. Sci.*, **200**, 401–4.

Wagner, J.C. and McCormick, J.N. (1967) Immunological investigations of coalworkers' disease. *J. R. Coll. Phys. (Lond.)*, **2**, 49–56.

Wagner, J.C., Burns, J., Munday, D.E. and McGee, J.D. (1982) Presence of fibronectin in pneumoconiotic lesions. *Thorax*, **37**, 54–6.

Wallaert, B., Lassalle, P., Fortin, F., Werts, C., Bart, F., Fournier, E. and Voisin, C. (1990) Superoxide anion generation by alveolar inflammatory cells in simple pneumoconiosis and in progressive massive fibrosis of non-smoking coal workers. *Am. Rev. Respir. Dis.*, **141**, 129–33.

Watson, A.J., Black, J., Doig, A.T. *et al.* (1959) Pneumoconiosis in carbon electrode workers. *Br. J. Ind. Med.*, **16**, 274–85.

12 Coal workers' pneumoconiosis in developing nations

Changqi Zou and Guo-An Shen

Coal workers' pneumoconiosis (CWP) has been an important health concern of miners, mining engineers, epidemiologists and medical researchers. As the energy needs of developing nations, such as the People's Republic of China, the Republic of Korea, and other countries have grown, the coal mining industry has expanded, and scientists have paid increasing attention to the respiratory health of miners. In China, as well as in other nations around the world, CWP remains an important cause of disability and premature mortality.

PREVALENCE OF CWP IN DEVELOPING COUNTRIES

The exact magnitude of CWP remains unclear in many parts of the world. Accurate and comparable surveillance data from developing countries are often inadequate. In addition, data collected for different purposes and from various programs such as compensation statistics, periodic surveys and voluntary screening programs, makes direct comparisons between nations difficult. Although some prevalence data are available from cross-sectional studies, incidence figures and risk estimates are usually lacking because of the absence of reliable information about the size of the population at risk and exposure measurements. In China, for example, a high proportion of the mining workforce work in small coal mines. These miners might not come under the jurisdiction of conventional legislation. Information regarding the prevalence of disease in these workers is often not available. Problems such as these contribute to serious under-reporting of pneumoconiosis (He and Mikheev, 1993). Furthermore, because diagnostic criteria may vary among countries, it is difficult to compare the prevalence and incidence figures of CWP from these countries. Due to these limitations, comparable global statistics are unavailable.

In highly industrialized countries, dust suppression and medical surveillance of the work force have resulted in a substantial reduction in the prevalence of CWP. However, even in the most highly industrialized countries where dust suppression has been systematically practiced for many years, new cases continue to occur (He and Mikheev, 1993). In newly industrialized and developing countries, even with limitations in data collection, crude estimates have shown that CWP remains a serious threat to the miners' health. In these countries, estimates of the incidence and prevalence of CWP are high and appear to be increasing. In most instances, working conditions do not meet commonly accepted standards and, at least in China, the rapid expansion of coal mines places many new workers at risk each year. Hygienically poor working conditions often expose workers to unacceptably high concentrations of respirable dust, both increasing the number and severity of cases.

Occupational Lung Disease. Edited by Daniel E. Banks and John E. Parker. Published in 1998 by Chapman & Hall. ISBN 0 412 73630 6

The World Health Organization has recognized that the number of new cases of CWP is increasing in Asia, Africa and South America (WHO Study Group, 1986). In China, the number of workers newly diagnosed with CWP has exceeded 5000 per year for the past several years (Zou *et al.*, 1993). In Brazil, although the absolute number of cases per working population approximates that of other countries, the short latency period points to a grave situation (Algranti *et al.*, 1992). When silica exposure is present as part of coal mine exposure, the health risk is increased due to the risk of tuberculosis (vanSprundel, 1990).

CWP IN CHINA

China is the largest and most populous of the developing nations. Rapid industrial development in China, with the use of coal as the main energy source for new industries, has driven increases in coal production. China currently leads the world in coal production. Since 1990, the global output of coal has approached 4600 million tons. Approximately one-fourth of this (1200 million tons) has been produced in China. During this recent industrialization of China, especially with the rapid development of village–township enterprises, many small coal mines with poorly controlled working conditions have been opened.

Over the past 20 years, numerous epidemiologic surveys have been performed. Different indices, such as determination of the incidence and prevalence of cases of CWP, mortality studies of miners, and studies focusing on dust exposure–clinical response relationships have been undertaken to investigate the importance of pneumoconiosis in many prefectures and in industrial systems in China.

Recent Ministry of Health data on pneumoconioses (CWP, silicosis and others) collected throughout the country from 1949 to 1986, identified over 150 000 miners with CWP among a population of 1.93 million in the coal mining industry. This approximated 39% of the total number of cases of pneumoconioses. Only silicosis exceeds CWP as a contributor to the burden of pneumoconiosis in China (Figure 12.1). Over 125 000 individuals with CWP are still living. Among them, over 95% have simple pneumoconiosis (Chinese classification system stages I and II) and about 4% have complicated disease (Chinese stage III) (Appendix 12A). More than 15 000 cases (10%) were complicated with pulmonary tuberculosis. Nearly 30% of miners with Chinese stage III disease were also diagnosed with tuberculosis, a condition which remains a serious complication in Chinese coal miners

Among all coal miners, the prevalence of pneumoconiosis was about 6%, with the prevalence greatest in Sichuan, Hunan, Liaoning, Sanxi and Jiangxi provinces (Figures 12.2 and 12.3). In mortality studies of miners with CWP, 43% died from respiratory diseases. Of those who died, 17% did so with tuberculosis. Other important causes of mortality included cardiovascular disease (25%) and cancer (12.3%) (Ministry of Public Health of the People's Republic of China, 1992). Between 1987 and 1992, the number of miners

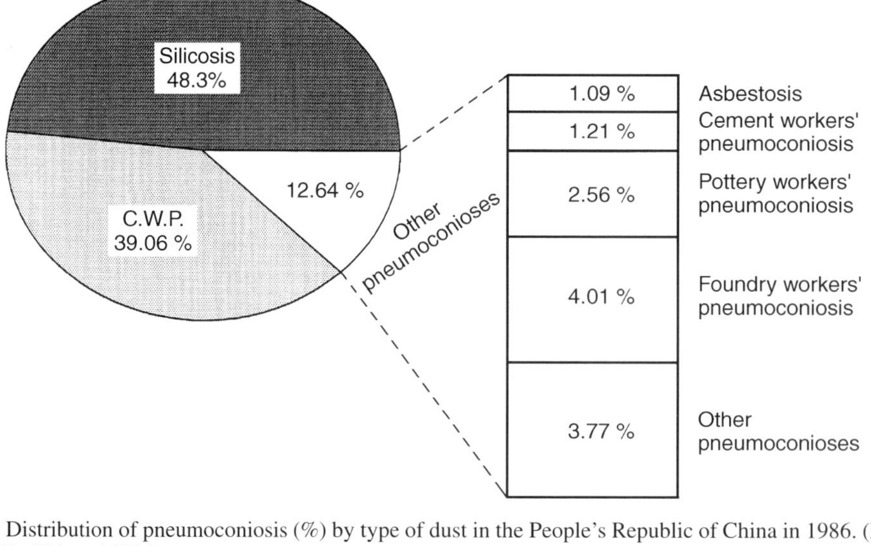

Figure 12.1 Distribution of pneumoconiosis (%) by type of dust in the People's Republic of China in 1986. (Ministry of Public Health of P. R. China, 1992.)

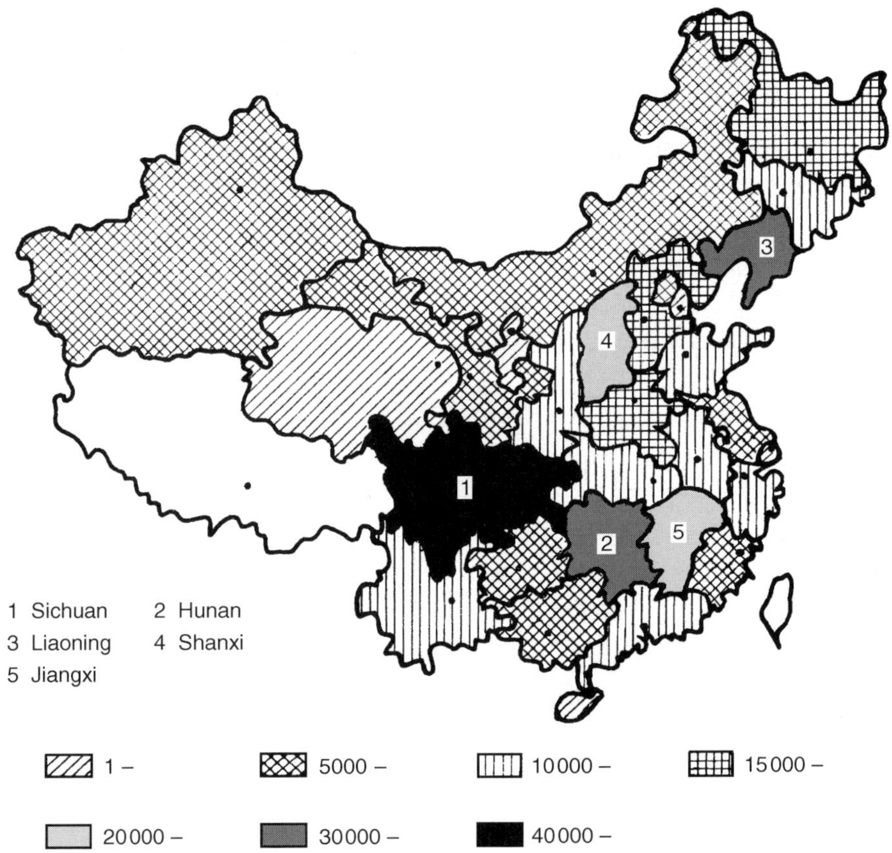

Figure 12.2 Prevalence of pneumoconiosis by province in China, 1986. (Reproduced with permission from Ministry of Public Health of P. R. China, 1992.)

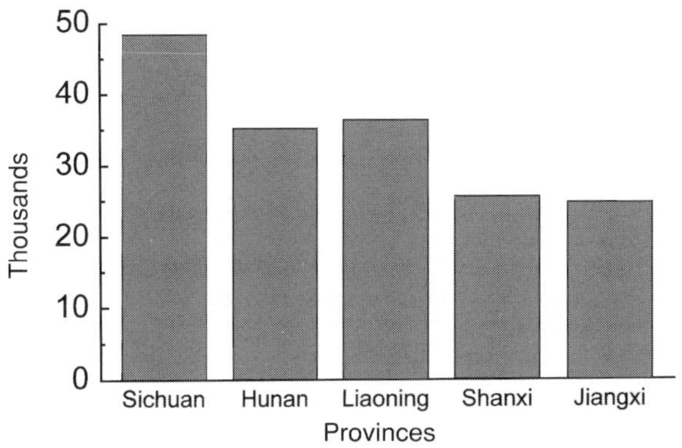

Figure 12.3 Distribution of pneumoconiosis in 1986 in five Chinese provinces where dust diseases are most prevalent. The total number of cases of pneumoconiosis in these five provinces was 170 180, reflecting 42% of the total number of cases in China. (Ministry of Public Health of P. R. China, 1992.)

diagnosed with CWP annually remained above 5,000 (Figure 12.4). It is likely that about 20% of the cases with pneumoconiosis are not reported. In addition, because workers with CWP may have developed the condition while working in small mines associated with village–township enterprises, these numbers were not included.

For more than 40 years, pneumoconioses in China has increased yearly, the increases being most obvious in the coal industry. Indeed, in 1992, the number of new cases in the coal industry were the greatest of any Chinese industry (Figure 12.5). However, results of epidemiologic surveys and the combined analysis of the relationship between dust concentration and pneumoconiosis in mines from the 1950s to the 1980s showed that the average age of workers diagnosed as having stage 1 CWP was increased from 40.3 to 50.5 years. The latency period between the first occupational exposure to dust and

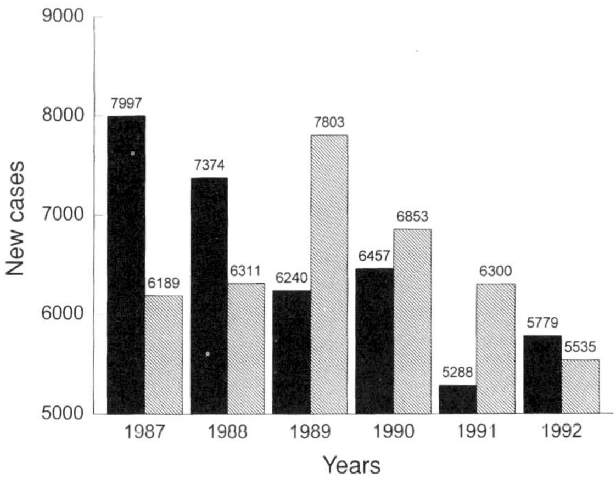

Figure 12.4 Incidence of CWP and silicosis in China from 1987–1992, by year. Solid bars represent the number of silicosis cases; and hatched bars represent the number of cases of CWP. (Redrawn with permission from Zou *et al.*, 1993.)

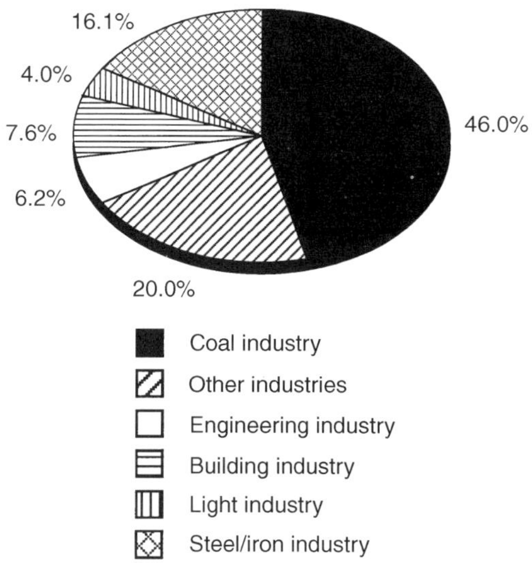

Figure 12.5 Percent of new cases of pneumoconiosis by industry in China for 1992. (Zou *et al.*, 1993.)

diagnosis of stage I CWP also increased from 15.5 to 24.0 years, while the average age at death increased from 44.8 to 61.4 years. In approximately half of those dying with CWP, the time between diagnosis of the condition and death increased by 7.2 years (Ministry of Public Health of P.R. China, 1992). These data indicate that the dust-suppression measures in effect for more than 40 years in the Chinese coal industry have indeed resulted in some improvement in miners' health.

CWP IN OTHER DEVELOPING COUNTRIES

vanSprundel (1990) summarized the occurrence of CWP in some developing countries. Prevalence data of CWP from countries in Africa, Asia and South America are shown in Table 12.1.

CWP was reported to affect 15% of coal miners in Columbia and 18.8% in India. Among 13 000 Korean anthracite miners, the prevalence of simple pneumoconiosis was 9.3% (7% category 1, 1.6% category 2 and 0.7% category 3), with progressive massive fibrosis (PMF) present in 0.01% of those examined. Among coal miners in Zimbabwe, chest radiographic changes of pneumoconiosis rarely occurred before 15 years of exposure and the prevalence of CWP among all miners was reported to be 20%. An epidemiologic study of the progression of CWP in an Indian coal mine showed that 12% of miners had the condition when studied after 16 years of exposure. In Brazil, CWP was present in 5.6% of coal miners after a mean exposure of 8.4 years, while among 3754 Chilean coal workers, the prevalence of CWP was 14.1% and that of PMF 0.22%.

In general, dust control measures remain inadequate in developing countries, with concentrations of respirable dust likely to exceed permissible standards and medical surveillance less likely to be effective (He and Mikheev, 1993). There are also suggestions that the cumulative attack rate of pneumoconioses in workers with dust exposure in these countries may be increasing, rather than decreasing (WHO Study Group, 1986).

Table 12.1 Prevalence of CWP in developing countries

Country	*Population*	*Prevalence (%)*	*Year*
Brazil	Coal miners	5.6	1988
Chile	Coal miners	14.1	1988
Columbia	Coal miners	15	1986
India	Coal miners	18.8	1986
	Coal miners	12	1988
Korea	Anthracite miners	9.3	1986
Zimbabwe	Coal miners	20	1988

From *Experimental Lung Research* (1990) **16**, 5–13, van Sprundel, M., Taylor and Francis, Inc., Washington, DC. Reproduced with permission.

Table 12.2 Classification of the extent of pulmonary function damage

Degree of damage	*FVC*	*FEV_1*	*MVV*	$\frac{FEV_1}{TLC}$ %	$\frac{RV}{TLC}$ %	*D_Lco*	*Pao_2 (kPa)*	*$Paco_2$ (kPa)*	*$(A\text{–}a)o_2$ (kPa)*
Normal	>80	>80	>80	>70	<35	>80	–	–	–
Mild	60–79	60–79	60–79	55–69	36–45	60–79	–	–	–
Medium	40–59	40–59	40–59	35–54	46–55	45–59	–	–	–
Serious	<40	<40	<40	<35	>55	<45	4–8	6–8	>9.3

The FVC, FEV_1, MVV and D_Lco measurements are the % predicted values quoted from the *Distinguished Criteria of Disability Resulting from Industrial Injury and Occupational Diseases*, p. 32, as formulated by the Ministry of Labour and the Ministry of Health, P. R. China.

PREVENTION AND CONTROL OF CWP

Despite the lack of systematic and precise data collection and unreliable statistical evidence, the fact that the incidence of CWP appears to be increasing in developing countries is not to be ignored. Experience from industrialized countries has shown that a significant reduction in the incidence and prevalence of CWP can only be achieved by primary prevention; that is, by reduction of the intensity and duration of exposure of workers (Lou and Zhou, 1989; He and Mikheev, 1993; Shen, 1994). Therapeutic interventions aimed at reducing the tissue reaction to inhaled dust have been disappointing, so that the main emphasis must be the prevention of dust inhalation (vanSprundel, 1990). The only way to prevent this unacceptably high figure from rising further is by implementing programs of CWP prevention for miners exposed to dust. These are essentially engineering programs which minimize the release of dust by lessening the amount of dust generated, by wetting the dust, by personal protective measures, and through medical surveillance and education (He and Mikheev, 1993).

China has developed methods to attack the high rate of CWP. Studies using different techniques for dust suppression have been conducted since 1960s and a maximum allowable concentration of coal dust exposure was formulated, with regular monitoring of dust exposures in workplaces being initiated. Radiographic criteria of pneumoconioses were revised to meet ILO international classification of radiographs of pneumoconiosis (1980). Pathologic diagnostic criteria for pneumoconiosis and classification of pulmonary function abnormalities attributable to pneumoconiosis were also drafted (Li *et al.*, 1989) (Table 12.2; Appendix 12B). Epidemiologic investigations have made clear the hazards of coal mining, the clinical characteristics of CWP, and the prevalence of disease. These data have provided a scientific basis for the government to draft measures of dust control. A series of regulations such as *The Regulation of Prevention and Control of Pneumoconiosis Disease in People's Republic of China* have been promulgated. Workers exposed to dust have been enrolled in medical surveillance programs and given drug therapy. Because of the high prevalence of pulmonary tuberculosis in many developing countries and the poor prognosis of tuberculosis in this mining population, priority should be given to the prevention and control of this disease.

The effectiveness of these measures will not be fully identified in the short term, as CWP is a disease with a long latent period. Indeed, the CWP population will likely remain unchanged for the next 20 years, and there will probably be newly diagnosed workers with CWP 30 years from now (Lou and Zhou, 1989). To prevent CWP, hygienic conditions at the workplaces must be improved and appropriate dust standards implemented. The health surveillance of exposed workers should be organized. Indeed, for the ultimate elimination or reduction of the burden of cases of CWP on a global basis, the health risks of miners must be given high priority in a world with accelerating socioeconomic development.

REFERENCES

Algranti, E., Mendonga, E.M.C., Silva, R.C. *et al.* (1992) Incidence, radiological progression and probability of coal workers pneumoconiosis in Brazilian coal workers, in *Abstracts of the Eighth International Conference on Occupational Lung Diseases*. ILO, p. 4.

He, F. and Mikheev, M.I. (1993) WHO strategies on global elimination of silicosis, in *Proceedings of Second International Symposium on Silica, Silicosis and Cancer*, Centers for Disease Control and Prevention, National Institute for Occupational Safety and Health, pp. 34–8.

Ministry of Public Health of P.R. China (1992) *Data collection of epidemiologic study on pneumoconioses in China (1949–1986)*, Combined Publishing House of Beijing Medical University and Chinese Xiehe Medical University, Beijing, pp. 16–49. [in Chinese]

Li, Y.R., Fu, S.C., Ding, M.B. *et al.* (1989) Achievements in research in the area of pneumoconiosis prevention and treatment in 40 years in China. *Chinese J. Ind. Hyg. Occup. Dis.*, **7**, 268–271 [in Chinese].

Lou, J.Z. and Zhou, C. (1989) The prevention of silicosis and prediction of its future prevalence in China. *Am. J. Public Health*, **79**(12), 1613–16.

Shen, G.A. (1994) Coal workers' pneumoconiosis, in *Contemporary Occupational Hygienics* (eds Y.L. Wang and B.Q. Gang), The People's Medical Publishing House, Beijing, pp. 125–32. [in Chinese].

van Sprundel, M.P.M. (1990) Pneumoconioses: the situation in developing countries. *Exp. Lung Res.*, **16**, 5–13.

WHO Study Group on Recommended Health. Based Limits in Occupational Exposure to selected mineral dusts (1986). Recommended health based limits in occupational exposure to selected mineral dusts (silica, coal report of a WHO study group [meeting held in Geneva 6–12 March 1984]).

Zou, C.Q., Liu, Z.Y., Cheng, Y.H. *et al.* (1993) Epidemiological study of pneumoconioses of China, in *Abstract Compilation of Papers in 5th Country's Academic Meeting of Labour Health and Occupational Disease*, Chinese Preventive Medical Association, Wuhan, pp. 1–6 [in Chinese].

APPENDIX 12A
THE 1986 ROENTGENOGRAPHIC DIAGNOSTIC CRITERIA OF PNEUMOCONIOSES AND PRINCIPLES OF MANAGEMENT

The 1986 roentgenographic diagnostic criteria of pneumoconioses are applicable to all prescribed types of pneumoconioses.

1. Principles of diagnosis

Radiologic examination of the chest is the basic approach used in the diagnosis and classification of pneumoconioses. Correct interpretation of the radiographic manifestations of pneumoconioses depends on a reliable, detailed occupational history followed by the careful scrutinization of standard size Posterior–Anterior (PA) chest film of optimal technical quality. To understand progression of the disease, previous chest films are invaluable and should be sought and reviewed. Data describing dust exposure monitoring is exceptionally important; confirmatory information, if available, should be consulted.

In addition to the roentgenologic diagnosis and classification of X-ray manifestations, the clinical diagnosis of pneumoconioses includes the early recognition of any complications and reasonable exclusion of other diseases which may give rise to similar X-ray appearances. Evaluation of the compensatory status is achieved by an integrated study of the history, symptoms and signs, physical examination, routine laboratory examination and, if indicated, examination of the sputum and pulmonary function testing.

2. Criteria for diagnosis and 'stages' of pneumoconiosis

2.1 No pneumoconiosis (Code 0)

a. 0: No roentgenologic manifestations of pneumoconiosis.

b. 0^+: With roentgenologic manifestations of pneumoconiosis up to the lower limit of 'stage I'.

2.2 Pneumoconiosis 'stage I' (Code 1)

a. I: Small rounded opacities, with profusion of category 1, and extent of one area in at least two zones of the lung, the diameter of each area being not less than 2 cm.

For small irregular opacities, with profusion of category 1, and extent of not less than two zones of the lung.

b. I^+: Small opacities, either rounded or irregular, with profusion of category 1, and extent of more than four zones of the lung; or with profusion of category 2, and extent of up to four zones of the lung.

2.3 Pneumoconiosis 'stage II' (Code II)

a. II: Small rounded or irregular opacities, with profusion of category 2, and extent of more than four zones of the lung; or with profusion of category 3, and extent up to four zones of the lung.

b. II^+: Small rounded or irregular opacities of profusion category 3, and extent of more than four zones of the lung; or with a large opacity less than 2 cm in length and 1 cm in width.

2.4 Pneumoconiosis 'stage III' (Code III)

a. III: With a large opacity of not less than 2 cm in length and not less than 1 cm in width.

b. III^+: With a large opacity or opacities of total area larger than the right upper zone of the lung.

3. Principles of management

The purposes of management of pneumoconioses are to delay the progress of the disease, to alleviate the symptoms, to adopt effective measures to prevent and treat the complications. Hence, the combined measures of management should include medical treatment, an educational program for personal hygiene, a well-scheduled life pattern, appropriately individualized exercise and rehabilitation, and preventive measures for pulmonary tuberculosis.

4. Evaluation of working ability

The working ability of a patient is ascertained in accordance with his/her stage of pneumoconiosis and status of functional compensation.

5. Medical surveillance

A pre-employment physical examination, including radiological examination, must be carried out before dust exposure for any worker, and must be repeated periodically during work exposure.

The aim of the periodic physical examination during exposure is to discover new cases of pneumoconiosis or to follow the progress of old cases. Examinations are carried out at intervals decided by the local office of health. Generally, the interval between two examinations is 1 year (maximum 2 years) for heavy exposure, 2–3 years for light exposure, and 3–5 years in special circumstances such as migratory workers, intermittent exposure or where facilities are not available.

6. Contraindications for dust exposure

a. Any severe chronic diseases of the upper respiratory tract.
b. Active tuberculosis of lung, or any extra-pulmonary active tuberculosis. Severe chronic pulmonary disease.
c. Diseases of the pleura, thoracic cage or vertebra, markedly influencing pulmonary function.
d. Severe cardiovascular disease.

APPENDIX 12B PATHOLOGIC DIAGNOSTIC CRITERIA OF PNEUMOCONIOSES

Pneumoconioses are a group of diseases resulting from the inhalation of dust and characterized by lung fibrosis.

1. Principle of diagnosis

The pathologic diagnosis of pneumoconiosis should be made on the bases of a detailed and confirmed history of occupational exposure and standard pathologic technical procedures. Relevant X-ray chest films, a clinical abstract or death record of the patient and the labor hygiene data of the patient's work places are essential references.

2. Criteria of pathologic diagnosis and categorization

2.1 No pneumoconiosis

Those cases in which only dust tissue reactions are demonstrated in the lungs and/or in the hilar lymph nodes (HLNs); or pneumoconiosis lesions demonstrated in the lungs and/or in HLNs, but not sufficient in extent or severity to diagnose stage I pneumoconiosis.

2.2 Stage I pneumoconiosis

A. Those cases in which the number of pneumoconiosis nodules (Nds) demonstrated on the lung cut surfaces macroscopically and/or confirmed microscopically ≥20; or where the number of pneumoconiosis Nds ≥10 in association with diffuse fibrosis (DF) close to fibrosis of profusion category 1/1; or
B. DF ≥ fibrosis of profusion category 1/1; or
C. Maculo-emphysematous areas ≥50% of the cut lung surfaces.

2.3 Stage II pneumoconiosis

A. Those cases in which the number of pneumoconiosis nodules (Nds) demonstrated of the lung cut surfaces macroscopically and/or confirmed microscopically ≥50, or where the number of pneumoconiosis Nds ≥20 plus DF ≥1/1; or
B. DF ≥ fibrosis of profusion category 2/2; or
C. Maculo-emphysematous areas exceed ≥75% of the cut lung surfaces.

2.4 Stage III pneumoconiosis

A. Massive fibrosis demonstrated in the lungs with a background of at least stage I pneumoconiosis; or
B. DF ≥ fibrosis of profusion category 3/3.

13 Quartz and silicosis

William G. B. Graham

INTRODUCTION

Silicosis is the term used to designate the occupational lung disease caused by inhaling crystalline silica (alpha-quartz or SiO_2) or its polymorphs, tridymite or cristobalite. These latter compounds occur uncommonly in nature, but may be produced by the calcining or flux calcining of amorphous silica, usually as diatomaceous earth. In fact, this definition as an 'occupational disease' may be too limited. Published reports (Nouh, 1989; al-Kassimi *et al.*, 1991; Norboo *et al.*, 1991) suggest that dusty non-occupational exposures, as in villages in the high Himalayas or in desert communities, may lead to abnormal chest radiographs or even histologic changes suggestive of silicosis. In a different context, workers exposed to very low levels of quartz who have no pulmonary symptoms and normal lung function and chest films have shown in bronchoalveolar lavage fluids done for research purposes not only dust particles, but also abnormal cellular and protein constituents (Christman *et al.*, 1985; Calhoun *et al.*, 1986), suggesting at least a reaction to inhaled particles. Whether these findings represent the presence of a disease process (alveolitis) is almost a subject for philosophical discussion, akin to asking whether tanning of the skin is a pathologic process or a normal response to an imposed stimulus.

This analogy is not too far-fetched, since exposure to both quartz and sunlight have been constant companions in the evolution of biologic systems. Silicon is one of the eight major elements making up the chemistry of the earth's crust and by weight is the second largest component after oxygen. It is combined in myriad forms and combinations with other elements to produce chemical structures which may or may not be harmful to exposed humans or animals. Amorphous silica, which lacks a crystalline structure, is essentially benign. Crystalline silica in any form is potentially toxic when absorbed or inhaled in sufficient quantities. Primitive art going back to the Stone Age suggests that creative persons not fully occupied with scrounging sustenance were exposing themselves to the same hazards as are ceramicists and sculptors today. The Bronze Age is so named because of the mining of metals and the creation of alloys. No written records exist for these eons of exposure to harmful dust until Hippocrates commented on the shortness of breath of a metal miner, and Pliny the elder (23–79 AD) spoke of respirators to prevent dust inhalation (Plinius). Recalling the Mt St Helen's catastrophic eruption in 1980, it is interesting to note that Pliny, an endlessly curious observer, suffocated when Vesuvius exploded and buried Pompeii in volcanic ash and magma. Georgius Agricola, the nom-de-plume of George Bauer, in 1557 advocated the ventilation of mines, and stated, 'other mines are very dry, and the constant dust enters the blood and lungs, producing that difficulty of breathing the Greeks call asthma. When the dust is corrosive it ulcerates the lungs and produces consumption, hence it is that in the Carpathian mountains there are women who have married seven husbands, all of whom this dreadful disease has brought to an early grave' (Agricola). Ramazzini quotes a description given by Diembrock of several stonecutters who died asthmatic, 'in whose lungs he found such heaps of sand that in running the knife through the pulmonary vesicles he thought he was cutting some sandy body' (Ramazzini, 1705). The term silicosis, derived from the Latin 'silex', or flint, was coined by Visconti in 1870, but in 1713, a patent granted to Thomas Benson of

Occupational Lung Disease. Edited by Daniel E. Banks and John E. Parker. Published in 1998 by Chapman & Hall. ISBN 0 412 73630 6

Newcastle-under-Lyme for grinding flints by a wet method states that pounding flints dry 'proved very destructive to mankind insomuch that any person, ever so healthful and strong, working in that business, cannot possibly survive over 2 years occasioned by the dust sucked into his body by the air he breathes' (Owen, 1901).

Silicosis presents itself in many different forms, varying from a relatively acute form called silicoproteinosis resulting from massive exposures to quartz, to an indolent and slowly progressive process which may cause negligible clinical manifestations. Chronic silicosis may take decades to develop and the radiographic changes which occur initially produce few if any symptoms and minor or no changes in pulmonary function. When the disease progresses to conglomerate radiographic opacities, patients often develop symptoms of dyspnea and cough, with severe limitation of physical activity, and in some cases respiratory failure and death. An association which appears to be almost unique to silicosis among occupational disorders occurs in its facilitation of tuberculosis, which before antituberculous therapy was almost invariably progressive and ultimately fatal.

With the advent of roentgenography during the past century, epidemiologic studies clarified what types of jobs and exposures cause silicosis and the dose–response relations; that is, how much dust, measured by intensity and duration of exposure, causes the onset and progression of radiographic disease. Engineering studies have taught industries how to measure environmental levels of dust and devised methods of control. Radiographic criteria have been standardized to assess in a uniform way the changes which occur as the fibrotic process evolves. Various governmental agencies have established or proposed exposure limits which will, if observed, protect exposed workers from developing clinically apparent disease. Current limits of dustiness are called the threshold limit values (TLV), the permissible exposure limit (PEL), or the maximum allowable concentrations (MAC). Despite all these decades of research, however, an ongoing debate concerns the total amount of inhaled quartz that is likely to cause the earliest stages of radiographic abnormalities (Muir *et al.*, 1989; Hnizdo *et al.*, 1993a; Hughes, 1995; Steenland and Brown, 1995; Hughes and Weill, 1995). Despite this controversy, which concerns the lowest dose of quartz which will produce the most minimal radiographic silicosis in one or a few workers, if current PELs are observed in the workplace, silicosis as a clinical disease, causing symptoms, disability and early death would almost as a certainty be eliminated. It will become clear that all the studies cited here dealing with silicosis and its complications are analyzing disease induced decades ago at high dust levels.

For a disease recognized and studied world-wide for decades, why does there continue to be such interest in silicosis? For the molecular biologist, quartz provides an excellent stimulus in animals for inflammation, allowing detailed examination of biologic mechanisms of injury and repair. Cellular reactions involving all cell types and various constituents of cells such as cytokines can be studied sequentially and allow observations to be made regarding the components and modifying factors involved in the initiation, course and final outcome of the scarring process. Most experiments with quartz have been made in cell cultures or animal models. However, similar experiments have been carried out in humans exposed to quartz, with or without radiographic evidence of disease, using analyses of both sera and bronchoalveolar fluids obtained by bronchoscopic lavage (Schuyler, 1977; Schuyler *et al.*, 1980; Christman *et al.*, 1991; Doll *et al.*, 1981; Rom *et al.*, 1987).

A second important concern involves public policy: whether the current governmental guidelines for exposure are in fact sufficiently low to prevent disease after a full working lifetime. NIOSH recommended in 1974 that the then current exposure limit of 0.1 mg/m^3 be lowered to 0.050 mg/m^3 (NIOSH, 1974). OSHA, the agency which must adopt the standard, inspect industry for compliance with regulations and levy penalties for violation, has so far not accepted the 0.050 mg/m^3 standard. Among the complex reasons for this inaction are doubts about the validity of some epidemiologic data on which the recommendations were based. The controversy is discussed more fully in later sections.

A third major concern is the question of whether quartz is a carcinogenic agent. Historically, quartz was not considered to be carcinogenic, but an influential paper called this almost universal assumption into question (Goldsmith *et al.*, 1982). Subsequently, an intense debate has developed, fuelled by animal studies which suggest the induction of neoplasms in certain species, and the pronouncements of IARC (International Agency for Research in Cancer, 1987) which has labeled quartz a definite carcinogen in animals, and a 'probable' carcinogen in human beings.

Defining the health effects of quartz has become increasingly difficult as exposure levels have been lowered. In the early 1900s, as advances in industrial techniques created more dust in workers' environments, silicosis occurred early and obviously, leading to indisputable radiographic changes and often to fatal tuberculosis. As dust levels have fallen as a result of environmental controls and governmental standards, the health effects have become more subtle: radiographic changes may be so slight that interpreting a film as 'abnormal' or 'normal' may be difficult and subject to disagreement, even by expert readers.

Likewise, when conglomerate silicosis was often the outcome of extremely high dust exposures, pulmonary function changes were certain and inevitable. Now, however, whether quartz exposure has any effect in the absence of radiographic changes is very much in doubt.

A controversy in the making is whether very slight exposures should be subject to regulatory standards. This concern is particularly germane to the question of carcinogenicity. If governmental agencies conclude without heavy qualifications that quartz is a carcinogen, surely the ensuing public panic would make the asbestos litigation industry seem trivial. The US Environmental Protection Agency is in the process of developing a position statement regarding non-cancer health effects due to ambient quartz levels in the environment. The report cites various sources of quartz within the environment such as quarrying, industrial manufacturing plants which exhaust dust into the environment, agricultural activities, forest fires and wind erosion. By extrapolating selected studies carried out on miners and other quartz-exposed workers – which in turn sometimes make assumptions regarding past exposure – the statement makes predictions regarding the possibility that environmental exposure to ambient air might result in silicosis. Ambient levels have been measured at 3 $\mu g/m^3$, with high levels being 8 $\mu g/m^3$. Adjusted exposure to ambient air for a 24-hour period would result in a total daily exposure of 8.4 to 22.4 $\mu g/m^3$. This dose over a 1-year period is called a $\mu g/m^3$-year. The amount of inhaled dust over the course of a 70-year lifetime would produce a total dust burden of 1.6 mg/m^3-year of quartz at the high level. In a certain low percentage, workers exposed to this quantity of dust in some studied cohorts are estimated to develop radiographic silicosis. The implications for cancer are not discussed.

How common is silicosis in this day and age? The judgment of many students of the disease is that the incidence has been declining for years, especially in large industries which have effective industrial hygiene. A 1994 report shows the mortality from silicosis declining from 1157 deaths in 1968 to 538 deaths in 1978 (Centers for Disease Control and Prevention, 1994). 'Cause' was defined as silicosis being listed on the death certificate as either underlying or contributing to death. Mortality data are neither a good reflection of the prevalence of disease, nor as informative as the induction of new disease due to current levels of exposure. Undoubtedly, the majority of men with silicosis do not die of it, particularly since tuberculosis combined with silicosis was the major cause of death before antituberculosis therapy. Much of the advanced silicosis leading to death was likely to have developed when dust exposures were uncontrolled. For example, during the decade 1968–1978, Vermont was listed as having 88 deaths due to silicosis (Centers for Disease Control and Prevention, 1994). However, mortality data related to initial date of employment in the granite industry (the major source of silicosis in the state) shows that no deaths occurred from this cause among workers first employed after 1940, when dust controls were fully operational (Costello and Graham, 1988). The 88 deaths were a tragic legacy from the days of poorly controlled environmental conditions.

Several states have attempted a systematic surveillance program. New Jersey identified 60 cases of silicosis annually between 1979–1987 by studying hospital discharge diagnoses, consulting the National Center for Health Statistics, and surveying all the physicians in the state (Valiante *et al.*, 1989). The workers were generally older, had begun work before 1960, and had worked for 15 years or more in potteries, construction, foundries and sand mines. The authors concluded that extrapolation of their state incidence to the national work force would yield many cases, and suggested that the prevalence of silicosis was markedly underestimated. More recently, the Ohio Department of Health, which identified workers with silicosis from compensation claims, physicians reports, death certificates and hospital discharge diagnoses found 562 cases from 1989–1994 (Anonymous, 1995); a large majority of these cases (404) were found through hospital reporting and 41% of these were confirmed. Other data compiled by NIOSH from four industrialized states (New Jersey, Michigan, Ohio and Wisconsin) counted 447 cases of silicosis diagnosed between 1988–1992, 259 of which were from primary metal industries (Centers for Disease Control and Prevention, 1994). These industries employ machine operators who grind, abrade, buff, polish, mold and cast. However, this type of information is crude in the sense that the data are not stratified by age, duration of exposure or severity of disease. However, a recent editorial comments on the persistence of silicosis, pointing out that sampling for airborne quartz by MSHA (Mine Safety and Health Administration) showed that about 30% of the samples were over the PEL and nearly 20% exceeded twice the PEL (Wagner, 1995).

That silicosis continues to occur is a reflection of uncontrolled and probably egregious exposures. For example, the Ohio case which led to surveillance was a sandblaster with advanced silicosis and a *Mycobacterium kansasii* infection who had worked in a metal preparation shop with no dust controls (Anonymous, 1995); a survey of the shop showed four additional cases in current or former workers. An investigation of two silica flour plants in Illinois (Banks *et al.*, 1981) showed that of 161 current or previous workers, 26%

had simple silicosis, and 11% progressive massive fibrosis. Quartz levels were many times over the PEL. Other reports of accelerated silicosis have occurred in church stonemasons working with no respiratory protection (Seaton *et al.*, 1991).

PATHOGENESIS OF SILICOSIS

During the past two decades enormous progress has been made in defining the events which occur in lung cells or tissues after the inhalation of quartz (de Shazo, 1982; Davis, 1986). These studies have utilized cell cultures exposed to quartz, and extensive animal exposures have been carried out after 'dusting' or injecting quartz into body cavities, with analysis of the inflammatory and scarring process over time. These exposures of course are qualitatively and quantitatively very different from those which occur in humans exposed in an occupational setting. Questions will always exist whether the cellular mechanisms initiated by the intensity of the stimulus are being altered by the comparatively large, though short, exposures. A typical example of experimental animal dose is 40 mg/m^3 of pure quartz, 6 hours per day for 8 days. After the exposure, the cellular and biochemical reactions are analyzed over the next few days, weeks or months. As 40 mg of quartz is 400 times the federal standard applied to humans in an occupational setting, several workers (Morrow, 1988; Holland, 1995; Mauderly, 1996) have suggested that such 'overloading' by dust may exceed the capacity of the cells involved in particle clearance such as alveolar macrophages. This in turn may produce in certain species cellular reactions which falsely appear to be neoplastic. Indeed, similar changes may occur after exposures to totally inert particles, suggesting that the changes are not specifically related to the toxic qualities of quartz.

Another serious concern in interpreting animal data is that species differences exist between animal models. Rats, a favored experimental model, appear to have particularly hyperreactive lung epithelial cells (Mauderly, 1996), while some histologic changes produced in rats are not reproducible in Syrian hamsters or in mice (Holland, 1995). These caveats apply to both the interpretation of the inflammatory process and the neoplastic response to quartz. Although animal studies require cautious interpretation, they have illustrated the complexity of the tissue response to quartz (and other substances such as asbestos).

The biologic effects of quartz have also been studied in humans, using the technique of bronchoalveolar lavage (BAL). This enables the removal via fibreoptic bronchoscopy of cells and fluid from human subjects exposed to quartz, with or without radiographic silicosis (Rom *et al.*, 1987; Schuyler *et al.*, 1980; Christman *et al.*, 1985; Calhoun *et al.*, 1986). Information from these diverse sources has allowed the formulation of a schema for the events which occur in the silicotic process. In Figure 13.1 this information is simplified in order to help in conceptualizing the pathogenetic events.

The physical features of inhaled silica are an important determinant of the ensuing tissue response. In both humans and experimental animals, the quartz particles which initiate the process must be small enough to reach the alveolar spaces. Particles less than 1 μm are probably the most dangerous, as larger particles become impacted in the nasopharynx, trachea and bronchi and are then removed by the movement of mucus up the respiratory tract. The inflammatory potency of the particle may be greater if recently fractured (Vallyathan *et al.*, 1988), though this enhanced reactivity decays fairly promptly. The surface of the quartz particle generates silicon-based radicals which in turn lead to the production of hydroxyl, hydrogen peroxide and other oxygen radicals, which may initiate lipid peroxidation of cell membranes and oxidatively inactivate essential cell proteins (Ghio *et al.*, 1990; Shi *et al.*, 1995). The potency will be reduced by coating the particles, for example by aluminum dust, which was used several decades ago in miners to prevent silicosis. The protective effect of aluminum dust has recently been thoroughly studied in a sheep model (Begin *et al.*, 1993). The phospholipids of alveolar surfactant may have a similar ameliorating effect. The surface properties of the quartz particle are conveyed by its crystalline structure. Amorphous silica does not have a regular structure and lacks the capacity to generate radicals; it is an essentially benign substance.

Alveolar macrophages, the earliest and most important cellular participants in the inflammatory process, ingest the particle. Previous studies done in cell cultures had suggested that the quartz is toxic, leading to the death of the macrophage with subsequent release of various harmful lysosomal cellular components. The ingested quartz was then recycled to continue the destructive process (Allison *et al.*, 1966). This effect however may have been due to the higher doses of quartz used in tissue culture. In contrast, studies of bronchoalveolar fluid in workers exposed to low quartz levels show that the macrophages (Rom *et al.*, 1987; Schuyler *et al.*, 1980; Christman, 1986) are activated rather than destroyed. These activated cells seem to have an enhanced ability to release various messengers such as cytokines and chemotatic factors which recruit other cells in the inflammatory process and damage resident structural cells and extracellular matrix (Davis, 1986).

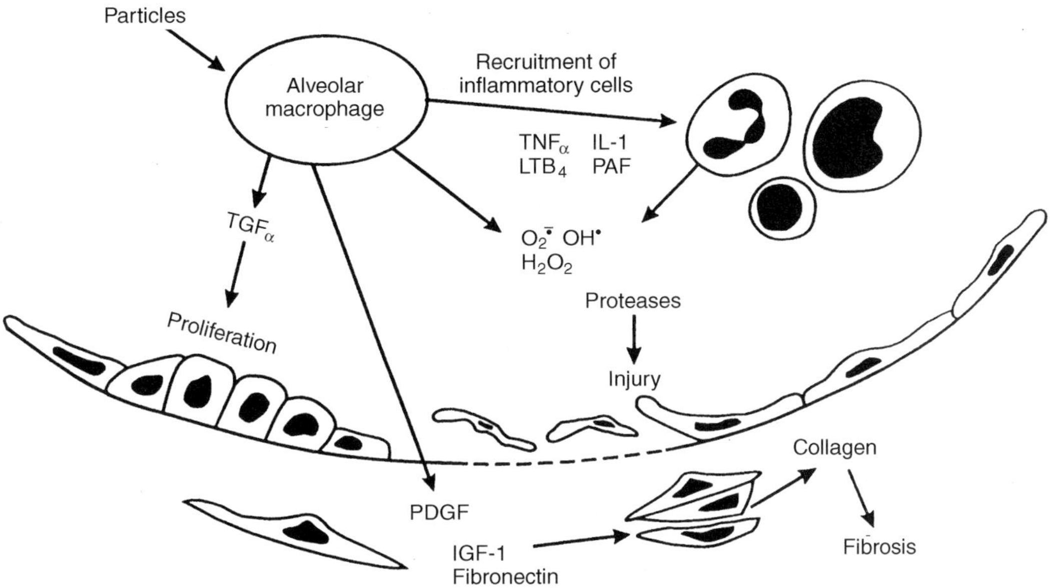

Figure 13.1 Alveolar macrophages, after ingesting particles, send cytokines to recruit inflammatory cells. The ensuing reactions release oxygen radicals and proteases which lead to epithelial cell damage. PDGF (platelet-derived growth factor), IGF-1 (insulin-like growth-factor) and fibronectin stimulate fibroblasts to produce collagen. Tumor growth factor (TGFα) leads to proliferation of Type 2 pneumocytes. Tumor necrosis factor (TNFα); IL-1, interleukin-1; LTB_4, leukotriene B-4; PAF, platelet activating factor. This list of reactants is selective.

When animals are exposed to dusting there is an active turnover of ingested particles so that most of the inhaled material is cleared within several months (Davis *et al.*, 1981; Davis, 1986). The uncleared particles remain in the lung and presumably remain biologically active, unless sequestered in calcified or fibrotic lesions. The removal of the quartz particles ingested by macrophages is largely by mucociliary clearance initially, as macrophages migrate upward to the ciliated bronchial epithelium. Macrophages within the alveoli apparently continue to contain particles, since BAL fluid washed from the lungs of retired granite workers not recently exposed to quartz continue to show quartz particles (Christman *et al.*, 1991). Macrophages remaining in the lung may also reside in the interstitial spaces or in regional lymph nodes, which are invariably involved in the inflammatory process. Resident quartz particles outlive the macrophages which originally ingest them and are repeatedly picked up by newly bred macrophages.

Cellular events after ingestion of the particles have been defined in animal models. The activated macrophage releases a wide variety of cytokines. These may recruit inflammatory cells such as leukocytes and lymphocytes (Davis *et al.*, 1981), as well as increasing the macrophage population in the lung. Transforming growth factor alpha (TGF-α) leads to the proliferation of Type II cells; other cytokines such as platelet-derived growth factor (PDGF) and insulin-like growth factor (IGF) stimulate fibroblasts to proliferate and produce collagen, a major component of the scarring process. Oxygen radicals such as hydrogen peroxide and superoxide, and proteases may lead to epithelial cell injury. Other biologic substances involved in the inflammatory process are tumour necrosis factor (TNF) and leukotriene B-4.

The end point of the inflammatory process in animal models is an increase in collagen in the lung, as measured by increases in collagen synthesis rates and in lung hydroxyproline contents; these changes parallel the formation of granulomas (Reiser *et al.*, 1982, 1983), and may occur within 1–2 weeks of instilling the silica. Parallel with these changes are histologic changes similar to those of alveolar proteinosis, which may be seen in humans exposed to extremely high dust levels. These elevated rates of collagen synthesis persist for up to a year afer the initial instillation, again suggesting that once initiated the process continues, probably as a result of quartz sequestered in the lung.

How are these cellular events, largely described in heavily dosed, short-term animal models, reflected in human broncheoalveolar lavage fluid? Non-smoking workers with established silicosis showed no differences from non-exposed subjects in cell number, viability, adherence or ability to phagocytose and kill intracellular organisms such as *Listeria monocytogenes* (Schuyler *et al.*, 1980). Significantly increased numbers of Type II pneumocytes were present in the retired silicotic subjects, reflecting the continuing presence of biologically active material, and a continuation of the inflammatory process. Christman *et al.* (1985) confirmed that there is no difference in phagocytic function or viability of macrophages from granite workers compared with non-exposed subjects. BAL fluids from non-smoking granite workers showed increased numbers of lymphocytes and increased concentrations of immunoglobulins in BAL, with normal albumin levels in lavage specimens and in serum.

The cellular and biochemical reactions which occur in the lung after inhalation of quartz are common to other types of inflammation, such as that seen in asbestosis or interstitial fibrosis. The differences between them in terms of their histology, clinical presentation and susceptibility to diseases are of great interest, and they will continue to be studied and clarified.

PATHOLOGY OF SILICOSIS

The pathologic diagnosis of silicosis requires that histologic changes consistent with the process initiated by quartz dust exposure be associated with the presence of quartz, demonstrated either by polarized light microscopy or by other evidence derived from more advanced technology. Techniques identifying quartz in tissue section include scanning electron microscopy or X-ray energy spectrometry. X-ray diffraction is useful for estimating quartz in air samples and in ashed tissue. The point of emphasizing both consistent histology and the presence of quartz is that some of the early cellular reactions (as opposed to mature silicotic granulomas or nodules) induced by quartz are not pathognomonic; nor, on the other hand, does the presence of quartz in the tissue prove conclusively that quartz is the causative agent for the histologic abnormalities. A good example of this diagnostic dilemma was the finding of non-specific fibrosis and inflammation in the lungs of granite workers who died of assorted diseases (Craighead and Vallyathan, 1980). Radiographs were negative for silicosis. Quartz was present in the tissues only by scanning electron microscopy with back-scatter imaging and X-ray spectrometry. Smoking history was incomplete, but was probably heavy since 10 of 15 cases died of lung cancer, cirrhosis or trauma. Did these subtle inflammatory and fibrotic changes indicate an early reaction to quartz or non-specific inflammatory changes from smoking or other causes?

The earliest histologic changes in humans as mentioned above are hard to interpret. Experimental animals dosed with quartz show accumulations of macrophages and neutrophils, soon joined by lymphocytes and other mononuclear cells, without evident structural change (Craighead and Kleinerman, 1988). This type of response is that of lung injury which may occur after many diverse insults. New cellular structures appear, arranged in chaotic or organized patterns. Occasionally, poorly formed granulomata and proteinaceous debris in the alveolar spaces are seen. Eventually these mature into well-formed collagenized nodules. There is no necrosis.

Human disease shows distinctive patterns, depending on whether the amount of dust inhaled was extremely large, or whether low-level exposure has taken place over decades. Some cases of heavy exposure show histologic features of both acute and chronic silicosis; these have been described clinically as 'accelerated' disease. Acute silicosis (or 'silicoproteinosis') has been described in patients exposed to massive amounts of quartz from sandblasting, tunnel drilling or exposure in silica flour plants. The disease develops in months to a few years, and at post-mortem or biopsy the alveoli are choked with lipoproteinaceous material which stains with periodic acid–Schiff (PAS) (Buechner and Ansari, 1969; Suratt *et al.*, 1977). The histologic picture is similar to the idiopathic lung disease alveolar proteinosis which appears to be related to a defect in clearance mechanisms. Type II pneumocyte hyperplasia and hypertrophy appear with increased production of phospholipid, specifically dipalmitoyl lecithin and neutral lipids (Craighead and Kleinerman, 1988). Loosely organized reticular nodules and fibrosis are present in the interstitium and lymph nodes. The intra-alveolar material is resistant to diastase digestion and non-reactive with traditional mucin stains. Unlike idiopathic alveolar proteinosis, interstitial inflammation and fibrosis are present, and in some cases the alveoli are filled with macrophages. The appearance can resemble that of desquamative interstitial pneumonitis (Seaton *et al.*, 1991). Necrosis, however, is not a component of the histologic picture, nor are multinucleated giant cells (Craighead and Kleinerman, 1988).

The essentially pathognomonic finding in mature silicosis is the silicotic nodule, generally solitary and less than 1 cm in diameter, with whorled, avascular collagen scattered preferentially in the upper lobes and in the regional lymph nodes (Figures 13.2 and

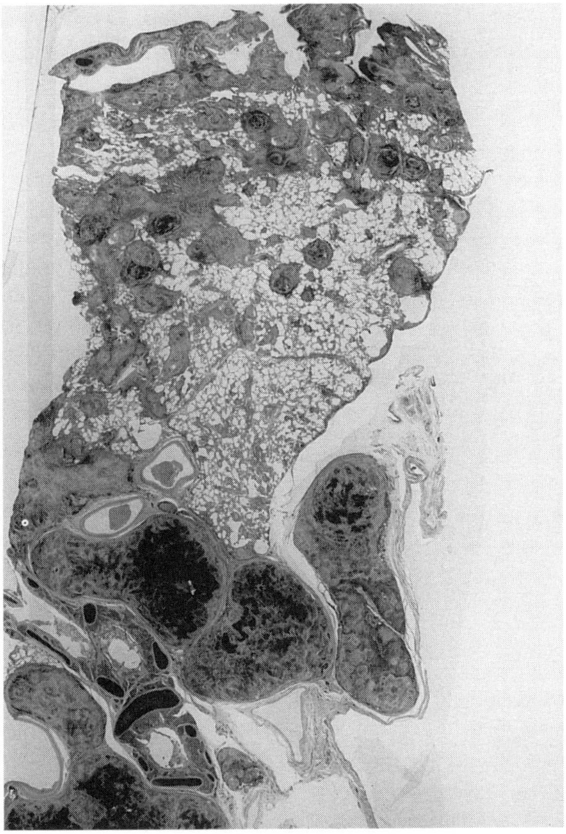

Figure 13.2 Thick section of lung and lymph node showing the nodular configuration of silicotic nodules in lung parenchyma. The darker areas are anthracotic pigment. Lymph node shows collagenized silicotic nodules. (Reproduced from W.G.B. Graham (1992) *Clinics in Chest Medicine,* **13**, (2), 253–67, by permission of W.B. Saunders, Philadelphia.)

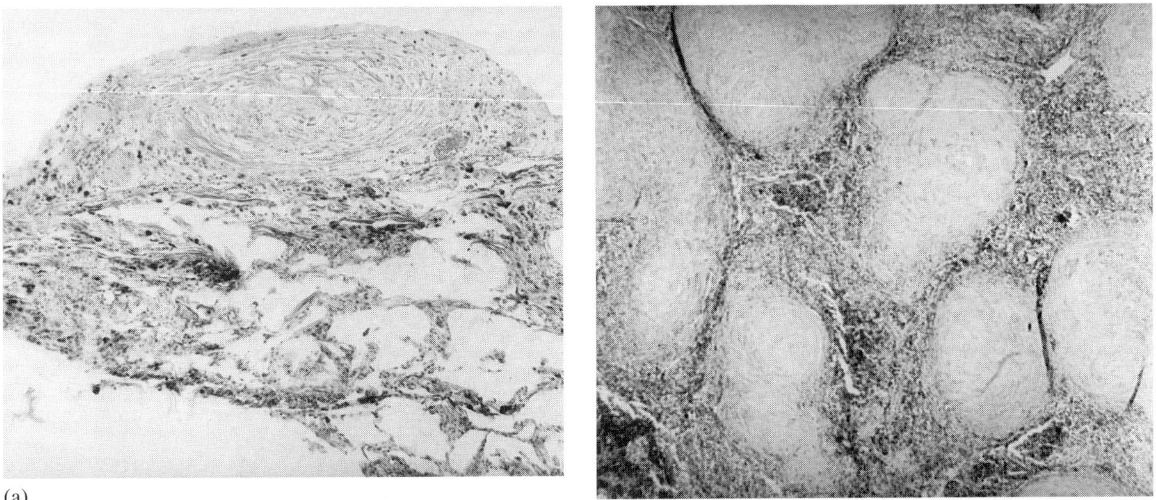

(a)

(b)

Figure 13.3(a,b) Hyalinized, collagenized silicotic nodules in a lymph node. (Reproduced from W.G.B. Graham (1992) *Clinics in Chest Medicine,* **13**, (2), 253–67, by permission of W.B. Saunders, Philadelphia.)

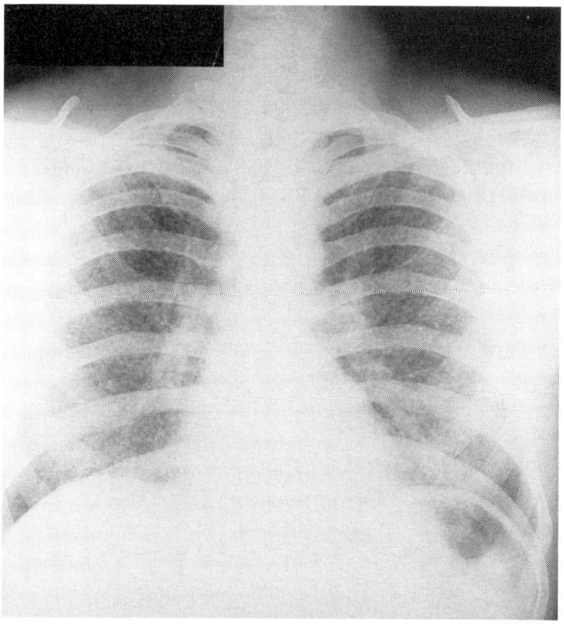

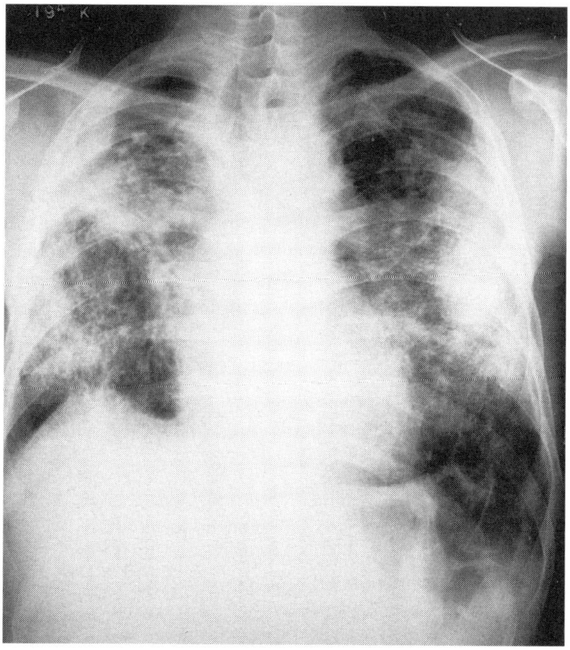

Figure 13.4 and 13.5 Progression of silicosis over an 11-year period, from finely nodular opacities to massive fibrosis. The second film shows bullae at the bases, tracheal deviation, elevation of the right diaphragm and cardiac enlargement from cor pulmonale. The patient worked outdoors in a stone crushing operation and died of respiratory failure. (Reproduced from W.G.B. Graham (1992) *Clinics in Chest Medicine,* **13**, (2), 253–67, by permission of W.B. Saunders, Philadelphia.)

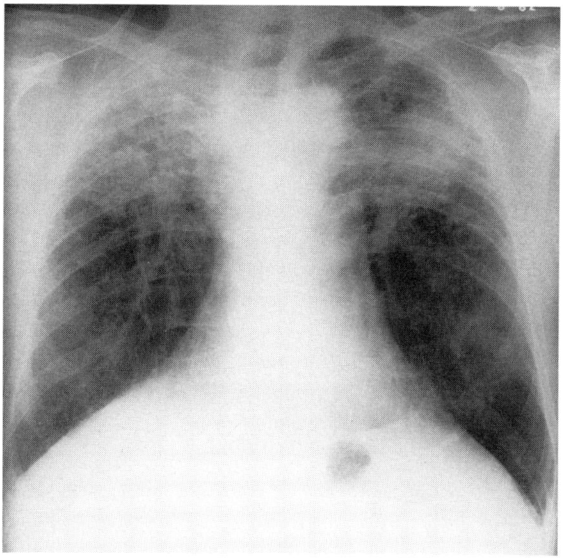

Figure 13.6 Extreme distortion of hilar structures from massive fibrosis. Note tenting of the left diaphragm and deviation of trachea. Pulmonary arteries are elevated and incorporated into fibrotic masses.

13.3). This lesion begins as a collection of dust-laden macrophages, with reticulin fibers, located within the interstitium near respiratory bronchioles and pulmonary vessels, as well as subpleurally. The central area becomes densely hyalinized fibrous tissue; the midzone shows a concentric arrangement of collagen fibers, having a laminated appearance. The central zone may calcify. Around the periphery are macrophages and lymphoid cells, and sometimes pigment. As the fibrotic process progresses, these nodules may fuse together, forming conglomerate masses of collagenized and hyalinized tissue which encroach on or replace the normal parenchyma (Figures 13.5–13.7). If a mycobacterial infection is engrafted onto the silicotic process, caseation necrosis and multinucleated giant cells will suggest the need for acid-fast staining. Traction produced by these masses on adjacent lung parenchyma lead to emphysema both in the vicinity of the nodulation and in more distant areas of the lung fields (Figures 13.6 and 13.7). Whether the development of conglomerate masses should be called 'progressive massive fibrosis' (PMF) is debatable, since this term has often been applied to coal workers' pneumoconiosis (CWP). As in tubercu-

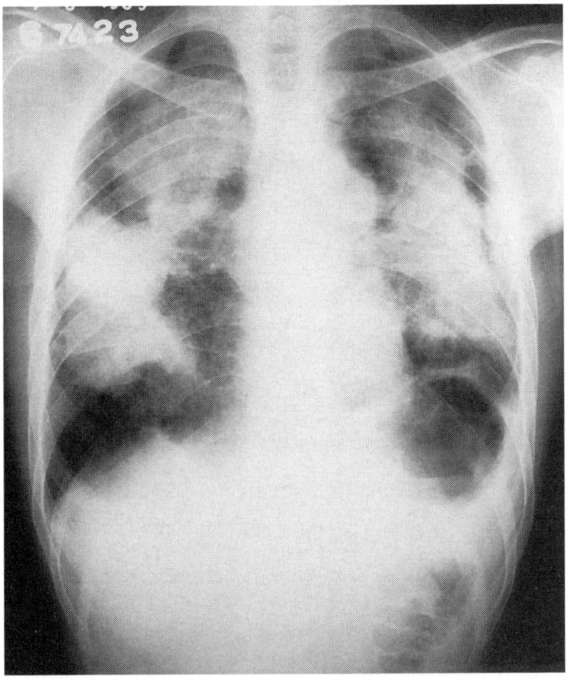

Figure 13.7 Massive conglomerate silicosis in a pumice worker. The appearance of the densely calcified masses has been called 'angel's wings'. The patient died in his forties of respiratory failure. (Reproduced from W.G.B. Graham (1992) *Clinics in Chest Medicine,* **13**, (2), 253–67, by permission of W.B. Saunders, Philadelphia.)

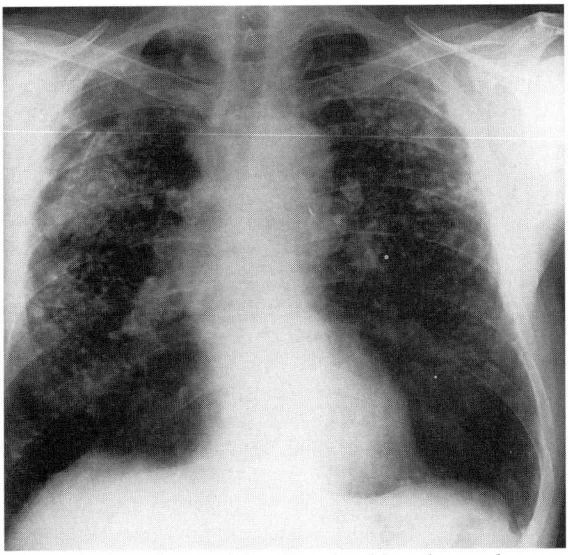

Figure 13.8 Diffuse nodular silicosis with early conglomeration. The nodules are heavily calcified. Note egg-shell calcification and apical bullae.

losis with contraction of the upper lobes from scarring, the hilar structures may be pulled up, producing rupture of the alveolar walls and emphysematous blebs in the lower lung fields (see Figures 13.6 and 13.7).

Identifying the quartz particles in the nodules or other areas of inflammation is not always easy. Birefringent crystals representing silica may be identified in the center of the nodule using polarized light. Brighter shapes seen in the periphery may be either silicates or calcium-containing crystals. Positive idenification may require scanning elecron microscopy with X-ray spectroscopy (Craighead and Kleinerman, 1988).

An unusual variant of silicosis is Caplan's rheumatoid pneumoconiosis, originally described in CWP. In workers with rheumatoid arthritis, nodules usually 5–10 mm in diameter may appear relatively quickly, and sometimes cavitate (Caplan *et al.*, 1962). This type of behavior is very different from that of usual silicotic nodules. The histologic picture resembles rheumatoid nodules with central necrosis and palisading of inflammatory cells at the periphery.

DIAGNOSIS OF SILICOSIS

A practical working diagnosis of silicosis is the presence of rounded opacities, or more advanced changes such as conglomerate shadows, in the chest radiograph in a worker with long-term exposure to dust containing quartz at least in some percentage. If the worker or patient has a plausible exposure history, the diagnosis is supportable, though not totally certain. For example, sarcoidosis in unusual cases may mimic the nodular appearance of silicosis, especially if there is some prominence of the hilar lymph nodes, and radiographic scars from granulomatous diseases such as histoplasmosis or tuberculosis may lead to apical or upper-lobe opacities. If uncertainty exists regarding the exposure history, and circumstances such as litigation require a definitive diagnosis, lung biopsy with careful examination of tissue for characteristic histologic findings may be necessary. In practical terms, most litigation never goes this far. BAL, if positive for quartz, will only indicate that some degree of exposure to respirable-size quartz particles has occurred; such findings may be present in exposed workers who have chest radiographs considered normal by certified readers. Occasionally, histologic findings of silicosis may occur unexpectedly in tissue resected for treatment of carcinoma, even in the face of a normal chest radiograph, or at autopsy. Such a finding in South Africa or Canada will entitle the patient to worker's compensation. The diagnosis has also been made by lung biopsy when silicosis was never considered as a diagnostic possibility; an example are the patients who have inhaled cleaning powders, inducing a puzzling clinical picture (Dumontet *et al.*, 1991).

DEVELOPMENT OF THE EXPOSURE STANDARD FOR CRYSTALLINE SILICA

An important governmental responsibility accepted by all industrialized societies is to provide guidelines for exposure to toxic or disease-producing substances which will protect worker health and provide a safe working environment. In the United States, this charge is embodied in the Occupational Safety and Health Act of 1970, which is administered by the Department of Labor. Setting these guidelines is hardly a simple task, depending as they do on animal toxicology as well as epidemiologic studies, either prospective or retrospective, of exposed occupational cohorts. Often, past exposures are uncertain because of fragmentary data and constantly changing work environments. Clinical data such as chest X-rays may never have been taken systematically and maintained. Predicting a safe exposure for a working lifetime involves the conundrum of protecting workers but not establishing standards which unfairly burden the industry, although the necessary tendency is clearly to err on the side of worker health. Although many epidemiologic investigations of silicosis have been carried in such industries as ceramics, glass, tin and mining, the Vermont granite industry ulimately provided the most complete epidemiologic data on the health effects of quartz through a series of landmark investigations (Russell *et al.*, 1929; Russell, 1941; Hosey *et al.*, 1957; Ashe and Bergstrom, 1964; Ayer, 1995). The nature of these studies is instructive, since they illustrate both points of occupational disease in general and silicosis in particular, and also show the problems and difficulties encountered in establishing scientifically valid standards of exposure.

The stimulus for the study of the Vermont granite industry was a tragic increase in the prevalence of silicotuberculosis after 1900 because of a huge increase in dust generated by the introduction of pneumatic chisels and surfacing machines. The classic study published in 1929 was exhaustive in its reach, including dust counting, chest radiographs, mortality rates and physical examinations (Russell *et al.*, 1929). It indicated that 100% of cutters (who used pneumatic tools) developed radiographic silicosis within 4 years; the average age at death was 46 years; 70% of granite workers died of silicotuberculosis. Few workers died of silicosis; tuberculosis was the killing disease when engrafted onto silicosis. Large differences in dust exposures in the manufacturing plants or 'stone sheds', where the granite was processed into monuments, existed in different job categories, enabling a dose–response relationship to be developed. Cutters who used pneumatic tools were exposed on average to 60 million parts/cubic foot of air (mppcf), general shed air was 20 mppcf, and some occupations such as wash-stand operators and boxers (who cleaned and crated the granite before shipment) less than 10 mppcf. The estimation of dust levels in the stone sheds was based on impinger samples collected in a liquid which was then diluted, put in a hemocytometer chamber and counted with an optical microscope. The quartz in the dust was 30%, determined by chemical analysis.

Based on the presence of radiographic silicosis and exposure history (length of employment in a given job category), the investigaors concluded that dust levels between 10 and 20 mppcf would prevent the development of silicosis. The ultimate standard of 10 mppcf was adopted by Vermont after a follow-up study carried out in 1937 (Russell, 1941) suggested that a higher level might not be protective. The objective of setting safe limits was not so much to prevent silicosis of minor extent, but to prevent the advanced silicosis which led to tuberculosis.

The Occupational Hygiene Division, set up at that time by the Health Department, then monitored dust levels, checked for proper performance of the suction devices which had been installed to reduce dust, and by state statute offered workers chest films annually, thereby providing the long-term follow-up necessary to assess the safety of the standard. This comprehensive follow-up continued through 1965 and determined that no cases of silicosis had appeared in workers first employed after 1938 (Hosey *et al.*, 1957; Ashe and Bergstrom, 1964). During this period, the number of silicosis cases declined as workers retired or died of silicosis or silicotuberculosis; it was projected that no cases of silicosis would exist in the industry after 1975.

The problem of setting a dust standard for Vermont involved dealing with only one level of quartz in inhaled air; that is, the level of quartz, estimated at 30% was more or less constant, whereas quartz percentages in all industries might vary from 1% to close to 100% in operations such as sand blasting. It was necessary to devise a standard which would take into account differing percentages of quartz. The American Conference of Governmental Industrial Hygienists (ACGIH) was particularly influential in devising a formula which would fulfill this requirement. The first standard for quartz, published in 1946, divided quartz levels into three broad categories based on percent quartz and prescribed a maximum allowable concentration (MAC). The high value was for dust containing over 50% quartz, with a MAC of 5 mppcf; medium quartz dust levels were between 5% and 50%, with a MAC of 20 mppcf; the low dust level was less than 5% quartz, with a MAC of 50 mppcf of dust.

The formula was revised in 1962, the TLV being set at 250 mppcf/% SiO_2 + 5. The constant of 5 assured that the total dust level would never go above 50 mppcf; if the percent quartz was 10%, the TLV would be 16.6 mppcf.

In 1968, a major change occurred in the methodology of estimating quartz exposures. The particle counting method was cumbersome and subject to error in all but the most careful hands, and did not provide information on particle size or the percentage of quartz in the respirable fraction of dust. The new development of technology enabled this to be done. However, a conversion factor had to be developed which would show the equivalence of particle counting versus gravimetric weighing. Significant studies of this problem were done in Vermont (Sutton and Reno, 1968; Ayer *et al.*, 1973; Ayer, 1995). The standard which was developed was:

$$\text{TLV} = 10\,\text{mg/m}^3 \text{ divided by } \%\text{ respirable free silica} + 2.$$

This formula was based on collection of dust in the breathing zone of the workers by cyclone filters which selected only particles small enough to penetrate into the gas-exchanging regions of the lung. The quartz percentage of the dust was determined by infrared spectroscopy. The constant '2' was added to prevent 'nuisance dusts' from becoming too large a burden as quartz percent fell; thus, the concentration of respirable dust with less than 1% silica would be limited to 5 mg/m^3. Cristobalite and tridymite were assigned limits half as great because of epidemiologic and animal data indicating that they were more fibrogenic. In essence, the above TLV became the standard in 1970 and was considered equivalent to 0.1 mg/m^3 of quartz in inspired air. A recent article describes the problems of establishing the basis for the conversion between millions of particles per cubic foot and mg/m^3 (Ayer, 1995).

In 1970, under the aegis of the National Institute for Occupational Safety and Health (NIOSH), a further comprehensive sudy of the Vermont sheds was begun to evaluate whether the low levels of quartz which had been maintained since 1938 had any adverse health effects. This study used the new techniques of dust sampling; chest radiographs of the entire work force were analyzed using the ILO classification system, and for the first time, both cross-sectional and longitudinal pulmonary function measurements were made using spirometry. The results concluded that a significant loss of lung function was occurring at the current dust levels (heretofore considered low enough to be eliminating silicosis and well below the TLV) and that perhaps even these low levels were producing radiographic changes (Theriault *et al.*, 1974a–c; Musk *et al.*, 1977). These conclusions became the basis of the recommendation of the criteria document published by NIOSH that the quartz permissible exposure limit (PEL) should be lowered to 0.05 mg/m^3, or half the previous value (NIOSH, 1974). These epidemiologic studies were however called into question because of technical and procedural problems, and the recommended exposure limit has not been accepted by the Department of Labor as the exposure limit at the present time. The current permissible exposure limit for crystalline silica is 0.1 mg/m^3; the value for tridymite and cristobalite is 0.05 mg/m^3. Most of the industrialized countries of the world have similar or higher values; none has a lower standard.

As mentioned previously, amorphous silica has no regular crystalline structure, and has little capacity to induce fibrogenesis. The current federal standard for this substance is therefore set high, at 80 mg/m^3 divided by the percent crystalline silica. A particular type of amorphous silica is diatomaceous earth (also called keiselguhr), a form of amorphous silica from sedimentary deposits of unicellular bivalve aquatic animals whose shells are siliceous. Processing this material involves heating (calcining or flux calcining), which can transform the amorphous quartz into tridymite or cristobalite, the so-called polymorphs of crystalline silica. Based on animal experiments, these two compounds are deemed to be more fibrogenic than quartz, and the current exposure standards are 0.05 mg/m^3.

A recent approach to the problem of setting safe exposure limits is estimating lifetime quartz exposures for individual workers, based on historical levels of dusts in mines, and attempting to set a lifetime limit for exposure which will prevent any silicosis, even of the earliest radiographic type (Muir *et al.*, 1989; Hnizdo *et al.*, 1993; Steenland and Brown, 1995). Unfortunately, estimates of this sort have diverged widely between hard rock miners in Canada and gold miners in South Africa (Hughes and Weill, 1995). Calculation of total lifetime dust inhalation is tricky since it depends on fragmentary or subjective estimates of historical dust levels. That the two most influential studies come to very different conclusions suggests that the methodology is difficult. A recent article evaluates the differing estimates, and concludes that the data from Canada are the more plausible (Hughes, 1995).

WHERE DOES ONE GET SILICOSIS?

Given that quartz is such a ubiquitous component of the earth's crust, it is hardly surprising that any perturbation of the earth's surface, from high winds blowing through the New Mexico mesas or the sandstone formations in the painted desert, from the erosion of building stone in cities, even blowing sand on beaches, will create some degree of free silica in the environment. As mentioned earlier, the EPA is preparing an analysis of ambient quartz levels and non-cancer health effects.

Historically, exposures occurring in industry have accounted for virtually all the silicosis with its deadly toll over the past several hundred years. A complete list of occupations where quartz exposure might occur would involve hundreds of listings. The important disruptions producing dust are mining, drilling, quarrying, crushing stone and foundry work, which employ constituents many of which contain quartz such as refractory bricks, sand, molds for casting and the ore from which the iron is derived. Sandblasting is particularly dangerous; even if sand is not used as the abrasive, the material being blasted may generate respirable-size particles. Quartz is used in ceramics, abrasives, glass and cement manufacturing. Silica flour is a component of fillers in various industrial products such as paint and paper. Batch plants use crushed stone as a component of the mix to repair potholes. Cases of silicosis have been reported in gemstone manufacturing (White *et al.*, 1991). The unusual case of silicosis induced by inhaling abrasive cleaning powder has been mentioned. (Dumontet *et al.*, 1991). The construction industry statistically represents a large source of silicosis cases (Centers for Disease Control and Prevention, 1994).

Radiology

The International Labour Organization has, over several decades, developed a radiographic classification scheme for judging the presence of opacities consistent with pneumoconiosis. The early versions were more concerned with silicosis and CWP, but later versions have been expanded to include radiographic changes related to asbestosis exposure. This is fully discussed in Chapter 7.

Exposure to quartz in sufficient quantities may lead to an extraordinary variety of radiographic opacities, from the subtlest markings of indeterminate shape through grotesque clumps of fibrocalcific scarring which lead to major distortions of normal anatomic boundaries. The earliest opacities are discernible when their size equals about 0.3 cm (Fraser and Pare, 1977), though when a profusion of smaller opacities exist, a summation of images may make the opacities discernible. Calcification probably enhances the recognition of lesions. With the advent of high-resolution computed tomography (CT), the power to define opacities has been greatly enhanced, making visible lesions only 0.1–0.2 mm in size (remy-Jardin *et al.*, 1992).

The classifical findings in early, or simple, silicosis are rounded opacities, generally located in the upper lung zones but in some cases, especially with progression, disseminated through all lung zones (Figure 13.9; see also Figures 13.4 and 13.8). Occasionally, the first abnormalities may be seen in a single lung zone. Lesions initially may be so subtle that expert readers disagree on their presence. For this reason, when research studies are being carried out, a panel of readers should judge the film in an attempt to reach a consensus. The most difficult films to judge are those on the borderline of normal, when repeated readings by the same reader may vary, or films may be judged to show regression in the opacities, which of course does not occur in pneumoconiosis. As the rounded opacities become more profuse, they may coalesce, forming conglomerate masses (see Figures 13.5–13.7). The individual rounded opacities may be gradually incorporated into large aggregations of tissue which exert traction on surrounding structures, leading to gross distortion of hilar structures, mediastinal deviation, the development of traction emphysema and bullae which may rupture and cause pneumothorax (see Figures 13.5–13.7).

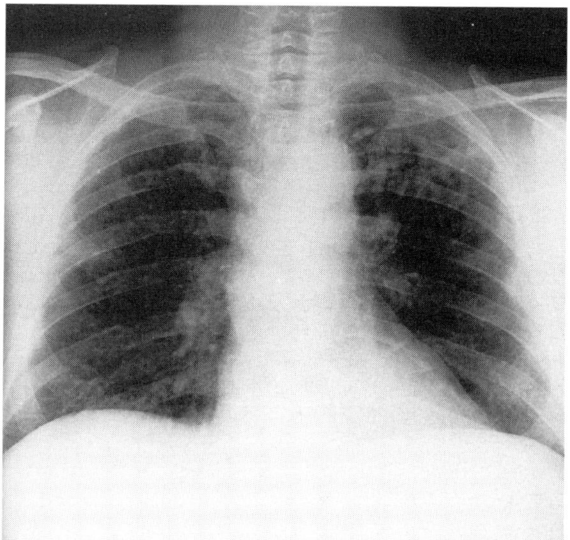

Figure 13.9 Diffuse nodular silicosis, with fibronodular concentration of opacities in left upper lobe. Active pulmonary tuberculosis superimposed on silicosis. (Reproduced from W.G.B. Graham (1992) *Clinics in Chest Medicine*, **13**, (2), 253–67, by permission of W.B. Saunders, Philadelphia.)

Although rounded opacities are the 'classical' findings, and are consistent with the lesions present in whole lung sections (see Figure 13.2), linear opacities were described by Russell *et al.* (1929) in their monograph on the Vermont granite industry. This appearance was said to be distinct from the nodular formations or 'mottling' which had been described earlier in South African gold miners. The linear markings tended to be in the lower lobe, perhaps representing enlarged lymphatics heading toward the hilar area, where lymph node drainage may be blocked with inflammatory hyperplasia. These changes were occurring in pneumatic tool operators who were exposed on average to 60 mppcf; workers would often show abnormal chest radiographs within 2 years of starting work and by 4 years, 100% of such workers had silicosis. These radiographic findings may in fact have been similar to those occurring in acute silicoproteinosis, though perhaps with a less exuberant inflammatory reaction. Advanced silicosis may also appear as very coarse reticular shadows which seem to overshadow the presence of nodular opacities (Figure 13.10).

As the pathology of silicosis involves the lymphatics and regional lymph nodes, it is no surprise that hilar enlargement may be an early sign of silicosis (Figure 13.11). Unless calcification occurs, such enlargment may be difficult to discern. An almost pathognomic radiographic finding indicating silicosis is the presence of 'egg-shell' calcification in hilar, intrapulmonary or mediastinal lymph nodes, although dense calcification may also occur (Figure 13.12; see also Figure 13.8).

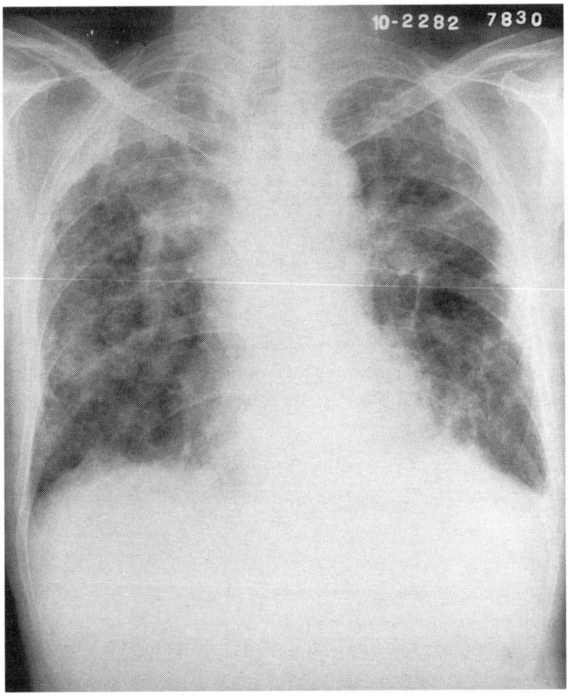

Figure 13.10 Diffuse coarse generalized fibrosis due to quartz exposure in a granite cutter. Spotty conglomerate disease bilaterally. Note the heart border is obscured.

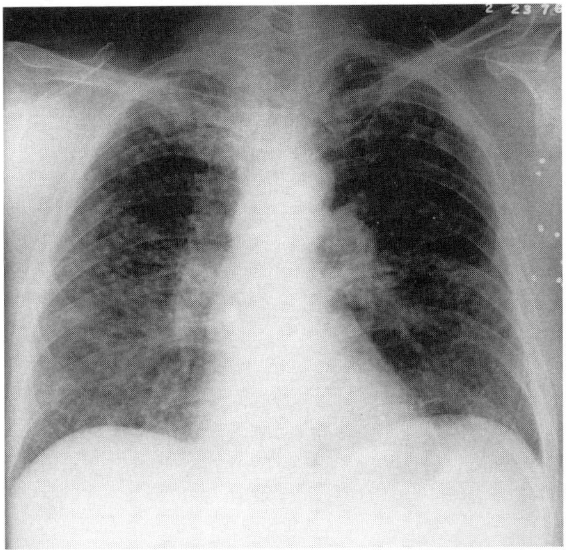

Figure 13.11 Nodular silicosis with enlarged hilar structures. Early conglomerate disease in right upper lobe. Calcifications in soft tissues on left are bird-shot. (Reproduced from W.G.B. Graham (1992) *Clinics in Chest Medicine,* **13**, (2), 253–67, by permission of W.B. Saunders, Philadelphia.)

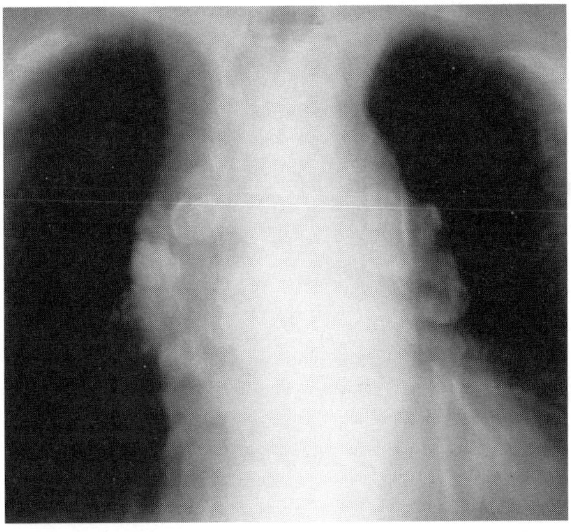

Figure 13.12 Tomograms showing egg-shell calcification. This radiographic finding is essentially pathognomonic for silicosis. Note nodulations in peripheral lung fields. (Reproduced from W.G.B. Graham (1992) *Clinics in Chest Medicine,* **13**, (2), 253–67, by permission of W.B. Saunders, Philadelphia.)

Similar shadows may occur rarely in sarcoidosis or histoplasmosis, but their presence in a chest film should always raise the question of occupational exposure to quartz, and will provide almost incontrovertible evidence that silicosis exists.

When dust exposures are extreme, as in sandblasting or tunneling without dust controls, the radiographic picture is that of acute silicosis, or silicoproteinosis. In this case, the shadows are similar to the ground-glass appearance of idiopathic alveolar proteinosis or irregular reticular opacities, resembling interstitial fibrosis. Diffuse nodular densities have also been described (Goodman *et al.*, 1992). Hilar lymph nodes may also be enlarged in acute disease but may be obscured by the reticular changes. In some histologically proven cases, the radiographic abnormalities are described as 'large opacities'.

When tuberculosis complicated silicosis, as was often the case in the past when silicosis became definitely advanced, discrete regions of the lung would show a localized increase in nodulations, coalescence or cavitation (Figure 13.13). Pleural effusions may still be due to coexistent tuberculosis, but as tuberculosis has declined in incidence, the majority of effusions are probably due to non-related conditions such as congestive heart failure or lung cancer (Figure 13.14).

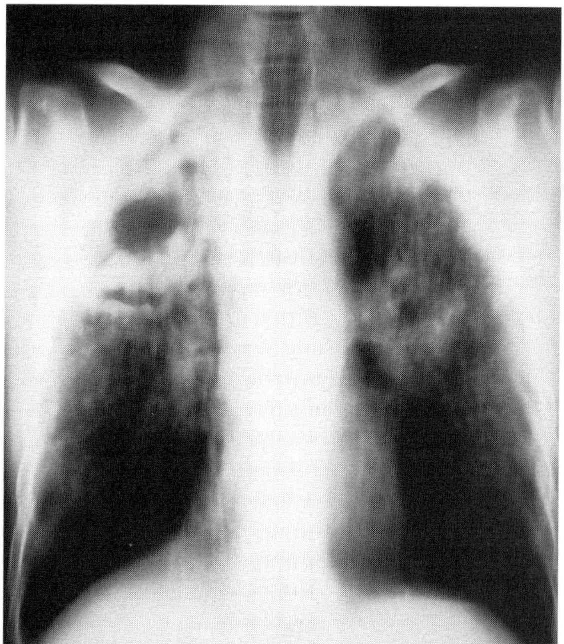

Figure 13.13 Tomogram of tombstone sandblaster who developed fatal *M. avium* infection. Huge cavity is present in right upper lobe. Note nodular opacities. Tomograms have been outmoded by computed tomography. (Reproduced from W.G.B. Graham (1992) *Clinics in Chest Medicine,* **13**, (2), 253–67, by permission of W.B. Saunders, Philadelphia.)

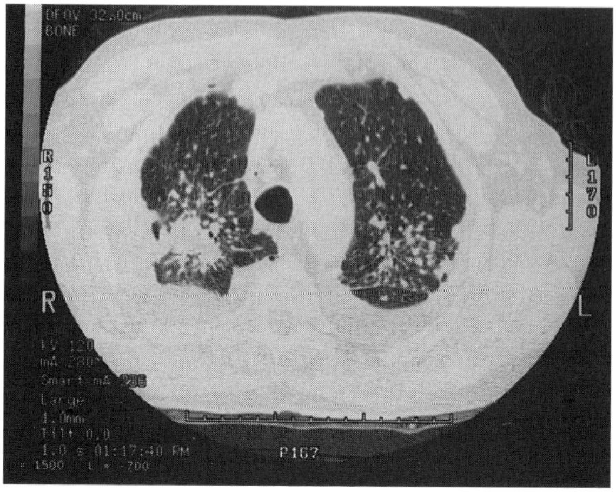

Figure 13.14 CT of the upper lobes, showing nodular opacities and conglomeration on the right. Note subpleural plaques. Pleural effusions were due to congestive heart failure.

Lung cancer developing in the lung parenchyma can easily be concealed early in its development by silicotic nodules. A reader seeing a new film on a worker with established silicosis might well assume that a new or enlarging shadow is the result of the underlying silicotic process. An inordinately rapid increase in the size of the opacity should raise the suspicion that a new process is evolving.

An unusual presentation of pneumoconiosis, described initially by Caplan in coal workers, was the occurrence of rapidly developing nodulations, sometimes with cavitation, in workers with seropositive rheumatoid arthritis (Caplan *et al.*, 1962). This syndrome is not common, and the pathogenesis is unclear. A peculiar feature is that the nodules often occur in workers who have relatively slight background radiographic abnormalities of pneumoconiosis. The same process has been described in workers with other collagen vascular diseases, such as scleroderma.

CT has changed the boundaries of radiographic evaluation of abnormalities due to pneumoconiosis because of its extraordinary power of resolution (see Figure 13.14). The detection of tiny opacities not visible with standard roentgenology is possible, very much as tissue examination from autopsy or lung biopsy has in the past been able to detect silicosis in negative chest radiographs (Hnizdo *et al.*, 1993). The extent of conglomerate disease can also be discerned earlier and more accurately than with standard radiographs (Begin *et al.*, 1987, 1991). The subtleties of anatomic relationships, such as the location of opacities near lymphatics or vessels, are clearly shown, as well as traction or scar emphysema (remy-Jardin *et al.*, 1992). Begin and colleagues (1991) studied workers exposed to quartz whose plain chest films, read under the ILO/UICC classification, were graded either 0/1 (meaning a negative film for pneumoconiosis) or 1/0. CT scans showed that nearly half of the normal films in fact had opacities consistent with silicosis.

The large increase in sensitivity afforded by CT scans raises the question of whether tiny opacities, hitherto not discernible on plain chest radiographs, will become the standard for making the diagnosis of work-related lung disease. The implications are manifold, including the question of whether future radiographic surveys of industries should include CT, whether the more sensitive detection of abnormalities will be important in litigation or disability evaluations, and whether arguments can be made that at least part of the responsibility for these changes rests with the general exposure to environmental particles, and is not therefore stricly related to employment.

Pleural changes are generally not seen in quartz exposure, unless there are mixed dust exposures, although the visceral pleural surfaces certainly contain nodules which are grossly visible. CT has demonstrated 'pseudoplaques' which are linear opacities underlying the visceral pleura (remy-Jardin *et al.*, 1992).

Pulmonary function

Pulmonary function abnormalities show the same variability as chest radiographic abnormalities, with few or no changes in lung function with early disease, but progressing in parallel as the disease progresses (Begin *et al.*, 1988). The consensus is that simple silicosis of the lower grades of profusion produce little change in the FEV_1, the forced vital capacity (FVC), lung volumes, diffusing capacity, or compliance. As the profusion of opacities increases, moderate restrictive impairment may occur. Serious symptomatic lung function changes emerge as large opacities, or conglomerate disease develops. These changes can produce a mixed picture of severe airway obstruction and restriction, that is limitation of flow rates in the face of small lung volumes. One would expect concomitant reductions in diffusing capacity. Eventually, gas exchange may become so abnormal that respiratory failure supervenes, though this outcome is usually precipitated by intercurrent infections initially. Some authors have suggested that workers exposed to quartz but with negative chest films demonstrate abnormalities of small airways, using tests such as closing volumes (Chia *et al.*, 1992). Given the uncertainty regarding the prognostic or clinical significance of this type of abnormality, and often the lack of smoking histories in the subjects who are tested, as well as the lack of histologic confirmation, such observations should be regarded with caution.

The lung function changes of acute silicoproteinosis are of course a special case, with a severe restrictive impairment of the type seen in any advanced parenchymal process such as interstitial fibrosis, alveolar proteinosis or congestive heart failure. Such patients have small lung volumes, no evidence of airflow limitation, marked hypoxemia and presumably very stiff, non-compliant lungs.

The most difficult problem in interpreting lung function changes in anyone with silicosis, or with quartz exposure, is the confounding fact of smoking. If clincially important lung dysfunction occurs in a quartz-exposed worker who does not have radiographic evidence of silicosis, or only simple silicosis involving the upper lung fields, the cause is likely to be due to the effects of smoking, the coexistence of bronchial asthma, or some other underlying cause of airway dysfunction. When radiographs showing silicosis are stable and the worker is not host to coexisting illnesses such as congestive heart failure, longitudinal losses approximate or are slightly larger than those associated with aging (Ng and Chan, 1992). In such workers, losses of FVC were 59 ml/year and of FEV_1 64 ml/year. In this study, however, nearly 80% of these workers were either current smokers (64%) or ex-smokers (12%).

An important question in terms of occupational safety and health is whether levels of quartz dust exposure below the current PEL leads to accelerated losses of lung function, independent of smoking. Becklake (1985) has argued forcefully that many industrial settings with diverse fumes or particles lead to industrial bronchitis with associated lung function losses. Although on average the losses are small, she states that some susceptible individuals will have severe declines in lung function, producing clinical illness or early demise. Certain exposures such as toluene diisocyanate never lead to radiographic changes but may lead to severe airway dysfunction. However, Becklake is discussing mixed exposures which may be irritating, leading to mucus hypersecretion, cough and sputum production.

Whether Becklake's premise applies to granite dust has been addressed in studies of the Vermont granite industry. Average quartz exposures have been below 0.1 mg/m^3 since 1940. The dust itself is composed of about 10% quartz, with the remainder being feldspar and mica; other materials such as tin and aluminum oxide used in processing are admixed. In 1970, the comprehensive study of the Vermont granite industry (Theriault *et al.*, 1974a–c) referred to earlier addressed this issue and concluded – based on a cross-sectional analysis of the working population – that losses in FVC were excessive: the then current rate of dust exposure (523 μg/m^3 per average working year) led to a 2 ml loss, in contrast to a 9 ml loss for smoking and a 30 ml loss from aging. The loss, trivial in clinical terms (Morgan, 1975) and so small as to beg confirmation, was nonetheless an important basis for the stricter standard for quartz inhalation proposed by the National Institute of Occupational Safety and Health (NIOSH, 1974). The existing standard at that time (one which continues to be the federal standard endorsed by OSHA) was equivalent to 0.1 mg/m^3; NIOSH proposed a limit of 0.050 mg/m^3. When data from four years of longitudinal observation were analyzed, the losses were estimated to be far larger: 50–70 ml/year for FEV_1, and 70–80 ml for FVC (Musk *et al.*, 1977). The results of this study were seriously undermined when subsequent testing showed that technical problems underestimated the FVC measurements (Graham *et al.*, 1981), a fact that the authors later confirmed (Wegman *et al.*, 1983).

A later longitudinal study of the Vermont granite population over an 8-year period concluded that loss of both FVC and FEV_1 were of the same order of magnitude as in non-exposed workers (Graham *et al.*, 1994). Gradients of loss were highest for smokers and ex-smokers. Multivariate analysis showed no correlation with years working in granite, which were used as a measure of exposure. During this study, the average quartz level was 0.060 mg/m^3, although about 10% of workers were exposed to levels above the current federal standard.

Eisen *et al.* (1983) discussed excessive losses of lung function in Vermont granite workers who failed to perform spirometries in a consistent manner, and made the assumption that this failure (and the estimated losses) were due to dust exposures. It is difficult to explain this result, since this subset of workers, after working an average of 15 years in the industry, had, at the beginning of the study, the best mean lung function values.

Other studies have attempted to estimate lung function losses in miners based on historical levels of dust exposures (Oxman *et al.*, 1993). The authors suggest that 8% of workers will have a significant loss of lung function from underground dust exposure independent of smoking, though these losses average less than 10 ml/year (Oxman *et al.*, 1993). The analysis was statistically complex, and it was not clear that dust levels were in conformance with current quartz exposure standards. It is also true that coal dust may impose a different type of irritant effect on the lung. Hnizdo (1992) reported small excessive losses in South African gold miners in relation to dust exposures; however, these occurred at average quartz levels estimated to be three times (3 mg/m^3) the current PEL. At this time, there are no convincing data which relate levels of quartz exposure below the current OSHA standard to excessive losses of lung function.

CLINICAL CONSIDERATIONS

The clinical picture of silicosis is quite variable, from the catastrophic illness of acute silicoproteinosis, the course of which is usually fatal in months or a few years, to a very indolent process diagnosed by chest radiography, with few pulmonary symptoms and minor or no changes in pulmonary physiology. Acute silicosis, occurring after massive exposures to dust in unregulated and illegal environments (Samimi *et al.*, 1978) leads to severe dyspnoea, hypoxemia, weight loss, cough, fever and pleuritic pain. Only one report has been published of improvement and survival after treatment with steroids (Goodman *et al.*, 1992). The course has often been complicated by superinfection with either typical tuberculosis or atypical mycobacteria such as *M. kansasii* or *M. avium-intracellulare*. A unique feature is the occurrence of antibodies usually associated with collagen–vascular diseases such as rheumatoid arthritis or lupus erythematosus. An unusual percentage of high titers for positive tests of rheumatoid factor and positive antinuclear antibody have been reported (Jones *et al.*, 1976; Doll *et al.*, 1981; Schuyler, 1977; Suratt *et al.*, 1977) as well as the coexistence of immunologic diseases such as glomerulitis. These immunologic abnormalities have been studied in humans who have undergone BAL, and in sandblasters with definite nodular silicosis, both with and without large opacities and conglomerate lesions, the rate of positive ANA tests was 44%, with a tendency for positive tests to be found in workers whose radiographs were more advanced. Among Suratt's four cases, two developed glomerulonephritis and rheumatoid arthritis (Suratt *et al.*, 1977). Interestingly, these cases bridged the histologic spectrum between classical nodular silicosis and acute silicoproteinosis in that both PAS-positive material and silicotic nodulation were present. Studies have also indicated that rheumatoid arthritis is more common in men with silicosis than in the general population (Klockars *et al.*, 1987), suggesting a general effect of quartz on the immune system.

The more typical clinical presentation of silicosis is an abnormal chest radiograph that may or may not be associated with symptoms of lung function changes early in the course, often developing after more than 10–20 years of exposure to quartz dust. If large opacities and later conglomerate masses appear, symptoms particularly of dyspnea and cough appear. If workers have smoked, an important part of the symptomatology may be related to smokers' bronchitis or emphysema, especially if only simple silicosis is present. With progression, severe airflow limitation may appear, with the eventual development of cor pulmonale and respiratory failure. At this juncture, standard therapy for obstructive lung disease such as bronchodilators and oxygen may be symptomatically helpful. Cessation of smoking may slow the development of airway obstruction.

A realistic concern is how rapidly silicosis will progress after it is first diagnosed, since the rapidity of progression is in a sense a prognosis regarding disability or death. Ordinarily, when silicosis is diagnosed, the worker is advised to work in a dust-free environment since any addition of dust to the overall lung burden might maintain or accelerate the process, while total avoidance of dust might slow progression. Various workers have given estimates of the rate of progression (Hughes, 1973; Ng *et al.*, 1987; Infante-Rivard *et al.*, 1991), but it seems intuitively true that silicosis which appears in very subtle form after 30 years of low-level exposure will have a very different prognosis than silicosis developing after 4 years of exposure, with the early appearance of conglomerate disease. The study by Infante-Rivard *et al.* suggests that silicotics without dyspnea or cough have a prognosis similar to that of average Quebecers. An unknown factor is individual suceptibility to develop quartz-induced disease or maintain the inflammatory reaction which is fuelled by the resident dust burden in the lung. It is probable that the rate of progression is a continuum, varying from months to a few years to death in acute silicoproteinosis, to progression over years with moderately aggressive silicosis, with gradual deterioration of lung function and increasing disability, to virtually undiscernible progression in radiographic findings over years, and little change in lung function or in symptoms.

In chronic silicosis, as in acute disease, mycobacterial infection either with *M. tuberculosis* var. *hominis* or with atypical mycobacteria may supervene (Snider, 1978; Althouse *et al.*, 1995). This statement has to be qualified for atypical mycobacteria, since few epidemiologic data germane to this question are published, although case reports exist. Therapy of tuberculosis must be guided by drug suceptibility tests and at least three effective drugs should be used (Hong Kong Chest Service, 1991); the odds of a successful outcome will improve if directly observed therapy is used. Silicotuberculosis continues to be a more difficult disease to treat than simple tuberculosis and probably requires longer therapy. Standard short-course therapy with isoniazid, rifampin and pyrazinamide (PZA), with PZA given for the first 2 months may have a high relapse rate (Lin *et al.*, 1987). Four-drug therapy with the addition of daily streptomycin will increase the efficacy of treatment, but may be difficult to administer and lead to side effects such as deafness. Our policy has been to use three drugs for at least the first 2 months and until we know the drug sucseptibility pattern, and then to continue treatment for a full 12 months. Treatment programs in the past have continued Isoniazid indefinitely, but given the current potent tuberculocidal drugs, there is little evidence to support this treatment.

Steroids have been apparently successful in a case of acute silicoproteinosis, as noted earlier. Whether long-term steroids could be used in chronic silicosis is unclear given the probable need for lifetime therapy. Sharma *et al.* (1991) reported results of a 6-month program of prednisolone treatment, using a maintenance dose of 0.25 mg/kg/day. Pulmonary function showed significant improvement in all physiological parameters, including diffusing capacity and arterial oxygen tension. BAL fluid showed a decrease in total cell count. The authors concluded that daily prednisolone therapy would suppress alveolitis and bring about an improvement in lung function in patients with chronic silicosis.

Given that most workers with silicosis have been smokers (80% in our cohort), the development of lung cancer should be suspected if the evolution or progression of radiographic opacities seems accelerated or becomes asymmetric. Early detection, however, may be delayed because neoplastic growth may be concealed in profuse nodulations or in conglomerate disease, or the appearance of a neoplasm may be mistakenly assumed to represent progression of silicosis. While non-metastatic cancers may be resectable, the operative procedure may be complicated by technical difficulties in closing the bronchus or in mobilizing the affected lobe. Surgeons at our institution have explored a number of cases of pulmonary malignancy where resection was not technically feasible, or the patients developed bronchopleural fistuli or tuberculosis effusions postoperatively.

Are other treatment options available? Perhaps the most important maneuver is the application of a tuberculin skin test. If positive, which in the context of silicosis means induration greater than 5 mm, isoniazid prophylaxis should be administered. Various short-term prophylactic treatment programs reduce the rate of subsequent tuberculosis (Anonymous, 1992b) but in my opinion therapy of at least a year should be prescribed. The development of silicotuberculosis undoubtedly varies depending on the local tuberculosis scene. If the area is endemic for tuberculosis or if a high prevalence of drug resistance is known to occur, the chances of seeing infectious tuberculosis are far greater. Perhaps it is an oversight that the recent Statement of the American Thoracic Society discussing therapy of tuberculosis did not mention isoniazid chemoprophylaxis in silicotics with positive tuberculin tests (American Thoracic Society, 1994). Whether isoniazid prophylaxis should be given to tuberculin-positive quartz-exposed workers without silicosis has not been studied.

Experimental treatment approaches have recently been reviewed (Banks *et al*., 1993), including whole-lung lavage to remove quartz from the lung. A few case reports have been published which suggested some symptomatic relief but no change in pulmonary function (Mason *et al*., 1982). This therapeutic approach is currently under investigation (Wilt *et al*., 1996). Lung transplant has also been attempted in one case (Anonymous, 1992a), though the technical difficulty of operating on advanced silicosis is likely to make this procedure a limited option compounded by the problems of finding donor organs.

CANCER AND QUARTZ

The question whether quartz is a carcinogenic agent became a hot issue only in the 1980s, prompted by an article which listed increased standardized mortality ratios from lung cancer in occupational groups exposed to silica (Goldsmith *et al*., 1982). Up to that time, an almost universal consensus stated that quartz did not cause cancer. In fact, the eminent researchers in silicosis in the 1930s (Vorwald and Karr, 1938; Gardner, 1940) are all on record against this idea; and at that time, there was a great amount of silicosis world-wide, suggesting that if quartz were carcinogenic, cancer would be occurring. For example, an autopsy series of 2100 workers with silicosis showed only three cases of lung cancer (Vorwald and Karr, 1938). A report from the Miner's Phthisis Medical Bureau in South Africa describes autopsy findings in 4510 workers separated into three groups: miners exposed to granite without silicosis; miners with silicosis; and workers with no underground experience. No difference in the percentage of lung cancer (which was less than 1%) was found among the groups. In the proceedings of the National Silicosis Conference (1938), only a few individual case reports in a bibliography of more than 1500 citations mention lung cancer. In fact, before 1940, lung cancer was a relatively uncommon cancer, because the habit of cigarette smoking had come in vogue only in the 1920s and 1930s.

The article by Goldsmith cited only positive studies, whereas a rebuttal by Heppleston (1985) cited only negative studies. Nonetheless, a torrent of mortality studies (see McDonald, 1995) has since been published, and the majority of these suggest at least an association between lung cancer and silicosis as indicated by standardized mortality studies, which compare age-stratified rates for observed cancers over the number expected on the basis of national mortality data. The International Association for Research on Cancer (IARC) stated that definite evidence exists for the carcinogenicity of quartz in animal species, and that quartz was 'probably carcinogenic to humans' (International Agency for Research on Cancer, 1987). More recently, the IARC has upgraded crystalline silica to a definite carcinogen in humans (IARC Monograph 68, 1997). The text states 'There is sufficient evidence in humans for the carcinogenicity of inhaled crystalline silica in the form of quartz or cristobalite from occupational sources'. This cautious wording suggests that the conclusion should not include all silica, such as ambient or environmental quartz. Another cautionary statement in the Overall Evaluation states, 'the Working Group noted that carcinogenicity in humans was not detected in all industrial circumstances studied. Carcinogenicity may be dependent on inherent characteristics of the crystalline silica or on external factors affecting its biological activity or distribution of its polymorphs'. The fact that some very good studies on carcinogenicity from silica have been negative tends to undermine the strength of the overall verdict, since one cardinal tenet of sound epidemiology is that results should be consistent, as they certainly have been in studies of cancer and tobacco smoke. Further, the Working Group rendered a split decision; one would think on a matter of such importance that scientific certainty would be a requirement. Of interest in the same IARC Monograph is an evaluation of coal dust, which of course contains significant crystalline silica; the Working Group concludes that there is inadequate evidence in humans for the carcinogenicity of coal dust.

Evaluating this problem is at the least complex. For example, in animal studies, the species most used to investigate the effect of quartz, the rat, appears to be unusually susceptible to epithelial cellular changes resembling neoplasia (Holland, 1995; Mauderly, 1996). Other species such as Syrian hamsters and mice do not develop similar neoplastic lesions after equivalent exposures to quartz. Whether rat neoplasia reflects comparable suceptibility in humans is another question. The differential effects of dioxin on animals and humans is a cautionary example.

Questions of confounding influences in the production of cancer also complicate the evaluation. For example, many cohorts studied have coincident exposures to known carcinogens: for underground miners radioactive gases (radon and daughters) and for foundry workers, polycyclic hydrocarbons. If the cohort is a registry of men with established silicosis or men receiving disability for silicosis, a selection bias may exist based on poor lung function related to chronic bronchitis and emphysema induced by smoking. Cigarette exposure is undoubtedly the most important confounder, and studies must evaluate the dose of tobacco as rigorously as they evaluate quartz exposures. Many published studies provide either no history of smoking or very rudimentary data regarding dose. There certainly appear to be gradients of pulmonary carcinoma depending on the number of cigarettes smoked, but such data are rarely cited in the silica literature. Our studies of mortality in the Vermont granite industry show an increase in pulmonary cancer in workers exposed before the 1930s, when dust levels were high (Costello and Graham, 1988). However, when smoking histories were available, every cancer occurred in a smoker. Some 80% of our cohort were smokers, a fact which appears to be true among other blue-collar workers. It is also important to show a gradient of pulmonary malignancy between those workers with silicosis and those exposed to quartz who do not have silicosis, the reason being that silicosis assuredly implies a much heavier exposure to quartz. Essentially no data exist at this point to show an increase in lung cancer in non-silicotic workers. The ideal cohort to settle the question would be non-smoking, quartz-exposed population – something difficult to locate given the world-wide use of tobacco.

If in fact the increased incidence of cancer in silicosis is borne out when careful correction has been made for cigarette use, the question is whether the altered anatomy produced by silicosis with its attendant inflammatory constitutents enhances the carcinogenic potential of cigarette smoke, much as asbestos seems to do. Because of the declining incidence of silicosis, we will now be dealing with cohorts who have been quartz-exposed but do not have silicosis. Studying such workers may help to resolve the issue.

For the interested reader, the most judicious reflection on quartz and cancer has recently been authored by McDonald (1995). After carefully reviewing the many mortality studies published since 1983, he concludes his discussion with this statement: '... there is fairly clear evidence of a systematic relationship between cumulative exposure to crystalline silica and pulmonary fibrosis detected radiologically. Secondly, there is little evidence of any increase in lung cancer in the absence of silicosis or at relatively low levels of exposure and disease prevalence ... Public health priority should not be deviated from the prevention of silicosis by concern for small or doubtful cancer risks'.

REFERENCES

al-Kassimi, F.A., al Majed, S.A., al Hajjaj, M.S.S. and Hawass, N.E. (1991) Silicosis in a Himalayan village population: role of environmental dust (letter, comment). *Thorax*, **46**, 861–2.

Agricola, G. (1557) De Re Metallica. Froben, Basel. (cited in Bulletin No. 21 National Silicosis Conference 1938, United States Department of Labor).

Allison, A.C., Harington, J.S. and Birbeck, M. (1966) An examination of the cytotoxic effects of silica on macrophages. *J. Exp. Med.*, **124**, 141–53.

Althouse, R.B., Bang, K.M. and Castellan, R.M. (1995) Tuberculosis comortality with silicosis – United States, 1979–1991. *Appl. Occup. Environ. Hyg.*, **10**(12), 1037–41.

American Thoracic Society (1994) Treatment of tuberculosis and tuberculosis infection in adults and children. *Am. J. Respir. Crit. Care Med.*, **149**, 1359–74.

Anonymous (1992a) Single lung transplantation for end-stage silicosis: report of a case. *J. Formosa Med. Assoc.*, **91**, 926–32.

Anonymous (1992b) A double-blind placebo-controlled clinical trial of three antituberculosis chemoprophylaxis regimens in patients with silicosis in Hong Kong. *Am. Rev. Respir. Dis.*, **145**, 36–41.

Anonymous (1995) From the Centers for Disease Control and Prevention. Occupational silicosis – Ohio, 1989–1994. *MMWR*, **44**, 62–4.

Ashe, H.B. and Bergstrom, D.E. (1964) Twenty-six years experience with dust control in the Vermont granite industry. *Ind. Med. Surg.*, **33**, 73–8.

Ayer, H.E. (1995) Origin of the U.S. respirable mass silica standard. *Appl. Occup. Environ. Hyg.*, **10**(12), 1027–30.

Ayer, H.E., Dement, J.M., Busch, K.A., Ashe, H.E., Levadie, B.T.H., Burgess, W.A. and DiBerardinis, L.J. (1973) A monumental study – reconstruction of a 1920 granite shed. *Am. Ind. Hyg. Assoc. J.*, **34**, 206–11.

Banks, D.E., Morring, K.L., Boehlecke, B.A., Althouse, R.B. and Merchant, J.A. (1981) Silicosis in silica flour workers. *Am. Rev. Respir. Dis.*, **124**, 445–50.

Banks, D.E., Cheng, Y.H., Weber, S.L. and Ma, J.K.H. (1993) Strategies for the treatment of pneumoconiosis. *Occupational Medicine: State of the Art Reviews*, **8**, 205–32.

Becklake, M.R. (1985) Chronic airflow limitation: its relationship to work in dusty occupations. *Chest*, **88**, 608–17.

Begin, R., Bergeron, D., Samson, L., Boctor, M. and Cantin, A. (1987) CT assessment of silicosis in exposed workers. *Am. J. Roentgenol.*, **148**, 509–14.

Begin, R., Ostiguy, G., Cantin, A. and Bergeron, D. (1988) Lung function in silica-exposed workers. A relationship to disease severity assessed by CT scan. *Chest*, **94**, 539–45.

Begin, R., Ostiguy, G., Fillion, R. and Colman, N. (1991) Computed tomography scan in the early detection of silicosis. *Am. Rev. Respir. Dis.*, **144**, 697–705.

Begin, R., Cantin, A., Masse, S., Dufresne, A., Perreault, G. and Sebastien, P. (1993) Aluminium inhalation in sheep silicosis. *Int. J. Exp. Pathol.*, **74**, 299–307.

Buechner, H.A. and Ansari, A. (1969) Acute silico-proteinosis. *Dis. Chest*, **55**, 274–84.

Calhoun, W.J., Christman, J.W., Ershler, W.B., Graham, W.G.B. and Davis, G.S. (1986) Raised immunoglobulin concentrations in bronchoalveolar lavage fluid of healthy granite workers. *Thorax*, **41**, 266–73.

Caplan, A., Payne, R.B. and Withley, J.L. (1962) A broadened concept of Caplan's syndrome related to rheumatoid factors. *Thorax*, **17**, 205–9.

Centers for Disease Control and Prevention (1994) *Work-related lung disease surveillance report 1994*, National Institute for Occupational Health and Safety, U.S. Department of Health and Human Services.

Chia, K.S., Ng, T.P. and Jeyaratnam, J. (1992) Small airways function in silica-exposed workers. *Am. J. Ind. Med.*, **22**, 155–62.

Christman, J.W., Emerson, R.J., Graham, W.G.B. and Davis, G.S. (1985) Mineral dust and cell recovery from the bronchoalveolar lavage of healthy Vermont granite workers. *Am. Rev. Respir. Dis.*, **132**, 393–9.

Christman, J.W., Emerson, R.J., Hemenway, D.R., Graham, W.G.B. and Davis, G.S. (1991) Effects of work exposure, retirement and smoking of bronchoalveolar lavage measurements of lung dust in Vermont granite workers. *Am. Rev. Respir. Dis.*, **144**, 1307–13.

Costello, J. and Graham, W.G.B. (1988) Vermont granite workers' mortality study. *Am. J. Ind. Med.*, **13**, 483–97.

Craighead, J.E. and Kleinerman, J. (1988) Diseases associated with exposure to silica and non-fibrous silicate minerals. *Arch. Pathol. Lab. Med.*, **112**, 673–720.

Craighead, J.E. and Vallyathan, N.V. (1980) Cryptic pulmonary lesions in workers occupationally exposed to dust containing silica. *JAMA*, **244**, 1939–41.

Davis, G.S. (1986) Pathogenesis of silicosis: current concepts and hypotheses. *Lung*, **164**, 139–54.

Davis, G.S., Hemenway, D.R., Evans, J.N., Lapenas, D.J. and Brody, A.R. (1981) Alveolar macrophage stimulation and population changes in silica-exposed rats. *Chest*, **80**, 8s–10s.

deShazo, R.D. (1982) Current concepts about the pathogenesis of silicosis and asbestosis. *J. Allergy Clin. Immunol.*, **70**, 41–9.

Doll, N.J., Stankus, R.P., Hughes, J., Weill, H., Gupta, R.C., Rodriguez, M., Jones, R.N., Alspaugh, M.A. and Salvaggio, J. (1981) Immune complexes and autoantibodies in silicosis. *J. Allergy Clin. Immunol.*, **68**, 281–5.

Dumontet, C., Biron, F., Vitrey, D. *et al.* (1991) Acute silicosis due to inhalation of a domestic product. *Am. Rev. Respir. Dis.*, **143**, 880–2.

Eisen, E.A., Wegman, D.H. and Louis, T.A. (1983) Effects of selection in a prospective study of forced expiratory volume in Vermont granite workers. *Am. Rev. Respir. Dis.*, **128**, 587–91.

Fraser, R.G. and Pare, J.A.P. (1977) *Diagnosis of Diseases of the Chest*, W.B. Saunders, Philadelphia.

Gardner, L.U. (1940) The pathology and roentgenographic manifestations of pneumoconiosis. *JAMA*, **114**, 535–45.

Ghio, A.J., Kennedy, T.P., Schapira, R.M., Crumbliss, A.L. and Hoidal, J.R. (1990) Hypothesis: is lung disease after silicate inhalation caused by oxidant generation? *Lancet*, **336**, 967–9.

Goldsmith, D.F., Guidotti, R.T. and Johnston, D.R. (1982) Does occupational exposure to silica cause lung cancer? *Am. J. Ind. Med.*, **3**, 423.

Goodman, G.B., Kaplan, P.D., Stachura, I., Castranova, V., Pailes, W.H. and Lapp, N.L. (1992) Acute silicosis responding to corticosteroid therapy. *Chest*, **101**, 366–70.

Graham, W.G.B., O'Grady, R.V. and Dubuc, B. (1981) Pulmonary function loss in Vermont granite workers. *Am. Rev. Respir. Dis.*, **123**, 25–8.

Graham, W.G.B., Weaver, S., Ashikaga, R. and O'Grady, R.V. (1994) Longitudinal pulmonary function losses in Vermont granite workers. *Chest*, **106**, 125–30.

Heppleston, A.G. (1985) Silica, pneumoconiosis, and carcinoma of the lung. *Am. Rev. Ind. Med.*, **7**, 285–94.

Hnizdo, E. (1992) Loss of lung function associated with exposure to silica dust and with smoking and its relation to disability and mortality in South Africa gold miners. *Br. J. Ind. Med.*, **49**, 472–9.

Hnizdo, E. and Sluis-Cremer, G.K. (1993a) Risk of silicosis in a cohort of white South African gold miners. *Am. J. Ind. Med.*, **24**, 447–57.

Hnizdo, E., Murray, J., Sluis-Cremer, G.K. and Rhomas, R.G. (1993) Correlation between radiological and pathological diagnosis of silicosis: an autopsy population based study. *Am. J. Ind. Med.*, **24**, 427–45.

Hosey, A.D., Trasko, V.M. and Ashe, H.B. (1957) *Control of silicosis in Vermont granite industry. Progress Report*, U.S. Department of Health, Education and Welfare, Public Health Service Publication No. 557, Washington, DC.

Holland, L.M. (1995) Animal studies of crystalline silica: results and uncertainties. *Appl. Occup. Environ. Hyg.*, **10**(12), 1099–103.

Hong Kong Chest Service/Tuberculosis Research Centre, Madras/British Medical Research Council (1991) A controlled clinical comparison of 6 and 8 months of antituberculosis chemotherapy in the treatment of patients with silicotuberculosis in Hong Kong. *Am. Rev. Respir. Dis.*, **143**, 262–7.

Hughes, J.M. (1995) Radiographic evidence of silicosis in relation to silica exposure. *Appl. Occup. Environ. Hyg.*, **10**(12), 1064–9.

Hughes, J.M. and Weill, H. (1995) Silicosis risk: Canadian and South African Miners (letter). *Am. J. Ind. Med.*, **27**, 617–22.

Hughes, J.M., Jones, R.V., Gilson, J.C. *et al.* (1973) Determinants of progression in sandblasters' silicosis. *Ann. Occup. Hyg.*, **26**, 701.

Infante-Rivard, C., Armstrong, R., Ernst, P., Petitclerc, M., Cloutier, L. and Theriault, G. (1991) Descriptive study of prognositic factors influencing survival of compensated silicotic patients. *Am. Rev. Respir. Dis.*, **144**, 1070–4.

International Agency for Research on Cancer (1987) Working group on the evaluation of carcinogenic risks to humans: overall evaluations of carcinogenicity. An updating of IARC monographs, Supplement 7, vols 1–42, World Health Organization, Lyon.

International Ageny for Research on Cancer, WHO (1997) Monograph on the evaluation of carcinogenic risks to humans: Silica, some silicates, coal dust and para-aramid fibrils, IARC, Lyon, France.

Jones, R.N., Turner-Warwick, M., Zinkind, M. and Weill, H. (1976) High prevalence of antinuclear antibodies in sandblasters' pneumoconiosis. *Am. Rev. Respir. Dis.*, **113**, 393–5.

Klockars, M., Koskela, R.S., Jarvinen, E., Kolari, P.J. and Rossi, A. (1987) Silica exposure and rheumatoid arthritis: a follow up study of granite workers 1940–81. *Br. Med. J.*, **294**, 997.

Lin, T.P., Suo, J., Lee, C.N. *et al.* (1987) Short course chemotherapy of pulmonary tuberculosis in pneumoconiotic patients. *Am. Rev. Respir. Dis.*, **136**, 808.

Mason, G.R., Abraham, J.L., Hoffman, L., Cole, S., Lippmann, M. and Wasserman, L. (1982) Treatment of mixed dust pneumoconiosis with whole lung lavage. *Am. Rev. Respir. Dis.*, **126**, 1102–7.

Mauderly, J.L. (1996) Lung overload: the dilemma and opportunities for resolution. *Inhalation Toxicol.*, **8** (suppl.), 1–28.

McDonald, C. (1995) Silica, silicosis, and lung cancer: an epidemiological update. *Appl. Occup. Environ. Hyg.*, **10**(12), 1056–63.

Morgan, W.K.C. (1975) The walrus and the carpenter or the silica criteria standard. *J. Occup. Med.*, **17**, 782–3.

Morrow, P.E. (1988) Possible mechanisms to explain dust overloading of the lungs. *Fundam. Appl. Toxicol.*, **10**, 369–84.

Muir, D.C.F., Julian, J.A., Shannon, H.S., Verma, D.K., Sebestyen, A. and Bernholz, C.D. (1989) Silica exposure and silicosis among Ontario hardrock miners: III. Analysis and risk estimates. *Am. J. Ind. Med.*, **16**, 29–43.

Musk, A.W., Peters, J.M., Wegman, D.H. and Fine, L.J. (1977) Pulmonary function in granite dust exposures: a four year follow-up. *Am. Rev. Respir. Dis.*, **115**, 769–76.

National Silicosis Conference (1938) Bulletin No. 21, United States Department of Labor, Washington, DC.

Ng, T.P. and Chan, S.L. (1992) Lung function in relation to silicosis and silica exposure in granite workers. *Eur. Respir. J.*, **5**, 986–91.

Ng, T.P., Chan, S.L. and Lam, K.P. (1987) Radiological progression and lung function in silicosis. *Br. Med. J.*, **295**, 164–8.

NIOSH (1974) *Occupational exposure to crystalline silica: criteria for a recommended standard*, HEW Publication No. (NIOSH), 75–120.

Norboo, T., Angchuck, P.T., Kamat, S.R. *et al.* (1991) Silicosis in a Himalayan village population: role of environmental dust. *Thorax*, **46**, 341–3.

Nouh, M.S. (1989) Is the desert lung syndrome (nonoccupational dust pneumoconiosis) a variant of pulmonary alveolar microlithiasis? Report of 4 cases and review of the literature. *Respiration*, **55**, 122–6.

Owen, H. (1901) *The Staffordshire Potter.* Grant Richards. Cited in Bulletin No. 21, National Silicosis Conference (1938), United States Department of Labor.

Oxman, A.D., Muir, D.C.F., Shannon, H.S., Stock, S.R., Hnizdo, E. and Lange, H.J. (1993) Occupational dust exposure and chronic obstructive pulmonary disease. *Am. Rev. Respir. Dis.*, **148**, 38–48.

Plinius Secundus, Caius. *Historiae Naturalis.* lib. xxxiii, sec xi.

Ramazzini, B. (1705) *De Morbis Artificum.* Translation by Dr James of London. Cited in Bulletin No. 21, National Silicocis Conference (1938), United States Department of Labor.

Reiser, K.M., Hesterberg, T.W., Haschek, W.M. and Last, J.A. (1982) Experimental silicosis: I. Acute effects of intratracheally instilled quartz on collagen metabolism and morphologic characteristics of rat lungs. *Am. J. Pathol.*, **107**, 176–85.

Reiser, K.M., Haschek, W.M., Hesterberg, T.W. and Last, J.A. (1983) Long-term effects of intratracheally instilled quartz on collagen metabolism and morphologic characteristics of rat lungs. *Am. J. Pathol.*, **110**, 30–41.

remy-Jardin, M., Remy, J., Farre, I. and Marquette, H. (1992) Computed tomographic evaluation of silicosis and coal workers' pneumoconiosis. *Radiol. Clin. North Am.*, **30**, 1155–76.

Rom, W.N., Bitterman, P.B., Rennard, S.I., Cantin, A. and Crystal, R. (1987) Characterization of the lower respiratory tract inflammation of nonsmoking individuals with interstitial lung disease associated with chronic inhalation of inorganic dusts. *Am. Rev. Respir. Dis.*, **136**, 1429–34.

Russell, A.E. (1941) The Health of workers in dusty trades. VII. Restudy of a group of granite workers. Public Health Bulletin No. 269, Washington, DC.

Russell, A.E., Britten, R.H., Thompson, L.R. *et al.* (1929) Health of workers in dusty trades. II. Exposure to siliceous dust (granite industry). Public Health Bulletin No. 187, US Public Health Service, Washington, DC.

Samimi, B., Ziskind, M. and Weill, H. (1978) The relation of silica dust to accelerated silicosis. *Ecotoxicol. Environ. Safety*, **1**, 429–36.

Schuyler, M.R., Gaumer, H.R., Stankus, R.P., Kaimal, J., Hoffman, E. and Salvaggio, J. (1980) Bronchoalveolar lavage in silicosis. *Lung*, **157**, 95–102.

Schuyler, M.R., Ziskind, M. and Salvaggio, J. (1977) Cell–mediated immunity in silicosis. *Am. Rev. Resp. Dis.*, **116**, 147–151.

Seaton, A., Legge, J.S., Henderson, J. and Kerr, K.M. (1991) Accelerated silicosis in Scottish stonemasons. *Lancet*, **337**, 341–4.

Sharma, S.K., Pande, J.N. and Verma, K. (1991) Effect of prednisolone treatment in chronic silicosis. *Am. Rev. Respir. Dis.*, **143**, 814–21.

Shi, X., Mao, Y., Daniel, L.N., Saffiotti, U., Dalal, N.S. and Vallyathan, V. (1995) Generation of reactive oxygen species by quartz particles and its implication for cellular damage. *Appl. Occup. Environ. Hyg.*, **10**(12), 1138–44.

Snider, D.E. (1978) The relationship between tuberculosis and silicosis. *Am. Rev. Respir. Dis.*, **118**, 455–9.

Steenland, K. and Brown, D. (1995) Silicosis among gold miners: exposure-response analyses and risk assessment. *Am. J. Public Health*, **85**, 1372–7.

Suratt, P.M., Winn, W.C. Brody, A.R., Bolton, W.K. and Giles, R.D. (1977) Acute silicosis in tombstone sandblasters. *Am. Rev. Respir. Dis.*, **115**, 521–9.

Sutton, G.W. and Reno, S.J. (1968) Respirable mass concentration equivalent in impinger count data: Barre, Vt granite sheds. Presented at 1968 annual meeting, American Industrial Hygiene Association, St. Louis, Mo.

Theriault, G.P., Burgess, W.A., DiBerardinis, L.J. *et al.* (1974a) Dust exposure in the Vermont granite sheds. *Arch. Environ. Health*, **28**, 12–17.

Theriault, G.P., Peters, J.M. and Fine, L.J. (1974b) Pulmonary function in granite shed workers of Vermont. *Arch. Environ. Health*, **28**, 18–22.

Theriault, G.P., Peters, J.M. and Johnson, W.M. (1974c) Pulmonary function and roentgenographic changes in granite dust exposure. *Arch. Environ. Health*, **28**, 23–7.

Valiante, D.J. and Rosenman, K.D. (1989) Does silicosis still occur? *JAMA*, **262**, 3003.

Vallyathan, V., Shi, X., Irr, W. *et al.* (1988) Generation of free radicals from freshly fractured silica dust: potential role in acute silica-induced lung injury. *Am. Rev. Respir. Dis.*, **138**, 1213–19.

Vorwald, A.J. and Karr, M.D. (1938) Pneumoconiosis and pulmonary carcinoma. *Am. J. Pathol.*, **14**, 49–57.

Wagner, G.R. (1995) Editorial: The inexcusable persistence of silicosis. *Am. J. Public Health*, **85**, 1346–7.

Wegman, E.H., Eisen, E. and Peters, J.M. (1983) Pulmonary function loss in Vermont granite workers. *Am. Rev. Respir. Dis.*, **128**, 776–7.

White, N.W., Chetty, R. and Bateman, E.D. (1991) Silicosis among gemstone workers in South Africa. *Am. J. Indust. Med.*, **9**, 205–13.

Wilt, J., Banks, D.E., Weissman, D.N. *et al.* (1996) Reduction of lung dust burden in pneumoconiosis by whole lung lavage. *J. Occup. Environ. Med.*, **38**, 619–24.

14 Silicosis in developing countries

Surinder K. Jindal and Jagdeep Whig

INTRODUCTION

Silicosis is an ancient disease and in India, at least, it is reflected in the history of workers who built stone structures. For example, in India, thousands of temples were built by cutting stone as far back as several centuries BC. These included the famous Ajanta and Ellora caves. In the Indian culture, the earliest reports of 'increasing physical disability' associated with rock cutting were vividly described by workers building pagodas on the sea coast of Mahabalipuram near Madras during the Pallava Dynasty in 7th century AD (Ghosh, 1964). As recently as 1969, silicosis was apparent in 35.2% of 227 stone cutters in Northern India (Gupta *et al.*, 1969).

Unlike the Western world where silicosis was diagnosed in the late 19th and early 20th century in rock crushers, gold miners and cutlery factory workers, most reports of silicosis in developing countries can be traced back to the past 50 years or so. This paucity of documented reports of silicosis in developing countries from earlier this century may be attributed to several reasons. Infectious diseases such as tuberculosis overshadowed most other respiratory problems. Because these diseases were so common, the presence of respiratory symptoms and opacities on chest radiographs led to a diagnosis of tuberculosis, regardless of the dust exposure history. The lack of an organized occupational health service structure was another important factor. There were no routine screening facilities for health hazards in most industrial units. Finally, and perhaps most importantly, the need of industrialization required for economic and social development was more urgent and outweighed all other considerations.

Even when silicosis was recognized, there was no uniform practice of medical reporting and documentation. It was only in more recent years, when investigators appreciated the occupations at risk for silicosis and attributed these illnesses to dust exposure, rather than to mycobacterial infection, that the gravity of the problem was recognized.

EPIDEMIOLOGY

It is difficult to assess the true prevalence of silicosis in the absence of a registry. There are several reports regarding the prevalence of silicosis in occupational groups from a number of developing countries, but there are few epidemiologic studies designed to identify its incidence. Further, methodology used for detection and criteria for diagnosis have been either inadequate or not standardized.

In India, the first case of silicosis in a modified form was reported in the underground workers in the Kolar gold fields of Mysore (Caplan and Burdon, 1947), although reports describing problems of pneumoconiosis in the Kolar Mines were published earlier (Jones, 1933). Sikand and Pamra (1949) reported radiographic evidence of silicosis in 28 of 42 stone cutters and in three of 24 stone breakers. Reports of silicosis in

Occupational Lung Disease. Edited by Daniel E. Banks and John E. Parker. Published in 1998 by Chapman & Hall. ISBN 0 412 73630 6

industrial workers soon followed (Banerjee, 1951; Methuna, 1953). Surveys which followed reported silicosis in 34.1% of miners, 16.7% of workers employed in silica brick making, and 22.8% of fire-clay brick makers (Gupta, 1954; Rao and Lundgren, 1955).

Of the numerous mining operations in India, mica mining is of special relevance in terms of the risk of silicosis because of the introduction of dry, compressed air drilling. Radiographic evidence of nodular or conglomerate silicosis was reported in 34.1% of mica miners and 45.2% of the supervisory staff. In one case, this disease developed within 18 months in a 27-year-old man working as a hand driller in a mica mine in Nellore (Thiruvengadam *et al.*, 1968).

The prevalence of silicosis reported in later studies has varied. Banerjee and Roy (1969) reported pneumoconiosis in 0.98% of iron and steel workers. This was low compared with figures subsequently reported in most studies. Pneumoconiosis was detected in seven out of 284 workers (of whom 100 were X-rayed) from 15 foundries of Kanpur (Aggarwal *et al.*, 1973).

In the steel industry (i.e. work in and around blast furnaces, sintering of iron ore, pig casting and coke ovens), there are several sources of silica exposure and several potential reasons which may explain the prevalence of silicosis in these workers (Grewal *et al.*, 1979). First, in the foundry process, the high melting point of steel causes this metal to penetrate the sand mold; second, the mixture for steel castings contains more free silica than the molds for iron castings; and finally, the use of pneumatic chisels on steel castings to remove burnt sand produces respirable-size particles.

Similarly, silicosis in abrasive wheel polish workers employed in the cycle industry in Ludhiana was attributed to the inhalation of respirable dust composed of emery (a material which contains 50–70% alumina and is thought to have little or no effect on the lungs) and silica coating the polishing wheel. This dust is generated when the cycle part to be polished is held tightly over the rim of the revolving polishing wheel (Malik *et al.*, 1985) (Figure 14.1 and Plates 9 and 10).

Silicosis has been reported in workers of other industries. Specifically, grinders employed in the pottery and ceramic industry, workers in the cutlery industry, and slate pencil workers are at risk. Respirable silica exposure in metal casting, boiler scaling, grinding and polishing, and in the manufacture of glass, paints and rubber can also be associated with this disease.

The Indian National Institute of Occupational Health (NIOH) has performed environmental and medical surveys in several industries and found the prevalence of silicosis to be very high in most instances (Table 14.1) (Kashyap, 1994). The prevalence of silicosis was attributed to a high concentration of total and respirable dust and the great percentage of free silica in the dust. The average duration of exposure before the development of silicosis in most of these surveys was approximately 10–12 years or less. The human cost was great, with approximately 55% of slate pencil industry workers having silicosis and nearly 18% developing progressive massive fibrosis. Of these, 22% died within 16 months of diagnosis.

Figure 14.1 Young men employed grinding bicycle sprockets.

In addition, silica flour mill workers are at risk for this aggressive accelerated form of silicosis (Natarajan *et al.*, 1992). Recently, workers handling silica powder bags were reported to have developed acute silicosis occurring within 1–3 years of exposure, and occasionally within months (Purohit *et al.*, 1992). In individuals with heavy exposure to freshly crushed silica, progression is rapid and a shortened life is the norm.

Environmental dust exposure causing silicosis has been described in residents in the Sahara and sub-Saharan desert. Similar cases have been reported from a mountainous Himalayan village (Norboo *et al.*, 1991).

Table 14.1 Prevalence of silicosis in select Indian workforces

	Dust level			*Prevalence*	
	Total (mg/m³)	*Respirable (mg/m³)*	*Free silica (%)*	*No. studied*	*No. with silicosis (%)*
Slate pencil industry (Madhya Pradesh)	46.5	10.4	56.5	593	324 (54.6)
Agate industry grinders (Gujarat)	14.75	2.35	60	353	135 (38.2)
Stone cutters (Uttar Pradesh)	22.4	1.6	–	125	28 (22)

CLINICAL FEATURES ASSOCIATED WITH SILICA DUST EXPOSURE

Unless detected incidentally on routine screening, silicosis often presents when the disease is either advanced or complicated. Based on clinical features and rapidity of development, silicosis is classified as simple (Plate 11), complicated (or progressive massive fibrosis), accelerated or acute. Definitions and descriptions of these differing presentations and patterns of progression of silicosis are described in detail elsewhere (see Chapter 13).

In addition to the development of pneumoconiosis in silica-exposed workers, investigators have reported increased respiratory symptoms and impaired lung function in industrial workers, miners, stone cutters and others in the absence of a definitive radiologic diagnosis of silicosis. This non-specific symptomatology is attributed to chronic dust exposure. Furthermore, the relationship between silica dust exposure, silicosis and mycobacterial infection remains an important problem in developing countries.

Non-specific respiratory morbidity attributable to dust exposure

Breathlessness and cough are frequent in industrial and mining workers, even when radiographic evidence of silicosis is not demonstrable. In stone quarry workers exposure to even low levels of respirable dust for 20 years or more caused significant effects on lung function (Ghotkar *et al.*, 1995). One or the other of these two respiratory symptoms was present in 20% of 500 cycle industry workers and the mean FEV_1 in both smoker and non-smoker silica-exposed workers was significantly lower than the control subjects not exposed to silica (Whig *et al.*, 1993).

Stone crushers around Chandigarh, India, were screened with a respiratory questionnaire, physical examination, chest radiograph and spirometry. Respiratory symptoms were present in 46.6% and approximately 10% had a forced vital capacity (FVC) <75% predicted (Jindal and Singh, 1992). Similar observations had been made in Hong Kong gem-stone workers exposed to silica (Ng *et al.*, 1987). The heavily exposed workers reported a greater prevalence of respiratory symptoms. Further, a decline in FVC was significantly related to increasing years of employment after allowing for the effects of age, height and smoking. These investigators concluded that there was an independent effect of respirable dust exposure in reducing FVC. This condition has been referred to as 'industrial bronchitis' caused by dust exposure, with the higher prevalence of respiratory symptoms explained by mucus hypersecretion.

Chronic airflow limitation has been described independently of mucus hypersecretion in dusty occupations (see Chapter 5). Since many blue-collar workers are often smokers, both mucus hypersecretion and airways obstruction could be at least partly attributed to tobacco smoking. Air pollution in areas where these workers live and work, as well as recurrent respiratory infections which are more likely to occur in these individuals, may also be responsible for this respiratory morbidity.

Silicotuberculosis

Tuberculosis is an important consideration in the diagnosis and management of silicosis, particularly in developing countries. It can either complicate silicosis or pose a problem in the differential diagnosis in view of the similarity of clinical symptoms and radiographic findings. In India, most silicotics with radiographic opacities in the upper zones are prescribed anti-tubercular treatment on empirical grounds. In one study of silica and copper miners >50% of the workers with silicosis had received anti-tuberculous treatment at one or the other time (Deshmukh, 1984).

Reports appearing in the past two decades continue to identify a clinically important rate of tuberculosis in silicotics (Table 14.2). Such illness in silicotics was reported in 18.6% of mica workers and 60% of workers in the pottery and ceramic industries. The prevalence of tuberculosis varied from 1.9 to 3.6% in reports describing pottery and ceramic workers, brick makers, metal workers and coal miners with silicosis. According to NIOH reports, tuberculosis was present in 4.8% of coal miners and 48% of stone cutters (Kashyap, 1994). In stone cutters, tuberculosis was nearly three times more common in silicotic than in non-silicotic workers (Gupta *et al.*, 1972).

In view of the high prevalence of tuberculosis in the Indian general population, as well as in the populations of other developing countries, some of the silica-exposed workers may suffer from tuberculosis independently of silicosis. For example, tuberculosis was reported in 8.75% of stone quarries workers exposed to silica dust, even though the prevalence of silicosis was not determined (Ghotkar *et al.*, 1995), yet the percentage of patients with these two diseases far exceeds the expected prevalence rates.

Several reasons have been put forward to explain the increased occurrence of tuberculosis. Free silica impairs macrophage function, thereby increasing the chances of mycobacterial infection (Snider, 1978; Morgan, 1979). The other important factors relate to social circumstances and to living and working conditions. Most of these workers live in crowded and unsanitary conditions, while their jobs are in dusty environments and silica exposures are not controlled. The chances of spread of infection are higher in such surroundings. Many reports of silicotuberculosis cited earlier do not necessarily attribute the presence of tuberculosis to underlying silicosis. However, when the conditions are present together, there is increased morbidity and mortality, with clear evidence of rapid silicosis progression (Figure 14.2). Studies from Singapore have reported the occurrence of tuberculosis as a significant factor associated with the development of progressive massive fibrosis and a strong predictor of mortality in patients with silicosis (Ng and Chan, 1991; Ng *et al.*, 1992).

Treatment of tuberculosis in these workers poses tremendous problems. Results of standard chemotherapeutic protocols for tuberculosis in the presence of silicosis are less effective than therapy for tuberculosis alone. Some of the earlier reports suggested indefinite treatment with isoniazid, streptomycin and para-aminosalicylic acid (Morgan, 1979). A partial explanation for the poor response has been attributed to impaired macrophage function and poor penetration of the drugs into fibrotic nodules (Snider 1978; Lowrie, 1982). Short-course chemotherapy with 2 months of streptomycin, isoniazid, rifampicin and pyrazinamide followed by 7 months of daily isoniazid and rifampicin was effective, with only a 5% failure rate in 16 previously treated silicotics and 56 anthrasilicotic

Table 14.2 Prevalence of tuberculosis complicating silicosis in different reports from India

Occupation	*Proportion with silicosis (%)*
Mica miners	18.6
Coal miners	4.8
Stone cutters	48.0
Pottery and ceramic industry	60.0

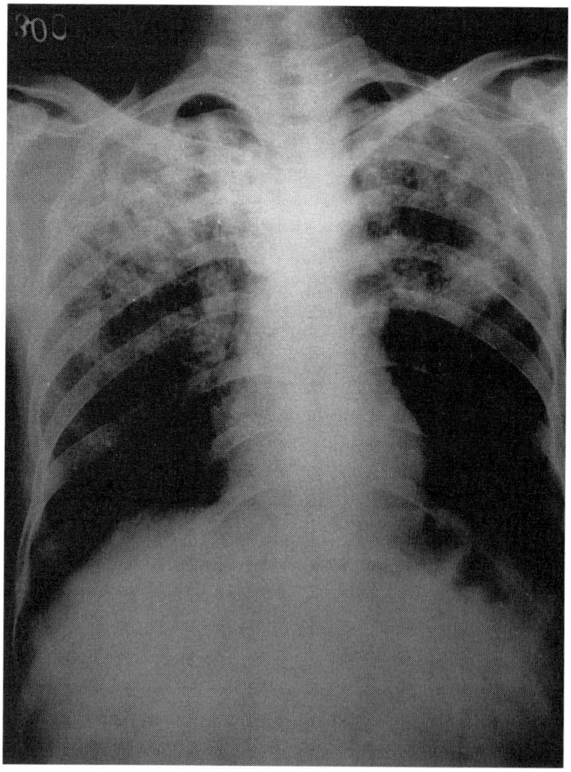

Figure 14.2 Chest radiograph showing bilateral upper zone opacities consistent with progressive massive fibrosis with a left upper zone cavitation in a patient with silicotuberculosis.

patients with mycobacterial positivity (Lin *et al.*, 1987). Clinical comparison was made of two anti-tuberculosis treatment regimens of four primary drugs administered thrice weekly for 6 and 8 months in 240 Chinese silicotics. Although the 6-month regimen was inadequate, since 22% of 41 patients relapsed bacteriologically during the first 3 years, the 8-month regimen was substantially more effective (Hong Kong Chest Service, 1991).

Chemoprophylaxis with rifampicin and isoniazid, both alone and together, was monitored in a double-blind design study in 679 Chinese men with silicosis. Although chemoprophylaxis halved the proportion of patients in whom tuberculosis developed, the proportion was still substantial (Hong Kong Chest Service, 1992). There was no evidence that chemoprophylaxis led to the selection of drug-resistant strains.

Overall, the efficacy of therapeutic as well as preventive interventions was proven in China on the 18-years cumulative incidence of silicosis in 26 603 dust-exposed workers (Lou and Zhou, 1989). From the 1950s to 1970s the 8-year cumulative incidence of tuberculosis decreased from 51.7% to 16.7% and the case fatality rate of silicosis fell from 53.9% to 18.3%.

Further discussion regarding possible therapeutic approaches to silicosis occurring in the absence of tuberculosis, as well as a more detailed discussion regarding prevention is included in Chapter 13.

SUMMARY

Silicosis remains an important health problem in workers exposed to silica dust. The endemic prevalence of tuberculosis in the population of developing countries increases the risk that the silicotic will develop tuberculosis infection. India, like other developing countries, is in a stage of development where utilization of natural resources by mining, tunneling and quarrying plays a great role in the progress that will be made, yet inadequate preventive measures put the workers at risk for disease. It is reasonable to conclude that the pace of industrialization in developing countries exceeds the current pace of such a transformation of society in developed countries. In developed countries, this process of industrialization has occurred in previous generations. Moreover, since industrialization is increasing in India as well as in other developing countries, silicosis will likely be more frequently recognized in the developing world than before.

Acknowledgments

The authors acknowledge the information supplied by Dr Shiela Yadav, Chandigarh; Dr V.K. Arora, Pondicherry; Dr P.R. Gupta, Udaipur; Dr M.L. Gupta, Jaipur; Dr Y.K. Meisheri, Mumbai; Dr S. Kashyap, Shimla; and Dr P.C. Manoria, Bhopal.

REFERENCES

Aggarwal, M.C., Bhargava, R.R.S. and Mittal, O.P. (1973) Health hazards in foundary workers. *Indian J. Chest Dis.*, **15**, 276–84.

Banerjee, D. and Roy, S. (1969) Respiratory morbidity in iron and steel workers at Burnpur. *Indian J. Ind. Med.*, **13**, 15.

Banerjee, L. (1951) Pneumoconiosis and its relation to tuberculosis. *Proceedings of the Society for the Study of Industrial Medicine*, **3**, 134.

Caplan, A. and Burdon, D.J. (1947) Pneumokoniosis on the Kolar Gold Field. Proceedings, Silicosis Conference, London, April 1947.

Deshmukh, P.A. (1984) Chest Diseases in industry. *Indian J. Tuberculosis*, **31**, 53–64.

Ghosh, P.K. (1964) Industrial pulmonary disease in India. *Ind. Med. Surg.*, **33**, 732–7.

Ghotkar, V.B., Maldhure, B.R. and Zodpey, S.P. (1995) Involvement of lung and lung function tests in stone quarry workers. *Indian J. Tuberculosis*, **42**, 155–60.

Grewal, K.S., Arora, V.K. and Gupta, S.P. (1979) Industrial lung diseases, in *Progress in Clinical Medicine in India*, 3rd series (ed. M.M.S. Ahuja), Arnold Heinemann, New Delhi, pp. 420–38.

Gupta, M.N. (1954) Vital capacity of the lungs of silica and fire clay brick workers in Bihar with special reference to dust exposure, exercise tolerance test and incidence of silicosis. Report No. 5, Office of the Chief Adviser Factories, New Delhi.

Gupta, S.P., Garg, A.K. and Gupta O.P. (1969) Silicosis amongst stone cutters. *J. Assoc. Physicians India* **17**, 163.

Gupta, S.P., Bajaj, A., Jain A.L. and Vasudeva, Y.L. (1972) Clinical and radiological studies in silicosis; based on a study of the disease amongst stone cutters. *Indian J. Med. Res.*, **60**, 1309–15.

Hong Kong Chest Service/Tuberculosis Research Centre, Madras/British Medical Research Council (1991) A control clinical comparison of 6 and 8 months of anti-tuberculosis chemotherapy in the treatment of patients with silicotuberculosis in Hong Kong. *Am. Rev. Respir. Dis.*, **143**(1), 262–7.

Hong Kong Chest Service/Tuberculosis Research Centre, Madras/British Medical Research Council (1992). A double-blind placebo-controlled clinical trial of three anti-tuberculosis chemoprophylaxis regimens in patients with silicosis in Hong Kong. *Am. Rev. Respir. Dis.*, **145**(1), 36–41.

Jindal, S.K. and Singh, K. (1992) Health effects of environmental pollution due to stone crushing in Haryana State. Report of Environment Department, Government of Haryana, Postgraduate Institute for Medical Education and Research, Chandigarh.

Jones, W.R. (1933) Silicotic Lungs: the minerals they contain. *J. Hyg.*, **33**, 307.

Kashyap, S.K. (1994) Occupational pneumoconiosis and tuberculosis. *Indian J. Tuberculosis*, **41**, 73–6.

Lin, T.P., Soo, J., Lee, C.M., Lee, J.J. and Yung, S.P. (1987) Short course chemotherapy of pulmonary tuberculosis in pneumoconiotic patients. *Am. Rev. Respir. Dis.*, **136**, 808–10.

Lou, J.Z. and Zhou, C. (1989) The prevention of silicosis and prediction of its future prevalence in China. *Am. J. Public Health*, **79**(12), 1613–16.

Lowrie, D.B. (1982) What goes wrong with the macrophage in silicosis? *Eur. J. Respir. Dis.*, **63**, 180–2.

Malik, S.K., Behera, D., Awasthi, G.K. and Singh, J.P. (1985) Pulmonary silicosis in emery polish workers. *Indian J. Chest Dis. Allied Sci.*, **27**(2), 116–21.

Methuna, N.C. (1953) Incidence of silicosis in the Mysore iron and steel works. *Proceedings of the Society of the Study of Industrial Medicine*, **5**, 131.

Morgan, E.J. (1979) Silicosis and tuberculosis. *Chest*, **75**, 202–3.

Natarajan, A.S., Gajalakshmi, L. and Karunakaran, S. (1992) Accelerated silicosis in a silica flour mill. *Lung India*, **10**(1), 33–7.

Ng, T.P. and Chan, S.L. (1991) Factors associated with massive fibrosis in silicosis. *Thorax*, **46**(4), 229–32.

Ng, T.P., Tsin, T.W., O'Kelly, F.J. and Chan, S.L. (1987) A survey of the respiratory health of silica exposed gem stone workers in Hong Kong. *Am. Rev. Respir. Dis.*, **135**, 1249–54.

Ng, T.P., Chan, S.L. and Lee, J. (1992) Predictors of mortality in silicosis. *Respir. Med.*, **86**(2), 115–19.

Norboo, T., Angchuk, P.T., Yahya, M. *et al.* (1991) Silicosis in a Himalayan village population: role of environmental dust. *Thorax*, **46**(5), 341–3.

Purohit, S.D., Gupta, P.R., Gupta, M.L. and Chauhan, A. (1992) Acute silicosis in handlers of silica powder bags. *Lung India*, **10**(4), 145–8.

Rao, M.N. and Lundgren, N.P.V. (1955) *A Review of Occupational Health Research in India*, Indian Council of Medical Research, Special Report Series No. 29, New Delhi.

Sikand, B.K. and Pamra, S.P. (1949) Preliminary report on the occurrence of silicosis amongst stone masons. Proceedings of Seventh Tuberculosis Workers' Conference, Bombay, November 1949, p. 106.

Snider, D.E. (1978) The relationship between tuberculosis and silicosis. *Am. Rev. Respir. Dis.*, **118**, 455–60.

Thiruvengadam, K.V., Anguli, V.C., Shetty, R., Sambanban, S. and Kosairam, R. (1968) Silicosis in a mica mine workers. *J. Indian Med. Assoc.*, **51**(5), 248–50.

Whig. J., Sood, A., Khurana, S.B. and Mahajan, R. (1993) Association of FEV_1 with minispirometer as a screening test in cycle industry workers in Ludhiana. *Lung India* **11**(3), 91–4.

15 Asbestos-related lung diseases

Raymond Bégin

HISTORICAL REVIEW

Asbestos, 'inextinguishable' in Greek, has been known since antiquity. This material was initially used for the wicks of torches in temples, and beginning in 2500 BC, was used in the production of fireproof pottery. In the time of the pharaohs, asbestos linens were used for the incineration of corpses.

Later, Charlemagne stunned his guests by throwing an asbestos-made tablecloth into the fire to clean it (Bégin *et al.*, 1993). Yet, until the recognition of vast deposits in Quebec and Russia at the end of the 19th century, asbestos usage remained small, and was principally related to crafts. In Quebec, mine exploration started in 1878 in ASBESTOS township, a location named by its product, and a site where chrysotile asbestos has been produced since. In Russia, commercial production of chrysotile began in 1885. In South Africa, crocidolite asbestos was initially produced in 1880, and amosite and anthophyllite asbestos became commercially available following the First World War. Asbestos-related diseases occur following exposure to all of the different types of asbestos fibers, but the incidence of various diseases and the intensity of the pathologic processes vary with the different fiber types.

Cumulative world production of asbestos reached 5×10^9 kg by 1930. Later, rapid acceleration of production resulted in a yearly production of 5×10^9 kg, responding to the demand for its use in some 3000 applications (Table 15.1). Compared with the cost of man-made mineral fibers, asbestos is very inexpensive.

Twenty years after the industrial use of asbestos began in Europe, Auribault (1906) described pulmonary fibrosis in an asbestos textile factory in Normandy, France. In England, Murray (1907) described the first case of asbestosis in a 30-year-old asbestos worker, the last survivor of 10 employed in the same workshop. In 1906, Marchand described asbestos bodies in the lung. After 1926, it became accepted that asbestos was toxic to the pleura, lung and airways. Yet, the extent of health problems posed by the industrial use of asbestos only became clear in the 1950s. In 1959, Jacob and Bohling reported calcified pleural plaques and in 1960, Wagner *et al.* described the unique relationship between malignant mesothelioma and asbestos. Of 33 cases of malignant mesothelioma, 28 had been exposed to crocidolite asbestos in South Africa. Later, epidemiologic studies of American and European teams were reported at international

Table 15.1 Most common asbestos uses

- Asbestos cement products, including pipes, shingles, clapboards, flat sheets, corrugated sheets and molded sheets for the construction industry
- Vinyl–asbestos floor tiles
- Asbestos paper for use in insulation and filtering products
- Friction materials for brake linings and clutch facings
- Asbestos textile products such as cloth, yarn, tape, felt, tubing, cord and rope
- Spray products for acoustical, thermal and fireproofing purposes

Occupational Lung Disease. Edited by Daniel E. Banks and John E. Parker. Published in 1998 by Chapman & Hall. ISBN 0 412 73630 6

conferences in New York in 1964, 1979 and 1990 (Selikoff and Churg, 1965; Selikoff and Hammond, 1979; Landrigan and Kazemi; 1991), at the International Agency for Research on Cancer (IARC) in 1972 and 1979 (Wagner, 1980), in Montreal in 1980 (Anonymous, 1982), in Cardiff, Wales in 1986, where the biologic effects of chrysotile was addressed (Wagner, 1986), and in Paris in 1991 and 1995, where the focus was malignant mesothelioma (Jaurand *et al.*, 1993. In the 1980s, studies of biological mechanisms increased our understanding of the pathogenesis of these asbestos-related disease processes (Bégin *et al.*, 1989, 1992a; Rom *et al.*, 1991).

ASBESTOS

Mineralogy

The term 'asbestos' refers to a family of naturally occurring fibrous hydrous silicate minerals that are relatively indestructible and heat resistant. Asbestos fibers are the result of fragmentation of larger and more massive crystals. They have a ratio of length to diameter (aspect ratio) exceeding 3 : 1, and, like organic fibers they are flexible and posses very thin units (the elementary fibril) arranged length-wise to form bundles. The fibrils can aggregate in quite organized hair-like bundles, or in less oriented aggregates, apparently based on variations in mineral content and characteristics of the component crystals.

Although numerous minerals possess characteristics similar to asbestos, only six asbestos fibers are of industrial use and will be considered here. The first type of fiber has a curved or serpiginous shape and is termed *chrysotile*. The remaining five are straight and categorized as *amphiboles*. These include amosite, anthophyllite, tremolite, actinolite and crocidolite (Table 15.2).

Table 15.2 Classification of asbestos minerals

Asbestiform	*Non-asbestiform*	*Chemical formula*
Actinolite	Actinolite	$Ca_2(Mg,Fe)_5(Si_8O_{22})(OH)_2$
Amosite	Cummingtonite-grunerite	$(Fe,Mg)_7(Si_8O_{22})(OH)_2$
Anthophyllite	Anthophyllite	$(Mg,Fe)_7(Si_8O_{22})(OH)_2$
Chrysotile	Antigorite, lizardite	$Mg_3(Si_2O_5)(OH)_4$
Crocidolite	Riebeckite	$Na_2Fe_5(Si_8O_{22})(OH)_2$
Tremolite	Tremolite	$Ca_2Mg_5(Si_8O_{22})(OH)_2$

Chrysotile

Chrysotile fibers result from a peculiar hydrothermic transformation (serpentinization) of ultrabasic rocks (dunites, pyroxenites and peridotites). The crystalline structure of chrysotile is complex and composed of hydrated magnesium silicate. Chrysotile fibers are relatively translucent with a silk sparkle. Depending on the mine of origin, they may have more or less flexibility and are found within veins of serpentine minerals. The elementary fibrils of chrysotile have an average diameter of 0.03 μm and are arranged as an open-core cylinder with layers of intertwined fibrils suggesting an onion-peel around the core.

Chrysotile is of commercial interest because of its mechanical properties of resistance to heat and traction, flexibility, adsorbancy, and chemical resistance to alkali. It has weak resistance to acid. It is the softest of the asbestos fibers and has been most used for spinning and weaving to produce asbestos cloth. It has the ability to resist heat over 100°C, but loses its structure above 575°C. Chrysotile represents 95% of the world production.

Amphiboles

Amphiboles, as a group, contain several minerals and are distinct in chemical composition but similar in structure. The structure of the five asbestos amphiboles differs slightly from that of chrysotile, as they are more rigid and do not form a cylinder, but are arranged in parallel to form a chain of silicate tetrahedra. This chain-like structure is stacked lightly, which permits good cleavage. Compared with chrysotile, the amphiboles are more rigid, form less aggregates, and release fibers more readily during handling. Amphiboles are more resistant to acids and heat compared with chrysotile. Of the five amphiboles, tremolite and actinolite are of little commercial value but may be contaminants of true talc and be associated with adverse respiratory effects (Zussman, 1979; Anonymous, 1991) (Table 15.3).

- *Crocidolite* is a sodium iron silicate formed in thermally metamorphosed banded ironstone. Crocidolite is referred to as 'blue asbestos' and has been mined in South Africa, Australia and Bolivia. It is less harsh than amosite and has been used in spinning and insulation. It is reputed to be the most dangerous, as evidenced by its strong association with mesothelioma.
- *Amosite* is an amphibole mined only in the Transvaal area of South Africa. It is an iron magnesium silicate, brownish in color, and is the longest of fibers. It has a harshness which renders it unsuitable for spinning but has excellent heat-insulating properties.

Table 15.3 Major sites of asbestos deposits

Fiber type	*Site of deposit*
Actinolite	Taiwan, South Africa
Amosite	South Africa, India
Anthophyllite	Finland, USA, South Africa, Bulgaria
Chrysotile	Canada (Quebec, British Colombia, Yukon, Newfoundland, Ontario), Siberia, Zimbabwe, Botswana, Swaziland, Australia, Cyprus, China, Brazil, USA (Vermont, Arizona, California)
Crocidolite	South Africa, Bolivia, Australia
Tremolite	Italy, South Africa, Pakistan, Korea

- *Anthophyllite* is a white amphibole and a magnesium silicate containing various quantities of iron. It occurs in fibrous masses with short fiber bundles. It was mined in Finland until 1970, but production has now ceased.

Asbestos – production and uses

World asbestos production dating from the late 19th century increased until about 1975–1980. At that time, production approximated 5 million tons material per year but since then annual production has declined and stabilized at about 4 million tons.

Asbestos mining is typically an open-pit operation where mechanical shovels and bulldozers mine and then load trucks or train cars for transport of the raw material to a processing mill. The raw material is further processed by fragmenting, sorting and screening to isolate and concentrate the asbestos and eliminate undesired ores. Milling permits further concentration of the fibers through elimination of undesired contaminants, drying of the fibers and separation of fibers into commercial grades.

The major end users of the past have been the construction industry, the shipbuilding industry and the automobile and railroad equipment industries. Today, much of the insulation and friction materials are made of non-asbestos replacement fiber materials.

The longest asbestos fibers are used mainly in the textile and insulation industries; intermediate-length fibers are used in asbestos cement production and friction and filter production; while the shortest fibers are used in the vinyl–asbestos tile industry and as an admixture to industrial paints. Mill tailings from the mining process have been used in road construction.

ASBESTOS FIBERS IN THE ENVIRONMENT AND IN HUMAN TISSUES

Asbestos in the environment

Control of asbestos dust in the workplace has proven to be the most effective mode of disease prevention. The current US standard is 0.1 fiber/ml of air. In the past, although peak asbestos fiber counts in the ambient workplace air may have exceeded 100 fibers/ml, the more typical exposures ranged between 5 to 20 fibers/ml.

To enforce fiber control legislation, methods for airborne dust sampling and fiber counting were necessary. Fibers are collected on a membrane filter. Those having a length >5 μm and length-to-width ratio >3 are counted under a light microscope at a magnification of 500×. Unfortunately, many chrysotile fibers are not seen by this method. Transmission electron microscopy (TEM) can magnify up to 100 000×, but specimen preparation is time consuming and expensive. Yet, TEM has made a significant impact in recognizing the level of environmental air pollution in neighborhoods near asbestos mills, plants and factories.

Analysis of lung tissue with identification of asbestos fibers supports the usual clinical methods of diagnosing asbestos-related diseases. Light microscopy allows visualization of ferruginous bodies. These are ferritin–protein-coated fibers (typically >10 μm in length) which appear as relatively long bead-like structures. These are formed by macrophages as they attempt to ingest and break down fibers. Although ferruginous bodies can be found in the lung tissue of virtually all of us, identifying these fibers in digested tissue, sputum and cells garnered by lung lavage may provide a fundamental index of asbestos exposure (Bignon *et al.*, 1979; Sebastien, 1982; Churg and Green, 1989). Although most often composed of amphibole asbestos fibers [and described as asbestos bodies (AB)], these ferruginous bodies are not specific for this exposure as other fibers, such as cotton or man-made mineral fibers, may be handled in this fashion.

Asbestos bodies in lung tissue (Churg and Green, 1989)

In a sputum cytology sample, an unexposed individual typically has no AB. The presence of any AB indicates a significant exposure. On a lung lavage sample, non-exposed subjects have less than 1 AB/ml of lavage effluent; more than 1 AB/ml suggests the presence of 1 AB/mg in dry lung, more than non-trivial exposure.

Study of non-exposed Quebec residents reported 808 fibers/mg dry lung and 0.142 asbestos bodies/mg dry

Table 15.4 Relationship between asbestos exposure and asbestos bodies (AB) (as identified by electron microscopy) in a lung tissue sample

- Non-exposed white-collar workers without exposure: <0.1 AB/mg
- Non-exposed blue-collar workers without exposure: <0.5 AB/mg
- Non-exposed blue-collar workers with minimal exposure: >0.5 AB/mg and <2 AB/mg
- Patients with pleural plaques have in the order of 1.7 AB/mg
- Patients with mesothelioma have a variable number of AB, ranging from non-exposed to asbestosis levels
- Long-term asbestos workers without asbestosis have up to 1300 AB/mg
- Long-term asbestos workers with category 1 asbestosis have about 8000 AB/mg
- Long-term asbestos workers with category 2 asbestosis have about 73 000 AB/mg

lung. No crocidolite or amosite fibers were identified. Values exceeding these represent a significant exposure (Dufresne *et al.*, 1995) (Table 15.4).

Deposition and clearance of asbestos

For non-fibrous minerals, it is rare that particles exceeding 10 μm in diameter reach the lung. Most particles deposited in the alveoli have a diameter of less than 5 μm. However, these considerations do not apply to fibrous minerals such as asbestos. Fiber deposition in the lung is largely ruled by fiber diameter, with fiber length being less important. Fibers 200–300 μm long can be found in the lung as their diameter is <5 μm. The typical fiber length of asbestos bodies is 20–50 μm, although many fibers <5 μm in length can also be found (Sebastien, 1982).

In animal experiments, inhaled fibers are primarily deposited in the bifurcations of the conducting airways and in the alveolar parenchyma (Rom *et al.*, 1991), but deposition of fibers can be diffuse and extend to subpleural lung tissue. Clearance of fibers occurs by the mucociliary escalator; translocation of fibers into the interstitium and then to the lymphatic system; and by dissolution, degradation, defibrillation and break down within the lung. Short fibers are cleared more readily, as reflected by the increase in the overall mean length of fibers over time. Similarly, chrysotile fibers are cleared more rapidly than amphiboles (Sebastien, 1982; Sebastien and Bégin, 1990). These considerations are relevant to the effects of fiber type, size and length-to-width ratio in the pathogenesis of the asbestos-related diseases.

Pathogenesis: the asbestos effects on tissues

Asbestos fibers (Bégin *et al.*, 1989, 1992; Rom *et al.*, 1991) induce fibrosis and cancers of the lung and pleural space. Fibrosis begins as an inflammatory reaction which evokes a repair process, resulting in persistent scarring. Cancers (Jaurand, 1991) begin as a multi-step process in which the target cell DNA suffers increasing amounts of damage (genetic mutations) through a variety of molecular injuries. Numerous experimental studies in animals and *in vitro* studies of cell cultures have shown that all types of asbestos fibers can produce all of the asbestos-related diseases.

In vitro, asbestos fibers (Figure 15.1) can cause cell damage which, if sufficiently severe, will result in cell death, gene mutation, chromosomal aberration, aneuploidies and malignant cell transformation. These fibers stimulate pulmonary macrophages to produce growth factors for fibroblasts and provoke other cells to take part in the development of disease. The mechanisms of pathogenesis are incompletely understood but the outlines are developed (Churg and Green, 1989; Jaurand, 1991).

Specificities of asbestos fibers: type, dimension and durability

The fiber dose needed to cause disease, the type, dimension, durability and the surface chemical characteristics influence the toxicity, carcinogenicity and fibrogenicity of the fibers to variable degrees. The

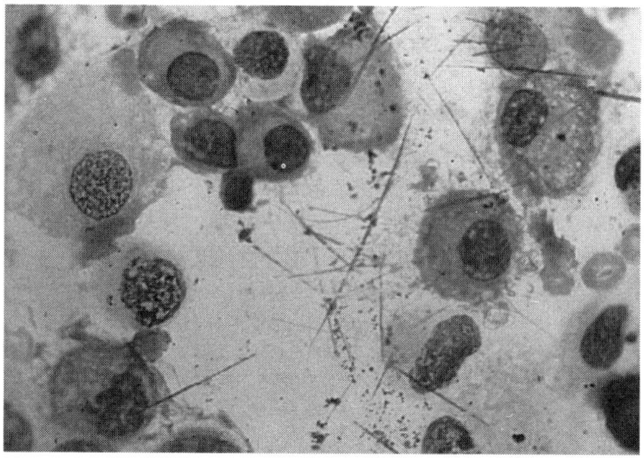

Figure 15.1 Cytospin preparation of a BAL specimen from an asbestos worker showing numerous fibers, many of these of length greater than the average macrophage diameter (8 μm) (Wright-Giemsa; original magnification ×1000).

dose–response is well documented epidemiologically and experimentally for the majority of asbestos-related diseases. *In vitro* and animal studies have shown that fiber toxicity depends on the fact that the material is a fiber, as indicated by the absence of toxicity following an exposure to pulverized asbestos or to non-fibrous analogues of asbestos. The diameter of the fiber is also important. Fibers with diameters >3 μm do not penetrate in the lower lung, but those <3 μm penetrate cell membranes, are translocated to the pulmonary interstitium and pleural space, and can result in asbestos-related disease. The length of the fiber is also important. Fibers <3 μm in length are phagocytosed or translocated to the lymphatics to be drained into the pleural space, whereas those longer than 5 μm are incompletely phagocytosed and continue to reside in the tissue, initiating, and then sustaining, the cascade of cellular and molecular events leading to asbestos-related diseases.

Amphibole fibers (particularly crocidolite) are about 10-fold more carcinogenic than chrysotile for mesothelial cells and are more prone to induce pleural fibrosis. Lung fibrosis, however, is equally affected by both types of asbestos fibers (Dufresne *et al.*, 1995).

Fiber composition can influence toxicity. The relative solubility of chrysotile attenuates its toxicity whereas its 'splittability' (multiplication effect) enhances it toxicity. Finally, surface properties, as well as electrical charge, influence fiber toxicity. Breakdown of the bonds between atoms on the fiber surface leads to an unstable situation which results in active sites for reactions between tissue and fiber. Adsorption of carcinogens such as benzene or polycyclic aromatic hydrocarbons (PAH) enhance the carcinogenicity of the fibers. Similarly, breakdown of chrysotile fibers liberate free radicals with established toxicity for cells.

ASBESTOSIS

Diagnostic criteria

The criteria for diagnosis of asbestosis have been the subject of several task force reports (Task Force on Occupational Respiratory Disease, 1979; America Thoracic Society, 1986; Harber and Smitherman, 1991). Briefly, all recognize that examination of histopathologic material coupled with mineralogic assessment is the most sensitive and specific method of diagnosing asbestosis. In the absence of pathologic material, which is most often the situation, the diagnosis of asbestosis is a medical judgment based on:

- a reliable and significant history of asbestos exposure;
- radiographic changes of diffuse lung fibrosis on chest radiograph or computed tomography (CT) scan;
- the absence of other medical illnesses which can mimic asbestosis:
- an appropriate time interval between exposure and detection (latency);
- a restrictive pattern of lung function;
- bilateral fixed inspiratory crackles; and
- clubbing of fingers and/or toes.

The first three criteria are essential, the others are confirmatory. The extent of profusion of opacities on the chest radiograph needed for the diagnosis of asbestosis has been debated, with ILO grade profusion category of 1/1 or greater being preferred as it increases the specificity of the diagnosis. Other illnesses – particularly idiopathic pulmonary fibrosis or lung fibrosis attributable to connective tissue disease – may be a part of the differential diagnosis. Most often, proving that the fibrosis is attributable to asbestos exposure can be determined by recognizing the extent of dust exposure and the absence of systemic illness.

Pathogenesis of asbestosis

(a) The initial lung injury

Inflammation begins almost immediately after exposure at the alveolar duct bifurcation, where the terminal airways divide into individual alveolar spaces. In animals, after a 1-hour exposure to asbestos, there is active uptake of fibers by Type I epithelial cells. Within 48 hours of such an exposure, alveolar macrophages accumulate at the alveolar duct bifurcations and whereas, one such exposure may not cause asbestosis, chronic exposure will cause this lesion to progress, forming first a localized peribronchiolar fibrosing alveolitis, followed by diffuse scarring. A cascade of events begins whereby activated macrophages flood the site of the injury and stimulate fibroblast proliferation. Increasing the asbestos dose amplifies the cellular response.

(b) Macrophage-derived cytokines regulate the disease process

Fibronectin is a glycoprotein produced by the macrophage which recruits fibroblasts and initiates fibroblast proliferation. Levels of this cytokine are increased significantly in the asbestos-exposed worker (Bégin *et al.*, 1985). Similarly, as a marker of inflammation and early fibrosis, increased procollagen 3 levels indicate collagen production. In asbestos workers without disease, but with asbestos-induced alveolitis, and in those with asbestosis, levels of fibronectin and procollagen 3 in bronchoalveolar lavage (BAL) are significantly elevated. In contrast, excessive levels of these two macrophage-derived substances are not increased after exposure to inert dusts.

In addition to fibronectin, alveolar macrophages recovered from workers with asbestosis release exaggerated quantities of growth factors, including platelet-derived growth factor (PDGF), insulin-like growth factor (IGF-1), and fibroblast growth factor (FGF) (Bégin *et al.*, 1989, 1991; Rom *et al.*, 1991). These cytokines attract fibroblasts to sites of injury and up-regulate fibroblast activity. Fibronectin, PDGF, IGF-1 and FGF interact synergistically in a complex cascade of events to induce fibroblast proliferation and form scar tissue, irreversibly altering the structure of the lung.

Damage to the initially injured area is further aggravated by asbestos-activated macrophages which release free oxygen radicals. These damage tissue and sustain the inflammatory process through direct cytotoxicity and peroxidation of cell membranes (Mossman *et al.*, 1990; Kamp *et al.*, 1992). Additional destructive substances come from macrophage release of plasminogen activator (Cantin *et al.*, 1989), a protease which converts plasminogen to plasmin and degrades interstitial matrix glycoproteins.

(c) Progression and latency of asbestosis

The chronic and progressive inflammation and injury produced by asbestos fibers continues from the time of exposure, through the subclinical phase, to the time when clinical disease is identifiable by the classic methods of chest X-ray and pulmonary function testing, as well as by more investigational approaches including BAL, electron microscopic evaluation, high-resolution CT (HRCT) scan and ^{67}Ga lung uptake (Bégin *et al.*, 1983, 1985, 1989, 1991; Rom *et al.*, 1991).

BAL studies have shown changes at the time of subclinical disease. HRCT scans show abnormalities in the presence of normal radiography in approximately one-third of long-term asbestos workers without a diagnosis of asbestosis. Similarly, ^{67}Ga scans are abnormal in a substantial portion of asbestos-exposed workers without radiographic evidence of asbestosis. Three-fourths of the asbestos-exposed workers who initially had a positive ^{67} Ga scan without clinical asbestosis, progressed to 'full-blown' clinical asbestosis over four years (Bégin *et al.*, 1985). Thus, during the latency period, before the overt clinical manifestations of the disease are detected by routine clinical tests, inflammation and injury is progressive and involves an increasingly greater numbers of macrophages and fibroblasts, with the formation of scar. This may eventually progress to clinically recognizable asbestosis.

(d) Worker susceptibility to asbestos fibers

When exposure to a toxic substance grossly exceeds the permissible exposure level, the disease appears in all exposed subjects, as was the case in the British workshop where Murray (1907) described his first cases of asbestosis. Today, typically only a fraction of the workforce develops an asbestos-related disease, although workers who develop the manifestations of asbestos-related diseases may not be recognized to have been more exposed than their coworkers (Becklake, 1976; Becklake *et al.*, 1983; Bégin *et al.*, 1983, 1985, 1987a). This is recognized as individual susceptibility and has been assessed through immunologic, anatomic and physiologic parameters.

Human immunohistocompatibility studies (Bégin *et al.*, 1987b) have been unable to find an immunologic marker of the HLA (Human leukocyte group A) system associated with individual susceptibility to asbestos-related diseases. However, some characteristics of the upper airways have also been linked to the susceptibility factor (Becklake *et al.*, 1983) and could influence the clearance rate of asbestos fibers. In humans it is not possible to directly measure the clearance rate of inhaled asbestos dust and to relate this to the development of asbestosis. However, several independent observations suggest that the rate of fiber clearance is linked to the risk of disease (Bégin and Sebastien, 1989). First, the lung tissue fiber burden is higher in asbestos-exposed workers than in the general population; second, individuals with a decrease in small airway flow rates attributable to asbestos have twice the fiber burden of asbestos-exposed workers without limitation of airway flow; third, patients with asbestosis have twice the fiber burden of patients with disease limited to the airways; fourth, analyses of BAL fiber content of workers with asbestosis are significantly higher than values found in exposed workers without asbestosis; and finally, those with more severe asbestosis have the highest levels of fiber retention. In this regard, studies in the sheep model have been particularly useful, documenting higher level of fiber retention in animals with asbestosis than in those without the disease after a comparable exposure (Bégin and Sebastien, 1989). The longer fibers were particularly associated with this effect. Thus, the rate of clearance appears to play a critical role in an individual's susceptibility to disease, and once disease is apparent, appears to be affected by the presence of disease.

The risk of developing asbestosis is primarily determined by the level of asbestos fiber exposure. Recent reports suggest a 1% risk for this disease after a cumulative dose of 10 fiber-year/m^3 of air. This finding contributed to lowering of the currently accepted level of exposure (threshold limit value or TLV) to less than 1 fiber/m^3 of air, given the expected 30–40 years of work for most asbestos workers.

(e) Dose–response relationship
Epidemiologic studies in asbestos-exposed individuals have repeatedly documented a higher incidence of asbestosis and lung cancer among workers with greater cumulative dose. This relationship is less valid for mesothelioma (Bégin *et al.*, 1989, 1992a, 1993a).

Pathology of asbestosis

The earliest change of asbestosis is scarring in and around the bronchioles (Bégin *et al.*, 1989) (i.e. peribronchiolar fibrosis)(Figure 15.2). Because of the relative lack of sensitivity of clinical tools, this lesion cannot usually be recognized by routine pulmonary function tests or chest radiograph unless nearly half of the airways are affected. Yet, even then, the specificity of this lesion as being asbestos-induced is in question. It is only when the peribronchiolar process becomes sufficiently diffuse and extends into the interstitium between airways (Figure 15.3) that it can be detectable radiographically, even though there may be only minimal functional impairment, and the clinician can be more certain regarding the diagnosis. Similarly, histologic studies of the lung in mildly symptomatic subjects complaining of dyspnea with normal chest radiographs may have pathologic evidence of fibrosis, but the lesions are not radiographically apparent because they are not sufficiently widespread.

Exposure of the lung to asbestos dust can initiate one of the following reactions (Bégin *et al.*, 1989).

1. A transient inflammatory reaction without a persistent lesion occurs when there is rapid clearance of inhaled fibers. This is a common situation in subjects occasionally exposed or chronically exposed to negligible doses of fibers who will never develop an asbestos-related disease, and in a large portion of long-term asbestos workers who have no detectable changes attributable to their asbestos exposure. This type of transient tissue reaction occurring without a histopathologic lesion has been reproduced in animal models (Bégin *et al.*, 1990).
2. A low-retention reaction occurs in the asbestos-exposed subject who seems to possess a particularly effective clearance mechanisms, and therefore has a low susceptibility to asbestos. The pulmonary response is limited to the site of fiber deposition, i.e. the bifurcation of conducting airways. The presence of fibers at the bifurcation stimulates an influx of macrophages, provokes an inflammatory reaction, and may evolve to fibrosis in the distal airway, but does not result in overt disease.
3. A high-retention reaction occurs in the most susceptible subjects who retain the most fibers. There is a dense accumulation of inflammatory cells in the tissue, an influx of activated macrophages and neutrophils, release of humoral proinflammatory and profibrotic mediators, and the development of fibrosing alveolitis, recognized as asbestosis (Figure 15.3). This intense secretory activity of macrophages and neutrophils at the site of fiber deposition is sustained by the inability to clear the longest fibers and by the multiplicative effect of fibers as the bundles break down into fibrils.

Pathologically, asbestosis is fibrosis associated with asbestos bodies (visualized by light microscopy) or by the recognition of a significant amount of asbestos fibers (by TEM). This definition of asbestosis by a

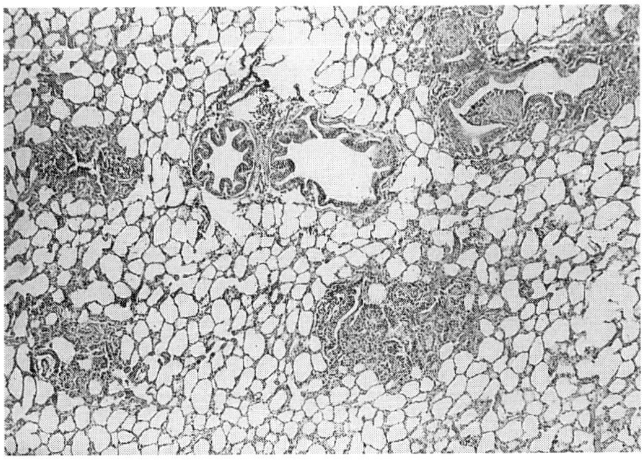

Figure 15.2 Asbestosis Grade 1 seen in a sheep as peribronchiolar inflammation and fibrosis limited to some of the surroundings of some of the peripheral airways. (Hematoxylin–eosin; ×40.)

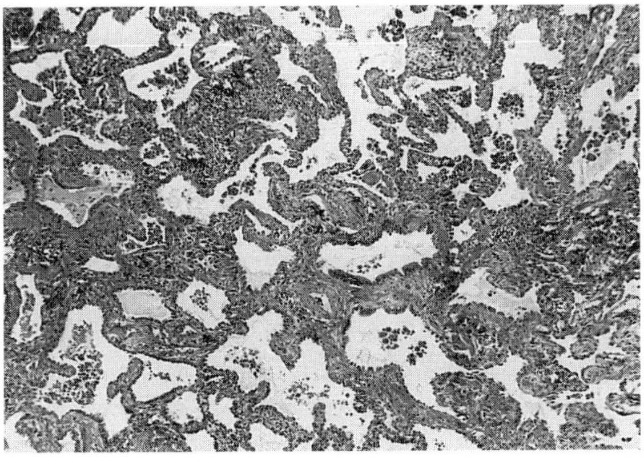

Figure 15.3 Asbestosis Grade 3 is seen in a more fibrotic stage with fibrosis around the peripheral airways and extension of the fibrosis in the interstitium between airways. (Hematoxylin–eosin; ×40.)

Table 15.5 American College of Pathologists scheme for assessing the severity of asbestosis

Grade 1:	Fibrosis involving the wall of at least one respiratory bronchiole with or without extension into the septa of the immediately adjacent layer of alveoli; there is no fibrosis in more distant alveoli. This is the asbestos airway disease of Churg and Green (1989)
Grade 2:	Fibrosis as in Grade 1, plus involvement of alveolar ducts or two or more layers of adjacent alveoli; there still must be a zone of nonfibrotic alveolar septa between adjacent bronchioles (Figure 15.3).
Grade 3:	Fibrosis as in Grade 2, but with coalescence of fibrotic change such that all alveoli between at least two adjacent bronchi have thickened, fibrotic septa; some alveoli may be obliterated completely.
Grade 4:	Fibrosis as in Grade 3, but with formation of new spaces of a size larger than alveoli, ranging up to 1 cm; this lesion has been termed honeycombing. Spaces may or may not be lined by epithelium.

The pathologic description also includes three extension grades based on the proportion of respiratory bronchioles involved by the disease process. These are:

Grade A:	Only occasional bronchioles are involved; most show no lesion.
Grade B:	More than occasional involvement is seen, but less than half of all bronchioles are involved.
Grade C:	More than half of all bronchioles are involved.

committee of the College of American Pathologists recognizes the severity and the extent of the fibrotic process (Craighead *et al.*, 1982; Churg and Green, 1989) (Table 15.5). This approach allows for the quantification of disease in any given specimen so that correlation studies with other parameters of disease severity, such as the radiograph and lung function tests, can be performed.

Macroscopically, the gross features depend on the severity of the disease. Early on, the visceral pleura loses its transparent appearance and the parenchyma becomes fibrotic in the interlobar and interlobular septa. With time, this extends further into the lung tissue. In more advanced disease, the pleural surface becomes nodular, similar to liver cirrhosis, and the lung parenchyma is characterized by volume loss, scar and cyst formations, typically most prominent in the lower zones.

Clinical findings in asbestosis

The symptoms and signs are like those of other interstitial lung diseases, i.e. dyspnea, dry cough and non-specific chest discomfort. Dyspnea on exertion is the usual symptom of presentation, which worsens as the disease progresses and lung function declines. Non-productive cough and chest pain are present in some cases, and occur late in the disease. When the cough is productive, it should be considered due to bronchitis or pneumonia. Chest tightness and chest discomfort can be attributed to muscle pain appearing only when the cough and the dyspnea become severe. Hemoptysis is not an usual complaint and should be investigated as it suggests lung carcinoma. All of these symptoms may occur during the working years or begin after exposure has ceased, becoming clinically apparent only after retirement.

Fine, crisp, persistent crackles are the most important physical finding. They are an indicator that disease is present and occasionally precede changes in lung volumes and airflow parameters. In most cases, however, their appearance coincides with detectable changes in lung function and abnormalities on the chest radiograph. They are first heard in the lower lateral lung fields in the late-inspiratory phase and, as the disease progresses, increase in number to become audible in the lower posterior lung fields in the mid- and late phases of inspiration. In severe disease, these crackles have a crisper, higher tone and are audible throughout the entire inspiratory phase. Other adventitious sounds are usually absent.

Finger clubbing may be present but does not necessarily relate to asbestosis severity. Cyanosis and reduced chest expansion are late manifestations of the disease.

Early on, lung function tests show a restrictive change associated with mild end-expiratory airflow limitation, as reflected by a decline in the maximal mid-expiratory flow rate. This is a non-specific finding and is also a frequent feature of early airways obstruction, most commonly attributable to cigarette smoking. Restriction progresses as the disease worsens, but airflow obstruction does not worsen as increasing lung rigidity enhances the peripheral airway airflow conductance.

Once recognized, asbestosis can be a chronic and progressive disease. The evolution of asbestosis has changed significantly since the original observation by Murray (1907) who reported 10 workers with asbestosis who died before age 30. Viallat *et al.* (1983) studied Corsican asbestos workers when almost 80% of the workers had no radiographic evidence of disease. Later, despite no further exposure, only 34% had no radiographic evidence of disease. Other studies have also documented disease progression after exposure has ceased. Currently, with improved industrial hygiene, the long latency period, and early disease recognition at time of a minimally abnormal chest radiograph (ILO category 1 pneumoconiosis), most cases are recognized after age 50 and fewer than 20% of cases so recognized progress.

Radiology

Radiologic tools used in the diagnosis of asbestos-related disorders are the posteroanterior (PA), lateral and oblique chest radiographs, and CT scans of the chest in conventional and high-resolution (HRCT) modes.

The PA and lateral chest radiographs provide basic radiologic information. On the PA radiograph, asbestosis is usually manifested as diffuse reticulonodular infiltrates at the lung bases (Becklake, 1976; Bégin *et al.*, 1993a) (Figure 15.4). Early on, a fine increase in the thickening of the lung markings occurs in the lower two thirds of the lung; later, the markings become more clearly reticular, noticeably in the lung periphery, and then extend into the other lung fields. The irregular markings of asbestosis do not have the clarity of the silicotic nodule and sometimes suggest vascular congestion due to left heart failure. In the advanced stage of asbestosis, honeycombing with coarse and more exaggerated interstitial markings appears and is associated with tissue destruction and distortion. The radiographic appearance of the parenchyma is non-specific and can be recognized in other lung diseases, most notably idiopathic pulmonary fibrosis and fibrosis attributable to collagen vascular diseases. When the lung changes are accompanied by pleural manifestations of asbestos exposure, the radiologic diagnosis of asbestosis can be made with more certainty. Oblique films can facilitate recognition of pleural-based abnormalities.

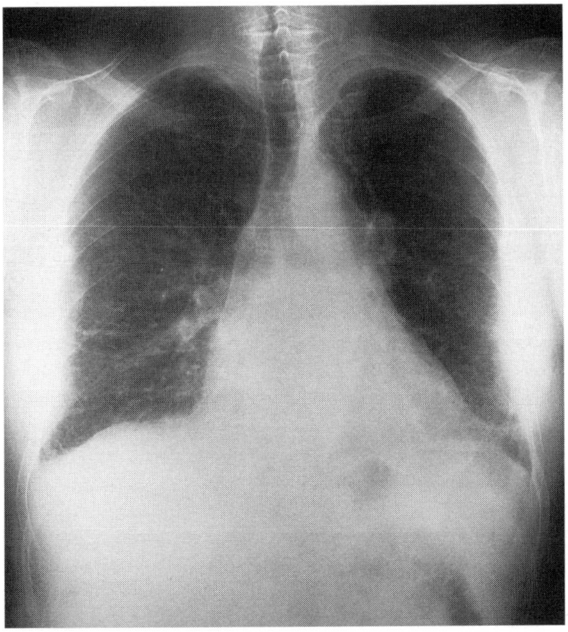

Figure 15.4 Classic radiograph of an asbestos worker with asbestosis and bilateral diffuse pleural thickening and a calcified pleural plaque on the right diaphragm.

The early, small opacities, the exaggerated septal lines, the coarse or fine linear opacities, the ground-glass appearance and honeycombing are part of the spectrum of changes seen in asbestosis. Lung size may become smaller, but only when advanced disease is apparent. Progressive massive fibrosis can occur, but it happens much less frequently than in the 'nodular' pneumoconioses, and it appears in the lower zone. In view of the risk of malignancy with asbestosis and the infrequency of progressive massive fibrosis, focal masses should be investigated as malignancies until proven otherwise.

The International Labor Office (ILO) is an organization that has promoted an internationally standardized guideline on the classification of chest radiographs for pneumoconiosis (ILO, 1980). The goal of the classification is to code the abnormalities in a simple and reproducible matter. The ILO provides standard reference chest radiographs which are used in the classification of the radiographs.

The use of the CT scan is not justified in the periodic examination of asbestos-exposed workers, but it has a place in the clinical investigation of these individuals. It can better define pleural-based abnormalities. When lung cancer or mesothelioma is considered, the CT scan better defines the anatomic extension of disease.

Compared with the conventional CT scan, HRCT scan has dramatically improved our assessment of the interstitium and allows us to better appreciate interstitial and emphysematous changes (Staples *et al.*, 1989; Bégin *et al.*, 1992a, 1993b). Although the pulmonary abnormalities seen on HRCT are non-specific and cannot be attributed to asbestosis without characteristic pleural plaques and/or a history of significant asbestos exposure, the HRCT scan in asbestosis characteristically shows bilateral:

- thickening of the interlobar and interlobular septa in the lung periphery;
- fibrotic bands extending from the pleural surface into the parenchyma (Figure 15.5);
- honeycombing with small, thick-walled cystic zones of destroyed lung, mostly located in posterior and non-dependent areas (Figure 15.6);

- curvilinear lines in non-dependent areas. These are parallel to but approximately 1 cm distant from the pleural surface. These opacities are non-specific and may be explained by fibrotic changes which precede honeycombing, atelectasis caused by local mechanical changes associated with adjacent pleural abnormalities, or to an inflammatory component of the disease process; and
- non-dependent subpleural densities which are non-specific indicators of interstitial lung disease (Figure 15.7).

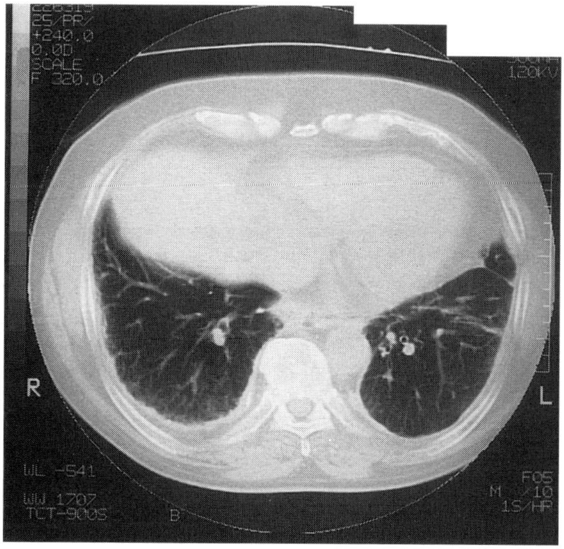

Figure 15.5 CT scan of an asbestos worker with asbestosis. Note the subpleural densities and a few bands.

Early detection of asbestosis

Diffuse fibrosing alveolitis due to asbestos fiber inhalation, the lesion recognizable on the chest radiograph, is the overt manifestation of an inflammatory process typically requiring decades of exposure. Diagnosing the illness early may alter the outcome by allowing the physician to advise the worker to cease exposure. In order to use the available clinical tools appropriately, the clinician must know the sensitivity and specificity of the tool.

(a) Crackles

End-inspiratory crackles in the axillary lung fields in an asbestos-exposed worker are an early indicator of asbestosis. This information is easily obtained and correlates well with the radiographic findings and lung function changes. In asbestos workers with asbestosis, crackles are usually present; however, in exposed workers not yet diagnosed to have asbestosis, these adventitious sounds constitute the only abnormality in less than 5% of the cases (Bégin *et al.*, 1985).

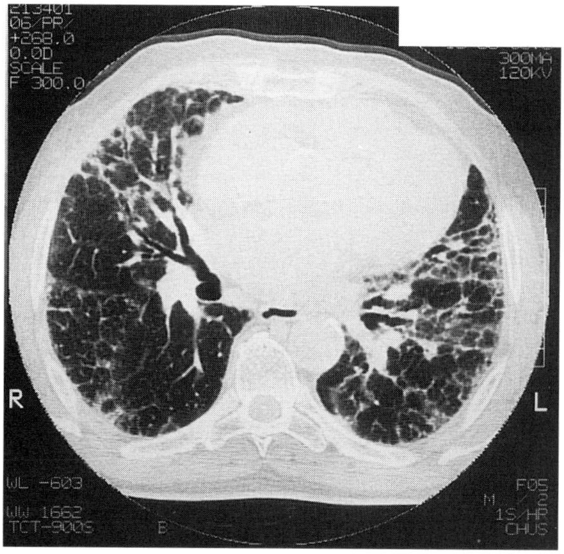

Figure 15.6 CT scan of an asbestos worker with asbestosis, showing advanced interstitial reticulation with honeycombing.

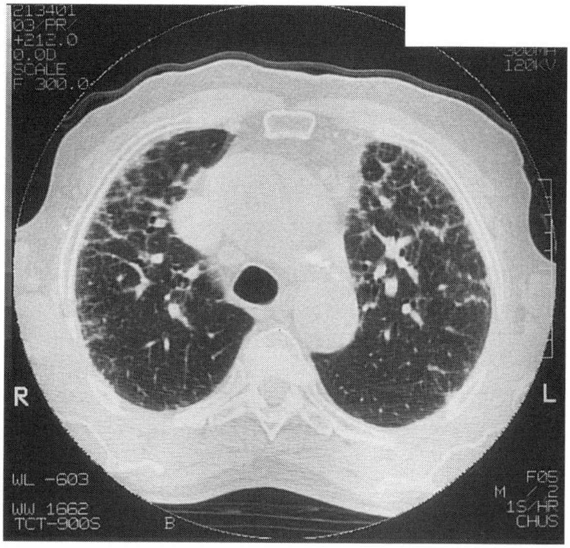

Figure 15.7 CT scan of an asbestos worker with asbestosis and bilateral minimal pleural thickening. Parenchymal changes of reticular nature are well appreciated.

(b) Chest radiograph
The standard PA chest radiograph is the main tool in the detection of asbestos-related diseases. It is specific for interstitial changes, but not always sensitive. In a fraction of symptomatic subjects with biopsy-proven interstitial lung disease, the chest radiograph is normal. In asbestos-exposed workers, similar observations have been reported. Furthermore, it is not uncommon to find pathologic changes of asbestosis in workers where asbestosis was not recognized on clinical grounds.

(c) Chest CT scan and ^{67}Ga scan
Investigation of workers at risk for asbestosis with the conventional CT scan does not show more abnormalities than the standard chest radiograph, but permits a better appreciation of the pleura. With the HRCT scan, detection of parenchymal abnormalities is increased. The HRCT scan can detect early interstitial lung disease in 10–20% of symptomatic asbestos workers with normal chest radiographs (Staples *et al.*, 1989; Bégin *et al.*, 1993b). The ^{67}Ga lung scan has been used for over 20 years to detect occult tumors and infections and, more recently, to quantify lung inflammation. In asbestosis, the ^{67}Ga lung scan permits the detection of inflammation which has been correlated pathologically (Bégin *et al.*, 1983, 1985). Although providing an index of inflammation, the ^{67}Ga scan has been replaced by the HRCT scan because the etiology of the measured inflammation is non-specific, the test requires injection with a considerable dose of radioactive nucleotides, and it does not provide sophisticated anatomic information seen on the HRCT scan.

(d) Bronchoalveolar lavage
In the study of diffuse interstitial lung diseases, BAL has provided information on ways that effector cells and humoral agents mediate the disease process. In asbestos-exposed workers, BAL is of interest because it can document the specific dust exposure and provide information on fiber burden; aid in determining whether infectious processes such as tuberculosis, or fungal illnesses, are a part of the active process; support other clinical data suggestive of alveolitis; and study the biologic mechanisms of the disease. Unfortunately, BAL is not an accurate method of diagnosing asbestosis or of predicting disease progression.

(e) Pulmonary function tests
Abnormalities in lung function, such as the diminution of the lung volume and the diffusion capacity, appear at about same time as changes in the chest radiograph (Becklake, 1976; Bégin *et al.*, 1983, 1985; American Thoracic Society, 1986). Identification of a decline in small airway flow rates is a sensitive but non-specific measurement of early indicator of asbestosis (Bégin *et al.*, 1983).

Medical intervention

Although life expectancy after the diagnosis remains less than normal, only 20% will progress. There is now experimental evidence to support that ceasing exposure reduces the rate of progression (Bégin *et al.*, 1991; Sebastien *et al.*, 1994). Once diagnosed, the worker must avoid further exposure to asbestos dust.

Although the diagnosis of asbestosis is most often made on the basis of the chest radiograph, the degree of functional impairment should be most important when it comes to structuring appropriate levels of compensation. Serial radiographs may identify progression and the need to reassess the compensation.

At the time of diagnosis, the worker with asbestosis will require an assessment of the extent of impairment. Follow-up may require treatment of intercurrent respiratory infections and serial chest radiographs may be necessary to address symptoms which may signal the onset of chest malignancy, or to identify the need for treatment of hypoxemia and right heart failure, late complications of this disease. Symptomatic relief of cough and immunizations to lessen the likelihood of influenza or pneumococcal infections are indicated. Corticosteroid or immunosuppressive therapy do not favorably alter the natural history of asbestosis. In fact, the latter treatment worsens experimental asbestosis. End-stage respiratory failure remains an occasional cause of death.

OTHER ASBESTOS-RELATED DISEASES

Benign pleurisy

(a) Pathogenesis and pathology
How benign asbestos pleurisy develops is largely unknown (Hillerdal, 1980; Schwartz, 1991). Inhaled asbestos fibers are cleared in part through the lung lymphatics into the interstitium, the pleural cavities, and lymph nodes (Rey, 1990). Direct contact between asbestos fibers and cells in the pleural space appears to be an early event and the release of chemotactic factors for neutrophils and other inflammatory cells has been stressed as a means of initiating and perpetuating inflammation (Rey, 1990). The exudative non-specific inflammatory reaction of benign pleurisy is characterized by the cellular and humoral characteristics of the pleural fluid and, rarely, by biopsy. Within the pleural space, an inflammatory reaction with fibrin deposition is accompanied by mesothelial cells, giving

the impression of pseudo-organization (Churg and Green, 1989). The presence of exudative fluid in the pleural space is most often transient, but very often scars and results in pleural thickening of the costo-diaphragmatic and costophrenic angles, and occasionally an adhesive fibrothorax. Although this process affects the pleural space, it is not a precursor of mesothelioma.

(b) Clinical findings

The criteria for the diagnosis of benign pleurisy are somewhat arbitrary, but practical. They include (Epler *et al.*, 1982):

- asbestos exposurc;
- radiographic or thoracentesis confirmation of effusion;
- absence of another explanation for the effusion; and
- absence of tumor during a follow-up period of at least 3 years.

Most with asbestos-induced benign pleurisy are asymptomatic, with the pleural effusion recognized at the time of an incidental chest radiograph. About one-third have symptoms of shortness of breath and/or chest pain, with or without fever. After diagnosis, approximately one-fourth of the pleural effusions recur. The fluid is exudative, it is often bloody, but it is usually less than 500 ml. It may occur with or without asbestosis, and can be unilateral or bilateral. The features on physical examination are localized to the area of inflammation and include dullness to percussion and a rub upon auscultation. Lung function tests show a restrictive pattern proportional to the severity of the disease (effusion + pain), but only in rare instances is the worker sufficiently affected to show a clinically important decrease in lung function. At the time of thoracentesis a pleural biopsy may be needed in order to exclude other causes of pleural effusion.

The incidence of benign pleurisy in the asbestos-exposed population approximates 3%, with a dose–response relationship. The doses need not be excessive, as white-collar workers in the asbestos industries may be affected. According to Epler *et al.* (1982), benign pleurisy is the most frequent asbestos-related disorders in the 20 years following initial asbestos exposure and usually the only one to be seen within 10 years of the first exposure to asbestos. Yet, the average latency time can be longer, and at times overlap that of mesothelioma, leading to the recognition that pleural effusions in those with significant asbestos exposure must be fully investigated to exclude malignancy. The diagnosis of benign pleurisy then becomes a diagnosis of exclusion.

(c) Radiology

Radiographically, the effusion presents as blunting of the costophrenic angles. The amount of fluid is usually small, it is free within the chest, and it is transient, as opposed to the progressive nature of a malignant tumor. HRCT can help to exclude mesothelioma, a process which presents with pleural effusion and nodularities. Benign pleurisy typically resolves leaving a thickened pleura at the affected costophrenic angle, appearing as diffuse pleural thickening and/or rounded atelectasis.

(d) Medical intervention

Benign pleurisy may cause the worker to take temporary medical leave. Permanent compensation is not necessary as, in most instances, the process leaves a minimal scar and has no relation to whether other asbestos-related diseases develop. Treatment is relief of symptoms. Follow-up for at least 3 years is reasonable to rule out a malignancy affecting the pleural space.

Pleural plaques

Pleural plaques are focal, irregular thickenings adjacent to the exterior aspect of the parietal pleura. They are, in reality, extrapleural (Becklake, 1976; Bégin *et al.*, 1993a). These plaques can also form adjacent to the visceral pleura, uncommonly in the interlobar fissures. They are the most frequent manifestation of asbestos exposure and, if bilateral, an accurate indicator of asbestos exposure. Latency time for the production of plaques is on the average 30 years, but has been reported to be as short as 3 years and as long as 57 years. The development and progression of pleural plaques appears to be in direct relation to amphibole exposure, since chrysotile has a limited tendency to produce plaques.

(a) Pathogenesis

Two pathogenic theories for the development of pleural plaques have been suggested (Hillerdal, 1980). The first, the direct contact theory, is the most plausible. Short and thin fibers are found in the pleural space soon after inhalation exposure, with the most likely means of transport via the lymphatics with probable additional contributions of gravity and respiratory motion (Parkes, 1973). The thin and pointed nature of the fibers in the pleural space facilitate tissue penetration, scratching of the pleural tissue, and initiation of an inflammatory reaction, followed by organization, all leading to the pleural plaque.

The second, or indirect effect, is in part based on the observation of Epler *et al.* (1982) who reported 44 subjects who developed pleural plaques following transient pleural effusions. Churg and Green (1989) have also described the occasional transformation of a chronic inflammatory and cellular pleural reaction into an acellular plaque. Yet, exudative pleural effusions attributable to asbestos exposure are thought more likely to result in diffuse pleural thickening of the costophrenic angle and lower one-third of the lateral chest wall pleura.

(b) Pathology

Pleural plaques consist of hyaline fibrosis in the submesothelial layers of the parietal pleura at the level of the costal margins, the diaphragms, and the paraspinal areas. They can be found in the pericardium and, less often, in other mediastinal pleura. Visceral pleural plaques occur much less frequently, but can extend in the interlobar fissures. Microscopically, plaques are composed of layers of virtually acellular collagen, the surface being covered with a thin layer of mesothelial cells. Calcifications are of dystrophic nature and may be present in plaques which have persisted for a number of years. Generally, neither asbestos bodies nor asbestos fibers are found in the plaques.

(c) Clinical findings

Workers with pleural plaques are without symptoms or signs of disease. The plaques are usually located on both hemithoraces, although to variable extent. The plaques in the retrocardiac area are not easily seen on the chest radiograph. Plaques typically have limited progression and are often found in the absence of asbestosis. Although not specific for asbestos disease, the appearance of bilateral pleural plaques is so often seen in asbestos-exposed workers to be reasonably attributed to asbestos exposure.

(d) Radiology

Plaques rarely extend more than four intercostal spaces. About one-third contain some calcification, not necessarily seen on standard radiography. Calcification affects mainly the parietal plaques and is characteristically located along the diaphragm and the posterolateral chest wall. Calcifications also can be found in paravertebral areas, an area poorly visualized on the PA film.

Plaques begin as thin, linear, raised opacities with sharp margins; early detection depends on the thickness of the plaques and optimal radiographic technique. Plaques can have a smooth or uneven surface. With time, the margins become more rounded and better defined. Oblique films of the chest increase the radiographic visibility of plaques by 50% over the standard PA and lateral chest films.

The conventional CT scan recognizes plaques much earlier and at a less well-defined stage than the chest radiograph. The paravertebral and pericardiac plaques are particularly better seen. The diaphragm plaques which were not always well appreciated with the conventional CT scan are better evaluated with the HRCT. CT scans can clearly differentiate plaques from extrapleural fat pads, a sometimes difficult distinction on the plain chest radiograph. Furthermore, in the presence of extensive and calcified pleural plaques, the CT scan permits a clearer appreciation of the lung parenchyma than the plain radiograph.

(e) Significance

Pleural plaques in asbestos workers are only markers of asbestos exposure. In the absence of diffuse pleural thickening, pleural plaques do not significantly affect lung function. Pleural plaques can enlarge and/or calcify and become confluent. They do not become malignant or transform into mesothelioma, although sometimes mesothelioma can be found at the site of pleural plaques.

Diffuse pleural thickening

(a) Pathogenesis and pathology

Diffuse pleural thickening is a disease of the visceral pleura. It is not specific for asbestos exposure and can be associated with a residual inflammatory reaction from tuberculous infection, surgery, hemorrhagic chest trauma or drug reaction. The mechanisms differentiating asbestos-induced diffuse pleural thickening from circumscribed pleural plaques are not defined, but an irritative mechanism of pleural tissue is likely and may be caused by fibers deposited in the parenchymal subpleural areas, reflecting the association of diffuse pleural thickening with asbestosis (Hillerdal, 1980; Rey, 1990). The mechanisms which cause diffuse pleural thickening are unknown, but this appears to be the residual from exudative benign pleurisy in a large percentage of cases. In other cases, where there is no history of a previous episode of benign pleurisy, authors have speculated that confluence of pleural plaques or extension of subpleural parenchymal fibrosis into the visceral pleura has occurred to cause diffuse thickening. In the most aggressive cases, diffuse pleural thickening can extend into the interlobular spaces of the lung and affect the fissures, leading to restriction of lung expansion, even in the absence of interstitial lung fibrosis (Hillerdal, 1980).

(b) Clinical findings

Diffuse pleural thickening may be responsible for dyspnea on exertion and perhaps dry cough associated with a significant restriction of lung function due to a 'trapped lung'. Because of the relative thinness of the pleura even when pleural thickening is present, pleural thickening cannot typically be detected by physical examination, although dullness may be present if this thickening is extensive. Decreased respiratory excursion may be noted if the lung is 'trapped' and moves little on the affected side. Diffuse pleural thickening can be limited to one side or involve both sides.

(c) Radiology

On the PA radiograph, diffuse pleural thickening appears as a continuous, smooth pleural opacity extending more than 25% of the pleural surface with blunting of the costophrenic angle. Bilateral diffuse pleural thickening may be the best predictor for asbestos exposure as an etiology for this process. It is typically made on the basis of the chest radiographic features and rarely requires histologic confirmation unless a diagnosis of malignant pleural disease is considered.

By CT scan, diffuse pleural thickening is defined as pleural thickening >5 cm wide, >8 cm long and >3 mm thick. Because this primarily affects the visceral pleura in the posterior and posterolateral lower zones, the CT scan provides a better visualization than the standard chest radiograph. Diffuse pleural thickening can be complicated by extension of fibrosis in the interlobar and interlobular fissures (Figure 15.8) to form the crow's foot and rounded atelectasis (Figure 15.9).

(d) Significance

Asbestos-related diffuse pleural thickening may be associated with interstitial fibrosis. Because these pleural changes can mimic a pleural-based malignant disease, workers should have serial radiologic investigations and medical follow-up as needed. Early diagnosis may allow for a better understanding for a decline in lung function. Diffuse pleural thickening may restrict lung function modestly if limited, but may cause respiratory insufficiency and failure in some cases if it is extensive and bilateral. The evaluation of functional impairment may lead to consideration of compensation for loss of lung function.

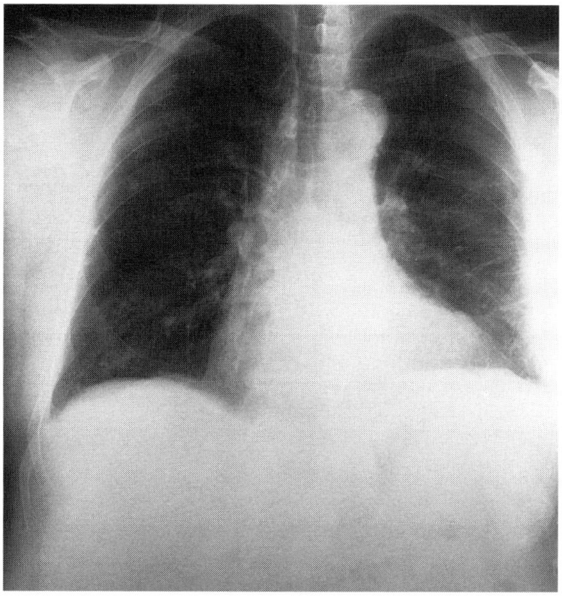

Figure 15.8 Chest radiograph of an asbestos worker showing left side diffuse pleural thickening, with extension of the fibrosis in the interlobular fissures to form a 'crow's foot'.

Rounded atelectasis

(a) Pathogenesis and pathology

Rounded atelectasis is an asbestos-induced effect on the peripheral lung and pleura caused by focal scarring, with retraction of the scar and partial collapse of adjacent lung. It is relatively uncommon and the mechanisms causing this pleuropulmonary process are not known beyond those causing the pleural disease. This condition has been more often recognized with the advent of the CT scan.

(b) Clinical findings

Rounded atelectasis is usually asymptomatic and may be detected on the chest radiograph as a pleural-based opacity which is suspected of being a malignancy. Occasionally, localized chest pain may be the presenting symptom. In the past, these pseudotumors were often resected, but with the CT scan, the true nature of the rounded atelectasis can be recognized without surgery.

(c) Radiology

Extension of visceral pleura fibrosis into the interlobar and interlobular fissures typically occurs years after ceasing exposure and, through retraction of scarred tissue, cause a torsion of the adjacent lung, and give rise to 'rounded atelectasis', 'folded lung', 'Bleskovsky's syndrome' (Figure 15.9), or 'pseudotumor'. This mimics carcinoma or a confluence of opacities sometimes seen in advanced pneumoconiosis. The characteristic 'comet's tail', curling of the bronchovascular markings toward the center of the mass from the hilum, can be better appreciated on an oblique chest radiograph or best seen on the CT scan.

The diagnosis of rounded atelectasis can be made on the chest radiograph. Radiologic criteria include:

- a pleural-based rounded lesion of 2–7 cm in diameter;
- curvilinear shadows extending from the hilum (comet's tail);
- an intrapulmonary location (acute angle between pleura and lesion) with pleural thickening adjacent to lesion;

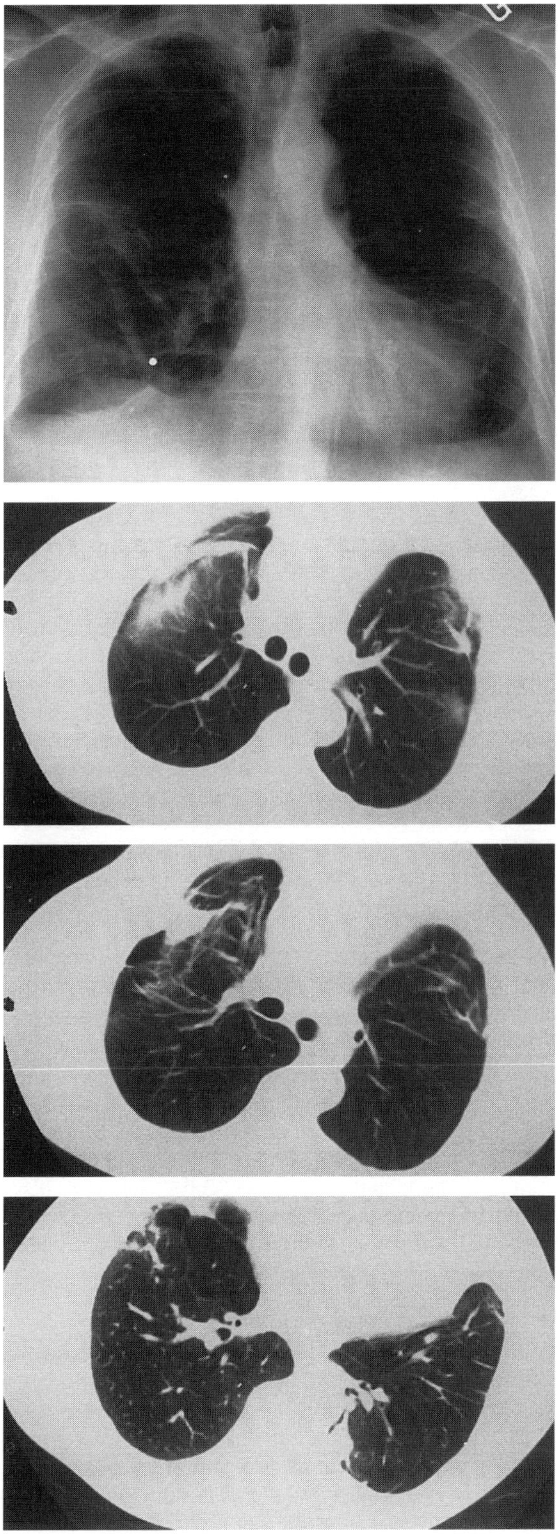

- thickening of the interlobar fissure; and
- a slowly progressive process (as opposed to malignant lesions).

(d) Significance

Rounded atelectasis usually occurs long after diffuse pleural thickening is first noted. Recognition of this entity is often a clinical problem that necessitates use of CT scan, bronchoscopy and transthoracic needle aspiration biopsy to rule out a malignant process.

Lung carcinoma

(a) Pathogenesis and pathology

Our understanding of the way that asbestos induces lung cancer is incomplete (Antman and Aisner, 1987). Some consider asbestos to be a tumor promoter, rather than an initiator, on the basis of its weak activity in standard tests addressing a carcinogen's potential (i.e. its mutagenicity and ability to produce chromosomal abnormalities). Based on this information, asbestos fibers increase the susceptibility of the lung to other carcinogens. Others consider asbestos to be a tumor initiator in view of the excess number of lung cancers in animals not exposed to carcinogens other than asbestos and the excess of lung malignancies in lifetime non-smokers. A third point of debate reflects whether asbestosis is necessary for the genesis of lung carcinoma. This position is supported by epidemiologic studies which have recorded an excess number of lung cancers only in the most heavily exposed workers with the highest incidence of radiographic opacities consistent with asbestosis. However, opposite views contend that bronchogenic carcinoma are found in lifetime non-smoking asbestos workers and that these lesions can be produced in animals in the absence of fibrosis.

Our understanding of asbestos-related lung cancers in the absence of asbestosis is further complicated by the multiplicative risk of smoking. The 10-fold increased risk of lung cancer in smokers appears to be increased to 50-fold in smokers with asbestos exposure. This risk estimate is largely a function of the amount of tobacco smoked and the amount of asbestos exposure. The fiber type appears irrelevant, although it has been suggested that crocidolite fiber exposure may represent a higher risk. Other risk factors in the workplace which can increase the risk of lung cancer in

Figure 15.9 Upper panel: chest radiograph of an asbestos worker showing bilateral diffuse pleural thickening, with extension of fibrosis in the interlobar and interlobular fissures to form rounded atelectasis. Lower panel: findings are more clearly appreciated on CT images.

asbestos-exposed workers include contaminant metals, ionizing radiation and chemicals such as benzopyrene and polycyclic aromatic hydrocarbons.

Asbestos-related cancers are not distinct in type, nature or their location within the lung from those solely associated with cigarette smoking. The histologic distribution of asbestos-related cancers approximates 35% epidermoid, 25% small cell, 30% adenocarcinoma and 10% large cell carcinoma, a distribution similar to the cell types of cancer in those who smoke without asbestos exposure.

(b) Clinical significance

Asbestos-related lung tumors present in the same manner as lung tumors caused by other carcinogens, except when the symptoms of asbestosis are present. In presence of asbestosis, it is universally accepted that lung cancers should be compensated. In the absence of asbestosis, compensation has been debated, some refusing compensation for all cases, others favoring compensation for cases of long exposure (about in excess of 20 years). A detailed discussion of this point is beyond the scope of this review and the interested reader is referred to text by Churg and Green (1989).

Malignant mesothelioma

(a) Pathogenesis and pathology

In the absence of asbestos exposure, malignant mesothelioma is a rare disease with an incidence of one to two cases per year per million people in the general population of North America. In the asbestos-exposed populations, the rates can be 5- to 20-fold higher, with a gradient of 10 to 1 for crocidolite to chrysotile exposures (Bégin *et al.*, 1992b).

The clear association of asbestos with mesothelioma was initially established by Wagner and colleagues (1960, 1986) and has been confirmed repeatedly. After gathering careful occupational and environmental exposure histories and undertaking sophisticated mineralogic analysis of lung tissue from those with mesothelioma, most cases are found in subjects exposed to asbestos fibers (Parkes, 1973; Preger, 1978; Selikoff and Lee, 1978). In animals, asbestos acts as an initiator and a promoter of mesothelioma, and is considered a complete carcinogen. In women, where the incidence is two- to 10-fold less common than in men, tissue analysis for mineral fibers revealed that 98% of 117 cases in the United Kingdom had lung amphibole counts greater than controls (Dawson *et al.*, 1993) In general, the geometric properties (a greater diameter, more rigidity) of the amphiboles favor deeper penetration and the resistance of these fibers to degradation favor their persistence in lung tissues, leading this fiber type to be most often implicated as the cause of mesothelioma. Yet, cases associated with chrysotile exposure alone have also been recorded. In some instances, the nearly equal number of the different types of fibers in the lung makes it impossible to establish a causal relationship for either fiber type. It is universally accepted that mesothelioma has no association with cigarette smoke exposure.

In animal studies (Davis, 1974), the development of mesothelioma begins soon after fiber injection into the pleural space. Granulomatous lesions form with subsequent deposition of acellular dense collagen. These lesions coalesce and are surrounded by a layer of mesenchymal cells and a surface of normal mesothelium. Later, neoplastic transformation with extensive surface growth occurs, showing differentiation into mesenchymal and epithelial cells. Experimentally, all asbestos fiber types have produced mesothelioma.

(b) Pathology

Early on, mesothelioma appears as multiple small grayish nodules on the visceral and parietal pleura. These coalesce and form larger tumor masses, accompanied by a pleural effusion. This malignancy spreads by direct extension into adjacent structures of the chest wall, interlobar fissure, lung parenchyma, mediastinum, pericardium, diaphragm, esophagus, large vessels of the mediastinum, contralateral pleura and the peritoneal cavity. Death is usually caused by restriction of one or more of these vital structures. Malignant mesothelioma is of peritoneal origin in less than 25% cases. In the abdomen, the gross appearance of the tumor is quite similar with direct invasion of the abdominal structures.

(c) Histopathology

Malignant mesotheliomas are histologically classified as epithelial in approximately 50% of the cases, sarcomatoid in 16% of the cases, and mixed in 34% cases. The epithelial type expresses one or more growth patterns such as tubular, papillar, microcystic and solid, which are often mixed within the tumor and vary from area to area. The sarcomatoid type has spindle-shaped and elongated cells; nuclear pleomorphism and mitoses are common. In the mixed form, epithelial and sarcomatous forms are present in variable portions.

(d) Histochemistry

It is often difficult to separate mesothelioma from adenocarcinoma, a primary pulmonary malignancy which may grow around the outside of the lung and mimic the pattern of mesothelioma growth. Considerable expertise by an interested pathologist and immunochemistry laboratory workers is sometimes necessary to be certain about the final diagnosis. Malignant mesothelioma elaborate acid mucous substances rich in hyaluronic acid. These provide a histochemical profile distinct from that of adenocarcinoma. Malignant mesothelioma are usually Periodic acid–Schiff–diastase (PAS–D) and mucicarmine-negative and hyaluronidase–alcian blue-positive. Adenocarcinoma are PAS–D and mucicarmine-positive, but hyaluronidase–alcian blue-negative.

The carcinoembrionic antigen (CEA) is negative in 90% cases of malignant mesothelioma (and when positive it is very weakly so) and positive in 90% cases of adenocarcinoma. Cytokeratin immunochemical staining is usually positive and the pattern of staining is distinct from that of adenocarcinoma, being more diffuse in mesothelioma and exhibiting a predominantly peripheral location in the adenocarcinoma. Vimentin is usually negative in both type of tumors, but may be positive in 20% cases of mesothelioma.

(e) Clinical findings

The chief complaint is usually a dull, persistent chest pain, occasionally worsened by inspiration, and requiring progressively higher doses of analgesics. Cough and dyspnea are frequent and attributed to the pleural effusion. Weight loss, fever and general malaise are present at least in the late phase of this disease. Shortness of breath progresses as the tumor extends within the chest, encases the lung and infiltrates the mediastinal structures. The symptoms of osteoarthropathy are uncommon. Regardless of the most aggressive therapy, all with malignant mesothelioma die within 24 months of diagnosis.

Mesothelioma can present in the pleural, pericardial or abdominal cavities, the latter being more often seen with high amphibole exposure. The reason for these preferences is unknown. One of the peculiarities of mesothelioma is the apparent 20–50-year latency period, which is unexplained but not unique, as other cancers also have similar dormancy period after exposure (e.g. radiation-induced cancers).

(f) Radiology

The chest radiograph shows a dense pleural abnormality appearing as a diffusely circumscribed thickening of the pleura with an irregular nodular surface (Figure 15.10). These features are attributed to an pleural effusion superimposed over multiple pleural nodules. As the disease progresses, the radiograph shows further involvement of the lung parenchyma, reduction in the size of the involved hemithorax, mediastinal and/or hilar invasion, pericardial thickening and/or effusion, abdominal extension, and chest wall invasion. A great many of these changes can be appreciated on the plain chest radiograph, but the CT scan may add precision and clarity to the observations.

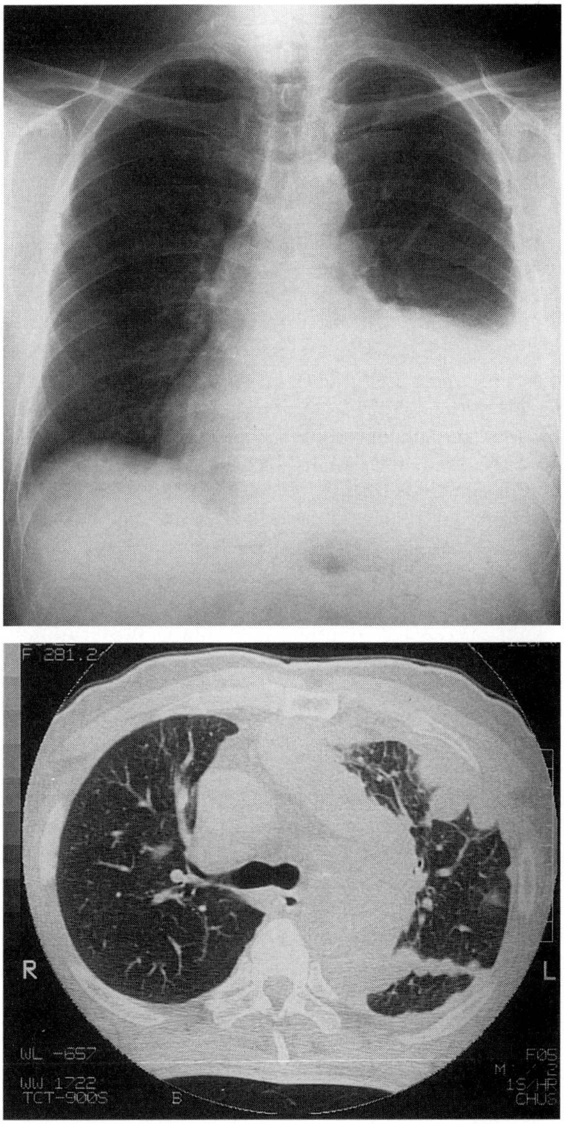

Figure 15.10 Upper panel: chest radiograph of an asbestos-exposed worker. The left lung field shows obliteration of the costophrenic angle suggestive of effusion. Lower panel: CT scan of patient shows diffuse irregular thickenings with invasion of the fissure in a 'rosary-like appearance', highly suggestive of mesothelioma, which was later biopsy-proven.

(g) Significance

There is no method for the early recognition of mesothelioma. Most patients with mesothelioma cite chest pain as the reason that they consult a physician. An open thoracotomy with pleural biopsy or thoracoscopic biopsy are usually needed to provide adequate amounts of tissue to establish a certain diagnosis.

Surgery, radiotherapy, chemotherapy and, of late, immunotherapy or cytokine therapy – or a combination of the above – have been equally unsuccessful in curing this disease. At best, extension of a few months of survival after the initial diagnosis can be demonstrated in selected studies (Jaurand *et al.*, 1993).

PREVENTION OF ASBESTOS-RELATED DISEASES

Public health practices protecting workers against job-related health risks are legislated in each country. These laws establish the standards of permissible exposure levels and health surveillance of the workers by periodical medical examinations. Interested readers are referred to a detailed ILO publication (International Labor Bureau, 1974).

Industrial hygiene is based on traditional techniques of reduction of dust exposure by enclosure of the sources of dust, adequate ventilation, coating, sealing of products, use of specialized tools, and frequent cleaning of the workplace. In some particularly dusty circumstances, individual protection is recommended. Hygienists use the threshold limit values (TLV) for specific dusts as standards to regulate the exposures in the workplace. Recognizing the difference in fiber fibrogenicity and carcinogenicity, the TLV for asbestos is currently at 1 fiber/ml of air for chrysotile and 0.2 fiber/ml for amphiboles in most industrialized countries.

Despite the most sophisticated monitoring program, excessive asbestos exposures can occur. Perhaps the best protection for the worker is education and a firm resolve by both the worker and the management to minimize exposures. This may be achieved by engineering controls, but often requires protective clothing and respiratory protection.

DEFINITIONS

Among the diseases considered to be associated with asbestos exposure, some have been proven to be definitely associated, whereas the relationship between asbestos exposure and the other illnesses remains uncertain. Listed are diseases unequivocally recognized to be attributable to asbestos fiber inhalation:

- **Asbestosis**: interstitial fibrosis of the lung parenchyma;
- **Benign asbestos pleurisy** (benign pleural effusion): an exudative and transient inflammation of the pleura;
- **Pleural plaques**: an accumulation of collagen fibers which form avascular, acellular and circumscribed hyaline masses usually limited to the parietal pleura;
- **Diffuse pleural thickening**: the result of an accumulation of collagen fibers which form avascular, acellular and diffuse hyaline masses affecting both the parietal and visceral pleura. This may occasionally invade the interlobular spaces of the lung and be recognized as 'crow's feet'. This is not to be confused with asbestosis.
- **Rounded atelectasis**: the result of asbestos-induced pleural disease caused by scarring of the pleura and adjacent lung tissue with retraction of the scarred lung and partial collapse of adjacent lung tissue.
- **Bronchogenic carcinoma**: a malignant bronchopulmonary tumor, no different from those attributed to exposures to cigarette smoke and other lung carcinogens.
- **Mesothelioma**: a malignant tumor of the mesothelial cells of the pleura. It is a rapidly growing tumor which is usually fatal 12–24 months after clinical diagnosis.

Illnesses which have not been unequivocally accepted as attributable to asbestos exposure are not discussed and readers are referred to an exhaustive review (Antman and Aisner, 1987). Of these illnesses where the attributability to asbestos is in question, perhaps the most hotly debated is laryngeal carcinoma. The association between laryngeal cancer and asbestos exposure remains controversial, with opposite conclusions regarding attributability formed by independent reviewers.

REFERENCES

American Thoracic Society (1986) The diagnosis of nonmalignant diseases related to asbestos. *Am. Rev. Respir. Dis.*, **134**, 363–8.

Anonymous, (1982) Proceedings of the World Symposium on Asbestos. Canadian asbestos information centre, Montreal, Canada.

Anonymous (1991) *Health Effects Institute-Asbestos Research: Asbestos in public and commercial Buildings: a literature review and synthesis of current knowledge* (ed. A.C. Upton, chairman of the review panel), HEI-AR, Cambridge, MA.

Antman, K. and Aisner, J. (1987) *Asbestos Related Malignancy*, Grune & Stratton, New York.

Auribault, M. (1906) Note sur l'hygiene et la securite des ouvriers dans les filatures et tissages d'amiante. *Bull. Insp. Trav. Paris*, **14**, 120.

Becklake, M.R. (1976) Asbestos-related diseases of the lung and other organs: their epidemiology and implications for clinical practice. *Am. Rev. Respir. Dis.*, **144**, 187–227.

Becklake, M.R., Toyota, B., Stewart, M. *et al.* (1983) Lung structure as a risk factor in adverse pulmonary response to asbestos. *Am. Rev. Respir. Dis.*, **128**, 385–8.

Bégin, R. and Sebastien, P. (1989) Alveolar dust clearance capacity as determinant of individual susceptibility to asbestosis: experimental observations. *Ann. Occup. Hyg.*, **33**, 279–82.

Bégin, R., Boileau, R. and Peloquin, S. (1987a) Asbestos exposure, cigarette smoking, and airflow limitation in long-term Canadian chrysotile miners and millers. *Am. J. Ind. Med.*, **11**, 55–66.

Bégin, R., Menard, H., Decarie, F. and St-Sauveur, A. (1987b) Immunogenetic factors as determinants of asbestosis. *Lung*, **165**, 159–63.

Bégin, R., Cantin, A. and Masse, S. (1989) Recent advances in the clinical assessment and pathogenesis of the mineral dust pneumoconiosis. *Eur. Respir. J.*, **2**, 988–1001.

Bégin, R., Cantin, A. and Masse, S. (1991) Influence of continued asbestos exposure on the outcome of asbestosis in sheep. *Exp. Lung Res.*, **17**, 971–84.

Bégin, R., Cantin, A. and Sebastien, P. (1990) Chrysotile asbestos exposure can produce an alveolitis with limited fibrosing activity in a subset of susceptible sheep. *Eur. Respir. J.*, **3**, 81–90.

Bégin, R., Cantin, A., Berthiaume, Y. *et al.* (1985) Clinical features to stage alveolitis in asbestos workers. *Am. J. Ind. Med.*, **8**, 521–36.

Bégin, R., Cantin, A., Drapeau, G. *et al.* (1983) Pulmonary uptake of Gallium-67 in asbestos exposed humans and sheep. *Am. Rev. Respir. Dis.*, **127**, 623–30.

Bégin, R., Ostiguy, R., Fillion, R. and Groleau, S. (1992a) Recent advances in the early diagnosis of asbestosis. *Semin. Roentgenol.*, **27**, 121–39.

Bégin, R., Gauthier, J.J., Desmeules, M. and Ostiguy, G. (1992b) Work related mesothelioma in Quebec 1967–1990. *Am. J. Ind. Med.*, **22**, 531–42.

Bégin, R., Masse, S., Groleau, S. and Sebastien, P. (1993a) Asbestose et atteintes pleurales de l'asbeste. *Encyclopedie Medico-Chirurgicale*, **6–039–10**, 1–22.

Bégin, R., Ostiguy, G., Filion, R., Colman, N. and Bertrand P. (1993b) Computed tomography in the early detection of asbestosis. *Br. J. Ind. Med.*, **50**, 689–98.

Bignon, J., Sebastien, P. and Gaudichet, A. (1979) Measurement of asbestos retention in human respiratory system related to health effects, in *Proceedings of the Workshop on Asbestos, Gaitherburg, Maryland.* Special publication #506, National Bureau of Standards, Washington, DC, pp. 95–119.

Cantin, A., Allard, C. and Bégin, R. (1989) Increased alveolar plasminogen activator in early asbestosis. *Am. Rev. Respir. Dis.*, **139**, 604–9.

Churg, A. and Green, F.H.Y. (1989) *Pathology of Occupational Lung Disease.* 1st edn, IGAKU-SHOIN, New York.

Craighead, J.E., Abraham, J.L., Churg, A. *et al.* (1982) The pathology of asbestos-associated diseases of the lung and pleural cavities: diagnostic criteria and proposed grading scheme. *Arch. Pathol. Lab. Invest.*, **106**, 544–95.

Davis, J.M. (1974) Histogenesis and fine structure of peritoneal tumors produced in animals by injection of asbestos. *J. Natl. Cancer Inst.*, **52**, 1823–33.

Dawson, A., Gibbs, A.R. and Pooley, F.D. (1993) Malignant mesothelioma in women. *Thorax*, **48**, 269–74.

Dufresne, A., Bégin, R., Churg, A. and Masse, S. (1996) Mineral fibre content of lungs in mesothelioma cases seeking compensation in Quebec. *Am. J. Respir. Crit. Care Med.*, **153**, 711–18.

Epler, G.R., McLoud, T.C. and Gaensler, E.A. (1982) Prevalence and incidence of benign asbestos pleural effusion in a working population. *JAMA*, **247**, 617–22.

Harber, P. and Smitherman, J (1991) Asbestosis: diagnostic dilution. *J. Occup. Med.*, **33**, 786–93.

Hillerdal, G. (1980) The pathogenesis of pleural plaques and pulmonary asbestosis: possibilities and impossibilities. *Eur. J. Respir. Dis.*, **61**, 129–38.

International Labor Bureau (1974) *Asbestos: Health Risk and Prevention*, Publication # 30 du ILB, International Labor Bureau, CH-1211 Geneve 22, Suisse 16.

International Labor Office/University of Cincinnati (1980) *International classification of radiographs of pneumoconiosis*, 1980, no. 22 (revised), Occupational Safety and Health Series, International Labor Office, Geneva.

Jacob, G. and Bohlig, H. (1959) Lungenkrebserwartung der deutschen asbestarbeiter. *Krebsforresch Krebsbekampf*, **3**, 130.

Jaurand, M.C. (1991) Observations on the carcinogenicity of asbestos fibers. *Ann. N.Y. Acad. Sci.*, **643**, 258–70.

Jaurand, M.C., Bignon, J. and Brochard, P. (eds) (1993) International conference on: 'Mesothelial cell and mesothelioma Past, present and future'. *Eur. Respir. Rev.*, **3**, 1–237.

Kamp, D.W., Graceppa, P., Pryor, W.A. and Weitzman, S.A. (1992) The role of free radicals in asbestos-induced diseases. *Free Radical Biol. Med.*, **12**, 293–315.

Landrigan, P.J. and Kazemi, H, (eds) (1991) The third wave of asbestos disease: exposure to asbestos in place. Public Health Control. *Ann. N.Y. Acad. Sci.*, **643**, 1–628.

Marchand, F. (1906) Uber eigentuemliche pigment kristalle in den lungen. *Ver. Deutsch. Gesell.*, **17**, 223.

Mossman, B.T., Marsh, J.P., Sesko, A. *et al.* (1990) Inhibition of lung injury, inflammation, and interstitial pulmonary fibrosis by polyethylene-conjugated catalase in a rapid inhalation model of asbestosis. *Am. Rev. Respir. Dis.*, **141**, 1266–71.

Murray, H. (1907) Report of the Commission on compensation of industrial disease (1907) Her Majesty's Stationary Office, London, **3496**, 127–8.

Parkes, W.R. (1973) Asbestos related disorders. *Br. J. Dis. Chest*, **67**, 261–300.

Preger, L. (ed.) (1978) *Asbestos Related Disease*, Grune & Stratton, New York.

Rey, F. (1990) La migration pleurale des fibres d'amiante. A propos d'une etude experimentale chez le rat. Memoire du diplome d'etudes et de recherche en biologie humaine sous la direction de C. Boutin, Faculte de medecine de Marseille, France.

Rom, W.N., Travis, W.D. and Brody, A.R. (1991) Cellular and molecular basis of the asbestos-related diseases. *Am. Rev. Respir. Dis.*, **143**, 408–22.

Schwartz, D.A. (1991) New developments in asbestos induced pleural disease. *Chest*, **99**, 191–8.

Sebastien, P. (1982) *La biometrologie des fibres inhalees*, PhD Thesis, Paris XII University, France.

Sebastien, P. and Begin, R. (1990) Mass, number and size of fibres in the pathogenesis of asbestosis in sheep. *Br. J. Exp. Path.*, **71**, 1–10.

Sebastien, P., Dufresne, A., Masse, S. and Begin, R. (1994) Asbestos fibres lung retention and the outcome of asbestosis with or without exposure cessation. *Ann Occup. Hyg.*, **18**(suppl. 1),672–82.

Selikoff, I.J. and Churg, J. (eds) (1965) Biological effects of asbestos. *Ann. N.Y. Acad. Sci.*, **132**, 1–76.

Selikoff, I.J. and Hammond, E.C. (eds) (1979) Health hazards of asbestos exposure. *Ann. N.Y. Acad. Sci.*, **330**, 1–814.

Selikoff, I.J. and Lee, D.H.K. (eds) (1978) *Asbestos and Disease*, Academic Press New York.

Staples, C.A., Gamsu, G., Ray, C.S. and Webb, W.R. (1989) High resolution computed tomography and lung function in asbestos-exposed workers with normal chest radiograph. *Am. Rev. Respir. Dis.*, **139**, 1502–8.

Task Force on Occupational Respiratory Disease (G. Ostiguy, chairman) (1979) Health and Welfare Canada. Canadian Government Publication, Ottawa: pp. 35–48.

Viallat, J.R., Boutin, C., Pietri, J.F. and Fondarai, J. (1983) Late progression of radiographic changes in Canari chrysotile mine and mill exworkers. *Arch. Environ. Health*, **38**, 54–8.

Wagner, J.C. (ed.) (1980) *Biological effects of mineral fibres*, WHO IARC Scientific Publication # 30, Lyons, France.

Wagner, J.C. (ed.) (1986) The biological effects of chrysotile, in *Accomplishment in Oncology*, vol. 1, #2, JB Lippincott Co., Philadelphia, pp. 1–162.

Wagner, J.C., Sleggs, C.A. and Marchand, P. (1960) Diffuse pleural mesothelioma and asbestos exposure in North Wertern Cape Province. *Br. J. Ind. Med.*, **17**, 260–71.

Zussman, J. (1979) The mineralogy of asbestos, in *Asbestos: Properties, Applications, and Hazards* (eds L. Michaels and S.S. Chissick), John Wiley & Sons, Chichester, UK, pp. 45–65.

16 Changing patterns of asbestos-related pulmonary diseases

Colin R. Woolf

The pattern of asbestos-related pulmonary diseases has not changed but new information about these conditions has become available and altered our perspectives regarding these illnesses. This information has been used to make changes in allowable asbestos exposure, to improve our ability to diagnose these illnesses, and to determine the best approach for the evaluation of respiratory impairment and long term follow-up. It is these changing patterns which will be considered.

When appraising potentially toxic materials, Woolf (1991–92) emphasized that the recognition of changing patterns has been part of a pendulum effect. First, there is an intense denial of any health danger. This is followed by a reaction and a major revision in thinking, the pendulum swings to the opposite side, and the material is damned even when there is exposure to the smallest quantities. Finally, as scientific evidence accumulates, good sense prevails and the pendulum comes to a midpoint. The relationship between exposure to asbestos and disease has certainly gone through the first two stages and the reasoned third stage is becoming a reality.

CHANGES IN EXPOSURE, RECOGNITION AND APPROACH

Exposure to asbestos has occurred not only in miners, but also in workers employed in the manufacture of asbestos products, particularly cement-based products, cloth and brake and clutch linings. There has also been asbestos exposure when this material was sprayed on boilers in ships and on the steel beams of buildings. Initially, these workers took no precautions to avoid inhalation of fibers, but later, ineffective simple gauze face masks were used. About 20 years ago, truly effective respirators were issued and dust control by vacuum cleaning and exhaust ventilation was initiated. There is now little danger of significant exposure in industries where asbestos is mined or used when respiratory protection is used correctly.

Sporadic cases of pulmonary fibrosis, as a possible result of exposure to asbestos, were reported in the early 1900s. It was only in 1930, after the report of Wood and Gloyne (1930), that asbestosis was accepted as a serious disease, yet for many years the warnings were not heeded. The latent period of about 20 years allows control measures, instituted about 20 years ago, to be effective and new cases are now seen only rarely as a result of industrial exposure. It is likely that asbestosis will become an 'extinct disease' after workers who presently have this condition die.

Although urban dwellers usually have asbestos fibers in their lungs from atmospheric pollution, the amount is unlikely to be sufficient to cause asbestosis. However, Rudd (1990) has reviewed some non-industrial unusual situations. Endemic pleural mesothelioma occurs in central Turkey, where there have been deaths in middle aged and even in teen-aged residents.

Occupational Lung Disease. Edited by Daniel E. Banks and John E. Parker. Published in 1998 by Chapman & Hall. ISBN 0 412 73630 6

The agent causing the problem was fibrous zeolite and possibly other asbestos materials such as tremolite. These fibers were present in quarried volcanic material and exposure also occurred in the construction of buildings. Mesothelioma has also been reported in Cyprus and Greece where there is naturally occurring asbestos such as tremolite. Pleural plaques have also been reported, probably due to environmental exposure to geologic deposits of tremolite in the Kabul area of Afghanistan (Voisin *et al.*, 1994) and in north-east Corsica (Rey *et al.*, 1993).

Mesothelioma was related to asbestos exposure in a report from South Africa by Wagner and his co-workers (1960). In most places in the world, this inevitably fatal disease has become the most feared outcome of asbestos exposure. Although mesothelioma remains an uncommon condition, the increase in risk which can occur with relatively low exposures and a latency period of up to 40 years means that the incidence of mesothelioma will increase, as control of asbestos exposure became effective only in the past 20 years. There is difficulty in deciding the amount of exposure which might result in a mesothelioma. There is a dose–response effect (although it is much less straightforward compared with the relationship between exposure and asbestosis or lung cancer) although the dose might occur over a short time (even 2 years) if the intensity of the exposure is high (Morgan, 1991–92). There is also the risk associated with long-term exposure where the intensity of exposure is light but a significant accumulated dosage occurs. This has led unjustifiably to concern about occupants of buildings, particularly children in schools, where asbestos has been used in the construction. The exposure is so slight that the chances of developing asbestosis or lung cancer is infinitesimal. The risk of developing mesothelioma is also negligible, especially as chrysotile was virtually the only type of asbestos used in most buildings. Morgan (1991–92) writes that there is a much greater health risk to school children from high-school football, and certainly from starting cigarette smoking.

Morgan (1991–92) has emphasized that legal proceedings against companies that have used asbestos material in buildings has become a major industry particularly benefiting lawyers in the United States. There have been settlements involving millions of dollars without evidence that definite harm has been done to a worker's health. In Canada the situation is different. A recent court ruling (as reported in the *Toronto Globe & Mail* newspaper on September 29, 1995) affirmed that there was no proof that the building in question, where an asbestos product had been used to fireproof the skeleton of the building, posed a health hazard to the occupants. The judgment noted that a product should be judged by the safety standards in effect at the time (in this case during the 1970s), and the past should not be judged by the standards of the present.

Churg (1993) wrote that exposure of the general population to the very low levels of chrysotile found in some public buildings (levels not greatly different from the ambient air) will never produce mesothelioma, asbestosis or lung cancer because these diseases all require quite high-level chrysotile exposure. Gaensler (1992) emphasized that the assumption of a linear dose–response down to very low levels of exposure with no threshold, is unproven. He strongly criticizes some of the legislative and regulatory mandates, promotional activities of abatement companies, adverse court decisions placing the onus of repairs on asbestos manufacturers, and a 'pandemic of media-genic disease'. Provided that asbestos in buildings is well covered, there seems no significant risk. If asbestos-containing materials are exposed and are flaking then there may be risk and these exposed materials should be well covered or removed. Removal of asbestos-containing material is associated with risk and the workers must be fully protected by positive-pressure respirators and special clothing, which is then decontaminated.

The association of lung cancer and exposure to asbestos has also been subject to the pendulum effect. Doll, in 1955, was the first to report such a relationship, which caused considerable concern. However, a better understanding of the relationship between asbestos exposure and cigarette smoking in causing lung cancer has become increasingly evident. The risk of lung cancer in an asbestos-exposed 'never-smoker' is now considered not great, yet approximates five times that of non-smokers (Rudd, 1990), although Morgan (1991–92) considers an increased risk in the absence of smoking to be doubtful. Rudd (1990) has presented evidence that, in the presence of smoking, additional exposure to asbestos increases the risk for lung cancer to more that 50 times that of non-smokers. A case has been reported of simultaneous development of a mesothelioma and adenocarcinoma of the lung (Cagle *et al.*, 1993). This association is of such rarity that it is considered unique and suggests that the mechanism by which asbestos fibers induce lung cancer is different from the way in which they induce mesothelioma.

There has also been controversy regarding whether radiographic evidence of asbestosis is necessary to attribute lung cancer to asbestos exposure. Churg (1993) wrote that the only association of asbestos exposure and lung cancer is the association between asbestosis and lung cancer. Thus, lung cancer should only be attributed to asbestos exposure where asbestosis is present on clinical or pathologic grounds.

Notwithstanding this opinion, some Workers' Compensation Boards allow compensation if there is lung cancer and a history of asbestos exposure, even if the chest radiograph shows no evidence of asbestosis.

Discrete pleural plaques have been a prominent feature on the radiographs of individuals with asbestos-related disease. Plaques were originally considered important in causing shortness of breath, but with more detailed pulmonary function testing this has now come into question and it is accepted that isolated plaques seldom cause restriction in lung function or respiratory symptoms. Yet, an asbestos-exposed worker may come to the physician's attention because of pleural pain and a pleural effusion be recognized. The concern may be for a malignant pleural process, but this is typically found to be a benign exudative pleural effusion which gradually undergoes spontaneous resolution. Not infrequently, the result of this pleural inflammatory process is residual diffuse extensive pleural thickening involving both the lateral walls of the chest and the costophrenic angles. This may cause a restrictive ventilatory defect and dyspnea.

In the absence of asbestosis, Weiss (1993) concluded that persons with asbestos-related pleural plaques do not have an increased risk of lung cancer.

DIAGNOSIS

Usually, the diagnosis of asbestos-related pulmonary diseases is based on a history of exposure to asbestos, clinical findings, a chest radiograph compatible with asbestos-related health effects, and the absence of other illness(es) which may be considered to be responsible for this overall picture. However, the amount of exposure or even the exposure itself may be questionable and the interstitial changes are not diagnostic of asbestosis. The radiographic appearance of asbestosis is identical to that of cryptogenic fibrosing alveolitis. New considerations in distinguishing these two illnesses have included bronchoalveolar lavage (BAL) and newer methods of examining biopsy material.

In BAL washings, a few asbestos (ferruginous) bodies may be found in individuals not known to be exposed to asbestos. However, the presence of many such bodies strongly suggests that the lung condition being investigated is due to asbestos, although this finding is not absolutely diagnostic. It is now possible to be more precise, as previously uncounted fibers may be counted using electron microscopy techniques.

Transbronchial needle biopsy does not provide sufficient material and even a trephine biopsy is only marginally sufficient for diagnostic purposes. For more certainty, open-lung biopsy is usually required. Churg (1993) is of the opinion that cases without asbestos bodies visible in routine or iron-stained tissues sections almost never turn out to be asbestosis. He emphasized that workers with asbestos exposure also develop all of the interstitial lung diseases to which the remainder of the population is subject. Others believe that the histology alone is not diagnostic and it is necessary to perform a fiber count on digested tissue. The number of uncoated fibers (per gram of dry lung tissue) is considered of particular importance. Fibers may be counted using light microscopy and phase contrast microscopy, but electron microscopy yields a higher fiber count as it also counts uncoated fibers.

There is now emphasis on the appearance of the fibers. There has been a tendency to give chrysotile a 'good press' and it has been considered less dangerous than other fibers. However, it does produce significant amounts of disease and must continue to be of concern. Sturm *et al.* (1995) recorded that the former German Democratic Republic imported 1.4 million tons of asbestos from 1960 to 1990, most of this chrysotile. During this period there were 843 mesotheliomas, 600 cases of asbestosis, and 2700 workers with pleural disease attributed to asbestos exposure. Amphibole fibers, most notably crocidolite and amosite, are more dangerous in terms of their potential to cause adverse health effects.

Analysis of a biopsy specimen becomes even more difficult because of the possibility that chrysotile has leeched out of the lung over time. Therefore, a pulmonary condition may be attributable to asbestos yet the analysis of the tissue biopsy shows little asbestos. However, chrysotile is often contaminated by an amphibole, tremolite – perhaps the most commonly recognized contaminant – and this will remain in the lung and be detected on analysis of the lung biopsy. It is unlikely that, in the absence of asbestos bodies and uncoated fibers, pulmonary fibrosis is attributable to asbestos.

Data reviewed by Wylie *et al.* (1993) suggest that the specific width of fibers is important. Fibers with widths greater than 1 μm are not implicated in the occurrence of lung cancer or mesothelioma. At the other end of the scale, fibers thinner than a few tenths of a micrometer must be abundant in order for them to be a causation of mesothelioma. The most dangerous fibers have a width between these two extremes. Fiber definition used in public policy should include this new information related to specific fiber width.

An open-lung or pleural biopsy may occasionally be required to distinguish a unilateral pleural plaque associated with pleural effusion and chest pain from mesothelioma or secondary adenocarcinoma. Histologically, the plaque shows hyalinized fibrous tissue, and calcification is common. This is easily distinguished from a malignancy. The challenge is to distinguish mesothelioma from adenocarcinoma of the lung, a difficult task using usual histologic techniques. Immunohistochemical methods are now more specific. The presence or absence of intracellular mucin (using the periodic acid–Schiff technique) is not reliable in differentiating these two conditions. There is now emphasis on mesothelioma showing a negative carcinoembryonic antigen (CEA) and positive human milk fat globule, 1 and 2, surface epithelial antigens. Adenocarcinoma is positive for both these tests.

When are computed tomographic (CT) scans indicated in asbestos-related pulmonary diseases? They are of particular value in the early diagnosis of asbestosis. Staples (1992) described curvilinear subpleural lines of variable length in high-resolution (slices 1–2 mm) CT scans. These represent peribronchiolar collapse of alveoli due to fibrosis and occur in the dependent parts of the lungs. These may be present on high-resolution CT scan when the conventional film and even the conventional CT (slices of 8–10 mm) scan are within normal limits.

If there is extensive pleural disease, this conceals the appearance of the underlying lung. A CT scan is useful to show if there is additional pulmonary fibrosis. The CT scan may show if emphysema is present in addition to pulmonary fibrosis.

Gevernois and co-workers (1994) emphasized the greater value of high-resolution CT compared with conventional CT in detecting both plaques and fibrosis. A CT scan may show the extent of pleural thickening better than an ordinary radiograph. The use of conventional radiographs, even with oblique views, often results in false positives for pleural fibrosis due to significant subpleural fat (Ameille *et al.*, 1993). The presence of pleural plaques does not predict the presence or absence of asbestosis, even when high-resolution CT is used (Ren *et al.*, 1991).

EVALUATION OF RESPIRATORY IMPAIRMENT, AND COMPENSATION

A complaint of dyspnea and the recognition of an abnormal chest radiograph has been sufficient, in the past, to award compensation for asbestos-related pulmonary diseases. The availability of elegant pulmonary function testing has provided an additional measure to determine respiratory impairment. Now, some Compensation Boards take the view that compensation should be made for respiratory impairment alone and not for an abnormal radiograph. The American Medical Association produces a book of *Guidelines to the Evaluation of Permanent Impairment* (now in its 4th edition) and describes how to assess respiratory impairment. Impairment ratings are in ranges and no distinction is made between a restrictive or obstructive ventilatory defect. Woolf *et al.* (1992) developed specific percentage impairment ratings which allows for recommendations to be easily made and takes into account the type of lung function abnormalities (restrictive or obstructive). These guidelines are shown in Table 16.1, which has been slightly modified from the original report. Use of this table may produce determinations of respiratory impairment that are more consistent and rational than has otherwise been possible. They are best described to be of an interim nature and should be replaced by others as better techniques for assessing respiratory impairment become available.

Even after 25 years, controversy continues to exist about the role of obstruction in the pattern of pulmonary function abnormality in asbestosis. Kilburn and Warshaw (1995) claim that small airways obstruction is a major physiologic defect in asbestosis. Jones *et al.* (1995) reviewed such results and concluded that the data proved the opposite, namely that there is no significant obstruction in small airways in asbestosis. Miller (1993) suggests that there is an interaction between the effects of smoking and asbestos fibers, and that there is a greater frequency of obstructive airway disease in asbestos workers who smoke than in other smokers. In medicine, when a situation remains controversial over 25 years or more, the subject of the controversy is probably unimportant for the clinical aspect of the condition under consideration.

Based on work by Nakadate (1995), there appears to be no adverse effect of continued low-level exposure to chrysotile in the yearly decline in FEV_1. However, mild pulmonary fibrosis, even when radiographically stable, is associated with an accelerated decline of this index. Pulmonary fibrosis acts as an independent risk factor in the deterioration of pulmonary function.

LONG-TERM FOLLOW-UP

Physicians who are responsible for the care of workers with asbestos-related pulmonary disease may wish to arrange long-term follow-up as part of good medical care. A second important reason for follow-up is to detect changes that might result in a revised assessment of the degree of respiratory impairment. If worsening of pulmonary function is noted at a follow-up visit, then a change in impairment assessment should be made in keeping with the recommendations in Table 16.1. This may result in an increase in compensation benefits.

Follow-up visits increase the cost of care of these workers. In the past, follow-up was often done at a predetermined time, typically yearly, but should be

Table 16.1 Pulmonary function guidelines to permanent respiratory impairment

Impairment (%)	*RE/OB*	*FVC**	*FEV_1**	*FEV_1/ FVC†*	*TLC**	*$D_L$co**
0	No abnormality	>80	>80	>70	>80	>70
10	RE	70–79	70–79		70–79	60–69
	OB		65–79	65–69		
20	RE	65–69	65–69		65–69	
	OB		55–64	60–64		55–59
30	RE	60–64	60–64		60–64	
	OB		45–54	55–59		50–54
45	RE	50–59	50–59		55–59	
	OB		35–44	45–54		35–49
70	RE	40–49	40–49		50–54	
	OB		25–34	30–44		20–34
100	RE	<40	<40		<50	
	OB		<25	<30		<20

RE, restrictive ventilatory pattern.
OB, obstructive ventilatory pattern.

* Expressed as % of predicted; † = FEV_1 as % of forced vital capacity (FVC).

Ventilatory tests and lung volumes normal predictions are based on reference values by Crapo, R.O. *et al.* (1981) *Am. Rev. Resp. Dis.*, **123**, 659–64.

Diffusing capacity normal predictions include relation to alveolar volume and are based on reference values by Gaensler, E.A. and Smith, A.A. (1973) *Chest*, **63**, 136–45.

rather at rational intervals based on the natural history of the disease. Unpublished data by Woolf show that pleural plaques and asbestosis do not show inevitable progression either on the radiograph or on pulmonary function tests. Similar observations have been made by Jones *et al.* (1989). Even in asbestosis, only 12–13% of cases show progression. With the decrease in exposure to asbestos in the past 20 years, progression can be expected to become uncommon. If there is no progression, follow-up is unnecessary.

How long does it take to determine whether progression is on-going? Workers who have pleural plaques and/or asbestosis and no radiographic change and no significant pulmonary function changes over 4 years need have no other regular follow-up. Further evaluations might be done at the request of the worker if there is a change in respiratory symptoms. If there has been radiographic progression or a significant worsening of pulmonary function, then follow-up at 2-year intervals is reasonable, as changes in one year are unlikely. If there is concern about the possibility of developing lung cancer or mesothelioma then review might be done by the family physician as required, especially if new symptoms such as change in cough, sputum production or chest pain occur.

TREATMENT

There is no effective treatment for asbestosis other than prevention by ending exposure to asbestos. The use of lung lavage to remove asbestos fibers has not been attempted, although removal of silica-containing dust is being explored in the treatment of silicosis. Pleural plaques need no treatment. Diffuse pleural thickening has occasionally required decortication. Lung cancer associated with asbestosis is treated in the same way as lung cancer alone, except for a need for further consideration of the amount of pulmonary function that will be available after lung resection or irradiation.

There are some aspects of the treatment of workers with asbestosis which have not been fully considered. Although it is not possible to alter the underlying condition, it is possible to improve shortness of breath using respiratory rehabilitation methods. These techniques have been described in numerous publications including a book by Woolf (1986) where the approaches are described in detail and an analysis of outcome is determined in 1000 patients with lung disease. Patients with pulmonary fibrosis do not do as well as patients with chronic obstructive lung disease or asthma, but there is short-term benefit. There is also the long-term value of supplemental oxygen therapy from portable and home-based oxygen equipment.

Some workers develop respiratory symptoms from other conditions present in addition to asbestos-related pulmonary disease. After an occupational condition has been diagnosed, other coexisting conditions may not be sought or may be neglected. Woolf (1990) undertook a prospective survey of 100 claimants for pneumoconiosis seen sequentially by the Occupational Chest Disease Review Committee of the Ontario Workers' Compensation Board. In this series, there were 44 workers with asbestos-related respiratory disease. Of these, 21 would have benefited from other diagnostic investigations and/or treatment directed to respiratory conditions other than those due to asbestos exposure, such as asthma and chronic obstructive lung disease. There is an opportunity to help these individuals which should not be neglected.

Changes in patterns have occurred in the treatment of mesothelioma (Aisner, 1995). Local removal has been of no value. Major surgery is involved in the use of extrapleural pneumonectomy. This has resulted in a few 5-year survivors but the removal of the lung with the pleura, pericardium and diaphragm makes this unlikely to be acceptable as a regular procedure. Modern approaches with combinations of radiation therapy, chemotherapy and biologic response modifiers have shown no striking improvement in survival time. The use of intracavitary photoactivated protoporphyrin derivatives is to be explored. It seems not yet possible to alter this pattern of inevitable progress and death. Unfortunately, mesothelioma is likely to continue to increase in incidence as a long-delayed legacy of the high exposures to asbestos 20 or more years ago.

QUESTIONS WHICH REMAIN TO BE ANSWERED

Numerous questions remain to be addressed and are of considerable concern especially to physicians who are responsible for making decisions for compensation and other medicolegal reasons.

- Are pleural or lung abnormalities accepted as due to asbestos if there is an abnormal CT scan but no symptoms, a normal chest radiograph, and normal pulmonary function tests?
- Where there is pulmonary fibrosis but only low asbestos exposure, at what level of exposure would the two conditions be considered related?
- What findings in lung biopsy are definitively acceptable in deciding whether or not there is a relationship between pulmonary fibrosis and asbestos exposure?
- Where there is cancer of the lung and a history of heavy smoking and of uranium mining exposure, and a relatively minor exposure to asbestos, may one assume that the smoking and uranium played a major role in the causation of the cancer?
- Should one accept a relationship between a lung cancer and asbestos exposure if there is no pleural or parenchymal evidence of asbestos-related pulmonary disease?
- What is the least level of asbestos exposure to allow acceptance that a mesothelioma is asbestos-related?
- Is there truly a threshold of exposure to asbestos below which there is no risk of any adverse pulmonary effects from asbestos?

These are a few of the questions that remain unanswered and should encourage further research. However, if there continues to be effective control of exposure to asbestos, it may well be that asbestos-related diseases will become extinct within 50 years and this chapter will be considered only of historical value.

REFERENCES

Aisner, J. (1995) Current approach to malignant mesothelioma of the pleura. *Chest*, **107** (suppl. 6), S332–44.

Ameille, J., Brochard, P., Brechot, J.M. *et al.* (1993) Pleural thickening: a comparison of oblique chest radiographs and high-resolution computed tomography in subjects exposed to low levels of asbestos pollution. *Int. Arch. Occup. Environ. Health*, **64**(8), 545–8.

Cagle, P.T., Wessels, R. and Greenberg, S.D. (1993) Concurrent mesothelioma and adenocarcinoma of the lung in a patient with asbestosis. *Modern Pathol*, **6**(4), 438–41.

Churg, A. (1993) Asbestos-related disease in the workplace and the environment: controversial issues. *Monographs Pathol.*, **36**, 54–77.

Doll, R. (1955) Mortality from lung cancer in asbestos workers. *Br. J. Ind. Med.*, **42**, 81–6.

Gaenster, E.A. (1992) Asbestos exposure in buildings. *Clin. Chest Med.*, **13**, 231–42.

Gevenois, P.A., DeVuyst, P., Dedeire, S. *et al.* (1994) Conventional and high-resolution CT in asymptomatic asbestos-exposed workers. *Acta Radiol* **35**(3), 226–9.

Jones, R.N., Diem, J.E., Hughes, J.M. *et al.* (1989) Progression of asbestos effects: a prospective longitudinal study of chest radiographs and lung function. *Br. J. Ind. Med.*, **46**, 97–105.

Jones, R.N., Glindmeyer, H., III and Weill, H. (1995) Airways obstruction from asbestos exposure. *Chest*, **107**(6), 1727–9.

Kilburn, K.H. and Warshaw, R.H. (1995) Airways obstruction from asbestos exposure and asbestosis revisited. *Chest*, **107**(6), 1730–1.

Miller, A. (1993) Pulmonary function in asbestosis and asbestos-related pleural disease. *Environ Res*, **61**(1), 1–18.

Morgan, W.K.C. (1991–92) On chrysotile, crocidolite, and cupidity. *Ontario Thoracic Rev.*, **4**(3), 1–4.

Nakadate, T. (1995) Decline in annual lung function in workers exposed to asbestos with or without preexisting fibrotic changes on chest radiograph. *Occup. Environ. Med.*, **52**(6), 368–73.

Ren, H., Lee, D.R., Hruban, R.H. *et al.* (1991) Pleural plaques don't predict asbestosis: high-resolution computed tomography and pathology study. *Modern Pathol.*, **4**(2), 201–9.

Rey, F., Boutin, C., Steinbauer, J. *et al.* (1993) Environmental pleural plaques in an asbestos exposed population of northeast Corsica. *Eur. Respir. J.*, **6**(7), 978–82.

Rudd, R.M. (1990) *Asbestos-Related Disease in Respiratory medicine*, 1st edn (eds R.A.L. Brewis, G.J. Gibson and D.M. Geddes), Ballière Tindall, London, pp. 1241–65.

Staples, C.A. (1992) Computed tomography in the evaluation of benign asbestos-related disorders. *Radiol. Clin. North Am.*, **30**(6), 191–207.

Sturm, W., Menze, B., Krause, J. and Thriene, T. (1995) Asbestos related diseases and asbestos types used in the former GDR. *Exp. Toxicol. Pathol.*, **47**(2–3), 173–8.

Voisin, C., Marin, T., Brochard, P. and Pairon, J.C. (1994) Environmental airborne tremolite asbestos pollution and pleural plaques in Afghanistan. *Chest*, **106**(3), 974–6.

Wagner, J.C., Sleggs, C.A. and Marchand, P. (1960) Diffuse pleural mesothelioma and asbestos exposure in the North Western Cape Province. *Br. J. Ind. Med.*, **7**, 260–71.

Weiss, W. (1993) Asbestos-related pleural plaques and lung cancer. *Chest*, **103**(6), 1854–9.

Wood, W.B. and Gloyne, S. (1930) Pulmonary asbestosis. *Lancet*, **1**, 445–8.

Woolf, C.R. (1986) *Respiratory Rehabilitation: a Practical Guide*, Tower Litho Company, Toronto, Ontario.

Woolf, C.R. (1990) Investigations of occupational diseases. *Occup. Health in Ontario*, **11**(4), 193–8.

Woolf, C.R. (1991–2) Editorial comments. *Ontario Thorac. Rev.*, **4**(3), 2.

Woolf, C.R., Davies, G.M., Gray, C.C. *et al.* (1992) Guidelines for the assessment of respiratory impairment in silica and asbestos exposure. *Occup. Health in Ontario*, **13**, 43–54.

Wylie, A.G., Railey, K.F., Kelse, J.W. *et al.* (1993) The importance of width in asbestos fiber carcinogenicity and its implications for public policy. *Am. Ind. Hyg. J.*, **54**(5), 239–52.

17 Asbestos-related diseases in Russia

Favel M. Kogan

In Russia, large-scale mining and milling of chrysotile asbestos began nearly a century ago. Since the birth of this industry, the production of asbestos fibers and the manufacture of asbestos-containing materials has dramatically increased. Yet, in Russia, as in other countries which have mined and marketed this mineral, there have been painful consequences for workers and their families associated with this industry and exposed to asbestos-containing dust.

AIRBORNE DUST LEVELS

Asbestos in mined in an open-pit process by excavation of exploded rock. The ore is transported by truck and railcar to adjacent mills for processing. Exposure limits in this Russian industry are determined by the maximal allowable concentration (MAC), i.e. the maximal level of exposure recognized to cause no adverse health effects. The current MAC for dust in these mines is based on a gravimetric measurement. The MAC for dust exposures in mines which contain a relatively low content of asbestos fibers (2–5%) has been established at 4 mg/m^3. Typically, in the Ural mountain asbestos mines, the present dust levels approximate 2–6 mg/m^3, and the number of asbestos fibers in the dust are recognized to be within the low content range.

Potential workplace exposures to airborne asbestos-containing dusts occur at mines, mills and enrichment plants, as well as in textile and other manufacturing operations. Between the 1950 and 1970s, dust exposures among asbestos miners and millers were excessive, ranging from 6 to 66 mg/m^3. Declines in asbestos exposures at these worksites began in the 1960s when an extensive program of dust suppression was initiated. For example, by isolating and enclosing long belt transporters and other equipment, gravimetric measurements of the dust levels in the mills or enrichment plants decreased from very high levels of 150–250 mg/m^3 in the 1950s to only 3–4 mg/m^3 in the 1990s (Figure 17.1). Concurrent with reducing exposures within the workplace, the filtering of asbestos-containing dust emissions in the milling and the further processing of asbestos has almost ended the environmental contamination of adjacent communities.

The ambient air environment in the asbestos textile plants has also improved. After modernization and installation of dust-suppressive engineering measures, such plants in the Urals – as well as in similar plants in other Russian regions (i.e. Egorievsk and Moscow) – now average dust levels which approximate 3–5 mg/m^3. Although these values exceed the current

Occupational Lung Disease. Edited by Daniel E. Banks and John E. Parker. Published in 1998 by Chapman & Hall.
ISBN 0 412 73630 6

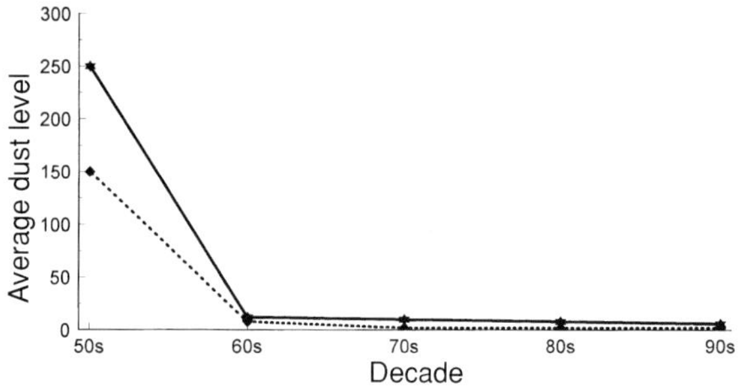

Figure 17.1 Dust levels (mg/m^3) in asbestos mills during the 1950s, 1960s, 1970s, 1980s and 1990s. •••••, minimal values; ———, maximal values.

MAC, they are ten and hundreds of times less than those encountered in the 1950s and 1960s. For example, exposures in asbestos textile weaving shops and carding shops range from two to eight times the MAC.

CLINICAL FEATURES OF ASBESTOSIS – RUSSIAN OBSERVATIONS

The features of asbestosis and its complications were recognized in Russia before the Second World War. Although the clinical features have been well documented by numerous investigators in Western literature, important observations have been proposed by Russian investigators and clinicians. We have recognized the classically described features of asbestosis which include dyspnea, weakness on exertion, dry cough and chest findings of crackles. In the early stages, hypoxemia may be present only with exertion, but in advanced disease, hypoxemia can occur at rest. With progressive pulmonary fibrosis comes pulmonary artery hypertension and right ventricular hypertrophy. We have reported more subtle features of asbestosis in the Russian literature which include dyspepsia, albuminuria, a decrease in hepatic function and a diminished skin response to ultraviolet light. We have also reported a reduction in myocardial contractility and stroke volume in the later stages of asbestosis. More than half of our patients with advanced disease had myocardial dysfunction and decreased cardiac output. We have found little correlation between hypoxemia and ventilatory rate. In our experience, finger clubbing is unusual.

The diagnosis of asbestosis is based mainly on the chest radiograph. The chest radiographs are best categorized by the international classification of pneumoconioses (ILO, 1980). In asbestosis, the chest radiograph shows diffuse linear (or 'ground-glass') interstitial markings in the lower and mid zones, and sometimes smaller lung volumes, depending on disease severity. Pleural plaques and pleural thickening with obliteration of the costophrenic angles are inconstant features. As a rule, as the profusion of opacities increases, lung function decreases.

The natural history of asbestosis may be hastened by accompanying chest infectious diseases. In our experience tuberculosis is an important and problematic complication. In Russia, asbestosis progression occurs, as a rule, during the 5 years after the diagnosis, and outcomes are worse in those who then continue to work in a dusty environment.

Asbestos workers exposed to airborne concentrations of dust which are two to four-fold greater than the MAC had respiratory complaints twice as frequent than in those working in an atmosphere of low exposures. Rarely, pneumoconiosis has been recognized in workers exposed to asbestos cement dust (Balashov *et al.*, 1983). Asbestosis occurred less frequently among workers in areas with intermittent dust exposures (i.e. repair and maintenance workers) compared with those continuously exposed.

THE OCCURRENCE OF ASBESTOSIS

The prevalence of asbestosis in chrysotile asbestos production workers has declined during an observation period of more than 40 years. Based on examinations of workers in the early years after the Second World War 29% were recognized to have asbestosis. This high percentage is not only attributed to excessive dust exposures, but also to an accumulation of cases, since no examinations were performed during the war. In 1950, the number of asbestotics approximated 21%. This decline is likely to be attributable to the removal of those with serious disease (through death or retirement) and in association with lower dust exposures (Figure 17.2). In the 1960s, 2% were found to have asbestosis while in the 1970s, the prevalence was approximately 0.7% (Kogan *et al.*, 1982). Most recently (1992–1994), examinations of millers and miners showed the majority of cases of asbestosis to be in retired workers (Table 17.1). This

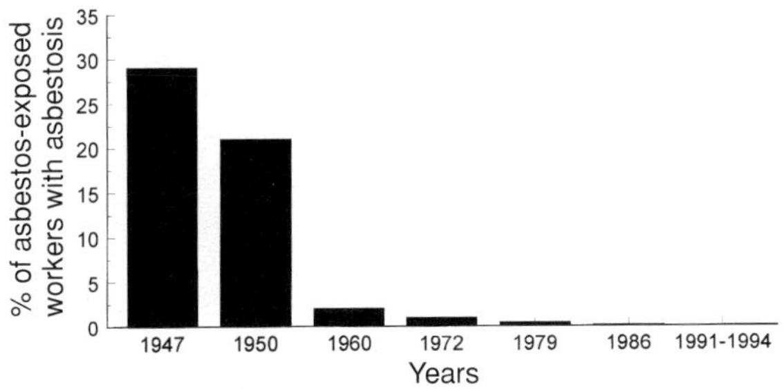

Figure 17.2 Incidence of asbestosis in asbestos mill workers (expressed as a percentage of the number of workers tested) from 1947 to 1994.

Table 17.1 Incidence of asbestosis in 1992, 1993 and 1994 in current employed asbestos workers and in pensioners

	1992	*1993*	*1994*
No. of workers examined	12 019	12 817	8 074
Current workers with asbestosis	15	19	9
Pensioners with asbestosis	30	21	28
Incidence (%)	0.003	0.003	0.005
Mean (range) of exposure (years) for current workers	25 (8–38)	24 (9–35)	26 (14–36)
Mean (range) of exposure (years) for pensioners	19 (8–20)	25 (10–32)	27 (10–40)

may be related to legislation passed in the early 1990s which provided generous compensation for industrial injuries, particularly occupational diseases. As a consequence, former asbestos workers who were not considered to have asbestosis in the past, were re-examined and identified to have this disease. Many of these retired workers were diagnosed to have minimal asbestosis (Category I), apparently attributable to progression since retirement, implying that medical examinations for these exposed workers should continue for life (Aksenova, 1992). Among those examined in 1992, the ratio of pensioners to current workers with asbestosis was 2:1. In 1993, the ratio approximated 1:1; and in 1994, it was 3:1 (Table 17.1). The average time from first exposure to diagnosis for the groups of current workers and pensioners ranged from 20 to 25 years. Because the pensioners are no longer exposed, this suggests that the asbestos workplace exposures for the pensioners was greater and that radiographic progression occurred, even in the absence of dust.

In our experience, a diagnosis of asbestosis is made after the accumulated amount of dust in the lung exceeds a threshold. Depending on the intensity of exposure, the time from first exposure to diagnosis commonly ranges from 9 to 35 years, although a typical latency period is 20–27 years. We believe that, in addition to dust burden, the risk for asbestosis depends on: (i) social factors (lifestyle, smoking or alcohol habits); (ii) concurrent illnesses (bronchitis, pneumonia, tuberculosis, rheumatism); and (iii) biologic parameters (gender, age of first dust exposure). Thus, in epidemiologic studies when controlling for these descriptive parameters, a more accurate prediction of individuals who would develop disease was possible (Polzik *et al.*, 1989).

Further investigations into factors that predict the development of asbestosis included 'hepatoglobin (Hp I)', genetic types S and M_3, and rare variants of α_1-antitrypsin phenotypes Rh-negative blood group (Gladkova, 1993). In contrast, individuals with significant asbestos exposure but without asbestosis have a predominance of Hp_2, M_1 and M_2, typical α_1-antitrypsin phenotypes, and Rh-positive blood group. Using these determinants, screening of workers was implemented soon after such differences were recognized and the workers' medical data at the time of their starting work has been computerized (Asbest-1) to define each individual's risk for disease.

Asbestotics have a nearly two-fold higher risk for cancer compared with those exposed but without asbestosis. In our populations of asbestos-exposed workers, asbestosis, as well as asbestos dust-induced bronchitis, are considered premalignant conditions (Kogan *et al.*, 1972). These observations have been confirmed by others (Gladkova, 1993; Weill, 1993). Histologic studies (Browne, 1993) identified the presence of premalignant airway changes such as epithelial metaplasia and dysplasia. The probability of developing asbestosis increases when α_1-antitrypsin deficiency is combined with a series of transferrin forms in the presence of a homozygous form of haptoglobin. These workers deserve a more complete evaluation as they may be at particularly high risk for asbestosis development (Gladkova, 1993). We also use a combination of 15 characteristic skin features or dermatoglyphic signs to identify workers whom we consider to be predisposed to develop asbestosis (Polzik *et al.*, 1989).

ASBESTOS EXPOSURE AND ASBESTOS-RELATED MALIGNANCIES

Since the early 1980s neither advanced asbestosis (Stage II radiographic disease) nor asbestosis complicated by tuberculosis have been recognized in our populations under study. In most cases of asbestosis (ILO radiographic category 1/1 or greater), the average exposure approximated 20–25 years. As early as the 1960s, even in the absence of clinical or radiographic asbestosis, lung surgery in asbestos exposed workers occasionally revealed incidental interstitial fibrosis, highlighting the relative insensitivity of the chest radiograph for the early diagnosis of asbestosis. Such minimal fibrotic changes may become a focus for the development of pulmonary neoplasms.

We have observed a reduction of lung cancer in asbestos-exposed workers and a steep decline in new cases of asbestosis over time (Tables 17.2 and 17.3; Figure 17.2). This is in contrast to work by both Nicholson *et al.* (1993) and Peto *et al.* (1977), who reported that reduction of asbestosis was not accompanied by a simultaneous lowering of lung cancer risk.

Between 1947 and 1979, the mortality due to lung, stomach, intestine and uterine cancer among asbestos mine and mill workers was several times greater than that of the general population in the adjacent Ural Mountain community of Asbest city. This city, which is named after the nearby vast asbestos mineral deposit, has over 100 000 residents, many employed in the asbestos mining, milling or textile product industries. Although little change was observed for a decade, since 1987 this mortality trend has decreased (Table 17.2).

Such a downward trend in mortality has been slow, partly because of the failure to reduce asbestos exposures below the MAC level (despite an aggressive dust reduction campaign), and partly perhaps reflecting the long latency from exposure to neoplasia for these diseases. Furthermore, we cannot be certain whether the large number of cases are attributable to the chronic excessive exposures or whether short-term peaks of intermittent dust emissions have played a role in increasing cancer risks.

Table 17.2 Relative risk for malignancy in Asbest City residents for 1948–1979 and 1987–1991. Asbestos-exposed workers are compared with non-exposed workers

Site of malignancy	*Men*		*Women*	
	1948–1979	*1987–1991*	*1948–1979*	*1987–1991*
Lung	3.7	1.5	6.2	1.2
Stomach	2.4	1.2	4.3	1.7
Intestine	4.0	0.6	4.4	2.1
Uterus	N/A	N/A	3.4	1.5
Total	2.7	1.6	3.4	1.5

N/A, not applicable.

We consider asbestos dust a systemic carcinogen. Our data show that asbestos exposure increased not only the lung cancer risk but also risks for gastrointestinal, gynecologic and hematopoetic malignancies (Frash *et al.*, 1988). For example, the average risk for all cancers between 1948 and 1979 in men with asbestos exposure was 2.7 but between 1987 and 1991 it was 1.6 (Table 17.2). In women, the rates for 1987–1991 were 3.4 and 1.5, respectively. More sharply drawn differences are apparent when comparing the rates for lung and stomach cancer. The two-fold reduction of the cancer rate in lung, stomach, intestinal and uterine malignancies during the years 1987–1991 reflect the gradual decline in exposures. The exceedingly high baseline rates likely reflect the very high exposure levels occurring during the Sec-

Table 17.3 Standardized mortality rate (SMR) for malignancy in asbestos textile workers by gender

Site of malignancy	*Men*		*Women*	
	Observed/ Expected	*Ratio*	*Observed/ Expected*	*Ratio*
Lung	7/4.9	1.4	1/1.4	0.7
Stomach	4/5.6	0.7	12/7	1.7
Uterus	N/A	N/A	4/1.6	2.5
Total	21/16.9	1.2	34/25.2	1.3

N/A, not applicable.

ond World War and in the immediate post-war years. With evidence of a continued decrease in asbestos exposure levels, the projected increase in occupationally induced cancers made in 1979 appears exaggerated (Bridbord *et al.*, 1978).

In addition to asbestos mine and mill workers, the cancer rate for workers in the asbestos textile, asbestos cement and asbestos friction products industries have been studied. The highest lung cancer risk was seen in the asbestos textile industry with a standardized mortality rate (SMR) of 1.3, possibly as a result of exposure to long, thin asbestos fibers (Gurvich, 1992). Lower lung cancer risks were observed for the asbestos cement industry and the friction products industry, possibly due to lower exposures occurring as a result of wet mixing processes. We also postulate that fiber toxicity may be altered by the coating of asbestos cement dust by hydrated cement, and coating by asbestos–bakelite and asbestos–rubber mixtures in the friction industry (Kerimova, 1991).

We have observed substantial differences of the SMR in 'closed' workforces (cohorts without newly hired workers) and 'open' (cohort in which new persons enter in the original group in the following years). The SMR of lung cancer was 1.3 in the 'open' and 1.7 in the 'closed' cohort when the mortality rates of two asbestos-exposed industrial cohorts were compared with the general population of the Moscow region (Gurvich, 1992).

The risk of lung cancer in workers in the Yegorievsk asbestos textile plant was 1.7 times the risk of residents in the Moscow region and 1.8 times that of residents of the Yegorievsk district without asbestos exposure. Among workers of a Moscow asbestos textile plant, the SMR for all cancers was 1.7 and for lung cancer, stomach cancer, uterine cancer the risks were 1.6, 2.6 and 1.6, respectively (Kogan *et al.*, 1987). In the Asbest City asbestos textile plant, the mortality rate from lung cancer for men in the 1960s and 1970s was three times higher than for men residing in the adjacent district (Kogan and Guselnikova, 1981). It may well be that the direct carcinogenic effect of asbestos fibers and the metabolic derangement resulting from the pulmonary fibrosis of asbestosis play a role in the development of pulmonary malignancies. We have also observed in other studies that the rate of stomach cancer increased at a rate nearly parallel to the rate of increase in lung cancer. This observation might be explained by the process whereby exposures to the lung and stomach are greatest as a result of these organs serving as the sites of 'entry and exit' for asbestos fibers.

The rate of cancer mortality of workers in friction product shops increased with latency from first exposure. In the first two decades after initiating asbestos exposure only two cases of lung cancer were detected, the first after 15 years of exposure to dust. Yet, in the group with the longest latency period of 30 years, the SMR dramatically increased to 3.2, emphasizing the importance of the latency effect in this illness. The most important competing causes of deaths in these workers included tuberculosis (18%), lung cancer (8%) and cardiovascular diseases (16%) (Kogan *et al.*, 1972).

Exposure to dust levels of 2 mg/m^3 in Moscow asbestos textile plants for more than 20 years (1958–1978) was associated with no new cases of asbestosis (Kogan *et al.*, 1987). However, this dust level did not eliminate new cases of lung cancer which occurred in excess of those expected in the general population and were thought to be attributable to asbestos. In order to assess the dose–response relationship between asbestos exposure and the development of lung cancer, we addressed both the intensity and duration of the workers' asbestos exposure during all work activity, i.e. a worker's cumulative exposure. Based on this, it appears that an average workplace asbestos dust level of 0.5 mg/m^3 or less is necessary to reduce the lung cancer mortality to a level equal to the general population (Kogan, 1988, 1992; Gurvich, 1992).

MESOTHELIOMA OF THE PLEURA

In numerous studies, the increased mortality from mesothelioma of the pleura was observed only in the cohorts exposed to amphibole asbestos (mainly crocidolite) (Dunnigan, 1988; Weill, 1993). In our experience, this does not appear to be the case. In the Middle Ural mountains, the main asbestos mining region in Russia, only chrysotile asbestos is produced. In the 50 districts of this region, the mortality from mesothelioma over a 10-year period was six-fold higher than the average rate in the Sverdovsk region, an area of negligible asbestos mining. Most with mesothelioma had worked at the asbestos mining and milling plants, or had lived in an adjacent town near old and very 'dusty' mills (Kogan and Berzin, 1986).

CONCLUSIONS

Based on our findings and on-going investigations we can conclude that:

- Serial assessments of the dust levels collected over the past 40 years in asbestos production shops in Russia have shown a dramatic lowering of the asbestos levels from 150–250 mg/m^3 to 3–5 mg/m^3. This reduction in output of asbestos fibers has resulted from mechanization, isolation of the processes and containment of many of the dusty operations.
- As a consequence of lessening workplace asbestos exposure, the number of new cases of asbestosis among exposed workers was reduced from 29% to less than 1%. During the past few years no cases of Category II asbestosis or asbestotuberculosis were detected. Due to changes in the compensation law, which provided a financial supplement to those with occupational disease, workers and pensioners without a previously recognized diagnosis of asbestosis during their employment in the asbestos plants were examined. Although a number of pensioners and some workers examined were found to have asbestosis (Category I), the proportion of those with the condition was rather small.
- During the 1950s, 1960s and 1970s, the cancer risk among the asbestos production and asbestos textile workers exceeded that of the general population by three- to five-fold. During the past 20 years, there has been a steady and substantial decline in such risk, though this has yet to achieve the rate identified in the general population.
- We consider asbestos to be a systemic carcinogen. The continuing excess rate of cancer mortality in Russian asbestos production workers appears to be attributable to the on-going excessive airborne dust level in our workplaces.
- In addition to implementing measures to lessen dust exposures, we have recognized that the use of specific medical parameters aids in identifying those workers who may have a particular sensitivity to asbestos dust. Our system uses 16 different criteria which include immunologic, economic, genetic, social and other parameters, in order to identify workers at high risk for the adverse effects of asbestos exposure.

REFERENCES

Aksenova, A.O. (1992) *Remote consequences of asbestos-effect and their prevention*, MD dissertation, Moscow.

Balashov, V.A., Kogan, F.M., Gusel'nikova, N.A., Rozenberg, I.A. and Serov, A.N. (1983) Gigiena truda i sostoiane zdorovia, rabotaiushchikh v prizvodste asbestotsementnykh izdelii. *Gig. Truda. Prof. Zabol.*, **Sep(9)**, 8–12.

Bridbord K., Decoufle, P. *et al.* (1978) *Estimates of the fraction of cancer in the US related to occupations*, US Dept of H.E.W., Washington, DC., September 15.

Browne, K. (1993) Proceedings of 8th Bienial Conference of Asbestos, Intern. Assoc. Paris, 11–12, May.

Dunnigan, J. (1988) Linking chrysotile asbestos with mesothelioma. *Am. J. Ind. Med.*, **14**, 205–9.

Frash, V.N., Gusel'nikova, N.A., Vanchugova, N.N., Sibiriakov, A.G. and Varzina, N.V. (1988) Immunologic changes as an index of biological effect of asbestos-containing dust. *Gig. Truda. Prof. Zabol*, **Jan**(1), 27–30.

Gladkova, E.V. (1993) *Occupational diseases of the broncho-pulmonary apparatus in the oncodangerous industries and their prevention*, MD dissertation (in Russian).

Gurvich, E.B. (1992) *Epidemiologic and hygienic evaluation of oncological risk in the industry*, MD dissertation, Moscow (in Russian).

ILO (1980) Guidelines for the use of the ILO international classification of radiographs of pneumoconiosis. No. 22 (Rev.) Occupational Safety & Health Series, International Labor Office, Geneva.

Kerimova, T.T. (1991) *Comparative hygienic evaluation of the asbestos-containing dusts, impact of the workers in asbestos-cement and asbestos products industry*, MD dissertation, Moscow (in Russian.).

Kogan, F.M. (1989) Research on asbestos health problems in the USSR, in *Sourcebook on Asbestos Diseases, Vol. 4*, Butterworth, p. 544.

Kogan, F.M. (1991) The approaches to the determining of the MAC of asbestos containing dust, in *Sourcebook on Asbestos Diseases, Vol. 6*, Butterworth, New York.

Kogan, F.M. and Berzin, S.A. (1986) The frequency of pleural mesothelioma under the impact of chrysotile asbestos dust. *Gig. Truda. Prof. Zabol.*, **Sept**(9), 9–12.

Kogan, F.M. and Guselinikova, N.A. (1981) *Proc. Occup. Dis. Dust Ethiology*, **2**, 24–9.

Kogan, F.M., Gusel'nikova, N.A. and Gulevskaia, M.R. (1972) Smernost' ot raka sredi rabochikh asbestovoi promyshennosti Urala. *Gigirna sanitaria*, **37**, 29–32.

Kogan, F.M., Deminov, A.G. *et al.* (1982) *Arch. Immunol. Ther. Exper.*, **30**, 277–83.

Kogan, F.M., Gusel'nikova, N.A., Gladkova, E.V. and Gurvich, E.B. (1987) Ob epidemiologii kroniche-skikh fonovykh, predopukholevykh i pnkololiche-skikh zabolevanii v rabot aiuschikh pri hizkikh kontsentrat siiakh. *Gig. Truda. Prof. Zabol.*, **March**(3), 37–40.

Nicholson, W. and Pooley, F. (1993) Workshop of the Health Risks associated with Chrysotile Asbestos, Jersey, UK November 14–17.

Peto, J., Doll, R., Howard, S.V., Kinlen, L.J. and Lewinsohn, H.C. (1977) A mortality study among workers in an English asbestos factory. *Br. J. Ind. Med.*, **34**, 169–73.

Polzik, E.V., Kogan, F.M., Kochnevba, M.Ju. and Potanin, N.I. (1989) Risk factors for asbestosis in workers in asbestos mills. *Med. del Lavoro*, **80**, 363–9.

Weill, H. (1993) Proc. 8th Bienniali Conference of Asbestos, Intern. Association. Paris, 11–12, May.

18 A national program to reduce the risk of asbestos-related disease in Finland

Antti Tossavainen and
Matti S. Huuskonen

A HISTORY OF ASBESTOS IN FINLAND

Finland has been the world's main producer of anthophyllite asbestos. Of the 350 000 tons of anthophyllite mined, 230 000 tons were exported. If one includes both the asbestos mined in Finland and all imported asbestos, 300 000 tons of asbestos were used in Finland between 1905 and 1988. Of this, 175 000 tons were chrysotile, 120 000 tons were anthophyllite and 5000 tons were crocidolite and amosite. The total production of asbestos was greatest at the beginning of the 1970s when more than 10 000 tons were produced per year (2.5 kg/per capita). Thereafter the use diminished when the mining and milling of anthophyllite was discontinued in 1975, and the manufacturing of asbestos-cement products ceased in 1988. The current annual use of asbestos products is less than 300 tons. This amount consists mainly of imported packing and friction materials (Huuskonen *et al.*, 1980; Ministry of Labour, 1990).

In 1992, the State Councill banned, with few exceptions, the manufacture, import and use of all asbestos-containing products including chemicals, vehicle parts, consumer appliances and building materials after January 1, 1993. Although exposure to asbestos was minimized, this restriction did not mean that asbestos exposure ceased. There are approximately one million buildings in Finland, and one-third of these (built or renovated between 1920 and 1980) contain about 200 000 tons of asbestos. Every year about 5000–10 000 asbestos-containing buildings are renovated. Intact asbestos surfaces are marked and removed during normal renovation or maintenance. Information to identify asbestos which should be removed and insight into the health risks of asbestos exposure has been provided. By the end of 1991, more than 4000 persons had attended a four-day course on asbestos removal, and about 300 companies were licensed to remove asbestos by the National Board of Labor Protection. Training and information, as well as effective sanctions, keep the exposures under control (Ministry of Labour, 1990).

Our best estimates suggest that more than 200 000 active or retired workers in Finland have been exposed to asbestos (at least 2 fibers/ml for at least 2 months). About 50 000 workers, mostly over 40 years of age, have had more than 10 years of exposure at work, most often in the construction or shipbuilding industries

Occupational Lung Disease. Edited by Daniel E. Banks and John E. Parker. Published in 1998 by Chapman & Hall.
ISBN 0 412 73630 6

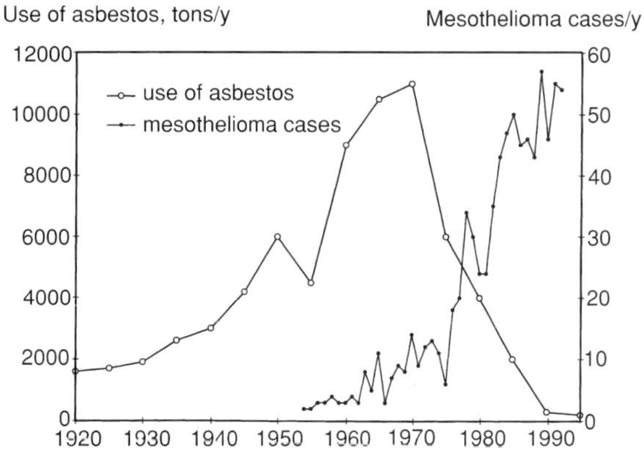

Figure 18.1 Use of asbestos and number of annual mesothelioma cases in Finland.

(Huuskonen *et al.*, 1987). Today around 10 000 construction and maintenance workers are exposed to asbestos. Thus, from what is known about workplace exposures in Finland, the incidence of asbestos-related diseases can be expected to reach its peak after the year 2000 (Huuskonen *et al.*, 1987; Ministry of Labour, 1990).

The first worker with asbestosis received financial compensation for an occupational disease in 1938. At that time the association between asbestos and the risk of lung cancer was unclear (Huuskonen and Tossavainen, 1992). About 50 years ago the radiologic and histopathologic findings in workers with asbestosis were known from case reports (Noro, 1945, 1946; Wegelius, 1947). Later, in the 1960s, a high prevalence of non-occupational pleural calcifications were found in the residents of communities around the Paakkila anthophyllite mine and their causal relationship to environmental exposure was established (Kiviluoto, 1960; Meurman, 1966; Raunio, 1966). In the general population of this area, the frequency of cancer was not higher than in other rural populations. However, when asbestos exposure occurred in the workplace, the frequency of lung cancer was particularly high, regardless of the presence of asbestosis (Meurman, 1966).

In the 1970 s, mortality studies of Paakkila miners and millers confirmed the observations relating workplace anthophyllite exposure to lung cancer. At the same time, pleural mesothelioma was shown to occur as a result of exposure to various types of asbestos in several occupations and industries (Nurminen, 1972, 1975; Meurman *et al.*, 1974). Before 1991, four cases of mesothelioma had been reported in the cohort of 999 anthophyllite miners and millers (Karjalainen *et al.*, 1994a; Meurman *et al.*, 1994).

Trends relating asbestos use to the incidence of mesothelioma in Finland are shown in Figure 18.1. Some historical perspectives of asbestos and disease issues in Finland in the 20th century are summarized in Figure 18.2.

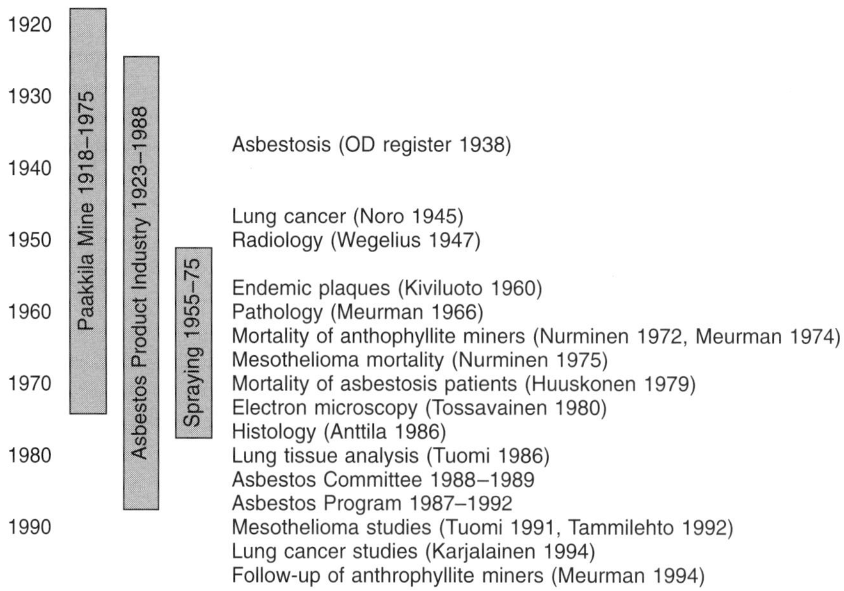

Figure 18.2 Historic perspectives of asbestos exposure and disease in Finland. (Names and years refer to the reference list, academic dissertations, or to the introduction of research methods.)

ASBESTOS-RELATED PULMONARY DISEASE IN FINLAND

About 1600 asbestosis cases were registered in Finland between 1938 and 1994. Of the 396 persons with asbestosis diagnosed before 1977, 132 were asbestos miners, 144 asbestos product manufacturers, 100 were insulators, eight had sprayed asbestos and 12 others had been exposed at work (Huuskonen *et al.*, 1980). During the past 10 years asbestos-related pulmonary abnormalities have most frequently been found among construction workers. Up to the mid-1980s, asbestosis was the only asbestos-related disease which was reported to the Register of Occupational Diseases. Considered in 5-year periods, the number of new asbestos-related diseases registered was 117 between the years 1976 and 1980, 164 during the years 1981 to 1985, and 1145 between 1986 and 1990. The number of individuals with pleural diseases registered has increased heavily in recent years (Table 18.1).

ASBESTOS-RELATED LUNG CANCER

A large percentage of those who develop asbestosis die of cancer. Among 174 men diagnosed to have asbestosis, about 40% died of lung cancer, 5–10% of mesothelioma, and 10% have died of other cancers (Huuskonen *et al.*, 1980). The causes of mortality in a second cohort of 115 asbestosis patients were compared with 144 silicosis patients and followed from 1980 to 1987. Twenty-one of the asbestosis patients (18%) contracted lung cancer and two were diagnosed to have mesothelioma. Three lung cancers (2%) and no mesothelioma occurred among the silicosis patients. The mean age at death was 62 years for the asbestosis cases and 67 years for the silicosis cases. The total cancer mortality rate was higher among asbestosis than among silicosis patients (17% versus 4%) (Oksa *et al.*, 1989).

In a series of 65 surgically treated lung cancer patients, past exposure to asbestos was evaluated by personal interview and by scanning electron microscopy (SEM) analyses of mineral fibers in the lung. Lung tissue samples of 17 autopsied male office workers served as a comparison group. Based on the occupational history, 37% of the lung cancer patients had definite or probable exposure to asbestos, 31% had possible exposure to asbestos, and 32% were unlikely to have had asbestos exposure. The fiber concentration in lung tissue ranged from less than 0.1 to 0.8 million fibers per gram dry tissue in the reference group. In 26% of the lung cancer patients, but none in the comparison group, the fiber concentration exceeded 1 million fibers per gram dry tissue. Anthophyllite, amosite and crocidolite were found in the tissues of the lung cancer patients. Most of those exposed patients had been employed in various construction jobs. Mineral analyses of lung tissue showed a good correlation between the exposure and the lung fiber burden (Karjalainen *et al.*, 1993a). In a series of 300 urban autopsy cases, pulmonary fiber concentrations exceeding 1 million fibers per gram (as determined by SEM) were highly indicative of occupational exposure to asbestos. When the limit of 1 million fibers per gram with SEM and 3 million fibres per gram with transmission electron microscopy (TEM) was used to indicate occupational exposure to asbestos, a consistent relationship was obtained in 93% of the patients (Karjalainen *et al.*, 1994b,c).

Asbestos body (AB) counts over 1 AB/ml in bronchoalveolar lavage (BAL) fluid were also indicative of past occupational exposure to asbestos. In 107 Finnish lung cancer patients, a significant exposure to asbestos was found by personal interview in one-third of the patients and in 27% of the patients by counting AB in the BAL (Karjalainen *et al.*, 1994d). In 25 individuals from the Tampere area with lung cancer who underwent lung resection, the lungs from six patients (24%) contained at least 1 million fibers/g dry tissue by SEM analysis. Anthophyllite was the main fiber type (Vilkman *et al.*, 1993). Lung tissue from 113 men who received surgical treatment for lung cancer was compared with autopsied lung specimens from 297

Table 18.1 Registered asbestos diseases in Finland (1964–1994)

Diagnosis	*1964–87*	*1988*	*1989*	*1990*	*1991*	*1992*	*1993*	*1994*
Lung cancer	14	12	33	29	56	64	34	97
Mesothelioma	24	14	25	20	12	19	28	25
Asbestosis	520	49	88	89	142	133	201	195
Pleural changes	NA	NA	179	439	809	1390	888	842

NA, Statistics not available.

autopsied men who resided in the Helsinki area. Some 19% of the lung cancer cases were attributed to occupational asbestos exposure (Karjalainen *et al.*, 1994e). The risk of lung cancer depended on the fiber dose. If the fiber concentration in the lung ranged from 1.0 to 4.99 million fibres/g, the risk of lung cancer increased to 1.7 that of those without asbestos exposure. When the asbestos fiber count exceeded 5.0 million fibres/g, the risk increased to 5.3 times that of the reference group. In this population, the risk of lung cancer was increased among asbestos-exposed patients regardless of the presence of diffuse interstitial fibrosis (Anttila *et al.*, 1993; Karjalainen *et al.*, 1993b; Nurminen and Tossavainen, 1994).

Meurman followed a cohort of 736 male anthophyllite miners from 1953 to 1991. Compared with the general Finnish population, the miners had a increased incidence of total cancer [standard incidence ratio (SIR) 1.7], mainly attributable to an excess in lung cancer (SIR 2.9). The risk of lung cancer was highest among those with at least 5 years of exposure (SIR 4.9). The authors concluded that anthophyllite is a more potent carcinogen for lung cancer than chrysotile or amosite and is nearly as potent as crocidolite (Meurman *et al.*, 1994).

According to Finnish insurance regulations (Karjalainen *et al.*, 1992), the criteria for 'significant asbestos exposure' to be used when lung cancer is compensated as an occupational disease include:

- at least one year of heavy exposure to asbestos (spraying, insulating) or at least 10 years of moderate exposure (construction work);
- at least 10 years of latency; and
- asbestos-related pleural or parenchymal findings.

The numbers of registered lung cancers and mesotheliomas caused by occupational exposure to asbestos are given in Table 18.1.

INCIDENCE OF MESOTHELIOMA

According to data in the Finnish Cancer Registry, the annual number of mesothelioma cases has increased 15-fold among men and 10-fold among women since the 1950s. In the 1960s, the number of new cases detected annually among men and women was roughly the same, but in the early 1990s there was an average of 34 new cases among men and 17 new cases among women each year. Standardized by age, the annual incidence of mesothelioma in the late 1980s was 11 per million among men and 4 per million among women. Between 1985 and 1989, the incidence of mesothelioma was greatest in urban men (16 cases per million men) and the least in men residing in rural areas (seven cases per million men). Women show less regional variation. The findings that men develop mesothelioma more frequently than women and that mesothelioma are more common in cities than in rural areas reflect the distribution of asbestos exposure among the Finnish population (Huuskonen *et al.*, 1993a,b).

Two-thirds of Finnish patients with mesothelioma have had some occupational exposure to asbestos. A more than 10-fold relative risk of mesothelioma was associated with fiber concentrations exceeding 1 million fibers per gram of dry lung tissue. In 29 patients with mesothelioma, crocidolite or amosite fibers were found in the lungs of 19 patients (66%), anthophyllite in 13 patients (45%), chrysotile in six patients (21%) and tremolite in five patients (17%) (Tuomi *et al.*, 1989, 1991a,b; Tuomi, 1992).

Among 999 anthophyllite miners and millers, 533 had died before 1991. Of these 533, only four died from mesothelioma, a clear excess. All four deaths (three from pleural, one from peritoneal mesothelioma) were the result of long-term exposure to anthophyllite and had high concentrations of anthophyllite fibers in their lung tissue (Karjalainen *et al.*, 1994a).

A NATIONAL PROGRAM TO REDUCE THE RISK OF ASBESTOS-RELATED DISEASE

A surveillance program for asbestos-exposed workers

In the past few decades, over 200 000 Finnish workers have been exposed to asbestos. Of these, over 150 000 have worked in the building industry, 20 000 in shipyards, 20 000 in car repair shops, and 10 000 in the asbestos industry (Ministry of Labour, 1990). In 1990, there were 87 000 workers in the construction industry and about 9000 in shipyards. Investigations by the Finnish Institute of Occupational Health (FIOH) showed that health surveillance of asbestos workers was not performed in a systematic manner. We charted the prevalence of parenchymal and pleural changes among construction workers in 1988. Current and retired workers who had started work before 1967 were selected from the local trade unions. Questionnaires detailing asbestos exposure were sent to 993 workers, of whom 453 (51%) participated in the health examination. Some 26% (116) had chest radiographic evidence of fibrosis or pleural changes that could be attributable to asbestos. Only 10 of the 116 people knew before this assessment that they had an asbestos-induced changes on their radiograph (Oska *et al.*, 1992).

A larger screening program for asbestos-exposed Finnish workers was carried out in 1990–1992 as a part of the Asbestos Program of the FIOH. Examination of active or retired workers included a personal interview detailing the work history and asbestos exposure, and chest radiographs. The target group for screening was workers under 70 years of age with at least for 10 years' employment in construction, or one year of shipyard employment or in the manufacture of asbestos products (Huuskonen *et al.*, 1995; Koskinen *et al.*, in press).

Workers were included as having asbestos-related health effects if they had radiographic findings:

- clear-cut lung fibrosis (at least ILO category 1/1);
- mild lung fibrosis (ILO category 1/0) with unilateral or bilateral pleural plaques;
- diffuse pleural thickening; or
- bilateral pleural and/or diaphragmatic plaques.

In addition to the screening, workers were educated to recognize asbestos in the work environment, which measures to take to prevent dust exposure, and the health effects of asbestos exposure and smoking. The screening prompted further national follow-up of asbestos-induced diseases among all workers who had been exposed to asbestos. It also aimed to minimize future asbestos exposure by identifying workers unaware of current exposure to asbestos and improving the diagnosis of asbestos-induced diseases. This investigation was necessary because general medical care was ineffective in identifying those with asbestos-related diseases. Occupational health care was insufficient, especially in the construction trades, and there was no organized follow-up of earlier exposed workers. One additional benefit of this approach was the opportunity to register exposed workers for future studies.

The target group comprised 54 409 people. Of these, 24 589 were invited to the screenings. 10 062 did not participate, and 1657 reported no exposure to asbestos. A total of 18 943 workers participated in the interview and chest radiograph examination. The mean age was 53 years. Most who participated were men (96%) and 65% were currently employed. About 28% were current smokers, 41% were ex-smokers and 31% were non-smokers. The great majority (95%) were employed in construction, while 2% worked in shipyards and 3% in the asbestos industry. Some 22% of those assessed were positive for screening. Of this group, bilateral pleural plaques occurred in 17%, diffuse pleural thickening was present in 7%, and lung fibrosis was reported in 30%. The effectiveness of screening increased with age. Asbestos-induced health effects were most common in carpenters (25% of the population), electricians (12%) and plumbers (12%). On average, the duration of employment in one of these occupations was 26 years (Huuskonen *et al.*, 1995; Koskinen *et al.*, 1996).

A surveillance program to identify asbestos-related health effects in a sample of the Finnish population

The presence of pleural plaques is a common and relatively specific indication of past asbestos exposure. The plaques may confer a greater risk for developing other asbestos-related sequela or be associated with abnormal lung function (Karjalainen *et al.*, 1994f; Valkila *et al.*, 1995). In the general population of the Paakkila mining area, the role of asbestos exposure as a cause of lung cancer was so low that the excess among the persons with radiographic plaques was not demonstrable (Partanen *et al.*, 1992).

In order to estimate the role of past asbestos exposure as a cause of radiographic small lung opacities and pleural abnormalities in the Finnish adult population, an investigation was conducted as a part of a large Mini-Finland health survey. The study was based on a population sample ($n = 8000$) representative of the Finnish population aged 30 years or over. Full-size chest radiographs and a complete job title history were available for 3811 women and 3274 men. Radiographs were classified according to the ILO 1980 classification of radiographs of pneumoconiosis and the work histories according to the probability of occupational asbestos exposure. Age- and smoking-adjusted relative risks for radiographic parenchymal and pleural abnormalities were calculated (Zitting, 1995; Zitting *et al.*, 1995).

About 13% of the men and 0.8% of the women were classified as probably exposed to asbestos. The risk of small lung opacities was significantly increased in the probably exposed men (RR = 1.7 for ILO profusion category 1/1 or more, and RR = 1.6 for profusion category 1/0). The risk of pleural plaques increased among men (RR = 3.0) and women (RR = 4.8) with probable exposure. The risk of a thickened horizontal interlobar fissure increased among probably exposed men (RR = 1.7). Among men, occupational asbestos exposure was responsible for approximately 30% of the small lung opacities, about 40% of pleural plaques, and about 20% of thickened horizontal fissures. Among women, asbestos attributability was similar for pleural, but was much lower for parenchymal, abnormalities. These figures probably underestimate the true effect of occupational asbestos exposure. At least 80 000 Finnish men have radiologically detectable bilateral pleural plaques and over 35 000 of these have had asbestos exposure at work. Occupational asbestos exposure has been common among Finnish men and it plays a significant role in the etiology of both pleural and parenchymal abnormalities (Zitting, 1995; Zitting *et al.*, 1995).

SUMMARY

Primary prevention carried out today can reduce the disease incidence in future decades. The present manifestations of asbestos-related health effects are the consequence of past asbestos exposure, mainly before the 1970s. The number of asbestos-related cancer deaths is approximately 150 per year, a value which exceeds the number of fatal work accidents. Asbestos-related cancer will continue to increase in Finland for 15 or 20 more years and reach a maximum of about 300 cases in 2010, but then begin to decline. About 20 000 asbestos-exposed workers have participated in medical screening and follow-up. Ceasing exposure to asbestos, introduction of anti-smoking campaigns, improved diagnostic evaluations, and careful attention to compensation issues have been the central issues of our national program.

Finally, a history of asbestos exposure and specific changes on the chest radiograph measure the adverse effects of asbestos long after the processes which lead to carcinogenesis or fibrogeneis are initiated. More sensitive diagnostic criteria, specifically the use of biomarkers, would be welcome and may allow us to limit the future damage attributable to asbestos.

REFERENCES

Anttila, S., Karjalainen, A., Taikina-Aho, O., Kyyrönen, P. and Vainio, H. (1993) Lung cancer in the lower lobe is associated with pulmonary asbestos fiber count and fiber size. *Environ. Health Perspect.*, **101**, 166–70.

Huuskonen, M.S. and Tossavainen, A. (1992) Fifty-year experience in diagnosing asbestos related cancers in Finland. Progress and detours. *Am. J. Ind. Med.*, **22**, 259–61.

Huuskonen, M.S., Ahlman, K., Mattsson, T. and Tossavainen, A. (1980) Asbestos disease in Finland. *J. Occup. Med*, **22**, 751–4.

Huuskonen, M.S., Karjalainen, A., Koskinen, K., Rinne, J.-P., Tossavainen, A. and Rantanen, J. (1993a) *Asbestos Prograrn 1987–92 Final Report*, Finnish Institute of Occupational Health, Helsinki.

Huuskonen, M.S., Karjalainen, A., Koskinen, K., Tossavainen, A., Tuomi, T. and Rantanen, J. (1993b) The Institute's action program against the health risks of asbestos, in: *Health Risks from Exposure to Mineral Fibres: An International Perspective*: (eds G.W. Gibbs J. Dunnigan M. Kido, T. Higashi. Captus Press Inc., Ontario, pp. 317–25.

Huuskonen, M.S., Koskilen, H., Roto, P., Tuomi, T., Heikkilä, P. and Rantanen, J. (1987) Asbestos will still be a significalt health hazard after the year 2000 in Finland. *Suomen Lääkärilehti*, **42**, 1669–72. (in Finnish).

Huuskonen, M.S., Koskinen, K., Tossavainen, A., Karjalainen, A., Rinne, J.-P. and Rantanen, J. (1995) Finnish Institute of Occupational Health Asbestos Program 1987–1992. *Am. J. Ind. Med.*, **28**, 123–42.

Karjalainen, A., Anttila, S., Heikkil, L., Karhunen, P. and Vainio, H. (1993a) Asbestos exposure among Finnish lung cancer patients: occupational history and fibre concentration in lung tissue. *Am. J. Ind. Med.*, **23**, 461–71.

Karjalainen, A., Anttila, S., Heikkil, L., Kyyrijnen, P. and Vainio, H. (1993b) Lobe of origin of lung cancer among asbestos-exposed patients with or without diffuse interstitial fibrosis. *Scand. J. Work Environ. Health*, **19**, 102–7.

Karjalainen, A., Huuskonen, M.S., Koskinen, K., Mattson, K., Riala, R. and Tossavainen, A. (1992) *Keuhkosyövän ja mesoteliooman ammattitautidiagnostiikka* (*Guidebook for exposure assessment and evaluation in the diagnostics of asbestos-related lung cancer and mesothelioma*), Institute of Occupational Health and Helsinki University Central Hospital, Helsinki, (in Finnish; English abstract).

Karjalainen, A., Meurman, L. and Pukkala, E. (1994a) Four cases of mesothelioma in miners. *Occup. Environ. Med.*, **51**, 212–15.

Karjalainen, A., Vanhala, E., Karhunen, P., Lalu, K., Penttil, A., Kyyrinen, P. and Tossavainen, A. (1994b) Asbestos exposure and pulmonary fibre concentration of 300 Finnish urban men. *Scand. J. Work Environ. Health*, **20**, 34–41.

Karjalainen, A., Taikina-aho, O., Anttila, S., Heikkil, L., Tossavainen, A. and Vainio, H. (1994c) Asbestos exposure among Finnish lung cancer patients – comparison of scanning and transmission electron microscopy in the analysis of lung burden, in *Inhaled Particles VII* (eds J. Dodgson and R.I. McCallum), Pergamon, Oxford, pp. 657–63.

Karjalainen, A., Anttila, S., Mäntylä, T., Taskinen, E., Kyyrönen, P. and Tukiainen, P. (1994d) Asbestos bodies in bronchoalveolar lavage fluid in relation to occupational history. *Am. J. Ind. Med.*, **26**, 655–54.

Karjalainen, A., Anttila, S., Vanhala, E. and Vainio, H. (1994e). Asbestos exposure and the risk of lung cancer in a general urban population. *Scand. J. Work Environ. Health*, **20**, 243–50.

Karjalainen, A., Karhunen, P.J., Lalu, K., Vanhala, E., Penttil, A., Kyyrinen, P. and Tossavainen, A. (1994f) Pleural plaques and exposure to mineral fibres in a male urban necropsy population. *Occup. Environ Med.*, **51**, 456–60.

Kiviluoto, R. (1960) Pleural calcification as a roentgenologic sign of non-occupational endemic anthophyllite asbestosis. *Acta Radiol. (Stockholm), (Suppl.)*, **194**, 1–67.

Koskinen, K., Rinne, J.-P., Zitting, A., Tossavainen, A., Kiveks, J., Reula, K., Roto, P. and Huuskonen, M.S. (1996) Screening of asbestos-induced diseases in Finland. *Am. J. Ind. Med.*, **30**, 241–51.

Meurman, L. (1966) Asbestos bodies and pleural plaques in a Finnish series of autopsy cases. *Acta Pathol. Microbiol. Scand. (Suppl.)*, **181**, 1–107.

Meurman, L. Kiviluoto, R. and Hakama, M. (1974) Mortality and morbidity among the working population of anthophyllite asbestos miners in Finland. *Br. J. Ind. Med.*, **31**, 105–12.

Meurman, L., Pukkala, E. and Hakama, M. (1994) Incidence of cancer among anthophyllite asbestos miners in Finland. *Occup. Environ. Med*, **51**, 42–5.

Ministry of Labour (1990) State Asbestos Committee. Final report 66/1989, Ministry of Labour, Helsilki. (in Finnish).

Noro, L. (1945) *Occupational Diseases and their Prevention*, Otava, Helsinki, (in Finnish).

Noro, L. (1946) On the histology of asbestosis. *Acta Pathol Microbiol. Scand.*, **23**, 53–9.

Nurminen, M. (1972) A study of the mortality of workers in an anthophyllite asbestos factory in Finland. *Work Environ. Health*, **9**, 112–18.

Nurminen, M. (1975) The epidemiological relationship between pleural mesothelioma and asbestos exposure. *Scand. J. Work Environ. Health*, **1**, 128–37.

Nurminen, M. and Tossavainen, A. (1994) Is there an association between pleural plaques and lung cancer without asbestosis? *Scand. J. Work Environ. Health*, **21**, 62–4.

Oksa, P., Huuskonen, M.S. Juntunen, J. *et al.* (1989) Follow-up study of asbestosis and silicosis patients. Final Report to Finnish Work Environment Fund, Helsinki, 13 p. (in Finnish).

Oksa, P., Koskinen, H., Roto, P., Rinne, J.-P., Zitting, A. and Huuskonen, M.S. (1992) Parenchymal and pleural fibrosis in construction workers. *Am. J. Ind. Med.*, **21**, 561–7.

Partanen, T., Nurminen, M., Zitting, A., Koskinen, H., Wiikeri, M. and Ahlman, K. (1992) Localized pleural plaques and lung cancer. *Am. J. Ind. Med.*, **22**, 185–92.

Raunio, V. (1966) Occurrence of unusual pleural calcifications in Finland: studies on atmospheric pollution caused by asbestos. *Ann. Med. Intern. Fenn. (Suppl.)*, **47**, 1–61

Tuomi, T. (1992) Fibrous minerals in the lungs of mesothelioma patients: comparison between data on SEM, TEM and personal interview information. *Am. J. Ind. Med.*, **21**, 155–6.

Tuomi, T., Huuskonen, M.S., Virtamo, M. *et al.* (1991a) Relative risk of mesothelioma associated with different levels of exposure to asbestos. *Scand. J. Work Environ. Health*, **17**, 404–8.

Tuomi, T., Huuskonen, M.S., Vanhala, E. and Virtamo, M. (1991b) Occupational exposure to asbestos as evaluated from work histories and lung tissue analysis of mesothelioma patients. *Br. J. Ind. Med.*, **48**, 48–52.

Tuomi, T., Segerberg-Konttinen, M., Tammilehto, L., Tossavainen, A. and Vanhala, E. (1989) Mineral fibre concentration in lung tissue of mesothelioma patients in Finland. *Am. J. Ind. Med.*, **16**, 247–54.

Valkila, E., Nieminen, M., Moilanen, A., Kuusisto, P., Lahdensuo, A. and Karvonen, J. (1995) Asbestos-induced visceral pleural fibrosis reduces pulmonary compliance. *Am. J. Ind. Med.*, **28**, 363–72.

Vilkman, S., Lahdensuo, A., Mattila, J., Tossavainen, A. and Tuomi, T. (1993) Asbestos exposure according to different exposure indices among Finnish lung cancer patients. *Int. Arch. Occup. Environ. Health*, **65**, 269–74.

Wegelius, C. (1947) Changes in the lungs in 126 cases of asbestos is observed in Finland. *Acta Radiol.*, **28**, 139–52.

Zitting, A.J. (1995) Prevalence of radiographic small lung opacities and pleural abnormalities in a representative adult population sample. *Chest*, **107**, 126–31.

Zitting, A.T., Karjalainen, A., Impivaara, O., Tossavainen, A., Kuusela, T., Mäki, J. and Huuskonen, M.S. (1995) Radiographic small lung opacities and pleural abnormalities as a consequence of asbestos exposure in an adult population. *Scand. J. Work Environ. Health*, **21**, 470–7.

19 Man-made vitreous fibers and respiratory health effects

Jacques Ameille, J.C. Pairon and P. De Vuyst

INTRODUCTION

Glass wool, rock wool, slag wool (all three are referred to as insulation wools), glass filaments and microfibers, and refractory ceramic fibers constitute the group of man-made mineral (MMMF) or vitreous (MMVF) fibers.

The term 'mineral fiber' evokes similarities with natural fibers (asbestos, erionite), which are the cause of various benign and malignant diseases. As the fibrous configuration of asbestos is an important parameter in toxicity, the synthetic fibers are also suspected of inducing serious adverse respiratory effects. Moreover, some of the uses and applications of MMVF and asbestos are very similar.

GENERAL POINTS

In this chapter, we will first define various aspects of MMVF classification and properties that may be relevant when examining their potential health effects. The factors influencing the fate of those fibers in a biologic environment (biopersistence) are also reviewed. General occupational and environmental hygiene data are reported.

Classification and composition

The Thermal Insulation Manufacturers' Association (TIMA) classification (Nomenclature Committee of TIMA Inc., 1991) is limited to vitreous silicate fibers, including the common purpose insulation wools (glass wool, rock wool, slag wool), glass continuous filaments, glass microfibers and vitreous silicate ceramic high-temperature (refractory) insulation fibers. Although all ceramic fibers are used in high-temperature insulation, TIMA proposed to exclude crystalline and non-silicate fibers from the definition of RCF (refractory ceramic fibers). Non-silicate and crystalline fibers represent small amounts of the total production of synthetic fibers and cover very specialized applications. Populations exposed to these fibers are small in number, while data concerning the fibers' potential toxicity are also scarce.

From an industrial point of view, essentially related to their manufacturing process, fiber diameter and final use, MMVF are also broadly categorized according to the following subdivisions: continuous filaments; insulation wools; special purpose glass microfibers; and refractory ceramic fibers.

MMVF are manufactured from molten glass, rock, slag, kaolin clay or combinations of silicon and aluminum oxide. Depending on the raw material used, the melting temperature ranges from 1000–1500°C

Occupational Lung Disease. Edited by Daniel E. Banks and John E. Parker. Published in 1998 by Chapman & Hall. ISBN 0 412 73630 6

for glass fibers, rock and slag wools, and 1800–2000°C for the most common RCF. During the fiberization process, several additives such as dust-suppressing agents or binders may be sprayed on the raw fibers (Klingholz, 1977).

The chemical composition of the various types of MMVF depends directly on the raw material used to manufacture them, and from the elements eventually added to meet some required properties. Continuous-filament glass fibers, glass wool and glass microfibers are made from silicated glass generally containing boron oxide. Rockwool is made from igneous (basalt) or metamorphic (granite, slate and limestone) rocks (Öhberg, 1987). Slags formed during the extraction of various metals from ore are used as raw material to produce slag wool. At present, only slag derived from iron blast furnaces is used, but in the past slags from various origins such as copper smelters were used.

RCF are manufactured from kaolin clay or from mixtures of pure silica and aluminum oxide.

In these different processes, variation in chemical composition can occur within the same type of production for fibers made of natural (sand, rock) or industrial (slag, reprocessed glass) materials.

Several additives are usually sprayed on the fibers during the fiberization process. Anti-static, anti-fungal, fire-retardant, hydrophobic or wetting agents may be added. These additives may represent as much as 25% of the weight of the final product. Mineral oil and polypropylene glycol are used as dust-suppressing agents. Typical binders include formaldehyde resins, urea, epoxy resins and bitumens.

The presence of several airborne contaminants or 'by-products' has been observed during the production of mineral wools, especially in the early production phases. In a realistic full-scale simulation of the working conditions prevailing at that time in a slag wool production plant, Fallentin and Kamstrup (1993) measured the emissions of various air pollutants. They reported the release of significant levels of PAH and airborne particulates during MMVF production.

The presence of asbestos fibers either as a contaminant of raw materials used in rockwool production or as a result of the parallel processing of asbestos-containing products in some MMVF production plants (Marsh *et al.*, 1990; Kilburn and Warshaw, 1991) is also described.

Uses

Continuous-filament glass fibers account for 10–15% of the total MMVF production. They are used as a reinforcement of cement, plastics and resinous materials, in paper and rubber products, in the form of textiles, and in electrical insulation. Insulation wools contribute more than 80% to the total MMVF production. They are mainly used for thermal and acoustical insulation, for fire protection (rock and slag wool), in acoustic ceiling tiles and panels, in the manufacture of air conditioning and ventilation ducts and as growing media for horticulture. RCF represent 1–2% of the total MMVF production; their main applications are in high-temperature insulation of furnaces and kilns. Special purpose glass fibers represent less than 1% of total fiber production.

MMVF properties

(a) Size characteristics of MMVF

The dimensional parameters of fibers are of utmost importance when considering fiber respirability and toxicity.

Diameter distribution and mean diameter of MMVF products depends on the composition of the fiber and the production process. Table 19.1 gives an overview of the 'mean' diameters reported in the literature for each type of product.

For continuous glass filaments, the diameter distribution is very narrow, all the fibers having approximately the same diameter. As a result of the relatively wide diameter of these fibers and the fabrication process (drawing), very few respirable fibers are produced. Insulation wool, RCF and special purpose fibers have a wider diameter distribution around the mean (Klingholz, 1977). Some of these products fall in the range of respirable fibers.

Due to production changes of MMVF between the 1930s and the present time, the average diameter of the products has decreased over the years. The fiber distribution of the products has become concentrated about the mean as a result of the reduction in the largest fiber diameters, but with no change in the percentage of the smaller-diameter fibers.

Table 19.1 Mean fiber diameters for MMVF

Type of fibers	*Mean diameter (μm)*
Continous glass filaments	5–16
Glass wool	4–6
Glass microfibers	0.1–3
Rock wool	3–8
Slag wool	3–8
Refractory ceramic fibers	1.2–3.5

Source: De Vuyst *et al.* (1995).

(b) Biopersistence of MMMF

It is accepted that the pathogenic properties of mineral fibers are related to their accumulation and to their persistence as fibers in target organs. This is actually a major field of concern and research in the evaluation of MMVF toxicity, since it is unlikely that fibers that are rapidly cleared or dissolved in the lung would have time enough to exert the potential toxicity linked to their dimensions and fibrous shape.

Because of their vitreous microstructure, MMVF do not split longitudinally like asbestos fibers, but break transversely. From this transverse breakage, fibers become shorter with the same diameter as the original ones. This is quite different from numerous fibers with smaller diameters but without changes in length that occur with defibrillation of asbestos fibers (Klingholz, 1977).

MMVF breakage may be enhanced by the fiber chemistry alteration that occurs *in vivo*.

Biopersistence of MMVF has been assessed after intratracheal injection and inhalation of MMVF in *in vivo* animal experiments, globally taking into account both the clearance mechanisms of the lung and fiber durability. After intratracheal injection in rats, rock wool exhibited limited solubility as compared with glass fibers after 18 months (Morgan and Holmes, 1984). RCF and some special purpose microfibers showed increased biopersistence, at a level similar to that of crocidolite, when compared to the current glass or rock fibers (Bellmann *et al.*, 1987).

Short-term (1-week) exposure of rats to aerosols of respirable fractions of commercial MMVF and crocidolite showed the following gradient for fiber biopersistence: crocidolite > rock wool > slag and glass wools (Musselman, *et al.*, 1994).

Fiber length plays an important role. Usually, for durable fibers like amphiboles, longer fibers are cleared more slowly than shorter ones (Morgan and Holmes, 1986). The same is observed with RCF (Yamato *et al.*, 1994). In contrast, long glass fibers are cleared more rapidly than shorter ones (Morgan *et al.*, 1982; Holmes *et al.*, 1983).

Morphologic aspects of *in vitro* and *in vivo* MMVF fiber alteration, confirming their dissolution, were observed by electron microscopy.

Biometrology in human lung tissues

Only three studies dealing with the quantification and characterization of MMVF by analytical electron microscopy in lung samples of exposed workers are available (McDonald *et al.*, 1990; Roggli, 1994; Sebastien *et al.*, 1994). In the largest of these studies (McDonald *et al.*, 1990), the fiber content of lung tissue samples from 131 deceased MMVF (glass, rock and slag wool) workers from a large US cohort and from 112 matched controls was analyzed. The most notable finding in this paper was the absence of any MMVF in the lung tissue of most exposed workers. The explanation is either that production workers were essentially exposed to non-respirable fibers, or more probably, that inhaled fibers did not 'survive' in the pulmonary environment. In addition, asbestos fibers were more commonly observed than MMVF fibers, both in workers and in controls.

The two other studies deal with workers exposed to RCF. Sebastien *et al.* (1994) examined the fiber content in the bronchoalveolar lavage (BAL) fluid of seven healthy workers currently employed in a RCF production plant. Si-Al fibers similar to ceramic fibers were recovered in the BAL samples from all workers but one. Roggli (1994) examined lung tissue from three ceramic fiber workers, including one with adenocarcinoma of the lung and parietal pleural plaques. In both studies, aluminium silicate fibers consistent with RCF were found. Electron microscopy demonstrated alteration of the chemical composition and the morphology of some of those fibers. The formation of ferruginous bodies on RCF mimicking asbestos bodies was reported.

However, the available data were too restricted to draw general conclusions about the biopersistence of all MMVF in the human lung.

Industrial and environmental hygiene data

Accurate estimation of past airborne fiber levels is of great importance for the interpretation of the data from the large-scale epidemiologic studies. Fiber levels in the workplace have been measured for only the past two decades. In production plants, airborne fiber concentrations increase when the mean diameter of the fiber under production decreases (Esmen *et al.*, 1979; Cherrie *et al.*, 1986). Average respirable fiber concentrations recorded in the production of continuous filaments range from 0.003 to 0.5 fibers/ml (Davies and Cherrie, 1992), in insulation wool from 0.01 to 2.0 fibers/ml (most values being lower than 1.0 fibers/ml), and for RCF and special purpose microfibers from 0.05 to 10 fibers/ml (exceptional values around 20) (Hori *et al.*, 1993).

With regard to users of MMVF, although some high peak values have been detected, levels are commonly lower than 1 fiber/ml (IPCS-WHO, 1988; Jaffrey *et al.*, 1990; Van Der Wal *et al.*, 1987; Schneider *et al.*, 1993). Highest levels are reached during blowing of insulation wools (Fowler *et al.*, 1971; Schneider *et al.*, 1993). There is a potential risk for higher exposure in secondary production processes and in end-users in the absence of environmental control measures and proper training in work practices (Lockey and Wiese, 1992).

From the limited concentration measurements available on MMVF airborne fiber levels in public buildings and homes it can be inferred that values in the range 10^{-4} to 10^{-3} fibers/ml are expected (Gaudichet *et al.*, 1989; Schneider *et al.*, 1990). Airborne fiber levels measured during major disturbances of in place loft MMVF insulation ranged from less than 0.002 to 0.3 fibers/ml (Jaffrey, 1990).

EXPERIMENTAL DATA

In vitro studies

In vitro cell studies provide valuable information on the mechanisms of fiber-related toxicity. These short-term tests allow screening for the effects of various physicochemical parameters of a given fiber, such as length, diameter, chemical composition and oxidant release. This cannot be systematically assessed in long-term inhalation studies.

The contribution of these tests in mechanistic studies and their possible use as screening assays has been emphasized in recent reviews (Hesterberg *et al.*, 1993a; Achard-Ellouk and Jaurand, 1994). Nevertheless, these *in vitro* cell culture models also have limitations. They bypass the natural mechanisms of deposition and clearance of particles which occur in the respiratory tract; they do not allow assessment of fiber durability and dissolution, nor do they define immunogenic cellular interactions. Therefore some authors have suggested that a battery of short-term tests should also include short-term inhalation tests and *in vitro* fiber solubility tests (Hesterberg *et al.*, 1993a; Bauer *et al.*, 1994). Moreover, attention should be paid to other mechanisms such as transfection of DNA by some fibers or oncogene activation which might also be involved in fiber-induced adverse responses.

In vitro cellular studies used for MMVF testing have been performed on various systems, mainly on rodent cells and human cells. They are aimed at assessing three main biologic end-points: cytotoxicity; inflammatory and/or fibrogenic response; and DNA or chromosome damage, or cellular transformation.

Comparative interpretation of results reported by authors of the different studies is generally difficult because information on fiber characteristics is often poor and doses employed are not expressed in similar units. Ideally, characterization of the fibers studied should indicate their chemistry, specific surface area, bivariate size distribution or at least mean, median and range values of length and diameter. Moreover, as fibers generally sediment in the culture device, dose should be expressed using a per surface area unit (μg per cm^2 of culture dish). A review of available data published on these cellular tests has recently been published (De Vuyst *et al.*, 1995). Studies of cytotoxicity and inflammatory/fibrogenic responses mainly concerned glass fibers. It was shown that the role of fiber dimension was prominent in the biologic response, the thin and long fibers exhibiting the highest potency on a per mass basis (Pickrell *et al.*, 1983; Gormley *et al.*, 1985; Brown *et al.*, 1986; Forget *et al.*, 1986). Some authors concluded that glass fibers having a distribution of length and diameter similar to that of asbestos fibers were comparable in toxicity on guinea-pig lung macrophages (Tilkes and Beck, 1983a, b).

Results of DNA and chromosome damage tests and of cellular transformation tests are available almost exclusively for glass microfibers. Genotoxic and/or transforming activity was observed for some MMVF in rodent cells (Sincock, 1977; Brown *et al.*, 1979; Sincock *et al.*, 1982; Woodworth *et al.*, 1983; Hesterberg and Bameit, 1984; Oshimura *et al.*, 1984; Hesterberg *et al.*, 1986; Mikalsen *et al.*, 1988; Fischer, 1989; Hart *et al.*, 1992), and less frequently in human cells (Linnainmaa, *et al.*, 1991; Pelin *et al.*, 1995). It should be emphasized that few laboratories used specific respiratory cells such as human mesothelial cells or airway epithelial cells (Linnainmaa *et al.*, 1991; Riebe-Imre *et al.*, 1994; Pelin *et al.*, 1995). The different genotoxic response of rodent cells and human cells might be linked to a better capacity for DNA repair and/or to a lower phagocytic capacity of human cells (Sincock *et al.*, 1982; Casey, 1983). Phagocytosis of fibers has been suggested to be essential for the induction of chromosomal damage, perhaps because of interference of fibers with chromosomes during mitosis (Oshimura *et al.*, 1984; Hesterberg *et al.*, 1986). Long fibers, because they are better phagocytosed than short fibers, would thus exhibit higher genotoxic potency. However, it is not possible to evaluate precisely the quantity of fibers which have been internalized in cells in these short-term tests. Alternatively, DNA damage could also be attributed to the production of hydroxyl radicals which results from the interaction of MMVF with the cell plasma membrane (Hansen and Mossman, 1987; Maples and Johnson, 1992).

In vivo studies

Prediction of the potential health effects of MMVF in humans using animal studies requires models which take into account the different steps following fiber exposure. Different events include deposition, epuration, retention of a fraction of the fibers inhaled, and translocation to the target cell. Rodents have been widely used to evaluate the *in vivo* toxicity of MMVF, using different routes of administration.

Studies can be divided into three main groups, each having a different purpose.

(a) Intracavitary studies

These studies allow the direct contact of a high dose of fibers with mesothelial cells after injection into the pleural or peritoneal cavity. This indicates that intracavitary studies are a very sensitive means for detecting the carcinogenic potency of different dusts, including fibers (Pott *et al.*, 1987, 1991). The intrapleural route of injection seems to provide results which fit better to human mesothelioma risk than results obtained with the intraperitoneal model (Carthew *et al.*, 1992). However, some recent data suggest that a more or less similar mechanism occurs in the peritoneal cavity and in the pleural space after injection of MMVF (Collier *et al.*, 1994). The intrapleural model allowed demonstration of the Stanton hypothesis concerning fibers (Stanton *et al.*, 1981): i.e. the incidence of pleural tumors was the highest with fibers having a length greater than 8 μm and a diameter less than 0.25 μm. Nevertheless, tumors also occurred with shorter and thicker fibers. Besides dimension, it has recently been emphasized that other parameters were also important in this model, i.e. the chemical composition and the dose injected (De Vuyst *et al.*, 1995). The main results obtained by intraperitoneal or intrapleural injection of MMVF are summarized in Figure 19.1.

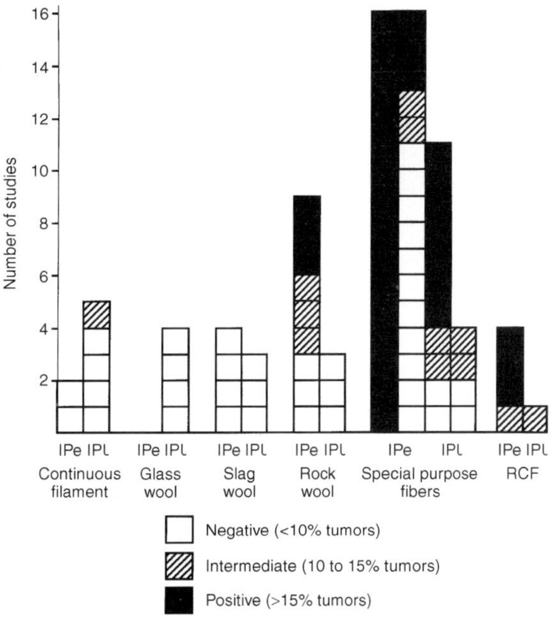

Figure 19.1 Results obtained by intraperitoneal and intrapleural injection of MMVF in rats. Each square represents result of one study. IPe, intraperitoneal injection; IPl, intrapleural injection; RCF, refractory ceramic fiber. Special purpose fibers include John Manville fine fibers, Owens Corning fine E-fibers and Bayer fine fibers. Highest rate of tumours in control rats: intraperitoneal experiments, 6.2%; intrapleural experiments, 2.8%. (Adapted from De Vuyst *et al.*, 1995.)

Although they allow exploration of the intrinsic carcinogenic effect of a given fiber, such intracavitary studies have some limitations which make extrapolation of results to humans hazardous: first, they bypass the phenomenon of translocation of fibers from the alveoli to the pleura and the different clearance mechanisms which occur following inhalation. Moreover, some non-durable fibers will probably reach the pleural cavity after significant surface changes, while the test only uses 'native' fibers. Secondly, although results may indicate some serosal carcinogenic potency of a given fiber, extrapolation of the results to the lung epithelial carcinogenic or fibrotic risk remains questionable.

(b) Short-term inhalation studies and intratracheal studies

These studies have been proposed for the assessment of the biopersistence of various MMVF in the rat lung (Morgan *et al.*, 1982; Bellman *et al.*, 1994; Bernstein *et al.*, 1994; Musselman *et al.*, 1994).

(c) Long-term inhalation studies

The inhalation route is generally considered as the most suitable model to assess the fibrogenic and carcinogenic effect of the fibers. Recommendations were established recently by the World Health Organization for this model (1992), and were taken into account in recent studies (Bernstein *et al.*, 1995). Thus, a detailed description of the bivariate dimensions (length, diameter) and chemical composition of the fibers tested is required. Only fibers having a diameter less than 1 μm are respirable in rodents. Consequently, size-selected fibers have to be used to allow deep penetration into the rodent airways. An adequate range of dose is required, i.e. up to the maximum tolerated dose, but less than the dose leading to an overload effect. Asbestos is generally used as a positive control, but protocols used do not allow a fiber to fiber comparison (asbestos versus MMVF) according to size. Results of the most recent studies in rats are summarized in Figure 19.2.

No significant excess of fibrosis or tumors was observed in rats following inhalation of glass wool and slag wool at the doses used [up to 1400 fibers/ml in the study of Wagner *et al.* (1984) and up to 246 fibers/ml in the recent study of Bunn *et al.* (1993)]. The

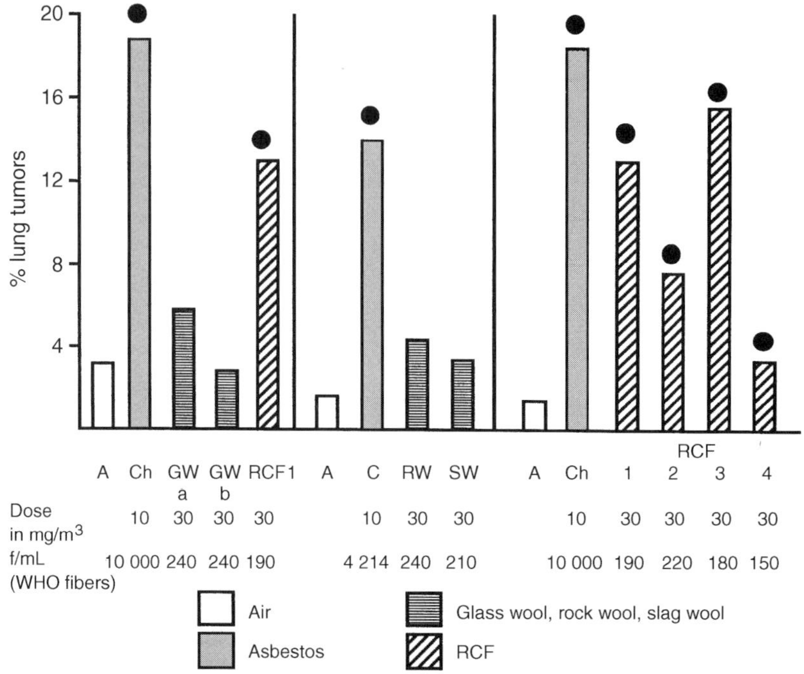

Figure 19.2 Summary of recent inhalation studies in the rat (Hesterberg *et al.*, 1993b; Bunn *et al.*, 1993; McConnell *et al.*, 1994; Mast *et al.*, 1995). Only the highest dose employed for each type of fiber is indicated. Rats were exposed 6 hours/day, 5 days/week for 24 months. Lung tumors include adenomas and carcinomas. A, air; Ch, chrysotile; C, crocidolite; GWa, glass fiber MMVF10; GWb, glass fiber MMVF11 (a frequency of 7.5% tumors was observed for the intermediate dose, i.e. 16 mg/m^3, for this fiber); RW, rock wool fiber; SW, slag wool fiber; RCF, refractory ceramic fiber. ● indicates occurrence of mesothelioma in the group considered.

incidence of lung tumors did not exceed 7.5% for the intermediate dose of glass fibers (16 mg/m^3, i.e. 150 fibers/ml), without any clear dose–response relationship. However, it has been reported that a significant excess of lung tumors was observed after pooling of all published inhalational data (Achard-Ellouk and Jaurand, 1994).

Lung fibrosis, exceeding Grade 4 in the Wagner scale, was reported in rats treated with rock wool fibers at the two highest doses employed (16 mg/m^3; 145 fibers/ml, and 30 mg/m^3; 247 fibers/ml), although there was no excess of tumors (Bunn *et al.*, 1993).

Results obtained for glass microfibers were negative, contrasting with the results of intracavitary tests, possibly due to variations in the chemical composition of these fibers.

A strong fibrotic and carcinogenic effect was demonstrated for refractory ceramic fibers (RCF) in both rat and hamster, although these species exhibited different pattern of response. Exposure to 30 mg/m^3 (200 fibers/ml) of a RCF (RCF1) led to an incidence of mesothelioma of 42% in hamsters and no lung tumors; the same exposure led to mesothelioma in only 1.6% of rats, but 13% had a lung tumor. Moreover, a significant fibrotic effect was observed within 6 months of beginning the experiment in the lungs of rats with this dose (Bunn *et al.*, 1993). There was a clear dose–response relationship for the fibrotic response in this study. Results of this carcinogenicity study contrasted with those obtained previously by Smith *et al.* (1987), who used a similar level of exposure (about 200 fibers/ml) but employed fibers having a thicker diameter (mean diameter 1.8 μm instead of 0.98 μm): no tumors were observed in rats and only one mesothelioma in hamsters.

HUMAN DATA

Non-malignant respiratory diseases

(a) Lung fibrosis (pneumoconiosis)
There is no firm evidence of an association between exposure to MMVF and small parenchymal opacities (Wright, 1968; Nasr *et al.*, 1971; Weill *et al.*, 1983; Hill *et al.*,1984).

Two cross-sectional studies on the same group of workers from glass, slag and rockwool production plants (Hughes *et al.*, 1993; Weill *et al.*, 1983) have been carried out. In the first study, on 941 X-ray films coded according to the ILO 1980 classification, small irregular opacities were detected in a small percentage of workers, all with low-grade profusion: 7% with Grade 0/1; 3% of 1/0 and 0.6% with 1/1 (Weill, 1983). The prevalence of abnormalities (> 0/1) increased with age and cigarette smoking, as well as with exposure to fine fibers among current smokers. In a recent follow-up study, the same authors did not find a significant effect of MMVF exposure on the X-ray findings when compared with a group of local blue-collar workers (Hughes *et al.*, 1993). Interestingly, there was no evidence of radiographic progression between the two studies, a supplementary argument against the existence of a MMVF-induced pneumoconiosis.

It must be borne in mind that a low profusion of small opacities must not be considered evidence for pneumoconiosis. In those with mild profusion categories (up to 1/1), cigarette smoking and age influence ILO gradings when emphysema is present (Castellan *et al.*, 1985; Dick *et al.*, 1992). CT scan studies clearly confirm that small opacities do not necessary correspond to pneumoconiotic lesions (Gevenois *et al.*, 1994). This misleading interpretation appears in a study on 284 MMVF appliance workers, in whom Kilburn *et al.* (1992) found 15% of the workers had evidence of 'pneumoconiosis', including category 0/1. One must be careful ascribing X-ray abnormalities to pneumoconiosis attributable to MMVF. In the same study, 17 of 43 workers with X-ray abnormalities had previous asbestos exposure.

(b) Pleural lesions

There is no evidence of a relationship between glass, rock and slag wool exposure and pleural abnormalities. In the first Weill study (1983), the prevalence of pleural thickening was 1.6% [(2.3% in the general US population (Rogan *et al.*, 1987)] and that of pleural plaques was 0.2%. Interestingly, in a series of workers involved in the manufacture of refractory ceramic fibers, 3.4% had pleural changes considered to be pleural plaques. The prevalence reached 11.4% in the group with more than 20 years since beginning their first RCF production job (Lemasters *et al.*, 1994).

(c) Non-specific respiratory diseases

Several studies have been undertaken to evaluate the relationship between MMVF exposure and chronic obstructive pulmonary disease (COPD). These studies are difficult to compare with each other, since different parameters have been used. Nevertheless, the results of most studies do not support an increase in non-malignant respiratory disease due to exposure to glass wool and rock wool in the production industry.

In a population of 2024 workers from glass and rock wool plants in France, correlations were found between respiratory symptoms and age or smoking (Moulin *et al.*, 1987), but not with exposure.

More recent studies (Chiazze *et al.*, 1992; Hughes *et al.*, 1993) confirmed the absence of a relationship between respiratory symptoms or loss of pulmonary function and exposure to MMVF. The last follow-up of the US cohort study (Marsh *et al.*, 1990) found no evidence of a relationship between MMVF exposure and deaths from non-malignant respiratory disease.

Some data indicate that there may be an interaction between smoking and fiber exposure in the causation of obstructive airway disease in ceramic production workers. Cumulative exposure to RCF could cause airways obstruction by promoting the effects of cigarette smoke (Trethovan *et al.*, 1995).

(d) Discussion

At the present time, there is no evidence of a pneumoconiosis risk in workers exposed in either glass, rock or slag wool production plants. This is probably due to the low respirability and biopersistence of these fibers when compared with asbestos. Nor is there clear evidence of a relationship between MMVF exposure and non-specific respiratory disease observed in association with smoking. In this regard, studies of bronchial hyper-reactivity in workers exposed to fine respirable fibers are required. The possibility that chronic exposure to fine MMVF could induce inflammation or irritation of small airways – and thus create bronchial hyper-responsiveness – should be examined.

The possible role of ceramic fibers in the development of pleural plaques should also be confirmed.

Exposure to several contaminants in the work place, such as inorganic dust (asbestos, talc, silica), formaldehyde, asphalt fumes or resins can be the cause of several respiratory diseases not directly related to MMVF. Of particular environmental interest is the possible contamination of the work place by microorganisms, due to high levels of humidity and high temperature. Cases of extrinsic allergic alveolitis (proven by serologic studies, BAL and lung biopsy) have been observed in glass filament workers who were exposed to various microscopic fungi contaminating the air conditioning system of the factory (De Vuyst *et al.*, 1992).

Malignant respiratory diseases

Most of the current knowledge concerning the risk of malignant respiratory disease in MMVF production workers is provided by two large historical cohort studies, one conducted in the United States (Enterline, *et al.*, 1987; Marsh *et al.*, 1990), and one conducted in Europe (Simonato *et al.*, 1987). Two smaller studies from Ontario, Canada gave additional information concerning the risk of lung cancer in glass wool (Shannon *et al.*, 1987) and glass continuous filament (Shannon *et al.*, 1990) production workers.

The US cohort consisted of 16 661 MMVF workers with 6 months or more of work experience in production or maintenance during the years 1945 to 1963 (1940 to 1963 for one plant) at one or more of 17 manufacturing plants (11 glass wool and six mineral wool plants).

The European study, conducted by the IARC, consisted of 21 967 workers producing MMVF in seven European countries (Denmark, Finland, Norway, Sweden, United Kingdom, Germany and Italy). These workers were followed for death from the beginning of MMVF production, which ranged from 1933 to 1961.

The two Canadian studies were conducted by the same investigators. The first one included 2557 men employed in the production of glass wool, who were followed-up from 1955 to 1984 (Shannon *et al.*, 1987). The second one consisted of 1465 men and women employed at least one year at a glass filament plant between 1951 and 1986 (Shannon *et al.*, 1990).

All four cohorts succeeded in tracing over 95% of the defined population. The US and Canadian studies gave information on airborne fiber concentrations and cumulative exposure. In the European cohort, no hygiene measurements from the past were available, and airborne concentration levels could not be quantified. However, technological phases which reflect different levels of exposure to fibers at different time periods were identified. Airborne levels of MMVF were estimated to be highest when no dust-suppressing agent was used (early technological phase). The periods when oil and resin binders were in use with modern mechanized production methods (late technological phase) was judged to correspond to periods of lowest airborne levels of fibers. The intermediate technological phase (use of processes both with and without dust suppressors) was considered to reflect intermediate levels of exposure.

(a) Lung cancer

European study (Simonato, 1987)

Mortality from all malignant neoplasms in the total exposed cohort, using national comparison rates, is statistically significantly elevated (SMR = 111; 95% C.I. 102–119), the excess being essentially related to lung cancer. The SMR value for lung cancer is 125 when compared with national rates ($P < 0.05$) and 110 (non-significant) when compared with local rates. Using national rates for comparison, SMR values are respectiveley 124 for the rock/slag wool subcohort (non-significant), 127 ($P < 0.05$) for the glass wool subcohort, and 120 (non-significant) for the continuous-filament subcohort. Using local rates for comparison, respective SMR values are 124, 103 and 97.

For the rock/slag wool and glass wool subcohorts, non-significant increasing trends for the SMR values with time since first exposure are observed. No such pattern was apparent for the continuous-filament subcohort. For none of the three processes was there any association between lung cancer mortality and length of employment. A significant excess of lung cancer is found among rock/slag wool workers employed during the early technological phase (SMR = 257; $P < 0.05$). There is no excess in workers employed in the intermediate and late technological phases.

The European study has been recently updated (Boffetta *et al.*, 1994). Using national rates for comparison, the SMRs are 137 for the rock/slag wool subcohort ($P < 0.05$), 127 for glass wool subcohort ($P < 0.05$) and 127 for continuous filament. Using local rates, the respective SMRs are 139 ($P < 0.05$), 112 and 107.

US study (Marsh, 1990)

Using local death rates to estimate expected deaths, there is a small but significant excess for all malignant neoplasms (SMR = 108.3; $P < 0.05$) and for lung cancer (SMR = 112.5; $P < 0.05$) for the period 1946 to 1985. The mortality from lung cancer is highest in rock/slag wool production (SMR = 135.6; $P < 0.05$), intermediate in glass wool production (SMR = 106.7), and lowest in glass filament production (SMR = 98.4).

For none of the subcohorts does there appear any relationship between lung cancer and duration of employment or time since first exposure. Furthermore, for each of the subcohorts the majority of lung cancer deaths occurred among workers employed for less than 10 years, and there is no evidence of a direct relationship between MMVF exposure and lung cancer, using either cumulative fiber exposure or average intensity of exposure. However, the SMRs for lung cancer in rock/slag wool, glass wool and continuous-filament subcohorts seem to run parallel to the average fiber exposure: 0.351 fiber/ml for the rock/slag wool subcohort; 0.033 fiber/ml for the glass wool subcohort; and 0.011 fiber/ml for the continuous filament subcohort.

In addition to SMR analyses, the US study used Poisson regression to investigate the dependance of SMRs on combinations of fibers exposure and/or potential confounding variables. The modeling showed that although the overall SMR for lung cancer is significantly elevated in the total MMVF cohort, there is no statistically significant pattern of increased risk associated with any of the indicators of exposure including the surrogate measures (e.g. process, plant, duration of employment), and the measures based on individual worker fiber exposure estimates.

Canadian studies (Shannon *et al.*, 1987, 1990)

In the study of glass wool production workers, there is a statistically significant excess of mortality due to lung cancer (SMR = 199; *P* <0.05) using regional rates for comparison. However, the lung cancer excess is not related to time since first exposure, or to duration of exposure. Moreover, SMRs show a decreasing trend with years employed and no relationship with cumulative exposure.

In the study of continuous-filament production workers (Shannon *et al.*, 1990) overall mortality is below that expected. There are 11 deaths due to lung cancer (SMR = 136; *P* = 0.31). Analysis by both duration of employment and cumulative exposure shows this increase to be inconsistent with an occupational etiology.

Pooled results

The results from the US (Enterline *et al.*, 1987), European (Simonato *et al.*, 1987) and Canadian (Shannon *et al.*, 1987) cohorts have been pooled by Doll (1987), providing information about a total of 7862 deaths in 41 185 workers. For the sake of comparability the data on persons employed for less than 1 year were excluded.

The pooled estimates of the three studies are given in Table 19.2. The mortality from lung cancer (SMR = 121) was significantly raised. On the other hand, the mortality from all other causes of death was close to that expected. The mortality from lung cancer is highest in the rock/slag wool production. For rock/slag wool workers there is a trend in SMRs by time since first exposure: the lowest SMRs are observed within 10 years after first exposure and the highest 30 or more years after first exposure.

According to the most recent review by Hughes and Weill (1994), the pooled data of the US, European and Canadian studies show SMRs significantly elevated for glass wool (SMR: 114.9; *P* <0.05) and rock/slag wool (SMR = 136.2; *P* <0.01) as compared with local rates.

(b) Mesothelioma

The European and US studies also provide information on mortality from mesothelioma. A total of five mesothelioma cases occurred in the two cohorts, one in the European study (Simonato *et al.*, 1987), and four in the US study (Marsh *et al.*, 1990). Given the size of the cohorts, these numbers are not in excess of the number which may be expected. The general consensus on this issue is that no increase in risk of mesothelioma has been demonstrated in the MMVF production industry. However, it must be borne in mind that epidemiologic studies have limitations in discriminating between the lack of any risk and the possibility of small risk.

(c) Discussion

An excess of mortality from lung cancer, but not from mesothelioma, has been demonstrated in the MMVF production industry. The risk is mainly for workers exposed in the rock/slag wool industry during the early technological phase, and no excess is observed in

Table 19.2 Mortality from lung cancer by type of exposure (number of deaths in parentheses)

	SMR			*Total*
	Rock/slag wool	*Glass wool*	*Continuous filament*	
Europe	124 (81)	103 (93)	97 (15)	128 (154)**
US	133 (60)	109 (267)	92 (64)	116 (370)**
Canada		174 (19)*		174 (19)*
All cohorts	128 (141)**	110 (379)	93 (79)	121 (543)***

SMR, standardized mortality ratio; national rates were used as reference for US and Europe cohorts, provincial rates for the Canadian cohorts.

* *P* <0.05; ** *P* <0.01; *** *P* <0.001.

Source: Doll (1987).

workers producing continuous filaments. No epidemiologic data are available for workers exposed to ceramic fibers.

As stated by Doll (1987), the increased mortality is not large in relative terms, but lung cancer is so common that even a 10 or 20% increase in mortality is a matter of concern.

Agreement between the US and European findings is quite good, so it has been postulated that it is unlikely that chance alone has played a significant role in the excess of lung cancer (Infante *et al.*, 1994). However, the question arises as to whether the excess mortality from lung cancer is due to the fibers themselves or to other agents. The evidence in favour of and against a causal role of MMVF in lung cancer has recently been summarized (De Vuyst *et al.*, 1995) in a review devoted to the respiratory health effects of MMVF (Table 19.3).

Several facts support a possible link between exposure to MMVF and lung cancer risk. For rock/slag wool workers, lung cancer mortality increases with time since first exposure to a maximum after 30 years or more (Doll, 1987). Moreover, the gradient of lung cancer is consistent with the gradient of fiber concentrations across sectors of the industry. However, as it has been pointed out by Enterline (1990, 1991), one of the difficulties in attributing a causal relationship between respiratory cancer and MMVF is that the estimated fiber concentrations to which workers were exposed were extremely low. Given our knowledge of the effects of asbestos fibers, similar levels of exposure to asbestos would probably have not produced a detectable excess of lung cancer. It may be however that the concentration of MMVF to which employees had been exposed in the early technological phase was notably underestimated (Doll, 1987).

Table 19.3 Summary of the evidence in favor of and against a casual role of MMVF in lung cancer

Evidence in favor

1. Similarities of SMRs for lung cancer in the main cohort studies.
2. Elevated SMRs are found for lung cancer only.
3. Parallel between the lung cancer SMRs and the airborne concentration of the fibers.

Evidence against

1. No increase of lung cancer risk with increase of exposure duration.
2. The increase in lung cancer occurred mostly in workers producing slag wool in the early production phase, which contained traces of arsenic, a well-established human lung carcinogen.
3. Documented exposures to other agents in the work place known to be capable of inducing lung cancer, including asbestos and polycyclic aromatic hydrocarbon.
4. Lack of risk of lung cancer in case-control studies adjusted for smoking.

SMR, standardized mortality ratio.
Source: De Vuyst *et al.*, (1995).

Evidence against a causal relationship between MMVF exposure and lung cancer include a lack of relationship between risk and duration of employment, used as a surrogate for dose. Yet, short duration of employment does not necessarily mean a low cumulative exposure. Thus, a short duration of employment in the early technological phase may have carried more risk than a long duration of employment in the late technological phase.

The increase in mortality from lung cancer in persons employed in the MMVF industry is also difficult to interpret because of the inability to control adequately for socioeconomic status, cigarette smoking and exposure to carcinogenic contaminants in the manufacturing process.

Exposed workers in MMVF production are mostly unskilled laborers of a lower socioeconomic status. This is possibly associated, for reasons that go beyond smoking, to an elevated SMR for lung cancer (Miettinen and Rossiter, 1990).

The possible confounding effect of cigarette smoking has been recently investigated in two case-control studies (Wong *et al.*, 1991; Chiazze *et al.*, 1992, 1993). Wong *et al.* (1991) carried out a nested case-control study in 55 workers with lung cancer and 98 controls from nine mineral plants, four of which were included in the US study. Cigarette smoking was found to be responsible for the observed increase in mortality from lung cancer and the risk increased with increasing pack-years of cigarette smoking. No increased risk of lung cancer was associated with exposure to MMVF and analysis by cumulative fiber exposure did not indicate any trend of increasing risk of lung cancer. The other case-control study included 144 lung cancer cases and more than 200 controls among employees of the largest fibrous glass plant in the US. The results indicated that smoking was the most important factor for lung cancer and that respirable fibers were not a significant factor (Chiazze *et al.*, 1992, 1993).

The possibility of confounding by exposures to human carcinogens must also be seriously considered. Several carcinogens may have been present in the industrial environment including asbestos, polycyclic aromatic hydrocarbons and arsenic (Cherrie *et al.*, 1986; Öhberg, 1987).

In several plants included in the US and European studies evidence has surfaced of concomitant exposure to asbestos. A study of lung tissue from workers in the

US study showed the presence of amosite in two rock/slag wool plants (McDonald *et al.*, 1990). However, in both the US and European studies, the lung cancer SMR for plants in which asbestos might have been used was about the same as in the plants in which it was believed asbestos was not used (Enterline, 1991). Consequently, the use of asbestos seems an unlikely candidate to account for the excess of lung cancer, particularly in a population in whom no apparent excess in mesothelioma cases has been detected (De Vuyst *et al.*, 1995).

Among rock/slag wool workers, the highest increased risk for lung cancer was found in the workers exposed to slag wool. In the US study, four of the six mineral wool plants used slag almost exclusively. In these plants the SMR for lung cancer, 20 years from first exposure, was double that expected. Two plants were originally rock wool plants and estimated fiber exposure levels in these plants did not differ from those in slag wool plants. No excess of respiratory cancer was observed in these two plants (Enterline, 1990). Similar facts were noted in the European study. Some types of copper slag are known to contain significant amount of arsenic, a human carcinogen, and this could have played a role in the excess of lung cancer.

CONCLUSIONS

Because of similarities to asbestos fibers, the concern about health hazards associated with MMVF exposure is great, even though MMVF differs from asbestos fibers in several ways. It has been demonstrated that most commercial MMVF are more soluble, less biopersistent and coarser than asbestos fibers. The assessment of the respiratory health effects of MMVF is a complex issue. The term MMVF includes very large non-respirable glass filaments as well as very fine refractory ceramic fibers or glass microfibers. It is also difficult to extrapolate the results from cellular and animal studies to humans.

To date, there is insufficient evidence to conclude that occupational exposure to MMVF entails a mesothelioma risk. Nor is there firm evidence that exposure to glass filaments, glass wool, rock wool and slag wool is associated with lung fibrosis, pleural disease or COPD.

An elevated SMR for lung cancer has been demonstrated for workers employed in the early technological phase of rock/slag wool production. At that time, carcinogenic agents including asbestos, arsenic and polycyclic aromatic hydrocarbons were also present at the workplace, and quantitative data about smoking are lacking. For these reasons it is not possible to ascribe the elevated risk of lung cancer only to MMVF themselves. Moreover, animal data provide insufficient evidence of a carcinogenic risk of mineral wools.

Nevertheless, insulation wools have been categorized as Grade 2b, 'possibly carcinogenic to humans', by the International Agency for Research on Cancer (IARC, 1988).

At the present time few data are available concerning neoplastic diseases in ceramic fiber production workers. However, it must be borne in mind that, in chronic inhalation studies, ceramic fibers produce an increase in mesothelioma incidence in rats and hamsters, and an increased incidence of lung tumors in rats. Due to persistent uncertainties, occupational exposures to MMVF should be lowered as much as possible with proper workers' training and by safe work practices.

REFERENCES

Achard-Ellouk, S. and Jaurand, M.C. (1994) Review on animal/in vitro data on biological effects of man made fibers. *Environ. Health Perspect.*, **102**(Suppl. 2), 47–61.

Bauer, J.F., Law, B.D. and Hesterberg, T.W. (1994) Dual pH durability studies of man-made vitreous fiber (MMVF). *Environ. Health Perspect.*, **102**(Suppl.5), 61–5.

Bellmann, B., Muhle, H., Pott, F., König, H., Klöppel, H. and Spurny, K. (1987) Persistence of man-made mineral fibres (MMMF) and asbestos in rat lungs. *Ann. Occup. Hyg.*, **31**, 693–709.

Bellmann, B., Muhle, H., Kamstrup, O. and Draeger, U.F. (1994) Investigation on the durability of man-made vitreous fibers in rat lungs. *Environ. Health Perspect.*, **102**(Suppl.5), 185–9.

Bernstein, D.M., Mast, R., Anderson, R., Hesterberg, T.W., Musselman, R., Kamstrup, O. and Hadley, J. (1994) An experimental approach to the evaluation of the biopersistence of respirable synthetic fibers and minerals. *Environ Health Perspect.*, **102**(Suppl.5), 15–18.

Bernstein, D.M., Thevenaz, Ph., Fleissner, H., Anderson, R., Hesterberg, T.W. and Mast, R. (1995) Evaluation of the oncogenic potential of man-made vitreous fibres: the inhalation model. *Ann. Occup. Hyg.*, **39**, 661–72.

Boffetta, P., Saracci, R., Andersen, A. *et al.* (1994) The IARC historical cohort study of man-made mineral fiber production workers in seven European countries: preliminary results of an extension of the mortality follow-up. Abstract; 10th International Symposium of Epidemiology in Occupational Health. ISEOH 1994, Como, Italy.

Brown, G.M., Cowie, H., Davis, J.M.G. and Donaldson, K. (1986) *In vitro* assays for detecting carcinogenic mineral fibres: a comparison of two assays and the role of fibre size. *Carcinogenesis*, **7**, 1971–4.

Brown, R.C., Chamberlain, M., Davies, R., Gaffen, J. and Skidmore, J.W. (1979) *In vitro* biological effects of glass fibers. *J. Environ. Pathol. Toxicol.*, **2**, 1369–83.

Bunn, W.B., Bender, J.R., Hesterberg, T.W., Chase, G.R. and Konzen, J.L. (1993) Recent studies of man-made vitreous fibers. Chronic inhalation studies. *J. Occup. Med.*, **35**, 101–13.

Carthew, P., Hill, R.J., Edward, R.E. and Lee, P.N. (1992) Intrapleural administration of fibres induces mesothelioma in rats in the same relative order of hazard as occurs in man exposure. *Hum. Exp. Tox.*, **11**, 530–4.

Casey, G. (1983) Sister-chromatid exchange and cell kinetics in CHO-K1 cells, human fibroblasts and lymphoblastoid cells exposed *in vitro* to asbestos and glass fibre. *Mutat. Res.*, **116**, 369–77.

Castellan, R., Sanderson, W. and Petersen, M. (1985) Prevalence of radiographic appearance of pneumoconiosis in an unexpected blue collar population. *Am. Rev. Respir. Dis.*, **131**, 684–6.

Cherrie, J., Dodgson, J., Groat, S. and MacLaren, W. (1986) Environmental surveys in the European man-made mineral fibre production industry. *Scand. J. Work Environ. Health*, **12**, 18–25.

Chiazze, L., Watkins, D. and Fryar, C. (1992) A case-control study of malignant and non-malignant respiratory disease among employers of a fibreglass manufacturing facility. *Br. J. Ind. Med.*, **49**, 326–31.

Chiazze, L., Watkins, D.K., Fryar, C. and Kozono, J. (1993) A case-control study of malignant and non-malignant respiratory disease among employees of a fibreglass manufacturing facility. II Exposure assessment. *Br. J. Ind. Med.*, **50**, 717–25.

Collier, C.G., Morris, K.J., Launder, K.A., Humphreys, J.A., Morgan, A., Eastes, W. and Townsend, S. (1994) The behaviour of glass fibers in the rat following intraperitoneal injection. *Regul. Tox. Pharmacol.*, **20**, S89–103.

Davies, L. and Cherrie, J.W. (1992) Airborne fibre concentrations in a glass continuous filament factory. *Ann. Occup. Hyg.*, **36**, 609–27.

De Vuyst, P., Symoens, F., Saint Remy, P., Jacobovitz, D., Van de Weyer, R. and Nolard, N. (1992) Alvéolite allergique extrinsèque chez un ouvrier de fabrication de fibres de verre. *Rev. Fr. Mal. Resp.* (suppl. 3), 150.

De Vuyst, P., Dumortier, P., Swaen, G.M.H., Pairon, J.C. and Brochard, P. (1995) Respiratory health effects of man-made vitreous (mineral fibres). *Eur. Respir. J.*, **8**, 2149–73.

Dick, J., Morgan, K., Muir, D., Reger, R. and Sargent, N. (1992) The significance of irregular opacities on the chest roentgenogram. *Chest*, **102**, 251–60.

Doll, R. (1987) Symposium on MMMF, Copenhagen, October 1986: overview and conclusions. *Ann Occup Hyg.*, **31**, 805–19.

Enterline, P. (1990) Role of man made mineral fibers in the causation of cancer. *Br. J. Ind. Med.*, **47**, 145–6.

Enterline, P.E. (1991) Carcinogenic effects of man-made vitreous fibers. *Annu. Rev. Publ. Health*, **12**, 459–80.

Enterline, P.E., Marsh, G.M., Henderson, V. and Callahan, C. (1987) Mortality update of a cohort of U.S. man-made mineral fibre workers. *Ann. Occup. Hyg.*, **31**, 625–56.

Esmen, N., Corn, M., Hammad, Y., Whittier, D. and Kotsko, N. (1979) Summary of measurements of employee exposure to airborne dust and fiber in sixteen facilities producing man-made mineral fibers. *Am. Ind. Hyg. Assoc. J.*, **40**, 108.

Fallentin, B. and Kamstrup, O. (1993) Simulation of past exposure in slag wool production. *Ann. Occup. Hyg.*, **37**, 419–33.

Fischer, A.B. (1989) Induction of sister chromatid exchanges by fibrous dusts alone and in combination with other xenobiotics in Chinese hamster cells, in *Effects of Mineral Dusts on Cells*, (eds B.T. Mossman and R.O. Bégin), NATO ASI Series, Berlin, H30, pp. 149–56.

Forget, G., Lacroix, M.J., Brown, R.C., Evans, P.H. and Sirois, P. (1986) Response of perifused alveolar macrophages to glass fibers: effect of exposure duration and fiber length. *Environ. Res.*, **39**, 124–35.

Fowler, D.P., Balzer, J.L. and Cooper, W.C. (1971) Exposure of insulation workers to airborne fibrous glass. *Am. Ind. Hyg. Assoc. J.*, **32**, 86.

Gaudichet, A., Petit, G., Billon-Galland, M.A. and Dufour, G. (1989) Levels of atmospheric pollution by man-made mineral fibres in buildings, in *Non-Occupational Exposure to Mineral Fibres*, (eds J. Bignon, J. Peto and R. Saracci), IARC Scientific Publications 90, Lyon, pp. 291–8.

Gevenois, P.A., Pichot, E., Dargent, F., Dedeire, S., Van de Weyer, R. and De Vuyst, P. (1994) Low grade coal worker's pneumoconiosis comparison of CT and chest radiograph. *Acta Radiol.*, **35**, 351–6.

Gormley, I.P., Brown, G.M., Cowie, H., Wright, A. and Davis, J.M.G. (1985) The effects of fiber length on the *in vitro* cytotoxicity of asbestos samples in three different assay systems, in *In vitro Effects of Mineral Dusts*, (eds E.G. Beck and J. Bignon), NATO ASI Series, Berlin, G3, pp. 397–404.

Hansen, K. and Mossman, B.T. (1987) Generation of superoxide ($O_2^{\bullet -}$) from alveolar macrophages exposed to asbestiform and nonfibrous particles. *Cancer Res.*, **47**, 1681–6.

Hart, G.A., Newman, M.M., Bunn, W.B. and Hesterberg, T.W. (1992) Cytotoxicity of refractory ceramic fibres to Chinese hamster ovary cells in culture. *Toxic. In Vitro*, **6**, 317–26.

Hesterberg, T.W. and Barrett, J.C. (1984) Dependence of asbestos- and mineral dust-induced transformation of mammalian cells in culture on fiber dimension. *Cancer Res.*, **44**, 2170–80.

Hesterberg, T.W., Butterick, C.J., Oshimura, M., Brody, A.R. and Barrett, J.C. (1986) Role of phagocytosis in Syrian hamster cell transformation and cytogenetic effects induced by asbestos and short and long glass fibers. *Cancer Res.*, **46**, 5795–802.

Hesterberg, T.W., Hart, G.A. and Bunn, W.B. (1993a) *In vitro* toxicology of fibers: mechanistic studies and possible use for screening assays, in *Fiber Toxicology* (ed. D.B. Warheit), Academic Press Inc., San Diego, pp. 139–70.

Hesterberg, T.W., Miller, W.C., McConnell, E.E., Chevalier, J., Hadley, J.G., Bernstein, D.M., Thevenaz, P. and Anderson, R. (1993b) Chronic inhalation toxicity of size-separated glass fibers in Fischer 344 rats. *Fund. Appl. Toxicol.*, **20**, 464–76.

Hill, J.W., Rossiter, C.E. and Foden, D.W. (1984) A pilot respiratory morbidity study of workers in a MMMF plant in the United Kingdom, in *Biological Effects of Man-Made Mineral Fibres*, Proceedings of a WHO/IARC Conference, Copenhagen, Denmark, 20–22 April 1982, Copenhagen. World Health Organization, Regional Office for Europe, Volume 1, pp. 413–26.

Holmes, A., Morgan, A. and Davidson, W. (1983) Formation of pseudoasbestos bodies on sized glass fibers in the hamster lung. *Ann. Occup. Hyg.* **27**, 301–13.

Hori, H., Higashi, T., Fujino, A., Yamato, H., Ishimatsu, S. and Oyabu, T. (1993) Measurement of airborne ceramic fibres in manufacturing and processing factories. *Ann. Occup. Hyg.*, **37**, 623–9.

Hughes, J. and Weill, H. (1994) Asbestos and man made mineral fibers, in *Epidemiology of Lung Cancer* (ed. J. Samet), Lung Biology in Health and Disease, pp. 185–205.

Hughes, J., Jones, R., Glindmeyer, H., Hammad, Y. and Weill, H. (1993) Follow-up study of workers exposed to man made mineral fibres. *Br. J. Ind. Med.*, **50**, 658–67.

IARC (1988) *IARC Monographs on the Evaluation of the Carcinogenic Risk of Chemicals to Humans, Vol. 43, Man-made Mineral Fibers and Radon.* International Agency for Research on Cancer, Lyon.

Infante, P.F., Schuman, L.D., Dement, J. and Huff, J. (1994) Fibrous glass and cancer. *Am. J. Ind. Med.*, **26**, 559–84.

IPCS-WHO (1988) *Environmental Health Criteria 77: Man-made Mineral Fibers*. World Health Organization, Geneva.

Jaffrey, S.A. M.T. (1990) Levels of airborne man-made mineral fibers in U.K. dwellings. I Fiber levels during and after installation of insulation. *Atmos. Environ.*, **24A**, 133–41.

Jaffrey, S.A. M.T., Rood, A.P., Llewellyn, J.W. and Wilson, A.J. (1990) Levels of airborne man-made mineral fibers in UK dwellings: II Fiber levels during and after some disturbance of loft insulation. *Atmos. Environ.*, **24A**, 143–6.

Kilburn, K.H. and Warshaw, R.H. (1991) Difficulties of attribution of effect in workers exposed to fiberglass and asbestos. *Am. J. Ind. Med.*, **20**, 745–51.

Kilburn, K., Powers, D. and Warshaw, R. (1992) Pulmonary effects of exposure to fine fibreglass: irregular opacities and small airways obstruction. *Br. J. Ind. Med.*, **49**, 714–20.

Klingholz, R. (1977) Technology and production of man-made mineral fibers. *Ann. Occup. Hyg.*, **20**, 153–9.

Lemasters, G., Lockey, J., Rice, C., McKay, R., Hansen, K., Lu, J., Levin, L. and Gartside, P. (1994) Radiographic changes among workers manufacturing refractory ceramic fibre and products. *Ann. Occup. Hyg.*, **38**(suppl. 1), 745–51.

Linnainmaa, K., Pelin-Enlund, K., Jantunen, K., Vanhala, E., Tuomi, T., Fitzgerald, J. and Yamasaki, H. (1991) Chromosomal damage and gap junctional intercellular communication in mesothelioma cell lines and cultured human primary mesothelial cells treated with MMMF, asbestos and erionite, in *Mechanisms in Fibres Carcinogenesis* (eds R.C. Brown, J.A. Hoskins and N.F. Johnson), Plenum Press, New York, pp. 327–34.

Lockey, J. and Wiese, N. (1992) Health effects of synthetic vitreous fibers, in *Clinics in Chest Medicine. Occupational Lung Diseases*, (ed. G. Epler), W.B. Saunders, pp. 329–38.

Maples, K.R. and Johnson, N.F. (1992) Fiber-induced hydroxyl radical formation: correlation with mesothelioma induction in rats and humans. *Carcinogenesis*, **13**, 2035–9.

Marsh, G.M., Enterline, P.E., Stone, R.A. and Henderson, V.L. (1990) Mortality among a cohort of U.S. man-made mineral fiber workers: 1985 follow-up. *J. Occup. Med.*, **32**, 594–604.

Mast, R.W., McConnell, E.E., Anderson, R., Chevalier, J., Kotin, P., Bernstein, D.M., Thevenaz, P., Glass,

L.R., Miiller, W.C. and Hesterberg, T.W. (1995) Studies on the chronic toxicity (inhalation) of four types of refractory ceramic fiber in male Fischer 344 rats. *Inhal. Toxicol.*, **7**, 425–67.

McConnell, E.E., Kamstrup, O., Musselman, R., Hesterberg, T.W., Chevalier, J., Miiller, W.C. and Thevenaz, P. (1994) Chronic inhalation study of size-separated rock and slag wool insulation fibers in Fischer 344/N rats. *Inhal. Toxicol.*, **6**, 571–614.

McDonald, J.C., Case, B.W., Enterline, E., Henderson, V., McDonald, A.D., Plourde, M. and Sebastien, P. (1990). Lung dust analyses in the assessment of past exposure of man-made mineral fibre workers. *Ann. Occup. Health*, **34**, 427–41.

Miettinen, O.S. and Rossiter, C.E. (1990) Man-made mineral fibers and lung cancer. *Scand. J. Work Environ. Health*, **16**, 221–31.

Mikalsen, S.O., Rivedal, E. and Sanner, T. (1988) Morphological transformation of Syrian hamster embryo cells induced by mineral fibres and the alleged enhancement of benzo(a)pyrene. *Carcinogenesis*, **9**, 891–9.

Morgan, A. and Holmes, A. (1984) Solubility of rockwool fibres *in vivo* and the formation of pseudo-asbestos bodies. *Ann. Occup. Hyg.*, **28**, 307–14.

Morgan, A. and Holmes, A. (1986) Solubility of asbestos and man-made mineral fibers *in vitro* and *in vivo*: its significance in lung disease. *Environ Res.*, **39**, 475–84.

Morgan, A., Holmes, A. and Davison, W. (1982) Clearance of sized glass fibers from the rat lung and their solubility *in vivo*. *Ann. Occup. Hyg.*, **25**, 317.

Moulin, J.J., Pham, Q.T., Mur, J.M. *et al.* (1987) Enquête épidémiologiques dans deux usines productrices de fibres minérales artificielles. II. Symptômes respiratoires et fonction pulmonaire. *Arch. Mal. Prof.*, **48**, 7–16.

Musselman, R.P., Miller, W.C., Eastes, W., Hadley, J.G., Karmstrup, O., Thevenaz, P. and Hesterberg, T.W. (1994) Biopersistences of man-made vitreous fibers and crocidolite fibers in rat lungs following short term exposures. *Environ. Health Perspect.*, **102** (suppl. 5), 139–44.

Nasr, A., Ditchek, T. and Scholtens, P. (1971) The prevalence of radiographic abnormalities in the chests of fiber glass workers. *J. Occup. Med.*, **13**, 371–6.

Nomenclature Committee of TIMA Inc. (1991) *Man-made Vitreous Fibers: Nomenclature, Chemical and Physical Properties* (ed. W. Eastes), Owens-Corning Fiberglass, Stamford, CT.

Öhberg, I. (1987) Technological development of the mineral wool industry in Europe. *Ann. Occup. Hyg.*, **31**, 529–45.

Oshimura, M., Hesterberg, T.W., Tsutsui, T. and Barrett, J.C. (1984) Correlation of asbestos-induced cytogenetic effects with cell transformation of Syrian hamster embryo cells in culture. *Cancer Res.*, **44**, 5017–22.

Pelin, K., Kivipensas, P. and Linnainmaa, K. (1995) Effects of asbestos and man-made vitreous fibers on cell division in cultured human mesothelial cells in comparison to rodent cells. *Environ. Mol. Mutagen.*, **25**, 118–25.

Pickrell, J.A., Hill, J.O., Carpenter, R.L., Hahn, F.F. and Rebar, A.H. (1983) *In vitro* and *in vivo* response after exposure to man-made mineral and asbestos insulation fibers. *Am. Ind. Hyg. Assoc. J.*, **44**, 557–61.

Pott, F., Ziem, U., Reiffer, F.J., Huth, F., Ernst, H. and Mohr, U. (1987) Carcinogenicity studies on fibres, metal compounds, and some other dusts in rats. *Exp. Pathol.*, **32**, 129–52.

Pott, F., Roller, M., Rippe, R.M., Germann, P.G. and Bellmann, B. (1991) Tumours by the intraperitoneal and intrapleural routes and their significance for the classification of mineral fibres, in *Mechanisms in Fibre Carcinogenesis* (ed R.C. Brown *et al.*), Plenum Press, New York, pp. 547–65.

Riebe-Imre, M., Aufderheide, M., Emura, M., Straub, M., Roller, M., Mohr, U. and Pott, F. (1994) Comparative studies with natural and man-made fibres *in vitro* and *in vivo*, in *Cellular and Molecular Effects of Mineral and Synthetic Dusts and Fibres* (eds J.M.G. Davis and M. Jaurand), NATO ASI Series vol. H85, Springer-Verlag, Berlin, Heidelberg, pp. 273–84.

Rogan, W., Gladen, B. and Ragan, N. (1987) US prevalence of occupational pleural thickening: a look at chest x rays from the first national health and nutrition examination survey. *Am. J. Epidemiol.*, **126**, 893–900.

Roggli, V.L. (1994) Rare pneumoconiosis: metalloconiosis, in *Pathology of Pulmonary Disease*, (ed. M.S. Saldana), J.B. Lippincott, Philadelphia, pp. 411–22.

Schneider, T., Nielsen, O., Bredsdorff, P. and Linde, P. (1990) Dust in buildings with man-made mineral fiber ceiling boards. *Scand. J. Work Environ. Health*, **16**, 434–9.

Schneider, T., Husemoen, T., Olsen, E., Christensen, V. and Kamstrup, O. (1993) Airborne fibre concentration during standardized building insulation with bonded man-made vitreous fibre insulation material having different nominal diameters and oil content. *Ann. Occup. Hyg.*, **37**, 631–44.

Sebastien, P., Vergnon, J.M., Blanchard, O., Wastiaux, A. and Emonot, A. (1994) Durability of ceramic fibres in the human lung. Preliminary results. *Ann. Occup. Hyg.*, **38** (suppl. 1), 723–30.

Shannon, H.S., Jamieson, E., Julian, J.A., Muir, D.C.F. and Walsh, C. (1987) Mortality experience of Ontario glass fibre workers – extended follow-up. *Ann. Occup. Hyg.*, **31**, 657–62.

Shannon, H.S., Jamieson, E., Julian, J.A. and Muir, D.C.F. (1990) Mortality of glass filament (textile) workers. *Br. J. Ind. Med.*, **47**, 533–6.

Simonato, L., Fletcher, A.C., Cherrie, J.W., Andersen, A. and Bertazzi, P. (1987) The International Agency for Research on Cancer historical cohort study of MMMF production workers in seven European countries: extension of the follow-up. *Ann. Occup. Hyg.*, **31**, 603–23.

Sincock, A.M. (1977) Preliminary studies of the in vitro cellular effects of asbestos and fine glass dusts, in *Origins of Human Cancer*, Cold Spring Harbor Laboratory Press, Cold Spring Harbor, New York, pp. 941–54.

Sincock, A.M., Delhanty, J.D.A. and Casey, G. (1982) A comparison of the cytogenetic response to asbestos and glass fibre in Chinese hamster and human cell lines. *Mutat Res.*, **101**, 257–68.

Smith, D.M., Ortiz, L.W., Archuleta, R.F. and Johnson, N.F. (1987) Long-term health effects in hamsters and rats exposed chronically to man-made vitreous fibres. *Ann. Occup. Hyg.*, **48**, 731–54.

Stanton, M.F., Layard, M., Tegeris, A., Miller, E., May, M. and Kent, E. (1981) Carcinogenicity of fibrous glass: pleural response in the rat in relation to fibre dimension. *J. Natl Cancer Inst.*, **58**, 587–603.

Tilkes, F. and Beck, E.G. (1983a) Macrophage functions after exposure to mineral fibers. *Environ. Health Perspect.*, **51**, 67–72.

Tilkes, F. and Beck, E.G. (1983b) Influence of well-defined mineral fibers on proliferating cells. *Environ. Health Perspect.*, **51**, 275–9.

Trethowan, W., Burge, P.S., Rossiter, C.E., Harrington, J.M. and Calvert, I. (1995) A study of the respiratory health of employers in 7 European ceramic fibre manufacturing plants. *Occup. Environ. Med.*, **52**, 97–104.

Van der Wal, J.F., Ebens, R. and Tempelman, J. (1987) Man-made mineral fibres in homes caused by thermal insulation. *Atmos Environ.*, **21**, 13–19.

Wagner, J.C., Berry, G., Hill, R.J., Munday, D.E. and Skidmore, J.W. (1984) Animal experiments with MMVF, in *Biological Effects of Mineral Fibres*, Vol 2. WHO, Copenhagen, pp. 209–33.

Weill, H., Hughes, J., Hammad, Y., Glindmeyer, H., Sharon, G. and Jones, R. (1983) Respiratory health in workers exposed to man-made vitreous fibers. *Am. Rev. Respir. Dis.*, **128**, 104–12.

Wong, O., Foliart, D. and Trent, L.S. (1991) A case-control study of lung cancer in a cohort of workers potentially exposed to slag wool fibres. *Br. J. Ind. Med.*, **48**, 818–24.

Woodworth, C.D., Mossman, B.T. and Craighead, J.E. (1983) Induction of squamous metaplasia in organ cultures of hamster trachea by naturally occurring and synthetic fibers. *Cancer Res.*, **43**, 4906–12.

World Health Organization (1992) Validity of methods for assessing the carcinogenicity of man-made fibres. Executive summary of a WHO consultation, 19–20 May, 1992. World Health Organization, Copenhagen.

Wright, G. (1968) Airbone fibrous glass particles: chest roentgenograms of persons with prolonged exposure. *Arch. Environ. Health*, **16**, 175–81.

Yamato, H., Hori, H., Tanaka, I., Higashi, T., Morimoto, Y. and Kido, M. (1994) Retention and clearance of inhaled ceramic fibres in rat lungs and development of a dissolution model. *Occup. Environ. Med.*, **51**, 275–80.

20 Lung disease from metal exposure

Benoit Nemery

INTRODUCTION

Definitions

For centuries metals have been known to be capable of causing human disease, including pulmonary disease (Hammond and Beliles, 1980; Parkes, 1982; Sakula, 1983; Bisetti, 1988; Friberg *et al.*, 1990a,b; Rom, 1992). Elements may be defined as metals on the basis of their physical or chemical properties, but these definitions do not allow a sharp distinction to be made between metals and non-metals, because a number of elements, e.g. selenium (Se), tellurium (Te), arsenic (As), antimony (Sb) and germanium (Ge) – often called metalloids – share properties of both. From a technical point it is customary to divide metals and their alloys into ferrous metals (iron and various types of steels) and non-ferrous metals (John, 1994). From a toxicological viewpoint, a metal can be defined as 'an element which under biologically significant conditions may react by losing one or more electrons to form cations' (Vouk, 1986). In this sense, carbon (C) and silicon (Si) do not behave as metals and this chapter will, therefore, not deal with the lung disorders caused by C and Si, i.e. coal pneumoconiosis, silicosis, asbestos-related conditions and other silicatoses, which are covered elsewhere, although it must be realized that recent research has pointed to a possibly important role of metals in the toxic and carcinogenic mechanisms of, for example, asbestos fibers. Metallic impurities are also increasingly implicated in the toxicity of the fine or ultrafine particles of (urban) air pollution, so that it is fair to say that the toxicology of metals for the respiratory tract encompasses more than what will be treated here, i.e. the lung diseases caused by *occupational* exposure to metal compounds.

Summarizing the pulmonary effects caused by occupational exposures to metallic compounds is not an easy task, not only because there are so many different agents to be considered, but also because the conditions caused are quite diverse, ranging from acute inhalation injury to lung cancer, from airway disease to parenchymal disorders. In general, lung disorders caused by metallic compounds concern fewer people than is the case for the classical pneumoconioses, and often the total number of individuals exposed to some agents is very small. Consequently, the clinical experience and epidemiology concerning metal-induced pulmonary disease is much less extensive than is the case for silicosis, coal pneumoconiosis or asbestos-related disease. In its general lines, this chapter is an updated version of an earlier review article on the subject (Nemery, 1990). In general, common knowledge or historical data are not referred to specifically, and priority has been given to more recent or less publicized aspects of metal toxicology.

General toxicity of metals

(a) Physicochemical aspects of metallic compounds

It is important to realize that 'metals' are a very diverse and rather non-descript category of substances covering not only a wide range of elements, but even many forms and derivatives for a single metallic element. Failure to take this into account may lead to serious

Occupational Lung Disease. Edited by Daniel E. Banks and John E. Parker. Published in 1998 by Chapman & Hall. ISBN 0 412 73630 6

mistakes, because the toxicity of a metallic compound is heavily dependent on its physicochemical characteristics. People are rarely exposed, in a toxicologically significant way, to the pure metallic form (zero oxidation state) of metals or metalloids, as is the case, for example, with mercury (Hg) vapors. More usually, there is exposure to oxides or other binary metal compounds, such as sulfides, halides, hydrides, carbides, or to multi-element compounds, mainly salts. Metals, most notably transition metals, can also form coordination complexes with various ligands such as ammonia (NH_3), carbon monoxide (CO), cyanogen (CN-), organic nitrogen or sulfur molecules. Some metals may also form organometallic compounds in which the metal is bound to the carbon atom of an organic group.

The chemical form of a metal, also called *metal speciation*, and the resulting physicochemical properties have important consequences in terms of toxicokinetics and biologic effects. Thus, the biologic availability and absorption of metals will be greatly influenced by their solubility in water and lipids, but more importantly by their actual solubility in biologic fluids, which contain a variety of organic ligands. Thus, the more insoluble the metal compound deposited in the airways, the more likely it is that it will be cleared by the mucociliary escalator. Conversely, soluble salts will readily dissociate, thus facilitating their transport as metal ions into lung cells or into the blood circulation.

(b) Mechanisms of the biologic effects of metals

The strong attraction between metal ions and organic ligands situated within tissue molecules may result in the binding of the metal ion to those molecules. This underlies a number of the biologic effects and toxic actions of metals. Thus, a number of metals, e.g. iron (Fe), copper (Cu), manganese (Mn), cobalt (Co), zinc (Zn), chromium (Cr) are essential as coenzymes for many enzymes. The transport and accumulation of metals is often the result of their ability to interact with ligands. Thus, the availability and abundance of transport and storage proteins for Fe and Cu (transferrin, ferritin, albumin, ceruloplasmin), probably constitute an important safeguard against the toxicity of free iron and copper ions. Similarly, cadmium (Cd) and other divalent metals bind to metallothionein, a low-molecular weight protein rich in sulfhydryl (SH) groups. This process plays a role in the defense against Cd, but it eventually leads to the retention and progressive accumulation of Cd in various tissues, including probably the pulmonary tissue.

The interaction of metals with functional groups on macromolecules is an important mechanism for their toxicity and their carcinogenicity. Thus, several metal ions react avidly with free SH groups, thereby possibly inhibiting active centers of enzymes, coenzymes or membrane-bound receptors. Direct interaction of metals with deoxyribonucleic acid (DNA) is one of the possible mechanisms for metal carcinogenesis (Jennette, 1981; Martell, 1981). Another toxicologically relevant consequence of metal binding to proteins is the possible acquisition of antigenicity. Thus, platinum (Pt), chromium, nickel and cobalt are sensitizers, probably by mechanisms similar to those of other reactive organic molecules of low molecular weight which may function as haptens.

The biologic activity and toxicity of some metals is also greatly influenced by their ability to change their oxidation state by oxidation (loss of electrons) and reduction (gain of electrons). Transition metals are electronically stable in more than one oxidation state. As a result of this property, transition metals play important roles in catalyzing biologic oxidation reactions. Of the transition metals, Fe and to a lesser extent Cu have been extensively studied because of their implication in many pulmonary and non-pulmonary disease processes by virtue of their ability to enhance the production of toxic (free radical) species of oxygen. Free radical oxygen toxicity, which seems always to involve metal catalysis, is considered to be a mechanism for the effects of many toxic agents (Aust *et al.*, 1985).

Important preliminary points

Before discussing the various lung disorders caused by metals it is worth emphasizing some important points.

1. The respiratory system is not necessarily the only, nor the principal, target for the toxicity of many metal compounds, even when their entry into the body is by inhalation.

   Thus, the chronic inhalation of lead (Pb), mercury (Hg) or manganese (Mn) may lead to systemic effects, such as neurologic damage, without leading to serious respiratory damage. Other metals, such as cadmium (Cd), are capable of causing both lung and kidney alterations, depending in part upon the route of exposure. It is, therefore, important to be aware of possible extrapulmonary manifestations in metal toxicology.
2. Lung disease in people who are occupationally exposed to metals, is not necessarily due to these metals.

   The most obvious example is the presence of smoking-related lung disease in industrial workers. On the other hand, cigarette smoking often obscures, or interferes with, the possible occupational origin of chronic obstructive pulmonary disease or lung cancer in the individual patient. Other examples of mixed exposures to metals and non-metals concern silicosis and asbestos-related lung disease in iron or other metal foundries. In metal smelting, cutting or burning, as well as during welding, toxic gases are produced in addition to metal oxides. In these instances sulfur dioxide (SO_2), ozone (O_3), or nitrogen dioxide (NO_2) and other volatile compounds released from the pyrolysis of coating materials or solvents and degreasers, rather than the metals themselves, may be responsible for acute or chronic airway or lung disease. Organic compounds are also present in metal industries; for example, isocyanate asthma has been described by several groups in iron and steel plants, in processes where synthetic resin binders are used (Zammit-Tabona *et al.*, 1983; Johnson *et al.*, 1985; Low and Mitchell, 1985; Venables *et al.*, 1985).

3. Exposure to metals is not confined to workers involved in metal mining or metallurgy.

   Metal compounds are indeed used in almost every sector of industry and even in agriculture. Thus, cobalt-induced fibrosing alveolitis and bronchial asthma have been described in diamond polishers who used cobalt-containing polishing discs (Demedts *et al.*, 1984; Gheysens *et al.*, 1985). Dental technicians are at risk of various types of pneumoconiosis (Rom *et al.*, 1984; Anonymous, 1985). Several processes in the electronics industry involve the use of metals or metalloids, e.g. gallium (Ga) or germanium, the toxicology of which is often poorly known. Numerous metal compounds are used as pigments in paints or ceramics (Raffn *et al.*, 1988), as catalysts in the chemical industry (Hery *et al.*, 1994), or as additives for various purposes in plastics (Pillière *et al.*, 1990). The respiratory effects of these exposures are not always well characterized.

4. Pulmonary diseases from occupational exposure to metals involve more than 'pneumoconioses'.

   Because the surveillance and diagnosis of work-related lung disease has long been almost exclusively confined to the chest radiograph, occupational lung disease has been traditionally viewed as consisting essentially of various forms of pneumoconioses, such as those caused by silica, coal dust or (fibrous) silicates. Thus, in older times, lung diseases caused by other minerals tended to be labeled according to a similar terminology and they were often lumped into a category of 'other' pneumoconioses, such as siderosis, 'berylliosis', 'hard-metal pneumoconiosis' and other rare conditions. Another consequence was that agents or processes not associated with manifest radiographic changes were not considered to be of serious concern (hence, for instance, the rather dubious concept of 'inert' dust). However, as this chapter will show, metal exposure may be implicated in a wide variety of respiratory diseases, many of which do not at all exhibit the features of the classic mineral pneumoconioses.

5. Significant exposure to metals is not confined to the work environment.

   Hobbies and domestic activities may lead to significant exposure. A classical example of para-occupational exposure is that of berylliosis in housewives and family contacts of beryllium-workers who brought factory-dust to their homes (Knishkowy and Baker, 1986). The role of exposures to various metals through environmental pollution is only beginning to be studied.

The pulmonary disorders due to inhaled metallic compounds are quite diverse. Two approaches may be adopted to describe them. Either the metals can be listed and their possible pulmonary effects enumerated, or different disease entities can be grouped and their possible metal etiologies discussed. The second option has been adopted here, but the main toxicity of the individual metals is given in a summary form in Table 20.1.

PULMONARY DISEASES CAUSED BY EXPOSURE TO METALS

Acute toxic effects

(a) Metal fume fever

A well-known, relatively benign condition following exposure to high concentrations of metal fumes, mainly zinc oxide (ZnO), is metal fume fever (Blanc and Boushey, 1993; Gordon and Fine, 1993). This condition, of which there are several synonyms, is an influenza-like or malaria-like reaction consisting of fever, chills and malaise with relatively mild respiratory symptoms, and classically little or no X-ray or functional abnormalities, although this is not always the case (Nemery and Demedts, 1991). The symptoms, often accompanied by a sweet metallic taste in the mouth, usually begin at home a few hours after a heavy exposure to metal oxides, e.g. after welding on galvanized steel in a confined space, and they then subside spontaneously. Peripheral leukocytosis is present during the acute illness, and recent studies using bronchoalveolar lavage have showed marked neutrophil infiltration and cytokine release (Blanc *et al.*,

Table 20.1 Summary of pulmonary toxicity of metals

Metal (1)	*Exposures and uses (2)*	*Acute toxicity (3)*	*Airways (4,5)*	*Lung parenchyma (5)*	*Lung cancer (5,6)*	*Extrapulmonary toxicity (7)*	*Comments*
Aluminum Al 13	bauxite, cryolite, many alloys, electrical engineering, transport and building industry, packaging, abrasives (corundum = Al_2O_3)		'potroom asthma' bronchial hyperreactivity (fluorides?, some salts), COLD	fibrosis (*aluminosis*) [granulomatous fibrosis] [alveolar proteinosis]	Al production	CNS	pulmonary disease rare in view of widespread use
Antimony Sb 51	alloys (with Sn, Pb, Cu), chemical & textile industry, pigments, ceramics, glass	pneumonitis ($SbCl_3$, $SbCl_5$), hemolysis (SbH_3)		pneumoconiosis (*antimoniosis*)	possible	heart	often concomitant exposure to As
Arsenic As 33	smelting of Cu, Pb, Au, Zn, Co alloys, glass, pigments, tanning, electronics, pesticides	metal fume fever hemolysis (AsH_3)	nasal septum perforation		definite	skin, PNS, C-V system, blood	monitoring As-U (! total As influenced by As of marine origin)
Barium Ba 56	alloys, paints, glass, ceramics, contrast material			pneumoconiosis (*baritosis*)		heart	
Beryllium Be 4	alloys (with Cu, Al, Ni), ceramics, electronics, nuclear, space, aircraft, dental technicians, metal reclaiming	pneumonitis		granulomatous pneumonitis/ fibrosis (*chronic beryllium disease, berylliosis*)	probable	skin, liver and other organs	cell-mediated immune response (Be-specific lymphocyte stimulation)
Cadmium Cd 48	in association with Zn and Pb alloys, electroplating, batteries, pigments, plastics, welding electrodes and Ag solders	pneumonitis, metal fume fever	COLD	emphysema (fibrosis in animals)	probable	kidney	cumulative in man and environment
Chromium Cr 52	alloys (stainless steel, stellite), chrome plating, pigments, tanning, welding, traces in cement		allergic asthma (rare?), nasal septum perforation		definite (CrVI)	contact dermatitis (frequent)	strong sensitizer
Cobalt Co 27	alloys, hard metal (with WC), diamond polishing, magnets, pigments, catalysts, animal feed	pneumonitis	allergic asthma COLD	fibrosing alveolitis (DIP, GIP) (*hard metal lung disease, cobalt-lung*)	possible	heart, contact dermatitis	'bizarre' multinucleated giant cells in BAL and lung
Copper Cu 29	electrical equipment, alloys (brass = Cu+Zn, bronze = Cu+Sn), pesticides, pigments	metal fume fever		fibrosis (*vineyard sprayer's lung*)		skin	Cu smelters: exposure to As, SO_2
Gold	jewellery, therapeutic			pneumonitis (gold salt		skin	gold refining: exposure

Iron Fe 26	mining, iron and steelmaking, alloys, welding	metal fume fever	COLD	pneumoconiosis (*siderosis*)	iron ore mining, foundries		concomitant exposures to SiO_2, asbestos, PAH
Lanthanides (rare earths) Cerium Ce 58	photoengraving glass			pneumoconiosis (*rare earth or cerium pneumoconiosis*)			
Lead Pb 82	lead smelting, alloys, batteries, pigments, paints, welding	metal fume fever				G-I tract, PNS, CNS, kidney, blood	contamination with Cd and As
Lithium Li 3	alloys, ceramics, electronics, catalysts	pneumonitis (LiH)					
Magnesium Mg 12	alloys	metal fume fever					
Manganese Mn 25	steel, alloys, metal slags (fertilizers), batteries, pigments, welding, chemical industry	pneumonitis, metal fume fever	COLD	pneumonia		CNS	
Mercury Hg 80	chlor-alkali industry, Au and Ag refining, electrical, paints, measuring devices, pesticides, drugs dentistry	pneumonitis, metal fume fever		(Hg embolism)		CNS, kidney, dermatitis	saturation of air with Hg rapidly achieved
Nickel Ni 28	steel, alloys, nickel plating, coins, batteries, electronics, catalysts, jewellery	metal fume fever, pneumonitis ($Ni(CO)_4$)	allergic asthma (rare)		definite (mainly soluble compounds), also sinonasal cancer	contact dermatitis (frequent)	
Platinum Pt 78	catalyst, photography, jewellery, anticancer drugs			allergic rhinitis and asthma			urticaria
Osmium Os 76	electron microscopy	pneumonitis					
Selenium Se 34	alloys, pigments, electronics, rubber, glass, Cu refining	pneumonitis					garlic odor breath
Silver Ag 47	alloys, jewellery, soldering, photography, catalyst					discoloration of skin and mucosa (*argyria*)	
Tin Sn 50	tin plating, alloys, welding	metal fume fever		pneumoconiosis (*stannosis*)			
Titanium Ti 22	alloys, white pigment (TiO_2)	pneumonitis ($TiCl_4$)	COLD	[granulomatous pneumonitis]			
Tungsten W 74	alloys, hard metal (WC + Co), cutting, drilling, polishing tools			co-factor in hard metal lung disease			

Table 20.1 Continued

Metal (1)	*Exposures and uses (2)*	*Acute toxicity (3)*	*Airways (4,5)*	*Lung parenchyma (5)*	*Lung cancer (5,6)*	*Extrapulmonary toxicity (7)*	*Comments*
Uranium U 92	nuclear fuel, ceramics	pneumonitis (UF_6)			definite	kidney	
Vanadium V 23	steel, alloys, catalysts, pigments, fuel ash and metal slags	tracheobronchitis (*boilermaker's bronchitis*)	irritant-induced asthma	pneumonia		green tongue	
Zinc Zn 30	alloys, brass, galvanizing, pigments, pesticides, smoke bombs ($ZnCl_2$)	metal fume fever (ZnO), pneumonitis ($ZnCl_2$)	asthma				contamination with Cd
Zirconium Zr 40	alloys, nuclear industry, pigments, catalysts, abrasives	pneumonitis ($ZrCl_4$)		granulomatous pneumonitis/fibrosis		skin granulomas	

* Summary compiled from general sources and ***only intended for use as a quick reference***. Not all species of a metal cause the indicated disorder; the reader is referred to the text and original references for a complete description of exposure conditions and diseases. Metals not indicated in the table have not been associated with pulmonary disease in humans. Conditions in square brackets concern isolated case reports.

(1) Metal, symbol and atomic number.
(2) Only typical or noteworthy uses have been indicated.
(3) Metal fume fever and chemical pneumonitis are not always easy to distinguish.
(4) Chronic excessive exposure to dusts may lead to industrial bronchitis and/or chronic obstructive lung disease (COLD).
(5) Underground mining of metal ores may lead to exposure to free silica (and silicosis) and/or radioactive radon daughters (and lung cancer).
(6) High-temperature treatment of metals during refining, smelting, cutting or welding may lead to exposure to toxic or irritant gases and/or carcinogenic compounds (e.g. polycyclic aromatic hydrocarbons, PAH).
(7) Only selected target organs are indicated, particularly when relevant to chronic exposure by inhalation,

C-V system, cardiovascular system; PNS, peripheral nervous system; CNS, central nervous system.

1991, 1993). A strange feature of metal fume fever (and also of other inhalation fevers) is the occurrence of tolerance: symptoms only appear when exposure takes place after a period of days without exposure and they do not appear on subsequent days.

The exact pathogenesis of metal fume fever is poorly understood. In some instances allergic mechanisms may be involved (Farrell, 1987), but then metal fume fever may be a misnomer or may be superimposed on bronchial asthma or hypersensitivity pneumonitis. There is a striking resemblance between metal fume fever and the organic dust toxic syndrome (ODTS), which occurs after heavy exposure to organic dust contaminated with microorganisms. Both syndromes have a similar clinical course with fever, leukocytosis, acute transient neutrophilic alveolitis and occurrence of tolerance. These similarities indicate common pathogenic mechanisms, which in fact still remain obscure, with regard to both the activation of the cells (presumably macrophages) and the self-limited nature of the inflammatory reaction.

With appropriate environmental control and individual preventive measures, metal fume fever should be relatively easy to prevent, but the disease has certainly not disappeared. In fact, workers at risk usually are quite familiar with the condition and its relatively benign course; a high proportion of welders mention having ever had a bout of metal fume fever, but few cases come to the attention of physicians, who presumably often overlook the syndrome as a simple viral infection.

Metal fume fever is said not to lead to sequelae, but this has not been adequately investigated, and there are data (Cotes *et al.*, 1989; Chinn *et al.*, 1990) suggesting that subjects who have had metal fume fever are more at risk of subsequent respiratory symptoms, although this may simply reflect a higher past exposure.

(b) Acute inhalation injury

The inhalation of various metallic compounds can cause severe inhalation injury, involving the upper and/or lower respiratory tract. The older literature, and also many animal experiments, show that exposure to fumes or dusts containing high quantities of various metals may lead to damage to the airways and lungs. Depending on the agent and the intensity of exposure, the injury may manifest itself as rhinitis, laryngitis, tracheobronchitis or chemical pneumonitis and non-cardiogenic pulmonary edema. Symptoms will, therefore, possibly include lacrimation, nose irritation with sneezing and bleeding, sore throat, hoarseness, cough, wheezing, chest pain or tightness, and dyspnea. In severe cases the clinical picture may be that of the adult respiratory distress syndrome (ARDS), with refractory arterial hypoxemia and eventual death.

It is important to realize that when the deep lung is involved, respiratory distress is often delayed for several hours until severe non-cardiogenic pulmonary edema develops. Such delay is a particularly dangerous feature of the inhalation of water-insoluble agents (such as NO_2 or phosgene), which cause little or no sensory irritation to the eyes or upper airways and, hence, have poor warning properties. The fact that the exposure was well tolerated and the long delay before the occurrence of the first symptoms may be possible reasons for not recognizing the toxic etiology of the 'pneumonia'.

Cadmium-induced pneumonitis

Cadmium is a by-product of the zinc and lead industry; it is used in metal plating and in special alloys; it is also used in the production of batteries, pigments and plastic stabilizers. Cadmium-pneumonitis is perhaps the best documented example of metal-induced acute pneumonitis and occasional cases still occur (Barnhart and Rosenstock, 1984; Taylor *et al.*, 1984; Yates and Goldman, 1990; Seidal *et al.*, 1993; Fernandez *et al.*, 1996). From a practical point of view, the important points to be aware of in relation to the risk of cadmium pneumonitis are: (i) that cadmium oxides may be liberated, often unknowingly to the worker, from the welding or burning of cadmium-containing alloys and cadmium-plated metal, from the use of hard solders, or from the smelting of zinc or lead (or scrap metal), which often contain significant levels of contaminating cadmium; and (ii) that exposure to toxic levels of cadmium fumes does not necessarily lead to immediate respiratory symptoms.

Mercury vapors

Although the target organ for inhaled mercury vapors is mainly the central nervous system, severe chemical pneumonitis may result from exposure to high levels of mercury vapors. This may occur during the refining of gold or silver (using amalgams) in confined spaces, as is often the case in the cottage industry, where several fatalities or severe pulmonary involvement resulting from such activities have been described (Liles *et al.*, 1985; Levin *et al.*, 1988; Rowens *et al.*, 1991). In a recent interesting report, 'smelter disease' which included respiratory, gastrointestinal, dermal and general symptoms, among workers replacing pipes in plants producing sulfuric acid, was attributed to mercury poisoning via skin absorption (Koizumi *et al.*, 1994).

The embolization of droplets of metallic mercury in the pulmonary circulation following the deliberate injection or accidental inoculation of mercury leads to a peculiar pattern of small very radiodense dots on the chest radiograph. This condition does not appear to be associated with either pulmonary or systemic manifestations of disease (Shaffer and Schmidt-Nowara, 1989; dell'Omo *et al.*, 1996).

Vanadium pentoxide and other metal oxides

Vanadium pentoxide (V_2O_5) may be present in significant quantities in slags from the steel industry (ferrovanadium) and, because some fuel oils contain high quantities of vanadium, in furnace residues from oil refineries or in soot from oil-fired boilers. Dust containing V_2O_5 may cause upper and lower airway irritation: rhinitis with sneezing and nose bleeds, pharyngitis, acute tracheobronchitis, with cough, wheeze and (possibly) airway hyper-reactivity ('boilermakers' bronchitis'), as well as possibly bronchopneumonia (Lees, 1980; Musk and Tees, 1982; Levy *et al.*, 1984; Pistelli *et al.*, 1991; Hauser *et al.*, 1995a,b).

The older literature also indicates that exposure to high levels of oxides of beryllium, cobalt, manganese and osmium may cause airway irritation and even bronchopneumonia. However, new technologies, such as those involving the thermal spraying of metals may also prove to be particularly hazardous, as is shown by a recently described fatal case caused by spraying nickel (Rendall *et al.*, 1994)

Metal halides, carbonyls and hydrides

Cases of adult respiratory distress syndrome, some with a protracted course, have been reported in military or civilian personnel accidentally exposed to smoke bombs which liberate zinc chloride ($ZnCl_2$) (Hjortso *et al.*, 1988; Matarese and Matthews, 1988; Allen *et al.*, 1992). Accidental exposure, e.g. as a result of explosions, burst pipes or leaks in chemical plants, to antimony trichloride ($SbCl_3$) and pentachloride ($SbCl_5$) (Cordasco and Stone, 1973), zirconium tetrachloride ($ZrCl_4$), titanium tetrachloride ($TiCl_4$) (Park *et al.*, 1984) and uranium hexafluoride (UF_6) (Brooks *et al.*, 1985) may also lead to inhalation injury.

Nickel carbonyl [$Ni(CO)_4$] is a volatile liquid of very high toxicity for the lungs and brain. Acute inhalation may cause hemorrhagic pulmonary edema (Zhicheng, 1986).

Lithium hydride (LiH), phosphine (hydrogen phosphide, PH_3, used as a doping agent for the manufacture of silicon crystals, or released from aluminum phosphide grain fumigants or zinc phosphide rodenticides), hydrogen selenide (SeH_3), and diborane (B_2H_5, used as high-energy fuel) have also been reported to cause acute inhalation injury with, possibly, pulmonary edema (Cordasco and Stone, 1973). (Boron and phosphorus are not really metals).

It is worth mentioning that the inhalation of the hydrogenated forms of arsenic (arsine, AsH_3) or antimony (stibine, SbH_3) can also be lethal as a result of fulminant hemolysis, which may sometimes manifest itself initially as dyspnea, besides abdominal and lumbar pain with hemoglobinuria.

(c) Possible sequelae of acute inhalation injury

Following acute inhalation injury, there is often complete recovery. However, this is certainly not always the case and various chronic sequelae such as lung fibrosis, bronchiolitis, bronchiectasis and other bronchial lesions may result from the acute injury, depending on the severity of the initial damage and perhaps also depending on treatment modalities, although very little hard data, let alone controlled studies, are available regarding the latter issue (Nemery, 1996).

Moreover, even in the absence of structural sequelae, which may be identified by imaging studies or through bronchoscopy, or in the absence of significant functional defects, a state of permanent non-specific bronchial hyper-reactivity may be observed. This condition of adult-onset, non-allergic asthma has been named 'reactive airways dysfunction syndrome' (RADS) (Brooks *et al.*, 1985) and occurs in a number of survivors of (severe) airway injury, by mechanisms that still remain to be elucidated (Nemery, 1996).

Chronic obstructive lung disease

(a) Chronic obstructive lung disease (COLD) and occupation

There is no doubt that the main exogenous cause of chronic obstructive lung disease and emphysema in the general population is cigarette smoking. However, clinically important airflow limitation occurs in less than 15% of smokers (Fletcher *et al.*, 1976). This concept underscores two important points with regard to occupational factors. Firstly, since the basis for the susceptibility to cigarette smoke is still largely unknown, there is no *a priori* reason why subjects who are (or would be) particularly susceptible to smoking-induced lung and airway disease, will not be equally susceptible to respiratory insults from the occupational environment, including metallic compounds which are capable of causing tissue destruction and inflammation. Secondly, even for environmental substances which would carry a risk similar to that of cigarette smoking (i.e. affecting 'only' 15% of the exposed subjects), the chances of conclusively demonstrating such a risk are small because of:

- the relatively small number of subjects that are generally exposed;

- the generally lower duration of exposure in the case of occupational exposures (as opposed to that of smoking);
- the attempts at and possibilities of avoiding occupational exposure (as opposed to the intentionality of smoking);
- the complexity of the exposures (in terms of their nature as well of as their intensity and duration);
- the high prevalence of smoking in industrial populations (Covey and Wynder, 1981); and
- selection factors collectively known as 'the healthy worker effect'.

Despite these methodologic difficulties, several studies have shown that occupational factors may cause not only symptoms of industrial bronchitis, but also loss of ventilatory function, which may be of similar magnitude to that associated with cigarette smoking (Becklake, 1985; Burge, 1994). This view, however, is not universally accepted and the issue of the real impact of occupational exposures on ventilatory function is still controversial. In practice, it remains impossible to determine the contribution of occupation in the causation of chronic obstructive lung disease in the individual smoking patient.

(b) Metal workers and COLD

The longitudinal studies that have led to the conclusion that significant airflow limitation may result from dust exposure, have essentially involved coal miners, grain dust-exposed workers or workers exposed to poorly defined 'industrial dust, fumes or gases' (Becklake, 1985; Burge, 1994). No such studies have been performed with workers exposed to specific metallic compounds and the available information regarding metals mainly stems from less powerful cross-sectional observations.

Cadmium

Cadmium is the only agent, besides cigarette smoke, for which there is good evidence that it causes pulmonary emphysema in humans. The relationship between occupational exposure to cadmium and the development of emphysema had long been suggested, and disputed, but a study of a large group of 101 workers and ex-workers from a cadmium alloy factory in England, clearly settled the issue, at least with regard to high levels of exposure (Davison *et al.*, 1988). In that study a clear excess of functional (ventilatory function and diffusing capacity) and radiologic signs of emphysema was demonstrated in the exposed subjects compared with appropriate controls. The causal role of cadmium was strengthened by the existence of a positive relationship between effects and dose, with the latter being estimated both by past hygiene measurements and by the internal (liver) cadmium burden. Moreover, the significance of these findings has been borne out by the demonstration of an increased mortality from non-malignant respiratory disease in cadmium-exposed workers (Armstrong and Kazantzis, 1985; Kazantzis *et al.*, 1988. A recent case report has also described a rapid evolution of emphysema in a cadmium-exposed furnace worker (Leduc *et al.*, 1993).

The mechanisms for cadmium-induced emphysema are not elucidated. In fact, animal studies suggest that fibrosis, rather than alveolar wall destruction, precede the development of emphysematous lesions (Niewoehner and Hoidal, 1982; Snider *et al.*, 1988).

Iron and steel industry and welders

There have been many surveys of the prevalence of chronic obstructive lung disease in workers from the iron and steel industry, as well as among metal welders. Although exposure in the iron and steel industry is mainly to iron-dust, many other metallic and non-metallic particulates (including silica and asbestos), as well as gases may be inhaled. During welding, even of steel, the plume is not so much composed of the metals being welded together, but of the materials which make up the electrode, the filler wire and the fluxes used.

Despite the generally consistent finding of increases in the prevalence of chronic bronchitis, as defined by questionnaire, among welders and steelworkers, the results regarding ventilatory function have been largely negative, inconclusive or showing only small effects (Sferlazza and Beckett, 1991; Billings and Howard, 1993). However, because of the healthy worker effect and the other methodologic reasons alluded to above, one should not conclude from these prevalence studies that there are no specific work processes within these broad categories which entail a risk of significant obstructive respiratory impairment in susceptible subjects.

Thus, population studies indicate that there is an excess mortality (Coggon *et al.*, 1994) and morbidity (Valentin and Smidt, 1978) from non-malignant respiratory disease among steelworkers and welders. A recent longitudinal study of shipyard workers has shown a faster annual decline in FEV_1 in welders and caulker/burners than in controls (Chinn *et al.*, 1990). In any case, the absence of certainty regarding the magnitude of the risk of COLD in welders and steelworkers should not lead to complacency about the need for adequate surveillance and prevention of dust exposure in these jobs, also because COLD is not the only risk incurred.

Other metals
Cross-sectional studies have also suggested an increased prevalence of chronic bronchitis and a loss of ventilatory function, sometimes mainly of forced vital capacity (FVC), associated with chronic exposure to beryllium (Kriebel *et al.*, 1988b), aluminum (Chan-Yeung *et al.*, 1983; Townsend *et al.*, 1985; Abramson *et al.*, 1989), cobalt (or hard metal) (Sprince *et al.*, 1984; Kusaka *et al.*, 1986a), manganese (Roels *et al.*, 1988), or titanium dioxide (TiO_2) (Garabrant *et al.*, 1987), apparently independently from the overt forms of other respiratory diseases seen with some of these metals (see below). However, other surveys of workers exposed to these compounds do not always reach the same conclusions (Roto, 1980; Cotes *et al.*, 1983; Chen and Fayerweather, 1988; Sprince *et al.*, 1988), possibly because of differences in total dust, in concomitant exposures, or in population characteristics and study designs. Here too, the absence of demonstrable effects does not exclude the existence of a preventable health risk.

Bronchial asthma

(a) General considerations
Several metals are known to be capable of causing bronchial asthma (Bernstein and Brooks, 1993). Metal-induced asthma is a particular case of occupational asthma caused by agents of low molecular weight (see Chapter 23). In essence, the clinical characteristics of metal-induced asthma do not appear to be dissimilar from occupational asthma caused by other agents.

Some forms of metal-induced occupational asthma clearly belong to the category of 'occupational asthma with a latency' and they are clearly due to immunologic sensitization to the offending agent, thus resulting in asthmatic reactions to even minute amounts of the material. Nevertheless, the exact molecular mechanisms of sensitization to metallic compounds have not been elucidated. It is generally believed that (in genetically susceptible individuals?) some metals may function as haptens and thereby cause specific immunologic sensitization. For some metals (platinum salts) there is good evidence for an IgE-mediated hypersensitivity reaction, whereas in many other instances IgE mechanisms have not been documented and are even questionable, and cell-mediated immune mechanisms should be envisaged. With some metals (as in fact with some organic molecules), alveolitis and asthma may coexist, as has been described with cobalt (Van Cutsem *et al.*, 1987). However, it is fair to say that, in general, the immunology of metal-induced respiratory reactions has not yet been thoroughly investigated.

Other forms of metal-related occupational asthma are more likely to belong to the category of non-immunologic asthma, i.e. 'irritant-induced asthma' or 'occupational asthma without latency'. It is also possible that other syndromes may occasionally be mistaken for or be superimposed on occupational asthma: thus, the clinical history (and even the sequential peakflow measurements) of patients who experience repeated bouts of metal fume fever may be suggestive of occupational asthma.

The diagnosis, management and prognosis of metal-induced occupational asthma do not appear to differ from occupational asthma caused by other agents. Thus, the golden standard for the diagnosis of occupational asthma is also the bronchial provocation testing with non-irritant quantities of the suspected agent, which in the case of metals must include considerations of the species of metal.

(b) Platinum salts
Occupational asthma caused by complex platinum salts has mainly been described in workers from precious-metal refineries, where platinum is obtained from the ore or from scrap metal through solubilization in concentrated hydrochloric acid, followed by precipitation, thus yielding ammonium hexachloroplatinate $[(NH_4)_2PtCl_6]$. Exposures to complex salts of platinum may also occur in the manufacture of catalysts, in photographic applications, and in electroplating.

Platinum salts appear to be very potent sensitizers. In fact, platinum asthma provides a unique example of a situation where a high proportion of exposed subjects may become sensitized (rather than the traditional 'minority' of subjects being affected in other forms of occupational asthma). Thus, prevalences of symptomatic subjects of up to 60% of the workforce have been reported (Hughes, 1980) and surveys of exposed populations have often pointed to the high turnover in platinum refineries because of medical terminations for allergic symptoms. In general, workers become allergic and/or symptomatic during the first year of employment (Venables *et al.*, 1989; Calverley *et al.*, 1995). The high sensitization potential of platinum salts also means that workers not directly involved in the refining process, but exposed in the vicinity or through maintenance or cleaning work, may also become sensitized. Thus, we have come across a case of platinum asthma in a worker who only cleaned the changing rooms used by the refinery workers (unpublished results).

There is good evidence for an IgE-mediated mechanism in platinum salt asthma. Although Pt-specific antibodies have been demonstrated by radioallergosorbent test (RAST) (Biagini *et al.*, 1985), the consensus is that skin testing is more appropriate for the clinical diagnosis of Pt-hypersensitivity. Skin testing should be performed by the prick-test method using freshly prepared solutions of ammonium or sodium hexachloroplatinate. Although positive reactions can occur with concentrations as low as 10^{-6} g/l (Biagini *et al.*, 1985) and (mild) systemic reactions may occur following skin testing (Merget *et al.*, 1991), a concentration of 10^{-2} mol/l (i.e. approximately 4.5 g/l or 0.45%) has been proposed for practical purposes (Merget *et al.*, 1991); others (Venables *et al.*, 1989) have used twice as much (10 g/l) for 3- to 6-monthly testing. A positive skin-prick test against a platinum salt may (in the appropriate clinical context) be considered diagnostic for platinum-induced rhinitis or asthma, i.e. it has a 100% specificity. However, there is not a perfect coincidence between the occurrence of respiratory symptoms and a positive skin test, since symptoms may precede a positive skin test by more than one year (Venables *et al.*, 1989; Brooks *et al.*, 1990), i.e. skin testing is not 100% sensitive. Those with a positive test ultimately develop symptoms of allergy (dermatitis, conjunctivitis, rhinitis, asthma) if exposure continues (Venables *et al.*, 1989; Calverley *et al.*, 1995) and this justifies the strategy of using skin testing as a surveillance tool in exposed subjects, in order to identify allergic subjects and remove them early from further exposure. Bronchial provocation using an aerosol of hexachloroplatinate [successive inhalation of 10 breaths of 10^{-8} to 10^{-2} mol/l, i.e. approximately 4.5×10^{-6} to 4.5 g/l of $(NH_4)_2PtCl_6$] has also been done and it was found that skin test-positive subjects generally also have a positive bronchial provocation test, thus demonstrating the good specificity of the skin test; on the other hand, the bronchial provocation test may occasionally be positive in skin test-negative subjects, thus demonstrating the relative lack of sensitivity of the skin test (Merget *et al.*, 1991). In the latter study, except for one dual reaction, all asthmatic reactions were of the immediate type, although clinical experience suggests that nocturnal respiratory problems are common in refinery workers (Hughes, 1980). Interestingly, there is some relation between the degree of skin and bronchial responsiveness to platinum, but not between the bronchial response to platinum and the non-specific bronchial hyperreactivity (Merget *et al.*, 1996). An important recent notion is that, although sensitization to platinum salts apparently occurs via IgE mechanisms, the risk of sensitization to platinum salts is, perhaps surprisingly, not or only poorly associated with atopy (defined as allergy to common aeroallergens), but very significantly with smoking (Venables *et al.*, 1989; Baker *et al.*, 1990; Calverley *et al.*, 1995). This indicates that selection of employees for work in precious-metal refineries on the basis of atopy is no longer warranted. Another feature of platinum asthma, which is also common to other types of occupational asthma, is that a considerable proportion of affected subjects remain symptomatic, even after removal from exposure, certainly if allergen avoidance is not complete (Baker *et al.*, 1990; Merget *et al.*, 1994).

Platinum is not the only metal of the platinum group that is capable of inducing sensitization and allergy to salts of related metals (palladium, rhodium, iridium, ruthenium) has been shown to coexist occasionally with platinum sensitization, but isolated allergy to these metals was not found (Biagini *et al.*, 1985; Murdoch and Pepys, 1987). However, we have recently investigated a worker with occupational asthma which was clearly related to exposure to fumes evolving from a metallization bath containing palladium salts; the subject exhibited an isolated positive skin test to tetra-ammine palladium (II) chloride (1 μg/ml) as well as a positive bronchial provocation test to this palladium salt (10 μg/ml in aerosol for a total of 3.5 min) Daenen *et al.*, 1997).

(c) Nickel, chromium and cobalt

These three metals are grouped here because they seem to share a number of characteristics with regard to their potential to cause sensitization and asthma. The number of reported cases of asthma caused by Ni, Cr and Co is not high and most data about this come from case reports or small series of patients. The apparent rarity of asthma caused by these metals is rather surprising, because these widely used compounds are very well known as sensitizers by dermatologists, who consider them among the chief causes of allergic contact dermatitis. Contact dermatitis caused by Cr is indeed the most prevalent occupational dermatitis in men, and Ni is the most prevalent contact allergen in women (probably because of exposure to jewellery), with cobalt allergy frequently being associated (Dooms-Goossens *et al.*, 1980; Fisher, 1986). Aerogenic contact dermatitis has even been documented with these agents (Dooms-Goossens *et al.*, 1986; Dooms-Goossens and Deleu, 1991). The relative rarity of respiratory rather than dermal manifestations of metal sensitization may result in part from under-diagnosis and under-reporting, but it is unlikely that this is the only explanation. Differences related to route of exposure and mechanisms of sensitization must be involved. Allergic contact dermatitis is not an IgE-mediated type of allergy, but a 'cellular' allergy, though this in itself is not sufficient to explain the (apparent) rarity of respiratory disease caused by these metals.

Asthma caused by nickel has been described in case reports of subjects exposed to nickel dust during metal polishing (Block and Yeung, 1982) or to nickel salts used for metal-plating (Malo *et al.*, 1982, 1985; Novey *et al.*, 1983) or as catalysts for the food industry (Davies, 1986). A number, but not all, of these individuals also had skin symptoms (contact dermatitis or urticarial rash). In some of these cases skin-prick tests to $NiSO_4$ were positive and Ni-reactive IgE antibodies could be demonstrated in serum (Malo *et al.*, 1982; Novey *et al.*, 1983; Dolovich *et al.*, 1989). The diagnosis of nickel-induced asthma can be verified by bronchial provocation testing using an aerosol of $NiSO_4$ (10 mg/ml for up to 30 min and up to 100 mg/ml for 4 min have been administered).

The relevant species of chromium with regard to toxicity, sensitization and carcinogenicity is the hexavalent form (CrVI). Significant exposure to hexavalent chromium occurs mainly in chromate production industries, in chrome plating, during stainless steel welding, but also in the building industry because cement generally contains small amounts of chromates, which is a frequent source of allergic contact dermatitis. Although probably unrelated to asthma, it is worth mentioning here that exposure to chromic acid mist in chrome plating causes nasal septum ulcers and perforation, even at relatively low levels of exposure (Lindberg and Hedenstierna, 1983; Lin *et al.*, 1994).

Despite the many opportunities for aerial exposure to chromium, asthma caused by chromium has not been described frequently. Although this has not been studied systematically, it appears from the reported cases that there is a rather close relation between signs of dermal and respiratory allergy in chrome asthma. Thus, patients often have preceding or concomitant skin symptoms (urticaria or dermatitis), and bronchial provocation testing may also elicit pruritus or a skin rash. One subject, a welder, experienced a delayed anaphylactoid reaction (generalized urticaria, angioedema and bronchospasm) upon bronchial challenge with a sodium chromate aerosol (Moller *et al.*, 1986). Olaguibel and Basomba (1989) have described five subjects (two from the concrete industry, one from a tannery, one metal plater and one construction worker) with contact dermatitis and positive patch tests, but negative skin-prick tests to potassium dichromate, who also had occupational asthma, which was confirmed by bronchial provocation testing with an aerosol of potassium dichromate ($K_2Cr_2O_7$, 0.1 mg/ml for 5 min). Park *et al.*, (1994) have reported chromate asthma in four subjects (two metal platers, one worker from the cement industry and one construction worker), of whom two had an immediate and two a delayed skin response to chromium sulfate [$Cr_2(SO_4)_3$]; bronchial provocation testing with an aerosol of $Cr_2(SO_4)_3$ (0.1 to 10 mg/ml, five inhalations) was positive with an early asthmatic reaction in one subject and a dual reaction in the other three. A brief bronchospastic reaction upon a bronchial provocation test with bichromate has also been reported in a worker from a metal-plating plant (Shirakawa and Morimoto, 1996). Occupational asthma due to stainless steel welding has also been attributed to either chromium or nickel in the welding fumes (Keskinen *et al.*, 1980). Cement asthma has also been described in building workers (Lob, 1985). From observations in a floorer who had both occupational dermatitis and occupational asthma, it has been suggested that the inhalatory exposure to chromates was via the smoking of (rolled) cigarettes contaminated with cement (Nemery *et al.*, 1995).

Cobalt has been more frequently reported as a cause of occupational asthma (Cirla, 1994). Most cases of cobalt asthma have been described in relation to the manufacture or grinding of sintered hard metal, in which cobalt is used as a binder for tungsten carbide (WC) (Davison *et al.*, 1983; Kusaka *et al.*, 1986b; Shirakawa *et al.*, 1989). It should be noted that coolants used in wet grinding of hard metal tools may contain a high concentration of dissolved (i.e. ionized) Co, thus leading to the paradox (in terms of prevention of pneumoconiosis) that wet grinding may be more 'harmful' than dry grinding (Sjögren *et al.*, 1980). Cobalt asthma has also been found in relation to the production of cobalt (Roto, 1980), the manufacture or use of cobalt pigments or additives (Pillière *et al.*, 1990), and in diamond polishers (Gheysens *et al.*, 1985). Bronchial provocation tests have been performed with cobalt powder and with cobalt salts (up to 1% $CoCl_2$ or $CoSO_4$), and late reactions appear to be common (Cirla, 1994). Several studies have been devoted to the pathogenesis of cobalt sensitization by Shirakawa and co-workers (Shirakawa *et al.*, 1990, 1992; Kusaka *et al.*, 1991), who found evidence for both humoral pathways (IgE against Co-conjugate with human serum albumin) and cellular pathways (lymphocyte transformation tests) in some, but not all of their patients with cobalt-induced asthma. However, other groups have been unsuccessful in obtaining similar results in their patients (Cirla, 1994). It is also unclear if and how cobalt-induced asthma relates to the interstitial lung disease associated with cobalt (see below). It is possible that cobalt asthma is but an 'airway variant' of hard-metal disease and there are occasional subjects who exhibit both asthmatic reactions and parenchymal involvement (Davison *et al.*, 1983; Van Cutsem *et al.*, 1987; Rivolta *et al.*, 1994).

(d) Potroom asthma

A peculiar form of occupational asthma has been described in aluminum smelters, who are exposed to the fumes evolving from the 'pots' in which aluminum ore is molten after addition of fluoride salts. The existence of this form of occupational asthma has been disputed, mainly because no positive bronchial provocation tests have been reported, but clinical and epidemiologic studies carried out not only in the Norwegian aluminum industry (Kongerud *et al.*, 1994), but also elsewhere (Abramson *et al.*, 1989; Desjardins *et al.*, 1994; Sorgdrager *et al.*, 1995), have clearly indicated that 'potroom asthma' does occur and that it shares many features of occupational asthma. However, the exact causal agent of this form of asthma is not known and the condition is not felt to be due to allergic mechanisms. It seems more likely that potroom asthma is a form of irritant-induced asthma, with the irritants being most probably the fluorides present in the fumes (Abramson *et al.*, 1989; Soyseth *et al.*, 1994). Occupational asthma has also been reported as a result of exposure to aluminum fluoride salts (Simonsson *et al.*, 1985; Hjortsberg *et al.*, 1986).

(e) Zinc

There are rare reports (Malo and Cartier, 1987) of asthma in subjects welding galvanized metal, with sensitization to zinc being postulated on the grounds of a positive bronchial challenge. Similarly, occupational asthma due to soldering fluxes containing zinc chloride has been described (Weir *et al.*, 1989). However, no immunologic support for sensitisation to zinc has been provided. It is conceivable that (some) individuals with non-specific bronchial hyper-reactivity exhibit asthmatic reactions when they have a bout of metal fume fever. The possible presence of metals other than zinc should also be considered in such instances.

(f) Vanadium

Exposure to vanadium, usually as vanadium pentoxide (V_2O_5), is often mentioned as a possible cause of occupational asthma. This exposure has been described in workers who maintain, clean or dismantle oil-fired boilers, because the fly ash produced by some types of fuel-oil may be very rich in vanadium (see above). From the available evidence it appears that this form of asthma essentially results from the strong irritation produced by V_2O_5 which may cause nasal ulceration and acute bronchitis (see above), with more or less lasting non-specific bronchial hyperreactivity as a result.

(g) Welding in general

In addition to the specific causes of occupational asthma described above, welding is often mentioned as a cause of occupational asthma or as a risk factor of asthma (Meredith *et al.*, 1991; Meredith, 1993; Wang *et al.*, 1994; Vandenplas *et al.*, 1995; Beach *et al.*, 1996). It is not always clear whether these are cases of asthma from metal sensitization, 'irritant-induced' asthmas due to irritant gases, or work-related asthmas.

Interstitial lung disorders

(a) General considerations and terminology

According to a leading authority (Parkes, 1994) 'pneumoconiosis may be defined as the non-neoplastic reaction of the lungs to inhaled mineral or organic dust and the resultant alteration in their structure but excluding asthma, bronchitis and emphysema'. Thus, all the occupational interstitial lung disorders, including for example hypersensitivity pneumonitis, may be considered to be 'pneumoconioses', which almost literally means 'dust-induced lung disease'. The specific origin of the disease is then indicated either by assigning the suffix '-osis' to the agent (e.g. silicosis, asbestosis, siderosis, berylliosis), or by qualifying the pneumoconiosis as such (e.g. coal worker's pneumoconiosis, hard-metal pneumoconiosis). While this terminology is adequate and widely used for the disorders in which the total dust burden in the lung plays a paramount role, the term pneumoconiosis is somewhat misleading for some other forms of interstitial lung diseases, where the occurrence of lung disease is not evidently related to the cumulative dust burden but is more likely due to individual susceptibility. The latter is clearly the case for hypersensitivity pneumonitis, which few authors would now label as a pneumoconiosis, but also for lung disease caused by beryllium or cobalt, which may be found in young subjects after only a brief exposure period or after exposures which are very low in comparison with those that lead to the traditional pneumoconioses. For this reason, it does not seem desirable to use a terminology that is too reminiscent of the other pneumoconioses for such diseases, and it is advised to abandon terms such as berylliosis or hard-metal pneumoconiosis, and to replace them by beryllium (lung) disease or hard-metal (lung) disease, as is in fact done by most authors in the field.

Metallic dusts deposited in the lungs may give rise to more or less marked pulmonary fibrosis, depending on the intrinsic properties and amount of the inhaled agent, as well as on hitherto poorly understood host factors. The fibrogenic potential of inhaled substances is presumably determined by their ability to interfere with the pulmonary immunoinflammatory system, either via immunologically mediated sensitization (e.g. Be), or indirectly via injury to alveolar macrophages and/or epithelial cells. Inflammation then sets in motion cytokine networks leading to fibroblast activation and proliferation. In other words, as in other forms of interstitial lung disease, the fibrotic process is probably dependent on the occurrence of alveolitis with an abnormal release of mediators by some cells. It must be recognized, however, that the exact pathogenic mechanisms of lung fibrosis, whether caused by metals or by other agents, have not been elucidated despite intense research efforts. Recent epidemiologic surveys have revealed that subjects with 'idiopathic' lung fibrosis are more likely than controls to have been occupationally exposed to metals (and also to wood) (Hubbard *et al.*, 1996). This means either that many patients receive a diagnosis of idiopathic lung disease because no good occupational and environmental history has been taken, or that some metals may play a hitherto unrecognized role in the causation of interstitial lung disease. It is reasonable to assume that both explanations are valid to some extent.

Traditionally, the mineral pneumoconioses are classified into pneumoconioses with fibrosis (e.g. silicosis) and pneumoconioses without fibrosis. The latter are sometimes defined as 'benign' pneumoconioses. The absence of fibrosis has often only been inferred from the fact that no confluence of (micro)noduli was found to occur on chest X-ray and that no significant functional deficits were observed in affected subjects. While this may be true, the labeling of a pneumoconiosis as 'benign' is not acceptable from an occupational health viewpoint and we should abandon this qualifier.

(b) Pneumoconioses without fibrosis

Of these pneumoconioses the most frequent and best (although still poorly) studied is siderosis, which is caused by the inhalation of iron compounds. More than 90% of the total world production of metallic materials is in the form of steels and cast irons (John, 1994), and occupational exposure to iron is, therefore, extremely widespread. It occurs during iron mining and related operations, during iron refining and at various stages in steelmaking, during welding, cutting and abrading of iron-containing materials, as well as during the manufacture or use of iron-containing abrasives (such as emery). Despite the large number of workers who are exposed to iron-containing dust, it must be admitted that little research has been carried out in recent years regarding the pulmonary effects of occupational exposure to iron dust. Yet, in the past decade there has been a considerable interest in the possible role of iron in other pulmonary diseases, because ferrous/ferric ions catalyze the Fenton reaction, whereby the very toxic hydroxyl radical ($OH^{\bullet}$) is produced. Thus, the presence of iron on crocidolite has been implicated as a determinant of this fiber's toxicity and carcinogenicity. A possibly important of iron has also been evoked with regard to the oxidant effects of cigarette smoke-induced lung disease. In view of the known effects of iron, it would be highly desirable to investigate further, in people who are occupationally exposed to iron, what the fate is of the inhaled iron in the lungs, what biochemical defense mechanisms operate against this high iron burden, what degree of inflammation is caused by chronic exposure to iron, etc.

According to most textbooks, siderosis is only a 'radiological disorder' which manifests itself by the presence of small, very radiodense opacities with uniform distribution throughout the lungs, but without formation of conglomerates. With cessation of exposure the radiographic opacities may gradually disappear. Pure siderosis is not associated with respiratory symptoms or functional impairment, and does not predispose to tuberculosis. However, it is important to realize that exposure to silica or asbestos is not uncommon in many jobs that involve exposure to iron, thus giving rise to mixed dust fibrosis ('siderosilicosis') or to asbestosis, which do have associated morbidity and complications. Moreover, the view that the symptomatic interstitial fibrosis, which is sometimes found in welders ('welder's pneumoconiosis'), is simply siderosis with coexisting silicosis has been recently challenged on the grounds that the pulmonary silicon content of such cases did not differ from that of control lungs (Funahashi *et al.*, 1988). Occasional case reports of siderosis with significant fibrosis are also being reported regularly in welders (Lasfargues *et al.*, 1991; Billings and Howard, 1993).

Other rare 'benign' pneumoconioses include those caused by tin (stannosis), barium (baritosis) and antimony (antimoniosis). The descriptions of these pneumoconioses are to be found in the older literature and little or no recent clinical and epidemiologic data are available on these pneumoconioses. It is generally considered that, during mining operations, the greatest danger for the lungs is that caused by the presence of free silica in the mine dust, rather than the specific metal ore, which is often only present in relatively small quantities (Seaton *et al.*, 1986; Masjedi *et al.*, 1989). Significant exposure to the metal dust, therefore, mainly occurs during procedures downstream of the ore mining, such as refining, metallurgical processes and further uses (Sluis-Cremer *et al.*, 1989).

Thus antimoniosis, described as dense, mainly p-type opacities in the middle to lower lung fields, has been reported in 51 Yugoslavian workers exposed for at least 9 years to high concentrations (17 to 86 mg/m^3) of antimony trioxide (Sb_2O_3) in an antimony smelting plant (Potkonjak and Pavlovich, 1983). Confluence of the lesions was never observed and no pure restrictive impairment with reduced diffusing capacity was found, but emphysematous changes (and hyperinflation) were frequent (34.5%). Although no comparison group was available and smoking data were not reported, it should be noted that seven workers (16%) reported 'permanent breathlessness in effort' and that nine (18%) and five (10%) subjects exhibited an obstructive or a mixed obstructive–restrictive impairment, respectively. From these data it may be concluded that antimoniosis resembles other forms of simple pneumoconiosis, but also that it is probably inappropriate to label it as 'benign'. Collagenous fibrosis has been described around dust areas in surgical lung biopsies of two subjects with antimoniosis (Le Bouffant *et al.*, 1987). On the other hand, recent mortality studies have not identified pneumoconiosis as a significant cause of death in antimony smelter workers (Jones, 1994; Schnorr *et al.*, 1995).

(c) Beryllium disease and other interstitial lung disease with granuloma formation

Beryllium disease

Berylliosis or chronic beryllium disease (CBD) is well known for its striking histologic and clinical resemblance to sarcoidosis. The clinical, epidemiologic and experimental aspects of the disease have been extensively reviewed (Eisenbud and Lisson, 1983; Cullen *et al.*, 1987; Kriebel *et al.*, 1988a; Meyer, 1994; Newman, 1995).

Apart from the extraction (as beryl ore) and primary refining industry, beryllium exposure is an occupational risk in several sectors of modern technological industries, such as aircraft and aerospace, electronics, computers and communications, where beryllium is utilized for its light weight, stability, hardness, lack of magnetic properties and good electrical conductivity. Beryllium is mainly found in alloys (often with copper) or in ceramics. Beryllium exposure also occurs in the nuclear power industry, because it facilitates fission reactions of uranium and plutonium. Beryllium used to be present in fluorescent lights, but this is no longer the case. However, it is important to realize that occult or unsuspected exposures to beryllium may take place during the refining of scrap metal, the machining or welding of non-ferrous alloys, in dental laboratories (Kotloff *et al.*, 1993), etc., and that the first manifestations of CBD may only occur long after exposure has ceased. Para-occupational exposure has also been described as a significant risk, not only in the older literature, when beryllium disease occurred in residents in the vicinity of beryllium refineries, but even very recently, when beryllium disease was diagnosed in the spouse of a beryllium production worker (Newman and Kreiss, 1992).

Classically, exposure to high concentrations of beryllium fumes may cause a chemical pneumonitis, which in some instances seems to progress toward chronic beryllium disease (unless this acute event represents already a manifestation of the hypersensitivity disease). However, modern exposures are mainly associated with chronic beryllium disease. This is characterized by diffuse pulmonary inflammation with formation of epithelioid non-caseating granulomas in the lung interstitium, mainly along the bronchovascular bundles, but also in the bronchial submucosa and subpleural regions, and in the intrathoracic lymph nodes. The pathology may also suggest extrinsic allergic alveolitis with more diffuse mononuclear cell infiltration. End-stage lung fibrosis with honeycombing may also be found, as well as conglomerate masses of granulomas and fibrosis. As described by Newman (1995) the presentation and clinical course of the disease are similar to those of sarcoidosis, with exertional dyspnea of gradual onset, cough and fatigue. However, asymptomatic hilar lymphadenopathy and pulmonary infiltrates, or ocular or skin lesions are not presenting features of CBD. Beryllium disease may also affect extrapulmonary organs, such as the lymph nodes, the skin (but there is no erythema nodosum), the salivary glands and other organs, although this is less frequently the case than in classic sarcoidosis, and no ocular or neurologic manifestations have been reported. Sarcoidosis and CBD cannot be distinguished from each other by radiology, pulmonary function testing, ordinary laboratory tests or bronchoalveolar lavage (BAL). The chest X-ray reveals diffuse parenchymal infiltrates of various types (reticular to nodular) and often also hilar lymphadenopathy. Functional impairment is variable with restrictive and obstructive patterns, as well as isolated reductions in $D_{L}CO$ and alterations in gas exchange on exercise, even years before radiologic abnormalities are seen. Serum ACE may also be elevated. In the BAL, lymphocyte numbers are increased and the CD4/CD8 ratio is markedly elevated. In principle, the treatment and prognosis of chronic beryllium disease are comparable with that of sarcoidosis, with some patients going into remission, but continuous steroid therapy is usually required (without always being effective) and progression to endstage lung fibrosis is more frequent in CBD than in sarcoidosis.

Most recent advances in the field of chronic beryllium disease have been made with regard to its immunopathology. Beryllium has become *the* example of a small-molecular weight antigen which is capable of leading, in susceptible individuals, to a cell-mediated, delayed-type hypersensitivity reaction in the lung. Although patch testing with Be had been shown in the 1950s to be positive in patients with CBD, a new impetus was given to the field by the use of *in vitro* tests, such as the beryllium lymphocyte transformation test (LTT) or beryllium-specific lymphocyte proliferation test (Rossman *et al.*, 1988). This test may be carried out in cells obtained by BAL or in peripheral blood lymphocytes. In this test, the proliferation of lymphocytes cultured in the presence of a beryllium salt (e.g. 10^{-5} M $BeSO_4$) is assessed, e.g. by measuring the incorporation of radiolabeled thymidine. This *ex vivo* test is considered to be highly specific, since lymphocyte proliferation does not occur in subjects without the disease, such as in ordinary sarcoidosis or extrinsic allergic alveolitis, or only rarely in exposed but non-diseased subjects. The test is also considered to be sensitive, particularly if it is performed on BAL lymphocytes, although technical improvements have also rendered it sensitive when applied to blood lymphocytes (Mroz *et al.*, 1991). The good positive and negative predictive value of the beryllium-specific lymphocyte proliferation test has led to its being proposed as a medical surveillance tool in beryllium-exposed subjects (Kreiss *et al.*, 1993b). Saltini *et al.* (1989) have shown that major histocompatibility complex class II antigens and functional IL-2 receptors are necessary for CD4+ lung T cells to proliferate *in vitro* upon stimulation by Be salts. That genetic factors determine, at least in part, the occurrence of this cell-mediated immunologic response is borne out by the much higher prevalence of subjects (32/33 patients) with the HLA-DPB1 glu 69 allele among patients with chronic beryllium disease, than in control beryllium-exposed workers (30%) (Richeldi *et al.*, 1993), as well as by the pronounced strain specificity of granuloma formation upon treatment of experimental animals with Be.

Finding a relationship between exposure levels and attack rates of sensitization to Be has been a difficult exercise, as with most immune-mediated lung diseases. Certain processes have been shown to carry a higher risk, possibly because of overexposures or because of particle characteristics, but cases of CBD have been shown to occur following very low exposures, e.g. in clerical jobs (Kreiss *et al.*, 1993a). A recent study (Kreiss *et al.*, 1996) has shown that sensitization occurred in workers exposed, during the machining of beryllia ceramics, to median levels of Be below $2\,\mu g/m^3$, thus leading the authors to suggest that beryllium exposure should be kept as low as reasonably achievable.

Other possible causes of lung disease with granuloma formation

It is possible, and conceivable, that beryllium is not the sole metal involved in causing sarcoid-like lung disease.

Thus, zirconium (Zr) may cause granulomas in human skin and exposure to Zr during the manufacture of ceramic tiles has been recently associated with granulomatous lung disease with a fulminant course (Liippo *et al.*, 1993). Similarly, hypersensitivity pneumonitis with granulomas has been reported in a welder on nuclear fuel rods containing zirconium (Schneider *et al.*, 1994) and a case of slowly progressive interstitial disease (without granulomas, but with high levels of zirconium being found in the lung) has also been described in a lens-grinder (Bartter *et al.*, 1991). On the other hand, a limited survey of a zirconium-exposed workforce has not found any significant risk of pulmonary disease (Hadji Michael and Brubaker, 1986).

Titanium, otherwise considered as virtually non-toxic (except when present as very fine particles), has been suggested as an etiologic agent in a single case of granulomatous lung disease based on a positive blood LTT to titanium chloride, and not to the other metals tested, including beryllium (Redline *et al.*, 1986). On the basis of similar reasoning, aluminum exposure has also been suggested to have led to sarcoid-like lung granulomatosis in a patient who had apparently not been exposed to beryllium (De Vuyst *et al.*, 1987).

Exposure to rare earth metals (or lanthanides), of which cerium (Ce) is the most abundant element, has also been associated with interstitial fibrosis in a small number of subjects (Husain *et al.*, 1980; Sinico *et al.*, 1982; Vocaturo *et al.*, 1983; Sulotto *et al.*, 1986; Vogt *et al.*, 1986; Haley, 1991; Pairon *et al.*, 1995). Rare earths are essential components of carbon arc lamps used for photoengraving, and the majority of cases of this pneumoconiosis have been described in photo-engravers. Deposits of rare earths have also been found, without radiologic abnormalities, in the lung of a movie projectionist (Waring and Watling, 1990). Rare earths are also used in the fabrication and polishing of glass, where they have been associated with pneumoconiosis (McDonald *et al.*, 1995; Pairon *et al.*, 1995). The presence of granulomatous interstitial alterations is mentioned occasionally (Husain *et al.*, 1980), but this is not the case in most available pathologic descriptions of cerium pneumoconiosis.

These cases confirm that occupational exposures (also to silicates, such as talc) should always be considered in cases of lung fibrosis and 'sarcoidosis.'

Aluminum

The question of the pulmonary effects of occupational exposure to aluminum dust and the very existence of 'aluminum-lung' has been the subject of some controversy. Indeed, in view of the extensive industrial use of aluminum, parenchymal lung disease caused by exposure to this metal appears to be very uncommon, both at the stage of its production and during its use. It should be mentioned that, on the basis of experimental data showing that aluminum counteracts the toxic effects of silica, inhalation of aluminum has been proposed and even implemented (without real success, but also without apparent pulmonary adverse effects) for the prevention of silicosis and even for its therapy

(Rifat *et al.*, 1990). On reviewing past literature, Dinman (1987,1988) concluded that fibrosis only occurred in: (i) workers who were heavily exposed to submicron-sized aluminum plates lubricated with an easily removed lubricant during the production of fireworks and explosives; and (ii) in workers involved in the smelting of bauxite for the production of Al_2O_3 (corundum) abrasive (Shaver's disease), but who were perhaps also exposed to crystalline silica. An extensive review also concluded that pulmonary fibrosis was not a significant problem in aluminum smelter workers (Abramson *et al.*, 1989). On this basis it may be stated that exposure to aluminum dust does not represent a common or potent risk of interstitial lung disease.

Nevertheless, this risk is not absent, as suggested by the description of severe interstitial fibrosis (of the UIP type) in three workers (out of a workforce of about 1000 workers), who had been heavily exposed, for 19 to 33 years, mainly, but not exclusively, to Al_2O_3 in the production of abrasives (Jederlinic *et al.*, 1990). A survey has also shown an increased prevalence of evenly distributed, small-sized irregular opacities in the lungs of aluminum refinery workers (Kilburn and Warshaw, 1992). Moreover, one should not discount the reports of isolated cases of granulomatous lung disease (De Vuyst *et al.*, 1987), fibrosis (Herbert *et al.*, 1982; Vallyathan *et al.*, 1982; De Vuyst *et al.*, 1986a; Akira, 1995) or alveolar proteinosis (Miller *et al.*, 1984) in aluminum welders or polishers. The physical characteristics of the aluminum particles, notably their surface area, or even their possibly fibrous nature (Gilks and Churg, 1987), have been suggested as important determinants of their bioreactivity and hence fibrogenicity.

Although the synthetic abrasive silicon carbide (SiC) or carborundum is not a metallic compound – in contrast to some other abrasives such as corundum (Al_2O_3) or emery (corundum with iron oxides) – it is worth mentioning that respiratory disease, including pneumoconiosis with fibrosis, has been associated with exposure to SiC, during its manufacture or use (Peters *et al.*, 1984; Funahashi *et al.*, 1984).

(d) Hard metal lung disease – 'cobalt lung'

Hard metal (not to be confused with 'heavy metals', which is a vague term used mainly in relation to environmental pollution by toxic metals such as lead, cadmium, mercury, etc.) is a composite material composed mainly of tungsten carbide (WC) and cobalt (Co). In contrast to other alloys, hard metal is not produced by melting together two or more metals, but it is obtained by a process of powder metallurgy, whereby the components are pressed together under specific conditions of pressure and temperature. The process which converts a powder compact into a solid polycrystalline material is called sintering, hence the other names sometimes given to hard metals: sintered carbides or cemented carbides (John, 1994). In hard metal, cobalt functions as a binder and it is present in proportions of 5% to 25% (generally about 10%), depending on the applications. Besides tungsten carbide, other carbides (e.g. with tantalum, titanium, niobium) may also be used and small amounts of other metals (e.g. nickel, chromium) may also be present. As its name indicates, hard metal is mainly used for its hardness, which may be about 90% that of diamond, hence one of its common commercial names 'Widia' (from the German 'wie Diamant' meaning 'like diamond'). Hard metals have numerous industrial applications: hard metal is mainly used to make tools for drilling, sawing, cutting, grinding or polishing various materials, such as stones, concrete, metals, wood, etc. In recent years, diamond itself has also been increasingly used to make grinding tools and here too, the binder used for the microdiamond powder is cobalt. In these applications cobalt makes up to 90% of the grinding surface. Another hard material used, e.g. for the tips of saw blades is Stellite, which is a super-alloy of cobalt (50–60%), chromium (30–35%) and other elements (Kennedy *et al.*, 1995).

The most significant exposures to cobalt dust take place during the various stages (weighing, mixing – milling, shaping, sintering, finishing) of the manufacture of hard metal tools, where very high levels of cobalt (in the hundreds to thousands of $\mu g/m^3$) may be found (Ferdenzi *et al.*, 1994; Mosconi *et al.*, 1994; Sala *et al.*, 1994). Significant exposure to cobalt may also occur during the filing and re-sharpening of hard metal tools, particularly during wet grinding (Sjögren *et al.*, 1980; Cereda *et al.*, 1994; Imbrogno and Alborghetti, 1994; Sala *et al.*, 1994; Kennedy *et al.*, 1995). Such activities often take place in small or medium-sized factories, where hygiene conditions may be far from ideal, and even in large plants, tool sharpening is often not recognized as a significant health risk. Coating with hard metal by the detonation gun process and related operations may also give rise to significant exposure (Figueroa *et al.*, 1992). Although exposure to cobalt may also conceivably occur during the use of hard metal-tipped tools during tunneling, construction, metal and wood working or oil drilling, it is not sure that this type of low exposure results in lung disease. However, the use of diamond–cobalt polishing disks has been associated with significant disease, at least in diamond polishers (see below).

Hard metal (lung) disease is a form of interstitial lung disease which may be encountered in subjects exposed to hard metal dust. (Hard metal-induced bronchial asthma, which may coexist with interstitial disease in the same patient, is sometimes included in the clinical entity of hard metal lung disease, but usually the term hard metal lung disease is restricted to the parenchymal form of the disease.) Several reviews of hard metal lung disease are available (Balmes, 1987; Van Den Eeckhout *et al.*, 1988; Cugell *et al.*, 1990) and a whole congress has even been devoted to it (Seghizzi *et al.*, 1994). From a clinical point, hard metal lung disease shares many features with hypersensitivity pneumonitis: in some subjects the presentation is that of a subacute alveolitis with fever, cough, dyspnea and a clear exposure-related clinical course, while in others a more insidious course with progressive diffuse fibrosis may be found. Overt disease only occurs in a minority of exposed workers and those affected may be young subjects with only a short exposure. These features clearly point to a condition caused mainly by a specific, perhaps immunologic, susceptibility, as opposed to a simple process of chronic accumulation of dust in the lungs, as in the classic mineral pneumoconioses. On the other hand, hard metal lung disease also exhibits characteristics which make it distinct from hypersensitivity pneumonitis caused by organic antigens. Thus, no precipitating antibodies are found in the serum and the lung pathology and cellular findings in the BAL generally differ from those of hypersensitivity pneumonitis. Thus, although granulomas have been occasionally described in hard metal lung disease, they are certainly not typical and the most characteristic finding in the lungs, as well as in bronchoalveolar lavage, is the presence of numerous so-called 'bizarre' giant multinucleated cells with 'cell in cell' or 'cannibalistic' features. The finding of the latter cells leads to a pathologic diagnosis of giant cell interstitial pneumonitis (GIP), and it has been proposed that GIP is pathognomonic for hard metal exposure (Abraham, 1987). Other pathologic features of hard metal lung disease include desquamation of numerous cells in the alveoli (desquamative interstitial pneumonitis or DIP), lymphoplasmocyte infiltration, and varying degrees of interstitial, mainly centrilobular fibrosis. However, the pattern of lung pathology in subjects with hard metal dust exposure appears to be very diverse, and it is possible that there are not only differences with respect to the stage of the disease (with alveolitis evolving towards end-stage fibrosis), but also with respect to mechanisms of disease, with some patients exhibiting features of mixed-dust pneumoconiosis rather than the hypersensitivity type of disease. Even a case of Goodpasture syndrome has been attributed to exposure to hard metal dust (Lechleitner *et al.*, 1993).

The pleomorphic character of the disease is also reflected in the fact that nearly all possible cellular patterns may be found in BAL, with all cell types – macrophages, lymphocytes, neutrophils or eosinophils, and even mast cells – being possibly increased. In fact, an increased overall number of cells has been proposed as an important reflection of the desquamative pneumonia (Forni, 1994). The ratio of T4 to T8 lymphocytes is generally found to be low and an increased number of eosinophils seems to be associated with more severe disease. In the BAL or in the lung tissue, increased concentrations of the constituents of hard metal may be detected by destructive analysis or by, for example, energy-dispersive X-ray spectrometry or neutron activation analysis, but often only tungsten is detected and not cobalt, which is more soluble and rapidly cleared. The finding of the constituents of hard metal in the BAL is a useful indicator of (past) exposure, but there is no quantitative relation with disease.

A turning point in the field of hard metal lung disease was reached in the early 1980s, when an outbreak of interstitial lung disease (as well as of asthma) occurred in diamond polishers following the introduction of a novel type of high-speed polishing disks (Demedts *et al.*, 1984; Lahaye *et al.*, 1984) The interstitial lung disease found in these polishers had all the features of hard metal lung disease, notably the presence of 'bizarre' multinucleated giant cells in the lung tissue or BAL fluid, although no exposure to cemented carbides could be documented, neither in the working environment, nor in the tissues of affected subjects. However, cobalt was present as a binder for the microdiamonds in the new polishing wheels, and it was detected both by hygiene measurements and in biologic samples of patients and workers (Nemery *et al.*, 1990, 1992; Vanden Oever *et al.*, 1990; Dinsdale *et al.*, 1991). The finding of 'hard metal lung disease', without exposure to hard metal but with exposure to cobalt was considered as evidence for a hypothesis, which had been formulated many years previously on the basis of animal experiments, that it is cobalt and not tungsten carbide which is responsible for hard metal lung disease. This led to the proposal to label the condition 'cobalt lung' (Lahaye *et al.*, 1984), rather than hard metal lung.

While the role of cobalt in the causation of interstitial lung disease in hard metal workers and diamond polishers is undisputed, exposure to cobalt alone does not appear to be sufficient to lead to parenchymal disease and fibrosis (Lison, 1996; Lison *et al.*, 1996). A series of both *in vivo* and *in vitro* experimental studies have indeed shown that the cellular toxicity and pro-oxidant effects of cobalt particles is greatly enhanced by the presence of carbides, most notably tungsten carbide. This is reflected by epidemiologic findings which indicate that occupational exposure to even high levels of 'pure' cobalt dust is apparently not associated with interstitial lung disease (although pure cobalt and cobalt salts may cause asthma) (Swennen *et al.*, 1993).

The mechanisms for the pulmonary toxicity of cobalt have not yet been entirely elucidated. The relative rarity of the condition and the fact that it may arise after short and apparently low exposures plead for an important role of individual susceptibility. The nature of this susceptibility is not yet known, but a number of indirect arguments, such as the known dermal and respiratory sensitizing properties of cobalt and some of the clinical features of the disease, plead for immunologic sensitization as a possible mechanism. Of note in the context of (auto)immune mechanisms is the recurrence of GIP in a (single) transplanted lung despite cessation of exposure (Frost *et al.*, 1993). However, unlike the situation with beryllium, no hard evidence for immunologic sensitization to cobalt as a cause for 'cobalt lung' is available: thus, lymphocyte transformation tests using cobalt have never convincingly proven positive in patients with the disease. Besides immunologic sensitization, the fact that cobalt may, in the right circumstances, lead to the production of toxic oxygen species may also be relevant in terms of susceptibility, in subjects with for example a genetically determined poor defense against oxidant toxicity. The plausibility that such effects are relevant also comes from the observation that oxygen administration may have led to a fulminant progression of 'cobalt lung' in a diamond polisher (Nemery *et al.*, 1990).

The evolution of hard metal lung disease is variable, but cessation of exposure usually results in clinical improvement and good recovery, though some patients do progress to end-stage fibrosis. Although corticosteroids have been given, no proof of their efficacy has been given.

As indicated above, besides its use in hard metal, cobalt is also used to make various special alloys, magnets, special corrosion-resistant steels, and some salts are also used as pigments in the ceramics industry or as additives in plastics. No cases of interstitial lung disease have been reported to occur in these circumstances. A special mention must be made here of the presence of cobalt in Vitallium, an alloy consisting of chromium, cobalt and molybdenum, which is commonly used in dental applications. Traces of this alloy may be found in the lungs of dental technicians with pneumoconiosis, but this does not necessarily mean that it is the cobalt which is responsible for 'dental technician's pneumoconiosis' (De Vuyst *et al.*, 1986b; Seldén *et al.*, 1996). From the published reports, this fibrosis is manifested by dense interstitial fibrosis around dust deposits, without giant cells, and it does therefore not exhibit the characteristics of hard metal disease; in addition, other agents, including silica and beryllium, to which dental technicians may be exposed must also be considered (Rom *et al.*, 1984).

LUNG CANCER

Radioactive metals

Several metallic compounds are proven lung carcinogens in humans (Cone, 1987; Norseth, 1988); these include radioactive metals (and their decay products) and non-radioactive metals. The increased incidence of lung cancer observed in uranium miners has been causally linked with the inhalation of radon daughters (Samet *et al.*, 1984). However, the underground mining of other compounds may also be associated with significant exposure to radioactivity, if there is insufficient ventilation of the radon which leaks from igneous rocks (Archer, 1988). This factor has been implicated in the higher incidence of lung cancer seen in various groups of mineworkers, such as Swedish iron ore miners (Radford and Renard, 1984), although it does not seem play a role in the similarly increased lung cancer incidence of French iron ore miners (Mur *et al.*, 1987). Domestic radon gas exposure is also implicated in the causation of bronchial cancer (Samet *et al.*, 1988).

Non-radioactive metals

Epidemiologic and experimental studies have clearly established the carcinogenic risk of exposure to arsenic, chrome and nickel, at least to some of their chemical forms (Peters *et al.*, 1986; Cone, 1987; Norseth, 1988). Thus, the relationship of arsenic to increased lung cancer risk in copper-smelting workers is unequivocal. This is also the case for other occupational exposures to arsenic, such as the manufacture or spraying of arsenical pesticides. A greatly increased risk of lung cancer has also been demonstrated for workers in the primary chromate production and in the chromate pigment industry. Epidemiologic studies of the carcinogenic risk of exposure to chromium during metal plating or during stainless steel welding have been considered inconclusive. However, the exposure in these jobs is to the carcinogenic form of chromium, i.e. hexavalent chromium, and there are several studies showing an increased lung cancer mortality in welders or platers. Occupational exposure to nickel in nickel smelters and refineries is also unequivocally associated with an increase in cancer of the lung and the nasal sinuses. Again the situation in nickel-using industries is less clear, but a cancer hazard has not been excluded.

There are also epidemiologic or experimental indications that antimony, beryllium, cadmium, cobalt and iron are carcinogenic for the human lung, as are occupations associated with these metals. Thus, studies of iron and steel foundry workers have consistently found an increased risk of lung cancer, but this may be due to the emission of polycyclic aromatic hydrocarbons as pyrolysis products of organic materials used. An increased mortality from lung cancer attributable to cadmium has also been shown in some, but not all, recent studies of cadmium-exposed workers. Clearly, the study of occupational exposures in the causation of lung cancer is hampered by the effect of cigarette smoking, which has been found to act in any fashion from less than additive to multiplicative with occupational exposure, depending on the agent, but also on characteristics of the study population. Since there are no indications that specific histologic types of lung cancer are associated with any specific environmental carcinomas, it is not possible to attribute a particular lung cancer to a particular etiology.

Various studies have shown that the proportion of lung cancer attributable to occupation is far from negligible, being around 15% (Doll and Peto, 1981), but possibly up to 47% in some populations (Simonato *et al.*, 1988). In all of these studies, most of the incriminated exposures are related to asbestos, underground mining, and metals or metal industries. It is therefore worthwhile, even in the individual patient with lung cancer, to search the past occupational history thoroughly for exposure to carcinogens rather than to limit the etiologic 'investigation' to cigarette smoking. Not only can this have implications in terms of compensation, where legislation provides for such compensation, but it may also help to discover or further strengthen the role of hitherto unproven carcinogenic exposures. particularly when clusters of cancers are found.

CONCLUSIONS

This chapter has attempted to show that the toxic effects of metallic compounds may be manifested in almost any form of pulmonary disease. The most important step towards uncovering a possible occupational or environmental cause for an illness in the individual patient is a careful and complete history taking. Inquiring about present and past jobs and understanding the patient's occupational history generally require a good knowledge of existing or past work practices.

Documentation of exposure to metals may be obtained from the analysis of metal concentrations in blood or urine (or even hair or nails) taken for biologic monitoring (Lauwerys, 1987). In addition, elemental analysis may be carried out on BAL, on biopsy tissue or on autopsy material. Both macroanalytical (or bulk) and microanalytical techniques may be applied. The former are destructive techniques which allow the detection, quantitation and/or characterization of the crystalline structure of inorganic elements. The latter techniques allow *in situ* analysis of individual cells and particles. Several sophisticated microanalytical techniques, coupled to scanning or transmission electron microscopy, exist: energy dispersive X-ray analysis (EDXA), particle-induced X-ray emission (PIXE), electron energy loss spectrometry (EELS) and laser microprobe mass analysis (LAMMA). These techniques have been successfully applied in both BAL fluid and lung biopsies, and databases are being constructed to obtain 'normal' values of inorganic particles (Dumortier *et al.*, 1989; Abraham *et al.*, 1991; Corhay *et al.*, 1992). Of course, the presence of a particular element, even in abnormally high quantities, does not in itself constitute proof that it is causally involved in the disease, but this information may be extremely useful to discover unsuspected past exposures or to document such exposures.

Various immunologic assays may be helpful for evaluating immunologically mediated lung disease, including that caused by metals, although these tests are generally still to be regarded as adjuncts to clinical diagnosis, and not as independent proof of causation or of diagnosis, except perhaps in the case of the beryllium-specific lymphocyte stimulation test. When hypersensitivity is suspected, bronchial challenge testing may be justified, but must obviously be carried out with the greatest care.

Finally in the field of environmental disease, be it caused by metals or other agents, it is essential for the practicing physician to keep abreast of the relevant contemporary literature, as well as to follow as much as possible the advances in technology. This is often done best by listening to workers, engineers and other non-medically qualified people and also by taking the time to visit factories and production sites. However, the recognition of occupational disease may in addition sometimes require an alertness for 'strange' occurrences and some degree of scepticism against 'idiopathic' or 'intrinsic' origins. This attitude must always be backed by elementary scientific principles of toxicology and epidemiology.

REFERENCES

Abraham, J.L. (1987) Lung pathology in 22 cases of giant cell intertsitial pneumonia (GIP) suggests GIP is pathognomonic of cobalt (hard metal) disease. (Abstract). *Chest*, **91**, 312.

Abraham, J.L., Burnett, B.R. and Hunt, A. (1991) Development and use of a pneumoconiosis database of human pulmonary inorganic particulate burden in over 400 lungs. *Scann. Microsc.*, **5**, 95–108.

Abramson, M.J., Wlodarczyk, J.H., Saunders, N.A. and Hensley, M.J. (1989) Does aluminum smelting cause lung disease? *Am. Rev. Respir. Dis.*, **139**, 1042–57.

Akira, M. (1995) Uncommon pneumoconioses: CT and pathologic findings. *Radiology*, **197**, 403–9.

Allen, M.B., Crisp, A., Snook, N. and Page, R.L. (1992) 'Smoke-bomb' pneumonitis. *Respir. Med.*, **86**, 165–6.

Anonymous (1985) Lung disease in dental laboratory technicians. *Lancet*, **i**, 1200–1.

Archer, V.E. (1988) Lung cancer risks of underground miners; cohort and case-control studies. *Yale J. Biol. Med.*, **61**, 183–94.

Armstrong, B.G. and Kazantzis, G. (1985) Prostatic cancer and chronic respiratory and renal disease in British cadmium workers: a case control study. *Br. J. Ind. Med.*, **42**, 540–5.

Aust, S.D., Morehouse, L.A. and Thomas, C.E. (1985) Role of metals in oxygen radical reactions. *J. Free Rad. Biol. Med.*, **1**, 3–25.

Baker, D.B., Gann, P.H., Brooks, S.M., Gallagher, J. and Bernstein, I.L. (1990) Cross-sectional study of platinum salts sensitization among precious metals refinery workers. *Am. J. Ind. Med.*, **18**, 653–64.

Balmes, J.R. (1987) Respiratory effects of hard-metal dust exposure. *Occup. Med.*, **2**, 327–44.

Barnhart, S. and Rosenstock, L. (1984) Cadmium chemical pneumonitis. *Chest*, **86**, 789–91.

Bartter, T., Irwin, R.S., Abraham, J.L. *et al.* (1991) Zirconium compound-induced pulmonary fibrosis. *Arch. Intern. Med.*, **151**, 1197–201.

Beach, J.R., Dennis, J.H., Avery, A.J. *et al.* (1996) An epidemiologic investigation of asthma in welders. *Am. J. Respir. Crit. Care Med.*, **154**, 1394–400.

Becklake, M.R. (1985) Chronic airflow limitation: its relationship to work in dusty occupations. *Chest*, **88**, 608–17.

Bernstein, I.L. and Brooks, S.M. (1993) Metals, in *Asthma in the Workplace* (eds I.L. Bernstein, M. Chan-Yeung, J.L. Malo and D.I. Bernstein), Dekker, New York, pp. 459–79.

Biagini, R.E., Bernstein, I.L., Gallagher, J.S., Moorman, W.J., Brooks, S. and Gann, P.H. (1985) The diversity of reaginic immune responses to platinum and palladium metallic salts. *J. Allergy Clin. Immunol.*, **76**, 794–802.

Billings, C.G. and Howard, P. (1993) Occupational siderosis and welders' lung: a review. *Monaldi Arch. Chest Dis.*, **48**, 304–14.

Bisetti, A.A. (1988) Bernardino Ramazzini and occupational lung medicine. *Ann. N.Y. Acad. Sci.*, **534**, 1029–37.

Blanc, P. and Boushey, H.A. (1993) The lung in metal fume fever. *Semin. Respir. Med.*, **14**, 212–25.

Blanc, P., Wong, H., Bernstein, M.S. and Boushey, H.A. (1991) An experimental human model of metal fume fever. *Ann. Intern. Med.*, **114**, 930–6.

Blanc, P.D., Boushey, H.A., Wong, H., Wintermeyer, S.F. and Bernstein, M.S. (1993) Cytokines in metal fume fever. *Am. Rev. Respir. Dis.*, **147**, 134–8.

Block, G.T. and Yeung, M. (1982) Asthma induced by nickel. *JAMA*, **247**, 1600–2.

Brooks, S.M., Weiss, M.A. and Bernstein, I.L. (1985) Reactive airways dysfunction syndrome (RADS). Persistent asthma syndrome after high level irritant exposures. *Chest*, **88**, 376–84.

Brooks, S.M., Baker, D.B., Gann, P.H. *et al.* (1990) Cold air challenge and platinum skin reactivity in platinum refinery workers. *Chest*, **97**, 1401–7.

Burge, P.S. (1994) Occupation and chronic obstructive pulmonary disease (COPD). *Eur. Respir. J.*, **7**, 1032–4.

Calverley, A.E., Rees, D., Dowdeswell, R.J., Linnett, P.J. and Kielkowski, D. (1995) Platinum salt sensitivity in refinery workers: incidence and effects of smoking and exposure. *Occup. Environ. Med.*, **52**, 661–6.

Cereda, C., Redaelli, M.L., Canesi, M., Carniti, A. and Bianchi, S. (1994) Widia tool grinding: the importance of primary prevention measures in reducing occupational exposure to cobalt. *Sci. Tot. Environ.*, **150**, 249–51.

Chan-Yeung, M., Wong, R., MacLean, L. *et al.* (1983) Epidemiologic health study of workers in an aluminium smelter in British Columbia. *Am. Rev. Respir. Dis.*, **127**, 465–9.

Chen, J.L. and Fayerweather, W.E. (1988) Epidemiologic study of workers exposed to titanium dioxide. *J. Occup. Med.*, **30**, 937–42.

Chinn, D.J., Stevenson, I.C. and Cotes, J.E. (1990) Longitudinal respiratory survey of shipyard workers: effects of trade and atopic status. *Br. J. Ind. Med.*, **47**, 83–90.

Cirla, A.M. (1994) Cobalt-related asthma: clinical and immunological aspects. *Sci. Tot. Environ.*, **150**, 85–94.

Coggon, D., Inskip, H., Winter, P. and Pannett, B. (1994) Lobar pneumonia: an occupational disease in welders. *Lancet*, **344**, 41–3.

Cone, J.E. (1987) Occupational lung cancer. *Occup. Med.*, **2**, 273–95.

Cordasco, E.M. and Stone, F.D. (1973) Pulmonary edema of environmental origin. *Chest*, **64**, 182–5.

Corhay, J.L., Weber, G., Bury, Th., Mariz, S., Roelandts, I. and Radermecker, M.F. (1992) Iron content in human alveolar macrophages. *Eur. Respir. J.*, **5**, 804–9.

Cotes, J.E., Gilson, J.C., McKerrow, C.B. and Oldham, P.D. (1983) A long-term follow-up of workers exposed to beryllium. *Br. J. Ind. Med.*, **40**, 13–21.

Cotes, J.E., Feinmann, E.L., Male, V.J., Rennie, F.S. and Wickham, C.A.C. (1989) Respiratory symptoms and impairment in shipyard welders and caulker/burners. *Br. J. Ind. Med.*, **46**, 292–301.

Covey, L.S. and Wynder, E.L. (1981) Smoking habits and occupational status. *J. Occup. Med.*, **23**, 537–42.

Cugell, D.W., Morgan, W.K.C., Perkins, D.G. and Rubin, A. (1990) The respiratory effects of cobalt. *Arch. Intern. Med.*, **150**, 177–83.

Cullen, M.R., Cherniack, M.G. and Kominsky, J.R. (1987) Chronic beryllium disease in the United States. *Semin. Respir. Med.*, **7**, 203–9.

Daenen, M., Rogiers, P., Van De Walle, C. *et al.* (1997) Occupational asthma caused by exposure to palladium. *Eur. Respir. J.*, **10**, 165s.

Davies, J.E. (1986) Occupational asthma caused by nickel salts. *J. Soc. Occup. Med.*, **36**, 29–31.

Davison, A.G., Haslam, P.L., Corrin, B. *et al.* (1983) Interstitial lung disease and asthma in hard-metal workers: bronchoalveolar lavage, ultrastructural, and analytical findings and results of bronchial provocation tests. *Thorax*, **38**, 119–28.

Davison, A.G., Newman Taylor, A.J., Darbyshire, J. *et al.* (1988) Cadmium fume inhalation and emphysema. *Lancet*, **i**, 663–7.

dell'Omo, M., Muzi, G., Bernard, A., Lauwerys, R.R. and Abbritti, G. (1996) Long-term toxicity of intravenous mercury injection. *Lancet*, **348**, 64.

Demedts, M., Gheysens, B., Nagels, J. *et al.* (1984) Cobalt lung in diamond polishers. *Am. Rev. Respir. Dis.*, **130**, 130–5.

Desjardins, A., Bergeron, J.P., Ghezzo, H., Cartier, A. and Malo, J.L. (1994) Aluminium potroom asthma confirmed by monitoring of forced expiratory volume in one second. *Am. J. Respir. Crit. Care. Med.*, **150**, 1714–17.

De Vuyst, P., Dumortier, P., Rickaert, F., Van de Weyer, R., Lenclud, C. and Yernault, J.C. (1986a) Occupational lung fibrosis in an aluminium polisher. *Eur. J. Respir. Dis.*, **68**, 131–40.

De Vuyst, P., Van de Weyer, R., Decoster, A. *et al.* (1986b) Dental technician's pneumoconiosis. A report of two cases. *Am. Rev. Respir. Dis.*, **133**, 316–20.

De Vuyst, P., Dumortier, P., Schandené, L., Estenne, M., Verhest, A. and Yernault, J.C. (1987) Sarcoidlike lung granulomatosis induced by aluminium dust. *Am. Rev. Respir. Dis.*, **135**, 493–7.

Dinman, B.D. (1987) Aluminium in the lung: the pyropowder conundrum. *J. Occup. Med.*, **29**, 869–76.

Dinman, B.D. (1988) Alumina-related pulmonary disease. *J. Occup. Med.*, **30**, 328–35.

Dinsdale, D., Verbeken, E.K., Demedts, M. and Nemery, B. (1991) Cobalt particles, identified by energy-dispersive X-ray analysis, in diamond polisher's lung. *Arch. Toxicol.*, **Suppl.14**, 92–95.

Doll, R. and Peto, R. (1981) The causes of cancer: quantitative estimates of avoidable risks of cancer in the US today. *J. Natl Cancer Inst.*, **66**, 1191–308.

Dolovich, J., Evans, S.L. and Nieboer, E. (1989) Occupational asthma from nickel sensitivity: I. Human serum albumin in the antigenic determinant. *Chest*, **95**, 29–37.

Dooms-Goossens, A. and Deleu, H. (1991) Airborne contact dermatitis; an update. *Contact Dermatitis*, **25**, 211–17.

Dooms-Goossens, A., Ceuterick, A., Vanmaele, N. and Degreef, H. (1980) Follow-up study of patients with contact dermatitis caused by chromates, nickel and cobalt. *Dermatologica*, **160**, 249–60.

Dooms-Goossens, A., Debusschere, K., Gevers. D.M. *et al.* (1986) Contact dermatitis caused by airborne agents. A review and case reports. *J. Am. Acad. Dermatol.*, **15**, 1–10.

Dumortier, P., De Vuyst, P. and Yernault, J.C. (1989) Non-fibrous inorganic particles in human bronchoalveolar lavage fluids. *Scann. Microsc.*, **3**, 1207–18.

Eisenbud, M. and Lisson, J. (1983) Epidemiologiocal aspects of beryllium-induced non malignant lung disease: a 30 year update. *J. Occup. Med.*, **25**, 196–202.

Farrell, F.J. (1987) Angioedema and urticaria as acute and late phase reactions to zinc fume exposure, with associated metal fume fever-like symptoms. *Am. J. Ind. Med.*, **12**, 331–7.

Ferdenzi, P., Giaroli, C., Mori, P. *et al.* (1994) Cobalt powder sintering industry (stone cutting diamond wheels): a study of environmental–biological monitoring, workplace improvement and health surveillance. *Sci. Tot. Environ.*, **150**, 245–8.

Fernandez, M.A., Sanz, P., Palomar, M., Serra, J. and Gadea, E. (1996) Fatal chemical pneumonitis due to cadmium fumes. *Occup. Med.*, **46**, 372–4.

Figueroa, S., Gerstenhaber, B., Welch, L., Klimstra, D., Smith, G.J. and Beckett, W. (1992) Hard metal interstitial pulmonary disease associated with a form of welding in a metal parts coating plant. *Am. J. Ind. Med.*, **21**, 363–73.

Fisher, A.A. (ed.) (1986) Dermatitis and discoloration from metals, in *Contact Dermatitis*, 3rd edn, Lea & Febiger, Philadelphia, pp. 710–44.

Fletcher, Ch., Peto, R., Tinker, C. and Speizer, F.F. (1976) *The Natural History of Chronic Bronchitis and Emphysema: an eight-year study of early chronic obstructive lung disease in working men in London*, Oxford University Press, Oxford.

Forni, A. (1994) Bronchoalveolar lavage in the diagnosis of hard metal disease. *Sci. Tot. Environ.*, **150**, 69–76.

Friberg, L., Nordberg, G.F. and Vouk, V.B. (1990a) *Handbook on the Toxicology of Metals, Volume I: General Aspects*, Elsevier, Amsterdam.

Friberg, L., Nordberg, G.F. and Vouk, V.B. (1990b) *Handbook on the Toxicology of Metals, Volume II: Specific Metals*, Elsevier, Amsterdam.

Frost, A.E., Keller, C.A., Brown, R.W. *et al.* (1993) Giant cell interstitial pneumonitis. Disease recurrence in the transplanted lung. *Am. Rev. Respir. Dis.*, **148**, 1401–4.

Funahashi, A., Schlueter, D.P., Pintar, K., Siegesmund, K.A., Mandel, G.S. and Mandel, N.S. (1984) Pneumoconiosis in workers exposed to silicon carbide. *Am. Rev. Respir. Dis.*, **129**, 635–40.

Funahashi, A., Schlueter, D.P., Pintar, K., Bemis, E.L. and Siegesmund, K.A. (1988) Welders' pneumoconiosis: tissue elemental microanalysis by energy dispersive X-ray analysis. *Br. J. Ind. Med.*, **45**, 14–18.

Garabrant, D.H., Fine, L.J., Oliver, C., Bernstein, L. and Peters, J.M. (1987) Abnormalities of pulmonary function and pleural disease among titanium metal production workers exposed to beryllium. *Scand. J. Work Environ. Health*, **13**, 47–51.

Gheysens, B., Auwerx, J., Van Den Eeckhout, A. and Demedts, M. (1985) Cobalt-induced bronchial asthma in diamond polishers. *Chest*, **88**, 740–4.

Gilks, B. and Churg, A. (1987) Aluminium-induced pulmonary fibrosis; do fibers play a role? *Am. Rev. Respir. Dis.*, **136**, 176–9.

Gordon, T. and Fine, J.M. (1993) Metal fume fever. *Occup. Med.*, **8**, 505–17.

Hadji Michael, O.C. and Brubaker, R.E. (1986) Evaluation of an occupational respiratory exposure to a zirconium-containing dust. *J. Occup. Med.*, **23**, 543–7.

Haley, P.J. (1991) Pulmonary toxicity of stable and radioactive lanthanides. *Health Physics*, **61**, 809–20.

Hammond, P.B. and Beliles, R.P. (1980) Metals, in *Casarett and Doull's Toxicology. The Basic Science of Poisons*, 2nd edn (eds J. Doull, C.D. Klaassen and M.O. Amdur), MacMillan, New York, pp. 409–67.

Hauser, R., Elreedy, S., Hoppin, J.A. and Christiani, D.C. (1995a) Airway obstruction in boilermakers exposed to fuel oil ash. *Am. J. Respir. Crit. Care Med.*, **152**, 1478–84.

Hauser, R., Elreedy, S., Hoppin, J.A. and Christiani, D.C. (1995b) Upper airway response in workers exposed to fuel oil ash: nasal lavage analysis. *Occup. Environ. Med.*, **52**, 353–8.

Herbert, A., Sterling, G., Abraham, J. and Corrin, B. (1982) Desquamative interstitial pneumonia in an aluminium welder. *Human Pathology*, **13**, 694–9.

Hery, M., Gerber, J.M., Hubert, G. *et al.* (1994) Exposure to metallic catalyst dust: manufacturing and handling of catalysts in the chemical industry. *Ann. Occup. Hyg.*, **38**, 119–35.

Hjortsberg, U., Nise, G., Orbaek, P., Soes-Petersen, U. and Arborelius, M. Jr (1986) Bronchial asthma due to exposure to potassium aluminium tetrafluoride (letter). *Scand J. Work Environ. Health*, **12**, 223.

Hjortso, E., Qvist, J., Bud, M.I. *et al.* (1988) ARDS after accidental inhalation of zinc chloride smoke. *Intensive Care Med.*, **14**, 17–24.

Hubbard, R., Lewis, S., Richards, K., Johnston, I. and Britton, J. (1996) Occupational exposure to metal or wood dust and aetiology of cryptogenic fibrosing alveolitis. *Lancet*, **347**, 284–9.

Hughes, E.G. (1980) Medical surveillance of platinum refinery workers. *J. Soc. Occup. Med.*, **30**, 27–30.

Husain, M.H., Dick, J.A. and Kaplan, Y.S. (1980) Rare-earth pneumoconiosis. *J. Soc. Occup. Med.*, **30**, 15–19.

Imbrogno, P. and Alborghetti, F. (1994) Evaluation and comparison of the levels of occupational exposure to cobalt during dry and/or wet hard metal sharpening. Environmental and biological monitoring. *Sci. Tot. Environ.*, **150**, 259–62.

Jederlinic, P.J., Abraham, J.L., Churg, A., Himmelstein, J.S., Epler, G.R. and Gaensler, E.A. (1990) Pulmonary fibrosis in aluminum oxide workers. Investigation of nine workers, with pathologic examination and microanalysis in three of them. *Am. Rev. Respir. Dis.*, **142**, 1179–84.

Jennette, K.W. (1981) The role of metals in carcinogenesis: biochemistry and metabolism. *Environ. Health Perspect.*, **40**, 233–52.

John, V.B. (1994) *Introduction to Engineering Materials*, The Macmillan Press Ltd, Houndsmills, Hampshire, London.

Johnson, A., Chan-Yeung, M., MacLean, L. *et al.* (1985) Respiratory abnormalities among workers in an iron and steel foundry. *Br. J. Ind. Med.*, **42**, 94–100.

Jones, R.D. (1994) Survey of antimony workers: mortality 1961–1992. *Occup. Environ. Med.*, **51**, 772–6.

Kazantzis, G., Lam, T.H. and Sullivan, K.R. (1988) Mortality of cadmium-exposed workers. *Scand. J. Work Environ, Health*, **14**, 220–3.

Kennedy, S.M., Chan-Yeung, M., Marion, S., Lea, J. and Teschle, K. (1995) Maintenance of stellite and tungsten carbide saw tips: respiratory health and exposure–response relationships. *Occup. Environ. Med.*, **52**, 185–91.

Keskinen, H., Kalliomäki, P.L. and Alanko, K. (1980) Occupational asthma due to stainless steel welding fumes. *Clin. Allergy*, **10**, 151–9.

Kilburn, K.H. and Warshaw, R.H. (1992) Irregular opacities in the lung, occupational asthma and airways dysfunction in aluminum workers. *Am. J. Ind. Med.*, **21**, 845–53.

Knishkowy, B. and Baker, E.L. (1986) Transmission of occupational disease to family contacts. *Am. J. Ind. Med.*, **9**, 543–50.

Koizumi, A., Aoki, T., Tsukada, M., Naruse, M. and Saitoh, N. (1994) Mercury, not sulphur dioxide, poisoning as cause of smelter disease in industrial plants producing sulphuric acid. *Lancet*, **343**, 1411–12.

Kongerud, J., Boe, J., Soyseth, V., Naalsund, A. and Magnus, P. (1994) Aluminium potroom asthma: the Norwegian experience. *Eur. Respir. J.*, **7**, 165–72.

Kotloff, R.M., Richman, P.S., Greenacre, J.K. and Rossman, M.D. (1993) Chronic beryllium disease in a dental laboratory technician. *Am. Rev. Respir. Dis.*, **147**, 205–7.

Kreiss, K., Mroz, M.M., Zhen, B., Martyny, J.W. and Newman, L.S. (1993a) Epidemiology of Beryllium sensitization and disease in nuclear workers. *Am. Rev. Respir. Dis.*, **148**, 985–91.

Kreiss, K., Wasserman, S., Mroz, M.M. and Newman, L.S. (1993b) Beryllium disease screening in the ceramics industry. *J. Occup. Med.*, **35**, 267–74.

Kreiss, K., Mroz, M.M., Newman, L.S., Martyny, J. and Zhen, B. (1996) Machining risk of beryllium disease and sensitization with median exposures below 2 $\mu g/m^3$. *Am. J. Ind. Med.*, **30**, 16–25.

Kriebel, D., Brain, J.D., Sprince, N.L. and Kazemi, H. (1988a) The pulmonary toxicity of beryllium. *Am. Rev. Respir. Dis.*, **137**, 464–73.

Kriebel, D., Sprince, N.L., Eisen, E.A., Greaves, I.A., Feldman, H.A. and Greene, R.E. (1988b) Beryllium exposure and pulmonary function. A cross-sectional study of beryllium workers. *Br. J. Ind. Med.*, **45**, 167–73.

Kusaka, Y., Ichikawa, Y., Shirakawa, T. and Goto, S. (1986a) Effect of hard metal dust on ventilatory function. *Br. J. Ind. Med.*, **43**, 486–9.

Kusaka, Y., Yokoyama, K., Sera, Y. *et al.* (1986b) Respiratory diseases in hard metal workers: an occupational hygiene study in a factory. *Br. J. Ind. Med.*, **43**, 474–85.

Kusaka, Y., Nakano, Y., Shirakawa, T., Fujimura, N., Kato, M. and Heki, S. (1991) Lymphocyte transformation test with nickel in hard metal asthma: another sensitizing component of hard metal. *Ind. Health.*, **29**, 153–60.

Lahaye, D., Demedts, M., Vanden Oever, R. and Roosels, D. (1984) Lung diseases among diamond polishers due to cobalt? *Lancet*, **1**, 156–7.

Lasfargues, G., Phan Van, J., Lavandier, M. *et al.* (1991) Sidérose pulmonaire et risques respiratoires à long terme du soudage à l'arc. *Rev. Mal. Resp.*, **8**, 304–6.

Lauwerys, R.R. (1987) Biological monitoring of metals: an overview, in *Genotoxic and Carcinogenic Metals: Environmental and Occupational Occurrence and Exposure*. Vol.XI, *Advances in Modern Environmental Toxicology*, (eds L. Fishbein, A. Furst and M.A. Mehlman), Princeton Scientific Publ., Princeton, pp. 265–78.

Le Bouffant, L., Giroulle, H., Guerrin, F., Normand, C., Cornu, L. and Tichoux, G. (1987) Pneumoconiose à l'antimoine. Etude anatomo-physique, in *Silicose et Pneumoconiose à Poussières Mixtes* (ed. L. Le Bouffant), INSERM, Paris, pp. 447–56.

Lechleitner, P., Defregger, M., Lhotta, K., Totsch, M. and Fend, F. (1993) Goodpasture's syndrome. Unusual presentation after exposure to hard metal dust. *Chest*, **103**, 956–7.

Leduc, D., de Francquen, P., Jacobovitz, D., Vandeweyer, R., Lauwerys, R. and De Vuyst, P. (1993) Association of cadmium exposure with rapidly progressive emphysema in a smoker. *Thorax*, **48**, 570–1.

Lees, R.E.M. (1980) Changes in lung function after exposure to vanadium compounds in fuel oil ash. *Br. J. Ind. Med.*, **37**, 253–6.

Levin, M., Jacob, J. and Polos, P.G. (1988) Acute mercury poisoning and mercurial pneumonitis from gold ore purification. *Chest*, **94**, 554–6.

Levy, B.S., Hoffman, L. and Gottsegen, S. (1984) Boilermakers' bronchitis. Respiratory tract irritation associated with vanadium pentoxide exposure during oil-to-coal conversion of a power plant. *J. Occup. Med.*, **26**, 567–70.

Liippo, K.K., Anttila, S.L., Taikina-Aho, O., Ruokonen, E.L., Toivonen, S.T. and Tuomi, T. (1993) Hypersensitivity pneumonitis and exposure to zirconium silicate in a young ceramic tile worker. *Am. Rev. Respir. Dis.*, **148**, 1089–92.

Liles, R., Miller, A. and Lerman, Y. (1985) Acute mercury poisoning with severe chronic pulmonary manifestations. *Chest*, **88**, 306–9.

Lin, S.C., Tai, C.C., Chan, C.C. and Wang, J.D. (1994) Nasal septum lesions caused by chromium exposure among chromium electroplating workers. *Am. J. Ind. Med.*, **26**, 221–8.

Lindberg, E. and Hedenstierna, G. (1983) Chrome plating: symptoms, findings in the upper airways, and effects on lung function. *Arch. Environ. Health*,**38**, 367–74.

Lison, D. (1996) Human toxicity of cobalt-containing dust and experimental studies on the mechanism of interstitial lung disease (hard metal disease). *Crit. Rev. Toxicol.*, **26**, 585–616.

Lison, D., Lauwerys, R., Demedts, M. and Nemery, B. (1996) Experimental research into the pathogenesis of cobalt/hard metal lung disease. *Eur. Respir. J.*, **9**, 1024–8.

Lob, M. (1985) Allergie respiratoire au ciment. *Z Unfallchir. Versicherungsmed. Berufskr.*, **78**, 47–50.

Low, I. and Mitchell, C. (1985) Respiratory disease in foundry workers. *Br. J. Ind. Med.*, **42**, 101–5.

Malo, J.L. and Cartier, A. (1987) Occupational asthma due to fumes of galvanized metal. *Chest*, **92**, 375–7.

Malo, J., Cartier, A., Doepner, M., Nieboer, E., Evans, S. and Dolovich, J. (1982) Occupational asthma caused by nickel sulfate. *J. Allergy Clin. Immunol.*, **69**, 55–9.

Malo, J.L., Cartier, A., Gagnon, G., Evans, S. and Dolovich, J. (1985) Isolated late asthmatic reaction due to nickel sulphate without antibodies to nickel. *Clin. Allergy*, **15**, 95–9.

Martell, A.E. (1981) Chemistry and metabolism of metals relevant to their carcinogenicity. *Environ. Health Perspect.*, **40**, 27–34.

Masjedi, M.R., Estineh, N., Bahadori, M., Alavi, M. and Sprince, N.L. (1989) Pulmonary complications in lead miners. *Chest*, **96**, 18–21.

Matarese, S.L. and Matthews, J.I. (1988) Zinc chloride (smoke bomb) inhalational lung injury. *Chest*, **89**, 308–9.

McDonald, J.W., Ghio, A.J., Sheehan, C.E., Bernhardt, P.F. and Roggli, V.L. (1995) Rare earth (cerium oxide) pneumoconiosis: analytical scanning electron microscopy and literature review. *Modern Pathol.*, **8**, 859–65.

Meredith, S. (1993) Reported incidence of occupational asthma in the United Kingdom, 1989–90. *J. Epidemiol. Community Health*, **47**, 459–63.

Meredith, S.K., Taylor, V.M. and McDonald, J.C. (1991) Occupational respiratory disease in the United Kingdom 1989: a report to the British Thoracic Society and the Society of Occupational Medicine by the SWORD project group. *Br. J. Ind. Med.*, **48**, 292–8.

Merget, R., Schultze-Werninghaus, G., Bode, F., Bergmann, E.M., Zachgo, W. and Meier-Sydow, J. (1991) Quantitative skin prick and bronchial provocation tests with platinum salt. *Br. J. Ind. Med.*, **48**, 830–7.

Merget, R., Reineke, M., Reuckmann, A., Bergmann, E.M. and Schultze-Werninghaus, G. (1994) Nonspecific and specific bronchial responsiveness in occupational asthma caused by platinum salts after allergen avoidance. *Am. J. Respir. Crit. Care Med.*, **1504**, 1146–9.

Merget, R., Dierkes, A., Rueckmann, A., Bergmann, E.M. and Schultze-Werninghaus, G. (1996) Absence of relationship between degree of nonspecific and specific bronchial responsiveness in occupational asthma due to platinum salts. *Eur. Respir. J.*, **9**, 211–16.

Meyer, K.C. (1994) Beryllium and lung disease. *Chest*, **106**, 942–6.

Miller, R.R., Churg, A.M., Hutcheon, M. and Lam, S. (1984) Pulmonary alveolar proteinosis and aluminium dust exposure. *Am. Rev. Respir. Dis.*, **130**, 312–14.

Moller, D.R., Brooks, S.M., Bernstein, D.I., Cassedy, K., Enrione, M. and Bernstein, I.L. (1986) Delayed anaphylactoid reaction in a worker exposed to chromium. *J. Allergy Clin. Immunol.*, **77**, 451–6.

Mosconi, G., Bacis, M., Leghissa, P. *et al.* (1994) Occupational exposure to metallic cobalt in the Province of Bergamo. Results of a 1991 survey. *Sci. Tot. Environ.*, **150**, 121–8.

Mroz, M.M., Kreiss, K., Lezotto, D.C., Campbell, P.A. and Newman, L.S. (1991) Reexamination of the blood lymphocyte transformation test in the diagnosis of chronic beryllium disease. *J. Allergy Clin. Immunol.*, **88**, 54–60.

Mur, J.M., Meyer-Bisch, C., Pham, Q.T. *et al.* (1987) Risk of lung cancer among iron ore miners; a porportional mortality study of 1,075 deceased miners in Lorraine, France. *J. Occup. Med.*, **29**, 762–8.

Murdoch, R.D. and Pepys, J. (1987) Platinum group metal sensitivity: reactivity to platinum group metal salts in platinum halide salt-sensitive workers. *Ann. Allergy*, **59**, 464–9.

Musk, A.W. and Tees, J.G. (1982) Asthma caused by occupational exposure to vanadium compounds. *Med. J. Aust.*, **1**, 183–4.

Nemery, B. (1990) Metal toxicity in the respiratory tract. *Eur. Respir. J.*, **3**, 202–19.

Nemery, B. (1996) Late consequences of accidental exposure to inhaled irritants: RADS and the Bhopal disaster. *Eur Respir J.*, **9**, 1973–6.

Nemery, B. and Demedts, M. (1991) Respiratory involvement in metal fume fever. *Eur. Respir. J.*, **4**, 764–5.

Nemery, B., Nagels, J., Verbeken, E., Dinsdale, D. and Demedts, M. (1990) Rapidly fatal progression of cobalt-lung in a diamond polisher. *Am. Rev. Respir. Dis.*, **141**, 1373–8.

Nemery, B., Casier, P., Roosels, D., Lahaye, D. and Demedts, M. (1992) Survey of cobalt exposure and respiratory health in diamond polishers. *Am. Rev. Respir. Dis.*, **145**, 610–16.

Nemery, B., De Raeve, H. and Demedts, M. (1995)

Dermal and respiratory sensitization to chromate in a floorer. *Eur. Respir. J.*, **8**,(suppl. 19), 222S.

Newman, L.S. (1995) Beryllium disease and sarcoidosis: clinical and laboratory links. *Sarcoidosis*, **12**, 7–19.

Newman, L.S. and Kreiss, K. (1992) Nonoccupational beryllium disease masquerading as sarcoidosis: identification by blood lymphocyte proliferative response to beryllium. *Am. Rev. Respir. Dis.*, **145**, 1212–14.

Niewoehner, D.E. and Hoidal, J.R. (1982) Lung fibrosis and emphysema: divergent responses to a common injury. *Science*, **217**, 359–60.

Norseth, T. (1988) Metal carcinogenesis. *Ann. N.Y. Acad. Sci.*, **534**, 377–86.

Novey, H.S., Habib, M. and Well, I.D. (1983) Asthma and IgE antibodies induced by chromium and nickel salts. *J. Allergy Clin. Immunol.*, **72**, 407–12.

Olaguibel, J.M. and Basomba, A. (1989) Occupational asthma induced by chromium salts. *Allergol. Immunopathol.*, **17**, 133–6.

Pairon, J.C., Roos, F., Sébastien, P. *et al.* (1995) Biopersistence of cerium in the human respiratory tract and ultrastructural findings. *Am. J. Ind. Med.*, **27**, 349–58.

Park, H.S., Yu, H.J. and Jung, K.S. (1994) Occupational asthma caused by chromium. *Clin. Exp. Allergy*, **24**, 676–81.

Park, T., Di Benedetto, R., Morgan, K., Colmers, R. and Sherman, E. (1984) Diffuse endobronchial polyposis following a titanium tetrachloride inhalation injury. *Am. Rev. Respir. Dis.*, **130**, 315–17.

Parkes, W.R. (1982) *Occupational Lung Disorders*, 2nd edn, Butterworths, London.

Parkes, W.R. (1994) *Occupational Lung Disorders*, 3rd edn, Butterworth-Heinemann, Oxford.

Peters, J.M., Smith, T.J., Bernstein, L., Wright, W.E. and Hammond, S.K. (1984) Pulmonary effects of exposures in silicon carbide manufacturing. *Br. J. Ind. Med.*, **41**, 109–14.

Peters, J.M., Thomas, D., Falk, H., Oberdörster, G. and Smith, T.J. (1986) Contribution of metals to respiratory cancer. *Environ. Health Perspect.*, **70**, 71–83.

Pillière, F., Garnier, R., Rousselin, X., Dimerman, S., Rosenberg, N. and Efthymiou, M.L. (1990) Asthme aux sels de cobalt. A propos d'un cas dû au résinate de cobalt. *Arch. Mal. Prof.*, **51**, 413–17.

Pistelli, R., Pupp, N., Forastiere, F. *et al.* (1991) Aumento della reattività bronchiale aspecifica dopo esposizione professionale a vanadio. *Med. Lav.*, **82**, 270–5.

Potkonjak, V. and Pavlovich, M. (1983) Antimoniosis: a particular form of pneumoconiosis. I. Etiology, clinical and X-ray findings. *Int. Arch. Allergy Appl. Immunol.*, **51**, 199–207.

Radford, E.P. and Renard, St. C. K.G. (1984) Lung cancer in Swedish iron miners exposed to low doses of radon daughters. *N. Engl. J. Med.*, **310**, 1485–94.

Raffn, E., Mikkelsen, S., Altman, G., Christensen, J.M. and Groth, S. (1988) Health effects due to occupational exposure to cobalt blue dye among plate painters in a porcelain factory in Denmark. *Scand. J. Work Environ. Health*, **14**, 378–84.

Redline, S., Barna, B.P., Tomashefski, J.F., Jr and Abraham, J.L. (1986) Granulomatous disease associated with pulmonary deposition of titanium. *Br. J. Ind. Med.*, **43**, 652–6.

Rendall, R.E.G., Phillips, J.I. and Renton, K.A. (1994) Death following exposure to fine particulate nickel from a metal arc process. *Ann. Occup. Hyg.*, **38**, 921–30.

Richeldi, L., Sorrentino, R. and Saltini, C. (1993) HLA-DPB1 Glutamate 69: a genetic marker of beryllium disease. *Science*, **262**, 242–4.

Rifat, S.L., Eastwood, M.R., Crapper McLachlan, D.R. and Corly, P.N. (1990) Effect of exposure of miners to aluminum powder. *Lancet*, **336**, 1162–5.

Rivolta, G., Nicoli, E., Ferretti, G. and Tomasini, M. (1994) Hard metal lung disorders: analysis of a group of exposed workers. *Sci. Tot. Environ.*, **150**, 161–5.

Roels, H., Lauwerys, R., Genet, P. *et al.* (1988) Epidemiological survey among workers exposed to manganese: effects on lung, central nervous system, and some biological indices. *Am. J. Ind. Med.*, **11**, 307–27.

Rom, W.N. (1992) *Environmental and Occupational Medicine*, Little, Brown and Company, Boston.

Rom, W.N., Lockey, J.E., Jeffrey, L.S. *et al.* (1984) Pneumoconiosis and exposure of dental laboratory technicians. *Am. J. Publ. Health.*, **74**, 1252–7.

Rossman, M.D., Kern, J.A., Elias, J.A. *et al.* (1988) Proliferative response of bronchoalveolar lymphocytes to beryllium. A test for chronic beryllium disease. *Ann. Intern. Med.*, **108**, 687–93.

Roto, P. (1980) Asthma, symptoms of chronic bronchitis and ventilatory capacity among cobalt and zinc production workers. *Scand. J. Work Environ. Health*, **6** (suppl.), 11–36.

Rowens, B., Guerrero-Betancourt, D., Gottlieb, C.A., Boyes, R.J. and Eichenhorn, M.S. (1991) Respiratory failure and death following acute inhalation of mercury vapor. A clinical and histologic perspective. *Chest*, **99**, 185–90.

Sakula, A. (1983) Ramazzini's de Morbis Artificum and occupational lung disease. *Br. J. Dis. Chest*, **77**, 349–61.

Sala, C., Mosconi, G., Bacis, M., Bernabeo, F., Bay, A. and Sala, O. (1994) Cobalt in 'hard metal' and diamonds grinding tools manufacturing and in grinding processes. *Sci. Tot. Environ.*, **150**, 111–16.

Saltini, C., Winestock, K., Kirby, M., Pinkston, P. and Crystal, R.G. (1989) Maintenance of alveolitis in patients with chronic beryllium disease by beryllium-specific helper T cells. *N. Engl. J. Med.*, **320**, 1103–9.

Samet, J.M., Marbury, M.C. and Spengler, J.D. (1988) Health effects and sources of indoor pollution. Part 2. *Am. Rev. Respir. Dis.*, **137**, 221–42.

Samet, J.M., Kutvirt, D.M., Waxweiler, R.J. and Key, C.R. (1984) Uranium mining and lung cancer in Navajo men. *N. Engl. J. Med.*, **310**, 1481–4.

Schneider, J., Freitag, F. and Rödelsperger, K. (1994) Durch Zirkonium-Einwirkung am Arbeitzplatz verusachte exogen-allergische Alveolitis (Nr.4201 BeKV). *Arbeitsmedizin Sozialmed Umweltmedizin*, **29**, 382–5.

Schnorr, T.M., Steenland, K., Thun, M.J. and Rinsky, R.A. (1995) Mortality in a cohort of antimony smelter workers. *Am. J. Ind. Med.*, **27**, 759–70.

Seaton, A., Ruckley, V.A., Addison, J. and Brown, W.R. (1986) Silicosis in barium miners. *Thorax*, **41**, 591–5.

Seghizzi, P., Mosconi, G., Sabbioni, E. and Minoia, C. (1994) Cobalt and hard metal disease (Special issue). *Sci. Tot. Environ.*, **150**, 1–278.

Seidal, K., Jörgensen, N., Elinder, C.G., Sjögren, B. and Vahter, M. (1993) Fatal cadmium-induced pneumonitis. *Scand. J. Work Environ. Health*, **19**, 429–31.

Seldén, A., Sahle, W., Sörensen, S. and Persson, B. (1996) Three cases of dental technicians's pneumoconiosis related to cobalt-chromium molybdenum dust exposure. Diagnosis and follow-up. *Chest*, **109**, 837–42.

Sferlazza, S.J. and Beckett, W.S. (1991) The respiratory health of welders. *Am. Rev. Respir. Dis.*, **143**, 1134–48.

Shaffer, B.A. and Schmidt-Nowara, W.W. (1989) Multiple small opacities of metallic density in the lung. *Chest*, **96**, 1179–81.

Shirakawa, T. and Morimoto, K. (1996) Brief reversible bronchospasm resulting from bichromate exposure. *Arch. Environ. Health*, **51**, 221–6.

Shirakawa, T., Kusaka, Y., Fujimura, N. *et al.* (1989) Occupational asthma from cobalt sensitivity in workers exposed to hard metal dust. *Chest*, **95**, 29–37.

Shirakawa, T., Kusaka, Y., Fujimura, N., Kato, M., Heki, S. and Morimoto, K. (1990) Hard metal asthma: cross immunological and respiratory reactivity between cobalt and nickel? *Thorax*, **45**, 267–71.

Shirakawa, T., Kusaka, Y. and Morimoto, K. (1992) Specific IgE antibodies to nickel in workers with known reactivity to cobalt. *Clin. Exp. Allergy*, **22**, 213–18.

Simonato, L., Vineis, P. and Fletcher, A.C. (1988) Estimates of the proportion of lung cancer attributable to occupational exposure. *Carcinogenesis*, **9**, 1159–65.

Simonsson, B.G., Sjöberg, A., Rolf, C. and Haeger-Aronson, B. (1985) Acute and long-term airway hyperreactivity in aluminum-salt exposed workers with nocturnal asthma. *Eur. J. Respir. Dis.*, **66**, 105–18.

Sinico, M., Le Bouffant, L., Paillas, J., Fabre, M. and Trincard, M.D. (1982) Pneumoconiose due au cérium. Documents anatomopathologiques. *Arch. Mal. Prof*, **43**, 249–52.

Sjögren, I., Hillerdal, G., Andersson, A. and Zetterström, O. (1980) Hard metal lung disease: importance of cobalt in coolants. *Thorax*, **35**, 653–9.

Sluis-Cremer, G.K., Thomas, R.G., Goldstein, B. and Solomon, A. (1989) Stannosis. A report of 2 cases. *S. Afr. Med. J.*, **75**, 124–6.

Snider, G.L., Lucey, E.C., Faris, B., Jung-Legg, Y., Stone, P.J. and Franzblau, C. (1988) Cadmium chloride-induced airspace enlargement with interstitial pulmonary fibrosis is not associated with the destruction of lung elastin. Implications for the pathogenesis of human emphysema. *Am. Rev. Respir. Dis*, **137**, 918–23.

Sorgdrager, B., Pal, T.M., deLooff, A.J., Dubois, A.E. and deMonchy, J.G. (1995) Occupational asthma in aluminium potroom workers related to pre-employment eosinophil counts. *Eur. Respir. J.* **8**, 1520–4.

Soyseth, V., Kongerud, J., Ekstrand, J. and Boe, J. (1994) Relation between exposure to fluoride and bronchial responsiveness in aluminium potroom workers with work-related asthma-like symptoms. *Thorax*, **49**, 984–9.

Sprince, N.L., Chamberlin, R.I., Hales, C.A., Weber, A.L. and Kazemi, H. (1984) Respiratory disease in tungsten carbide production workers. *Chest*, **86**, 549–57.

Sprince, N.L., Oliver, L.C., Eisen, E.A., Greene, R.E. and Chamberlin, R.I. (1988) Cobalt exposure and lung disease in tungsten carbide workers: a cross-sectional study of current workers. *Am. Rev. Respir. Dis.*, **138**, 1220–6.

Sulotto, F., Romano, C., Berra, A. *et al.* (1986) Rare earth pneumoconiosis: a new case. *Am. J. Ind. Med.*, **9**, 567–75.

Swennen, B., Buchet, J.P., Stanescu, D., Lison, D. and Lauwerys, R. (1993) Epidemiological survey of workers exposed to cobalt oxides, cobalt salts, and cobalt metal. *Br. J. Ind. Med.*, **50**, 835–42.

Taylor, A., Jackson, M.A., Burston, J. and Lee, H.A. (1984) Poisoning with cadmium fumes after smelting lead. *Br. Med. J.*, **288**, 1270–1.

Townsend, M.C., Enterline, P.E., Sussman, N.B., Bonney, T.B. and Rippcy, L.L. (1985) Pulmonary function in relation to total dust exposure at a bauxite refinery and aluminium-based chemical products plant. *Am. Rev. Respir. Dis.*, **132**, 1174–80.

Valentin, H. and Smidt, U. (1978) *Research project: chronic bronchitis and occupational dust exposure*, Harald Boldt Verlag KG, Boppard.

Vallyathan, V., Bergeron, W.N., Robichaux, P.A. and Craighead, J.E. (1982) Pulmonary fibrosis in an aluminium arc welder. *Chest*, **81**, 372–4.

Van Cutsem, E.J., Ceuppens, J.L., Lacquet, L.M. and Demedts, M. (1987) Combined asthma and alveolitis induced by cobalt in a diamond polisher. *Eur. J. Respir. Dis.*, **70**, 54–61.

Van Den Eeckhout, A., Verbeken, E. and Demedts, M. (1988) La pathologie pulmonaire due au cobalt et aux métaux durs. *Rev. Mal. Resp.*, **5**, 201–7.

Vanden Oever, R., Roosels, D., Douwen, M., Vanderkeel, J. and Lahaye, D. (1990) Exposure of diamond polishers to cobalt. *Ann. Occup. Hyg.*, **34**, 609–14.

Vandenplas, O., Dargent, F., Auverdin, J.J. *et al.* (1995) Occupational asthma due to gas metal arc welding on mild steel. *Thorax*, **50**, 587–8.

Venables, K.M., Dally, M.B., Burge, P.S., Pickering, C.A.C. and Newman-Taylor, A.J. (1985) Occupational asthma in a steel coating plant. *Br. J. Ind. Med.*, **42**, 517–24.

Venables, K.M., Dally, M.B., Nunn, A.J. *et al.* (1989) Smoking and occupational allergy in workers in a platinum refinery. *Br. Med. J.*, **299**, 939–42.

Vocaturo, G., Colombo, F., Zanoni, M. *et al.* (1983) Human exposure to heavy metals. Rare earth pneumoconiosis in occupational workers. *Chest*, **83**, 780–3.

Vogt, P., Spycher, M.A. and Rüttner, J.R. (1986) Pneumokoniose durch 'seltene' 'Erden' (Cer-Pneumokoniose). *Schweiz. Med. Wochenschr.*, **116**, 1303–8.

Vouk, V. (1986) General chemistry of metals, in *Handbook on the Toxicology of Metals. Vol.I General Aspects* (eds L. Friberg, G.F. Nordberg and V.B. Vouk), Elsevier, North-Holland, pp. 15–35.

Wang, Z.P., Larsson, K., Malmberg, P., Sjögren, B., Hallberg, B.O. and Wrangskog, K. (1944) Asthma lung function and bronchial responsiveness in welders. *Am. J. Ind. Med.*, **26**, 741–54.

Waring, P.M. and Watling, R.J. (1990) Rare earth deposits in a deceased movie projectionist. A new case of rare earth pneumoconiosis? *Med. J. Aust.*, **153**, 726–30.

Weir, D.C., Robertson, A.S., Jones, S. and Burge, P.S. (1989) Occupational asthma due to soft corrosive soldering fluxes contaning zinc chloride and ammonium chloride. *Thorax*, **44**, 220–3.

Yates, D.H. and Goldman, K.P. (1990) Acute cadmium poisoning in a foreman plater welder. *Br. J. Ind. Med.*, **47**, 429–31.

Zammit-Tabona, M., Sherkin, M., Kijck, K., Chan, H. and Chan-Yeung, M. (1983) Asthma caused by diphenylmethane disocyanate in foundry workers. *Am. Rev. Respir. Dis.*, **128**, 226–30.

Zhicheng, S. (1986) Acute nickel carbonyl poisoning: a report of 179 cases. *Br. J. Ind. Med.*, **43**, 422–4.

Part Three Occupational asthma, respiratory illness due to dust exposures, and organic dust diseases

21 Assessment of airway responsiveness relevant to occupational exposures

Susan M. Tarlo,
Neil Alexis and
Frances Silverman

INTRODUCTION

Occupational asthma is now the most common form of occupational lung disease in many industrialized countries such as the United Kingdom (Meredith *et al.*, 1991), France (Gervais and Rosenberg, 1986) and parts of Canada (Malo, 1990; Liss and Elinson, 1994). It is defined as a disease characterized by variable airflow limitation and/or bronchial hyper-responsiveness due to causes and conditions attributable to a particular working environment and not to stimuli encountered outside the workplace (Bernstein *et al.*, 1993). It may be caused by a high-level respiratory irritant exposure without a latent period (i.e. irritant-induced asthma; Tarlo and Broder, 1989) or may be caused by a specific workplace sensitizer, either through an immunologic response or by unknown mechanisms (Bernstein *et al.*, 1993).

A diagnosis of occupational asthma is likely to be associated with significant medicolegal implications or workers' compensation issues for which a high degree of certainty about the diagnosis is desirable. In addition, patients with occupational asthma due to sensitization to a workplace substance will often have to change their job description or their workplace, and may suffer significant socioeconomic difficulties. Therefore it is in the best interests of the patient to have the diagnosis made as definitively as possible, in order to prevent unnecessary work changes and to clarify medicolegal or compensation issues. This chapter will focus on the modalities used to assess airway hyper-responsiveness in patients suspected to have occupational asthma, however many of the approaches are relevant to the study of airway effects of other environmental exposures. Other aspects of the diagnostic process such as details of the exposure history, review of exposure conditions with use of safety data sheets from the workplace and/or worksite visits, assessment of contributing non-occupational asthma triggers, and assessment of other medical contributing factors will not be covered in this chapter, but are extremely important to assess and have been addressed in the recent American College of Chest Physicians consensus statement on assessment of asthma in the workplace (Chan-Yeung *et al.*, 1995). Likewise, assessment of specific immunologic responses to a

Occupational Lung Disease. Edited by Daniel E. Banks and John E. Parker. Published in 1998 by Chapman & Hall. ISBN 0 412 73630 6

Table 21.1 Triggers associated with increased airway reactivity

	Direct non-specific	*Direct specific*
Non-occupational	• Ozone • Viral infection • ? tobacco smoke	• Environmental allergens
Occupational	• Very high level respiratory irritants (spills, accidental high level exposures)	• Workplace allergen • Workplace sensitizer

workplace allergen by *in vivo* or *in vitro* methods is also important in the diagnosis of occupational asthma when feasible, but will not be discussed in detail in this chapter.

One of the hallmarks of asthma, both non-occupational and occupational, is the presence of increased responsiveness of the airways to environmental stimuli that can cause acute airway narrowing. This hyper-responsiveness is reflected by varying symptoms of dry cough, wheeze, shortness of breath and chest tightness in asthmatic patients. Such symptoms occur in response to exercise and to respiratory irritant agents such as dust, tobacco smoke, fumes and sprays to a varying extent in almost all symptomatic asthmatics (Table 21.1). In addition, there may be 'specific' hyper-responsiveness to common environmental allergens or to workplace allergens or low-molecular weight sensitizers which occur in a proportion of asthmatics. These 'specific' stimuli are also capable of increasing or inducing hyper-responsiveness of the airways for days to weeks (Boulet *et al.*, 1983; Mapp *et al.*, 1986), particularly in association with late asthmatic responses. This increase in airway responsiveness is generally also associated with an increase in asthma symptoms. Similarly, an increase in airway responsiveness often occurs with viral respiratory infections and can occur with agents such as ozone, possibly on the basis of associated airway inflammation.

HISTORICAL ASSESSMENT OF AIRWAY RESPONSIVENESS

In assessing a patient with possible occupational asthma, non-occupational factors relevant to airway hyper-responsiveness, such as viral respiratory infections, allergen exposure, beta-blocking medication use and smoking need to be considered in the medical history in addition to workplace factors, as inciting factors or exacerbating factors. The onset of symptoms suggestive of airway hyper-responsiveness within 24 hours of workplace exposure to a high concentration of a respiratory irritant agent, in the absence of previous symptoms of asthma or documented previous respiratory disease is suggestive of irritant-induced asthma. This diagnosis is circumstantial, relying on the history of relevant exposure and the objective documentation of asthma on the basis of pulmonary function tests showing significant airflow limitation responsive to bronchodilator, or increased airway responsiveness to histamine or methacholine inhalation challenge testing. Pre-existing significant respiratory disease should be excluded as far as possible by review of previous medical records.

In patients who have suspected occupational asthma due to a sensitizer and not related to a high-level irritant, the history of symptoms in keeping with airway hyper-responsiveness, beginning after a latent period of exposure at work, ranging from weeks to years is consistent with both occupational and non-occupational asthma. Improvement in symptoms on weekends and holidays away from work supports the likelihood of an improvement in airway responsiveness away from a workplace sensitizer. This information from the history has been shown to be sensitive but of low specificity in the diagnosis of occupational asthma in a study by Malo *et al.* (1991) in Quebec patients. Similarly, in our review of Ontario Workers' Compensation claims (Tarlo *et al.*, 1995) we found that 70% of patients with accepted claims for occupational asthma had described improvement in symptoms of weekends off work and 88% on holidays, statistically higher percentages than those with unrelated asthma. However, such improvement was also described in workers with unrelated asthma. Forty-one per cent improved on weekends off work and 54% improved on holidays.

Such findings emphasize the need for objective assessment of airway responsiveness to complement the history. Such documentation can be achieved in various ways (Table 21.2) which will be discussed in more detail subsequently. Spirometry pre- and post-bronchodilator can provide evidence of asthma and increased responsiveness to a bronchodilator stimulus but is usually unhelpful if baseline spirometry is normal. Serial spirometry performed several times a day using a small, portable spirometer may be used to assess work-related changes in airflow, but equipment is expensive to replace if lost by the patient. Serial peak flow recordings with simple meters are a less expensive alternative. They can provide evidence of

Table 21.2 Airway reactivity assessment techniques

	Serial peak flow (portable meters)	*Serial spirometry (portable and stationary)*	*Methacholine/ histamine challenge*	*Cold air/hyper/hypo saline/exercise challenge*	*Specific occupational challenges*
Advantages	• Inexpensive • High-frequency use • Assesses work-related changes • Portable/easy • Good correlation with other objective tests of occupational asthma	• Can use many times a day • Assesses work-related changes in airflow limitation	• Quantitative assessment of airway responsiveness • Useful even if spirometry is normal • Assesses work-related changes	• Assesses increased airway responsiveness	• Assesses specific trigger • Controlled environment • Patient performance can be assessed
Disadvantages	• Effort-dependent • ? Patient integrity • Frequent use of diary creates a compliance issue • Published data on compliance is poor • False +/– can occur in diagnosing occupational asthma	• Expensive if electronic • ? Patient integrity if basic type spirometer is used • Published data shows little value for single pre/ post-shift monitoring • No published data on electronic-type spirometer	• Requires technician to run • Requires physician to be present	• No quantitative assessment in a serial fashion to assess work-related changes	• Expensive • Time-consuming

the presence or absence of airway hyper-responsiveness and be used to assess workplace influences on this, but such readings have the disadvantage of being effort-dependent. Methacholine or histamine challenges have the advantage of being able to assess airway responsiveness even when baseline spirometry is normal and, if performed during a work week and at the end of a holiday period, will indicate whether there are transient changes in responsiveness related to workplace exposure. Cold air, hyper- or hypotonic saline challenges or exercise challenges can also be used to assess the presence or absence of airway hyper-responsiveness, but unlike histamine or methacholine challenges they do not provide a quantitative assessment which can be assessed on a serial basis during periods at work and off work. Finally, specific occupational challenges can be used to assess directly the airway responsiveness to a suspected workplace sensitizer.

SPIROMETRY IN ASSESSING WORK-RELATED AIRWAY HYPER-RESPONSIVENESS

Pulmonary function tests in a laboratory setting, including spirometry, are useful in determining the presence of airflow limitation, hyperinflation and gas trapping at the time of testing, and responsiveness to an inhaled bronchodilator. These tests have been well standardized; reproducibility of effort and technical performance by the patient can also be readily assessed (European Respiratory Society, 1993; American Thoracic Society, 1991, 1995). The finding of airflow limitation with a significant (at least 12%) improvement in FEV_1 following a bronchodilator confirms a diagnosis of asthma in a patient whose history is consistent, but normal pulmonary function results do not exclude asthma. Even when asthma is confirmed, a work relationship requires further investigation.

Serial spirometry is useful in investigating the variable airflow changes in asthma; in response to a bronchodilator as mentioned; in response to a bronchoconstrictor stimulus to assess airway responsiveness (e.g. in methacholine, histamine or exercise testing); or in response to exposure to a suspected workplace sensitizer, in the workplace or in the laboratory setting in a challenge chamber.

Previously, investigation of occupational asthma using measures of spirometry pre and post shift at work have utilized stationary spirometers in the workplace. There has been little value shown from such use (Burge, 1982) with assessment of pre to post shift changes in FEV_1, or other flow measures. However, in recent years electronic spirometers have become available which are small, portable and require little patient training for self use. These portable spirometers, if correctly used can provide assessment of airflow changes at multiple times during the work shift and after leaving work. Stored data can later be downloaded using a computer interface and also allow the interpreting physician to view the reproducibility and adequacy of patient effort for repeat spirometric maneuvers at each time of performance. To date however, there are few published data comparing this method of assessment for diagnosis of occupational asthma with other investigations. In the experience of the authors, simple peak flow meters which have been lent to patients for monitoring purposes in investigations of occupational asthma have frequently been lost or broken. While this is acceptable for a relatively inexpensive peak flow meter, it does provide a limitation to the usefulness of expensive portable electronic spirometers. In a case where frequent spirometric monitoring was performed by a trained technologist with and without exposure to the suspected work agent in a hospital setting, it has provided useful diagnostic information (Chan-Yeung *et al.*, 1993). However, large studies of the self-use by patients of portable spirometers in an occupational setting have not been reported to date.

PEAK FLOW MONITORING

Peak flow monitoring can provide a useful record of the presence or absence of changes in air flow rates in relationship to the workplace. Peak flow meters are relatively inexpensive, and have the advantage of being small enough to transport readily to and from work. Patients require careful instruction in the correct use of the peak flow meter. Hegewald and colleagues (1995) found that three efforts were adequate for most people to obtain a maximum peak flow reading. Patients can be asked to record the results of maximal peak expiratory efforts performed in triplicate at least four times a day (Malo *et al.*, 1993). A day-shift worker would typically be asked to record this before work in the morning, at the lunch break, immediately after work and at bedtime. Readings should be performed at the same times on days off work and additional recordings made at the time of any respiratory symptoms, preferably before use of a bronchodilator. Medication use and symptoms should also be recorded and notes made of the working hours each day and any unusual environmental exposures. Such records should be maintained for at least 2 weeks at work and 10 days off work unless contraindicated by the severity of asthma (Chan-Yeung *et al.*, 1995). Patients should be asked to write the actual peak flow result on a diary sheet rather than plotting it graphically, since this is considered to minimize any temptation to fill in any appropriate missing values. Nevertheless, such recordings do depend on good peak flow technique by the patient, with maximal inspiratory effort before the maximal expiratory effort. In addition, such recording requires good compliance by the patient in performing such repeat maneuvers, and honesty in the transcription of results.

In regard to feasibility of such tests, we have found that peak flow monitoring was feasible in only 35% of 154 consecutive referrals for assessment of occupational asthma (Tarlo and Broder, 1991). Workplace studies including peak flow monitoring had been considered advisable for diagnosis in 99 of these patients when assessed, but only 54 of these were willing and able to undergo these studies, and only 47 had adequate quantitative peak flow records. Among these, eight graphed recordings showed borderline changes which could not be clearly visually interpreted. Thus, only 39 (25% of the total population and 39% of those in whom serial peak flow monitoring was indicated) actually had useful results. Similarly, Henneberger and colleagues (1991) attempted to obtain serial peak flow monitoring in suspected cases of occupational asthma reported to the SENSOR surveillance program for occupational asthma in New Jersey. Among 70 cases, successful serial peak flow monitoring was feasible in only 20%, while a further 9% had readings which could not be interpreted.

Recent studies have questioned the validity of such self-reported monitoring even when patients maintain records (Malo *et al.*, 1995; Quirce *et al.*, 1995). Malo and co-workers (1995) have observed that when patients were not aware they were using a peak flow device which stored data electronically, the results recorded by the patients on diary sheets differed from electronically stored data on 48% of occasions. However, this study had required more frequent performance of peak flow readings than usually requested of

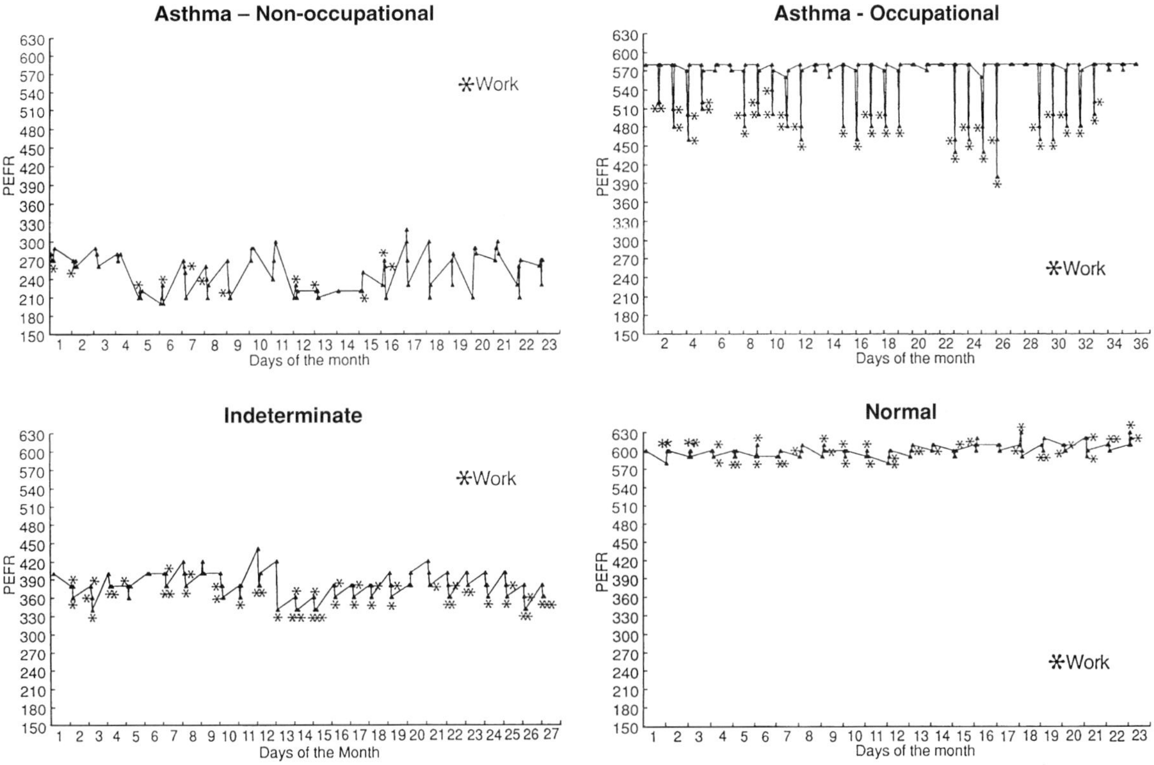

Figure 21.1 Examples of peak flow diagnostic interpretation. (Reprinted with permission from Liss, G. M. and Tarlo, S. M. (1991) Peak expiratory flow rates in possible occupational asthma. *Chest*, **100**, 63–9.)

patients in clinical practice – 2-hourly versus four times a day – and this may have contributed to poor compliance. Similar results were reported by Quirce and colleagues (1995); among peak flow recordings six times a day, 45% of recordings were inaccurate in time or value, or were fabricated. However, in neither study do the authors indicate how often the diagnostic conclusions were changed by the inaccurate data.

Previous studies from the same centers had shown a good correlation between results of non-electronic peak flow monitoring and specific occupational challenges (Côtè *et al.*, 1990a; Perrin *et al.*, 1992). They found the sensitivity of peak flow records to be 86% and 87%, and specificity 89% and 84%. More information is therefore needed to better evaluate concerns about the implications of poor compliance in these tests.

Interpretation of peak flow readings is usually performed by visual inspection of the graphed record. Some studies have advocated the use of the best of three peak flow efforts on each occasion (Sherwood-Burge *et al.*, 1979; Côtè *et al.*, 1990a; Perrin *et al.*, 1992) to be plotted on a single graph, while others have suggested using the daily mean, maximum and minimum as three separate graphs (Burge, 1982). In either event, these are usually plotted with peak flow (l/min) on the ordinate, and days on the abscissa. Indication should be made on the graph as to days at work and days off work (Figure 21.1). Visual interpretation of such graphs by physicians experienced in such interpretation have shown good interobserver correlation. In addition, interpretation in this manner has shown good correlation of peak flow monitoring results with more objective tests of occupational asthma such as specific chamber challenge testing in cedar workers (Côtè *et al.*, 1990).

Nevertheless, attempts have been made to interpret serial peak flow records using objective criteria. Quackenbush *et al.* (1991) have shown that the upper limits of diurnal fall of peak flow readings (maximum – minimum/mean × 100) in non-asthmatic adult subjects is 19% as identified by the ninety-fifth percentile. Perrin *et al.* (1992) found that visual interpretation by experts was more satisfactory than use of arithmetic indices. Conversely, we have found some

Table 21.3 Objective peak flow criteria

Asthma category	*< 20% Diurnal variation*	*>20% Diurnal variation*	*Relative frequency or severity of peak flow worsens more at work than off work*
Non-occupational asthma	No	Yes	No difference
Occupational asthma	No	Yes	Worse
Indeterminate	All days	1 day only	Borderline
No asthma	All days	None	No worse

value to objective indices (Liss and Tarlo, 1991a,b). In our study, 50 subjects who had other objective data to make or exclude a diagnosis of occupational asthma, recorded peak flow recordings four times a day at home and at work for at least 2 weeks and if possible 2 weeks off work. Two physicians who were unaware of the other test results used the following criteria for peak flow interpretation (Table 21.3), using the highest of three peak flow efforts at each recorded time:

1. Non-occupational asthma: ≥20% diurnal variation (defined as maximum – minimum PEFR/maximum PEFR × 100) on more than one day with no difference in the relative frequency or severity of diurnal variation on working days versus days off work, taking into account the proportion of working days to non-working days for which recordings were performed.
2. Occupational asthma: ≥20% diurnal variation in PEFR occurring relatively more frequently or with greater diurnal variation on working versus non-working days, accounting for the proportion of work/non-work days.
3. Indeterminate readings: ≥20% diurnal variation in PEFR on one day only, or <20% longitudinally across days with the lowest recordings on working days.
4. No asthma: none of the above.

Examples of these interpretations are shown in Figure 21.1. Using these criteria, 13 (26%) of recordings were indeterminate, of which six had no other objective evidence of asthma, four had objective evidence of occupational asthma, two non-occupational asthma and one had questionable occupational asthma. In the remaining 74% of recordings which were not indeterminate the sensitivity of the objective peak flow criteria for diagnosing occupational asthma (using the other objective tests as the gold standard) was 93% and sensitivity 90%. These were quite comparable with results obtained by visual inspection of peak flow readings in cedar workers and interpretation by consensus among at least two of three interpreting physicians as reported by Côtè *et al.* (1990b). This suggests that criteria can be developed to assist in diagnosis when experienced interpreters of such recordings are not available, or to provide a more objective interpretation.

Although a good correlation has been found between the result of serial peak flow recordings and other more objective tests for occupational asthma, false positive and false negative results can occur. False positive results may be due to: (i) coincidental worsening of asthma during the work period of monitoring, e.g. due to a concurrent upper respiratory infection or non-occupational allergen or other exposure; (ii) reduced patient effort at the end of a tiring working day; and (iii) failure to record accurately the results of peak flow readings (Malo *et al.*, 1995). Therefore, it is useful to obtain additional support for results of positive peak flow recordings for the diagnosis of occupational asthma. This includes serial methacholine challenges at the end of a working week and after a period off work, specific work allergen skin-prick tests, radioallergosorbent tests for IgE antibodies, or specific chamber challenges with the suspected causative work agent.

False negative peak flow results can similarly occur if: (i) the patient has concerns about loss of work and does not put consistent full effort into the monitoring or does not accurately record results; (ii) airflow limitation occurs mainly at low lung volumes; (iii) coincidental worsening of asthma occurs during the period of recordings off work; (iv) concurrent asthma medications prevent worsening at work; (v) work conditions are not typical during the monitoring time; or (vi) monitoring time off work is not long enough to observe an improvement. Finally, it is possible that after a prolonged period of work with occupational asthma there may no longer be a significant improvement away from exposure to the sensitizer.

If some of those causes are suspected, further investigations may be helpful. If medications may be masking a response by preventing diurnal peak flow changes, it is reasonable to reduce asthma medications carefully, with instructions to the patient as to an action plan using warning levels of peak flows or symptoms. If there is concern about a concurrent unrelated asthma trigger during the time off work, or atypical work exposure during the monitoring time at work, then repeat recordings should be performed. If compliance is suspect, then electronic peak flow devices may be helpful. If the time off work was short, then a longer period of observation off work may be needed.

While, as noted (Liss and Tarlo, 1991a), a significant proportion of peak flow recordings cannot be adequately interpreted and additional tests are needed for diagnosis, peak flow monitoring nevertheless remains a useful modality in the assessment of work-related airway hyper-responsiveness (Chan-Yeung *et al.*, 1995; Moscato *et al.*, 1995).

METHACHOLINE OR HISTAMINE CHALLENGE TESTING

Agents and implications of responses

The airway hyper-responsiveness which is a hallmark feature of asthma can be more directly and objectively assessed by determining airflow limitation in response to a physical stimulus (including hyper- or hypotonic saline, cold air, eucapnic hyperventilation of dry air, and exercise) or to a pharmacologic stimulus (e.g. histamine, methacholine, prostglandin D_2 or F_2-alpha, leukotriene C_4 or D_4). The subject with current asthma is generally hyper-responsive to some extent to all of the above stimuli even when baseline spirometry is normal. Therefore a challenge with one of these agents can be used to make the diagnosis of asthma in a patient with a consistent history with no initial airflow limitation and in whom assessment of bronchodilator responsiveness may not be diagnostic. Histamine and methacholine challenges, the most commonly used pharmacologic stimuli, are more sensitive but less specific in the diagnosis of asthma than physiologic stimuli such as cold air and exercise. However, in general, the level of responsiveness to histamine or methacholine challenge tends to parallel the response to physiologic stimuli. A quantitative dose–response curve can more easily be obtained with the pharmacologic challenges. The degree of reactivity to histamine correlates well with methacholine reactivity. Response to histamine or methacholine has been found in several studies to correlate with the severity of asthma as assessed by symptoms and medication requirements (Townley *et al.*, 1975, 1979; Cockcroft *et al.*, 1977a), diurnal variability of peak flow readings (Ryan *et al.*, 1982), and response to exercise or cold air (Anderton *et al.*, 1979; O'Byrne *et al.*, 1982). The degree of airway responsiveness to methacholine or histamine has also been shown to influence the degree of response to an inhaled allergen in a sensitized individual, as well as the degree of sensitization as demonstrated by specific skin tests responses (Killian *et al.*, 1976; Cockcroft *et al.*, 1987). A predictive formula (Cockcroft *et al.*, 1987) can be used to estimate the degree of airway allergen responsiveness based on the skin-prick test response to that allergen and airway response to methacholine or histamine.

In the diagnosis of suspected occupational asthma, a histamine or methacholine challenge can therefore be useful in four ways:

1. In providing an objective indication of the presence or absence of asthma.
2. If the suspect workplace sensitizer is a high-molecular weight allergen, the methacholine or histamine response can be useful in conjunction with the skin test response to the allergen in predicting the likelihood of airway hyper-responsiveness to the allergen.
3. Methacholine or histamine challenge tests performed in conjunction with serial peak flow monitoring, towards the end of a typical work week and after a period off work can be helpful as a more objective measure of the presence or absence of change in asthma severity in relation to workplace exposure (Fig. 21.2).
4. Methacholine or histamine challenge before and after a specific laboratory challenge with the suspected workplace agent can add to the results of spirometric monitoring by indicating whether the exposure induced an increase in airway responsiveness; this would be expected particularly in conjunction with a late asthmatic response to an occupational respiratory sensitizer.

Methods

The same well-standardized methods for methacholine and histamine challenges have been described in detail for the tidal breathing method (Juniper *et al.*, 1991) and for the dosimeter method (Ryan *et al.*, 1981). The results are expressed as provocation concentration causing a 20% fall in FEV_1 (PC_{20}) and appear to be comparable. The Yan method (Yan *et al.*, 1983) is a more rapid assessment method using a hand-operated nebulizer for delivery of the aerosol during inspiration. Results in this method are expressed as the cumulative dose causing a 20% fall in FEV_1 (PD_{20}). There is good agreement in results between the Yan method and the results of the other methods, taking into account the difference in the units used for reporting (Britton *et al.*, 1986).

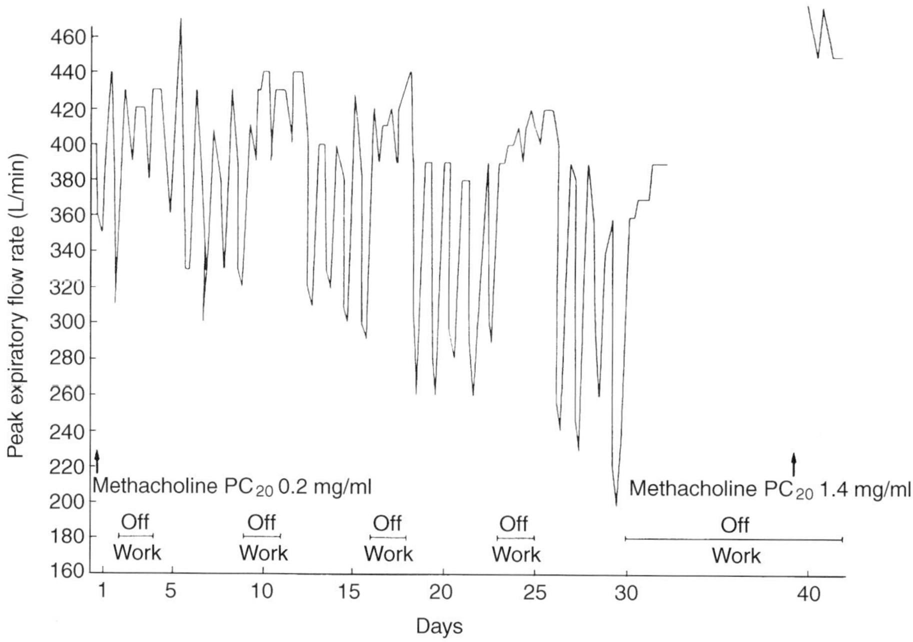

Figure 21.2 Peak flow readings and methacholine responsiveness in a surgical latex glove inspector with occupational asthma. (Reprinted with permission from Tarlo, S. M. *et al.* (1990) Occupational asthma caused by latex in a surgical glove manufacturing plant. *J. Allergy Clin. Immunol.*, **85**, 626–31.)

Careful attention to methodologic detail is essential; results can be affected by changes in storage of the solutions, aerosol generation and delivery (nebulizer output, particle size, modifications to the face mask or mouth piece, connecting tubing or valve box), and solution temperature (Sterk *et al.*, 1993). Therefore if comparison is to be made between repeated tests in the same patient they would best be performed in the same laboratory. It is essential that the patient complies with instructions regarding their medications before the test. Failure to discontinue inhaled or oral bronchodilators (and, for histamine challenges, antihistamines) for the appropriate time before the test will result in a reduced measure of airway responsiveness.

In assessing whether there is an improvement in airway responsiveness away from work, the methacholine or histamine challenge 'at work' should be a test obtained within 24 hours of being at work, towards the end of a typical working week during which asthma symptoms have occurred. Ideally this should be performed after several consecutive work weeks. The comparison test 'off work' should be arranged towards the end of a period of time off work which should be as long as is practicable for the patient to take, at least 10 days. If the patient is initially seen at a time off work, e.g. on short-term disability insurance, then the initial methacholine or histamine challenge may be performed just before a trial of return to work, and the repeat 'at work' challenge performed after a period back at work, after 3–4 weeks or sooner, depending on the severity of asthma.

If the patient is being sent back to work on a trial basis for monitoring of peak flow responses and methacholine or histamine challenges, careful explanation to them is needed so that they will immediately contact the monitoring physician if symptoms, peak flow recordings or medication requirements indicate a significant worsening of asthma, before the need for emergency treatment or the need to stop work. In that event, the patient will need reassessment and earlier scheduling of the histamine or methacholine challenge. Adherence to a rigid time schedule may be dangerous for the patient, or may lead to the finding that they have stopped work for several days or weeks before the 'at work' test, so that the results are uninterpretable. There is a significant risk of asthma worsening during a trial of return to usual work in a patient with suspected occupational asthma who has been off work due to their symptoms; thus, it is advisable to obtain informed consent from the patient. The employer will often require a note from the physician requesting and explaining such a trial.

Factors affecting results and interpretation

Non-occupational factors can affect the degree of airway hyper-responsiveness and may act as confounding factors which must be considered in interpreting the results in a patient with suspected occupational asthma (Table 21.4). Among some of the known effects, viral respiratory infections can increase airway responsiveness, even in non-asthmatic subjects, for up to 8 weeks (Empey *et al.*, 1976). Concurrent non-occupational specific allergen exposure in sensitized asthmatics can also increase responsiveness up to several weeks after exposure (Cartier *et al.*, 1982, 1986), especially in association with late allergic airway responses. Ozone at summer ambient levels with prolonged outdoor exercise can also transiently increase airway responsiveness (Juniper *et al.*, 1978). A careful history taken on the day of testing is therefore necessary to avoid incorrect interpretation of the results.

An additional factor to consider when interpreting changes in methacholine or histamine responsiveness is the comparability of baseline FEV_1. Since the end-point of the challenge is a 20% fall in FEV_1, a smaller absolute change is needed in patients when the baseline FEV_1 is lower. Therefore, to compare PC_{20} in repeat histamine or methacholine challenges in the same patient, it is accepted practice that the baseline FEV_1 measurements should vary by less than 10%. The other factor relevant to assessing changes in methacholine responses within the same patient is the reproducibility of the test. In clinically stable patients undergoing repeat histamine or methacholine challenges within 1 week, the 95% confidence interval of PC_{20} or PD_{20} is less than a two-fold concentration difference (Juniper *et al.*, 1978). However, for longer intervals in a less controlled setting, the 95% confidence interval increases towards a two- to three-fold doubling dose difference (Dehaut *et al.*, 1983).

Table 21.4 Sources of variability in methacholine challenge test results

Methodology-based	*Patient-based*
• Different methods of assessing airway reactivity, i.e. tidal breathing or metered dose delivery	• Prior medication use before test
• Solution storage, i.e. temperature, age of solution	• Concurrent exposures to non-workplace allergens
• Reproducibility of repeat testing – decreases as testing intervals increase	• Exposure to outdoor pollutant episodes, i.e. ozone
• Published data on sensitivity (62%) and specificity (78%) is relatively poor versus serial peak flow monitoring	• Inherent variability of baseline FEV_1

Comparison with other tests for occupational asthma

The sensitivity of work-associated changes in methacholine or histamine responsiveness for the diagnosis of occupational asthma has been reported to be less than that for peak flow monitoring (Côtè *et al.*, 1990; Perrin *et al.*, 1992). In a study of specific challenges, Côtè and colleagues (1990) considered the results of specific challenges with plicatic acid as the 'gold standard' for the diagnosis in 23 patients with suspected occupational asthma due to red cedar. These patients also underwent serial peak flow monitoring six times a day for 3 weeks at work and 2 weeks off work. Methacholine challenges were performed at the end of three working weeks and after 2 weeks off work. A two-fold improvement in PC_{20} was considered significant. The sensitivity and specificity of peak flow monitoring was 86% and 89%, and for repeat methacholine challenges was 62% and 78%, respectively. Combining peak flow results with methacholine responses did not improve diagnostic accuracy in this study. There was one additional patient excluded from analyses due to lack of performance of peak flow monitoring on holiday, but there was no apparent lack of compliance or inconclusive interpretation of peak flow graphs in the 23 other subjects. This suggested that the study population may have been more compliant and produced better records than in other reports. (Henneberger *et al.*, 1991; Liss and Tarlo, 1991a; Tarlo and Broder, 1991).

It is possible that a longer period away from work may be required before conclusions can be drawn from methacholine or histamine responses after work and holiday periods. It has been reported that hyper-responsiveness after exposure to a sensitizer may persist for several weeks (Cartier *et al.*, 1986); removal from work for 10 days to 2 weeks, even soon after the onset of occupational asthma, may not allow a reversal to be demonstrated. Therefore, if borderline improvement is seen during this time, a longer trial period off work with repeat testing may be helpful. Lastly, it must be noted that maximum improvement in occupational asthma after leaving the relevant workplace exposure may take up to 1–2 years (Malo *et al.*, 1988) and improvement may not occur at all (Merget *et al.*, 1994).

Conversely, a normal methacholine or histamine response in a patient who is initially tested after being away from the workplace exposure for over 24 hours does not exclude occupational asthma. In such a patient, a 3–4-week trial of return to work with peak flow monitoring should be arranged before repeating the histamine or methacholine challenge (unless symptoms or peak flow changes are marked earlier); as measurable changes in peak flows and PC_{20} into the asthmatic range may only occur after several days of exposure to the sensitizer (Hargreave *et al.*, 1984). In addition, a few cases of true isocyanate-induced asthma have been reported with normal methacholine responsiveness even after a positive isocyanate chamber challenge exposure (Banks *et al.*, 1986; Mapp *et al.*, 1986). Therefore, use of airway responsiveness as a diagnostic test is best combined with other investigations such as peak flow monitoring, specific skin testing and/or chamber challenges.

In the absence of a confounding factor such as a viral infection or non-occupational allergen exposure, the finding of a significant improvement in methacholine or histamine responsiveness while away from the workplace exposure suggests that the workplace asthmatic response relates to a specific sensitizer, rather than to an airway irritant. Therefore it potentially provides some information as to the mechanism of work-related asthma, thereby supplementing peak flow recordings. These can change similarly in response to a workplace irritant which aggravates underlying airway responsiveness, or in response to a workplace sensitizer which causes asthma. The degree of airway responsiveness to histamine or methacholine in patients with asthma appears to relate to the degree of airway inflammation as reflected by the proportion of eosinophils in sputum (Pin *et al.*, 1992) or bronchoalveolar lavage fluid (Wardlow *et al.*, 1988). It is particularly increased after exposure to sensitizing agents which induce late asthmatic responses (Cockcroft *et al.*, 1977b), with some exceptions (Malo *et al.*, 1989). Such late responses have been similarly associated with increases in airway inflammation (Fabbri *et al.*, 1987; Malo *et al.*, 1989); the increase in histamine or methacholine response has been documented to occur even before measurable changes of airflow limitation during the late asthmatic response to a workplace sensitizer (Durham *et al.*, 1987). Occasionally, an increase in histamine or methacholine response may be the sole manifestation of a positive specific challenge test to a workplace sensitizer (Butcher *et al.*, 1982), without significant immediate or late airflow limitation.

In summary, assessment of methacholine or histamine responsiveness supplements serial peak flow monitoring or serial self-performed spirometry by being a more objective test, and it may help to distinguish between a workplace response to a sensitizer with associated inflammation and an airway response to an irritant in a patient with underlying airway hyper-responsiveness.

FURTHER INVESTIGATIONS

There are other investigations which can serve as adjuncts to serial peak flow and methacholine or histamine challenge tests. The demonstration of a worsening of airway responsiveness during periods of work as compared with periods off work by peak flow monitoring and histamine or methacholine challenge tests can exclude or diagnose occupational asthma or aggravation of underlying airway hyper-responsiveness. However, additional information is needed to identify the responsible workplace agent(s). These may be suspected from the patient's history of work exposure and review of relevant WHMIS sheets (safety data sheets) from the workplace, particularly if there are known respiratory sensitizers or irritant agents.

Further assistance may be obtained from specific skin-prick testing with workplace high-molecular weight allergens and with certain low-molecular weight allergens such as complex platinum salts. *In vitro*, radioallergosorbent tests or enzyme-linked immunosorbent assays can also identify specific serum IgE antibodies to work allergens. Although the identification of specific IgE antibodies to such an allergen is not by itself diagnostic for occupational asthma, in the presence of documented airway hyper-responsiveness, it is highly likely that exposure to that allergen will trigger asthma (Cockcroft *et al.*, 1987) and in a patient with documented occupational asthma, this is likely to be the responsible sensitizer. Unfortunately, however, such tests are not useful for most common low-molecular weight sensitizers such isocyanates or plicatic acid, even when linked to a carrier protein, as specific IgE antibodies have only been demonstrated in a minority of sensitized patients and have poor specificity for occupational asthma.

SPECIFIC OCCUPATIONAL CHALLENGE TESTS

The assessment of serial peak flow readings or serial spirometry and methacholine or histamine challenges at work and off work are often sufficient to diagnose or exclude occupational asthma. However, such 'work-related studies' have some disadvantages as compared with specific chamber challenges:

- the patient may already have left work and may not be willing or able to return to the same work;

Table 21.5 Indications for specific chamber challenge tests

- Patient unable to return to work environment
- Work environment or process has changed since development of symptoms
- Safety concerns are high for reintroducing patient into work environment for study purposes without on-site physician
- Need to identify workplace sensitizer
- More than one sensitizer present in the workplace and specific differentiation is clinically important

Table 21.6 Factors leading to false positive and false negative chamber results

False positive	*False negative*
• Patient has severe airway reactivity and chamber exposure agent is a low-level irritant to patient • Coincidental worsening of pulmonary function on exposure day • Progressive worsening of airflow limitation due to medication withdrawal • Blinding of patient to exposure agent not successful, i.e. suggestive bias	• Incorrect exposure agent is used • Did not achieve identical chemical state of agent as per workplace • Exposure level in chamber is too low • Exposure duration in chamber is too short • Failure to perform methacholine/histamine challenge pre/post exposure • Lack of appropriate follow-up period post exposure – must be at least 6–8 hours for late asthmatic reaction • Follow-up period failed to use spirometry for assessment – peak flow not as sensitive as FEV_1 to monitor late asthmatic reaction

- the work environment or process may have changed since the onset of the patient's symptoms;
- the patient's asthma may be severe, raising concern about return to the environment without direct physician supervision;
- workplace studies may have given inconclusive results;
- workplace studies may have supported occupational asthma but there is no known respiratory sensitizer in the workplace;
- workplace studies may have supported occupational asthma but there is more than one chemical respiratory sensitizer in the workplace environment and it would be important for the patient to identify the relevant sensitizer to assist in future work placement (Table 21.5).

In these situations, a specific laboratory chamber challenge with the suspected work sensitizer(s) under direct physician supervision and informed consent, can clarify the diagnosis.

Nevertheless, since chamber challenge requires specialized knowledge, and equipment to generate and monitor appropriate exposure conditions, such tests are very time-consuming and expensive. They are not indicated in all patients with suspected occupational asthma.

Even in centers with specialized facilities for specific occupational challenges, there are potential false positive and false negative challenge results (Table 21.6). False positive challenges may occur for four reasons:

1. Some patients with severe airway hyper-responsiveness to histamine or methacholine (as assessed before the occupational challenge) may develop airflow limitation with exposure to relatively low levels of respiratory irritant agents, below levels which trigger a response in non-asthmatic individuals, potentially below the industrial TLV for that substance. Thus, exposure to a workplace agent may cause respiratory irritation in that individual.
2. Asthma is characterized by variable airflow limitation; there may be coincidental worsening of pulmonary function on the day of the exposure to the workplace agent, spontaneously or due to a non-occupational exposure, e.g. to a common relevant allergen or a respiratory irritant.
3. If asthma medications have been withdrawn for the challenge to prevent masking of an immediate or late response to the suspected sensitizer, then there may be subsequent worsening of asthma. If the active exposure day follows a control placebo day, or if several days of increasing exposure to the work agent are undertaken, then there may be a progressive worsening of airflow limitation which may be incorrectly attributed to the exposure agent.
4. If the patient is not blinded to the exposure, i.e. there is no placebo exposure or the placebo is not sufficiently similar in appearance and odor to the workplace agent, there may be a false positive response on the active exposure day due to suggestion.

By comparison, false negative chamber challenges may occur due to:

1. Exposure may be made to the wrong work agent.
2. Exposure to the correct agent but in the incorrect chemical state, i.e. there may be chemical interactions at the workplace, or different isomeric forms, or different particle sizes.
3. Levels in the challenge may be too low.
4. The exposure time may be too short. A single exposure day may not be sufficient to induce measurable changes in flow rates or airway responsiveness if the patient has been off work for a period of time before the challenge, and particularly if airway hyper-responsiveness has cleared (Hargreave *et al.*, 1984). In addition, specific reactivity to the work sensitizer may be lost after some years away from work (Butcher *et al.*, 1982); although this has been described only in relation to a single challenge exposure day and not on the basis of a resumption of workplace exposure to the sensitizer.
5. Histamine or methacholine challenges not obtained before and after the occupational challenge may also result in a false negative result, as occasionally a positive response may be manifest only by an increase in airway responsiveness without early or late airflow limitation on the exposure day (Cartier *et al.*, 1986).
6. A lack of an appropriate follow-up period after the occupational challenge may also lead to a false negative result. Isolated late asthmatic responses are common, especially related to low-molecular weight sensitizers such as isocyanates and plicatic acid. These can occur up to 6–8 hours after exposure; therefore, spirometry needs to be monitored for several hours after exposure.
7. Self-monitoring of peak flow by the patient has been used for further evaluation of airflow limitation beyond 6–8 hours post-exposure, but the peak expiratory flow rate has been shown to be less sensitive than FEV_1 in assessment of the late asthmatic response (Berube *et al.*, 1991) and may miss a small, true isolated late response, resulting in a false negative interpretation.

Therefore, rather than being the 'gold standard' for the diagnosis of occupational asthma (Sterk *et al.*, 1993), the occupational chamber challenge may rather be a silver or bronze standard! Unlike spirometry and methacholine or histamine challenges, standardized protocols have not been agreed upon for occupational challenges and attempts to assess sensitivity, specificity and reproducibility have been undertaken for challenges with only a few workplace sensitizers such as isocyanates and colophony in limited numbers of patients (Gerblich *et al.*, 1979; Sherwood Burge *et al.*, 1979; Vandenplas *et al.*, 1992). In the absence of an alternative 'gold standard' it is important to evaluate results with a combination of other tests such as peak flow monitoring, methacholine or histamine measures at and off work, as well as skin test responses when applicable. Another difficulty in trying to develop a standardized protocol for occupational challenges – particularly with low-molecular weight sensitizers – relates to the common occurrence of isolated late asthmatic responses; exposure can only be safely increased at daily intervals and the testing could potentially be very time-consuming. Therefore, the initial exposure level and duration has often been decided for these agents on an individual basis by the responsible physician. Factors taken into account include: the workplace exposure level; duration of exposure before the onset of symptoms; severity of work attributed symptoms and objective evidence of asthma severity from spirometry; and peak flows and/or histamine or methacholine response before chamber challenge. However, there is no systematic quantitative method currently used to predict response levels for low-molecular weight sensitizers; for high-molecular weight soluble allergens the predicted formula derived from knowledge of the specific skin test response and methacholine or histamine airway responsiveness (Cockcroft *et al.*, 1987) can give provide assistance in predicting a safe starting challenge exposure level to nebulized solutions of the allergen.

Methodologic issues

Factors to consider in occupational challenges include: (i) the exposure environment, i.e. chamber, or a partial or fully closed circuit device; (ii) method of generating and monitoring exposure concentrations; (iii) choice of a safe starting concentration, exposure duration, and appropriate incremental increases; (iv) choice of appropriate placebo substance; (v) frequency and duration of pulmonary function monitoring; and (vi) test environment, i.e. in a hospital setting or a separate unit, hospital admission and overnight observation or outpatient performance of tests. These will be considered individually.

(a) Exposure environment

Various exposure environments and methods of exposure generation have been used in different centers (Banks *et al.*, 1996) (Plate 12). For occupational challenges with dusts or vapours, the environment has generally consisted of a chamber large enough for the patient to sit inside comfortably, in order that they may be exposed to the relevant sensitizer without exposing other personnel. Such a chamber may have a static ventilation system turned off during short-term exposures with generation of appropriate levels of the sensitizer, i.e. no fresh air intake or exhaust. More commonly, a dynamic exposure chamber is used, with a relatively constant rate of generation of airborne levels of the sensitizer and a known rate of clean air intake and exhaust. This is more suitable for exposures lasting longer than 1 hour. It more readily allows for the maintenance of a constant concentration of the sensitizer, as well as control of temperature, humidity and carbon dioxide levels. In the chamber, the patient does not wear nose clips or a face mask during the exposure, thus more realistically simulating the workplace. In addition, if space permits there is the potential to reproduce workplace levels of exercise on a treadmill or stationary bicycle during the exposure. An alternative approach developed in recent years and now commercially available has been the use of small, closed-circuit particle and fume generation systems from which the patient breathes by face mask (Vandenplas *et al.*, 1992; Banks *et al.*, 1996). These have the advantage of requiring the generation of smaller amounts of the workplace agent and may provide less risk of accidental exposure to the laboratory personnel. The smaller volume into which the material is generated may also allow more complete mixing and more uniform concentrations than an exposure chamber, although exposure chambers are generally designed for appropriate mixing of the fumes or dusts. Monitoring of levels in the chamber should be performed in the breathing zone of the patient. Some patients may find it difficult to use a face mask for prolonged exposure periods and may find an exposure chamber to be less claustrophobic.

Materials used for the construction of an exposure chamber need to be chosen with care to avoid adsorption of exposure materials and, in the case of fumes, subsequent off-gassing. As an example, appropriate materials to use for reactive chemicals such as isocyanates are stainless steel, teflon and glass for any areas which will come into contact with the isocyanates. For safety reasons, there should be an ante-chamber with an air-lock and separate ventilation and exhaust for the chamber and ante-chamber. The chamber should be at a negative pressure relative to the ante-chamber, and the ante-chamber relative to the external environment to prevent dissemination of the exposure material into the surrounding environment. Appropriate scrubbing of exhausted air is needed also to protect the external environment.

(b) Generation systems

Soluble high-molecular weight allergens often are nebulized through a face mask or nebulizer with one-way inspiratory and expiratory valves. The appropriate filters for the exhaled gas can be incorporated into the outflow tubing. Such exposure is still best performed inside an exposure chamber to reduce staff exposure to the aerosolized agent.

In the case of solutions or emulsions of chemicals for which skin tests cannot be usefully performed, and for which the workplace exposure is in the form of an aerosol, there are often concerns about direct inhalation from a nebulizer through a face mask. A safer starting procedure is to reproduce the type of workplace exposure by nebulization into a closed-circuit apparatus (Vandenplas *et al.*, 1992) or a chamber environment (Banks *et al.*, 1996). Knowledge of workplace exposure levels and mandated short-term exposure limits is useful in determining a safe starting exposure level and exposure time for the individual patient. For other substances, fume generation at room temperature or with heat or by mixing with other work agents may be appropriate, thus reproducing the work procedure where feasible. Industrial hygiene input is essential in developing the appropriate exposure conditions and in monitoring of levels.

Dust generation and appropriate exposure may be achieved by use of a sophisticated commercial system (Cloutier *et al.*, 1989). Simpler but less controlled measures include use of a simple blender, or having the patient tip the relevant powder or dust from one container to another in a chamber. Such simpler methods (Pepys *et al.*, 1972) may be adequate for clinical diagnosis if exposure levels are established on a previous day with monitoring of respiratory dust levels in the area where the patient's breathing zone will be, to develop appropriate dust levels, and if similar monitoring of exposure is performed during the actual challenges.

(c) Exposure conditions and incremental increases

The principle guiding the exposure level is that it should not exceed the level of the substance to which the patient would likely have had exposure in the workplace, and it should not exceed allowable TLVs. Considerably lower starting levels may be required, depending on the history of the patient, the documented severity of asthma, and specific skin tests responses if available. The duration of exposure is similarly guided by the history of timing and severity of asthma at work and severity of asthma immediately before the challenge. Depending on these factors, starting exposures ranging from a few breaths or 1 minute up to 30 minutes may be appropriate, with subsequent increases being dependent on the response throughout the follow-up. Similarly, the time interval between increases in exposures will depend on the patient's history and the likelihood that responses will be immediate, late or dual. As mentioned, although high-molecular

Table 21.7 Assessment techniques for airway reactivity due to occupational asthma

Analysis strategy	*Technique*	*Purpose*
First: presence or absence of asthma	• Peak flow • Spirometry • Methacholine/histamine challenge	• Establish baseline reactivity – rule out other diagnoses
Second: analyze for specific suspect agent	• Serial peak flow • Serial spirometry • Serial methacholine/histamine – perform the above with and without workplace exposure	• Diagnose or exclude occupational asthma
	• Specific chamber challenge	• More direct form of assessing airway reactivity to suspected workplace agent

weight sensitizers are likely to induce immediate or dual responses and exposures may be increased on a single day, low-molecular weight sensitizers commonly induce isolated late responses which preclude more than one exposure per day for these agents.

(d) Choice of appropriate placebo

It goes without saying that in order to provide single blind exposure conditions, an appropriate placebo needs to be chosen for a control exposure day. This should have the same appearance as the exposure agent, but have no respiratory effects. In the case of nebulized solutions, it is relatively easy to select the diluent as a placebo; if there is a difference in color from the active agent, an opaque nebulizer or covered nebulizer can be used. In the case of wood dust, a different wood which is not suspected to cause an effect may be selected for the placebo. If a workplace procedure is reproduced, then the suspect agent may be omitted from a mixture of other workplace substances or replaced with an innocuous agent on the placebo day. In some cases use of nose-clips may be necessary to disguise any differences in odor on placebo and active challenge days.

(e) Frequency and duration of pulmonary function monitoring

It is useful to obtain a methacholine or histamine challenge the day before and day after specific occupational challenge in order to assess changes in airway responsiveness. Spirometry is most commonly used to monitor changes in airflow during the actual occupational challenge days, although airway resistance or conductance may be necessary in patients who cannot perform reproducible spirometry. Baseline, pre-exposure spirometry should be reproducible with FEV_1 variability within 10% on each exposure day. During all challenges patients should be advised to notify the staff of any symptoms and spirometry should be repeated at that time; at intervals (e.g. every 30 minutes) during long exposures and immediately following exposure. Further spirometry should be repeated every 10–15 minutes for 1 hour then hourly for 6–8 hours, or at the time of any symptoms. Exposure should be discontinued if the FEV_1 falls by 15%; if such a fall occurs on follow-up, monitoring is performed more frequently until improvement occurs. After leaving the laboratory, the patient should continue regular spirometry or peak flow recordings while awake, and return the next day. A positive challenge response is a fall in FEV_1 of at least 15% from baseline and at least 15% greater than any fall on the placebo challenge day.

(f) Hospital admission or out-patient testing?

Hospital admission is often necessary for occupational challenges in patients who have severe asthma or have had a history of severe responses at work, particularly late asthmatic responses. All patients may have a late asthmatic response on challenge testing and should have easy, quick access to emergency treatment; if patients are not admitted to hospital for challenge testing they must have clear instructions as to when their symptoms or peak flow/spirometry changes indicate that they should seek such treatment. If exposure facilities are not located within a hospital, there must be resuscitation facilities available nearby, along with ready access to a hospital. Challenges with nebulized high-molecular weight allergens or agents which have been reported to induce anaphylaxis should only be performed in a hospital setting. In some centers, chemical or dust exposures with agents which are less likely to cause severe acute responses are performed outside a hospital main building, but close to emergency facilities.

CONCLUSIONS

In order to assess airway responsiveness relevant to occupational exposures the first step is an assessment of responsiveness which is not specific to the suspected responsible exposure agent, i.e. by peak flow or spirometry monitoring, and/or histamine or methacholine challenges (Table 21.7). Secondly, responsiveness to the suspected occupational/environmental agent can be assessed by serial measures of peak flow/spirometry recordings and methacholine or histamine challenges both in and out of the actual specific environment. Laboratory challenge tests with the suspect agent provide an alternative and supplemental form of investigation for more directly assessing specific airway responsiveness, but such tests are time-consuming and expensive as they generally require specialized facilities and personnel. A diagnosis of occupational asthma can often be made or excluded without occupational chamber challenges if suspected early while the patient is still at work (Tarlo and Broder, 1991).

REFERENCES

American Thoracic Society (1991) Lung function testing: selection of reference values and interpretative strategies. *Am. Rev. Respir. Dis.*, **144**, 1202–18.

American Thoracic Society (1995) Standardization of Spirometry 1994; Update. *Am. J. Respir. Crit. Care Med.*, **152**, 1107–36.

Anderton, R.C., Cuff, M.T., Frith, P.A. *et al.* (1979) Bronchial responsiveness to inhaled histamine and exercise. *J. Allergy Clin. Immunol.*, **63**, 315–20.

Banks, D.E., Barkman, W., Butcher, B.T., Hammad, Y.Y., Rando, R.J., Glindmeyer, H.W., Jones, R.N. and Weill, H. (1986) Absence of hyper-responsiveness to methacholine in a worker with methylene diphenyl diisocyanate (MDI)-induced asthma. *Chest*, **89**, 389–93.

Banks, D.E., Tarlo, S.M., Masri, F., Rando, R.J. and Weissman, D.N. (1996) Bronchoprovocation tests in isocyanate-induced asthma. *Chest*, **109**, 1370–9.

Bernstein, I.L., Chan-Yeung, M., Bernstein, D.I. and Malo, J.-L. (1993) Definition and classification of asthma, in *Asthma in the Workplace*, (eds I.L. Bernstein, M. Chan Yeung, J.-L. Malo and D.I. Bernstein), Marcel Dekker, New York, pp. 1–4.

Berube, D., Cartier, A., L'Archeveque, J., Ghezzo, H. and Malo, J.L. (1991) Comparison of peak expiratory flow rate and FEV_1 in assessing bronchomotor tone after challenges with occupational sensitizers. *Chest*, **99**, 831–6.

Boulet, L.P., Cartier, A., Thomson, N.C. *et al.* (1983) Asthma and increases in non-allergic bronchial responsiveness from seasonal pollen exposure. *J. Allergy Clin. Immunol.*, **71**, 399–406.

Britton, J., Mortagy, A. and Tattersfield, A.E. (1986) Histamine challenge testing: comparison of three methods. *Thorax*, **41**, 128–32.

Burge, P.S. (1982) Single and serial measurements of lung function in the diagnosis of occupational asthma. *Eur. J. Respir Dis.*, **63**(suppl. 123), 47–59.

Butcher, B.T., O'Neil, C.E., Reed, M.A., Salvaggio, J.E. and Weill, H. (1982) Development and loss of toluene diisocyanate reactivity; immunologic, pharmacologic, and provocation challenge studies. *J. Allergy Clin. Immunol.*, **70**, 231–5.

Cartier, A., L'Archeveque, J. and Malo, J.-L. (1986) Exposure to a sensitizing occupational agent can cause a long lasting increase in bronchial responsiveness to histamine in the absence of significant changes in airway caliber. *J. Allergy Clin. Immunol.*, **78**, 1185–9.

Cartier, A., Thomson, N.C. and Frith, P.A. (1982) Allergen-induced increase in bronchial responsiveness to histamine: relationship to the late asthmatic response and change in airway caliber. *J. Allergy Clin. Immunol.*, **70**, 170–7.

Chan-Yeung, M., McMurren, T., Catonio-Begley, F. and Lam, S. (1993) Occupational asthma in a technologist exposed to glutaraldehyde. *J. Allergy Clin. Immunol.*, **91**, 974–8.

Chan-Yeung, M., Brooks, S., Alberts, M. *et al.* (1995) ACCP Consensus Statement. Assessment of asthma in the workplace. *Chest*, **108**, 1084–117.

Cloutier, Y., Lagier, F., Lemieux, R. *et al.* (1989) New methodology for specific inhalation challenges with occupational agents in powder form. *Eur. Respir. J.*, **2**, 769–77.

Cockcroft, D.W., Killian, D.N., Mellon, J.J.A. and Hargreave, F.E. (1977a) Bronchial reactivity to histamine: a method and clinical survey. *Clin. Allergy*, **7**, 235–43.

Cockcroft, D.W., Ruffin, R.E., Dolovich, J. and Hargreave, F.E. (1977b) Allergen induced increase in nonallergic bronchial reactivity. *Clin. Allergy*, **7**, 503–13.

Cockcroft, D.W., Murdoch, K.Y., Kirby, J. and Hargreave, F.E. (1987) Prediction of airway responsiveness to allergen from skin sensitivity to allergen and airway responsiveness to histamine. *Am. Rev. Respir. Dis.*, **135**, 264–7.

Côtè, J., Kennedy, S.M. and Chan-Yeung, M. (1990a) Sensitivity and specificity of PC20 and peak expiratory flow rate in cedar asthma. *J. Allergy Clin. Immunol.*, **85**, 592–8.

Côtè, J., Kennedy, S.M. and Chan-Yeung, M. (1990b) Objective versus qualitative analysis of PEFR in the diagnosis of occupational asthma. *Am. Rev. Resp. Dis.*, **141S**, 82.

Dehaut, P., Rachiele, A., Martin, R.R. and Malo, J.-L. (1983) Histamine dose–response curves in asthma: reproducibility and sensitivity of different indices to assess response. *Thorax*, **38**, 516–22.

Durham, S.R., Graneck, B.J., Hawkins, R. and Newman Taylor, A.J. (1987) The temporal relationship between increases in airway responsiveness to histamine and late asthmatic responses induced by occupational agents. *J. Allergy Clin. Immunol.*, **79**, 398–406.

Empey, D.W., Laitinen, L.A., Jacobs, L. *et al.* (1976) Mechanisms of bronchial hyperreactivity in normal subjects after upper respiratory tract infections. *Am. Rev. Respir. Dis.*, **113**, 131–9.

European Respiratory Society (1993) Official Statement. Standardized lung function testing. *Eur. Resp. J.*, **6**(suppl. 16), 5–40.

Fabbri, L.M., Boschetto, P. Zocca, E. *et al.* (1987) Bronchoalveolar neutrophilia during late asthmatic reactions induced by toluene diisocyanate. *Am. Rev. Respir. Dis.*, **136**, 36–42.

Gerblich, A.A., Horowitz, J., Chester, E.H., Schwartz, H.J. and Fleming, G.M. (1979) A proposed standardized method for bronchoprovocation tests in toluene diisocyanate induced asthma. *J. Allergy Clin. Immunol.*, **64**, 658–61.

Gervais, P. and Rosenberg, N. (1986) Occupational respiratory allergy: epidemiological and medicolegal aspects, in *Proceedings of the 12th International Congress of Allergology and Clinical Immunology* (ed. C.E. Reed), C.V. Mosby Co., St Louis, pp. 480–5.

Hargreave, F.E., Ramsdale, E.H. and Pugsley, S.O. (1984) Occupational asthma without bronchial hyperresponsiveness. *Am. Rev. Respir. Dis.*, **130**, 513–15.

Hegewald, M.J., Crapo, R.O. and Jensen, R.L. (1995) Intraindividual peak flow variability. *Chest*, **107**, 156–61.

Henneberger, P.K., Stanbury, M.J., Trimbath, L.S. and Kipen, H.M. (1991) The use of portable peak flow meters in the surveillance of occupational asthma. *Chest*, **100**, 1515–21.

Juniper, E.F., Frith, P.A., Dunnett, C., Cockcroft, D.W. and Hargreave, F.E. (1978) Reproducibility and comparison of responses to inhaled histamine and methacholine. *Thorax*, **33**, 705–10.

Juniper, E.F., Cockcroft, D.W. and Hargreave, F.E. (1991) *Histamine and methacholine inhalation tests: tidal breathing method. Laboratory procedure and standardization*. A.B. Draco, Lund, Sweden.

Killian, D., Cockcroft, D.W., Hargreave, F.E. and Dolovich, J. (1976) Factors in allergen induced asthma: relevance of the intensity of the airways allergic reaction and nonspecific bronchial reactivity. *Clin. Allergy*, **6**, 219–25.

Liss, G.M. and Elinson, L. (1994) Recognition of occupational asthma. *Practical Allergy Immunol.*, **9**, 252–60.

Liss, G.M. and Tarlo, S.M. (1991a) Peak expiratory flow rates in possible occupational asthma. *Chest*, **100**, 63–9.

Liss, G.M. and Tarlo, S.M. (1991b) Peak expiratory flow rates in possible occupational asthma (letter). *Chest*, **100**, 1480.

Malo, J.-L. (1990) Compensation for occupational asthma in Quebec. *Chest*, **98**, S236–9.

Malo, J.L., Cartier, A., Ghezzo, H., Lafrance, M., McCants, M. and Lehrer, S.B. (1988) Patterns of improvement on spirometry, bronchial hyper-responsiveness and specific IgE antibody levels after cessation of exposure in occupational asthma caused by snow crab processing. *Am. Rev. Respir. Dis.*, **138**, 807–12.

Malo, J.L., L'Archeveque, J. and Cartier, A. (1989) Significant changes in nonspecific bronchial responses after isolated immediate bronchospecific reactions caused by plicatic acid. *J. Allergy Clin. Immunol.*, **83**, 159–65.

Malo, J.L., Ghezzo, H., L'Archeveque, J., Lagier, F., Perrin, B. and Cartier, A. (1991) Is the clinical history a satisfactory means of diagnosing occupational asthma? *Am. Rev. Respir. Dis.*, **143**, 528–32.

Malo, J.L., Côtè, J., Cartier, A., Boulet, L.P., L'Archeveque, J. and Chan-Yeung, M. (1993) How many times per day should peak expiratory flow rates be assessed when investigating occupational asthma? *Thorax*, **48**, 1211–17.

Malo, J.L., Trudeau, C., Ghezzo, H., L'Archeveque, J. and Cartier, A. (1995) Do subjects investigated for occupational asthma through serial peak expiratory flow measurements falsify their results? *J. Allergy Clin. Immunol.*, **96**, 601–7.

Mapp, C.E., Dal Vecchio, L., Boschetto, P., De Marzo, N. and Fabbri, L.M. (1986) Toluene diisocyanate induced asthma without airway hyper-responsiveness. *Eur. J. Respir. Dis.*, **68**, 89–95.

Mapp, C.E., Di Giacomo, G.R., Omino, C. *et al.* (1986) Late but not early asthmatic reactions induced by toluene di-isocyanates are associated with increased airway responsiveness to methacholine. *Eur. J. Respir. Dis.*, **69**, 276–84.

Meredith, S.K., Taylor, V.M. and McDonald, J.C. (1991) Occupational respiratory disease in the United Kingdom, 1989: a report to the British Thoracic Society and the Society of Occupational Medicine by the SWORD Project group. *Br. J. Ind. Med.*, **48**, 292–8.

Merget, R., Reineke, M., Rueckmann, A., Bergmann, E.M. and Schultze-Werninghaus, G. (1994) Nonspecific and specific bronchial responsiveness in occupational asthma caused by platinum salts after allergen avoidance. *Am. J. Respir. Crit. Care Med.*, **150**, 1146–9.

Moscato, G., Godric-Cvar, J., Maestrelli, P. for the subcommittee on occupational allergy of European Academy of Allergy and Clinical Immunology (1995) Statement on self-monitoring of peak expiratory flows in the investigation of occupational asthma. *J. Allergy Clin. Immunol.*, **96**, 295–301.

O'Byrne, P.M., Ryan, G., Morris, M., McCormack, D., Jones, N.L., Morse, J.L.C. and Hargreave, F.E. (1982) Asthma induced by cold air and its relationship to nonspecific bronchial responsiveness to methacholine. *Am. Rev. Respir. Dis.*, **125**, 281–5.

Pepys, J., Pickering, C.A.C. and Loudon, H.W.G. (1972) Asthma due to inhaled chemical agents – piperazine dihydrochloride. *Clin Allergy*, **2**, 189–96.

Perrin, B., Lagier, F., L'Archeveque, J. *et al.* (1992) Occupational asthma: validity of monitoring of peak expiratory flow rates and non-allergic bronchial responsiveness as compared to specific inhalation challenge. *Eur. Resp. J.*, **5**, 40–8.

Pin, I., Freitag, A.P., O'Byrne, P.M., Gergis-Gabardo, A., Watson, R.M., Dolovich, J., Denburg, J.A. and Hargreave, F.E. (1992) Changes in the cellular profile of induced sputum after allergen-induced asthmatic responses. *Am. Rev. Respir. Dis.*, **145**, 1265–9.

Quackenbush, J.J., Lebowicz, M.D. and Krzyzanowski, M. (1991) The normal range of diurnal changes in peak expiratory flow rates. *Am. Rev. Respir. Dis.*, **143**, 323–30.

Quirce, S., Contreras, G., DyBuncio, A. and Chan-Yeung, M. (1995) Is peak expiratory flow monitoring reliable in investigating patients with possible occupational asthma? *Am. J. Respir. Crit. Care Med.*, **151**, A418.

Ryan, G., Dolovich, M.B., Roberts, R.S., Frith, P.A., Juniper, E.F., Hargreave, F.E. and Newhouse, M.T. (1981) Standardization of inhalation provocation tests; two techniques of aerosol generation and inhalation compared. *Am. Rev. Respir. Dis.*, **123**, 195–9.

Ryan, G., Latimer, K.M., Dolovich, J. *et al.* (1982) Bronchial responsiveness to histamine: relationships to diurnal variation of peak flow rate, improvement after bronchodilator and airway calibre. *Thorax*, **37**, 423–9.

Sherwood Burge, P., O'Brien, I.M. and Harries, M.G. (1979) Peak flow rate records in the diagnosis of occupational asthma due to colophony. *Thorax*, **34**, 308–16.

Sterk, P.J., Fabbri, L.M., Quanjer, P.H., Cockcroft, D.W., O'Byrne, P.M., Anderson, S.D., Juniper, E.F. and Malo, J.-L. (1993) Airway responsiveness. Standardized challenge testing with pharmacological, physical and sensitizing stimuli in adults. Official statement of the European Respiratory Society. *Eur. Respir. J.*, **6**(suppl. 16), 53–83.

Tarlo, S.M. and Broder, I. (1989) Irritant induced asthma. *Chest*, **96**, 297–300.

Tarlo, S.M. and Broder, I. (1991) Outcome of assessments for occupational asthma. *Chest*, **100**, 329–35.

Tarlo, S.M., Liss, G., Corey, P. and Broder, I. (1995) Workers' compensation claims for occupational asthma: comparison of subgroups. *Chest*, **107**, 634–41.

Townley, R.G., Ryo, U.Y., Kilotkin, B.M. and Kang, B. (1975) Bronchial reactivity to methacholine in current and former asthmatic and allergic rhinitis patients and control subjects. *J. Allergy Clin. Immunol.*, **56**, 429–42.

Townley, R.G., Bewtra, A.K., Nair, N.M., Brodkey, F.D., Watt, G.D. and Burke, K.M. (1979) Methacholine inhalation challenge studies. *J. Allergy Clin. Immunol.*, **64**, 569–74.

Vandenplas, O., Malo, J.L., Cartier, A., Perreault, G. and Cloutier, Y. (1992) Closed circuit methodology for inhalation challenge tests with isocyanates. *Am. Rev. Respir. Dis.*, **145**, 582–7.

Wardlow, A.J., Dunnette, S., Gleich, G.J., Collins, J.V. and Kay, A.B. (1988) Eosinophils and mast cells in bronchoalveolar lavage in subjects with mild asthma. Relationship to bronchial hyperreactivity. *Am. Rev. Respir. Dis.*, **137**, 62–9.

Yan, K., Salome, C. and Woolcock, A.J. (1983) Rapid method for measurement of bronchial responsiveness. *Thorax*, **38**, 55–61.

22 Mechanisms of occupational asthma

Chris Stenton, Gavin Spickett and David Hendrick

INTRODUCTION

Although occupational asthma is often considered to be a distinct disease entity, the adjective 'occupational' merely recognizes that a specific environmental cause has been identified. The disease is no different in most respects from that experienced by the majority of asthma sufferers. There is no reason to suppose that the underlying mechanisms of occupational asthma differ in any fundamental way from those of asthma arising under other circumstances, and so insights derived from studies of one condition may be applied to the other.

Asthma itself can be defined as a state of enhanced responsiveness of the intrathoracic airways which is manifest clinically by variable degrees of airflow obstruction (American Thoracic Society, 1962). It is associated with a characteristic inflammatory process in the airways which includes lymphocyte and eosinophil infiltration, epithelial disruption, and basement membrane thickening (Barnes *et al.*, 1995). Immunologic investigations have identified a series of complex intercellular interactions, and a vast array of chemical mediators which are involved in maintaining the inflammatory process. However, with the exception of type I hypersensitivity involving the production of specific IgE antibodies, the mechanisms which cause asthma to develop in the first place are not clearly defined. Atopy is the inherited propensity to form IgE antibodies to common aeroallergens and is an important risk factor for asthma, particularly when the disease develops in childhood (Sibbald, 1992). However, not all asthmatics are atopic, and other mechanisms must often be involved in inducing the disease. The increasing incidence of asthma in Westernized societies and the marked geographical variations in its prevalence suggest that environmental factors are important. Occupational asthma provides an almost unique opportunity to study the effects of clearly defined environmental factors and has the potential to provide information of relevance to all forms of asthma.

Occupational asthma refers to the induction of asthma for the first time, or to a substantial increase in its severity, as a consequence of occupational exposures (Bernstein *et al.*, 1993). It is currently recognized to occur under two distinct circumstances. In the most common form of the disease, symptoms start after a latent interval of asymptomatic exposure, usually to a known asthmagenic agent. Specific IgE antibodies can often be identified, and the underlying mechanism is believed to involve immune hypersensitivity. In about 5–10% of cases asthma develops following an acute, accidental, and often massive exposure to a toxic or irritant agent. Hypersensitivity is unlikely to be involved and the disease probably arises as a direct consequence of acute airway inflammation. It is commonly referred to as the reactive airways dysfunction syndrome or RADS (Brooks, 1985), a term which unfortunately does nothing to

Occupational Lung Disease. Edited by Daniel E. Banks and John E. Parker. Published in 1998 by Chapman & Hall. ISBN 0 412 73630 6

emphasize that the disease is asthma. Whether milder forms of RADS occur following lower level exposures to toxic or irritant agents, and whether such exposures might eventually lead to asthma is as yet unknown.

Agents which induce occupational asthma following a latent period of asymptomatic exposure can be categorized as being of high or of low molecular weight. The former include flour, animal proteins and biological enzymes. They are mostly proteinaceous, naturally occurring substances. They often induce specific IgE antibodies, and they thus have many similarities to common environmental aeroallergens derived from mites, pollens, animal proteins, etc., which are associated with asthma in non-occupational settings. Low-molecular weight occupational asthmagens are usually reactive manufactured chemicals such as diisocyanates and acid anhydrides. Some occur naturally, most notably plicatic acid which is found in the sap of western red cedar, but there are few if any recognized examples of asthma caused by low-molecular weight agents outside occupational settings. This is not surprising because although asthma induced by low-molecular weight chemicals displays clinical features suggestive of hypersensitivity, specific IgE antibodies can often not be identified. Their asthmagenic nature is apparent because symptoms are closely related to intermittent occupational exposures. If the relevant exposures occurred in the non-occupational environment and were more continuous, the cause of this form of asthma would be difficult to identify.

Asthma has traditionally been categorized as 'extrinsic', i.e. associated with environmental inducing agents (and specific IgE antibodies), or 'intrinsic' where no environmental inducing factor can be identified. At an early stage in its development, occupational asthma displays the characteristics of 'extrinsic' disease, with symptoms being closely associated with occupational exposures. If exposures cease, the asthma may disappear completely. Conversely, after a variable period of activity, the disease may take on the characteristics of 'intrinsic' asthma, persisting and progressing even in the absence of further exposure to the inducing agent. This suggests that the underlying immunologic processes change from being environment-dependent to being autonomous. Reduction of the relevant environmental exposures before this change takes place can greatly influence the outcome of the disease. A similar window of opportunity might exist with other forms of asthma, allowing the natural history of the disease to be influenced. This is suggested by the observation that early treatment with inhaled corticosteroids has a beneficial effect which is not seen if treatment is delayed (Haahtela *et al.*, 1994).

ASTHMA AND AIRWAY RESPONSIVENESS

The basic physiologic abnormality in asthma, whatever its etiology, is a high level of airway responsiveness. This concept refers to the tendency of the airways to constrict in response to a wide variety of 'non-specific' stimuli such as cold air, smoky atmospheres, or the cooling and drying of the airways which occurs with exercise. In combination with these everyday stimuli, and with changes in endogenous bronchodilator hormone levels, the enhanced responsiveness of the airways gives rise to the variable airflow obstruction by which asthma is usually recognized.

Asthma is commonly thought of as a distinct disease entity which is either present or absent. Airway responsiveness measurements on the other hand are distributed unimodally in the population with no clear distinction between the normal and abnormal ranges (O'Connor *et al.*, 1987). The measurements are similar in this respect to other measures of asthma activity such as the degree of diurnal variation of ventilatory function. Thus physiologically, asthma appears to be quantitatively rather than qualitatively different from normality. The same is true for the relationships between many other disease states and their physiologic counterparts, for example diabetes and glucose tolerance. This suggests that, in epidemiologic terms, it can be misleading to consider simply whether increased levels of airway responsiveness are present or not. It is more useful to consider whether the levels are sufficiently high to make asthma probable, or sufficiently low to make it improbable. With intermediate levels of airway responsiveness, the likelihood of a clinical diagnosis of asthma being made depends on additional factors such as the potency of environmental stimuli, and the ability of the sufferer to recognize and report the subjective sensations associated with bronchoconstriction. Thus, an increase in apparent asthma prevalence in a population could be a consequence of a change in the distribution of airway responsiveness, or of a change in the likelihood of the associated symptoms being recognized and reported.

Airway responsiveness can be measured in the laboratory by quantifying the bronchoconstrictor response to stimuli such as cold air or exercise, but more precise measurements can be obtained using pharmacologic agents such as methacholine or histamine (Beach *et al.*, 1993). The measurements are objective and if properly carried out they are repeatable. In studies of asthmatic populations they correlate well with other measures of asthma activity such as symptom severity and the use of medication. In individuals they increase and decrease as asthma worsens and improves, whether in response to treatment or to environmental changes. Very marked

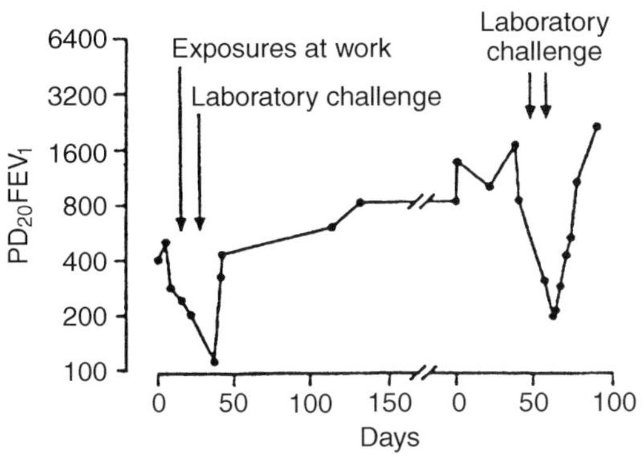

Figure 22.1 Marked changes in airway responsiveness following exposures to the detergent ingredient SINOS at work and in the laboratory during two series of inhalation challenge tests approximately 1 year apart. Note also the gradual decrease in airway responsiveness (increase in $PD_{20}FEV_1$) in the absence of occupational exposures.

changes in airway responsiveness can be observed with changes in occupational exposures in the early stages of occupational asthma (Stenton *et al.*, 1990) (Figure 22.1). It seems unlikely that such marked changes are specific to occupational asthma and similar changes might be observable in other emergent forms of asthma if the appropriate environmental or therapeutic manipulations could be carried out. This greater variability of airway responsiveness might be an important determinant of the greater responsiveness of asthma to inhaled corticosteroids in the early stages of the disease (Haahela *et al.*, 1994).

As asthma develops in an individual, airway responsiveness moves from the normal into the mildly abnormal and then into the clearly abnormal range. This increase in airway responsiveness can be rapid if there are exposures to a potent asthmagenic stimulus, but they can be gradual if the stimulus is not potent, or stepwise if there are exposures to a succession of mildly asthmagenic stimuli. Because the population distribution of airway responsiveness is unimodal with asthmatic subjects being represented at one end of that range, a small increase in airway responsiveness in everyone in an exposed population could shift some workers into the asthmatic range without the cause being readily apparent in any (see Figure 22.5). This phenomenon might help explain recent trends in asthma epidemiology in industrially developed countries.

PATHOLOGIC FEATURES OF ASTHMA

The precise anatomic and pathologic basis for the airway hyper-responsiveness of asthma is not yet established, but it seems clear that it is related to the characteristic inflammatory changes which can be observed in the airways (Barnes *et al.*, 1995). T lymphocytes are present in increased numbers in peripheral blood, bronchoalveolar lavage fluid and bronchial biopsies, and are in an activated state, as demonstrated by the expression of surface markers such as the interleukin-2 receptor (IL2R) and the HLA antigen HLA-DR (Weller *et al.*, 1995). Sputum, peripheral blood and airway eosinophilia are also prominent features (Venge, 1990). The eosinophils are in an activated state and secrete a range of pro-inflammatory mediators which have a variety of effects including causing toxic damage to epithelial cells. Epithelial disruption and thickening of the basement membrane are characteristic features of bronchial biopsies in asthmatics (Barnes *et al.*, 1995). The latter is due to collagen deposition in the subepithelial reticular layer rather than thickening of the true basement membrane (Saetta *et al.*, 1992a). It probably represents a response to fibroblast-stimulating growth factors and cytokines released as part of the inflammatory process. It develops at an early stage in the disease so is not necessarily related to the fixed airflow obstruction which sometimes develops with chronic disease (Saetta *et al.*, 1992a). Mast cell numbers are often increased in asthmatic airways and appear to be in a state of activation, showing features of degranulation (Barnes *et al.*, 1995). Other typical histologic features include, vascular dilatation, edema and smooth muscle and submucosal gland hypertrophy.

The pattern of airway inflammation in asthma appears to differ little between atopics and non-atopics, or between occupational and non-occupational disease (Fabbri *et al.*, 1994). A case has been reported of an isocyanate-sensitive worker who died during an acute asthma attack at work, presumably resulting from his occupational exposures (Fabbri *et al.*, 1988). His lungs showed the typical features of severe asthma with smooth muscle hypertrophy, basement membrane thickening, and an intense predominantly eosinophilic inflammatory infiltrate in the airways. Biopsy studies of workers with stable occupational asthma show similar but milder changes with lymphocyte and eosinophil infiltration and activation, basement membrane thickening, increased numbers of intraepithelial mast cells, and an increase in the intercellular spaces between basal epithelial cells. The latter might be associated with the epithelial shedding commonly seen in asthma (Saetta *et al.*, 1992a; Bentley *et al.*, 1994; Frew *et al.*, 1995). More mast cells have been found in the airways of those who develop occupational asthma to isocyanates shortly after initial exposure than of those who first develop asthma after many years, and it has been suggested that a high number of airway mast cells is a predisposing factor (di Stefano *et al.*, 1993).

Basement membrane thickening is reportedly greater in RADS than in other forms of asthma of similar severity (Gautrin *et al.*, 1994), but both these observations require confirmation. The thickening of the basement membrane is potentially reversible with cessation of exposure and it is likely that the other inflammatory changes are also reversible in those whose symptoms improve, but this is yet to be clearly established (Saetta *et al.*, 1992b).

ASTHMATIC REACTIONS

Immediate asthmatic reactions

Once asthma and airway hyper-responsiveness have developed, acute bronchoconstriction can be provoked by a variety of irritant exposures such as smoky or dusty atmospheres, cold air, or the cooling and drying of the airways which occurs with exercise. These immediate responses develop within minutes, reach their peak within 10–20 minutes, and usually resolve within 1 hour unless the exposure is prolonged. They occur in all asthmatics irrespective of the etiology of their disease, their severity being roughly proportionate to the level of airway responsiveness. The precise mechanism of these reactions is not established but is likely to include mast cell mediator release and sensory nerve stimulation in the airways. Immediate asthmatic reactions due to irritant exposures are not associated with any increase in airway responsiveness, and so this type of exposure will not of itself cause asthma to develop or worsen. The difference between causing acute bronchoconstriction and causing the disease is an important one, and it is useful to distinguish linguistically between 'provoking' the reaction and 'inducing' the disease. Work often involves physical exercise or exposure to adverse environments which can cause acute symptoms in asthmatic subjects, but this is not conventionally considered to be a form of occupational asthma as these exposures do not induce asthma in those not previously affected, and do not worsen its overall severity.

Immediate asthmatic reactions can also develop in hypersensitive subjects exposed to an appropriate allergen. These reactions have a similar pattern to the responses to 'non-specific' stimuli. They are mediated through an interaction between inhaled allergen and specific IgE antibodies on the surface of mast cells and basophils. As with 'non-specific' bronchoconstriction, isolated immediate reactions following exposure to aeroallergens are not associated with any increase in airway responsiveness. The reaction itself does not necessarily reflect any fundamental mechanism in the pathogenesis of asthma, though the evidence it provides of hypersensitivity does point to an agent with the ability to induce asthma. Thus, immediate asthmatic reactions following the inhalation of non-irritant quantities of occupational agents (i.e. those which cause no symptoms in the majority of asthmatic subjects) imply that the agent in question is a probable cause of asthma.

Late asthmatic reactions

Late asthmatic reactions first become evident some hours (1–10) after the relevant provoking exposure begins, and they usually persist for several hours before resolving spontaneously. When provoked by occupational exposures they often cause symptoms in the evenings or at nights after work and can provide important clues to the diagnosis. They do not occur following exposures to irritant stimuli and they thus appear to be consequences of specific hypersensitivity reactions in the airways. Late asthmatic reactions are closely associated with increases in airway responsiveness and so agents which provoke these reactions are also likely to be inducers of asthma. Because of similarities to naturally occurring severe asthma – a prolonged time course, airway inflammation, an association with increases in airway responsiveness, corticosteroid responsiveness, and relative beta-agonist resistance – allergen-provoked late reactions have been widely employed as model for studying the mechanisms of asthma.

Allergen challenge in atopic asthmatics results in a rapid reduction of T helper lymphocyte (CD4+) numbers in peripheral blood and a parallel increase in their numbers in the lung, with the lymphocytes showing features of activation (Weller *et al.*, 1995; Barnes *et al.*, 1995). Eosinophils show a similar pattern with an initial fall followed by a rise in their numbers in peripheral blood. Within the lungs there is evidence of eosinophil and neutrophil recruitment by 4–8 hours after allergen challenge and by 18–24 hours there is a marked eosinophilia which is detectable in bronchoalveolar lavage (BAL) fluid, and which persists up to 48 hours (Rossi *et al.*, 1991; Sedgewick *et al.*, 1992). Increased levels of albumin are also found in BAL fluid suggesting microvascular leakage, but lymphocyte numbers may be decreased suggesting that they are selectively retained within the airway, possibly because of increased expression of adhesion molecules (Gratziou *et al.*, 1996).

The changes are similar following inhalation challenge with occupational agents. Exposure to plicatic acid leads to increased numbers of eosinophils in the airways, increased amounts of albumin in BAL fluid, sloughing of epithelial cells, and mild neutrophilia (Lam *et al.*, 1987). Isocyanate exposure leads to increased numbers of eosinophils in induced sputum samples though curiously, the predominant cell type found in BAL fluid is the neutrophil (Fabbri *et al.*, 1987; Maestrelli *et al.*, 1994a). Changes in lymphocyte populations are observed, though at least in the case of isocyanate asthma, CD8+ cytotoxic rather than CD4+ helper T-cells appear to be activated.

In laboratory studies, late asthmatic reactions follow immediate reactions to common aeroallergens on about 50% of occasions. Their frequency increases with the challenge dose, and they are observed more commonly if the immediate reaction is inhibited pharmacologically, allowing greater amounts of allergen to be administered. Late asthmatic reactions should not be considered as 'all-or-nothing' phenomena. Rather, their magnitude and the extent of the associated inflammatory changes are likely to vary continuously (Durham, 1988). When they are of only modest magnitude they are difficult to detect, but because of their association with changes in airway responsiveness they are likely to be the more important reaction even in situations where an immediate reaction is more obvious clinically.

High-molecular weight occupational asthmagens behave like common aeroallergens in usually provoking isolated immediate or dual reactions. Low-molecular weight asthma inducers behave rather differently as they frequently provoke late asthmatic reactions without any preceding immediate reaction. Under these circumstances the late reaction cannot be a consequence of bronchoconstrictor and pro-inflammatory mediator release associated with an immediate reaction. It has also been demonstrated with occupational agents that isolated late asthmatic reactions can be provoked following challenge doses lower than those which provoke dual reactions. This suggests that even when immediate and late reactions occur following the same challenge, the latter is not necessarily a direct consequence of the former, and might possibly occur through a different pathogenetic mechanism.

The type of response (immediate, late or dual) is not agent-specific and probably depends on several factors including the amount of exposure, the level of 'non-specific' airway responsiveness, the amount of specific IgE antibody in the airways, and any medication used. Figure 22.2 illustrates two inhalation challenge tests with the antibiotic ceftazidime in a mildly symptomatic worker (Stenton *et al.*, 1995). The first challenge resulted in an isolated late reaction but the second challenge 1 week later using an identical dose and identical challenge technique resulted in a clear dual reaction. By then, both the background level of airway responsiveness and the specific immune response had probably been increased as a consequence of the initial challenge.

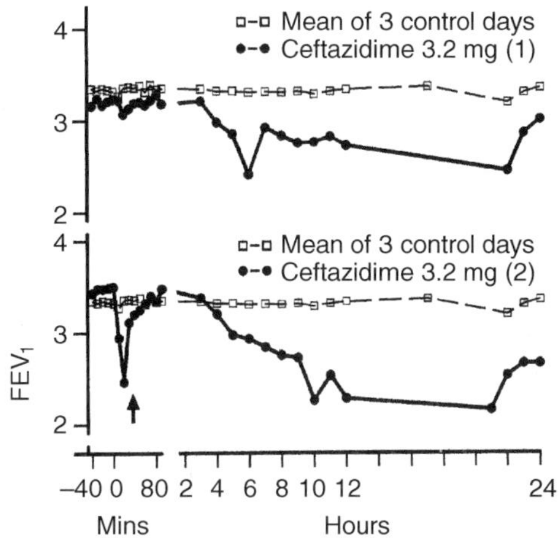

Figure 22.2 Responses to two inhalation challenge tests with 3.2 mg of the antibiotic ceftazidime performed 1 week apart. The first challenge was associated with an isolated late reaction but there was a dual reaction following the second challenge. Inhaled albuterol (↑) was required to relieve the acute bronchoconstriction following the second challenge. (modified with permission, from S. C. Stentan, J. H. Dennis and D. J. Hendrick (1995) Occupational asthma caused by ceftazidime. *Eur. Respir. J.*, **8**, 1421–3).

IMMUNOLOGIC MECHANISMS IN ASTHMA

The development of immune responses

The lungs are one of the major interfaces between humans and their external environment, and elaborate defense systems have evolved to deal with potentially harmful inhaled material (Figure 22.3). The first step in the generation of an immune response is the ingestion of foreign material by phagocytic cells such as macrophages, dendritic and epithelial cells in the airways. These cells are able to process foreign organic material, reduce it to short peptides, and bind it to molecules of the Major Histocompatibility Complex (MHC) which are then expressed on their cell surfaces. The cells travel to the bronchus-associated lymphoid tissue and the regional draining lymph nodes where the surface antigen–MHC complex is presented to T-helper lymphocytes with receptors which are able to recognize the short peptides bound to MHC. At the same time, costimulatory molecules on the T cell and antigen-presenting cell engage to provide a necessary second signal for T-cell responsiveness, and the cytokine interleukin-1 (IL-1) is secreted by the antigen presenting cells stimulating T-lymphocyte activation (Lanzavecchia, 1990; Picker and Butcher, 1992; Linsley and Ledbetter, 1993).

Mucosal immune reactions interact dynamically with the systemic immune system. Mature but naïve lymphocytes recirculate throughout all parts of the body. Once they are activated in a particular organ they undergo a phenotypic change which ensures that they subsequently recirculate to the site of primary activation (Picker and Butcher, 1992). Specific pair of ligands are

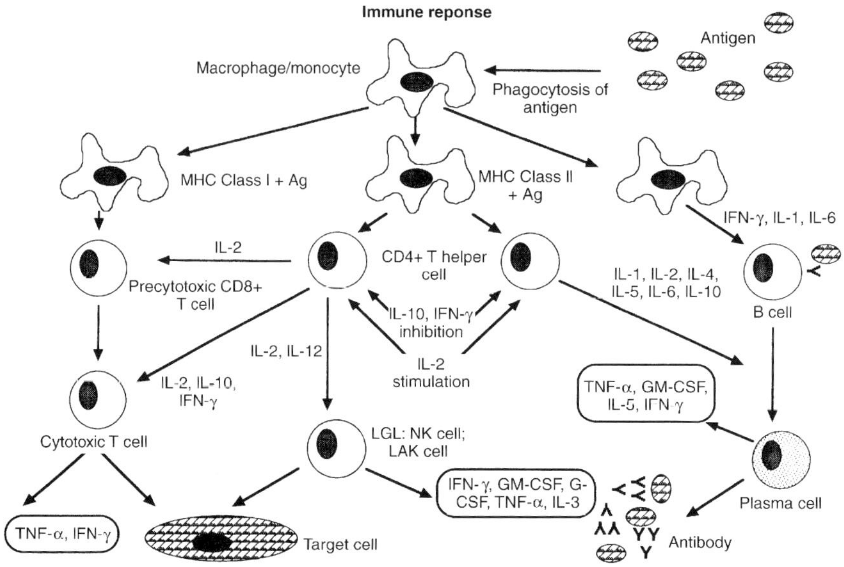

Figure 22.3 Overview of the generation of immune responses. MHC, major histocompatibility complex; IL, interleukin; IFN, interferon; TNF, tumor necrosis factor; GM-CSF, granulocyte-macrophage colony stimulating factor; LGL, large granular lymphocyte; NK, natural killer; LAK, lymphocyte-activated killer.

expressed on the target tissue and the lymphocytes to ensure the specificity of this reaction. Immune reactions thus often remain localized to a particular organ. The process has not been specifically demonstrated for the respiratory tract in the way that it has for the gut and the skin, but similar processes are likely to operate. It is possible, however, that lymphocytes sensitized in one part of the respiratory tract might lead to symptoms elsewhere, for example sensitization in the nose might lead to symptoms in the lower respiratory tract.

Once activated by antigen, T lymphocytes secrete a number of cytokines which attract, activate and promote the growth and differentiation of other leukocytes. Over 50 distinct cytokines have been identified and their relationships with each other, with their target cells, and with the cells from which they originate are complex. It is highly probable that these cytokines regulate the chronic inflammatory changes in asthma. One of their more recently described actions is to increase the expression of adhesion molecules which are important in attracting and maintaining cells such as eosinophils at the site of inflammation (Barnes *et al.*, 1995).

Studies in mice have identified at least two types of activated T helper (CD4+) lymphocytes (Th1 and Th2), and as there is some evidence that these populations also exist in humans they serve as a model for immunologic activation (Romagnani, 1991). Th1 lymphocytes secrete a variety of lymphokines including interleukin-2 (IL-2), tumor necrosis factor (TNFβ) and interferon (IFNγ) which favor the development of the cellular immune responses of delayed (type IV) hypersensitivity. Th2 lymphocytes on the other hand secrete interleukins IL-4, IL-5, IL-6 and IL-13 which favor humoral immune responses. The type of T-helper cell present at a site of immunologic activation is partly determined by hereditary factors, but is also influenced by the cell's cytokine milieu and both of these are important in determining the nature of the response. The cytokines produced by Th1 and Th2 cells are mutually antagonistic, so that Th1 cytokines tend to down-regulate Th2 responses and vice-versa. It is Th2-like cells which are of greatest relevance to asthma because of the range of cytokines which they produce; IL-4 is a potent stimulator of B-cell activity and of IgE production, while IL-5 promotes eosinophil activity.

Type I hypersensitivity

IL-4 secreted by activated lymphocytes and other cells plays a critical role in stimulating B lymphocytes to produce IgE, while other T lymphocyte-derived cytokines such as IL-6, IL-13 and TNFα have synergistic effects. IgE binds to high-affinity surface receptors on mast cells and basophils, and to lower-affinity receptors on platelets, macrophages, lymphocytes and eosinophils. Mast cells which are the most important effector cells of type I hypersensitivity are themselves under the control of T lymphocytes via the secretion of

IL-3 which is an obligatory growth factor in their development. A range of other cytokines such as IFN-γ, TNFα and IL-1 also influence their reactions by priming them and making them more likely to release their contents.

When surface IgE on mast cells is crosslinked by an appropriate allergen in the presence of extracellular calcium, mast cells are stimulated to secrete preformed mediators such as histamine from their granules, and to synthesize other proinflammatory metabolites including prostaglandin D_2 and leukotriene C_4. Together, these cause vasodilatation, mucus hypersecretion and the bronchoconstriction which are characteristic of immediate asthmatic reactions. Non-IgE-mediated release of mast cell products probably also has a role in the 'non-specific' responses to cold air, exercise and hyperventilation. Under these circumstances the stimulus for the mast cells to release their mediators probably involves changes in the osmotic pressure of the fluid lining the airways.

There is some evidence that mast cells and basophils are involved in the immediate asthmatic reactions to low-molecular weight occupational agents which are often noted even in the absence of detectable levels of specific IgE antibodies. Increased levels of mast cell-derived histamine and leukotrienes can be found in BAL fluid following exposure to plicatic acid (Chan-Yeung, 1989), and peripheral blood basophils from symptomatic workers have been shown to release histamine when exposed to plicatic acid *in vitro* (Chan-Yeung *et al.*, 1982). The latter reaction is usually regarded as evidence of antigen-specific IgE. Passive sensitization of basophils from unaffected individuals by incubation with serum from patients has not been possible and, as passive sensitization is possible in grass pollen allergy, this also suggests that the plicatic acid response is not IgE-mediated. The response appears to be specific to workers with occupational asthma, however, and is not due to a direct pharmacologic action as plicatic acid does not have the same effect on cells from control subjects (Frew *et al.*, 1993). Possible explanations include priming of basophils and mast cells *in vivo* by cytokines in those with occupational asthma, or a specific interaction between plicatic acid and lymphocytes or other cells leading to the release of cytokines which then induce histamine release from mast cells.

Mast cells might have a wider role in orchestrating chronic inflammatory changes in the airway. Late reactions to common aeroallergens can be inhibited by leukotriene antagonists, suggesting that these mast cell-derived products play some part. In allergic individuals, mast cell granules contain a range of cytokines more usually associated with T lymphocytes including TNFα, IL-4, IL-5 and IL-6, and these can be released following IgE-dependent activation (Bradding *et al.*, 1994). Mast cells might thus have a role in recruiting eosinophils to the airways (via IL-5), and in augmenting IgE production by B cells (via IL-4, TNFα and other mechanisms) (Gordon *et al.*, 1990). They also release enzymes such as tryptase which causes tissue damage, and together with TNFα and IL-4, up-regulates adhesion molecules.

Although the best characterized pulmonary manifestation of type I hypersensitivity, i.e. acute bronchoconstriction, is not associated with any increase in airway responsiveness and may not be an important mechanism in the induction of asthma, type I hypersensitivity clearly has an important role. A brief exposure to an aeroallergen in a sensitive subject can cause symptoms of worsened asthma which persist for several days. Pollen-sensitive individuals can experience a considerable worsening of their asthma during the appropriate season, or may suffer from asthma only during the pollen season. However, many asthmatic subjects display no features of type I hypersensitivity and it is clear that other mechanisms can be important in maintaining the disease.

Inflammatory changes in asthma

The increased numbers of activated lymphocytes seen in the airways in asthma suggests that immunologic mechanisms are important in maintaining the inflammatory changes which underlie the increased levels of airway responsiveness. The numbers of lymphocytes in the airway and their state of activation correlate with the degree of airway responsiveness (Walker *et al.*, 1992), and their numbers in peripheral blood are increased further during acute severe exacerbations of asthma (Kita *et al.*, 1993; Gratziou *et al.*, 1996). The predominant lymphocytes are CD4+ T-helper cells which from their patterns of intracellular cytokine messenger RNA expression correspond most closely with Th2 cells.

IL-5 is likely to be an important cytokine in asthma. It is specific in promoting the development, and activation of eosinophils and in increasing their adhesion to endothelial cells (Siracusa *et al.*, 1992; Barnes *et al.*, 1995). Messenger RNA coding for IL-5 can be found in bronchial biopsies from asthmatics (Hamid *et al.*, 1991), and IL-5 is the predominant eosinophil activating cytokine found in BAL during allergen-induced late asthmatic reactions (Kita *et al.*, 1993). There is a correlation between lymphocyte and eosinophil activation in both allergic and non-allergic asthma, further suggesting that eosinophil activation is under lymphocytic control (Walker *et al.*, 1991, 1992). The increased numbers of activated eosinophils in the airways parallels increases in the airway responsiveness both during natural exacerbations of asthma and following allergen challenge, suggesting that eosinophil-derived products are important regulators of airway physiology. Activated eosinophils secrete proinflammatory mediators such as LTC4, platelet activating factor (PAF), 15-HETE and granule proteins such as major basic protein, eosinophil cationic protein and eosinophil peroxidase, all of which are toxic to airway epithelium. Although PAF is a potent chemotactic agent for eosinophils and neutrophils, the use of specific inhibitors has shown that it plays little role in the asthmatic response.

Epithelial disruption is commonly reported in biopsies of asthmatic airways. The epithelium appears to have a protective or modulatory role in asthma, and its loss due to the effects of inflammatory mediators probably increases airway responsiveness. This might explain why agents such as ozone and other environmental pollutants can increase airway responsiveness. In addition to the loss of a physical barrier and the exposure of sensory nerve endings, disruption of the epithelium might lead to the loss of epithelium-derived relaxing factor (EDRP) which can protect against bronchoconstriction, and the release of proinflammatory mediators and including cytokines such as GM-CSF, IL-8 and RANTES which can assist the recruitment of inflammatory cells

T cell-derived cytokines also have an important regulatory role in the expression of adhesion molecules both on endothelial and epithelial cells. Increased expression of the adhesion molecules P-selectin, E-selectin and VCAM-1 on endothelial cells stimulated by IL-4 helps trap passing eosinophils which express the appropriate counter-receptors (Bochner *et al.*, 1994). IL-1 and TNFα increase the expression of ICAM-1 on airway epithelial cells and this might contribute to the adherence of leukocytes at the airway surface. It is likely that there are a wide variety of other direct and indirect cytokine actions in asthma, the significance of which remain to be clarified.

Allergic and non-allergic asthma

Some differences in have been observed in lymphocyte subpopulations between atopic and non-atopic asthma. Non-atopic asthmatics have higher numbers of cytotoxic CD8+ T lymphocytes in their BAL fluid and lower numbers in peripheral blood than atopic asthmatics, but they have higher numbers of CD4+ lymphocytes in bronchial biopsies than either atopic asthmatics or normal control subjects (Walker *et al.*, 1992; Bentley *et al.*, 1994). Little is known as yet about the role of cytotoxic T lymphocytes in asthma and so the significance of these observations is not clear.

The patterns of cytokine release from activated T lymphocytes are likely to be pivotal in influencing the expression of type I hypersensitivity and asthma and, as with lymphocyte populations, some differences have been observed between those with atopic and those with non-atopic asthma. IL-5 levels are increased in BAL fluid in both forms of asthma and correlate with the degree of eosinophilia (Walker *et al.*, 1992). IL-4, on the other hand, has been detected only in atopic asthmatics (Walker *et al.*, 1992). It could therefore be postulated that a T lymphocyte subpopulation might exist which secretes eosinophil stimulating interleukins such as IL-5, but not B cell-stimulating interleukins such as IL-4. Such cells could stimulate the airway eosinophilic inflammation which is characteristic of asthma without stimulating IgE production and features of type I hypersensitivity. They could be an important determinant of the differences between atopic and non-atopic asthma.

Occupational asthma

The differences which have been observed in the immunology of atopic and non-atopic asthma also might be of relevance to occupational asthma. CD8+ suppressor T cells appear to be increased in BAL fluid in diisocyanate asthma, a finding similar to that seen in non-atopic asthma (Walker *et al.*, 1992; Bentley *et al.*, 1994) but in contrast to most studies of non-occupational asthma which suggest that CD4+ T-helper cells play the major role (Hol *et al.*, 1992). This suggests that there might be similarities at the level of lymphocytic activation between low-molecular weight occupational asthma and non-atopic asthma. Cytokine patterns also show similar patterns. As with non-atopic asthmatics whose BAL fluid contains IL-5 but not IL-4, T cells cloned from subjects with diisocyanate asthma have been shown to secrete IFNγ and IL-5 with very little IL-4 (Maestrelli *et al.*, 1994b). Thus, just as non-occupational asthma can be considered to be either allergic ('extrinsic') or non-allergic ('intrinsic'), so occupational asthma can be considered to be IgE-associated or non-IgE associated. The occupational analogy does however suggest that the non-allergic asthma is just as 'extrinsic' as the allergic asthma and is merely caused by a different range of environmental agents.

THE DEVELOPMENT OF ASTHMA IN OCCUPATIONAL SETTINGS

Asthma associated with immune hypersensitivity

Occupational asthma usually displays features which are characteristic of immune hypersensitivity.

Symptoms develop after a latent interval of asymptomatic exposure lasting for weeks or months, and usually only a minority of exposed workers is affected. Those who are affected experience asthmatic symptoms following exposures which they previously tolerated, while others continue to be able to tolerate the same working environment without difficulty. Asthmatic subjects with similar levels of airway responsiveness to the affected workers show no adverse reactions when exposed to the causative agent or environment.

With most high-molecular weight asthmagens there is also direct evidence of hypersensitivity in the form of specific IgE antibodies. In some settings, for example among laboratory animal handlers, IgE antibodies can be found in all or almost all workers with occupational asthma. The development of type I hypersensitivity is therefore likely to be an important step in the pathogenesis of this type of asthma. It appears to precede the development of asthma as IgE antibodies can be found in asymptomatic exposed workers who later develop asthma, and other manifestations of type I hypersensitivity such as rhinitis often precede the development of asthma. The explanation for the frequently observed sequence of rhinitis preceding asthma is not clear. It might be because the allergic sensitization occurs in the upper respiratory tract and the asthma develops later as a consequence of appropriately sensitized lymphocytes travelling to the lung. It seems less likely that lymphocytes at both sites become sensitized at the same time and that the earlier onset of nasal symptoms is a reflection of greater exposures at that site, but it might be that the necessary increase of airway hyper-responsiveness into the asthmatic range takes longer to establish than the hyperemia and edema of rhinitis, thereby allowing nasal symptoms to appear first.

Once hypersensitivity has developed, further stimulation of the immune system from repeated exposures to the inducing agent are clearly important for the development of asthma. Those with the greatest degree of hypersensitivity, i.e. the highest levels of specific IgE antibodies, are at greatest risk of developing asthma (Block *et al.*, 1983), and removing workers from exposure after sensitization considerably reduces the risk of asthma developing. It is not possible to determine whether it is increasing levels of IgE resulting from continued exposures which are responsible for the development of the asthma or whether this involves stimulation of another arm of the immune system. It is also not clear whether all sensitized workers would eventually go on to develop asthma if they continued to be exposed for long enough, or whether some workers are protected.

While specific IgE antibodies can be sensitive markers for high-molecular weight occupational asthma it is clear that they are not diagnostic, as usually a greater proportion of the work force becomes sensitized than goes on to develop asthma. Evidence of specific IgE antibodies has been found in very high proportions of some working populations, and the presence of the antibodies often appears to be a better marker of exposure than of disease (Fabbri *et al.*, 1993). Some sensitized workers go on to develop occupational asthma, some develop other allergic diseases such as rhinitis, and some have no clinical manifestation of their sensitisation. This illustrates quite clearly that an additional or parallel step is necessary for the progression from immune sensitization to the development of occupational asthma.

Low-molecular weight chemicals are too small to stimulate an immune response directly and to do this they must first bind to a host protein or alter its structure so that it is recognized as foreign by the immune system. With many low-molecular weight asthmagens, the evidence that they stimulate the production of IgE antibodies is less clear than with high-molecular weight agents, as IgE antibodies often cannot be detected. There are several possible explanations for this including technical difficulties in creating the appropriate immunoreactive chemical–protein complex for *in vitro* or *in vivo* testing, the possibility that only very low concentrations of specific antibodies are present, or that they are localized only in the airway. However, its seems most likely that IgE-related mechanisms are not always involved in asthma caused by these agents.

Acid anhydrides provide the clearest evidence that low-molecular weight chemical agents can act by inducing IgE antibodies. The presence of specific antibodies to human serum albumin (HSA) conjugates with these agents correlates well with the presence of occupational asthma (Howe *et al.*, 1983). Occupational exposures to platinum salts also frequently result in the formation of IgE antibodies, and skin-prick tests to the salt itself are often positive (Venables *et al.*, 1989). Platinum workers are reported to have higher levels of total IgE antibodies than controls, suggesting that exposures might stimulate the production of antibodies to a variety of environmental agents, and not just to the salt itself (Baker *et al.*, 1990).

Asthma caused by other low-molecular weight inducers is less frequently associated with detectable levels of IgE antibodies. Only 10–20% of symptomatic diisocyanate workers have evidence of circulating IgE antibodies to haptenic conjugates (Kestinen *et al.*, 1988) and only about 30% of cedar workers with occupational asthma have antibodies to plicatic acid (Vedal *et al.*, 1986) – prevalences which are not much different from those found among exposed workers with no evidence of asthma. The clinical features of asthma caused by these agents are similar to those observed with high-molecular weight agents where there is evidence of type I hypersensitivity, but the mechanisms through which they induce asthma remain unknown.

Asthma following acute toxic/irritant exposures

The periods of minutes or hours over which asthma sometimes develops following high-level exposures to toxic or highly irritant agents does not allow sufficient time for a hypersensitivity response to develop and other mechanisms must be involved. This condition – RADS – has been described following the inhalation of agents such as chlorine, ammonia, acid aerosols and smoke from domestic fires (Brooks *et al.*, 1985). It is likely to be a response to acute inflammatory changes in the airways. Acute insults to the lung, for example from smoke inhalation, are characterized by epithelial disruption and inflammatory cell infiltration into the airways with prominent neutrophilia (Sheppard, 1992). The precise mechanism by which these changes give rise to asthma is, however, unclear. Increased numbers of neutrophils can often be detected in the airways at the time of late reactions following inhalation challenge with allergens and occupational agents. Neutrophil infiltration in the airways is an important cause of increased airway responsiveness in some animal models but not in others, and its importance to human asthma is unclear. A potentially more important inaugural event in RADS is the disruption of the airway epithelium which is likely to occur with massive inflammatory insults. As noted above, this can result in the release of cytokines and proinflammatory mediators with the potential to cause acute airway edema and bronchoconstriction, and to recruit other cells such as mast cells, eosinophils and lymphocytes which are relevant to the perpetuation of the asthmatic response. From Figure 22.4, it can be seen that there is very close integration of the neurogenic arm with the non-specific cellular arm and stimulation of sensory nerve endings due to disruption of the epithelium would be capable of contributing to an acute inflammatory response.

In RADS, the asthma persists indefinitely, but milder and temporary increases in airway responsiveness might be more common following acute irritant exposure to the airways. Temporary increases in airway responsiveness have been described in non-asthmatic subjects following mild inhalation injuries from smoke inhaled

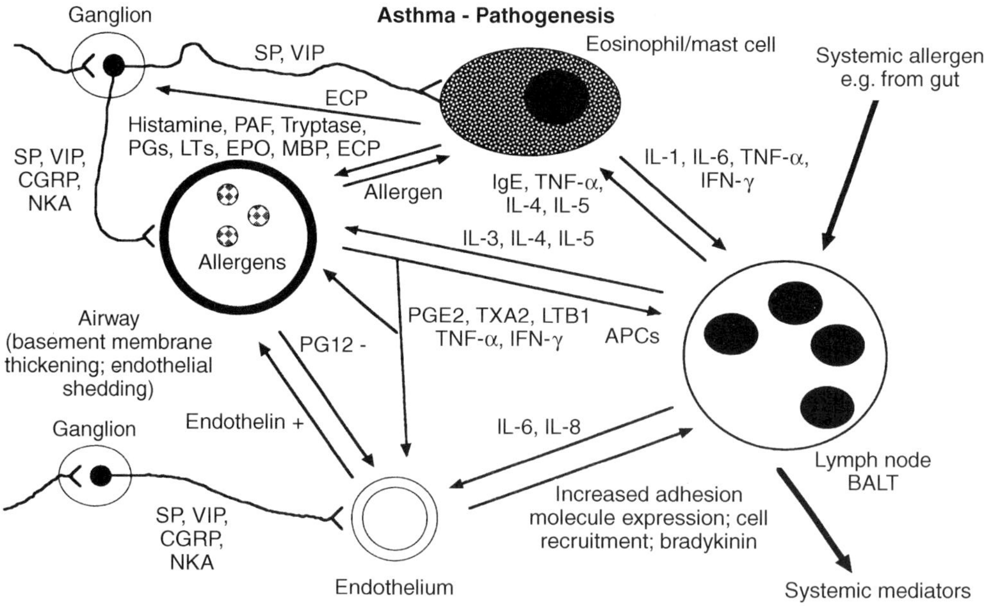

Figure 22.4 Interaction of innate and specific immune systems with neurogenic and local factors in the generation of asthmatic responses. SP, substance P; VIP, vasoactive intestinal polypeptide; NKA, neurokinin A; CGRP, calcitonin gene-related peptide; PAF, platelet activating factor; PG, prostaglandin; LT, leukotriene; TX, thromboxane; ECP, eosinophil cationic protein; MBP, major basic protein; EPO, eosinophil peroxidase; TNF, tumor necrosis factor; IFN, interferon; APCs, antigen-presenting cells; BALT, bronchus-associated lymphoid tissue.

from domestic fires and in experimental situations following exposure to ozone and nitrogen dioxide (Golden *et al.*, 1978; Stenton *et al.*, 1988). The persistence of the airway responsiveness in RADS appears to involve the triggering of an autonomous airway inflammatory process which is similar to that which develops after the cessation of exposures in other forms of occupational asthma. Once established, RADS shows typical clinical and pathologic features of asthma. Increased numbers of lymphocytes, eosinophils and neutrophils are seen in bronchial biopsies in those who have had their disease for 2–3 years (Gautrin *et al.*, 1994). A greater element of fixed airflow has been described in RADS than in other forms of asthma of similar duration and severity. This might be a consequence of additional obstructive bronchiolitis caused by the acute toxic event, but the marked degree of basement membrane thickening which has been observed might also be of some relevance.

Asthma following repeated submassive inflammatory exposures

The observation that a single massive exposure to an irritant agent can induce asthma in association with inflammatory changes in the airways raises the question of whether less massive exposures or exposures to less toxic agents might be capable of causing lesser but still significant adverse shifts in airway responsiveness. If that were the case, then prolonged exposures to these agents might lead to progressive and permanent increases in airway responsiveness, and asthma. Alternatively, small increases in airway responsiveness caused by occupational exposures could combine with other environmental factors such as viral infections and atmospheric pollutants which can also increase airway responsiveness, to shift the level in a stepwise manner into the asthmatic range. Under these circumstances there would be a latent interval of asymptomatic exposure during which airway responsiveness gradually increased, mimicking the latent interval associated with hypersensitivity. The only asthmatic response which would be observed in the workplace would be acute 'non-specific' bronchoconstriction following exposures to the irritant agent, and the cause of the asthma would be difficult to identify. It could only be detected epidemiologically by the demonstration of a greater than expected prevalence of asthma among a working population.

Hypersensitivity to common aeroallergens could contribute to asthma induced in this way. Inflammatory exposures can disrupt the integrity of the airway epithelium and increase the penetration of allergens. They also increase the recruitment of antigen-presenting cells and lymphocytes into the airway, leading to an increased risk of sensitization. Ozone, for example, can potentiate the responses to inhaled aeroallergens both in experimental animals and in humans (Molfino *et al.*, 1991). Such mechanisms might be relevant to situations such as aluminum smelting where there appears to be an increased prevalence of asthma but no single causative agent has been identified.

If irritant exposures do increase the risk of developing asthma then the response cannot be entirely non-specific. Cigarette smoking is the most common chronic irritant stimulus to the airways and causes marked increases in airway permeability and inflammation with increases in the numbers of neutrophils and other cells. Smoking does not cause asthma and it is therefore important to remember that, with the exception of RADS, the induction of asthma by toxic or irritant exposures is speculative.

Population changes in airway responsiveness

One final mechanism worth considering is that occupational exposures to irritant agents might cause a small shift in airway responsiveness in most or all of an exposed population. The increase in airway responsiveness in any individual might be too small to detect but the distribution of airway responsiveness would become shifted towards the asthmatic end of the spectrum (Figure 22.5). The pre-existing levels of airway responsiveness of some individuals will already be close to the threshold above which clinically active asthma becomes a likely consequence and a modest increase in the level of airway responsiveness might be sufficient to cause overt clinical asthma. By virtue of the distribution of airway responsiveness in the population, a small change affecting the whole population might have a marked effect on the proportion developing symptoms of asthma. We have recently observed a distribution of airway responsiveness among welders which appears significantly shifted compared with a control population and have speculated that this might be a consequence of small changes affecting the whole population (Beach *et al.*, 1996).

RISK FACTORS FOR THE DEVELOPMENT OF OCCUPATIONAL ASTHMA

Occupational asthma usually affects only a small proportion of an exposed work force, suggesting that some workers are more at risk than others. On the other hand, the very high prevalences of 50% or above which have been documented in some industrial settings suggest that any inherent predisposition is only partial and that the entire population has the potential to develop asthma given the appropriate stimulus.

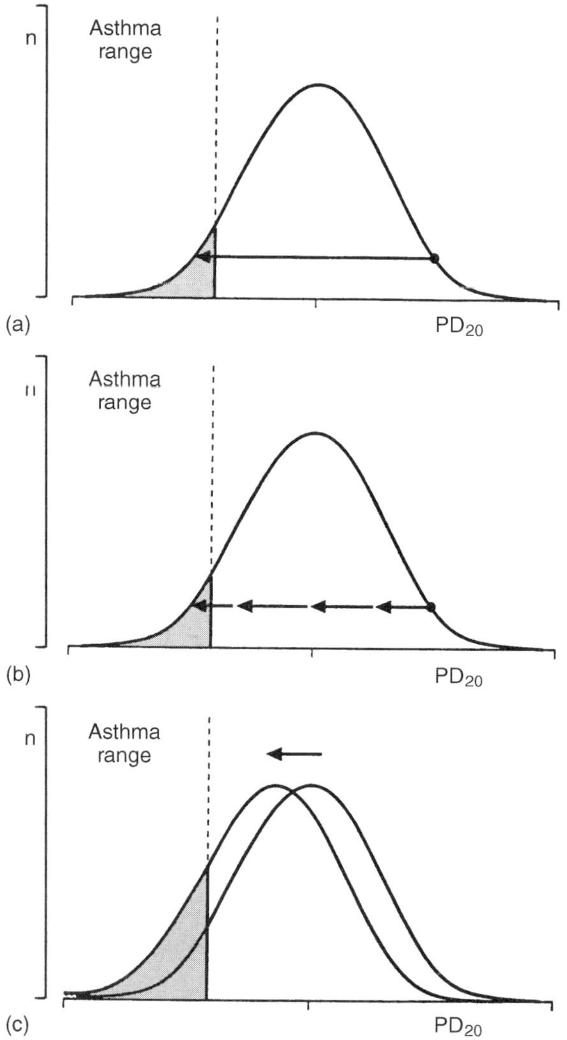

Figure 22.5 Models to explain the possible development of airway hyper-responsiveness and asthma. A subject can move into the asthmatic range of airway responsiveness following exposure to a potent asthma-inducing agent (a), as a result of a series of smaller increases in airway responsiveness (b), or due to a small shift in the distribution of airway responsiveness in the entire population (c).

Genetic/personal

Atopy is the most important personal predisposing factor for developing asthma, the risk being increased approximately four-fold among those with symptoms of atopy such as hay fever or eczema. Occupational asthma caused by high-molecular weight agents is also considerably more common among atopic subjects. Almost all psyllium workers who develop asthma are atopic (Malo *et al.*, 1990), atopic bakers are some 16 times more likely to become sensitized to occupational allergens than non-atopics (Musk *et al.*, 1989), and laboratory animal asthma is approximately twice as common in atopics as in non-atopics (Platt-Mills *et al.*, 1987). With laboratory animal asthma, atopy has been shown to increase the risk of developing symptomatic sensitization most during the first year of exposure, raising the possibility that its major effect is to bring forward the time of sensitization rather than to increase the overall risk (Botham *et al.*, 1987). It has also also been found that the levels of specific IgE antibodies are not greater in atopics than in non-atopics, suggesting that while atopy might predispose towards the development of sensitization, other factors are important in determining the magnitude of the response (Platt-Mills *et al.*, 1987).

Atopy does not appear to confer any increased risk for the development of asthma following exposure to low-molecular weight chemicals where there is no evidence of IgE involvement, e.g. with isocyanates and plicatic acid (Vedal *et al.*, 1986; Pisati *et al.*, 1993). It is, however, a risk factor with low-molecular weight agents such as platinum and acid anhydrides whose mode of action does appear to involve IgE; atopic workers are approximately twice as likely to become sensitized to platinum salts as non-atopics (Venables *et al.*, 1989)

Atopy is clearly under genetic control though its precise mode of inheritance remains unclear. A number of recent studies have identified possible candidate genes which might contribute to the development of atopy. These include a site at the 11q13 region of chromosome 11 (Cookson, 1995) which is close to, if not identical to, that of the gene coding for the beta unit of the high-affinity IgE receptor (Fcε RIb) normally expressed on mast cells. Genetic differences in IgE receptor density or affinity could clearly influence the expression of atopy. Another important genetic region controlling immune functions is found on chromosome 5 at the 5q31.1 or 5q31.2-q33 loci. This area contains genes coding for IL-4,-5,-9,-13 and other IgE-regulating cytokines and their receptors, and polymorphisms in this area have been related to IgE levels in family studies of atopics (Bleeker *et al.*, 1995; Marsh *et al.*, 1995). The regulation of these cytokines and their receptors is likely to be critical in determining whether an individual develops a preponderance of Th1 of Th2 cells and thus whether their immune response is weighted towards a humoral or cellular response.

Genes located on chromosome 6p encode MHC class II molecules which are essential to the process of presenting foreign antigenic material to T helper cells. Genetic variation at this site leads to configurational differences in HLA antigens which can influence their ability to bind foreign antigens or present them in an appropriate manner to lymphocytes. The possession of specific HLA antigens appears to be associated with risk of sensitization to a variety of allergens. The HLA DR4 haplotype has been associated with an overall increased risk of developing atopic asthma (Aron *et al.*, 1993) and HLA-DW2 is strongly associated with immune recognition of Amb tV and Amb aV ragweed proteins (Roebber *et al.*, 1984). Specific HLA class II alleles have also been shown to be associated with sensitivity to occupational agents. Young *et al.* (1993) found the HLA-DR3 haplotype to be six times more common among workers with positive RASTs to acid anhydrides compared with exposed controls with negative skin tests and negative RASTs. Among diisocyanate-exposed workers, those with occupational asthma were significantly less likely to express the HLA-DQB1*0501 allele and were more likely to express the HLA-DQB1*0503 allele than exposed but asymptomatic controls. The *0501 and the *0503 alleles differ by a single amino acid at position 57 – valine in the *0501 allele and aspartic acid in the *0503 allele (Fabbri *et al.*, 1994). The resulting change to the three-dimensional structure of the molecule might well be sufficient to affect its interaction with either the diisocyanate molecule itself or a diisocyanate protein hapten, and so influence its likelihood of stimulating the immune system.

Other potential genetic influences over the expression of atopy include polymorphisms of the T-cell receptor alpha chain located on chromosome 14 (Moffat *et al.*, 1994). However, as immune function is under the control of multiple genes the contributions of individual gene loci to allergic disease is likely to remain difficult to determine. None of the genetic influences which have so far been determined is thought to be of sufficient practical importance to allow it to be used as a basis for identifying workers who should be excluded from exposure to sensitizing agents.

Nature of causative agent and exposure characteristics

The nature of the inhaled material has an important influence on the type of response. Some agents such as diisocyanates induce occupational asthma commonly, with 5–15% of moderately exposed workers becoming affected (Diem *et al.*, 1982). Other chemical agents such as formaldehyde appear to induce asthma only rarely and following exceptionally heavy exposures. These differences in asthmagenic potency reflect differences in chemical structure and reactivity. Agents such as glues and polymer crosslinking agents are often potent modifiers of body proteins, an important step in forming haptens which are able to stimulate the immune system and induce asthma. Some agents such as proteolytic enzymes have other properties which might increase their penetration across the airway mucosal barrier and so increase their likelihood of stimulating an immune response.

The level of exposure is also an important determinant of sensitization. Quantitative relationships between exposure levels to western red cedar, flour and laboratory animal proteins and the risk of developing asthma have all been demonstrated (Musk *et al.*, 1989). Patterns of exposure have been less well characterized but are also likely to be important. Immunologic stimulation is best achieved by intermittent exposures to large amounts of the sensitizing material. The circumstances of many commercial operations predispose to such intermittent high exposures, e.g. during soldering or when emptying sacks of flour. These intermittent high-level exposures can be difficult to quantify and in some circumstances difficult to control.

Cigarette smoking has a number of effects on pulmonary defense mechanisms and on the immune system. It increases the permeability of the airway epithelium, depresses the function of antigen-presenting cells and suppresses mucosal lymphocyte-mediated immune reactions associated with the production of specific IgG antibodies. Smoking also increases total IgE levels, though this is not reflected in higher levels of specific IgE antibodies or higher prevalences of positive skin-prick tests. The effects of smoking on the risk of developing sensitization to occupational agents are variable. With platinum salts the prevalence of positive skin-prick tests is much higher in smokers. Conversely, with western red cedar, occupational asthma appears to be less common among smokers, but for most other agents no effect of smoking has been demonstrated. These differences might offer some clues to the mechanisms involved.

NATURAL HISTORY OF OCCUPATIONAL ASTHMA

Occupational asthma persists and often worsens so long as exposures to the causative agent continue. In one study, none of a group of 48 western red cedar workers who remained at work after their asthma was diagnosed became asymptomatic (Cote *et al.*, 1990). All were encouraged to reduce their exposures, and with this five showed some improvement in their asthma, but 25 remained unchanged and 18 deteriorated. This outcome is not surprising given that occupational asthma usually appears to involve a hypersensitivity mechanism. It is typical of an allergic condition that once the disease is established exposures to only very small amounts of the causative agent – much less than those necessary to induce sensitization in the first place – can perpetuate the disease.

Of considerably greater importance than the perpetuation of occupational asthma in the presence of continued exposures is the observation that occupational asthma does not always improve when exposures cease. This phenomenon was initially reported in the late 1970s and has been confirmed in a large number of follow-up studies. These have shown improvement rates ranging from 19% to 86% among those leaving work, with typically fewer than half of subjects recovering fully. An understanding of the factors which govern this outcome and the mechanisms involved would offer potentially valuable lessons for the management of asthma in general, and insight into the relationship between what are commonly thought of as 'extrinsic' and 'intrinsic' asthma.

The one variable which has most consistently been shown to be related to the outcome after leaving work is the duration of the period of exposure after symptoms have begun. Chan-Yeung *et al.*, (1982) demonstrated that cedar workers who became asymptomatic after leaving work had a shorter duration of symptomatic exposure before their asthma was diagnosed compared with those whose asthma persisted (17 versus 49 months). Pisati *et al.*, (1993) demonstrated similar findings for diisocyanate workers. Those who recovered after leaving work had a mean duration of symptomatic exposure of 1.6 years compared with 2.8 years for those whose asthma improved partially and 5.5 years for those who did not recover at all. It thus appears that, after a variable period from the time of sensitization, occupational asthma loses its dependence on environmental factors and becomes 'intrinsic.' Personal characteristics such as age, atopy or smoking status do not affect the outcome (Chan-Yeung *et al.*, 1982; Cote *et al.*, 1990).

Mechanisms involving a specific antigen–antibody interaction cannot be responsible for perpetuating the asthma once exposures to the causative agent cease. Once exposure stops, antibody levels fall slowly with time (Merget *et al.*, 1991) with a half-life of approximately 12 months for TCPA-HSA antibodies and 20 months for snow-crab IgE (Venables *et al.*, 1987; Malo *et al.*, 1988). Circulating antibodies can be detected in platinum workers for years after exposures cease and skin-prick tests can remain positive for even longer, suggesting that local tissue sensitivity might remain after systemic hypersensitivity has diminished. The levels of specific IgE antibodies continue to decline after any improvement in airway responsiveness has plateaued, supporting the view that they are not important in maintaining the asthma after exposures have ceased (Malo *et al.*, 1988).

Specific IgE antibodies cannot be used to study the mechanisms underlying the perpetuation of asthma following sensitization to most low-molecular weight agents. Studies using inhalation challenge tests suggest that specific hypersensitivity to these agents declines slowly and that this is not necessarily related to the perpetuation of the asthma. Paggiaro *et al.*, (1993) found that fewer than half of those who lost their specific hypersensitivity to isocyanates after leaving work also decreased their levels of airway responsiveness to methacholine into the normal range. Mapp *et al.*, (1988), on the other hand, found a rather closer association with seven of eight workers who lost their specific hypersensitivity to isocyanates becoming asymptomatic.

Inflammatory changes in the airway appear to persist after cessation of occupational exposures. Paggiaro *et al.*, (1990) showed increased numbers of lymphocytes, eosinophils and neutrophils in BAL fluid in 10 subjects with diisocyanate asthma who had left work. Bronchial biopsies also showed typical features of asthma with lymphocyte and eosinophil infiltration and basement membrane thickening. Half of the subjects had shown some recovery over the follow-up period and this was associated with some lessening of the BAL eosinophilia. Saetta *et al.*, (1992a,b, 1995) performed sequential bronchial biopsies on 10 subjects with diisocyanate asthma before and after stopping work. A decrease in the numbers of inflammatory cells was observed, though the changes were not statistically significant and the reduction in cell numbers showed no relationship to clinical improvement. The degree of basement membrane thickening and the numbers of subepithelial fibroblasts did decrease significantly, possibly reflecting a reduction in cytokine release at an early stage in the improvement of the inflammatory process. Chan-Yeung *et al.* (1988) studied a group of cedar workers after longer periods away from work (1–10 years) and remote from any inhalation challenge tests with the causative agent which might have induced acute changes in cellular profiles in the airways. The ex-workers with persisting symptoms had higher levels of albumin, and higher neutrophil, eosinophil and epithelial cell counts in their BAL fluid than either the asymptomatic ex-workers or control subjects. This suggests that airway inflammation persists in those whose asthma remains active after the cessation of occupational exposures but slowly improves in those who asthma disappears. The reduction in airway inflammation might lag behind clinical recovery.

Persisting occupational asthma can lead to the development of fixed airflow obstruction identical to that which develops with other forms of asthma. The mechanisms of this are poorly understood but they are presumed to involve remodeling of the airways and this might be related to the thickening of the basement membrane which is observed in the early stages of the disease. The end result – a form of chronic obstruction pulmonary disease (COPD) – is similar to the COPD which has been identified in a number of occupational settings, including industries where there is an increased risk of occupational asthma. This industrial COPD might even be the end result of a chronic airway inflammatory process with similar immunologic mechanisms to those seen in asthma.

REFERENCES

American Thoracic Society (1962) Definitions and classifications of chronic bronchitis, asthma, emphysema. *Am. Rev. Respir. Dis.*, **85**, 762–8.

Aron, Y., Bignon, J.S., Ju, L.Y., Lockhart, A., Charron, D. and Swierczewski, E. (1993) Haplotype HLA-DR is associated with asthma and atopic state. *Hum. Immunol.*, **36**(suppl II), 37–52.

Baker, O.B., Gann, P.H., Brooks, S.M. *et al.* (1990) Cross sectional study of platinum salts sensitization among precious metals refinery workers. *Am. J. Ind. Med.*, **18**, 653.

Barnes, P.J., Djukanovic, R. and Holgate, S. (1995) Asthma pathogenesis, In *Respiratory Medicine*, (eds R.A.L.B. Brewis, B. Corrin, D.M. Geddes and G.J. Gibson), W.B. Saunders, London. pp. 1108–53.

Beach, J.R., Young, C.L., Stenton, S.C., Dennis, J.H., Avery, A.J., Walters, E.H. and Hendrick, D.J. (1993) Measurement of airway responsiveness to methacholine: the relative importance of the precision of drug delivery and the method of assessing the response. *Thorax*, **48**, 133–5.

Beach, J.R., Dennis, J.H., Avery, A.J. *et al.* (1996) An epidemiological investigation of asthma in welders. *Am. J. Respir. Crit. Care Med.*, **154**, 1394–1400.

Bentley, A.M., Durham, S.R. and Kay, A.B. (1994) Comparison of the immunopathology of extrinsic, intrinsic, and occupational asthma. *J. Invest. Allergy Clin. Immunol.*, **4**, 222–32.

Bernstein, A.L., Chan-Yeung, M., Malo, J.-L. and Bernstein, D.I. (1993) Definition and classification of asthma, in *Asthma in the Workplace*, (eds A.L. Bernstein, M. Chan-Yeung, J.-L. Malo and D.I. Bernstein) Marcel Dekker, New York, pp. 1–4.

Bleecker, E.R., Amelung, P.J., Levitt, R.C., Postma, D.S. and Meyers, D.A. (1995) Evidence for linkage of total serum IgE and bronchial hyperresponsivness to chromosome 5q: a major regulatory locus important in asthma. *Clin. Exp. Allergy*, **25**(suppl. 2), 84–8.

Block, G., Tse, K.S., Kijek, K., Chan, H. and Chan-Yeung, M. (1983) Baker's asthma. *Clin. Allergy*, **13**, 359–70.

Bochner, B.S., Undem, B.J. and Lichtenstein, L.M. (1994) Immunological aspects of allergic asthma. *Annu. Rev. Immunol.*, **12**, 295–355.

Botham, P.A., Davies, G.E. and Teasdale, E.L. (1987) Allergy to laboratory animals: a prospective study of its incidence and of the influence of atopy on its development. *Br. J. Ind. Med.*, **44**, 627–32.

Bradding, P., Roberts, J.A., Britten, K.M., Montefort, S., Djukanovic, R., Heusser, C., Howarth, P.H. and Holgate, S.T. (1994) Interleukin-4,-5,-6 and TNFα in normal and asthmatic airways: evidence for the human mast cell as an important source of these cytokines. *Am. J. Respir. Cell Mol. Biol.*, **10**, 471–80.

Bradding, P., Feather, I.H., Wilson, S., Bardin, P.G., Heusser, C.H., Holgate, S.T. and Howarth, P.H. (1993) Immunolocalisation of cytokines in the nasal mucosa of normal and perennial rhinitic subjects: the mast cell as a source of IL-4, IL-5, and IL-6 in human allergic mucosal inflammation. *J. Immunol.*, **151**, 3853–65.

Brooks, S., Weiss, M.A. and Bernstein, I.L. (1985) Reactive airways dysfunction syndrome: persistent asthma after high level irritant exposure. *Chest*, **88**, 376–84.

Chan-Yeung, M., Lam, S. and Koener, S. (1982) Clinical features and natural history of occupational asthma due to Western Red Cedar. *Am. J. Med.*, **72**, 411–15.

Chan-Yeung, M., Leriche, J., Maclean, L. and Lam, S. (1988) Comparison of cellular and protein changes in bronchial lavage fluid of symptomatic and asymptomatic patients with red cedar asthma on follow-up examination. *Clin. Allergy*, **18**, 359–65.

Chan-Yeung, M., Chan, H., Tse, K.S., Salari, H. and Lam, S. (1989) Histamine and leukotrienes release in bronchoalveolar lavage fluid during plicatic acid induced bronchoconstriction. *J. Allergy Clin. Immunol*, **84**, 762–8.

Cookson, W.O.C.M. (1995) 11q and high affinity IgE receptor in asthma and allergy. *Clin. Exp. Allergy*, **25**(suppl. 2), 71–3.

Cote, J., Kennedy, S. and Chan-Yeung, M. (1990) Outcome of patients with cedar asthma with continued exposures. *Am. Rev. Respir. Dis.*, **141**, 373–6.

di Stefano, A., Saetta, M., Maestrelli, P., Milani, G., Pivrotto, F., Mapp, C.E. and Fabbri, L.M. (1993) Mast cells in the airway mucosa and rapid develoment of occupational asthma induced by toluene diisocyanate. *Am. Rev. Respir. Dis.*, **147**, 1005–9.

Diem, J.E., Jones, R.N., Hendrick, D.J. *et al.* (1982) Five year longitudinal study of workers employed in a new toluene diisocyanate manufacturing plant. *Am. Rev. Respir. Dis.*, **126**, 420–8.

Durham, S.R. (1988) The significance of the late responses in asthma. *Clin. Exp. Allergy*, **21**, 3–7.

Fabbri, L.M., Boschetto, P., Zocca, E. *et al.* (1987) Bronchoalveolar neutrophilia during late asthmatic reactions induced by toluene diisocyanate. *Am. Rev. Respir. Dis.*, **136**, 36–42.

Fabbri, L.M., Danieli, D., Crescioli, S. *et al.* (1988) Fatal asthma in a subject sensitized to toluene diisocyanate. *Am. Rev. Respir. Dis.*, **137**, 1494–8.

Fabbri, L.M., Ciaccia, A., Maestrelli, P., Saetta, M. and Mapp, C. (1993) Pathophysiology of occupational asthma, in *Asthma in the Workplace*, (eds A.L. Bernstein, M. Chan-Yeung, J.-L. Malo, D.I. Bernstein), Marcel Dekker, New York, pp. 61–92.

Fabbri, L.M., Maestrelli, P., Saetta, M. and Mapp, C.M. (1994) Mechanisms of occupational asthma. *Clin. Exp. Allergy*, **24**, 628–35.

Frew, A., Chan, H., Dryden, P., Salari, H., Lam, S. and Chan-Yeung, M. (1993) Immunological studies of the mechanism of occupational asthma. *J. Allergy Clin. Immunol.*, **92**, 466–78.

Frew, A.J., Chan, H., Lam, S. and Chan-Yeung, M. (1995) Bronchial inflammation in occupational asthma due to western red cedar. *Am. J. Respir. Crit. Care Med.*, **151**, 340–4.

Gautrin, D., Boulet, L.-P., Boutet, M. *et al.* (1994) Is reactive airways dysfunction syndrome a variant of occupational asthma. *J. Allergy Clin. Immunol.*, **93**, 12–22.

Golden, J., Nadel, J.A. and Boushey, H.A. (1978) Bronchial hyperirritibility in healthy subjects after exposure to ozone. *Am. Rev. Respir. Dis.*, **118**, 287–94.

Gordon, J.R., Burd, P.R. and Galli, S.J. (1990) Mast cells as a source for multifunctional cytokine. *Immunol. Today*, **11**, 458–63.

Gratziou, C., Carroll, M., Montefort, S., Teran, L., Howarth, P.H. and Holgate, S.T. (1996) Inflammatory and T-cell profile of asthmatic airways 6 hours after local allergen challenge. *Am. J. Respir. Crit. Care Med.*, **153**, 515–20.

Haahtela, T., Jarvinen, M., Kava, T. *et al.* (1994) Effects of reducing or discontinuing inhaled budesonide in patients with mild asthma. *N. Engl. J. Med.*, **331**, 700–5.

Hamid, Q., Azzawi, M., Moqbel, R. *et al.* (1991) Expression of mRNA for interleukin-5 in mucosal biopsies from asthma. *J. Clin. Invest.*, **87**, 1542–6.

Hol, B.E.A., Krouwels, F.H., Bruiner, B. *et al.* (1992) Cloning of T lymphocytes from bronchoalveolar lavage fluid. *Am. J. Respir. Cell Mol. Biol.*, **7**, 523–30.

Howe, W., Venables, K.M., Topping, M.D., Dally, M.B., Hawkins, R., Law, J.S. and Newman Taylor, A.J. (1983) Tetrachlorophthalic anhydride asthma: evidence for specific IgE antibody. *J. Allergy Clin. Immunol.*, **71**, 5.

Kestinen, H., Tupasela, O., Tiikkainen, U. and Normanden, H. (1988) Experiences of specific IgE in asthma due to diisocyanates. *Clin. Allergy*, **18**, 597–64.

Kita, O., Weiler, D., Sur, S. *et al.* (1993) Il-5 is the predominant eosinophil-active cytokine in the antigen-induced pulmonary late-phase reaction. *Am. Rev. Respir. Dis.*, **147**, 901–7.

Lam, S., LeRiche, J., Plillips, D. and Chan-Yeung, M. (1987) Cellular and protein changes in bronchoalveolar lavage fluid after late asthmatic reactions in patients with red cedar asthma. *J. Allergy. Clin. Immunol.*, **80**, 44–50.

Lanzavecchia, A. (1990) Receptor-mediated antigen uptake and its effect on antigen presentation to class II-restricted T lymphocytes. *Annu. Rev. Immunol.*, **8**, 8773–93.

Linsley, P.S. and Ledbetter, J.A. (1993) The role of the CD28 receptor during T cell responses to antigen. *Annu. Rev. Immunol.*, **11**, 11191–212.

Maestrelli, P., Callagni, P.G., Saetta, M., di Stefano, A., Hosslet, J.J., Santnastaso, A., Fabbri, L.M. and Mapp, C.E. (1994a) Sputum eosinophilia after asthmatic responses induced by isocyanates in sensitized subjects. *Clin. Exp. Allergy*, **24**, 29–34.

Maestrelli, P., Gianfranco, F., Carli, M. *et al.* (1994b) CD8 T-cell clones producing interleukin-5 and interferon-gamma in bronchial mucosa of patients with asthma induced by toluene diisocyanate. *Scand. J. Work Environ. Health*, **20**, 376–81.

Malo, J.L., Cartier, A., Ghezzo, H., Lafrance, M., Mocants, M. and Lehrer, S.B. (1988) Patterns of improvement in spirometry, bronchial hyper-responsiveness, and specific IgE antibody levels after cessation of exposure in occupational asthma caused by snow-crab processing. *Am. Rev. Respir. Dis.*, **138**, 807–12.

Malo, J.L., Cartier, A., L'Archeveque, J., Ghezzo, H., Lagier, F., Trudeau, C. and Dolovich, J. (1990) Prevalence of occupational asthma and immunological sensitization to psyllium among health care personnel in chronic care hospitals. *Am. Rev. Respir. Dis.*, **142**, 1359–66.

Mapp, C.E., Corona, P.C., de Marzo, N. and Fabbri, L. (1988) Persistent asthma due to isocyanates. *Am. Rev. Respir. Dis.*, **137**, 1326–9.

Marsh, D.G., Neely, J.D., Breazeale, D.R. *et al.* (1995) Total serum IgE levels and chromosome 5q. *Clin. Exp. Allergy*, **25**(suppl. 2), 79–83.

Merget, R., Schultze-Wernighaus, G., Bode, F., Bergmann, E.-M., Zachgo, W. and Weier-Sydow, J. (1991) Quantitative skin prick and bronchial provocation tests with platinum salt. *Br. J. Ind. Med.*, **48**, 830–7.

Moffat, M.F., Hill, M.R., Cornelis, F. *et al.* (1994) Genetic linkage on the T cell receptor a/d complex to specific IgE responses. *Lancet*, **343**, 1597–9.

Molfino, N.A., Wright, F.C., Katz, I. *et al.* (1991) Effect of low concentrations of ozone on inhaled allergen responses on asthmatic subjects. *Lancet*, **338**, 199–203.

Musk, A.W., Venables, K.M., Crook, B. *et al.* (1989) Respiratory symptoms, lung function, and sensitization to flour in a British bakery. *Br. J. Ind. Med.*, **46**, 636–42.

O'Connor, G., Sparrow, D., Taylor, D., Segal, M. and Weiss, S. (1987) Analysis of dose–response curves to methacholine: an approach suitable for population studies. *Am. Rev. Respir. Dis.*, **136**, 1412–17.

Paggiaro, P., Bacci, E., Paoletti, P., Bernard, P., Dente, F.L., Marchetti, G., Talini, D., Menconi, G.F. and Giuntini, C. (1990) Bronchoalveolar lavage and morphology of the airways after cessation of exposure in asthmatic subjects sensitized to toluene diisocyanate. *Chest*, **98**, 536–42.

Paggiaro, P.L., Vagaggini, B., Dente, F.L. *et al.* (1993) Bronchial hyper-responsiveness and toluene diisocyanate. Long term change in sensitized subjects. *Chest*, **103**, 1123–8.

Picker, L.J. and Butcher, E.C. (1992) Physiological and molecular mechanisms of lymphocyte homing. *Annu. Rev. Immunol.*, **10**, 561–91.

Pisati, G., Baruffini, A. and Zedda, S. (1993) Toluene diisocyanate induced asthma: outcome according to persistence of cessation of exposure. *Br. J. Ind. Med.*, **50**, 60–4.

Platt-Mills, T.A.E., Longbottom, J., Edwards, J. and Cockroft, A. (1987) Occupational asthma and rhinitis related to laboratory rats: serum IgE and IgE antibodies to the rat urinary allergen. *J. Allergy Clin. Immunol.*, **79**, 505–15.

Roebber, M., Klapper, D.G., Goodfriend, L., Bias, W.B., Hsu, S.H. and Marsh, D.G. (1984) Immunochemical and genetic studies of Amb iV (Ra5G): an Ra5 homologue from giant ragweed pollen. *J. Immunol.*, **134**, 3062–9.

Romagnani, S. (1991) Human Th1 and Th2 subsets doubt no more. *Immunol. Today*, **12**, 256–7.

Rossi, G.A., Crimi, E., Lantero, S. *et al.* (1991) Late phase asthmatic response to inhaled allergen is associated with early recruitment of eosinophils in the airways. *Am. Rev. Respir Dis.*, **144**, 379–83.

Saetta, M., di Stafano, A., Maestrelli, P. *et al.* (1992a) Airway mucosal inflammation in occupational asthma induced by toluene diisocyanate. *Am. Rev. Respir. Dis.*, **145**, 160–8.

Saetta, M., Maestrelli, P., di Stefano, A., de Marzo, N., Milani, G., Pivirotto, F., Mapp, C.E. and Fabbri, L.M. (1992b) Effect of cessation of exposure to toluene diisocyanate (TDI) on bronchial mucosa of subjects with TDI-induced asthma. *Am. Rev. Respir. Dis.*, **145**, 169–74.

Saetta, M., Maestrelli, P., Turato, G., Mapp, C.E., Milani, G., Pivirotto, F., Fabbri, L.M. and di Stefano, A. (1995) Airway wall remodelling after cessation of exposure to isocyanates in sensitized asthmatic subjects. *Am. J. Respir. Crit. Care Med.*, **151**, 489–94.

Sedgwick, J.B., Calhoun, W.J., Vertis, R.F., Bates, M.E., McAllister, P.K. and Busse, W.W. (1992) Comparison of airway and blood eosinophil function after *in vivo* antigen challenge. *J. Immunol.*, **149**, 3710–18.

Sheppard, D. (1992) Gases, in *Asthma: Basic Mechanisms and Clinical Management*, (eds P.J. Barnes, I.W. Rodger and N.C. Thompson), Academic Press, London pp 491–8.

Sibbald, B. (1992) Genetics, in *Asthma: Basic Mechanisms and Clinical Management* (eds P.J. Barnes, I.W. Rodger and N.C. Thompson), Academic Press, London pp 21–32.

Siracusa, A., Vecciarelli, A., Brugnami, G., Marabini, A., Felicioni, D. and Severini, C. (1992) Changes in interleukin-1 and tumor necrosis factor production by peripheral blood monocytes after specific bronchoprovocation test in occupational asthma. *Am. Rev. Respir. Dis.*, **146**, 408–12.

Stenton, S.C., Walters, E.H. and Hendrick, D.J. (1988) Effect of accidental smoke inhalation on bronchial responsiveness. *Br. J. Dis. Chest.*, **82**, 436–8.

Stenton, S., Dennis, J.H., Walters, E.H. and Hendrick, D.J. (1990) The asthmagenic properties of a newly developed detergent ingredient – sodium iso-nonoyl oxybenzene sulphonate. *Br. J. Ind. Med.*, **47**, 405–10.

Stenton, S.C., Dennis, J.H. and Hendrick, D.J. (1995) Occupational asthma caused by ceftazidime. *Eur. Respir. J.*, **8**, 1421–3.

Vedal, S., Chan-Yeung, M., Enarson, D.A., Chan, H., Dorken, E. and Tse, K. (1986) Plicatic acid-specific IgE and nonspecific bronchial hyper-responsiveness in western red-cedar workers. *J. Allergy Clin. Immunol.*, **78**, 1103–9.

Venables, K.M., Topping, M.D., Nunn, A.J. *et al.* (1987) Immunological and functional consequences of chemical (tetrachlorophthalic anhydride)-induced asthma after four years of avoidance of exposure. *J. Allergy Clin. Immunol.*, **80**, 212–18.

Venables, K.M., Dally, M.B., Nunn, A.J. *et al.* (1989) Smoking and occupational allergy in workers in a platinum refinery. *Br. Med. J.*, **299**, 939–42.

Venge, P. (1990) What is the role of the eosinophil? *Thorax*, **45**, 161–3.

Walker, C., Kaegi, M.K., Braun, P. and Blaser, K. (1991) Activated T cells and eosinophilia in bronchoalveolar lavages from subjects with asthma correlated with disease severity. *J. Allergy Clin. Immunol.*, **88**, 935–42.

Walker, C., Bode, E., Boer, L., Hansel, T.T., Blaser, K. and Virchow, J.C. (1992) Allergic and nonallergic asthmatics have distinct patterns of T-cell activation and cytokine production in peripheral blood and bronchoalveolar lavage. *Am. Rev. Respir. Dis.*, **146**, 109–15.

Weller, F.R., Jansen, H.M. and Kapsenberg, M.L. (1995) Lymphocyte markers and disease activity in asthma. *Clin. Exp. Allergy*, **23**, 678–9.

Young, R.P., Barker, R., Cookson, W.O.C.M. and Newman Taylor, A.J. (1993) HLA-DR and DP antigen frequencies and acid anhydride sensitization. *Am. Res. Respir. Dis.*, **147**, A113.

23 Occupational asthma due to low-molecular weight compounds

P. Maestrelli, M. Saetta, C.E. Mapp, L.M. Fabbri, J.E. Parker and D.E. Banks

INTRODUCTION

It is difficult to determine the overall prevalence of occupational asthma. Since many workers leave the industry once they become affected, identifying only current workers underestimates the prevalence of this disease. It may be that, of the 2–6% of the general adult population who have asthma, between 2 and 15% of this group have asthma induced by workplace agents (NIAID Task Force Report, 1979; Butcher *et al.*, 1977; Blanc, 1987). Gannon and Burge (1991) proposed the annual incidence for occupational asthma to range from 30 to 303 per million residents in the West Midlands area of England. Understandably, these figures are likely to vary considerably by country, industry, and the type of agent to which workers are exposed.

The importance of occupational asthma cannot be underestimated. It is now the most common work-related lung disease in developed nations (Keskinen *et al.*, 1978; Gannon and Burge, 1991 Contreras *et al.*, 1994; Meredith and McDonald, 1994; Meredith and Nordman, 1996; Reilly *et al.*, 1994). The importance of this illness lies not only in the recognition that it is associated with adverse respiratory effects, but also because of its social implications with its impact on a worker's quality of life (Gannon *et al.*, 1993; Malo *et al.*, 1993a). The primary treatment is removal of the worker from exposure to the causative agent.

Numerous chemicals, both man-made and natural, which are a found in a wide range of materials and industrial processes can act as sensitizers and induce occupational asthma. These agents are typically associated with a latency period (time from initial exposure to the development of symptoms) and can be divided into low- and high-molecular weight groups, and further subdivided into those that induce asthma through IgE-dependent and IgE-independent mechanisms. Examples of high-molecular weight agents associated with asthma include organic agents such as latex, cereals, sea foods, enzymes, gums and numerous animal allergens (Chan-Yeung and Malo, 1994).

Our interest in this chapter is the recognition and understanding of asthma provoked by agents of 'low molecular weight' (i.e. sensitizers which have a molecular weight less than 1000 Da). More than 140 small-molecular weight agents have been reported to induce occupational asthma (Davies and Blainey, 1984; Butcher and Salvaggio, 1986). As new industrial processes have been introduced into the workplace, this number has increased (Table 23.1). Examples of typical

Occupational Lung Disease. Edited by Daniel E. Banks and John E. Parker. Published in 1998 by Chapman & Hall.
ISBN 0 412 73630 6

Table 23.1 A selected series of manuscripts addressing small-molecular weight agents reported to induce occupational asthma

Therapeutic agents	
Antibiotics and related compounds	
Penicillins (including ampicillin)	Davies *et al.*, 1974; Moscato. *et al.*, 1995; Wuthrich and Hartmann, 1982; Estrada Rodriguez *et al.*, 1991
Cephalosporins	Stenton *et al.*, 1995; Coutts *et al.*, 1981
Tetracycline	Menon and Das, 1977; Fawcett and Pepys 1976; Steinbruegge and Judson, 1980
Phenylglycine acid chloride	Kammermeyer and Matthews, 1973
Spiramycin	Malo and Cartier, 1988; Moscato *et al.*, 1984; Paggiaro *et al.*, 1979; Davies and Pepys, 1975
Pharmaceuticals and related compounds	
Alpha-mcthyldopa	Harries *et al.*, 1979
Amprolium hydrochloride	Greene and Freedman, 1976
Cimetidine	Coutts *et al.*, 1984
Furan-based binder	Cockcroft *et al.*, 1980
Hydralazine	Perrin *et al.*, 1990
Opiates	Biagini *et al.*, 1990; Gorski and Ulinski, 1996
Piperazine	See text
Penicillamine	Lagier *et al.*, 1989
Chloramine	See text
Hexachlorophene	See text
Tannic acid	Johnston *et al.*, 1951
Inorganic chemicals	
Ammonium persulfate	Wrbitzky *et al.*, 1995; Fisher and Dooms-Goossens, 1976 Merget *et al.*, 1996
Metal fumes and salts	Ruegger, 1995; Rivolta *et al.*, 1994; Kusaka *et al.*, 1996a; Chiappino, 1994
Aluminum	Saric and Marelija, 1991; Kongerud and Samuelson, 1991
Chromium	Shirakawa and Morimoto, 1996; Joules, 1952
Cobalt	Shirakawa *et al.*, 1992; Sjogren *et al.*, 1980; Kusaka *et al.*, 1986; 1996b Davison *et al.*, 1983; Cirla, 1994; Lauwerys and Lison, 1994; Sala *et al.*, 1994; Sabbioni *et al.*, 1994
Fluoride	Soyseth and Kongerud, 1992
Nickel	McConnell *et al.*, 1973; Novey *et al.*, 1983; Malo *et al.*, 1982; Block and Chan-Yeung 1982; Shirakawa *et al.*, 1990
Platinum	See text
Stainless steel	Keskinen *et al.*, 1980
Tungsten carbide	Bruckner, 1967; Prince *et al.*, 1984
Welding fumes	Lee *et al.*, 1990; Beach *et al.*, 1996
Vanadium	Musk and Tees, 1982
Zinc	Malo and Cartier, 1987
Organic chemicals	
Abietic acid (colophony)	See text
Acrylates (methyl methacrylates, cyanoacrylates)	Lozewicz *et al.*, 1985; Mickelsen and Johnston, 1997; Raymond, 1996; Savonius *et al.*, 1993; Kopp *et al.*, 1985
Amines (diamines, ethanolamines, tetramines)	See text
Acid anhydrides (phthalic, tetrachlorophthalic, trimellitic)	See text
Azo dyes (azobisformamide, azodicarbonamide)	See text
Azodicarbonamide	Ferris *et al.*, 1977; Slovak, 1981
Formaldehyde and glutaraldehyde	See text
Insecticides	Senthilselvan *et al.*, 1992; Deschamps *et al.*, 1994; Bryant, 1985; Wax and Hoffman 1994; Weiner, 1961; Underner *et al.*, 1987
Isocyanates	See Chapter 24

Table 23.1 Continued

Organic chemicals *continued*	
Plicatic acid (western red cedar wood dust)	Chan-Yeung *et al.*, 1980, 1994; Salari *et al.*, 1994; Frew *et al.*, 1993; Malo *et al.*, 1992; Chan-Yeung and Desjardins, 1992; Cote *et al.*, 1990; Vedal *et al.*, 1986, 1988; Paggiaro and Chan-Yeung, 1987; Lam *et al.*, 1987; Tse *et al.*, 1982;
Polyvinyl chloride (decomposition products)	Gannon *et al.*, 1992; Moisan, 1991; Lee *et al.*, 1989; Nielsen, *et al.*, 1989; Markowitz, 1989; Eisen *et al.*, 1985; Butler and Culver, 1981; Brooks and Vandervort, 1977; Andrasch *et al.*, 1976; Polakoff *et al.*, 1975
Styrene	Becher *et al.*, 1996
Freon	Malo *et al.*, 1984

Adapted from Butcher and Banks (1992).

exposures associated with the development of asthma due to some of these agents are listed in Table 23.2.

Some occupational agents, particularly high-molecular weight sensitizers (although this has been increasingly recognized in small-molecular weight agents), act through an IgE-mediated mechanism. In these instances, the inhaled sensitizing agents bind to specific IgE on the surface of mast cells, basophils and probably macrophages, eosinophils and platelets. Occupational asthma induced by high-molecular weight compounds is more frequent in atopic subjects (Burrows *et al.*, 1989; Sporik *et al.*, 1990). High-molecular weight sensitizing agents behave as complete antigens. In contrast, small-molecular weight agents are of insufficient size, by themselves, to provoke an immune response. Low-molecular weight sensitizers (haptens) probably need to react with autologous or heterologous carrier proteins to produce a complete antigen, and they are thought to combine with native human macromolecules such as albumin. When IgE is formed in response to such hapten–carrier complexes, further exposure to the antigenic agent may result in a reaction between the antigen and IgE antibody attached to the surface of cells, inducing a cascade of events which activates inflammatory cells and initiates and continues the synthesis and/or release of a wide variety of preformed or newly formed inflammatory mediators. In the lung, the result is the inflammatory reaction which we recognize as asthma.

Table 23.2 Examples of low-molecular weight agents associated with asthma

Agent of exposure	*Occupations or workers at risk*
Isocyanates	Plastic, rubber, or foam manufactures, spray painters, and foam insulation installers
Acrylates	Adhesive handlers
Anhydrides	Plastic and epoxy resin workers
Amines	Lacquer and shellac workers, solderers
Dyes	Textile workers
Fluxes	Electronic workers
Glutaraldehyde, formaldehyde	Health care workers
Metals (platinum, nickel, aluminum, hard metals, etc.)	Refinery and potroom workers
Pharmaceuticals or drugs	Manufacturers of pharmaceuticals
Plicatic acid (wood dust)	Wood workers, foresters, artisans

The first conclusive evidence that small-molecular weight compounds in the work place could be immunogenic was the demonstration that a serum albumin conjugate of phthalic anhydride could bind specific IgE antibodies in sensitized individuals (Maccia *et al.*, 1976). Subsequently, similar findings were shown in workers exposed to another anhydride (Moller, *et al.*, 1985). In some instances, other classes of antibodies, in particular IgG, may play a role in the pathogenesis of occupational asthma (Akiyama *et al.*, 1984; Wenzel, 1994). The importance of the antibody-mediated immunity can be investigated in humans with *in vivo* and *in vitro* tests, most frequently with skin-prick tests and radioallergosorbent (RAST) or ELISA tests.

An increase in bronchial responsiveness (measured pharmacologically by inhalation of increasing amounts of methacholine or histamine) is a characteristic feature of occupational asthma. The extent of responsiveness may decrease with time away from exposure and may increase with re-exposure to the sensitizing agent or after specific challenge tests (Lam *et al.*, 1979; Banks and Rando, 1988; Banks *et al.*, 1990). Specific inhalation challenge tests may induce isolated early, isolated late, biphasic or continuous asthmatic reactions (Pepys and Hutchcroft, 1975). An immediate asthmatic reaction begins within a few minutes after an inhalation challenge, reaches maximal intensity within 30 minutes, and ends within 90 minutes. A late asthmatic reaction begins 4–6 hours after challenge, reaches maximal intensity within 8–10 hours, and ends after 24–48 hours. A biphasic reaction is an immediate reaction with a spontaneous recovery of lung function to baseline (in the absence of therapy), followed by a late reaction. A continuous asthmatic reaction shows no remission between the early and the late phases.

When one addresses the prevalence of occupational asthma in populations exposed to low-molecular weight agents, it appears that the higher the degree of exposure to an agent, the greater the prevalence of asthma. Although the role of smoking has not been well characterized for many agents, smoking appears to be more important than atopy in predisposing workers to sensitization to platinum salts (Venables *et al.*, 1989), while neither atopy nor smoking appear to be important determinants for agents that induce asthma through IgE-independent mechanisms (Chan-Yeung *et al.*, 1982).

The duration of exposure to the sensitizing agent before the development of occupational asthma varies. In general terms, about 40% of patients with occupational asthma develop symptoms within 1–2 years of exposure; yet in 20%, symptoms develop after more than 10 years of exposure (Malo *et al.*, 1992). Similarly, most cases of anhydride-related asthma develop in the first 2 years of employment (Howe *et al.*, 1983; Wernfors *et al.*, 1986; Venables, 1989). The duration of symptoms before removal from exposure is a prognostic indicator of the outcome of a worker's asthma, irrespective of the agent. Continuous exposure in affected workers is associated with a worsening of asthma (Cote *et al.*, 1990). Early removal from exposure increases the likelihood of recovery.

The diagnosis of occupational asthma requires the physician to diagnose asthma, to establish a temporal relationship between exposure in the workplace and episodes of asthma, and to identify a sensitizing agent in the workplace. The diagnosis of asthma is based on the presence of variable airflow limitation or, in the absence of airflow limitation, the presence of pharmacologically induced bronchial responsiveness. Because the therapy of occupational asthma requires removal of the worker from the workplace, a clear-cut relationship to work must be objectively demonstrated. For screening examinations in the workplace or for field epidemiologic surveys, less specific diagnostic requirements may increase the sensitivity of case identification.

Finally, although our interest is in identifying and reporting on the adverse effects of a number of low-molecular weight sensitizing agents, we will exclude isocyanates, agents widely used in many industries, including the manufacture of flexible and rigid polyurethane foam for a variety of industries, and as paints, adhesives and coatings. For the isocyanates, potent and frequently occurring inducers, there are intriguing insights regarding how asthma develops and is maintained, and these agents may be a general model for mechanisms which occur in asthma due to low-molecular weight agents. Yet, for many other small-molecular weight agents which are less potent inducers of asthma and which have been rarely recognized to induce asthma, there remains a paucity of understanding about the pathogenic mechanisms. Investigations into the mechanisms of asthma induced by isocyanate (and other low-molecular weight agents) will need to continue before specific treatments and therapies that block sensitization can be developed. Since the prevalence of isocyanate-induced asthma in exposed workers may approach 10%, the effect of exposures to this group of agents is of sufficient importance to merit their own chapter, yet because these agents have been the most frequently studied, some relevant data are cited below.

BIOCHEMICAL MECHANISMS OF 'LOW-MOLECULAR WEIGHT' ASTHMA

In a thoughtful review, Agius *et al.* (1991) attempted to define the biochemical features of a number of recognized sensitizers which can be separated into several widely disparate groups. These agents, at first glance, have no common threads linking their ability to induce asthma. Yet, there are several common themes within the groups and sometimes with overlapping groups regarding the chemical structure of these agents and their potential to provoke asthma.

First, binding sites between the metals and the ligand (molecules which share electrons with the metal) are most relevant. Platinum, nickel, cobalt, vanadium and chromium are metals which exist as complex salts, with ammonium hexochloroplatinate $[(NH_4)_2PtCl_6]$ being perhaps the most prominent example (Cleare *et al.*, 1976). It appears that it is the nature of the ligand–metal complex which determines whether a metal is capable of sensitizing a worker. Ligands with multiple binding sites (e.g. porphyrins, triethylenetetramine) can form chelate rings which hold the metal at several sites, a feature which adds stability to the complex and enhances the ability of the molecule to sensitize. In addition, the charge of the ligand–metal complex is important. When skin tests were performed with platinum bound to a halogen (primarily chlorine but also bromide) molecule with an ionic charge, reactivity was much greater than in skin tests where platinum was bound to neutral or less reactive ligands. Therefore, in this instance, it is both the charge of the compound, as well as its ability to interact with both the metal and protein (perhaps a receptor on the mast cell surface or an IgE antibody) with a resulting stable ligand–protein complex, which appears to predict platinum's ability to sensitize. This may also explain the low order of cross-reactivity to the compounds closely related to platinum as the antigenic determinant formed by the conjugation of the platinum salt with the protein carrier is different from that formed by the conjugation with closely related palladium and rhodium salts.

Second, the chemical characteristics of a variety of sensitizing molecules vary, but all appear to be quite reactive. Amines and azo dyes are strongly basic, and bind quickly to agents which normalize the pH. The azo dye molecule often contains a chromophobic end (the azo group), hydrophilic groups to improve solubility, and a 'reactive end' such as a heterocyclic halogen grouping or an aliphatic chain with reactive groups. Since analogs of these dyes with just one reactive site are not sensitizers, two available reactive sites may be necessary to combine with human protein. The increased reactivity associated with two available sites (typically at opposite ends of the molecule) is a common theme as many of these lesser molecular weight molecules form bridges joining larger molecules. Glutaraldehyde has an aldehyde (-NCO) group at each end of the molecule. Formaldehyde (-HCHO), when in an aqueous solution, takes on similar characteristics. An agent such as styrene may act as a sensitizer because of the basic nature of the agent, as well as the fact that the agent exists as an unsaturated hydrocarbon (Seaton *et al.*, 1988). Similarly, colophony, acrylates and cyanoacrylates are sensitizers which possess unsaturated double bonds, and are therefore quite reactive. When one addresses the explanation for some of the reactivity of abeitic acid, the sensitizing agent in colophony, hydrogenation of its unsaturated double bond lessens its reactivity and reduces contact allergy (Burge, 1984; Karsberg *et al.*, 1988). Yet Burge *et al.* (1980) have also noted that the unsaturated double bonds of this resin acid may be less important for protein binding than the carboxyl groups. The acid anhydrides, with the prototype being phthalic anhydride, are also sensitizers. They bind covalently to free amino groups, in particular to lysine residues of protein in the lung to form stable anhydride–protein complexes. These are able to stimulate both specific immunoglobulin production and responsive T cells (Zeiss *et al.*, 1977; Newman-Taylor *et al.*, 1987; Venables, 1989).

Wass *et al.* (1990) had limited success in predicting which agents might sensitize workers by addressing whether the agent reacts with proteins (or peptides). An *in vitro* system was designed to simulate *in vivo* protein interaction with sensitizing chemicals. Liquid chromatographic techniques showed that lysine-containing peptides reacted with different isocyanates, anhydrides and chloramine-T [a disinfectant of small molecular weight shown to induce sensitization (Kramps *et al.*, 1981)]. Metals were not tested while other recognized sensitizers such as amines, aldehydes and dyes failed to react.

IMMUNOLOGIC MECHANISMS OF 'LOW-MOLECULAR WEIGHT' ASTHMA

Occupational asthma shares many characteristics with IgE-mediated asthma: in both, the responsible agent is known, the clinical presentation and the response to inhalation challenge in the laboratory is identical, and the response to anti-asthma drugs is similar. The mechanisms of occupational asthma due to low-molecular weight agents are not straightforward and in many instances it is not possible to demonstrate an IgE-mediated mechanism (Chan-Yeung and Malo, 1995). Although non-atopic asthma has features of chronic cell-mediated disease which occur independently of an IgE-mediated immunologic mechanism, activated T lymphocytes, metachromatic cells and eosinophils are increased in the bronchial mucosa of both atopic and non-atopic asthmatics, as well as in subjects with occupational asthma induced by low-molecular weight agents, suggesting similar, probably immunologically mediated mechanisms for both non-occupational and occupational asthma (Saetta *et al.*, 1992; Boulet *et al.*, 1994; Frew *et al.*, 1995).

After activation by antigen, T lymphocytes secrete a number of lymphokines which attract, activate and promote the growth and the differentiation of other leukocytes. Based on murine studies, T helper (Th) clones may be subdivided on the basis of their cytokine profile (Romagnani, 1995). For example, Th1 clones produce interleukin-2 (IL-2) and interferon-gamma (IFNγ), but not interleukin-4 (IL-4) and interleukin-5 (IL-5), whereas Th2 clones produce IL-4, IL-5 and granulocyte-macrophage colony stimulating factor (GM-CSF). *In vivo* studies in humans confirm the existence of functionally distinct subsets of T lymphocytes (Romagnani, 1995).

In addition to their role as helper cells for the production of humoral antibodies by B cells, activated CD4+ lymphocytes may be considered to be inflammatory cells. Interleukin-8 (IL-8), a chemotactic cytokine for polymorphonuclear leukocytes, is produced by macrophages and lymphocytes (Maestrelli *et al.*, 1988), as well as stromal cells, such as epithelial and endothelial cells as a result of cytokine networking between these cells and cytokine-producing T cells. They are also an important source of the GM-CSF and IL-5. The first is important in eosinophil development and activation and plays a significant role in the amplification of eosinophilic inflammation. IL-5 appears to be specific in promoting the development, activation and survival of eosinophils, and it is the predominant eosinophil-active lymphokine present in bronchoalveolar lavage (BAL) fluids during allergen-induced late phase inflammation (Kita *et al.*, 1993).

Activated T cells may therefore initiate and propagate allergic inflammation in the airways and participate directly in the events responsible for asthma exacerbations (Robinson *et al.*, 1993a,b; Corrigan and Kay, 1990; Corrigan *et al.*, 1993). However, a number of cells produce IL-3, IL-4, IL-5, GM-CSF and other relevant cytokines (Kay *et al.*, 1991; Bradding *et al.*, 1992). T-lymphocyte activation and local accumulation of activated eosinophils in the bronchial wall occurs in asthma of diverse severity and etiology, i.e. atopic, non-atopic and occupational asthma of low-molecular weight agents (Azzawi *et al.*, 1990; Bradley *et al.*, 1991; Bentley *et al.*, 1992a,b).

Different subsets of CD8+ T cells can participate in B-cell suppression or act as cytotoxic cells for exogenous and endogenous antigens. These cells are MHC class I restricted, but can respond to soluble exogenous antigens or haptens (Kalish and Morimoto, 1988; Walker *et al.*, 1991).

Antibody-mediated hypersensitivity

Most studies in occupational asthma have concentrated on antibody-mediated immunity, i.e. skin-prick tests and circulating specific antibodies against extracts or conjugates of occupational sensitizers. Positive skin-prick tests and measurable levels of specific IgE and/or IgG antibodies are present mainly in atopic subjects sensitized to high-molecular weight sensitizing agents (Rasanen *et al.*, 1994), but also in some subjects sensitized to low-molecular weight sensitizers (e.g. anhydrides, platinum salts, nickel, plicatic acid, isocyanates) (Maccia, *et al.*, 1976; Biagini *et al.*, 1985; Chan-Yeung and Malo, 1994; Herd and Bernstein, 1994). In workers with asthma attributable to high-molecular weight compounds, positive skin tests and/or specific antibodies to occupational allergens are often present in subjects both with and without symptoms of asthma, suggesting that their presence reflects exposure rather than disease. These workers require monitoring with the aim of confirming sensitization. In occupational asthma due to a low-molecular weight compound such as isocyanates, specific IgG antibodies may be more important than IgE antibodies (Cartier *et al.*, 1989). Again, these antibodies may be present in some subjects with no history of asthma and a negative inhalation challenge (Frew *et al.*, 1993). Like isocyanate-induced asthma, specific antibodies against plicatic acid and against morphine have been found both in symptomatic and asymptomatic workers exposed to red cedar or to morphine, respectively (Chan-Yeung, 1994; Biagini *et al.*, 1990; Gorski and Olinski, 1996). Immunologic testing, without clinical correlation, is insufficient to make an accurate diagnosis of occupational asthma.

Cell-mediated hypersensitivity

Until recently, lymphocytes were considered to play a role in IgE-mediated immunity mainly through the induction and regulation of IgE by B lymphocytes (Holt, 1991; Harding, 1993). However, T lymphocytes may release potent cytokines which recruit and activate other inflammatory cells, and thereby directly cause inflammation, suggesting that T lymphocytes act as effector cells in atopic asthma through pathways distinct from B-cell regulation and IgE production. T lymphocytes are capable of modulating eosinophil adherence, chemotaxis, survival and activation, and of stimulating these cells to cause tissue damage. Cell-mediated immunity and its role in airway inflammation has not been extensively examined in occupational asthma, yet, like in the airways in atopic asthmatics, an increased number of activated CD25+ T lymphocytes (lymphocytes expressing the IL-2 receptor), activated eosinophils and mast cells have been observed in patients with occupational asthma induced by toluene diisocyanate (TDI) and other small-molecular weight sensitizers, suggesting that similar immunologic mechanisms may be involved (Saetta *et al.*, 1992a; Boulet *et al.*, 1994; Frew *et al.*, 1995). Activated lymphocytes and eosinophils in bronchial biopsies suggest that a T lymphocyte–eosinophil interaction is important in asthma of different origins, a hypothesis further supported by the findings of cells expressing IL-5 messenger RNA in bronchial biopsies of atopic asthmatics. IL-5 is, in fact, the most important eosinophil-regulating cytokine, and its concentration in the airway mucosa of asthmatics correlates with markers of T lymphocyte and eosinophil activation (Hamid *et al.*, 1991).

There have been a number of reports of alterations in T cell subsets either in blood or in BAL fluid following specific bronchoprovocation, all supporting an immune-mediated mechanism for the development of airflow obstruction (Finotto *et al.*, 1991). Circulating CD8+ cells and eosinophils increase significantly at 48–72 hours after a specific inhalation challenge with TDI, at a time when airflow obstruction has resolved, suggesting that, in sensitized subjects, exposure to TDI not only causes airflow obstruction but also the recruitment and activation of lymphocytes and eosinophils, probably through an immunologic mechanism. Similarly, airway hyper-responsiveness to histamine increases 3 hours after initiating exposure the sensitizing agent in a worker already sensitized, and before the development of a late reaction (Durham *et al.*, 1987). Nickel-specific T lymphocyte clones have been isolated in subjects with nickel-induced asthma, suggesting that T-cell-mediated immunity may be involved. Cobalt-sensitized lymphocytes proliferating after incubation *in vitro* with cobalt have been found in subjects with hard metal asthma (Kusaka *et al.*, 1989).

Although an immunologic mechanism has not yet been firmly established, occupational asthma induced by TDI shares several features with allergic asthma. For example, as previously mentioned, circulating CD8+ lymphocytes and eosinophils increase significantly after exposure to TDI and remain increased even when FEV_1 returns to baseline. This observation suggests that the events triggered by exposure to TDI in sensitized subjects include changes in lung function and systemic effects which last longer than bronchoconstriction, and involve suppressor/cytotoxic lymphocytes and eosinophils, i.e. an immunologic mechanism (Finotto *et al.*, 1991). In addition, isocyanate-induced asthma and atopic asthma have a similar pattern of inflammatory cell infiltrate, including mast cells, eosinophils and mononuclear cells. Mononuclear cells are probably activated lymphocytes, as shown by the expression of IL-2 receptor on the cell surface.

Yet the processes do not appear to be identical. In atopic asthma, the majority of T cell clones derived from the atopic patients are CD4+ with a Th2 pattern of cytokine production. These lymphocytes, because of their pattern of cytokine production, may play a central role in determining the nature of inflammatory response seen in bronchial mucosa of atopic patients through the induction of both allergen-specific IgE (via IL-4) and eosinophilia (via IL-5). In patients with TDI-induced asthma, the majority of T cell clones are CD8+ and are capable of producing IL-5, a cytokine which attracts, activates and increases survival of eosinophils in the bronchial mucosa (Del Prete *et al.*, 1993; Maestrelli *et al.*, 1994a). Interestingly, an increased percentage of CD8+ T cells has also been reported in the BAL fluid of intrinsic non-atopic asthmatics (Walker *et al.*, 1992).

Airway inflammation

Asthma is characterized by the presence of chronic airway inflammation which transiently increases during exacerbations (O'Byrne *et al.*, 1987; Drazen and Turino, 1995; Shelhamer, 1995). This has been perhaps best studied by inhalation challenge with the agent in question, followed by fiberoptic bronchoscopy where BAL fluid was aspirated and bronchial biopsies were performed (Fabbri and Ciaccia, 1992). A less well-characterized approach has been developed with analysis of cells in sputum (Pin *et al.*, 1992; Fahy *et al.*, 1994).

In subjects with occupational asthma, BAL fluid obtained between attacks and after cessation of exposure when the worker has recovered shows minor changes in the total count or distribution of inflammatory cells compared with control values (Fabbri *et al.*, 1987; Chan-Yeung *et al.*, 1988; Paggiaro *et al.*, 1990). Increased numbers of inflammatory cells have been shown with this technique in subjects who fail to recover after the cessation of exposure and during asthma attacks induced by occupational agents and allergens (De Monchy *et al.*, 1985; Lam *et al.*, 1987).

BAL fluid obtained during or after a late asthmatic reaction induced by an occupational sensitizer shows a significant increase of inflammatory cells, i.e. neutrophils and eosinophils, and a marked increase in albumin concentration. By contrast, although early asthmatic reactions induced by occupational agents are associated with increased levels of histamine, leukotrienes and other mediators in BAL fluid (Chan-Yeung *et al.*, 1989), they are not followed by a marked increase of inflammatory cells. After challenge with isocyanates, leukotriene B_4 levels are increased, which may be partially responsible for leukocyte infiltration in the airways (Zocca *et al.*, 1990). Macrophages, epithelial cells or neutrophils may be the source of this mediator. In the supernatant of BAL fluid obtained during late asthmatic reactions, there is an increased concentration of albumin compared with the findings obtained from controls or during an early asthmatic reaction. The increased concentration of albumin may reflect microvascular leakage and mucosal edema (Persson *et al.*, 1995b). Macromolecules and albumin leak between the gaps of endothelial cells in the postcapillary venule. They cross two membranes, the endothelial and the epithelial basement membrane, and accumulate in the airway lumen. The process of plasma exudation is associated with movement of water which causes mucosal edema and accumulation of exudate in the airway lumen. The plasma exudative process is not necessarily associated with airway edema, epithelial disruption or increased mucosal absorption, but rather it increases the hydrostatic pressure in the subepithelial space creating pathways for its luminal entry (Persson *et al.*, 1995a). If transient edema is induced by airway plasma exudation, it is unlikely to compromise airway function. When the exudative response is massive and prolonged, physical alterations of the wall and the lumen occur which may adversely affect the airways.

Like subjects who die of non-occupational asthma, subjects with occupational asthma who die after occupational re-exposure to the sensitizing agent have acutely inflamed airways with massive edema and exudate in the airway lumen (Fabbri *et al.*, 1988; Ehrlich, 1994; Carino *et al.*, 1997). In a sensitized subject who died in the workplace after re-exposure to isocyanates, histologic examination of his airways showed denudation of airway epithelium and thickening of the basement membrane with infiltration of the lamina propria by leukocytes (mainly eosinophils), and diffuse mucus plugging of bronchioles. Bronchial smooth muscle appeared hyperplastic and disarrayed, and the lung parenchyma showed focal areas of alveolar destruction adjacent to areas of intact alveolar walls. It is likely that bronchial walls thickened by persistent edema or by an increased amount of subepithelial collagen decrease the lumen diameter induced by a similar degree of smooth muscle shortening and provoke further clinical impairment.

In bronchial biopsies obtained from subjects with TDI-induced asthma, inflammatory cells are compartmentalized in the airway mucosa. Whereas eosinophils are increased in the epithelium and submucosa, CD45+ cells (mononuclear cells account for most of the CD45+ cells) are significantly increased only in the epithelium and in the most superficial layer of the submucosa, and mast cells are increased mainly in the epithelium (Saetta *et al.*, 1992a). In addition, there is persistent activation of these lymphocytes with the clonic expression of proinflammatory cytokines, such as tumor necrosis factor-alpha (TNFα) and interleukin-1β (IL-1β) (Maestrelli *et al.*, 1995a).

Although there is no difference between the percentage of the basement membrane covered by epithelial cells or in the width of the intercellular spaces between columnar cells in healthy asthmatics and controls, the intercellular spaces between the basal cells are wider in asthmatics (Saetta *et al.*, 1992a). As the attachment of the majority of the columnar cells to the basement membrane is mediated through basal cells, the wider spaces between basal cells may adversely affect adhesion of columnar cells to the basement membrane, contributing to the influx into epithelium of inflammatory cells and to the induction of epithelial fragility, as observed in non-occupational asthma (Laitinen *et al.*, 1985).

Asthmatics also exhibit a thicker subepithelial reticular layer of the basement membrane, whereas the width of the true basement membrane is similar in asthmatics and controls (Beasley *et al.*, 1989; Saetta *et al.*, 1989). This finding is a histopathologic feature of asthma, since in subjects with chronic bronchitis, a disease which causes airway inflammation and activation of T lymphocytes as does asthma, the thickness of the basement membrane does not differ from that in controls (Saetta *et al.*, 1993). Since a thickened reticular layer of the basement membrane has also been described in young asthmatics with mild asthma (Beasley *et al.*, 1989), it is likely that this histologic feature represents an early change of the disease.

Cessation of the exposure to an occupational sensitizer for approximately 6 months allows for the thickness of the reticular layer of the basement membrane to lessen, although the inflammatory cell infiltrate persists (Saetta *et al.*, 1992b). After a longer time away from exposure to the sensitizing agent (6–21 months), the subepithelial fibrosis, the number of mast cells and fibroblasts, and the sensitivity to sensitizing low-molecular weight agent is significantly reduced, yet the number of eosinophils and mononuclear cells and the degree of airway responsiveness to methacholine does not change (Saetta *et al.*, 1995). The association between the thickness of the reticular layer of the basement membrane and the number of mast cells and fibroblasts, suggests a role for these resident cells in the development of subepithelial fibrosis.

Interestingly, subjects who develop occupational asthma after a short period of exposure to TDI (average 2 years) have more mast cells in the airway mucosa than subjects who develop respiratory symptoms after a long period of exposure to TDI (average 22 years) (DiStefano *et al.*, 1993). This finding could be a 'predisposing factor' for a rapid onset of the disease, or the effect of different levels of exposure to the sensitizing agent.

Examination of induced sputum, although less well tested than evaluation of BAL or bronchial wall histology, has an advantage as it is a non-invasive method to investigate inflammatory cells in the airways of asthmatic subjects. The relative proportion of inflammatory cells is different in the samples obtained by BAL, bronchial biopsies and sputum. However, there is a fairly good agreement between the results of eosinophil counts with the three techniques (Maestrelli *et al.*, 1995b).

Asthmatic reactions induced by isocyanates in sensitized subjects are associated with a significant increase of eosinophils in induced sputum, a feature which lasts up to 24 hours after the exposure to isocyanates. Early, as well late, asthmatic reactions induced by isocyanates are associated with the increased number of sputum eosinophils (but not other intra-epithelial cells), indicating that development of eosinophilic inflammation occurs independently of the pattern of bronchoconstriction (Maestrelli *et al.*, 1994b). The time course of eosinophil accumulation in the sputum is compatible with the time necessary to release adhesion molecules and/or initiate cytokine activity.

In summary, with isocyanates frequently used as a model, there is increasing evidence that immunologic mechanisms play a role in the development of occupational asthma of some low-molecular weight agents. Additional work will be needed to prove that these same mechanisms are a part of the pathogenesis of sensitization and asthma of a number of other agents.

ASTHMA DUE TO SPECIFIC LOW-MOLECULAR WEIGHT AGENTS

An author is faced with difficult choices when describing the low-molecular agents which are encountered in the workplace and which have been shown to induce asthma. There are a great number of agents spanning a number of chemical groups. It is not possible to address all of the agents, but we have attempted to present an important 'flavor' of the breadth of the problem. Additional information regarding other agents which are only occasionally reported to induce asthma may be found in other sources, beginning with Table 23.1. Increasingly, the internet has become a very helpful source in learning about the effects of less frequently encountered chemicals.

Platinum

Once platinum is mined, it is ground to liberate mineral particles. The particles are recovered by flotation techniques and the concentrate is melted in an electric furnace. This smelting process yields a purer platinum product which is further separated from contaminating metals by leaching and/or magnetic separation and then refined. The process of refining involves precipitating platinum in the form of one of its complex salts, with ammonium tetrachloroplatinate [$(NH_4)_2PtCl_4$] and ammonium hexachloroplatinate [$(NH_4)_2PtCl_6$] being most commonly formed. Platinum and its salts are also encountered in jewelry making, electroplating, photography and in the manufacture of catalytic converters. These agents are very potent sensitizers, and studies from the 1970s have recognized a significant number of those exposed become sensitized.

Although Karasek and Karasek (1911) reported eight cases of profound irritation of the throat and nasal passages with skin cracking, bleeding and pain among workers in Chicago photo studios where a paper containing potassium chloroplatinate was used, it was not until Hunter *et al.* (1945) recorded the potent sensitizing effect of platinum with its associated asthma and dermatitis in a population-based report that the risk was better understood. They described the respiratory health of workers from four United Kingdom platinum refineries where workers were exposed to platinum complex salts as a dry dust or wet spray. Co-workers exposed to metallic platinum or complex salts of other precious metals such as palladium were not affected. Of 114 workers who participated, 91 were in contact with platinum salts. Of these, 52 had asthma, 13 had dermatitis and 43 had eosinophilia.

In another of the initial studies, Roberts (1951) observed 20 employees of a platinum works over a 5-year period. Although all developed scratch tests to platinum chloroplatinate, five of the 19 (26%) examined workers had overt asthma and another eight (42%) complained of chest tightness. Pepys *et al.*, (1972) described 16 platinum refinery workers with respiratory disease who were tested with solutions of complex salts of platinum. Ten of 16 gave immediate prick test reactions and seven of 11 showed an immediate nasal reaction to inhaled platinum. Laboratory-based inhalation tests performed in all 16 more than a year after the last exposure showed an immediate asthmatic reaction in seven, a dual reaction in one, and a late reaction in two. All but the two late reactors had an immediate prick test reaction. In all of the nine workers challenged with $(NH_4)_2PtCl_6$ and Na_2PtCL_6, the response was greater to the former complex. Calverley *et al.* (1995) later showed that workers with a positive skin test to platinum invariably developed asthma if they continued exposure.

Unlike some of the other small-molecular weight agents, atopy appears to be an important contributing factor in the development of asthma to this agent. This can be demonstrated by skin testing with the platinum salt and determination of specific IgE antibodies by RAST. The mechanism of platinum allergy was first recognized when Cromwell *et al.* (1979) showed that specific IgE antibodies to platinum were present in individuals with platinum sensitivity. Pepys *et al.* (1979) later showed that passive transfer of serum containing antibody to the complex salts of platinum from workers sensitized to $(NH_4)_2PtCl_6$ or $(NH_4)_2PtCl_4$ to monkeys resulted in a positive skin test to platinum in the monkeys.

Murdoch *et al.* (1986) collected sera from 306 South African platinum refinery workers to measure IgE levels and the workers were skin tested with salts of palladium, platinum, rhodium, iridium and ruthenium. The intent was to explore the extent of cross-reactivity to platinum with the other salts of the metals in this group – agents recognized to be less likely sensitizers. The question of cross-reactivity arose because of the complexity of the processes in the platinum refinery, where workers are exposed to more than one of the platinum group metal salts (Murdoch and Pepys, 1987). For example, platinum refinery workers are known to be exposed to comparable levels of salts of palladium and rhodium. No worker described asthma symptoms to common environmental agents at the time of hire. Twenty-eight workers developed skin test reactivity to the platinum salts, while 10 others reacted to the other metals singly and in combination, features implying minimal cross-reactivity.

Venables *et al.* (1989) concluded that smokers are at increased risk of sensitization by platinum salts. They studied 91 workers who started work in 1973 and 1974 and whose smoking habit and atopic state was documented at the time of hire. Skin-prick tests with several complex platinum salts were performed every 3–6 months and respiratory symptoms reported during the years of employment. Follow-up was until 1980 or until leaving refinery work, whichever occurred first. Results of the data reflect the very high sensitization risk associated with platinum exposure during the mid-1970s. [Ruff *et al.* (1979) estimated that 5–10% of platinum refinery workers each year became allergic to complex salts of platinum.] The risk for sensitization was greatest in the first year of exposure and declined with time. The risk of developing a positive skin test was four-to five-fold increased in smokers, and was more predictive than the presence of atopy for positive skin test. The number of cigarettes smoked per day, and not the presence of a positive skin tests to platinum, was the only significant predictor of respiratory symptoms. Similar results were reported by Niezborala and Garnier (1996). Perhaps the increased mucosal permeability induced by cigarette smoke allows antigen to cross the bronchial mucosa (Zetterstrom *et al.*, 1985). In this same cohort, Dally *et al.* (1980) showed that in those who became sensitive to platinum, conversion to a positive skin-prick test to soluble platinum salts was an important disease marker, but quite variable in its time course. Skin test conversion ranged from 6 months before to 14 months after the development of chest symptoms.

In summary, the salts of platinum are potent sensitizers, which unlike other small-molecular weight agents, appear to have IgE-mediated allergy and asthma. One part of the serial assessment of these workers is periodic skin testing with appropriate antigen. This appears to be important in identifying the population of workers at risk for developing platinum-induced asthma. Promptly removing such workers from the workplace is an important way to protect the worker's health. However, the chance for recovery once the worker becomes sensitized and develops overt asthma appears to be poor. Merget *et al.* (1994) showed that of 24 workers with platinum asthma, 17 continued to have both non-specific and specific airway hyper-responsiveness when re-evaluated from 1 to 77 months (mean 19 months) after removal from all exposure.

Colophony

Burge *et al.* (1978) noted that the soldering process has been in use since ancient times and had been described by Pliny in 77 AD. Ninth century reports describe a soldering cream, a mixture of two parts of axle grease and one part pine resin (also called rosin) with an equal dose of tin fillings. Soldering fluxes are needed to stop the oxidation of the heated metals at the solder site and to aid in the melting and spreading of the solder. Colophony, the solid material left after turpentine has been distilled from pine resin, has been the basis of soldering fluxes ever since and, beginning around 1930, additives have been added to the flux to hasten its action.

Although the 'typical' means of sensitization is through inhalation of heated fumes of colophony [containing the sensitizing agent 'abeitic acid' ($C_{20}H_{30}O_2$)], asthma can also occur after inhalation of untreated (and unheated) colophony as a dust (Burge *et al.*, 1986), to emulsified oil mist used to cool a cutting machine (in this instance, pine resin was a component of the coolant) (Hendy *et al.*, 1985), and to heated 'tall oil', an agent containing abeitic acid which has been used to coat strips of rubber in a tire plant (Tarlo, 1992). In other instances, the manufacture of solder is associated with sensitization even when the pine resin is heated to only 140°C, a temperature insufficient to cause resin breakdown products (Burge *et al.*, 1981). Furthermore, some workers sensitized to colophony complain of episodic asthma when walking among pine trees (Burge *et al.*, 1978). Heating the solder flux does not appear necessary to develop sensitization. A more complete discussion is provided by Burge *et al.* (1980).

The characteristics of colophony used in the soldering flux are dependent on the manner of extraction, as well as its country of origin, with rosins from different areas of the world having somewhat different properties (Smith *et al.*, 1996). Burge *et al.* (1980) showed that in 12 sensitized workers, Portuguese Y colophony caused slightly larger reactions than American WW colophony with similar exposures.

Specific antibodies to abeitic acid in the serum of workers sensitized to colophony have not been found and positive skin prick tests have not been reported. Although relatively little is known about the immunologic effects of exposure to abeitic acid, colophony can be immunogenic (Cullen *et al.*, 1992). Mice injected with abeitic acid bound to bovine serum albumin or human IgG produced antibodies to these agents. Unconjugated abeitic acid failed to produce specific antibody responses or dermal sensitization in mice. Spleen cells from the immunized mice proliferated when exposed to either of the conjugates.

The first report of challenge testing to colophony to prove the relationship between exposure to heated fumes of this agent and occupational asthma, was described by Fawcett *et al.* (1976). Here, four challenged workers gave immediate bronchial reactions to inhalation of the fumes, with exposures varying from one breath to 3 minutes of exposure. Two of the workers demonstrated immediate-type responses to bronchial challenge, one a prolonged (24-hour) response, and one a dual response. Burge *et al.* (1978) later proved asthma in 21 electronics industry workers with workplace respiratory complaints who were exposed to solder flux fumes containing colophony.

In the 1970s, the prevalence of asthma due to colophony exposure in English electronics factories was high. A study of respiratory symptoms in a factory mass-producing consumer electronics showed that 22% of the solders (compared with 6% of the office staff) had work-related wheeze or dyspnea or both. Although the levels of solder flux were reported to be below the threshold limit value during the survey, the authors postulated that previous exposures were excessive and led to the development of respiratory complaints (Burge *et al.*, 1979; Perks *et al.*, 1979). There is evidence that the figure would be greater, had not there been high labor turnover, as the mean duration from hire to sensitization was 4 years, and most workers had left the workplace by then. It should be noted that some of the convincing evidence leading to the use of a peak flow meter to aid in the diagnosis of occupational asthma came from the evaluation of these workers (Burge *et al.*, 1979).

Burge (1982) followed 39 electronic workers 1 to 4 years after being evaluated for asthma suspected to be due to exposure to colophony. The most important conclusion was the recognition that of the 20 with a diagnosis of occupational asthma attributable to colophony, only two were symptom-free after leaving the workplace, and most had a considerable decline in their quality of life following the diagnosis of asthma, a course recognized frequently among those with asthma attributable to low-molecular weight agents (Chan-Yeung and Malo, 1995).

Dyes

Reactive dyes are most important in industry, but a number of these agents are also used in personal services. In 1978, when the first report of allergy due to reactive dyes was reported in industry, azo-based reactive cotton dyes had been in use for approximately 20 years and more than 100 dyes had been developed. Their use had been welcomed in industry and they were widely used. Because of the ability of these dyes to produce strong covalent bonds to cellulose, protein and other textile fibers, they were colorfast and had the additional property of brightness.

In 1978, four workers who had weighed dyes for more than 2 years developed asthma and allergic rhinitis. Nasal and bronchial provocation tests produced immediate reactions (Alanko *et al.*, 1978). A high serum titer of specific IgE was demonstrated for at least one dye, indicating that the mechanism of hypersensitivity is immunologic, with reactive dyes apparently acting as haptens combining with body proteins which become allergens. The complexity of the immunologic features was well shown by Wass *et al.* (1990) when they demonstrated that RAST using directly dyed disks (i.e. disks without a carrier protein) provided an unreliable measurement of specific IgE in workers with reactive dye asthma, while disks with dye bound to human serum albumin showed very effective binding in workers. Interestingly, in one of the three workers tested, IgE was hapten-specific but not carrier-specific, in a second case, IgE was both carrier-specific and hapten-specific, and in the third case, intermediate between the two. This suggests that the interaction between the dye and the protein results in a new antigen which may be specific for each dye.

A population-based study in Sweden showed that 10 of 162 (6%) workers in dye houses reported work-related nasal or respiratory symptoms (Nilsson *et al.*, 1993). Although four of the five workers who underwent RAST were positive to a black dye recognized by these investigators to be a common sensitizer, cross-reactivity occurred between the dyes.

Docker *et al.* (1987) reported the health of a population of workers at dyehouses in Manchester and Leicester, in a study stimulated by the death of a worker with reactive dye asthma who was re-exposed to dyes. Respiratory questionnaires administered to 414 workers exposed to reactive dyes revealed that 53 had work-related respiratory or nasal symptoms. In the 49 workers with symptoms who were clinically evaluated, 21 had symptoms attributed to the reactive dyes (5% of the workforce) (18 of these 21 were atopic and 17 had serum IgE to dye–human serum albumin conjugate). Interestingly, half of the specific IgE measurements were to yellow and/or orange dye conjugates. Two patterns of airways response were described by the workers: the first, an immediate response of short duration, and the second, a longer-lasting response, usually of several hours, sometimes accompanied by nocturnal asthma.

In addition to the dyes reported above, there has been a series of selected cases reported where one or just a few individuals are described to have occupational asthma in association with an exposure to a dye. In cases where exposures to dyes are suspected of inducing asthma, a detailed review of the literature (beyond the scope of this chapter) would probably be in order to identify individual episodes of dye-related asthma. Examples of several instances are reported below.

Carmine is a natural red dye widely used as a food coloring agent, in cosmetics and as a histologic stain. It is extracted from the dried females of the insect *Dactylopius coccus* var. *Costa* (cochineal). These insects arc parasites which reside on cacti in the Americas and on the Canary Islands of Spain. Quirce *et al.* (1994) studied 10 current and former employees at a factory making natural dyes. Three workers with respiratory symptoms had positive skin-prick test reactions to both carmine and cochineal. Another employee had work-related symptoms of rhinitis and asthma attributable to the dye confirmed by bronchial provocation tests and specific IgE antibody to carmine and cochineal by RAST. Specific IgG antibodies against carmine and cochineal were found in the 10 subjects surveyed. These findings suggest that carmine may induce immunologic responses in workers with occupational asthma.

Pepys *et al.* (1976) reported a non-atopic hairdresser with immediate skin and bronchial reactions to an extract of henna dye, in addition to sensitivity to a persulphate-containing bleach in the form of a non-immediate episode of asthma following challenge. Henna is a dye isolated from the plant *Lawsonia inermis*, which is indigenous to Africa and Asia. The relationship between the development of asthma and the use of these two agents was not understood.

Miller *et al.* (1996) described a 55-year-old man who developed immediate-onset asthma and rhinitis immediately upon exposure to FD&C Blue Dye No. 2 (Indigotine) at work. He had worked for 10 years mixing and grinding a number of powdered synthetic dyes for use in foods. His symptoms had begun 2 years before evaluation and occurred only following exposure to Indigotine ($C_{16}H_8N_2Na_2O_8S_2$), a free-flowing blue powder. Prick testing to this agent was negative and ELISA to Indigotine–HSA conjugate showed no specific IgE, IgA, IgM or IgG antibodies. Bronchial challenge to diluted Indigotine powder provoked an immediate episode of asthma.

Finally, although the incentive to develop reactive dyes for wool has been less than for cotton-based materials (because of the need for a wide variety of pre-existing colors and rapid staining and drying), in 1966, dyes containing an acyl-amine reactive group which reacts to the protein constituent of wool fibers were introduced. Topping *et al.* (1989) reported work-related respiratory symptoms associated with exposure to these dyes in six employees in wool dyehouses. Five subjects had specific IgE to human serum albumin conjugates of one or more of the dyes, providing evidence of sensitization. In two subjects there was an association between symptoms to a particular dye and specific IgE to an albumin conjugate of that dye. Specific IgG antibodies were found in exposed subjects, irrespective of allergic symptoms, indicating that specific IgG reflects exposure rather than clinical sensitization. The relatively fewer number of cases of asthma in with exposure to wool dyes compared with cotton dyes led investigators to suggest that wool dyes might be relatively safer or that exposures might be less in this industry.

Amines

Over 40 different amines have been recognized to induce occupational asthma (Malo and Bernstein, 1993). A number of these are described below. Although there is an occasional instance where a workplace population with amine exposure has been identified and a prevalence of sensitization reported, for the most part, information regarding the asthmagenic properties of these agents are related on a case-by-case basis, without clear insight into disease mechanisms.

Silberman and Sorrell (1959) studied 80 referred workers during the years 1940 to 1955 who were employed in the fur industry with exposure to the dye paraphenylethylenediamine (PD) with skin or respiratory complaints (or both). Asthma occurred as the sole symptom in 30 (37%) and appeared alone or in combination with other allergic symptoms in 59 (74%). Thirty-one of 47 (65%) had positive patch tests to PD. These agents provoked asthma in 37/50 (74%) subjects, with 13 (26%) showing no airway hyper-responsiveness despite a history consistent with asthma to this agent. About two-thirds of the workers studied required 5 or more years of exposure to the dye before symptoms began to appear.

Soon after, Sterling (1967) – in one of the earliest reports of amine-induced asthma – described two patients with recent onset of late-onset dyspnea after inhalation of fumes from a new soldering flux recently developed for use in joining aluminum. Inhalation of one constituent of the flux, amino-ethyl ethanolamine, caused the same response as inhalation of the fumes from the soldering flux.

Of the numerous amines capable of inducing asthma, ethylenediamine (EDA), (NH_2CH_2-CH_2NH_2, molecular weight of 60) appears to be one of the most recognized, as it is widely used in the rubber, lacquer, shellac and beauty culture industries as a solvent and as an accelerator. Information of respiratory illness attributable to this agent is primarily available as a series of case reports, and systematic surveys of working populations with exposure to this agent are generally unavailable. Lam and Chan-Yeung (1980) showed amines to be capable of inducing asthma in non-atopic subjects. They described a 30-year-old man who developed asthma following exposures to a number of other chemicals used in developing color photographs for the past two-and-a-half years. Following exposures to the individual agents used in developing film, the worker developed a specific and reproducible late asthmatic reaction to EDA only. Results of skin tests to this agent were negative, and serum antibodies to this chemical were not found.

Nakazawa and Matsui (1990) reported two non-atopic chemical plant workers with EDA-induced asthma. The first was employed in a factory which synthesized EDA, while the second handled this agent. Inhalation challenge testing showed a late reaction in both. Although IgG and IgE antibodies and patch tests were negative in both, a positive reaction occurred when sera from the two patients were injected intracutaneously into a healthy host (a positive Prausnitz–Kustner reaction). Both recovered after transfers from EDA exposures. Vallieres *et al.* (1977) reported progressively severe sneezing, rhinorrhea, cough, wheezing and dyspnea in a spray-painter in relation to exposure to a particular spray paint. Laboratory-based challenge testing resulted in a dual asthmatic response to the whole paint (consisting of a 98% methyl methacrylate emulsion and 2% dimethyl ethanolamine solution), to a 2% solution of dimethyl ethanolamine alone, but not to a solution of methyl methacrylate emulsion, used as a paint thinner. Allergy skin tests with dimethyl ethanolamine and a mixture of dimethyl ethanolamine and human serum albumin were negative. Like many of the singular cases reported to induce asthma, the mechanism of this specific reactivity is not known.

Chloramine-T, a powder with a molecular weight of 281, is a water-soluble powder, which, when mixed with water, liberates chlorine. It has long been recognized as an irritant and has been used for more than 50 years as an effective sterilizing agent in the food and drink industry, as a water disinfectant, and as a topical antiseptic. Feinburg and Watson (1945) found occupational asthma and rhinitis in workers exposed to this agent based on wheal and flare reactions. Bourne *et al.* (1979) proved evidence that asthma due to this agent was immunologically mediated, when he reported seven brewery workers who developed asthma after using chloramine powder in the workplace. All had wheal and flare reactions to skin-prick tests of solutions of chloramine and an increased total serum IgE concentration, regardless of their atopic status. Ceasing exposure allowed for the resolution of symptoms and laboratory abnormalities. Kujala *et al.* (1995) recently reported additional evidence for immunologic mechanisms of chloramine-T-induced asthma. They described a 36-year-old woman who developed asthma with specific IgE antibodies to chloramine-T within several months of beginning a job which required her to aerosolize this agent to clean the municipal swimming pool.

Another amine, hexachlorophene (molecular weight 409), an agent which had been used as an antiseptic as either the powder or lotion, was shown by inhalation testing to induce asthma in a non-atopic children's nurse who had contact with hexachlorophene for 15 years (Nagy and Orosz, 1984). She had had symptoms of rhinitis for 10 years and reported symptoms of asthma upon immediate exposure to this agent for the past 2 years. Skin tests with hexachlorophene solution showed no effect, while non-specific bronchial reactivity to five breaths of inhaled acetylcholine was increased. The worker developed an immediate-type response to hexachlorophene with inhalation challenge testing.

Aldehydes

A number of cleaning fluids have been recognized to be irritants for many years, and more recently considered to be sensitizers which induce asthma. In addition to the amine-based liquids reported above, two aldehydes – formaldehyde and glutaraldehyde – have been reported in the past 20 years to be associated with asthma. Like many of the low-molecular weight agents which are thought to induce asthma, large-scale studies showing important numbers of cases in the workplace are absent. Again, the reports are anecdotal or case studies, often with less than optimal proof of diagnosis, implying that even though the exposures are widespread and frequently encountered by a large number of people, the association with asthma was quite infrequent.

(a) Formaldehyde

Exposure to formaldehyde is ubiquitous and, over the past 20 years, concerns have arisen regarding the effects of low-level exposures; in particular, the role of this agent as a respiratory irritant, as an inducer of workplace asthma, and as an irritant and potential sensitizer in homes with urea–formaldehyde insulation. The greatest exposures occur in the workplace (formaldehyde is a commonly used sterilizing agent with potentially important exposures in embalmers, health care workers who are employed in sterilizing material, and factory workers), with lesser exposures associated with the 'off-gassing' of fumes, i.e. the release of fumes in association with newly manufactured products. These latter exposures are typically a small fraction of industrial exposures. Most manufactured formaldehyde is used in the production of phenolic, urea, amelamine and acetal resins. These resins are then used extensively in the manufacture of textiles, floor covering, plywood, particle board, embalming fluid and some varieties of usual and carbon-less paper. Formaldehyde has a molecular weight of 30. It is very reactive and, for that reason, is marketed as formalin, a mixture of formaldehyde and alcohol. This combination keeps the formaldehyde from polymerizing.

Hendrick and Lane (1975) first used inhalation provocation tests to document the role of inhaled formalin fumes in the development of airways obstruction in two hospital staff members continually exposed to formalin in the course of their work. The first, a nurse who worked in the dialysis unit, had a late response to formalin, while the second, a pathologist with childhood asthma who complained of worsening breathing after a day's exposure to this agent, had no response. These investigators later showed that eight of 28 employees who worked around formalin in the dialysis unit reported bronchitis and five reported wheeze. Of these five, two nurses were proven to have late-onset asthma following formalin inhalation (Hendrick and Lane, 1977). Re-testing of these two nurses 5 years later (Hendrick *et al.*, 1982), showed that the first nurse, who had ceased formalin exposure soon after her initial complaints, had no additional respiratory complaints nor decline in lung function upon re-challenge. The second nurse, who was initially reactive to formaldehyde, and who had only had intermittent exposures to formaldehyde since the initial diagnosis, was again reactive, as a 5-minute exposure to 3.0 parts per million (ppm) of formaldehyde provoked a late reaction. In this case, proving a late episode of asthma provides convincing evidence for a sensitizing effect; however, it should be recognized that separating the irritant effect from the sensitizing effect can be difficult when an irritant effect can occur as low as 1.0 ppm (Bardana and Montanaro, 1991; National Research Council, 1981).

Over 6½ years, Nordman *et al.* (1985) reported on a select group of 230 individuals who were occupationally exposed to formaldehyde with respiratory symptoms in Finland and referred for assessment of occupationally induced asthma. An exposure period ranging from 1 month to 19 years preceded the onset of symptoms. All were tested with an unblinded specific bronchial provocation test for 30 minutes with 1 or 2 ppm of formalin. When a 15% decline in peak flow was considered to be a positive test, 12 (4%) were recognized to be positive. Three of these 12 sensitized individuals displayed no bronchial hyper-reactivity to histamine or methacholine. Another 71 of the 218 subjects who failed to react to formalin demonstrated bronchial hyper-reactivity. Removal from exposure was associated with an improvement in respiratory symptoms.

Specific IgE antibodies to formaldehyde have been recognized in association with intravenous exposure during dialysis (i.e. the use of formaldehyde-sterilized intravenous tubing) (Bousquet and Michel, 1991) and have been linked to asthma in a small number of patients. Formaldehyde-specific IgE (and in some instances specific IgG and IgM) has been recognized in workers exposed to formaldehyde at work, but this has not been linked to respiratory disease (Wilhelmsson and Holmstrom, 1987; Kramps *et al.*, 1989; Dykewicz *et al.*, 1991). Patterson *et al.* (1989) showed that specific IgG to formaldehyde was present in several patients who underwent dialysis with formaldehyde-sterilized tubing, but was not apparent in workers who had a history of gaseous exposures to formaldehyde. In a number of other reports, antibodies in workers exposed to formaldehyde were not found. Although neither specific antibodies nor immunologic abnormalities have been found when exposure has been to low-levels of formaldehyde in adults (Pross *et al.*, 1987), Wantke *et al.* (1996) identified a population of children who have formed specific antibodies when exposed to relatively low levels of this agent.

The question becomes, then, how important is formaldehyde as a cause of allergic respiratory disease? First, separating an irritant effect from the sensitizing action has been difficult. This is an example where a workplace challenge may not be sufficient to separate the sensitizing effect from the irritant effect. Specific challenge testing appears to be best performed in the laboratory and must be well controlled with accurate measurement of the concentration of exposure to provide convincing evidence of a sensitizing effect. This remains a critical issue in proving attributability. Second, even at (relatively) high exposure levels, inhaled formaldehyde rarely induces a specific immune response; and the relationship between this immunologic response and respiratory symptoms is poor. Co-workers without respiratory symptoms may develop antibodies. To date, environmental exposure to formaldehyde has not been proven as an important cause of immunologically mediated respiratory disease in adults. Since children may show an increased tendency to develop specific antibodies after environmental exposure to formaldehyde, they may turn out to be at greatest risk, although the clinical significance of these antibodies remains uncertain (Smedley, 1996).

(b) Glutaraldehyde

Alkaline glutaraldehyde has broad disinfectant properties and is currently the most widely used disinfectant for cold sterilization of fiberoptic endoscopes. It is also used as a histologic fixative, a biocide in cooling towers, an agent which aids in tanning leather, and a hardener for X-ray films. Like formaldehyde, It has a very low odor threshold and it is a potent irritant, with a number of studies describing eye irritation, rhinitis and complaints of wheeze and chest tightness (Abbott, 1995).

There have been a succession of reports of workers who were exposed to glutaraldehyde at work, typically in the role of sterilizing flexible fiberoptic scopes, leading investigators to conclude that this behaves as a sensitizer. Axon *et al.* (1981) were the first to write about the skin and respiratory difficulties with glutaraldehyde when they described 36 workers (37% of all employed) in 16 endoscopy units with complaints. In a second report, eight of nine employees associated with employment in the endoscopy unit complained of skin irritation, eye watering and nasal congestion in work areas where a 2% glutaraldehyde solution was in use (Jachuck *et al.*, 1989). Corrado *et al.* (1986) evaluated four nurses working in endoscopy units who complained of respiratory symptoms on exposure to alkaline glutaradehyde which was used as a disinfectant for fiberoptic endoscopes. Although an irritant effect could not be completely discounted, a provocation test was positive in two nurses showing a relation between exposure to glutaraldehyde and respiratory disease. All four nurses were atopic by history and three of the four had positive skin-prick test reactions to common allergens.

Chan-Yeung *et al.* (1993) documented asthma in an atopic respiratory technologist who had worsening of childhood asthma following employment in a bronchoscopy unit. In addition to helping with the procedures, it was her job to clean the bronchoscopes with Sporicidin® (Sporicidin International Ltd., Rockville, Md.), a commercial cleaning agent containing 3.6% glutaraldehyde, 7% phenol and 1.2% sodium phenolate. Workplace challenge testing showed a clinically significant decrement in FEV_1 following exposure to the commercial agent. The authors noted that the specific component causing asthma was likely glutaraldehyde, although tests with individual components of the commercial product were not undertaken. In addition, because levels of glutaraldehyde were not measured, it was not possible to determine whether the worker had been sensitized *de novo* to Sporicidin®, or whether this was asthma following an irritant exposure in an individual with underlying asthma. A worker employed in a similar role developed asthma with a 2% solution of glutaradehyde (Benson, 1984).

As we alluded to in testing for inhalational allergy to formaldehyde, 'challenge testing' to glutaraldehyde, a very potent irritant, can lead to difficulties in proving 'sensitivity'. Stenton *et al.* (1994) addressed this complex question in the evaluation of a 46-year-old endoscopy nurse who developed what was thought to be glutaradehyde-induced asthma. An initial inhalation challenge test at the endoscopy suite caused a dramatic immediate FEV_1 fall, yet a more detailed evaluation with double-blind exposures of 'placebo' and non-irritant doses of glutaraldehyde vapor in the laboratory suggested marked dual asthmatic reactions following challenges with 0.032 ppm glutaraldehyde, yet with similar 'declines' on placebo days, with increased airway responsiveness to methacholine. These data were potentially confounded by an intercurrent respiratory tract infection and by technically poor FEV_1 recordings, so the challenge was repeated several weeks later. The second series of double-blind inhalation challenges with carefully controlled exposures to glutaraldehyde showed no obvious asthmatic reactions, with only a slight increase in airway responsiveness. The authors concluded that these results illustrated the potential for misdiagnosis of occupational asthma when unblinded challenge tests are used and show that sophisticated investigatory techniques may not define the diagnosis. A sophisticated chamber has been developed for challenge testing with aldehydes (Lemiere *et al.*, 1996). It may not be enough to provoke asthma with exposure to aldehyde fumes; rather, to prove sensitivity one must be certain that the exposures are non-irritative and below the permissible exposure limit. Yet, Gannon *et al.* (1995) provided important convincing evidence for glutaraldehyde as a sensitizer when they showed that seven of eight hospital workers exposed to glutaraldehyde with respiratory complaints developed asthma following challenge tests where the mean level of glutaraldehyde in air was 0.068 mg/m^3, about one-tenth of the short-term occupational exposure standard of 0.7 mg/m^3.

Curran *et al.* (1996) compared the immunologic responses in subjects exposed to glutaraldehyde with occupational asthma, or work-related respiratory symptoms to unexposed workers and were the first to report evidence of immunologic sensitization in some workers exposed to glutaraldehyde. A series of glutaraldehyde-modified proteins were characterized, and a significant difference recognized between the level of glutaraldehyde-specific IgE antibodies in exposed subjects compared with the levels in unexposed subjects with normal levels of serum IgE. Approximately one-third of exposed workers with occupational asthma had antibody levels exceeding those in the unexposed population. Yet, false-positive results were obtained with serum from unexposed workers who had high total IgE levels. It appears that glutaraldehyde behaves like a number of other low-molecular weight chemicals in that specific antibodies are detected in only a small percentage

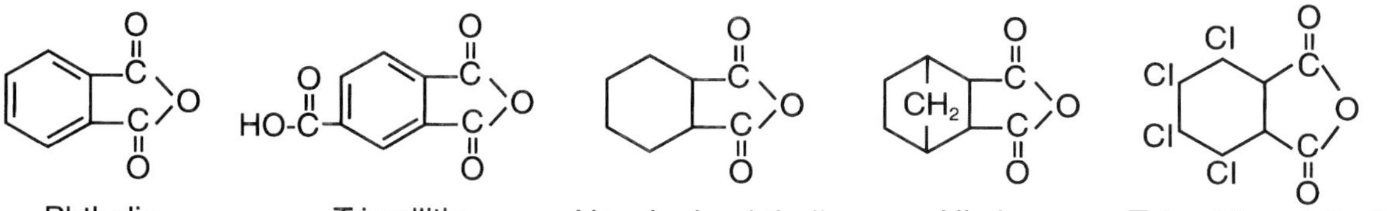

Figure 23.1 Commonly encountered acid anhydrides.

of exposed workers with work-related respiratory symptoms.

Like formaldehyde, glutaraldehyde appears able to sensitize workers, and those at greatest risk appear to be those employed in the health care field. In the workplace, using steam sterilization whenever possible, and minimizing exposures by providing appropriate ventilation in areas where this agent must be used as a disinfectant, appear to be the best ways to prevent sensitization. Again, like formaldehyde, there appears to be a poor correlation between specific antibodies and respiratory symptoms, making the importance of the presence of these antibodies – and the mechanism of asthma – unclear.

Anhydrides

The acid anhydrides are a group of reactive low-molecular weight chemicals which have been used in industry for more than 50 years (Figure 23.1). The major uses of the members of this chemical family are in epoxy and alkyd resins. These resin systems are long-chain polymers which become a hard solid with the addition of very reactive chemicals, known as curing agents or hardeners. Phthalic anhydride and maleic anhydride are common constituents of alkyd resins and form the base for paints, varnishes and reinforced plastics. The epoxy resins are the result of a combination of epichlorhydrin and bisphenol, which cure with the addition of a reactive chemical such as an amine or an acid anhydride. These agents have good adhesive strength and provide good electrical insulation and are widely used as coatings, sealants, inks, as fabric treatment and in dental and surgical applications. The primary adverse effects of these agents include mucosal and skin irritation, with sensitization of the respiratory tract and resultant asthma, with or without rhinitis, conjunctivitis or urticaria.

Like a number of the other low-molecular weight agents, these chemicals are antigenic and appear to behave as haptens and stimulate the formation of specific IgE antibody, a feature implicated in the mechanism of disease. Specific IgE antibody has been identified in workers with asthma attributable to phthalic (Maccia *et al.*, 1976), trimellitic (Zeiss *et al.*, 1977), tetrachlorophthalic (Howe *et al.*, 1983), himic (Bernstein *et al.*, 1984; Rosenman *et al.*, 1987), hexahydrophthallic (Bernstein *et al.*, 1984), and maleic anhydride (Topping *et al.*, 1986). Each anhydride appears to stimulate the formation of a distinct population of antibodies (Bernstein *et al.*, 1984; Topping *et al.*, 1986).

The most comprehensive summary of the health implications associated with exposure to the acid anhydrides was written by Venables (1989), where several important ideas were put forward. This work is most helpful because the author was able to make important conclusions based primarily upon numerous anecdotal reports. Although there are relatively few cases of asthma attributed to the anhydrides, it appears that phthalic anhydride may be the most effective sensitizer among this group of agents. When one reviews the numerous challenge tests reported to be positive in workers with exposure to these agents, the overwhelming majority are dual reactions. Of 24 cases with anhydride-induced asthma, 83% had elevations of serum IgE to anhydrides, and 81% had elevation of serum IgG to anhydrides. Yet, since not all have immunologic abnormalities, it is likely that there are probably non-immunologic mechanisms which are associated with anhydride-induced asthma. When one reviews immunologic data from populations of anhydride-exposed workers, the presence of an elevated level of serum IgE, rather than IgG, allowed for the better separation of cases with asthma from exposed workers without respiratory illness. As with many of these low-molecular weight agents, acute excessive exposures appear important in the sensitizing process, yet cases are reported where excessive exposures were not recognized but the worker has become sensitized. Frankly, like many of the other low-molecular weight agents, the mechanism of sensitization is frequently not known. Finally, although the decline in serum IgE antibody has been recognized to occur after ceasing exposure in workers sensitized to one of the phthalates, tetrachlorophthalic anhydride, asthma persisted despite being away from the workplace (Venables *et al.*, 1987).

Although a systematic study of outcomes to the other phthalates is not available, it remains prudent to remove workers from the sensitizing agent.

(a) Phthalic anhydride

Phthalic anhydride is a white crystalline solid which sublimes on heating, and the dust and fumes cause cough, sneeze and increased mucus secretion. Maccia *et al.* (1976) were the first to show a positive skin test, evidence for late asthma following a specific broncho-provocation test, and high serum titers of IgE (positive RAST) in a worker exposed to this dust. Soon after, Chester *et al.* (1977) and Fawcett *et al.* (1977) reported additional workers with asthma due to phthalic anhydride proven by specific inhalation tests. In a further series of workers with phthalic anhydride exposure, mean levels of total serum IgE were increased among those with exposure to this agent (compared with the non-exposed), although mean values for specific antibody to phthalic anhydride were no different between the exposed and control groups. This led the authors to speculate that exposure to this agent may facilitate the entry of common allergens (Nielsen *et al.*, 1991).

(b) Tetrachlorophthalic anhydride

Schleuter *et al.* (1978) reported the initial cases of asthma due to tetrachlorophthalic anhydride among employees in a plastics plant. The production procedure resulted in the excessive exposure of tetrachlorophthalic anhydride during the use of a large amount of powdered chemicals which were heated and then rolled into sheets of plastics. Five of 34 workers were shown to be sensitized by specific exposure testing, and 10 other untested employees voiced respiratory complaints. Specific IgE antibodies were negative in the five workers proven to be sensitized. Later, Howe *et al.* (1983) elicited positive skin tests and RAST to tetrachlorophthalic anhydride in seven referred workers [employed in a production facility where electronic parts were made and then 'dipped' in epoxy plastic (containing tetrachlorophthalic anhydride) to insulate and coat the wires] and provoked asthma in four of these employees (all were non-atopic). The workplace symptoms were reproduced by exposing the workers to less than one-tenth of the recommended exposure limit of this agent. Follow-up of these seven workers for 5 years showed the half-life of specific IgE to be approximately 1 year, although airway responsiveness persisted in all tested. A larger survey of the 300 workers of this factory showed 24 to have specific IgE antibody (none had asthma), with 20 of these 24 being current smokers. There was no clear predisposition to atopy within the group (Venables *et al.*, 1989). This high number of smokers led the authors to conclude that smoking (with its associated increase in lung permeability), rather than atopy, plays an important role in the sensitization process.

More recently, workers employed at a plant where solenoid coils were manufactured with the use of an epoxy resin developed respiratory difficulties (Liss *et al.*, 1993). Respiratory complaints affected approximately one-third of the workforce. Nearly a third of the 49 workers had elevated serum specific IgE levels and 39% had increased specific IgG, with the antibody levels greatest in those with the most exposure.

Venables *et al.* (1987) serially evaluated a number of workers with asthma to tetrachlorophthalic anhydride who had left the workplace. After avoidance of this agent, concentrations of specific IgE decreased (at a rate approximating 50% per year), and symptoms improved; however, both were present 4 years later, a feature consistent with the natural history of a number of low-molecular weight agents. It appears that the antibody is a sensitive marker of disease activity, as a marked rise in specific IgE occurred following brief sub-irritant inhalation exposures to this agent performed as a part of the on-going diagnostic investigations.

(c) Hexahydrophthalic anhydride

Hexahydrophthalic anhydride is a saturated analog of phthalic anhydride which was first shown to be a potent industrial sensitizer and to induce asthma by Moller *et al.* (1985). Twenty-seven workers with exposure to this agent were evaluated and seven were found to have symptoms of asthma and rhinitis (four were consistent with occupational asthma with significant levels of specific IgE) and 14 of the remaining 20 had nasal or ocular symptoms at work. Mean levels of specific IgE and IgG increased as exposures increased.

Grammer *et al.* (1993a) reported additional investigations in six workers exposed to hexahydrophthalic anhydride who developed rhinitis, nasal mucosal erosions and epistaxis (three also had asthma). All had high titers of specific serum IgE and IgG, implying an immunologic, and not solely an irritative mechanism. When these workers were removed from exposure, the nasal symptoms resolved spontaneously.

Later, Grammer *et al.* (1994) evaluated workers who were exposed to hexahydrophthalic anhydride in a workplace molding operation for the risk of developing immunologically mediated lung disease. There were two primary adverse immunologic respiratory outcomes; the development of a respiratory disease attributable to specific IgE antibody (asthma and/or rhinitis) or to specific IgG antibody (hypersensitivity pneumonitis or hemorrhagic rhinitis). Seven of the 57 workers had both IgE- and IgG-mediated disease, whereas nine displayed only IgE-mediated disease. Although smoking, age and race were not risk factors for these diseases, development of these illnesses was most closely associated with exposure level and development of specific IgE or IgG antibodies.

Twenty-eight workers with hexahydrophthalic anhydride immunologic respiratory disease who had been removed from exposure for at least 1 year were evaluated (Grammer and Shaughnessy, 1996). Seven had asthma, nine had hemorrhagic rhinitis, four had both, and eight had allergic rhinitis alone. Symptoms, signs and spirometry normalized in all but one employee and a decline in antibody titer for both IgE and IgG against hexahydrophthalic anhydride bound to human serum albumin, although specific antibody was still present in many. There were no chest roentgenograph findings at follow-up that could be attributed to these exposures.

In an attempt to understand the importance of immunologic parameters associated with hexahydrophthalic anhydride, Moller *et al.* (1985) evaluated 27 workers with occupational exposure to hexahydrophthalic anhydride; seven had asthma and rhinitis, four had occupational asthma, and 14 had nasal or ocular symptoms. Twelve workers had significant levels of specific IgE to hexahydrophthalic anhydride–human serum albumin complex; 11 had significant levels of specific IgG hexahydrophthalic anhydride–human serum albumin complex; all four with asthma had high levels of IgE to hexahydrophthalic anhydride–human serum albumin complex; three workers with asthma not clearly related to workplace did not have elevated specific IgE.

(d) Maleic anhydride

Relatively little is known about airway hyper-responsiveness to maleic anhydride. It is recognized to be a strong irritant to the skin and mucous membranes, but no systemic study of its ability to induce asthma has been performed. The accounts of asthma are anecdotal. Lee *et al.* (1991) described an atopic worker (who was also a heavy smoker) who was exposed to both phthalic anhydride and maleic anhydride, but was shown to react with a dual reaction only to maleic anhydride. Durham *et al.* (1987) identified another worker with a similar dual response to maleic anhydride.

(e) Trimellitic anhydride (TMA)

TMA differs from phthalic anhydride only by the absence of the carboxyl group, yet the presentations of illness with TMA are quite different (Bardana and Andrach, 1983). Three distinct syndromes were first observed following TMA sensitization (Zeiss *et al.*, 1977), and a later presentation, a pulmonary disease–anemia variant, has been recognized.

The first of these is a classic IgE-mediated 'asthma/rhinitis syndrome' occurring after a suitable latent or sensitizing period. Positive skin tests to TMA conjugated to human serum albumin were found in those sensitized, although total serum IgE levels are not clearly increased (Blackman *et al.*, 1996). Late asthma can also occur. There is also a IgG-mediated 'late respiratory systemic syndrome', which has also been termed 'TMA flu', an illness that the authors considered a variant of the late asthmatic reaction. This response was associated with late onset (beginning 4 to 12 hours after exposure) cough and wheezing, but included systemic symptoms of arthralgias, myalgias and fever. The greatest levels of total antibody to TMA–human serum proteins was found in this group. There is also a non-immunologic irritant response ('irritant syndrome') that appears acutely after the first short, heavy exposure to TMA dust or fumes and is characterized by epistaxis, rhinorrhea, cough, dyspnea and wheezing, and appears related to the chemical irritant properties of the anhydride, and is typically without measurable levels of serum TMA antibodies.

Further studies have identified a rarely occurring response categorized by hemoptysis, breathlessness, pulmonary infiltrates, restrictive ventilatory impairment and hemolytic anemia (designated the 'pulmonary disease–anemia syndrome'), typically occurring when heated metals are sprayed with materials containing TMA in order to make them resistant to corrosion. High levels of total antibody to TMA were present in serum and on erythrocytes (Turner *et al.*, 1980). A 5-year follow-up of 61 newly hired workers who were exposed to TMA fumes or dust in a TMA production plant showed six workers to have developed the asthma/rhinitis syndrome, five developed the late respiratory syndrome, and one worker had both (Zeiss *et al.*, 1983). Later Zeiss *et al.* (1992) undertook a 1-year cross-sectional survey of 474 employees at a large chemical manufacturing facility with TMA exposures, and provided insight into the prevalence of these problems in a working population. Thirty-two (7%) had a TMA immunologic syndrome. Twelve of 32 had asthma/rhinitis, 10 had 'current' late respiratory systemic syndrome, five had a 'distant' bout of late respiratory systemic syndrome, five had late-onset asthma, and one had late-onset arthralgia.

When workers with TMA-induced immunologic lung disease were removed to areas of lower exposure, those with diseases associated with high total antibody, late asthma and late respiratory systemic syndrome, did better uniformly (Grammer *et al.*, 1993b). In contrast, only about half of the workers with asthma/rhinitis improved, not unlike what is recognized in other instances of low-molecular weight agent-induced asthma (Banks *et al.*, 1990). With transfer, the mean serum IgE levels also decreased.

In summary, like a number of the agents discussed in this chapter, the acid anhydrides are a model for low-molecular weight agents which interact with proteins in the body (the specific anhydride bound to human serum albumin serves as a marker of this process), and induce asthma. With some of these agents, specifically trimellitic anhydride, the process is more sophisticated and less well explained, with a number of immunologic illnesses potentially occurring following exposure to this agent. Although the natural history of these unique immunologic responses is not well understood, it appears that the immunologic measures associated with the development of asthma resolve with ceasing exposure, while asthma induced by this agent is more likely to persist.

Two other important issues need to be addressed. First, it appears that the level of exposure to anhydrides is an important determinant of sensitization (Bernstein *et al.*, 1982; Grammer *et al.*, 1992). Second, in ways not yet well understood and apparently unlike some of the other small-molecular weight agents, cigarette smoking appears to increase the risk of sensitization. The fact that in some instances total serum IgE is measurably increased in those exposed to these anhydrides suggests the smoking effect may be a result of increased permeability.

CONCLUSIONS

The first step in treating workers with occupational asthma is removal from exposure. Sometimes, a worker might be able to be transferred to a job without exposure in the same company, but most often this is ineffective as sensitized workers develop an episode of asthma following such negligible exposures as the presence of the agent on a co-worker's shoes or clothes. Most people with occupational asthma have to be retrained for a job with another employer in a different field. When employees have irritant-induced asthma or work-aggravated asthma, employers should make every effort toward reasonable accommodation. Physicians should counsel patients concerning compensation; the specifics may vary from country to country and the appropriate public health agency should be notified. Such agencies should initiate surveillance programs when sentinel cases of occupational asthma have been identified. Pharmacologic treatment of patients with occupational asthma is similar to the treatment of patients with other forms of asthma. Continued monitoring is important if potential for exposure to the sensitizing agent remains.

Although removal from exposure generally results in improvement, a large number of patients with occupational asthma never recover, even after years away from exposure. Patients may continue to require medication and have airflow limitation or pharmacologically induced bronchial hyper-responsiveness for many months or years. The result is permanent respiratory impairment. Increased bronchial responsiveness is associated with chronic airway inflammation, which once initiated, perpetuates itself (Chan-Yeung *et al.*, 1988).

Many agents of low molecular weight are associated with occupational asthma and the list will likely continue to grow. Along with this greater recognition of etiologic agents, improved diagnostic methods and better understanding of pathogenesis and natural history are needed. Although the burden of the ancient occupational lung diseases such as asbestosis and silicosis are still poorly controlled in some parts of the world, the challenge to control occupational asthma may be even greater.

Acknowledgements

These studies were supported by the National Research Council (Grant no. FATMA 95.00820.PF41); MURST, Project ENFUMOSA, European Community (BIOMED1 BMHI-CT94–1521), ARCA, Padova; Ferrara Ricerche, Ferrara, Italy; and by National Institute for Occupational Safety & Health Grants U60/CCU306149–01–1 and U60/CCU313230.

REFERENCES

Abbott, L. (1995) The use and effects of glutaraldehyde: a review. *Occup Health (Lond).*, **47**, 238–9.

Agius, R.M., Nee, J., McGovern, B. *et al.* (1991) Structure activity hypotheses in occupational asthma caused by low molecular weight substances. *Ann. Occup. Hyg.*, **35**, 129–37.

Akiyama, K., Pruzansky, J.J. and Patterson, R. (1984) Hapten modified basophils: a model for human immediate hypersensitivity that can be elicited by IgG antibody. *J. Immunol.*, **133**, 3286–90.

Alanko, K., Keskinen, H., Bjorksten, F. *et al.* (1978) Immediate-type-hypersensitivity to reactive dyes. *Clin. Allergy*, **8**, 25–31.

Andrasch, R.H., Bardana, E.J., Jr, Koster, F. *et al.* (1976) Clinical and bronchial provocation studies in patients with meat wrappers asthma. *J. Allergy Clin. Immunol.*, **58**, 291–8.

Axon, A.T.R., Banks, J., Cockel, R. *et al.* (1981) Disinfection in upper-digestive-tract endoscopy in Britain. *Lancet*, **i**, 1093–4.

Azzawi, M., Bradley, B., Jeffery, P.K. *et al.* (1990) Identification of activated T lymphocytes and eosinophils in bronchial biopsies in stable atopic asthma. *Am. Rev. Respir. Dis.*, **142**, 1407–13.

Banks, D.E. and Rando, R.J. (1988) Recurrent asthma induced by toluene diisocyanate. *Thorax*, **43**, 660–2.

Banks, D.E., Rando, R.J. and Barkman, H.W., Jr (1990) Persistence of toluene diisocyanate-induced asthma despite negligible workplace exposures. *Chest*, **97**, 121–5.

Bardana, E.J., Jr, and Andrach, R.H. (1983) Occupational asthma secondary to low molecular weight agents used in the plastic and resin industries. *Eur. J. Respir. Dis.*, **64**, 241–51.

Bardana, E.J. and Montanaro, A. (1991) Formaldehyde: an analysis of its respiratory, cutaneous, and immunologic effects. *Ann Allergy*, **66**, 441–52.

Beach, J.R., Dennis, J.H., Avery, A.J. *et al.* (1996) An epidemiologic investigation of asthma in welders. *Am. J. Respir. Crit. Care Med.*, **154**, 1394–400.

Beasley, R., Roche, W.R., Roberts, J.A. *et al.* (1989) Cellular events in the bronchi in mild asthma and after bronchial provocation. *Am. Rev. Respir. Dis.*, **139**, 806–17.

Becher, R., Hongslo, J.K., Jantunen, M.J. *et al.* (1996) Environmental chemicals relevant for respiratory hypersensitivity: the indoor environment. *Toxicol. Lett.*, **86**, 155–62.

Benson, W.G. (1984) Case report: exposure to glutaraldehyde. *J. Soc. Occup. Med.*, **34**, 63–4.

Bentley, A.M., Maestrelli, P., Saetta, M. *et al.* (1992a) Activated T-lymphocytes and eosinophils in the bronchial mucosa in isocyanate-induced asthma. *J. Allergy Clin. Immunol.*, **89**, 821–9.

Bentley, A.M., Menz G., Storz, Chr. *et al.* (1992b) Identification of T-lymphocytes, macrophages, and activated eosinophils in the bronchial mucosa in intrinsic asthma: relationship to symptoms and bronchial responsiveness. *Am. Rev. Respir. Dis*, **146**, 500–6.

Bernstein, D.I., Patterson, R. and Zeiss, C.R. (1982) Clinical and immunologic evaluation of trimellitic anhydride-and phthalic anhydride-exposed workers using a questionnaire with comparative analysis of enzyme-linked immunosorbent and radioassay studies. *J. Allergy Clin. Immunol.*, **69**, 311–18.

Bernstein, D.I., Gallagher, J.S., D'Souza, L. *et al.* (1984) Heterogeneity of specific-IgE responses in workers sensitized to acid anhydride compounds. *J. Allergy Clin. Immunol.*, **74**, 794–801.

Biagini, R.E., Bernstein, I.L., Gallagher, J.S. *et al.* (1985) The diversity of reaginic immune responses to platinum and palladium metallic salts. *J. Allergy Clin. Immunol.*, **76**, 794–802.

Biagini, R.E., Klincewicz, S.L., Henninsen, G.M. *et al.* (1990) Antibodies to morphine in workers exposed to opiates at a narcotics manufacturing facility and evidence for similar antibodies in heroin abusers. *Life Sci.*, **47**, 897–908.

Blackman, K.S., Shaughnessy, M.A., Harris, K.E. *et al.* (1996) Total serum IgE in trimellitic anhydride-induced asthma. *J. Occup. Environ. Med.*, **38**, 347–51.

Blanc, P. (1987) Occupational asthma in a national disability survey. *Chest*, **92**, 613–17.

Block, G. and Chan-Yeung, M. (1982) Asthma induced by nickel. *JAMA*, **247**, 1600–2.

Boulet, L.P., Boutet, M., Laviolette, M. *et al.* (1994) Airway inflammation after removal from the causal agent in occupational asthma due to high and low molecular weight agents. *Eur. Respir. J.*, **7**, 1567–75.

Bourne, M.S., Flindt, M.L.H. and Walker, J.M. (1979) Asthma due to industrial use of chloramine. *Br. Med. J.*, **2**, 10–12.

Bousquet, J. and Michel, F.-B. (1991) Allergy to formaldehyde and ethylene oxide. *Clin. Rev. Allergy*, **9**, 357–70.

Bradding, P., Feather, I.H., Howarth, P.H. *et al.* (1992) Interleukin-4 is localised to and released by human mast cells. *J. Exp. Med.*, **176**, 1381–6.

Bradley, B.L. Azzawi, M., Jacobson, M. *et al.* (1991) Eosinophils, T-lymphocytes, mast cells, neutrophils and macrophages in bronchial biopsy specimens from atopic subjects with asthma: Comparison with biopsy specimens from atopic subjects without asthma and normal control subjects and relationship to bronchial hyper-responsiveness. *J. Allergy Clin. Immunol.*, **88**, 661–74.

Brooks, S.M. and Vandervort, R. (1977) Polyvinyl chloride film thermal decomposition products as an occupational illness. 2. Clinical studies. *J. Occup. Med.*, **19**, 192–6.

Brooks, S.M., Weiss, M.A. and Bernstein, I.L. (1985) Reactive airways dysfunction syndrome (RADS): persistent asthma syndrome after high level irritant exposures. *Chest*, **88**, 376–84.

Bruckner, H.C. (1967) Extrinsic asthma in a tungsten carbide worker. *J. Occup. Med.*, **9**, 518–19.

Bryant, D.H. (1985) Asthma due to insecticide sensitivity. *Aust.N.Z.J.Med.*, **15**, 66–8.

Burge, P.S. (1982) Occupational asthma in electronics workers caused by colophony fumes: follow-up of affected workers. *Thorax*, **37**, 348–52.

Burge, P.S. (1984) Occupational asthma, rhinitis, and alveolitis due to colophony. *Clin. Immunol. Allergy*, **4**, 55–81.

Burge, P.S., Harries, M.G., O'Brien, I.M. *et al.* (1978) Respiratory disease in workers exposed to solder flux fumes containing colophony (pine resin). *Clin. Allergy*, **8**, 1–14.

Burge, P.S., Perks, W., O'Brien, I.M. *et al.* (1979) Occupational asthma in an electronics factory. *Thorax*, **34**, 13–18.

Burge, P.S., Harries, M.G., O'Brien, I. *et al.* (1980) Bronchial provocation studies in workers exposed to the fumes of electronic soldering fluxes. *Clin. Allergy.*, **10**, 137–49.

Burge, P.S., Edge, G., Hawkins, R. *et al.* (1981) Occupational asthma in a factory making flux-cored solder containing colophony. *Thorax*, **36**, 828–34.

Burge, P.S., Wieland, A., Robertson, A.S. *et al.* (1986) Occupational asthma due to unheated colophony. *Br. J. Ind. Med.*, **43**, 559–60.

Burney, P.G.J., Laitinen, L.A., Perdrizet, S. *et al.* (1989) Validity and repeatability of the IUATLD (1984) Bronchial Symptoms Questionnaire: an international comparison. *Eur. Respir. J.*, **2**, 940–5.

Burrows, B., Martinez, F.D., Halonen, M. *et al.* (1989) Association of asthma with serum IgE levels and skin-tests reactivity to allergens. *N. Engl. J. Med.*, **320**, 271–7.

Butcher, B.T. and Banks, D.E. (1992) Immunologic and clinical features of occupational agents attributable to small molecular weight agents. *Immunol. Allergy Clin. North Am.*, **12**, 329–45.

Butcher, B.T. and Salvaggio, J.E. (1986) Occupational asthma. *J. Allergy Clin. Immunol.*, **78**, 547–56.

Butcher, B.T., Jones, R.N., O'Neil, C.E. *et al.* (1977) Longitudinal study of workers employed in the manufacture of toluene diisocyanate. *Am. Rev. Respir. Dis.*, **116**, 411–22.

Butler, J., Culver, B.H. and Robertson, H.T. (1981) Meat wrappers' asthma. *Chest*, **80**(1 supp.), 71–3.

Calverley, A.E., Rees, D., Dowdeswell, R.J. *et al.* (1995) Platinum salt sensitivity in refinery workers: incidence and effects of smoking and exposure. *Occup. Environ. Med.*, **52**, 661–6.

Carino, M., Aliani, M., Licitra, C. *et al.* (1997) Death due to asthma at workplace in a diphenylmethane diisocyanate-sensitized subject. *Respiration*, **64**, 111–13.

Cartier, A., Grammer, L., Malo, J.-L. *et al.* (1989) Specific serum antibodies against isocyanates: association with occupational asthma. *J. Allergy Clin. Immunol.*, **8**, 507–14.

Chan-Yeung, M. (1994) Mechanism of occupational asthma due to western red cedar (*Thuja plicata*). *Am. J. Ind. Med.*, **25**, 13–18.

Chan-Yeung, M. and Desjardins, A. (1992) Bronchial hyper-responsiveness and level of exposure in occupational asthma due to western red cedar (*Thuja plicata*). Serial observations before and after development of symptoms. *Am. Rev. Respir. Dis.*, **146**, 1606–9.

Chan-Yeung, M. and Malo, J.-L. (1994) Aetiologic agents in occupational asthma. *Eur. Respir. J.*, **7**, 346–71.

Chan-Yeung, M. and Malo, J.-L. (1995) Occupational asthma. *N. Engl. J. Med.*, **333**, 107–12.

Chan-Yeung, M., Giclas, P.C. and Henson, P.M. (1980) Activation of complement by plicatic acid, the chemical compound responsible for asthma due to western red cedar (*Thuja plicata*). *J. Allergy Clin. Immunol.*, **65**, 333–7.

Chan-Yeung, M., Lam, S. and Koener, S. (1982) Clinical features and natural history of occupational asthma due to western red cedar (*Thuja plicata*). *Am. J. Med.*, **72**, 411–15.

Chan-Yeung, M., Leriche, J. Maclean, L. *et al.* (1988) Comparison of cellular and protein changes in bronchial lavage fluid of symptomatic and asymptomatic patients with red cedar asthma on follow-up examination. *Clin. Allergy*, **18**, 359–65.

Chan-Yeung, M., Chan, H., Tse, K.S. *et al.* (1989) Histamine and leukotrienes release in bronchoalveolar lavage fluid during plicatic acid induced bronchoconstriction. *J. Allergy Clin. Immunol.*, **84**, 762–8.

Chan-Yeung, M., McMurren, T., Catonio-Begley, F. *et al.* (1993) Occupational asthma in a technologist exposed to glutaraldehyde. *J. Allergy Clin. Immunol.*, **91**, 974–8.

Chan-Yeung, M., Lam, S., Kennedy, S.M. *et al.* (1994) Persistent asthma after repeated exposure to high concentrations of gases in pulp mills. *Am. J. Respir. Crit. Care Med.*, **149**, 1676–80.

Chester, E.H. Schwartz, H.J., Payne, C.B. Jr *et al.* (1977) Phthalic anhydride asthma. *Clin. Allergy*, **7**, 15–20.

Chiappino, G. (1994) Hard metal disease: clinical aspects. *Sci. Total Environ.*, **150**, 65–8.

Cirla, A.M. (1994) Cobalt-related asthma: clinical and immunological aspects. *Sci. Total Environ.*, **150**, 85–94.

Cleare, M.J., Hughes, E.G., Jacoby, B. *et al.* (1976) Immediate (type I) allergic responses to platinum compounds. *Clin. Allergy*, **6**, 183–95.

Cockroft, D.W., Cartier, A., Jones, G. *et al.* (1980) Asthma caused by occupational exposure to a furan-based binder system. *J. Allergy Clin. Immunol.*, **66**, 458–63.

Contreras, G.R., Rousseau, R. and Chan-Yeung, M. (1994) Occupational respiratory diseases in British Columbia, Canada in 1991. *Occup. Environ. Med.*, **51**, 710–12.

Corrado, O.J., Osman, J. and Davies, R.J. (1986) Asthma and rhinitis after exposure to glutaraldehyde in endoscopy units. *Human Toxicol.*, **5**, 325–7.

Corrigan, C.J. and Kay, A.B. (1990) CD4+ T-lymphocyte activation in acute severe asthma. Relationship to disease severity and atopic status. *Am. Rev. Respir. Dis.*, **141**, 970–7.

Corrigan, C.J., Haczku, A., Gemou-Engesaeth, V. *et al.* (1993) CD4 T-lymphocyte activation in asthma is accompanied by increased serum concentration of interleukin-5. Effect of glucocorticoid therapy. *Am. Rev. Respir. Dis.*, **147**, 540–7.

Cote, J., Kennedy, S. and Chan-Yeung, M. (1990) Outcome of patients with cedar asthma with continuous exposure. *Am. Rev. Respir. Dis.*, **141**, 373–6.

Coutts, I.L., Dally, M.B., Taylor, A.J. *et al.* (1981) Asthma in workers manufacturing cephalosporins. *Br. Med. J.*, **283**, 950.

Coutts, I.L., Lozewicz, S., Dally, M.B. *et al.* (1984) Respiratory symptoms related to work in a factory manufacturing cimetidine tablets. *Br. Med. J.*, **288**(6428), 1418.

Cromwell, O., Pepys, J. Parish, W.E. *et al.* (1979) Specific antibodies to platinum salts in sensitized workers. *Clin. Allergy*, **9**, 109–17.

Cullen, R.T., Cherrie, B. and Soutar, C.A. (1992) Immune responses to colophony, an agent causing occupational asthma. *Thorax*, **47**, 1050–5.

Curran, A.D., Burge, P.S. and Wiley, K. (1996) Clinical and immunologic evaluation of workers exposed to glutaraldehyde. *Allergy*, **51**, 826–32.

Dally, M.B. Hunter, J.V. Hughes, E.G. *et al.* (1980) Hypersensitivity to platinum salts: a population study *Am. Rev. Respir. Dis.*, **121**, 230A.

Davies, R.J. and Blainey, A.D. (1984) Occupational asthma: classification and clinical aspects. *Semin. Respir. Med.*, **5**, 229–43.

Davies, R.J. and Pepys, J. (1975) Asthma due to inhaled chemical agents – the macrolide antibiotic Spiramycin. *Clin. Allergy*, **5**, 99–107.

Davies, R.J., Hendrick, D.J. and Pepys, J. (1974) Asthma due to inhaled chemical agents: ampicillin, benzyl penicillin, 6 amino penicillanic acid and related substances. *Clin. Allergy*, **4**, 227–47.

Davison, A.G., Haslam, P.L. Corrin, B. *et al.* (1983) Interstitial lung disease and asthma in hard metal workers: bronchoalveolar lavage, ultrastructure, and analytical findings and results of bronchial provocation tests. *Thorax*, **38**, 119–28.

Del Prete, G.F., DeCarli, M., Maestreli, P. *et al.* (1993) Allergen exposure induces the activation of allergen-specific Th2 cells in the airway mucosa of patients with allergic respiratory disorders. *Eur. J. Immunol.*, **23**, 1445–9.

De Monchy, G.R., Kauffman, H.F., Venge, P. *et al.* (1985) Bronchoalveolar eosinophilia following allergen-induced late-phase asthmatic reactions. *Am. Rev. Respir. Dis.*, **131**, 373–6.

Deschamps, D., Questel, F., Baud, F.J. *et al.* (1994) Persistent asthma after acute inhalation of organophosphate insecticide. *Lancet*, **344**(8938), 1712.

Di Stefano, A., Saetta, M., Maestrelli, P. *et al.* (1993) Mast cells in airway mucosa and rapid development of occupational asthma induced by toluene diisocyanate. *Am. Rev. Respir. Dis.*, **147**, 1005–9.

Docker, A., Wattie, J.M., Topping, M.D. *et al.* (1987) Clinical and immunological investigations of respiratory disease in workers using reactive dyes. *Br. J. Ind. Med.*, **44**, 534–41.

Drazen, J.M. and Turino, G. (1995) Progress at the interface of inflammation and asthma. Workshop report. *Am. J. Respir. Crit. Care Med.*, **152**, 385–424.

Durham, S.R., Graneek, B.J., Hawkins, R. *et al.* (1987) The temporal relationship between increases in airway responsiveness to histamine and late asthmatic responses induced by occupational agents. *J. Allergy Clin. Immunol.*, **79**, 398–406.

Dykewicz, M.S., Patterson, R., Cuggell, D.W. *et al.* (1991) Serum IgE and IgG to formaldehyde-human serum albumin; lack of relation to gaseous formaldehyde exposure. *J. Allergy Clin. Immunol.*, **87**, 48–57.

Ehrlich, R.I. (1994) Fatal asthma in a baker: a case report. *Am. J. Ind. Med.*, **26**, 799–802.

Eisen, E.A., Wegman, D.H. and Smith, T.J. (1985) Across-shift changes in the pulmonary function of meat-wrappers and other workers in the retail food industry. *Scand. J. Work Environ. Health*, **11**, 21–6.

Estrada Rodriguez, J.L., Lopez Serrano, C., Belchi Hernandez, J. *et al.* (1991) Hypocomplementemic urticarial vasculitic syndrome, asthma and anaphylactic reaction with ampicillin. *J. Investig. Allergol. Clin. Immunol.*, **1**, 69–71.

Fabbri, L.M. and Ciaccia, A. (1992) Investigative bronchoscopy in asthma and other airways diseases. *Eur. Respir. J.*, **5**, 8–11.

Fabbri, L.M., Boschetto, P., Zocca, E. *et al.* (1987) Bronchoalveolar neutrophilia during late asthmatic reactions induced by toluene diisocyanate (TDI). *Am. Rev. Respir. Dis.*, **136**, 36–42.

Fabbri, L.M., Danieli, D., Crescioli, S. *et al.* (1988) Fatal asthma in a subject sensitized to toluene diisocyanate. *Am. Rev. Respir. Dis.*, **137**, 1494–8.

Fahy, J.V., Liu, J., Wong, H. *et al.* (1994) Analysis of cellular and biochemical constituents in induced sputum after allergen challenge: a method for studying allergic airway inflammation. *J. Allergy Clin. Immunol.*, **93**, 1031–9.

Fawcett, I.W. and Pepys, J. (1976) Allergy to tetracycline preparation – a case report. *Clin. Allergy*, **6**, 301–4.

Fawcett, I.W., Newman Taylor, A.J. and Pepys, J. (1976) Asthma due to inhaled chemical agents, fumes from 'Multicore' soldering flux and colophony resin. *Clin. Allergy*, **6**, 577–85.

Fawcett, I.W., Newman Taylor, A.J. and Pepys, J. (1977) Asthma due to inhaled chemical agents – epoxy resin systems containing phthalic anhydride, trimellitic anhydride and triethylene tetramine. *Clin. Allergy*, **7**, 1–14.

Feinberg S.M. and Watson, R.M. (1945) Atopy to simple chemical compounds – sulfone-chloramides. *J. Allergy*, **16**, 209.

Ferris, B.G. Peters, J.M., Burgess, W.A. *et al.* (1977) Apparent effect of an azodicarbonamide on the lung: a preliminary report. *J. Occup. Med.*, **19**, 424–5.

Finotto, S., Fabbri, L.M., Rado, V. *et al.* (1991) Increase in numbers of CD8 positive lymphocytes and eosinophils in peripheral blood of subjects with late asthmatic reactions induced by toluene diisocyanate. *Br. J. Ind. Med.*, **48**, 116–21.

Fisher, A.A. and Dooms-Goossens, A. (1976) Persulfate hair bleach reactions. Cutaneous and respiratory manifestations. *Arch. Dermatol.*, **112**, 1407–9.

Forni, A. (1994) Bronchoalveolar lavage in the diagnosis of hard metal disease. *Sci. Total Environ.*, **150**, 69–76.

Frew, A.J., Chan, H., Dryden, P. *et al.* (1993) Immunologic studies of the mechanisms of occupational asthma caused by western red cedar. *J. Allergy Clin. Immunol.*, **92**, 466–78.

Frew, A.J., Chan, H., Lam, S. *et al.* (1995) Bronchial inflammation asthma due to western red cedar. *Am. J. Respir. Crit. Care Med.*, **151**, 340–4.

Gannon, P.F. and Burge, P.S. (1991) A preliminary report of a surveillance scheme of occupational asthma in the West Midlands. *Br. J. Ind. Med.*, **48**, 579–82.

Gannon, P.F., Burge, P.S. and Benfield, G.F. (1992) Occupational asthma due to polyethylene shrink wrapping (paper wrapper's asthma). *Thorax*, **47**, 759.

Gannon, P.F., Weir, D.C., Robertson, A.S. *et al.* (1993) Health, employment, and financial outcomes in workers with occupational asthma. *Br. J. Ind. Med.*, **50**, 491–6.

Gannon, P.F., Bright, P., Campbell M. *et al.* (1995) Occupational asthma due to glutaraldehyde and formaldehyde in endoscopy and X-ray departments. *Thorax*, **50**, 156–9.

Gorski, P. and Ulinski, S. (1996) Effect of occupational exposure to opiates on the respiratory system. *Int. J. Occup. Med. Environ. Health.*, **9**, 245–53.

Grammer, L.C., Harris, K.E., Sonenthal, K.R. *et al.* (1992) A cross-sectional survey of 46 employees exposed to trimellitic anhydride. *Allergy Proc.*, **13**, 139–42.

Grammer, L.C., Shaughnessy, M.A. and Lowenthal, M. (1993a) Hemorrhagic rhinitis. An immunologic disease due to hexahydrophthalic anhydride. *Chest*, **104**, 1792–4.

Grammer, L.C., Shaughnessy, M.A., Henderson, J. *et al.* (1993b) A clinical and immunologic study of workers with trimellitic-anhydride-induced immunologic lung disease after transfer to low exposure jobs. *Am. Rev. Respir. Dis.*, **148**, 54–7.

Grammer, L.C., Shaughnessy, M.A., Lowenthal, M. *et al.* (1994) Risk factors for immunologically mediated respiratory disease from hexahydrophthalic anhydride. *J. Occup. Med.*, **36**, 642–6.

Grammer, L.C. and Shaughnessy, M.A. (1996) Study of employees with anhydride-induced respiratory disease after removal from exposure. *J. Occup. Environ. Med.*, **38**, 771–4.

Greene, S.A. and Freedman, S. (1976) Asthma due to inhaled chemical agents – amprolium hydrochloride. *Clin. Allergy*, **6**, 105–8.

Hamid, Q., Azzawi, M., Ying, S. *et al.* (1991) Expression of MRNA for interleukin-5 in mucosal bronchial biopsies from asthma. *J. Clin. Invest.*, **87**, 1541–6.

Harding, C.V. (1993) Cellular and molecular aspects of antigen processing and the function of class II MHC molecules. *Am. J. Respir. Cell. Mol. Biol.*, **8**, 461–7.

Harries, M.G., Taylor, A.N., Wooden, J. *et al.* (1979) Bronchial asthma due to alpha-methyldopa. *Br. Med. J.*, **1**, 1461.

Hendrick, D.J. and Lane, D (1975) Formalin asthma in hospital staff. *Br. Med. J.*, **1**, 607–10.

Hendrick, D.J. and Lane, D.J. (1977) Occupational formalin asthma. *Br. J. Ind. Med.*, **34**, 11–18.

Hendrick, D.J., Rando, R.J., Lane, D.J. *et al.* (1982) Formaldehyde asthma: challenge exposure levels and fate after five years. *J. Occup. Med.*, **24**, 893–7.

Hendy, M.S., Beattie, B.E. and Burge, P.S. (1985) Occupational asthma due to an emulsified oil mist. *Br. J. Ind. Med.*, **42**, 51–4.

Herd, Z.L. and Bernstein, D.I. (1994) Antigen-specific stimulation of histamine releasing factors in diisocyanate-induced occupational asthma. *Am. J. Respir. Crit. Care Med.*, **150**, 988–94.

Holt, P.G., McMenamin, C., Schon-Hegrad, M.A. *et al.*, (1991) Immunoregulation of asthma: control of T-lymphocyte activation in the respiratory tract. *Eur Respir J Suppl.*, **13**, 6S–15S.

Howe, W., Venables, K.M., Topping, M.D. *et al.* (1983) Tetrachlorophthalic anhydride asthma: evidence for specific IgE antibody. *J. Allergy Clin. Immunol.*, **72**, 5–11.

Hunter, D., Milton, R. and Perry, K.M.A. (1945) Asthma caused by the complex salts of platinum. *Br. J. Ind. Med.*, **2**, 92–8.

Jachuck, S.J., Bound, C.L., Steel, J. *et al.* (1989) Occupational hazard in hospital staff exposed to 2 per cent glutaraldehyde in an endoscopy unit. *J. Soc. Occup. Med.*, **39**, 69–71.

Johnston, T.G., Cazort, A.G., Marvin, H.N. *et al.* (1951) Bronchial asthma, urticaria and allergic rhinitis from tannic acid. *J Allergy*, **22**, 494–9.

Joules, H. (1952) Asthma from sensitization to chromium. *Lancet*, **2**, 182–3.

Kalish, R.S. and Morimoto, C. (1988) Urushiol (poison ivy) triggered suppression T cell clone generated from peripheral blood. *J. Clin. Invest.*, **82**, 825–32.

Kammermeyer, J.K. and Matthews, K.P. (1973) Hypersensitivity to phenylglycine acid chloride. *J. Allergy Clin. Immunol.*, **52**, 73–84.

Karlsberg, A.-T., Boman, A. and Nilsson, L.G. (1988) Hydrogenisation reduces the allergenicity of colophony (rosin). *Contact Derm.*, **19**, 22–9.

Karasek, S.R. and Karasek, M. (1911) *The use of Platinum Paper*, Report of Illinois State Commission of Occupational Diseases, p. 97.

Kay, A.B., Corrigan, C.J. and Frew, A.J. (1991) The role of cellular immunology in asthma. *Eur. Respir. J.*, **4**, 105s–112s.

Keskinen, H., Alanko, K. and Saarinen, L. (1978) Occupational asthma in Finland. *Clin. Allergy*, **8**, 569–79.

Keskinen, H., Kalliomaki, P.-L. and Alanko, K. (1980) Occupational asthma due to stainless steel welding fumes. *Clin. Allergy*, **10**, 151–9.

Kita, O., Weiler, D., Sur, S. *et al.* (1993) IL-5 is the predominant eosinophil-active cytokine in the antigen-induced pulmonary late-phase reaction. *Am. Rev. Respir. Dis.*, **147**, 901–7.

Kongerud, J. and Samuelson, S.O. (1991) A longitudinal study of respiratory symptoms in aluminum potroom workers. *Am. Rev. Respir. Dis.*, **144**, 10–16.

Kopp, S.K., McKay, R.T., Moller, D.R. *et al.* (1985) Asthma and rhinitis due to ethylcyanoacrylate instant glue. *Ann. Intern. Med.*, **102**, 613–15.

Kramps, J.A., van Toorenenbergen, A.W., Voorhen, P.H. *et al.* (1981) Occupational asthma due to inhalation of chloramine-T: II. Demonstration of specific IgE antibodies. *Int. Arch. Appl. Immunol.*, **64**, 428–38.

Kramps, J.A., Peltenburg, L.T.C., Kerklaan, P.R.M. *et al.* (1989) Measurement of specific IgE antibodies in individuals exposed to formaldehyde. *Clin. Exp. Allergy*, **19**, 509–14.

Kujala, V.M., Reijula, K.E., Ruotsalainen, E.M. *et al.* (1995) Occupational asthma due to chloramine-T solution. *Respir. Med.*, **89**, 693–5.

Kusaka, Y., Yokoyama, K., Sera, Y. *et al.* (1986) Respiratory diseases in hard metal workers: an occupational hygiene study in a factory. *Br. J. Ind. Med.*, **43**, 474–85.

Kusaka, Y., Nakano, Y., Shirakawa, T. *et al.* (1989) Lymphocyte transformation with cobalt in hard metal asthma. *Industr. Health*, **27**, 155–63.

Kusaka, Y., Iki, M. and Kumagai, S.S. (1996a) Epidemiological study of hard metal asthma. *Occup. Environ. Med.*, **53**, 188–93.

Kusaka, Y., Iki, M., Kumagai, S. *et al.* (1996b) Decreased ventilatory function in hard metal workers. *Occup. Environ. Med.*, **53**, 194–9.

Laitinen, L.A., Heino, M., Laitinen, A. *et al.* (1985) Damage of the airway epithelium and bronchial reactivity in patients with asthma. *Am. Rev. Respir. Dis.*, **131**, 599–606.

Lagier, F., Cartier, A., Dolovich, J. *et al.* (1989) Occupational asthma in a pharmaceutical worker exposed to penicillamine. *Thorax*, **44**, 157–8.

Lam, S. and Chan-Yeung, M. (1980) Ethylenediamine induced asthma. *Am. Rev. Respir. Dis.*, **121**, 151–5.

Lam, S., Wong, R. and Chan-Yeung, M. (1979) Nonspecific bronchial reactivity in occupational asthma. *J. Allergy Clin. Immunol.*, **63**, 28–34.

Lam, S. LeRiche, J., Phillips, D. *et al.* (1987) Cellular and protein changes in bronchial lavage fluid after late asthmatic reaction in patients with red cedar asthma. *J. Allergy. Clin. Immunol.*, **80**, 44–50.

Lauwerys, R. and Lison, D. (1994) Health risks associated with cobalt exposure – an overview. *Sci. Total Environ.*, **150**, 1–6.

Lee, H.S., Yap, J., Wang, Y.T. *et al.* (1989) Occupational asthma due to unheated polyvinyl chloride resin dust. *Br. J. Ind. Med.*, **46**, 820–2.

Lee, H.S., Chia, S.E., Yap, J.C. *et al.* (1990) Occupational asthma due to spot-welding. *Singapore Med. J.*, **31**, 506–8.

Lee, H.S., Wang, Y.T., Cheong, T.H. *et al.* (1991) Occupational asthma due to maleic anhydride. *Br. J. Ind. Med.*, **48**, 283–5.

Lemiere, C., Cloutier, Y., Perrault, G. *et al.* (1996) Closed-circuit apparatus for specific inhalation challenges with an occupational agent, formaldehyde, in vapor form. *Chest*, **109**, 1631–5.

Liss, G.M., Bernstein, D., Genovese, L. *et al.* (1993) Assessment of risk factors for IgE-mediated sensitization to tetrachlorophthalic anhydride. *J. Allergy Clin. Immunol.*, **92**, 237–47.

Lozewicz, S., Davison, A., Hopkirk, A. *et al.* (1985) Occupational asthma due to methyl methacrylate and cyanoacrylates. *Thorax*, **40**, 836–9.

Maccia, C.A., Bernstein, I.L., Emmett, E.A. *et al.* (1976) *In vitro* demonstration of specific IgE in phthalic anhydride hypersensitivity. *Am. Rev. Respir. Dis.*, **113**, 701–4.

Maestrelli, P., O'Heir, R.E., Tsai, J.J. *et al.* (1988) Antigen-induced neutrophil chemotactic factor derived from cloned human T lymphocytes. *Immunology*, **65**, 605–9.

Maestrelli, P., Del Prete, G.F., De Carli, M. *et al.* (1994a) CD-8 T-cells producing interleukin-5 and interferon-gamma in bronchial mucosa of patients with asthma induced by toluene-diisocyanate. *Scand. J. Work Environ. Health*, **20**, 376–81.

Maestrelli, P., Calcagni, P.G., Saetta, M. *et al.* (1994b) Sputum eosinophilia after asthmatic responses induced by isocyanates in sensitized subjects. *Clin. Exp. Allergy*, **24**, 29–34.

Maestrelli, P., Di Stefano, A., Occari, P. *et al.* (1995a) Cytokines in the airway mucosa of subjects with asthma induced by toluene diisocyanate. *Am. J. Respir. Crit. Care Med.*, **151**, 607–12.

Maestrelli, P., Saetta, M., Di Stefano, A. *et al.* (1995b) Comparison of leukocyte counts in sputum, bronchial biopsies, and bronchoalveolar lavage. *Am. J. Respir. Crit. Care Med.*, **152**, 1926–31.

Malo, J.L. and Bernstein, I.L. (1993) Other chemical substances causing occupational asthma. In *Asthma in the Workplace* (eds I.L. Bernstein, J.L. Malo, M.C. Yeung, D.I. Bernstein), Marcel Dekker, New York, pp. 481–502.

Malo, J.L. and Cartier, A. (1987) Occupational asthma due to fumes of galvanized metal. *Chest*, **92**, 375–7.

Malo, J.L. and Cartier, A. (1988) Occupational asthma in workers of a pharmaceutical company processing spiramycin. *Thorax*, **43**, 371–7.

Malo, J.L., Cartier, A., Doepner, M. *et al.* (1982) Occupational asthma caused by nickel sulphate. *J. Allergy Clin. Immunol.*, **69**, 55–9.

Malo, J.L., Gagnon, G. and Cartier, A. (1984) Occupational asthma due to heated freon. *Thorax*, **39**, 628–9.

Malo, J.L., Ghezzo, H., D'Aquino, C. *et al.* (1992) Natural history of occupational asthma: relevance of type of agent and other factors in the rate of development of symptoms in affected subjects. *J. Allergy Clin. Immunol.*, **90**, 937–44.

Malo, J.L., Boulet, L.P., Dewitte, J.D. *et al.* (1993) Quality of life of subjects with occupational asthma. *J. Allergy Clin. Immunol.*, **91**, 1121–7.

Markowitz, J.S. (1989) Self-reported short-and long-term respiratory effects among PVC-exposed fire-fighters. *Arch. Environ. Health*, **44**, 30–3.

McConnell, L.H., Fink, J.N., Schleuter, D.P. *et al.* (1973) Asthma due to nickel sensitivity. *Ann. Intern. Med.*, **78**, 888–90.

Menon, M.P. and Das, A.K. (1977) Tetracycline asthma – a case report. *Clin. Allergy*, **7**, 285–90.

Meredith, S.K. and McDonald, J.C. (1994) Work-related respiratory disease in the United Kingdom, 1989–1992: report on the SWORD project. *Occup. Med.*, **44**, 183–9.

Meredith, S. and Nordman, H. (1996) Occupational asthma: measures of frequency from four countries. *Thorax*, **51**, 435–40.

Merget, R., Reineke, M., Rueckmann, A. *et al.* (1994) Nonspecific and specific bronchial responsiveness in occupational asthma caused by platinum salts after allergen avoidance. *Am. J. Respir. Crit. Care Med.*, **150**, 1146–9.

Merget, R., Buenemann, A., Kulzer, R. *et al.* (1996) A cross sectional study of chemical industry workers with occupational exposure to persulphates. *Occup. Environ. Med.*, **53**, 422–6.

Mickelsen, R.L. and Johnston, O.E. (1997) Control of ethyl methacrylate exposures during the application of artificial fingernails. *Am. Ind. Hyg. Assoc. J.*, **58**, 214–18.

Miller, M.E., Lummus, Z.L. and Bernstein, D.I. (1996) Occupational asthma caused by FD&C blue dye no. 2. *Allergy Asthma Proc.*, **17**, 31–4.

Moisan, T.C. (1991) Prolonged asthma after smoke inhalation: a report of three cases and a review of previous reports. *J. Occup. Med.*, **33**, 458–61.

Moller, D.R., Gallagher, J.S., Bernstein, D.I. *et al.* (1985) Detection of IgE-mediated respiratory sensitization in workers exposed to hexahydrophthalic anhydride. *J. Allergy Clin. Immunol.*, **75**, 663–72.

Moscato, G., Naldi, L. and Candura, F. (1984) Bronchial asthma due to spiramycin and adipic acid. *Clin. Allergy*, **14**, 355–61.

Moscato, G., Galdi, E., Scilbilia, J. *et al.* (1995) Occupational asthma, rhinitis, and urticaria due to piperacillin sodium in a pharmaceutical worker. *Eur. Respir. J.*, **8**, 467–9.

Murdoch, R.D. and Pepys, J. (1987) Platinum group metal sensitivity: reactivity to platinum group metal salts in platinum halide salt-sensitive workers. *Ann. Allergy*, **59**, 464–9.

Murdoch, R.D., Pepys, J. and Hughes, E.G. (1986) IgE antibody responses to platinum group metals: a large scale refinery study. *Br. J. Ind. Med.*, **43**, 37–43.

Musk, A.W. and Tees, J.G. (1982) Asthma caused by occupational exposure to vanadium compounds. *Med. J. Aust.*, **1**, 183–4.

Nagy, L. and Orosz, M. (1984) Occupational asthma due to hexachlorophene. *Thorax*, **39**, 630–1.

Nakazawa, T. and Matsui, S. (1990) Ethylenediamine-induced late asthmatic responses. *J. Asthma*, **27**, 207–12.

National Institute for Allergy and Infectious Diseases task force report, (1979) Occupational and environmental respiratory disease, In *Asthma and Other Allergic Diseases* (ed. J.E. Salvaggio), National Institute of Health publication 79–387.

National Research Council (1981) *Formaldehyde and Other Aldehydes*, National Academy of Sciences, Washington, D.C.

Newman Taylor, A.J., Venables, K.M., Durham, S.R. *et al.* (1987) Acid anhydrides and asthma. *Int. Arch. Allergy Appl. Immunol.*, **82**, 435–9.

Nielsen, J., Fahraeus, C., Bensryd, I. *et al.* (1989) Small airways function in workers processing polyvinyl chloride. *Int. Arch. Occup. Environ. Health*, **61**, 427–30.

Nielsen, J., Bensryd, I., Almquist, H. *et al.* (1991) Serum IgE and lung function in workers exposed to phthalic anhydride. *Int. Arch. Occup. Environ. Health*, **63**, 199–204.

Niezborala, M. and Garnier, R. (1996) Allergy to complex platinum salts: a historical prospective cohort study. *Occup. Environ. Med.*, **53**, 252–7.

Nilsson, R., Nordliner, R., Wass, U. *et al.* (1993) Asthma, rhinitis, and dermatitis in workers exposed to reactive dyes. *Br. J. Ind. Med.*, **50**, 65–70.

Nordman, H., Keskinen, H. and Tuppuranainen, M. (1985) Formaldehyde asthma – rare or overlooked? *J. Allergy Clin. Immunol.*, **75**, 91–9.

Novey, H.S., Habib, M. and Wells, I.D. (1983) Asthma and IgE antibodies induced by chromium and nickel salts. *J. Allergy Clin. Immunol.*, **72**, 407–12.

O'Byrne, P.M., Dolovich, J. and Hargreave, F.E. (1987) Late asthmatic responses. *Am. Rev. Respir. Dis.*, **136**, 740–51.

Paggiaro, P.L. and Chan Yeung, M. (1987) Pattern of specific airway response in asthma due to western red cedar (*Thuja plicata*): relationship with length of exposure and lung function measurements. *Clin. Allergy*, **17**, 333–9.

Paggiaro, P.L., Loi, A.M. and Toma, G. (1979) Bronchial asthma and dermatitis due to spiramycin in a chick breeder. *Clin. Allergy*, **9**, 571–4.

Paggiaro, P.L., Paoletti, P., Bacci, E. *et al.* (1990) Eosinophils in bronchoalveolar lavage (BAL) of patients with toluene diisocyanate (TDI) asthma after cessation of work. *Chest*, **98**, 536–42.

Patterson, R., Dykewicz, M.S., Evans, R. III *et al.* (1989) IgG antibody against formaldehyde human serum proteins: a comparison with other IgG antibodies against inhalant proteins and reactive chemicals. *J. Allergy Clin. Immunol.*, **84**, 359–66.

Pepys, J. and Hutchcroft, B.J. (1975) Bronchial provocation tests in etiologic diagnosis and analysis of asthma. *Am. Rev. Respir. Dis.*, **112**, 829–59.

Pepys, J., Pickering, C.A.C. and Hughes, E.G. (1972) Asthma due to inhaled chemical agents – complex salts of platinum. *Clin. Allergy*, **2**, 391–6.

Pepys, J., Hutchcroft, B.J. and Bresliln, A.B.X. (1976) Asthma due to inhaled chemical agents – persulphate salts and henna in hairdressing. *Clin. Allergy*, **2**, 399–404.

Pepys, J., Parish, W.E., Cromwell, O. *et al.* (1979) Passive transfer in man and monkey of type I allergy due to heat labile and heat stable antibody to complex salts of platinum. *Clin. Allergy*, **9**, 99–108.

Perks, W.H., Burge, P.S., Rehahn, M. *et al.* (1979) Work-related respiratory disease in employees leaving an electronics factory. *Thorax*, **34**, 19–22.

Perrin, B., Malo, J.L., Cartier, A. *et al.* (1990) Occupational asthma in a pharmaceutical worker exposed to hydralazine. *Thorax*, **45**, 980–1.

Persson, C.G.A., Andersson, M., Greiff, L. *et al.* (1995a) Airway permeability. *Clin. Exp. Allergy*, **25**, 807–14.

Persson, C.G.A., Greiff, L., Andersson, M. *et al.* (1995b) Mucosal exudation of plasma in asthma and rhinitis, in *Asthma and Rhinitis* (eds W.W. Busse and S.T. Holgate), Blackwell Scientific Publications, Boston, pp. 733–41.

Pin, I., Gibson, P.G., Kolendowicz, R. *et al.* (1992) Use of induced sputum cell counts to investigate airway inflammation in asthma. *Thorax*, **47**, 25–9.

Polakoff, P.L., Lapp, N.L. and Reger, R. (1975) Polyvinyl chloride pyrolysis products. A potential cause for respiratory impairment. *Arch. Environ. Health*, **30**, 269–71.

Prince, N.L., Chamberlin, R.I., Hales, C.A. *et al.* (1984) Respiratory disease in tungsten carbide production workers. *Chest*, **86**, 549–57.

Pross, H.F., Day, J.H., Clark, R.H. *et al.* (1987) Immunologic studies of subjects with asthma exposed to formaldehyde and urea-formaldehyde foam insulation (UFFI) off products. *J. Allergy Clin. Immunol.*, **79**, 797–810.

Quirce, S., Cuevas, M., Olaguibel, J.M. *et al.* (1994) Occupational asthma and immunologic responses induced by inhaled carmine among employees at a factory making natural dyes. *J. Allergy Clin. Immunol.*, **93**, 44–52.

Rasanen, L., Kuusisto, P., Penttila, M. *et al.* (1994) Comparison of immunologic tests in the diagnosis of occupational asthma and rhinitis. *Allergy*, **49**, 342–7.

Raymond, L.W. (1996) Pulmonary abnormalities and serum immunoglobulins in facsimile machine repair technicians exposed to butyl methacrylate fume. *Chest*, **109**, 1010–18.

Reilly, M.J., Rosenman, K.D., Watt, F.C. *et al.* (1994) Surveillance for occupational asthma – Michigan and New Jersey, 1988–1992. *MMWR*, **43**, 9–17.

Rivolta, G., Nicoli, E., Ferretti, G. *et al.* (1994) Hard metal lung disorders: analysis of a group of exposed workers. *Sci. Total Environ.*, **150**, 161–5.

Roberts, A.E. (1951) Platinosis. A five year study of the effects of soluble salts on employees in a platinum laboratory and refinery. *Arch. Ind. Hyg.*, **4**, 549–9.

Robinson, D.S., Hamid, Q., Bentley, A. *et al.* (1993a) Activation of CD4+ cells, increased Th2-type cytokine mRNA expression, and eosinophil recruitment in bronchoalveolar lavage after allergen inhalation challenge in patients with atopic asthma. *J. Allergy. Clin. Immunol.*, **92**, 313–24.

Robinson, D.S., Hamid, Q., Sun Ying, *et al.* (1993b) Prednisone treatment in bronchial asthma. Clinical improvement is accompanied by reduction in bronchoalveolar lavage eosinophilia and modulation of n–4, n–5 and IFN-gamma cytokine gene expression. *Am. Rev. Respir. Dis.*, **148**, 401–6.

Romagnani, S. (1995) Biology of human TH1 and TH2 cells. *J. Clin. Immunol.*, **15**, 121–9.

Rosenman, K.D., Bernstein, D.I., O'Leary, K. *et al.* (1987) Occupational asthma caused by himic anhydride. *Scand. J. Work Environ. Health*, **13**, 150–4.

Ruegger, M. (1995) Lung disorders due to metals. *Schweiz. Med. Wochenschr.*, **125**, 467–74.

Ruff, F., Matteo, G., Dupuis, J.P. *et al.* (1979) Reactions broncho-pulmonaires et asthma dus as platine. Incidences chez les ouvriers du platine en region parisienne. *Rev. Mal. Respir.*, **7**, 206–8.

Sabbioni, E., Minoia, C., Petra, R. *et al.* (1994) Metal determinations in biological specimens of diseased and non-diseased hard metal workers. *Sci. Total Environ.*, **150**, 41–54.

Saetta, M., Fabbri, L.M., Danieli, D. *et al.* (1989) Pathology of bronchial asthma and animal models of asthma. *Eur. Respir. J.*, **2** (suppl. 6), 477–82.

Saetta, M., Di Stefano, A., Maestrelli, P. *et al.* (1992a) Airway mucosal inflammation in occupational asthma induced by toluene diisocyanate. *Am. Rev. Respir. Dis.*, **145**, 160–8.

Saetta, M., Maestrelli, P., Di Stefano, A. *et al.* (1992b) Effect of cessation of exposure to toluene diisocyanate (TDI) in bronchial mucosa of subjects with TDI-induced asthma. *Am. Rev. Respir. Dis.*, **145**, 169–74.

Saetta, M., Di Stefano, A., Maestrelli, P. *et al.* (1993) Activated T-lymphocytes and macrophages in bronchial mucosa of subjects with chronic bronchitis. *Am. Rev. Respir. Dis.*, **147**, 301–6.

Saetta, M., Maestrelli, P., Mapp, C.E. *et al.* (1995) Airway remodeling in isocyanate induced asthma. *Am. J. Respir. Crit. Care Med.*, **151**, 489–94.

Sala, C., Mosconi, G., Bacis, M. *et al.* (1994) Cobalt exposure in hard metal and diamond grinding tools manufacturing and in grinding processes. *Sci. Total Environ.*, **150**, 111–16.

Salari, H., Howard, S., Chan, H. *et al.* (1994) Involvement of immunologic mechanisms in a guinea pig model of western red cedar asthma. *J. Allergy Clin. Immunol.*, **93**, 877–84.

Saltini, C., Vinestock, K., Kirby, M. *et al.* (1989) Maintenance of alveolitis in patients with chronic beryllium disease by beryilium-specific helper T cells. *N. Engl. J. Med.*, **320**, 1103–9.

Saric, M. and Marelja, J. (1991) Bronchial hypersensitivity in potroom workers and prognosis after stopping exposure. *Br. J. Ind. Med.*, **48**, 653–5.

Savonius, B., Keskinen, H., Tuppurainen, M. *et al.* (1993) Occupational respiratory disease caused by acrylates. *Clin. Exp. Allergy.*, **23**, 416–24.

Schlueter, D.P., Banaszak, R.F., Fink, G.N. *et al.* (1978) Occupational asthma due to tetrachlorophthalic anhydride. *J. Occup. Med.*, **20**, 183–8.

Seaton, A., Cherrie, B. and Turnbull, J. (1988) Rubber glove asthma. *Br. Med. J.*, **296** (6621), 531–2.

Senthilselvan, A., McDuffie, H.H. and Dosman, J.A. (1992) Association of asthma with use of pesticides. Results of a cross-sectional survey of farmers. *Am. Rev. Respir. Dis.*, **146**, 884–7.

Shelhamer, J.H. (ed.) (1995) Airway inflammation. *Ann. Intern. Med.*, **123**, 288–304.

Shirakawa, T. and Morimoto, K. (1996) A brief reversible bronchospasm resulting from bichromate exposure. *Arch. Environ. Health*, **51**, 221–6.

Shirakawa, T., Kusaka, Y., Fujimura, N. *et al.* (1990) Hard metal asthma: cross immunological and respiratory activity between cobalt and nickel? *Thorax*, **45**, 267–71.

Shirakawa, T., Kusaka, Y. and Morimoto, K. (1992) Specific IgE antibodies to nickel in workers with known reactivity to cobalt. *Clin. Exp. Allergy*, **22**, 213–18.

Silberman, D.E. and Sorrell, A.H. (1959) Allergy in fur workers with special reference to paraphenylenediamine. *J. Allergy*, **30**, 11–18.

Sjogren, I., Hillerdal, G., Anderson, A. *et al.* (1980) Hard metal lung disease: importance of cobalt in coolants. *Thorax*, **35**, 653–9.

Slovak, A.J.M. (1981) Occupational asthma caused by the plastics blowing agent, azodicarbonamide. *Thorax*, **36**, 906–9.

Smedley, J. (1996) Is formaldehyde an important cause of allergic respiratory disease? *Clin. Exp. Allergy.*, **26**, 247–9.

Smith, P.A., Son, P.S., Callaghan, P.M. *et al.* (1996) Sampling and analysis of airborne resin acids and solvent-soluble material derived from heated colophony (rosin) flux: a method to quantify exposure to sensitizing compounds liberated during electronics soldering. *Toxicology*, **111**, 225–38.

Soyseth, V. and Kongerud, J. (1992) Prevalence of respiratory disorders among aluminum potroom workers in relation to exposure to fluoride. *Br. J. Ind. Med.*, **49**, 125–30.

Sporik, R., Holgate, S.T., Platts-Mills, T.A.E. *et al.* (1990) Exposure to house-dust mite allergen (Der p I) and the development of asthma in childhood. *N. Engl. J. Med.*, **323**, 502–7.

Steinbruegge, J.M. and Judson, F.N. (1980) Type I allergic reaction to orally administered tetracycline hydrochloride: a case report. *Sex. Trans. Dis.*, **7**, 193–4.

Stenton, S.C., Beach, J.R., Dennis, J.H. *et al.* (1994) Glutaraldehyde, asthma and work – a cautionary tale. *Occup. Med. (Oxf.)*, **44**, 95–8.

Stenton, S.C., Dennis, J.H. and Hendrick, D.J. (1995) Occupational asthma due to ceftazidime. *Eur. Respir. J.* **8**, 1421–3.

Sterling, G.M. (1967) Asthma due to aluminum soldering flux. *Thorax*, **22**, 533–7.

Tarlo, S.M. (1992) Occupational asthma induced by tall oil in the rubber tyre industry. *Clin. Exp. Allergy* **22**, 99–101.

Topping, M.D., Venables, K.M., Luczynska, C.M. *et al.* (1986) Specificity of the IgE response to inhaled acid anhydrides. *J. Allergy Clin. Immunol.*, **77**, 834.

Topping, M.D., Forster, H.W., Ide, C.W. *et al.* (1989) Respiratory allergy and specific immunoglobulin E and immunoglobulin G antibodies to reactive dyes used in the wool industry. *J. Occup. Med.* **31**, 857–62.

Tse, K.S., Chan, H. and Chan-Yeung, M. (1982) Specific IgE antibodies in workers with occupational asthma due to western red cedar. *Clin. Allergy*, **12**, 249–58.

Turner, E., Pruzansky, J.J. Patterson, R. *et al.* (1980) Detection of antibodies in human serum using trimellityl-erythrocytes: direct and indirect hemagglutination and hemolysis. *J. Clin. Exp.*, **39**, 470.

Underner, M., Cazenave, F. and Patte, F. (1987) Occupational asthma in the occupational environment. *Rev. Pneumol. Clin.*, **43**, 26–35.

Vallieres, M., Cockroft, D.W., Taylor, D.M. *et al.* (1977) Dimethyl ethanolamine-induced asthma. *Am. Rev. Respir. Dis.*, **115**, 867–71.

Vedal, S., Chan-Yeung, M., Enarson, D.A. *et al.* (1986) Plicatic acid-specific IgE and nonspecific bronchial hyperresponsiveness in western red-cedar workers. *J. Allergy Clin. Immunol.*, **78**, 1103–9.

Vedal, S., Enarson, D.A., Chan, H. *et al.* (1988) A longitudinal study of the occurrence of bronchial hyperresponsiveness in western red cedar workers. *Am. Rev. Respir. Dis.*, **137**, 651–5.

Venables, K.M. (1989) Low molecular weight chemicals, hypersensitivity, and direct toxicity: the acid anhydrides. *Br. J. Ind. Med.*, **46**, 222–32.

Venables, K.M., Topping, M.D., Nunn, A.J. *et al.* (1987) Immunologic and functional consequences of chemical (tetrachlorophthalic anhydride)-induced asthma after four years of exposure. *J. Allergy Clin. Immunol.*, **80**, 212–18.

Venables, K.M., Dally, M.B., Nunn, A.J. *et al.* (1989) Smoking and occupational allergy in workers in a platinum refinery. *Br. Med. J.*, **299**, 939–42.

Walker, C., Bode, E., Boer, L. *et al.* (1992) Allergic and intrinsic asthmatics have distinct patterns of T-cell activation and cytokine production in peripheral blood and BAL. *Am. Rev. Respir. Dis.*, **148**, 109–15.

Walker, P.R., Fellowes, R., Hecht, E.M. *et al.* (1991) Characterization of streptococcal antigen-specific CD8+ II; MHC class I-restricted T cell clones that down-regulate *in vitro* antibody synthesis. *J. Immunol.*, **147**, 3370–80.

Wantke, F., Demmer, C.M., Tappler, P. *et al.* (1996) Exposure to gaseous formaldehyde includes IgE-mediated sensitization to formaldehyde in children. *Clin. Exp.*, **26**, 276–80.

Wass, U. and Belin, L. (1990) An *in vitro* method for predicting sensitizing properties of inhaled chemicals. *Scand. J. Work. Environ. Health.*, **16**, 208–14.

Wass, U., Nilsson, R., Nordlinder, R. *et al.* (1990) An optimized assay of specific antibodies to reactive dyes and studies of immunologic responses in exposed workers. *J. Allergy. Clin. Immunol.*, **85**, 642–8.

Wax, P.M. and Hoffman, R.S. (1994) Fatality associated with inhalation of a pyrethrin shampoo. *J. Toxicol. Clin. Toxicol.*, **32**, 457–60.

Weiner, A. (1961) Bronchial asthma due to organic phosphate insecticides. *Ann. Allergy*, **15**, 211–12.

Wenzel, S.E. (1994) Asthma as an inflammatory disease. *Semin. Respir. Med.*, **15**, 106–16.

Wernfors, M., Nielsen, J., Schutz, A. *et al.* (1986) Phthallic anhydride-induced occupational asthma. *Int. Arch. Allergy Appl. Immunol.*, **79**, 77–82.

Wilhelmsson, B. and Holmstrom, M. (1987) Positive formaldehyde-RAST after prolonged formaldehyde exposure by inhalation. *Lancet*, **ii**, 164.

Wrbitzky, R., Drexler, H. and Letzel, S. (1995) Early reaction type allergies and diseases of the respiratory passages in employees from persulphate production. *Int. Arch. Occup. Environ. Health*, **67**, 413–17.

Wuthrich, B. and Hartmann, A.L. (1982) Occupation-related bronchial asthma caused by ampicillin. Diagnostic significance of occupation-specific inhalation provocation tests. *Schweiz. Med. Wochenschr.*, **112**, 1046–8.

Zeiss, C.R., Patterson, R., Pruzansky J.J. *et al.* (1977) Trimellitic anhydride-induced airway syndromes: clinical and immunologic studies. *J. Allergy Clin. Immunol.*, **60**, 96–103.

Zeiss, C.R., Wolkonsky, P., Chcon, R. *et al.* (1983) Syndromes in workers exposed to trimellitic anhydride. A longitudinal clinical and immunologic study. *Ann. Intern. Med.*, **98**, 8–12.

Zeiss, C.R., Mitchell, J.H., VanPeenen, P.F. *et al.* (1992) A clinical and immunologic study of employees in a facility manufacturing trimellitic anhydride. *Allergy Proc.*, **13**, 193–8.

Zetterstrom, O., Nordvall, S.I., Bjorksten, B. *et al.* (1985) Increased IgE antibody responses in rats exposed to tobacco smoke. *J. Allergy Clin. Immunol.*, **75**, 594–8.

Zocca, E., Fabbri, L.M., Boschetto, P. *et al.* (1990) Leukotriene B_4 and late asthmatic reactions induced by toluene diisocyanate. *J. Appl. Physiol.*, **68**, 1576–89.

24 Isocyanate-induced asthma

Cristina E. Mapp

INTRODUCTION

Diisocyanates are being used extensively in a number of important industries all over the world. They are synthesized by the reaction of amines or their hydrochlorides with phosgene, all have an –N=C=O– group attached to a radical, and react exothermically with compounds containing active hydrogen atoms, to form a polyurethane. Isocyanates may be modified in many ways to give products with differing chemical and physical properties. Several aromatic and aliphatic isocyanates are available, but about 95% of all polyurethanes are based on two of them. These are toluene diisocyanate (TDI) and 4.4-diisocyanate-diphenlymethane (MDI). Most of the TDI used is a mixture of the two isomers 2.4 and 2.6 in an 80 : 20 mix. A 65 : 35 mix is also available. MDI is produced from aniline and formaldehyde, reacted together using hydrochloric acid as a catalyst. Other aromatic diisocyanates include 1.5-diisocyanato-naphthalene (NDI), and 1.4-diisocyanato-benzene (PPDI). The most widely available aliphatic diisocyanates, which are much less reactive than the aromatic diisocyanates, are 1-isocyanato-3-isocyanatemethyl-3,5.5-trimethylcyclohexane (IPDI), 1.6-diisocyanate-hexane (HDI), and 4.4-diisocyanate-dicyclohexylmethane (HMDI). TDI, NDI and MDI prepolymers are also available, and they have been shown to give rise to occupational asthma (Chan-Yueng and Malo, 1994).

HAZARDS AND SAFE PRACTICES FOR HANDLING ISOCYANATES

The hazards involved in making and using polyurethanes are those affecting the worker in the factory, the safety of the people outside the factory, and the customer using the final product. The threshold limit values (TLVs) and the ceiling levels (TLV-C) for a number of diisocyanates are respectively 0.02 and 0.005 ppm. Handling of isocyanates should only be done by trained personnel in positively vented enclosures. The main hazard in handling of isocyanates arises from the possible inhalation of isocyanate vapor, dust or aerosol droplets. The hazard is lowest when exposure is to polymeric MDI that has a low vapor pressure, and does not produce much vapor at room temperature. Emergency and safety equipment should be immediately available on the job. Isocyanate bulk storages must be kept separated from the workplace and from any inflammable materials. Any spillage of isocyanates should be decontaminated immediately. Aqueous liquid decontaminants are the best for minor spills and for treating contaminated clothing or equipment, whereas absorbent solid decontaminants are the best for larger spills of liquid isocyanates. An important consideration is that the only certain method of confirming the safe operations in a factory is the regular analysis of the air inside the factory. When discontinuous operations are the rule in the workplace, individual samples are

Occupational Lung Disease. Edited by Daniel E. Banks and John E. Parker. Published in 1998 by Chapman & Hall. ISBN 0 412 73630 6

essential to obtain a record of exposure. To measure low concentrations of isocyanate vapor in air, many methods are available, including chromatographic separation and dry and wet colorimetry.

Safe practices for handling isocyanates also include regular medical checks of employees, including lung function tests and measurement of non-specific airway responsiveness.

MEDICAL PROBLEMS

Medical problems have been reported following the first commercial production of toluene diisocyanate. In 1951, Fuchs and Valade described asthma in seven of nine subjects exposed to TDI. Asthma induced by exposure to TDI has been recognized as a leading cause of occupational asthma (Mapp *et al.*, 1988a), although in recent years, HDI and MDI seem to be more important as respiratory sensitizers (Vandenplas *et al.*, 1993a). The prevalence of TDI-induced asthma varies from 5% to 10% (Vandenplas *et al.*, 1993a), but estimates as high as 15% have been suggested (Luo *et al.*, 1990). The dose and length of exposure necessary to induce asthma are not known. There are numerous reports suggesting that a high level of exposure, e.g. a spill, may induce asthma (Luo *et al.*, 1990). A reactive airways dysfunction syndrome (RADS)-type of occupational asthma after a high exposure to isocyanates could not be excluded. In animal studies, synthesis of IgG_1 and of IgE antibodies has been reported in TDI hypersensitivity reactions (Karol, 1983; Mapp *et al.*, 1996). In guinea-pigs, TDI was absorbed mainly in the upper airways, and the uptake correlated with the concentration in inhaled air (Kennedy and Brown, 1989; Kennedy *et al.*, 1989). Labeled TDI was found in the epithelium and subepithelium from the nose to the terminal bronchioles. In the blood and in the airways, TDI was conjugated with a 70 kDa molecular weight protein, which was laminin in the airways.

When humans are exposed to high levels of isocyanates, for example 2 ppm, symptoms include acute inflammation, lacrimation, rhinitis, coughing, and burning in the throat and upper chest. Once an individual has been sensitized to TDI, levels of 1 ppb may initiate an attack of asthma. The pulmonary responses include isolated immediate, isolated late, biphasic or recurrent late asthmatic reactions (Cartier, 1994). Late asthmatic reactions are usually accompanied by an increase in non-specific airway responsiveness (Mapp *et al.*, 1986). Non-specific airway hyper-responsiveness is often associated with diisocyanate asthma, though the latter condition has frequently been reported in the absence of hyper-responsiveness. Once symptomatic, the persistence of asthma is variable, but symptoms remain in the majority of individuals even after several years of cessation of exposure (Mapp *et al.*, 1988b; Paggiaro *et al.*, 1994). The prognosis is better when both diagnosis and removal from exposure occur early after the diagnosis.

MECHANISMS OF ISOCYANATE ASTHMA

Although a number of mechanisms have been proposed to explain isocyanate-induced asthma e.g. immunologic, pharmacologic and neurogenic inflammation, the actual mechanism is not known (Mapp *et al.*, 1994a). Recently, a role for genetic factors in TDI-induced asthma has been shown, with certain HLA alleles which confer protection and others which convey susceptibility (Bignon *et al.*, 1994; Balboni *et al.*, 1996). Since asthma is likely to be a multifactorial disease, environmental factors may be necessary for its expression. It is likely that two factors such as exposure to the sensitizing agent and inheritance of the genetic marker are necessary for the expression of TDI asthma.

Although specific IgE antibodies are present in 10–20% of the subjects with a positive inhalation challenge with TDI, the immunologic mechanism seems to be the favorite mechanism in this type of asthma. In isocyanate-induced asthma, cell-mediated immunity may play a role. In fact, in subjects with TDI asthma, activated CD4 T-lymphocytes (CD25+), activated eosinophils and mast cells have been shown in the bronchial mucosa (Bentley *et al.*, 1992; Saetta *et al.*, 1992a). TDI-induced asthma has a cell infiltrate similar to that of non-occupational asthma, and T-lymphocytes and eosinophils are the key cells. Fragments of bronchial mucosa of subjects with TDI asthma, cultured in the presence of interleukin-2 (IL-2), with the aim to activate T-cells in the tissue and to clone T-blasts, produced clones which exhibited the CD8 phenotype (Maestrelli *et al.*, 1994a). All the CD8 clones produced interferon-gamma, and 44% produced interleukin-5 (IL-5). An increase of CD8+ cells and of eosinophils has been observed in the peripheral blood of subjects who developed a late asthmatic reaction after exposure to TDI (Finotto *et al.*, 1991). The inflammatory response of the airways in TDI asthma is characterized not only by increased numbers of mucosal eosinophils and mast cells, but also by activation of lymphocytes and by chronic expression of proinflammatory cytokines. In fact, subjects with TDI asthma had increased numbers of cells expressing CD25 and VLA-1 (markers of activation of lymphocytes), and tumor necrosis factor-alpha (TNFα) and interleukin-1beta (IL-1β) immunoreactivity was also increased (Maestrelli *et al.*, 1995). Airway inflammation in TDI asthma has also been documented with other techniques. An early increase in neutrophils followed by an increase in eosinophils, together with an increase in albumin concentration in bronchoalveolar lavage (BAL) fluid has been shown in TDI asthma (Fabbri *et al.*, 1987). After challenge with TDI, leukotriene B_4 (LTB_4) levels were also increased in BAL fluid (Zocca *et al.*, 1990). In sputum induced by inhalation of ultrasonically nebulized hypertonic saline, eosinophilia was observed both in early and late reactors 8 and 24 hours after TDI/MDI challenges (Maestrelli *et al.*, 1994b). In the peripheral blood of sensitized asthmatic subjects who developed a late asthmatic reaction after exposure to TDI, the serum eosinophil cationic protein (ECP) levels were increased (Mapp *et al.*, 1994b). Cessation of exposure to TDI has been shown to induce a reduction in the thickness of the reticular layer of the basement membrane, though the inflammatory cell infiltrate in the airways of sensitized subjects persists (Saetta *et al.*, 1992b). The development of subepithelial fibrosis, at least in TDI asthma, may be due to the presence of resident cells, since the thickness of the reticular layer of the basement membrane was correlated to the number of mast cells and fibroblasts (Di Stefano *et al.*, 1993). The occurrence of airway wall remodeling after cessation of exposure to isocyanates in sensitized asthmatic subjects has been confirmed by a recent study which showed a reduced thickness of subepithelial fibrosis and a reduced number of subepithelial fibroblasts, mast cells and lymphocytes in the bronchial mucosa with the avoidance of the sensitizing agent (Saetta *et al.*, 1995). The importance of airway inflammation in TDI asthma is further confirmed by the lung pathology of a TDI-sensitized asthmatic car painter who died at work after exposure to diisocyanates (Fabbri *et al.*, 1988). In this patient the picture was similar to that observed in the status asthmaticus of non-occupational asthma. The effects on health of MDI and HDI exposure have been studied to a lesser degree than those of TDI. In cases of occupational asthma resulting from exposure to HDI and MDI (Zamit-Tabona *et al.*, 1983; Vandenplas *et al.*, 1993b), both specific IgE and specific IgG antibodies to MDI and HDI protein conjugate were detected. However, whether such presence of antibodies results from exposure to both MDI and HDI is not clear, as hypersensitivity pneumonitis has been reported more frequently after exposure to MDI than to TDI (Vandenplas *et al.*, 1993a).

DIAGNOSIS

Diagnosis of diisocyanate asthma is difficult when symptoms occur several hours after exposure, and sometimes a specific bronchial inhalation challenge is required to identify the sensitizing agent. Medical history should include personal, family and occupational questions which can lead to an exact diagnosis. Determination of non-specific airway responsiveness, and peak expiratory flow (PEF) monitoring may be useful in the diagnostic procedure. Skin testing with diisocyanates has not proven useful as an indicator of diisocyanate asthma. The absence of specific antibodies does not refute or confirm the diagnosis of diisocyanate asthma (Cartier *et al.*, 1989). A specific bronchial challenge with diisocyanates remains the 'gold standard' in the confirmation of the diagnosis of occupational asthma, but its use is limited to specialized centers with trained personnel where resuscitative measures are available (Cartier, 1994). However, a negative test does not exclude the diagnosis of occupational asthma when a worker has not been exposed to the workplace for several months.

TREATMENT

When a worker develops diisocyanate asthma, he or she should be removed from the work environment, though on many occasions such a change in employment is not possible due to personal and financial reasons. In such cases, the worker must avoid or limit the exposure to diisocyanates in the workplace, and a pharmacologic treatment is indicated. The medical treatment of diisocyanate asthma is similar to that of non-occupational asthma and consists of inhaled steroids as controller medication plus inhaled short-acting beta$_2$-agonists as reliever medication. In the presence of exacerbations, repetitive administration of inhaled short-acting beta$_2$-agonists is usually necessary to relieve airflow limitation as quickly as possible. Oral corticosteroids should be used to speed resolution of exacerbation. In addition, the use of a peak flow meter is indicated, not only to monitor exposure to the sensitizing agent in the workplace, but also to follow the effects of therapy.

SUMMARY

In conclusion, diisocyanate-induced asthma forms a major proportion of cases of occupational asthma. The detection of asthma requires detailed inquiries to be made about the worker's occupation as part of the clinical history. Confirmation of asthma should be made with a specific inhalation challenge, or with peak flow monitoring at the subject's home and workplace. Once a diagnosis has been established, complete avoidance of exposure to diisocyanates is mandatory in order to permit remission of asthma, though in many cases the condition may not be completely reversible. Pharmacologic therapy for diisocyanate-induced asthma is identical to that used in other forms of asthma, but is less effective than avoidance of exposure. Adequate hygiene measures in the workplace are important in preventing sensitization to isocyanates, while atopy and tobacco smoking do not predispose workers to higher risk in jobs where these compounds are used.

REFERENCES

Balboni, A., Baricordi, O.R., Fabbri, L.M., Gandini, E., Ciaccia, A. and Mapp, C.E. (1996) Association between toluene diisocyanate-induced asthma and DQBI markers: a possible role for aspartic acid at position 57. *Eur. Respir. J.*, **9**, 207–10.

Bentley, A.M., Maestrelli, P., Saetta, M., Fabbri, L.M., Robinson, D.S., Bradley, B.L., Jeffery, P.K., Durham, S.R. and Kay, A.B. (1992) Activated T-lymphocytes and eosinophils in the bronchial mucosa in isocyanate-induced asthma. *J. Allergy. Clin. Immunol.*, **89**, 821–9.

Bignon, J.S., Aron, Y., Ju, L.Y. *et al.* (1994) HLA class II alleles in isocyanate-induced asthma. *Am. J. Respir. Crit. Care Med.*, **149**, 71–5.

Cartier, A. (1994) Definition and diagnosis of occupational asthma. 1994. *Eur. Respir. J.*, **7**, 153–60.

Cartier, A., Grammer, L. and Malo, J.L. (1989) Specific serum antibodies against isocyanates: association with occupational asthma. *J. Allergy Clin. Immunol.*, **84**, 507–14.

Chan-Yeung, M. and Malo, J.-L. (1994) Aetiological agents in occupational asthma. *Eur. Respir. J.*, **7**, 346–71.

Di Stefano, A., Maestrelli, P., Mapp, C.E., Fabbri, L.M. and Saetta, M. (1993) The reduction of airway subepithelial fibrosis after cessation of exposure to toluene diisocyanate (TDI) in sensitized subjects is associated with a reduction of mast cells and fibroblasts. *Am. Rev. Respir. Dis.*, **147**, A557.

Fabbri, L.M., Boschetto, P., Zocca, E., Milani, G.F., Pivirotto, F., Plebani, M., Burlina, A., Licata, B. and Mapp, C.E. (1987) Bronchoalveolar neutrophilia during late asthmatic reactions induced by toluene diisocyanate. *Am. Rev. Respir. Dis.*, **136**, 36–42.

Fabbri, L.M., Danieli, D., Crescioli, S., Bevilacqua, P., Meli, S., Saetta, M. and Mapp, C.E. (1988) Fatal asthma in a subject sensitized to toluene diisocyanate. *Am. Rev. Respir. Dis.*, **137**, 1494–8.

Finotto, S., Fabbri, L.M., Rado, V., Mapp, C.E. and Maestrelli, P. (1991) Increase in numbers of CD8 positive lymphocytes and eosinophils in peripheral blood of subjects with late asthmatic reactions induced by toluene diisocyanate. *Br. J. Ind. Med.*, **48**, 116–21.

Fuchs, S. and Valade, P. (1951) Etude clinique et experimentale sur quelques cas d'intoxication par le desmodur T (diisocyanate de tolylene 1–2–4 et 1–2–6). *Arch. Mal. Prof.*, **12**, 191–6.

Karol, M.H. (1983) Concentration-dependent immunologic response to toluene diisocyanate (TDI) following inhalation exposure. *Toxicol. Appl. Pharmacol.*, **68**, 229–41.

Kennedy, A.L. and Brown, W.E. (1989) Modification of airway proteins and induction of secondary responses by inhalation exposure to isocyanates. *Am. Rev. Respir. Dis.*, **139**, 387A.

Kennedy, A.L., Stock, M.F., Alarie, Y. and Broen, W.E. (1989) Uptake and distribution of 14C during and following inhalation exposure to radioactive toluene diisocyanate. *Toxicol. Appl. Pharmacol.*, **100**, 280–92.

Luo, J.C., Nelsen, K.G. and Fiscbein, A. (1990) Persistent reactive airway dysfunction syndrome after exposure to toluene diisocyanate. *Br. J. Ind. Med.*, **47**, 239–41.

Maestrelli, P., Del Prete, G.F., De Carli, M., D'Elios, M.M., Saetta, M., Di Stefano, A., Mapp, C.E., Romagnani, S. and Fabbri, L.M. (1994a) CD8 T-cell clones producing interleukin-5 and interferon-gamma in bronchial mucosa in toluene diisocyanate asthma. *Scand. J. Work. Environ. Health*, **20**, 376–81.

Maestrelli, P., Calcagni, P.G., Saetta, M., Di Stefano, A., Hosselet, J.J., Santonastaso, A., Fabbri, L.M. and Mapp, C.E. (1994b) Sputum eosinophilia after asthmatic responses induced by isocyanates in sensitized subjects. *Clin. Exp. Allergy*, **24**, 29–34.

Maestrelli, P., Di Stefano, P., Occari, P., Turato, G., Milani, G.F., Pivirotto, F., Mapp, C.E., Fabbri, L.M. and Saetta, M. (1995) Cytokines in the airway mucosa of subjects with asthma induced by toluene diisocyanate. *Am. J. Respir. Crit. Care. Med.*, **151**, 607–12.

Mapp, C.E., Boschetto, P., Dal Vecchio, L., Maestrelli, P. and Fabbri, L.M. (1988a) Occupational asthma due to isocyanates. *Eur. Respir. J.*, **1**, 273–9.

Mapp, C.E., Chiesura Corona, P., De Marzo, N. and Fabbri L.M. (1988b) Persistent asthma due to isocyanates. A follow-up study of subjects with occupational asthma due to toluene diisocyanate (TDI). *Am. Rev. Respir. Dis.*, **137**, 1326–9.

Mapp, C.E., Di Giacomo, G.R., Omini, C., Broseghini, C. and Fabbri, L.M. (1986) Late, but not early, asthmatic reactions induced by toluene diisocyanate are associated with increased airway responsiveness to methacholine. *Eur. J. Respir. Dis.*, **69**, 276–84.

Mapp, C.E., Lapa e Silva, J.R., Lucchini, R.E., Chitano, P., Rado, V., Saetta, M., Pretolani, M., Karol, M.H., Maestrelli, P. and Fabbri, L.M. (1996) Inflammatory events in the blood and in airways of guinea pigs immunized to toluene diisocyanate (TDI). *Am. J. Respir. Crit. Care. Med.*, **154**, in press.

Mapp, C.E., Saetta, M., Maestrelli, P., Di Stefano, A., Chitano, P., Boschetto, P., Ciaccia, A. and Fabbri, L.M. (1994a) Mechanisms and pathology of occupational asthma. *Eur. Respir. J.*, **7**, 544–54.

Mapp, C.E., Plebani, M., Faggian, D., Maestrelli, P., Saetta, M., Calcagni, P., Borghesan, F. and Fabbri, L.M. (1994b) Eosinophil cationic protein (ECP), histamine and tryptase in peripheral blood before and during inhalation challenge with toluene diisocyanate (TDI) in sensitized subjects. *Clin. Exp. Allergy*, **24**, 730–6.

Paggiaro, P.L., Vagaggini, B., Bacci, E. *et al.* (1994) Prognosis of occupational asthma. *Eur. Respir. J.*, **7**, 761–7.

Saetta, M., Di Stefano, A., Maestrelli, P., De Marzo, N., Milani, G.F., Pivirotto, F., Mapp, C.E. and Fabbri, L.M. (1992a) Airway mucosal inflammation in occupational asthma induced by toluene diisocyanate. *Am. Rev. Respir. Dis.*, **145**, 160–8.

Saetta, M., Maestrelli, P., Di Stefano, A., De Marzo, N., Milani, G.F., Pivirotto, F., Mapp, C.E. and Fabbri, L.M. (1992b) Effect of cessation of exposure to toluene diisocyanate (TDI) on bronchial mucosa of subjects with TDI-induced asthma. *Am. Rev. Respir. Dis.*, **145**, 169–74.

Saetta, M., Maestrelli, P., Turato, G., Mapp, C.E., Milani, G.F., Pivirotto, F., Fabbri, L.M. and Di Stefano, A. (1995) Airway wall remodeling after cessation of exposure to isocyanates in sensitized asthmatic subjects. *Am. J. Respir. Crit. Care Med.*, **151**, 489–94.

Vandenplas, O., Malo, J.-L., Saetta, M., Mapp, C.E. and Fabbri, L.M. (1993a) Occupational asthma and extrinsic alveolitis due to isocyanates: current status and perspectives. *Br. J. Ind. Med.*, **50**, 213–28.

Vandenplas, O., Malo, J.-L. and Dugas, M. (1993b) Hypersensitivity pneumonitis-like reaction among workers exposed to diphenylmethane diisocyanate (MDI). *Am. Rev. Respir. Dis.*, **147**, 338–46.

Zammit-Tabona, M., Sherkin, M., Kijek, K., Chan, H. and Chan-Yeung, M. (1983) Asthma caused by diphenylmethane diisocyanate in foundry workers. Clinical, bronchial provocation and immunologic studies. *Am. Rev. Respir. Dis.*, **128**, 226–31.

Zocca, E., Fabbri, L.M., Boschetto, P., Plebani, M., Masiero, M., Milani, G.F., Pivirotto, F. and Mapp, C.E. (1990) Leukotriene B_4 and late asthmatic reactions induced by toluene diisocyanate. *J. Appl. Physiol.*, **68**, 1576–89.

25 Occupational asthma due to high-molecular weight agents

Jeremy R. Beach,
Francis Thien and
E. Haydn Walters

INTRODUCTION

Physically large allergens of high molecular weight constitute a group of common and potent agents capable of causing occupational asthma. There have been numerous and increasingly frequent reports of these allergens causing disease, and despite some changes in work processes and environments, new cases continue to occur and new causes continue to be recognized. The SWORD (Surveillance of Work-related and Occupational Respiratory Disease) scheme has consistently reported flour and grains, laboratory animals and enzymes among the ten most frequently cited causes of occupational asthma in the UK (Meredith and McDonald, 1994; Sallie *et al.*, 1994). Organic agents, the group containing most large molecular weight antigens, accounted for approximately 25% of all new cases of occupational asthma in 1989–1991, and approximately 30% in 1993. Thus, not only are large-molecular weight antigens commonly cited as a cause of asthma, but they have been reported with increasing frequency, in contrast to the most frequently cited and best known cause of occupational asthma, isocyanate exposure.

A 1991 survey in British Columbia, Canada, reported that grain dust, wheat flour and animal products account for approximately 7% of new cases of occupational asthma (Contreras *et al.*, 1994). This lower estimate of incidence may in part reflect unusual patterns of employment opportunities within the locality, as >40% of all reported cases of occupational asthma were attributed to one agent, western red cedar dust. Another Canadian study noted a higher proportion of cases of occupational asthma due to large-molecular weight antigens. Approximately 12% of 469 cases of work-related asthma assessed by the Ontario Workers' Compensation Board were attributed to flour, grain, enzymes, latex or animals (Tarlo *et al.*, 1995).

Although not clearly defined, large-molecular weight allergens are usually accepted as those with a molecular weight >5–20 kDa (Novey *et al.*, 1989; Chan-Yeung and Malo, 1995). There are some similarities between the various agents, most of which are derived from animal, fungal and plant proteins, and it may be that they act via a similar IgE-mediated mechanism, although this is not yet clear. The main groups of allergens in this category are:

- laboratory animals and insects;
- flours and grains;
- enzymes;
- seafoods and other marine organisms;
- other food-derived products; and
- gums and latex.

In real-life situations there is often considerable overlap in exposure and sensitization to each of these

Occupational Lung Disease. Edited by Daniel E. Banks and John E. Parker. Published in 1998 by Chapman & Hall.
ISBN 0 412 73630 6

groups. For example, bakers may be exposed to flours and grains, insects and enzyme products simultaneously.

LABORATORY ANIMALS AND INSECTS

Laboratory animals

Keeping large numbers of animals in research laboratories involves close interaction between keeper and animal on a regular basis. Not only are the researchers themselves exposed, but inevitably so are others working in the same areas. While the number of people engaged in this work and regularly exposed to laboratory animals is not known, it is probably tens of thousands worldwide. In 1994, there were approximately 16 000 personal and 5000 project animal experimentation licence holders in the UK. Approximately 2 800 000 experiments involving animals were undertaken (Home Office, 1995). Most of the animals used are small rodents, especially rats and mice, but rabbits, guinea pigs and hamsters are also used in considerable numbers, and larger animals, such as cats, dogs, sheep, pigs and primates are used in some laboratories.

Although there have been reports of asthma related to animals since the 18th century (Salter 1860), it was not until more recently that much was known of this problem. A case report of occupationally induced allergy to mice (Sorrel and Gottesman, 1957) was followed by a case series of allergy to laboratory animals at the Karolinska Hospital (Rajka, 1961). However, the importance of the problem only began to become apparent as systematic surveys were published. Lincoln *et al.* (1974) reported laboratory animal allergy symptoms in 11.3% of 238 technicians, scientists and animal handlers with frequent exposure at a US laboratory. Thirteen of the 238 (5.5%) subjects noted asthma symptoms related to work. Ten subjects had modified their work pattern to reduce or eliminate exposure to animals because of the severity of their allergic symptoms. Positive prick tests to animal allergens occurred in 24.3% of subjects in a biology laboratory, suggesting that a higher proportion of subjects had become sensitized than had developed disease. Lutsky and Neuman (1975) reported laboratory animal allergy symptoms in approximately 15% of 1293 workers at 39 US research facilities. Asthma was present in 71% of those subjects with symptoms of laboratory animal allergy (i.e. approximately 10% overall).

Subsequently, cross-sectional studies suggest that symptoms of laboratory animal allergy occur in 20% to 30% of laboratory animal workers and work-related asthma occurs in 5% to 12% of this group (Table 25.1). One report described the incidence of symptoms of asthma since commencing employment as 3.1% among exposed workers, but only 0.7% among unexposed workers in the same laboratories (Davies and McArdle, 1981). Latency – that is, the time since commencing employment in the laboratories – was not stated. These studies were liable to both selection bias and a survivor effect, perhaps underestimating the true prevalence of disease in exposed populations. There have been few longitudinal studies to address these issues.

Botham *et al.* (1987) recruited a series of yearly cohorts of new employees working in a laboratory between 1979 and 1984, and followed each cohort for up to 3 years. The incidence of allergic symptoms in the first year of work decreased during the study (28% in 1979, 37% in 1980, 37% in 1981, 20% in 1982, 10% in 1983 and 12% in 1984) as the laboratory introduced workplace controls on exposure. The incidence of symptoms also decreased as the duration of employment increased, with two-thirds of subjects developing allergy symptoms during the first year. Chest symptoms were not comprehensively reported, but six subjects developed symptoms during the first year of exposure, representing an annual incidence of 1.6%. In another well-designed, longitudinal study, 225 students from five laboratory technician training schools were investigated at recruitment and approximately 4 years later towards the end of their training (Renstrom *et al.*, 1994). Of the 225 students, 168 had worked in a laboratory. Of these, 43 had been meaningfully exposed at work to laboratory animals and 38 participated in follow-up. No significant differences in pre-employment prevalence of respiratory symptoms, lung function or atopy were found between those who later worked with animals and those who did not, although subjects later working with animals had a non-significant trend to have a lower prevalence of positive skin-prick tests (21% versus 37%) to common allergens. Seven of these 38 subjects had developed a positive skin-prick test to one or more animal hair extracts, and six of these plus a further two subjects (21% in all) reported the development of allergic symptoms related to animal exposure. Six (16%) exposed subjects developed new symptoms of asthma during the follow-up. Total serum IgE levels at the outset were higher among those who subsequently developed symptoms related to exposure, with 7/8 having an IgE >100 kU/l. The 36 exposed subjects were matched with 36 unexposed subjects based on age, sex and smoking. In this paired analysis, atopy and allergic symptoms increased in both the animal-exposed group and the unexposed group, so that the differences between the two groups were not statistically significant. It remained unclear what had caused this increase in atopy and symptoms among

Table 25.1 Cross-sectional surveys of the prevalence of laboratory animal allergy and asthma

Authors	*Study population*	*Study methods*	*Prevalence of lab animal allergy*	*Prevalence of lab animal asthma*
Lincoln *et al.* (1974)	404 staff (238 exposed) Oak Ridge National Labs	Review of records, skin tests to common and animal allergens	27/238 (11.3%) with frequent exposure	13/238 (5.5%) with frequent exposure
Lutsky and Neuman (1975)	1293 exposed staff at 39 animal research facilities.	Questionnaire to lab director	191/1293 (14.7%) noted symptoms of allergy to animals	Of those with allergy: 71% noted asthma, 58% noted cough
Taylor *et al.* (1976)	474 animal workers at two labs	Questionnaire	23% noted symptoms within 12 h of animal contact	9% reported chest symptoms
Cockcroft *et al.* (1981)	213 (179 exposed) staff at three UK Medical Research Council labs	Questionnaire, spirometry, skin tests to common/animal allergens	49/179 (27.4%) exposed noted allergic symptoms related to work	21/179 (11.7%) exposed noted asthma symptoms related to work
Slovak and Hill (1981)	146 exposed staff at two pharmaceutical company labs	Questionnaire, spirometry, skin tests to common allergens	48/146 (32.9%) noted animal-related symptoms	15/146 (10.3%) with symptoms. FEV_1/FVC lower in asthmatics
Davies and McArdle (1981)	1487 staff (585 exposed) at four labs	Questionnaire	Symptoms developed in: 19.5% exposed 13.4% unexposed	Symptoms developed in: 3.1% exposed, 0.7% unexposed
Beeson *et al.* (1983)	390 (69 exposed) staff at a pharmaceutical research site	Questionnaire, skin tests to common/ animal allergens	15/69 (21.7%) of exposed subjects noted animal-related symptoms	3/69 (4.3%) noted animal related asthma symptoms
Agrup *et al.* (1986)	125 current and 79 previously exposed staff at university labs	Questionnaire, spirometry, skin tests to common/animal allergens	27.4% current and 35.4% previous staff noted animal-related symptoms	8.1% current and 7.7% previous staff noted animal related symptoms
Venables *et al.* (1988a)	138 exposed staff in a pharmaceutical lab	Questionnaire, skin tests to common/ animal allergens	60/138 (43%) noted animal-related symptoms	15/138 (10.9%) noted animal-related chest symptoms
Aoyama *et al.* (1992)	5641 exposed staff from 137 labs. 252 unexposed	Questionnaire	23.1% noted animal-related symptoms	6.4% noted animal-related chest symptoms
Bryant *et al.* (1995)	228 (130 exposed) staff at university/ college/hospital	Questionnaire, spirometry, skin tests – common/animal allergens	56% exposed 0% unexposed noted animal-related symptoms	22% exposed 5% unexposed noted animal-related symptoms

the unexposed group, and further analysis was not reported.

Relatively few papers have assessed the potency of different laboratory animals as causes of asthma. Lutsky and Neuman (1975) reported the prevalence of allergic symptoms attributed to specific animal exposures. In 191 subjects with allergic symptoms following animal exposure, 56% noted symptoms precipitated by rats, 37% by rabbits, 37% by mice, 24% by guinea pigs, 13% by cats, 10% by dogs and 2% by hamsters. The relative use, and hence exposure to each animal, was not reported. Many subjects reported allergic symptoms with exposure to more than one species. In 15 subjects with laboratory animal allergy, positive skin-prick tests to rat allergens were more common than reactions to mouse, guinea pig or rabbit allergens, but relative frequency of exposure to each type of animal was again not reported (Beeson *et al.*, 1983). Of those with laboratory animal asthma, 67% had a positive skin test to an animal allergen compared with just 2% of unexposed workers, and a positive skin test to a particular species was associated with symptoms precipitated by that animal. In another group of 101 laboratory technicians, the relative prevalence of positive skin-prick tests and exposure to various animal species was: rat, 15/83 exposed subjects (18%); mouse, 10/55 (18%); rabbit, 6/55 (11%); guinea pig, 9/35 (26%); cat 8/26, (31%); hamster, 1/9 (11%) (Agrup *et al.*, 1986). Many had positive reactions to more than one species. Botham *et al.* (1987) also reported that rat allergy was most commonly implicated in six subjects identified with work-related chest symptoms. Four subjects attributed symptoms to rat exposure, one to rabbit, and one to mouse, but again details of exposure were insufficient to determine whether this reflected differences in animal use. Venables *et al.* (1988a) attempted to assess the extent of exposure to different species, comparing both exposures with the current and previous employers with chest symptoms. Results suggested that cat exposure was most potent in causing exposure-related chest symptoms, occurring in 10/95 (11%) of the subjects, compared with 5/121 (4%) exposed to rats, 3/99 (3%) exposed to mice, 8/95 (8%) exposed to guinea pig, 5/105 (5%) exposed to rabbit, and 5/115 (4%) exposed to dog. Unfortunately, no results were presented for skin tests to cat or dog, but the most common reaction was to rat, whereas the most common positive radioallergosorbent test (RAST) was to mice. At present it remains difficult to know which species, if any, is most potent in terms of producing allergy, although all seem capable of doing so.

Although many of the first reports incriminated animal hair or dander (skin and hair fragments brushed from the animal's coat) as the source of the relevant allergens by using skin-prick tests (Rajka, 1961; Wilson, 1971; Lincoln *et al.*, 1974) similar positive skin reactions have also been reported to other animal tissues such as extracts of urine, saliva, and serum (Newman Taylor *et al.*, 1977; Cockcroft *et al.*, 1981; Beeson *et al.*, 1983). Two proteins in rat urine have been shown to be potent allergens, namely prealbumin, and α_2-euglobulin in male rats. There also appear to be other important allergens not derived from urine, which may be derived from skin, fur and saliva (Walls and Longbottom, 1983; Longbottom and Austwick, 1987). Mouse urine, too, contains significant amounts of proteins which can act as potent allergens, and allergens have also been identified in mouse pelt, one of which is probably the same protein as one of the major urine allergens (Longbottom and Price, 1987). Urinary and salivary allergens in guinea pigs, rabbits, dogs and cats have been identified and allergens have also been extracted from the pelts of the same animals (Walls *et al.*, 1985; Bardana, 1992). Allergens identified in urine, saliva and pelt may in some cases be common allergens originally derived from plasma proteins. Transfer of urinary proteins to pelt could occur during urination or from soiled bedding, and salivary proteins might be transferred during grooming. However, this does not appear to be the only source of pelt allergen. One study identified several rat allergens in laboratory dust not found in rat urine (Longbottom and Austwick, 1987). Similarly, there appears to be one major mouse allergen which is derived from hair follicles and skin rather than urine (Longbottom and Price, 1987). High concentrations of a major cat allergen have also been identified within sebaceous glands and are probably excreted directly onto fur and skin (Charpin *et al.*, 1991). The highest levels of allergen exposure seem to result from handling the animals or other close contact, such as cleaning out cages, and it appears to be activities such as this that are most commonly associated with symptoms. However, some exposure can result from simply entering an area containing animals: e.g. dust trapped in extraction systems contains significant amounts of allergens (Longbottom and Austwick, 1987), and so do static air samples collected in laboratories (Gordon *et al.*, 1994).

Atopy, whether defined as positive skin-prick tests or total serum IgE, is important in determining the development of laboratory animal asthma (Venables *et al.*, 1988a; Sjostedt and Willers, 1989; Renstrom *et al.*, 1994; Bryant *et al.*, 1995). Botham *et al.* (1987) reported that the relationship between atopy and the incidence of symptoms appears to be most apparent in the first year of exposure. Within one cohort, 93% of subjects who had developed symptoms during the first year of exposure were atopic, compared with an overall prevalence of atopy in the cohort of 45%. In the second year the prevalence of atopy among those developing symptoms was 73% and in the third year 43%. The authors speculated that eventually the prevalence of laboratory animal allergy may become similar in atopics and non-atopics, the effect of atopy being to precipitate symptoms within a shorter period after commencing exposure. There is evidence that atopy is particularly associated with respiratory symptoms rather than nasal or skin symptoms of allergy (Slovak and Hill, 1981; Cockroft *et al.*, 1981; Agrup *et al.*, 1986; Venables *et al.*, 1988a). It is worth noting that all studies have reported a considerable proportion of atopic subjects who do not develop laboratory animal asthma, and some subjects without atopy who have developed this disease. Positive skin test reactions or RAST to specific animal allergens seems to help to identify subjects with disease, or the species responsible. Yet, because all allergens causing symptoms have yet to be recognized, and an allergic response may be very specific, not all those with symptoms will have identifiable IgE antibodies against the antigens tested, and conversely not all those with antibodies will have symptoms. Venables *et al.* (1988a) noted 24% of exposed subjects with both positive skin tests and RAST to small-animal urine extracts had no symptoms of allergy, whereas 27% of subjects with negative skin tests and RAST had upper respiratory tract or skin-related allergic symptoms, and 4% had chest symptoms.

The effect of smoking on the development of laboratory animal asthma is unclear. A number of studies have failed to find any effect, or have suggested a weak and statistically non-significant relationship between smoking and the development of laboratory animal allergy (Agrup *et al.*, 1986; Sjostedt and Willers, 1989; Renstrom *et al.*, 1994). Venables *et al.* (1988b) addressed this question in some detail by combining results from three cross-sectional studies. Smoking was associated with a higher prevalence of animal-related allergic symptoms, but this association was not statistically significant. In addition, smoking was significantly associated with animal-related allergic chest symptoms, although the strength of this effect was not clear.

In summary, some features of laboratory animal asthma can be distilled from this information. It tends to occur most frequently within a year or two of commencing exposure, and chest symptoms are frequently preceded by rhinitis or other symptoms of allergy (Rajka, 1961; Cockcroft *et al.*, 1981; Slovak and Hill, 1981). It occurs more commonly in atopic workers (at least initially) and smokers, although neither effect is sufficiently specific or strong to act as an accurate predictor of who will subsequently develop asthma. Individuals who become sensitized are often sensitized to more than one of the species to which they are exposed (Agrup *et al.*, 1986; Venables *et al.*, 1988a).

There are a number of ways to reduce allergen levels within laboratories. Decreasing animal density and improving ventilation, either generally or locally around cages and handling areas, can reduce the levels of allergens (Swanson *et al.*, 1990; Gordon *et al.*, 1992). Filter-top cages, and substituting absorbent pads for wood-based litter such as sawdust or woodchips, also reduce airborne allergen levels (Gordon *et al.*, 1992; Newman Taylor *et al.*, 1994).

Insects

Insects are used in laboratory research for a variety of purposes. In addition, large numbers of insects are bred for many other purposes. For example: bees for the production of honey; silkworms for silk production; and common house fly maggots and other insects as fishing bait. They are also encountered coincidentally in a number of occupational settings where they have been implicated as a cause of asthma, for example grain weevils and storage mites in the grain industry, and fowl mites in poultry farmers and processors.

Although there is a considerable literature on asthma due to insects, there is less on asthma occurring in an occupational setting (Table 25.2). A cross-sectional survey by the US National Institute of Occupational Safety and Health (NIOSH) of employees of the Agricultural Research Service is probably the most comprehensive (NIOSH, 1984). A total of 753 employees at 85 insect-raising facilities responded, comprising 71% of all employees. Of these, 190 (25%) reported symptoms consistent with work-related allergy, and of these 38% reported cough, 26% wheeze, and 24% shortness of breath (i.e. approximately 6–8% overall). Symptoms were more commonly reported by workers with frequent contact rather than by those with little or no direct contact (33% versus 9%). The cause of symptoms most commonly identified was airborne insect material (83%) or direct contact with insects (50%), and the insect group most frequently cited was *Lepidoptera* (moths and butterflies).

Asthma has been described to result from exposure to a number of moths, butterflies and their larvae. An early report was of a single case of occupational asthma due to larvae of bee moths (*Galleria mellonella*) bred for use as fishing bait (Stevenson and Mathews, 1967). The Douglas fir tussock moth (*Oryia pseudotsugata*) has been described as a cause of asthma among forest workers (Perlman *et al.*, 1976). These insects inhabit the forests of Oregon and Washington states in the Pacific Northwestern US. In this study, 130 loggers, 39 forestry workers, 236 lumber mill workers and 23 controls (logging contractors from an uninfested area) were surveyed, and upper or lower respiratory tract symptoms were found in 58%, 21%, 33% and 13%, respectively. The prevalence of 'difficulty breathing' in these groups was 15%, 5%, 7% and 0%, respectively.

The silkworm has been incriminated as a cause of asthma in an occupational setting (Harindranath *et al.*, 1985), and sensitization to silkworm moth is common among asthmatics even without occupational exposure in some countries (Kino and Oshima, 1979). Harindranath *et al.* (1985) reported findings in 243 subjects randomly selected from 400 employed in silk processing at two Indian factories. Cocoons were sorted, steamed to kill the pupae, threads were then rolled over reels, skeined, and then either sent for weaving, or for further processing elsewhere. Of the 243 subjects, 36% suffered from asthma; in 17% this was perceived to be

Table 25.2 Surveys of the prevalence of insect-related allergy and asthma

Authors	*Insect type*	*Study population*	*Study methods*	*Prevalence of*
NIOSH (1984)	Various	753 employees in 83 research facilities in USA	Self administered postal questionnaire	25% reported work related allergic symptoms, 33% of those frequently exposed
Perlman *et al.* (1976)	Douglas Fir Tussock moth	405 forestry and mill workers from infested area, 23 loggers (controls) from uninfested	Questionnaire (skin tests to common/ insect allergens in 16 subjects only).	5–15% exposed workers noted 'difficulty breathing', 0% of controls
Harindranath *et al.* (1985)	Silkworm	243 employees at two silk processing factories	Questionnaire, skin tests to silkworm allergens	17% noted symptoms of occupational asthma, 29% positive skin test to silkworm extract
Burge *et al.* (1980)	Locust	118 laboratory and field workers (110 industrial workers-controls for skin tests only)	Questionnaire, spirometry, skin tests to locust allergens, IgE/IgG	26% lab workers noted work related respiratory symptoms. Positive skin tests to locust in: 19–43% lab workers versus 19% unexposed
Spieksma *et al.* (1986)	Fruit Fly	22 laboratory workers, 14 unexposed controls	Questionnaire, skin tests to common/ fruit fly allergens, IgE	7 (32%) lab workers noted work-related respiratory symptoms. Sensitization in 8/22 exposed versus 1/14 unexposed.

work-related. Some 29% of the employees had positive skin tests to extracts of cocoons and pupae, with a greater frequency in those with symptoms. IgE antibodies to the same constituents were identified at a higher concentration in a subgroup of 15 subjects with occupational asthma than in a group of 10 asymptomatic controls. This type of occupational asthma appears to be mediated via an IgE mechanism.

Poultry mites are a cause of occupational asthma (Lutsky *et al.*, 1984). Poultry are parasitised by a number of mites with the Northern fowl mite (*Ornithonyssus sylviarum*) being one of the most common in Europe and North America. In 16 symptomatic poultry workers (14 with asthma, two with rhinitis), 12 asymptomatic poultry veterinarians, and 27 non-exposed atopic clinic patients, positive skin tests to the Northern fowl mites were more frequent in the symptomatic group (10/16 versus 0/12 versus 2/27, respectively). Thirteen of 14 poultry workers who underwent RAST testing had specific IgE antibodies to the Northern fowl mite, and one subject who underwent a specific airway challenge had a 25% decrement in FEV_1, confirming this as a cause of their asthma. A single case of asthma and rhinoconjunctivitis due to red spider mite (*Tetranychus urticae*) infestation in carnations has been described in a farmer (Delgado *et al.*, 1994). The patient had a significant decrement in FEV_1 following a specific airway challenge, and demonstrated a specific IgE antibody against the mite.

Asthma due to locust allergen exposure was first reported by Frankland (1953). Subsequently, Burge *et al.* (1980) recorded work-related respiratory symptoms in 26% of insect handlers working with locusts at a research establishment. Sensitization to locusts was common, with 55% of atopic insect workers demonstrating a positive test to locust allergens. However, 44% of the unexposed atopics also had a positive test to locust allergens. The high prevalence of positive skin tests to locust allergens among unexposed atopics was surprising, and the authors suggested there may be some cross-reactivity with house dust mite allergens. Asthma due to crickets (*Acheta domestica*) was shown in two scientists (Bagenstose *et al.*, 1980). They were exposed to crickets bred to feed frogs used in the laboratories. Specific airway challenges with cricket extract confirmed the cause of their asthma. In addition, of 13 employees at the laboratory, four had a positive RAST and five positive skin tests to cricket extract.

Sewage filter flies were incriminated as a cause of

asthma in sewage workers by Ordman (1946). Gold *et al.* (1985) reported a case of asthma due to sewer flies (*Psychoda alternata*) confirmed by specific airway challenge to a sewer fly extract. Specific IgE to sewer fly extract was identified in the patient's serum. Interestingly, this patient had been previously described to have asthma due to bee moth larvae (Stevenson and Mathews, 1967). Seven of 22 laboratory scientists working with fruit flies (*Drosophila melanogaster*) reported work-related allergic symptoms, of which four had chest symptoms (Spieksma *et al.*, 1986). Three of these seven had a positive response to a specific airway challenge with fruit fly extract. Positive RAST and skin prick tests confirmed fruit fly as a cause of sensitization in these individuals, and both were associated with symptoms, suggesting an IgE mechanism to their asthma. There is also a report of asthma due to the Australian sheep blowfly (*Lucilia cuprina*) in a research worker (Kaufman *et al.*, 1986). The patient had positive skin-prick tests and RAST to blowfly and other insect extracts.

Other insects have been implicated as causes of occupational asthma. Beetle larvae (*Coleoptera*) were recognized to be a cause of asthma in a museum curator (Sheldon and Johnston, 1941). These larvae were used to remove flesh from skeletons of zoological specimens. Four workers breeding larvae of the beetle *Tenibrio molitor* (mealworms) for fish bait noted allergic symptoms and two developed asthma (Bernstein *et al.*, 1983). Mexican bean weevil (*Zabrotes subfasciatus*) caused occupational asthma in a group of bean sorters (Wittich, 1940); and honeybee exposure caused asthma in a worker in a honey packing plant (Ostrom *et al.*, 1986).

Although suggestions that cockroach exposure could cause asthma initially came from three researchers who had developed symptoms at work, most interest has focused on cockroach asthma in a non-occupational setting (Bernton *et al.*, 1972). In some populations, sensitization to cockroach allergens appears to be common, perhaps because they are a ubiquitous domestic pest (Lan *et al.*, 1988). A number of other insects can cause sensitization or asthma. These include the caddis fly (*Trichoptera*) (Osgood, 1957); the green Nimitti midge (Kay *et al.*, 1983); other moths (Wynn *et al.*, 1988); and mosquitoes (Gluck and Pacin, 1986). This list is not complete.

The source of the allergen is not as well defined for most insects as it is for laboratory animals. Generally, sensitization has been tested by exposure to whole body extracts rather than specific allergens. Silkworm moth wing extracts were more potent sensitizers than body parts (Kino and Oshima, 1979). Cockroach feces contain relevant allergens but so does whole body extract (Bernton and Brown, 1970; Bernton *et al.*, 1972). Scales and hair from wings and body, insect hemoglobin, or other body parts have also been suggested as sources of allergens. Where it has been tested, the mechanism underlying insect-induced asthma seems to be IgE-mediated, although in one instance an IgG antibody was also identified (Burge *et al.*, 1980).

Overall, the potency of insects as sensitizers in exposed populations remains unclear, perhaps partly because of the great variety of exposures encountered, but also because of the relative paucity of epidemiologic research in this area when compared with some other settings, such as work with laboratory animals. In most instances, the exact source of the allergens responsible for sensitization remains unknown.

GRAIN AND FLOUR

Grain and flour have been a staple part of the diet of many of the world's people for centuries, and the popularity of bread is growing even in areas where it is not a traditional food. There have been reports of respiratory disease among bakers and millers since the time of the Roman Empire. Ramazzini (1713) described respiratory diseases in 16th century bakers and millers. It is a sobering thought that although flour and grain have been known to have adverse effects for so many years, they are still responsible for illness in a considerable number of people each year. More recently, the process of sensitization and the development of allergic disease has been examined in considerable detail.

Van Dishoeck and Roux (1939) performed skin tests with extracts of wheat, barley, oats, buckwheat, rye, corn, rice and yeast using a superficial scratch method in 262 Amsterdam bakers and millers. Sixty-six (25%) of all the bakers and millers had a positive scratch test to at least one extract, and 49 (19%) reported symptoms of respiratory illness related to work, although not necessarily asthma. Positive skin tests to wheat extract were most frequent, and positive tests were more common among those with symptoms. Sensitization was more common in workers in small bakeries than large bakeries (44% versus 25%). This led to speculation that this might be due to improved hygiene in the large bakeries, or to a selection bias, with those employed in the small bakeries often being family members, and less likely to leave the job.

Frankland and Lunn (1965) identified the grain weevil (*Sitophilus granarius*) as a potential cause of asthma in two laboratory workers. Both had work-related asthma, and both had positive skin tests to grain weevil extract, one also to a flour extract. Both improved after reducing workplace exposure to insects. Lunn (1966) investigated sensitization to both flour and weevil extracts in 75 flour mill workers, and 100 non-exposed controls. Among the mill workers, positive skin tests to extracts of flour (68%) and weevil (57%) were more common than among the controls (17% and 34%, respectively). Fifteen (20%) of the mill workers reported work-related respiratory symptoms. In addition, three mill workers underwent specific airway challenge with flour and weevil extracts. Two had FEV_1 decrements with flour extract and two with weevil extract, confirming both flour and weevil allergens as causes of occupational asthma in mill workers. Although 25 of the controls underwent similar airway challenge to the weevil extract, none developed a significant decline in FEV_1.

Work-related asthma has occurred in farmers harvesting grain (Darke *et al.*, 1976). Eighteen farmers reported respiratory symptoms during harvesting. Dust sampling around the combine harvesters identified large numbers of fungal spores, and 36% of the symptomatic farmers had positive skin tests to fungal extracts. All five symptomatic farmers who took part in a specific airway challenge with aerosolized fungi identified by the air sampling developed significant changes in FEV_1. Thus, fungal allergens were added to the possible causes of asthma already identified in grain and flour. Work-related respiratory symptoms and decrements in lung function during the working week have been reported among grain elevator workers in British Columbia (Chan-Yeung *et al.*, 1980). Hendrick *et al.* (1976) reported two bakers with asthma caused by exposure to flour. Both wheat and rye flours were implicated as causes of occupational asthma by specific airway challenge in these two subjects.

Prichard *et al.* (1984) reported work-related chest symptoms in 176 bakers and 24 slicers and wrappers at bakeries in Western Australia. Some 19% of the bakers reported work-related chest symptoms while none of the slicers and wrappers had chest complaints. In addition, 15% of the bakers and 4% of the slicers and wrappers had a positive skin test to work-related allergens (in this case, wheat flour extract). Twenty (11.4%) of the bakers met the criteria for occupational asthma compared with none of the slicers and wrappers. Subsequently, 13% of 279 workers in a large bakery in the UK (Musk *et al.*, 1989) reported work-related chest symptoms. Respiratory and allergic symptoms, skin-prick tests to bakery antigen, and airway responsiveness to methacholine showed a relationship to workplace dust exposures.

The usefulness of immunotherapy in wheat flour allergy was evaluated in 1395 bakers and pastry cooks located in north-west Spain. Of the 139 referred for further evaluation, 35 (25.2%) demonstrated a specific allergy to wheat flour (Armentia *et al.*, 1990). Of these 35 with symptoms suggestive of allergy, 66% were atopic, and 57% had developed symptoms within 10 years of commencing exposure. An investigation of 226 bakers from 105 small bakeries in Italy reported work-related asthma in 4.9% and work-related bronchitis in 10.2% (De Zotti *et al.*, 1994). There is less information about the prevalence of work-related symptoms and asthma in other groups exposed to flour or grain dust, although the prevalence of wheeze and breathlessness among grain elevator workers approximated 10–15% for non-smokers, 25–30% for ex-smokers, and 40–50% for smokers (Chan-Yeung *et al.*, 1980).

A number of allergens have been identified in flour and dust in bakeries, flour mills and grain elevators. Protein allergens from grains, storage mites, other insects and fungal constituents arising both by contamination and through the deliberate addition to flour have been implicated. Most of these agents had already been clearly identified as possible allergens almost 50 years ago (Jimenez Diaz *et al.*, 1947). Early reports often used fairly crude extracts of flour or insects for skin testing and specific airway challenge (Frankland and Lunn, 1965; Hendrick *et al.*, 1976). As techniques have progressed it has become possible to identify more accurately the compounds responsible for inducing sensitization.

Wheat, rye, barley, triticale, corn, oats and rice flour can induce IgE antibodies in exposed individuals. Exposure to wheat flour may induce IgA and IgG antibodies, although their role, if any, in the etiology of symptoms remains unclear (Kolopp-Sarda *et al.*, 1994). Rye and wheat flours seem to be most frequently identified as causes of sensitization by skin-prick testing (Block *et al.*, 1984; Armentia *et al.*, 1990). Wheat, rye, barley and triticale are related species and share many proteins and hence allergens (Block *et al.*, 1984; Sandiford *et al.*, 1995). There is also some sharing of allergens with the less commonly used and taxonomically relatively distant grains such as oats, corn and rice, and even to some extent, soya.

It also appears that there are many potential allergens within each grain species. Pooled IgE from 35 allergic subjects was recognized to bind to several different constituents of wheat flour (Sutton *et al.*, 1984), and a number of wheat proteins with molecular weights between 12 and 69 kDa were identified as major allergens (Sandiford *et al.*, 1995). The proteins identified in this latter study were of similar molecular weights to a number of known flour enzymes.

There is additional evidence that much of the allergenic activity of flour lies in its enzyme content. A number of additives have been used to improve or standardize the quality of flour in modern manufacturing techniques. One of these is the enzyme α-amylase, which is derived from cultures of *Aspergillus oryzae*. This enzyme is also found naturally in flour in varying amounts. The fungal enzyme was initially identified as a cause of sensitization and asthma (Brisman and Belin, 1991; Quirce, 1992), but subsequently a grain-derived α-amylase and α-amylase inhibitor have also been identified as major allergens (Franken *et al.*, 1994; Sandiford *et al.*, 1994).

As well as intentional contamination of flour and grain by fungal products, inadvertent contamination is also an important source of allergens. Airborne *Aspergillus* and *Alternaria* species were identified as the cause of occupational asthma in two bakers by specific bronchial challenge (Klaustemeyer *et al.*, 1977), and *Cladosporium* and other fungal species in grain dust were also identified as a cause of sensitization among farmers harvesting grain (Darke *et al.*, 1976).

Insect contamination of flour and grain introduces a further group of allergens. The grain weevil (*Sitophilus granarius*) was identified as a relevant allergen among mill workers by Lunn (1966). A storage mite (*Lepidoglyphus destructor*) has also been identified as a cause of sensitization and occupational asthma in a grain handler (Warren *et al.*, 1983) and a farmer when shovelling or threshing grain (Sotillos *et al.*, 1991). Among North Italian bakers, 17.7% had a positive skin-prick test to storage mites, compared with 21.7% with a positive skin test to flour, suggesting sensitization to such insects is nearly as common among bakers as sensitization to flour itself (De Zotti *et al.*, 1994). Tee (1992) examined results from skin-prick testing and RAST to a number of grain storage mites (*L. destructor, A. siro, G. domesticus* and *T. putrescentiae*) among 279 bakery workers and 107 salt packers. Skin-prick testing showed the prevalence of atopy in the two groups to be similar, with 33% of the bakery workers skin test-positive to at least one storage mite compared with 30% of the salt packers who had never worked with flour. Positive RAST to storage mites was identified in a greater proportion of the salt packers than the bakery workers. The high prevalence of sensitization to storage mites among unexposed subjects could not be explained by workplace exposure. This raises the possibility that bakers, too, are significantly exposed to the relevant storage mite allergens outside work. The role of allergens derived from these insects remains to be clarified.

The mechanism of occupational asthma in grain and flour workers appears to be similar, regardless of the allergen responsible. Numerous investigators have reported positive skin-prick prick tests to a number of relevant antigens, suggesting sensitization capable of producing a type I, IgE-mediated immediate allergic reaction. A specific IgE response to a number of the allergens has also been identified, and these appear to be closely related to the development of allergic disease, including asthma (Block *et al.*, 1983; Sutton *et al.*, 1984; Quirce *et al.*, 1992; Valdivieso *et al.*, 1994). The relationship between the development of disease and atopy is unclear. Many case reports of occupational asthma have, not surprisingly, been in atopic individuals (Hendrick *et al.*, 1976; Klaustermeyer *et al.*, 1977; Sotillos *et al.*, 1991). However, while some studies have identified a link between pre-existing atopy and allergic sensitization to flour allergens or the presence of work-related symptoms (Musk *et al.*, 1989; Tee *et al.*, 1992; De Zotti *et al.*, 1994), this is not true for all studies (Chan-Yeung *et al.*, 1980; James *et al.*, 1990). These discrepancies may be due to selection biases removing atopic subjects disproportionately from the exposed population of cross-sectional studies as they develop allergic disease.

The role of smoking in flour and grain asthma has not been completely clarified. It has been reported in detail in only a few studies, and seems to have little influence on the development of occupational asthma in this group (Chan-Yeung *et al.*, 1980; Prichard *et al.*, 1984; De Zotti *et al.*, 1994). One study found that grain workers taking up smoking were at higher risk of developing positive skin-prick tests to common allergens than non-smokers (Lin *et al.*, 1994), but the significance of this finding remains unclear.

As might be expected, higher levels of exposure to dust and flour are associated with a higher prevalence of sensitisation and allergic disease (Musk *et al.*, 1989; Tee *et al.*, 1992). Peak concentrations of dust and flour appear to be associated with a limited number of tasks, particularly cleaning silos and bins, maintenance cleaning, and cleaning up spillages (Nieuwenhuijsen *et al.*, 1995). Among production workers, those in the dough-making area appear to have the greatest exposure (Burdorf *et al.*, 1994). It has been suggested that excluding individuals with pre-existing allergy from the workforce may also be effective in reducing the incidence of occupational asthma (De Zotti *et al.*, 1995). There may be some justification in excluding those with severe pre-existing respiratory disease on the grounds that a further respiratory insult would be proportionately more disabling than for others in the population. However, as with laboratory animals, neither a family history of allergy nor skin-prick tests to common allergens (both used as a marker for underlying atopy) is a sufficiently sensitive or specific predictor of the development of occupational asthma to be a useful screening test.

In addition to the commonly used grain flours, work-related asthma has been described in a worker in a jam factory exposed to carob bean flour used as a thickener (van der Brempt *et al.*, 1992), and in a flour mill worker processing peas into starch, protein and bran (Bhagat *et al.*, 1995).

ENZYMES

As discussed above, the enzyme content of grain and fungi has been identified as a cause of occupational asthma. A number of enzymes are used in other occupational settings, and a number of these have also been implicated as causes of occupational asthma. Such enzymes are often used for relatively specialized and obscure purposes by small groups of workers performing varied tasks.

Enzymes in detergent products

In 1963, an enzyme-containing cleaning product for pre-wash soaking was introduced in the Netherlands. In 1967, large-scale commercial production of a detergent containing the proteolytic enzyme 'alcalase' began in Britain and the US. 'Alcalase' was produced by a process using submerged fermentation of *Bacillus subtilis* and added to conventional detergent washing powder ingredients. 'Alcalase' powder, which contained approximately 5–10% active enzyme, was mixed with water and then sprayed onto phosphate granules and dried before being added to the other ingredients.

The first reports of alcalase-induced adverse respiratory effects followed shortly after its introduction. Flindt (1969) described breathlessness occurring approximately 8 hours after the commencement of exposure and associated with positive skin-prick tests to alcalase in detergent factory workers. The first symptoms had been reported within a few weeks of beginning full-scale production. Pepys *et al.*, (1969) described three cases of occupational asthma from the same factory, each having a greater than 20% decrement in FEV_1 following a specific airway challenge with an 'alcalase' preparation. Although all three workers had an immediate skin-prick test reaction to an extract of 'alcalase' and serum precipitins against 'alcalase', none was atopic.

A large proportion of exposed workers developed sensitization to 'alcalase' in another factory (Newhouse *et al.*, 1970). About 45% of workers who made the enzyme/phosphate complex or who mixed the powder had positive skin tests to 'alcalase', compared with 23% in packing, 12% of non-production workers, and none of the shipping and warehouse workers. Of 57 with positive skin tests, 42 (74%) reported acute respiratory symptoms compared with 75/214 (35%) workers with negative tests. Atopy was associated with positive skin tests to 'alcalase', as 82% of atopics had positive skin tests compared with 37% of non-atopics. However, no IgE antibodies to 'alcalase' could be identified to confirm the underlying mechanism.

Slavin and Lewis (1971) reported asthma in 26% of 238 workers in a US detergent factory. Atopy was statistically associated with symptoms, although nearly half of all subjects with atopy reported no symptoms. A positive skin-prick test to 'alcalase' was also associated with symptoms. In Australia, Mitchell and Gandevia (1971) reported symptoms suggestive of asthma in 50% of 98 workers in the detergent industry exposed to *B. subtilis* enzymes. The presence of respiratory symptoms was associated with atopy, although this was not statistically significant. Nasal symptoms of allergy showed a significant association with atopy, and a positive skin test to 'alcalase' (suggesting sensitization) was associated with positive skin tests to common allergens. Smoking did not influence the likelihood of developing work-related symptoms.

Juniper *et al.*, (1977) followed workers for 7 years employed at the factory where the initial cases of asthma due to 'alcalase' had been recognized by Flindt. Sensitization to 'alcalase' by skin tests was related to atopy and to the level of exposure, reaching 75% among atopics with high exposure, 40% in non-atopics with high exposure, but only 10% in non-atopics with medium exposure. Subjects with intermittent high exposure were less likely to become sensitized than subjects with continuous high exposure. In all groups, most subjects who became sensitized did so within 24 months of the first exposure. RAST identified 'alcalase'-specific IgE antibodies associated with positive skin-prick tests. Of 1642 workers, 53 (3.2%) developed work-related respiratory symptoms. Smoking appeared to have no effect on the risk of developing positive skin tests to 'alcalase'.

Because of fears regarding the risk of sensitization to *B. subtilis* enzymes in detergent washing powders in consumers, investigators have studied the prevalence of sensitization among subjects with non-occupational exposure. Bernstein (1972) found that 25% of 353 patients attending an allergy clinic had positive skin scratch tests to either 'alcalase' or a different detergent enzyme, 'amylase protease'. Fourteen patients with a history suggestive of symptoms following domestic exposure to detergent enzymes had more detailed testing and seven had specific airway challenge. These seven developed respiratory symptoms and a decrement in FEV_1 of 10% or more following inhalation of the detergent enzyme. In this report, it appeared that non-occupational exposures could induce sensitization and play a role in the development of asthma. In contrast, Pepys *et al.*, (1985) found little evidence of sensitization to either 'alcalase' or the antigenically identical enzyme 'maxatase' in 136 subjects attending an allergy clinic. Although a number of positive skin-prick tests were noted, the authors were not confident that these positive tests were not due to a direct irritant effect as they occurred only at high concentrations of allergen and they were not associated with a positive RAST for specific anti-enzyme IgE. The final picture, therefore, remains unclear.

Papain and bromelain

Papain is a proteolytic enzyme derived from the fruit of the paw-paw tree (*Caruca papaya*). It is used in laboratories for work such as the structural analysis of proteins, in the food industry as a meat tenderizer, as an ingredient in medicines, as a clarifying agent in brewing, and for other purposes. Bromelain is a proteolytic enzyme usually derived from pineapple (*Ananas comosus*) and similarly used in the food industry as a meat tenderizer, for clarifying beer after fermentation, and for other uses. It is also used in blood transfusion laboratories as a reaction booster.

Milne and Brand (1975) reported occupational asthma to papain in two chemists working in the same factory, although in different laboratories. Both had symptoms of asthma when exposed to papain and positive skin scratch tests to papain. Baur and Fruhmann (1979) studied 11 workers, nine from a large commercial kitchen where papain was used as a meat tenderizer, and two from pharmaceutical plants which produced digestive aids containing papain. Seven of the 11 workers reported work-related symptoms, of whom six had symptoms suggestive of asthma. All seven with symptoms had positive skin-prick tests to papain, whereas neither the four asymptomatic workers nor 40 unexposed asthmatics had positive tests. All six subjects with asthma had a positive RAST to papain. Six of the seven symptomatic workers also had positive skin tests to common allergens, suggesting an association between papain asthma and atopy. Five of the subjects underwent a specific airway challenge with papain and all had a greater than 50% decrement in specific airway conductance, confirming the papain as a cause of their asthma. Novey *et al.*, (1980) reported that 12/23 workers in a plant producing a new papain-containing product noted work-related respiratory symptoms. Levels of specific IgE to papain correlated with duration and intensity of exposure, symptoms and pulmonary function. Atopic subjects also developed symptoms after a shorter mean period of exposure than non-atopics.

Galleguillos and Rodriguez (1978) first reported two cases of occupational asthma due to bromelain, one in a non-atopic pharmaceutical worker, and one in an atopic delivery boy who worked for a pharmaceutical company and presumably had considerably lower exposure than the first. Both had a positive skin-prick test to a bromelain extract and a positive specific airway challenge. A further study reported four subjects from a blood grouping laboratory with bromelain-related asthma (Gailhofer *et al.*, 1988). All had a specific IgE to bromelain, and three had positive skin test to bromelain, although none was atopic. Among 17 others in the same laboratory, seven had specific IgE antibodies against bromelain compared with none of the 15 controls without occupational exposure.

Other enzymes

Occupational asthma has been reported in a subject processing crude trypsin from bovine and porcine pancreas (Zweiman *et al.*, 1967). The subject had a strongly positive skin test to a trypsin solution, and a strong positive specific airway challenge to trypsin. Another study of 14 workers who used trypsin in the production of polymer resins found that four had definite and two had equivocal work-related respiratory symptoms (Colten *et al.*, 1975). All four with definite symptoms had positive skin tests to trypsin, and all tested had a positive specific airway challenge to trypsin.

Asthma has been reported in a nurse who mixed pancreatic extract with formula for infants with cystic fibrosis (Hayes and Newman Taylor, 1991). Three pharmacists with asthma following exposure to 'flaviastase', a proteolytic extract of *Aspergillus fumigatus* used in digestive powders, have also been reported (Pauwels *et al.*, 1978). All were atopic and had a positive RAST for 'flaviastase'.

SEAFOODS AND OTHER MARINE ORGANISMS

Seafoods have been described to cause occupational asthma. A 1980 report described an outbreak of occupational asthma among employees in a Scottish prawn-processing factory (Gaddie *et al.*, 1980). Shortly before the outbreak the factory had converted from hand peeling the prawns to 'air blowing', a process by which the meat was blown out of the tail using compressed air and impacted against a partition. Of 50 workers, 13 had a positive skin-prick test to prawn extract, eight had a prawn-specific IgE by RAST, and 18 reported work-related respiratory symptoms. The symptomatic group had a higher prevalence of atopy and included most subjects with positive skin tests or RAST to prawn extract. Two symptomatic subjects were tested with specific airway challenge with purified prawn protein, and both had an asthmatic reaction confirming the prawn as a cause of their asthma.

There have been a series of papers describing occupational asthma in snow crab-processing workers in Canada (Cartier *et al.*, 1984, 1986; Hudson *et al.*, 1985). In 303 workers in two processing plants, 15.6% had objectively proven work-related asthma. Positive skin-prick tests to crab and the presence of specific IgE were significantly associated with occupational asthma. A weak association was also found with smoking as 81% of workers with occupational asthma were current smokers compared with 64% of those without asthma. Although the authors reported that no significant association between asthma and atopy was identified, 18% of those with occupational asthma were atopic compared with 9% of those without asthma, and positive skin-prick tests to common allergens were related to positive skin tests to crab.

Occupational asthma also occurs in fish processors. Five of eight Danish trout-processing factory workers working near an automatic gutting machine which produced an aerosol of fish proteins and endotoxin were found to have occupational asthma (Sherson *et al.*, 1989). Similarly, aerosols generated by automatic gutting machines during salmon processing were shown to induce asthma (Douglas *et al.*, 1995). Work-related asthma was associated with specific IgE to salmon serum and current smoking, but not atopy. Asthma has also been reported in processing workers to sea squirt, an organism commonly encountered on oyster shells (Jyo *et al.*, 1964), and in an atopic laboratory worker grinding dried sea sponge (*Dysidea herbacea*), who had become sensitized to a variety of sponges and soft corals (Baldo *et al.*, 1982).

OTHER FOOD-DERIVED PRODUCTS

Egg protein has been recognized for a number of years as a food allergen capable of causing or exacerbating urticaria, dermatitis, gastrointestinal complaints, asthma and anaphylaxis. IgE-mediated, work-related asthma to egg protein was described in 4/25 workers in a plant processing raw eggs into powdered yolk, powdered whole egg, and liquid egg-white (Bernstein *et al.*, 1987). Positive skin tests and IgE to egg products was associated with asthma, but not atopy. In a follow-up study at two additional plants, 58/188 (31%) workers reported work-related symptoms suggestive of asthma, and approximately 10% at each plant were estimated to have occupational asthma using a combination of symptoms, physician assessment and serial peak flow recordings (Smith *et al.*, 1990). Again, work-related asthma was associated with skin tests or specific IgE to egg products, although not all of the 34% of workers with a positive skin-prick test had asthma.

Egg protein contains a number of allergens, including ovalbumin, ovomucoid, conalbumin and the enzyme lysozyme. Egg-white derived lysozyme is used in the pharmaceutical industry as a preservative and enhancer of antibiotic activity. A case of occupational asthma in a worker extracting egg-white lysozyme was confirmed by specific airway challenge (Bernstein *et al.*, 1993). A specific IgE antibody was detected in the patient's serum.

Workers employed processing green coffee beans before roasting are at risk for occupational asthma (Figley and Rawling, 1950). When investigators realized that green coffee beans were often carried in bags that had previously been used to carry castor beans, it was thought that castor beans may be the source of the allergen, as these had previously been identified as a cause of asthma (Figley and Elrod, 1928; Figley and Rawling, 1950). However, allergens have been identified in green coffee beans distinct from those in castor bean dust. These antigens are specific to green coffee beans, being removed by processing (Karr *et al.*, 1978; Jones *et al.*, 1982).

The allergens in castor bean appears to lie in the 'pomace', i.e. organic material remaining after castor oil has been extracted (Figley and Elrod, 1928; Davison *et al.*, 1983). Epidemics of asthma were reported occurring around a linseed oil and castor bean mill in Toledo, Ohio, USA (Figley and Elrod, 1928).

Work-related asthma has been reported among workers in a variety of miscellaneous food-related processes. Tea fluff produced during the processing of tea leaves (Uragoda, 1970) and tea dust used for packing tea bags has been identified as a cause of occupational asthma (Roberts and Thomson, 1988). Scottish workers manufacturing mushroom soup from dried mushroom powder developed sensitization to a mushroom extract and occupational asthma (Symington, 1981). Four workers developed a greater than 30% decrement in FEV_1 when handling mushroom powder. Pectin has been described as a cause of asthma in a patient pouring pectin into a kettle before mixing jam (Cohen *et al.*, 1993). Ginseng was shown to be a cause of occupational asthma (Subiza *et al.*, 1991). There has also been a case report of asthma caused by exposure to two herb materials (*Dioscorea batatas*, *Pinellia ternata*) in an atopic worker in a herbal pharmacy in Korea (Park *et al.*, 1994). Again, an IgE mechanism was proposed as the patient had positive skin-prick tests to both materials and specific IgE antibodies to one of the herbs.

GUMS AND LATEX

Latex

In the late 1970s, latex exposure was recognized as a cause of immediate hypersensitivity reactions ranging from contact urticaria to anaphylaxis. Asthma due to latex exposure in an occupational setting was first reported in 1988 in a laboratory technician using rubber gloves (Seaton *et al.*, 1988). Since then, investigators have recognized latex to be a common sensitizing agent and an important cause of occupational asthma. Furthermore, the marked increase in the use of latex products, particularly in the health industry, has raised awareness that latex allergy has become a major occupational health problem.

Natural rubber latex is derived from the rubber tree, *Hevea brasiliensis*, with the basic unit being a low-molecular weight compound, *cis*-1,4-isoprene. Native latex coagulates by forming polymers of isoprene several thousand units in length, and is coated with a layer of protein, lipid and phospholipid which provides structural integrity. Proteins comprise only about 2% of the total content of raw latex. Although latex allergens have not yet been precisely characterized, most investigators concur that they are likely to be the proteins in both raw latex and in extracts of finished rubber products, rather than contaminants introduced by the manufacturing process.

Latex sensitization is common and occurs in 10% of atopic individuals undergoing routine skin testing for aeroallergens (Moneret-Vautrin *et al.*, 1993). The prevalence of latex allergy in health care workers, as assessed by a symptom questionnaire and skin-prick testing, ranged from 11% to 17% (Lagier *et al.*, 1992; Yassin *et al.*, 1994). Most sensitized subjects report contact urticaria from gloves, although allergic rhinitis and conjunctivitis are also common. Occupational asthma was diagnosed in 6% of 50 workers in a latex glove manufacturing plant on the basis of a positive latex skin-prick test associated with changes in spirometry and bronchial hyper-responsiveness related to workplace exposure (Tarlo *et al.*, 1990). A recently published Belgian study of hospital personnel (Vandenplas *et al.*, 1995) used a stepwise protocol with a standardized questionnaire, skin-prick tests, and assessment of non-specific bronchial hyper-responsiveness to identify nurses, cleaning staff and laboratory technologists with possible occupational asthma caused by latex. Thirteen of 273 (4.7%) subjects had positive skin reactivity to latex. Five of these subjects had a previous history of asthma, and seven had positive reactions to specific inhalation challenge with latex gloves, yielding a prevalence of 2.5%.

Using gel electrophoresis, latex allergens appear to be a mixture of proteins with molecular weight ranging from 10 to 67 kDa (Jaeger *et al.*, 1992). One group of investigators have reported that a protein called rubber elongation factor (REF) is a major allergen (Czuppon *et al.*, 1993). This 58 kDa protein facilitates the polymerization of isoprene units in the biosynthesis of natural rubber. Positive skin-prick wheal and flare reactions to REF were found in most sensitized subjects, and positive nasal and bronchial reactions were also induced on direct challenge. However, based on the results of immunoblotting tests to sera in allergic individuals, it is likely that there are multiple major allergens in addition to REF.

Risk factors for sensitization to latex include frequent use of disposable gloves, the presence of atopic disease, and prior or current hand dermatitis (Yunginger, 1995). It is likely that sensitization occurs through direct skin contact in predisposed atopic individuals.

There is an association between latex allergy and allergy to a variety of fruits, including banana, avocado and kiwi fruit (Ahlroth *et al.*, 1995), probably because latex proteins are structurally homologous with other plant proteins. It is possible that prior sensitization occurs from fruit ingestion, leading to subsequent latex allergy. However, it is the respirable particles which play the most important role in causing asthma following re-exposure in sensitized individuals.

Although latex proteins are the responsible allergens, cornstarch glove powder has an important role in precipitating asthma attacks in the occupational setting. Most cases of latex-associated asthma and rhinitis have been described in health care workers from inhalation of latex allergen-coated cornstarch particles from powdered gloves (Sussman and Beezhold, 1995). Latex allergens can be leached from rubber gloves by normal skin moisture, with subsequent adsorption to cornstarch powder inside the gloves. They can also be adsorbed directly to powder inside gloves that have not been worn. When the gloves are donned or removed, the latex allergen-coated cornstarch particles are then aerosolized at levels similar to those of other occupational respiratory allergens (Baur and Jaeger, 1990). Quantification of latex aeroallergen levels at a medical center confirmed that the highest levels occurred when powdered latex gloves were used frequently (Swanson *et al.*, 1994). Furthermore, latex allergen was airborne only when there was work activity in the area.

Apart from asthma in health care workers and in the surgical glove-manufacturing factory previously described (Tarlo *et al.*, 1990), there was also a report of asthma in a latex doll-manufacturing plant, where sanding or grinding of solid latex during the manufacturing process produced airborne latex proteins (Orfan *et al.*, 1994). There are no reports of latex-induced asthma among rubber plantation workers, which may reflect the lack of aerosolization in the initial processing of raw rubber.

Williams *et al.* (1995) recently put forward the intriguing hypothesis that sensitization to latex may also occur through inhalation of particulate air pollutants containing extractable latex antigens from rubber tire fragments. Using volumetric sampling devices and immunoassay, these investigators found a large proportion of the respirable particulate pollution in urban air contained latex. They hypothesized that these particles represented abraded tire fragments. This was supported by mass spectroscopy, which also indicated a spectrum of constituents similar to latex gloves. It is possible that these airborne particles could contribute through direct sensitization, or as specific adjuvants for IgE responses, to the increase in latex sensitization and asthma.

In patients demonstrated to have latex occupational asthma, the avoidance of direct contact by use of latex-free products is indicated. Synthetic latex-free gloves are available, although their degree of barrier protection may not be as great as with latex gloves (Yunginger, 1995). The use of powder-free latex gloves by co-workers will also help to provide a latex allergen-free airborne environment (Tarlo *et al.*, 1994). As with other forms of occupational asthma, respiratory disability may continue even after removal from exposure (Brugnami *et al.*, 1995).

Gums

Natural vegetable gums are high-molecular weight carbohydrates derived from various plants exudates. They are soluble in aqueous solutions, with stabilizing and suspending qualities, and are used as protective colloids and emulsifying agents in food and pharmaceutical products. Previously, they have also been used in the printing industry, to prevent freshly printed sheets from sticking to each other, and to stabilize ink to prevent running. Printer's asthma from exposure to gums in this occupational setting is well documented (Fowler, 1952), but is largely of historical interest, as they have been phased out of this industry. More recently, Lagier *et al.* (1990) reported three cases of asthma due to gums in a work-related setting: one worked in a pharmaceutical firm where gum was used as a tablet hardener, and two in carpet manufacturing plants where gum was used to fix colors. The diagnosis was confirmed by positive skin-prick tests and inhalation challenge studies. A follow-up study among employees of a carpet manufacturing plant, found a 5% prevalence of positive skin-prick test reactions to gums, and a 2% prevalence of gum-related occupational asthma (Malo *et al.*, 1990a).

Psyllium is a plant-derived gum used widely as a laxative. It can cause occupational asthma in employees of pharmaceutical companies manufacturing the laxatives, and in health care workers preparing the product for chronically ill patients. The prevalence of occupational asthma due to psyllium has been found to be 3.6% in 130 employees at a pharmaceutical company (Bardy *et al.*, 1987) and in 4% of 193 health care workers in chronic care hospitals (Malo *et al.*, 1990b).

SUMMARY AND CONCLUSIONS

There is a considerable literature on a wide variety of apparently diverse, large-molecular weight agents that can cause occupational asthma. Most reports concentrate on the definition of the problem in a particular group of workers or occupational setting, with a particular emphasis on confirming the diagnosis of occupational asthma and its etiology. This is inevitable given the importance of the clinical, occupational hygiene and medicolegal imperatives. However, it frequently means that the scale of the problem is rather poorly defined even within a particular work setting, and prevalence data, where they exist are largely limited to small cross-sectional studies with their inevitable sources of confounding. Little work has been performed on defining the antigens involved, and on whether they have common functional (e.g. enzymatic) or conformational or chemical characteristics.

However, a number of largely empirical and clinical observations can be made which tend to be common across this group of agents:

- Allergic problems tend to arise where exposure levels are high and repeatedly so; this has obvious implications for prevention and health/safety considerations in general.
- Clinical symptoms tend to arise early in the course of employment, frequently over the first few months.
- The mechanism is usually IgE-mediated.
- Atopic individuals are at particular risk, but the sensitivity and specificity of an atopic background for the development of specific sensitization and especially for development of occupational asthma is low. It would not be equitable or practicable to exclude all atopics from workplaces of known risk.
- Smoking is likely to be a risk factor in most circumstances, although the effect is not great. The mechanism of this effect may be related to the known induction of IgE production by smoking, or perhaps to an increase in permeability of the respiratory epithelium allowing greater penetration of the antigens.

REFERENCES

Agrup, G., Belin, L., Sjostedt, L. and Skerfving, S. (1986) Allergy to laboratory animals in laboratory technicians and animal keepers. *Br. J. Ind. Med.*, **43**, 192–8.

Ahlroth, M., Alenius, H., Turjanmaa, K., Makinen-Kiljunen, S., Reunala, T. and Palosuo, T. (1995) Cross reacting allergens in natural rubber latex and avocado. *J. Allergy Clin. Immunol.*, **96**, 167–73.

Aoyama, K., Ueda, A., Manda, F., Matsushita, T., Ueda, T. and Yamauchi, C. (1992) Allergy to laboratory animals: an epidemiological study. *Br. J. Ind. Med.*, **49**, 41–7.

Armentia, A., Martin-Santos, J.M., Quintero, A. *et al.* (1990) Bakers' asthma: prevalence and evaluation of immunotherapy with a wheat flour extract. *Ann. Allergy*, **65**, 265–72.

Bagenstose, A.H., III, Mathews, K.P., Homburger, H.A. and Saaveard-Delgado, A.P. (1980) Inhalant allergy due to crickets. *J. Allergy Clin. Immunol.*, **65**, 71–4.

Baldo, B.A., Krilis, S. and Taylor, K.M. (1982) IgE-mediated acute asthma following inhalation of a powdered marine sponge. *Clin. Allergy*, **12**, 179–86.

Bardana, E.J. (1992) Occupational asthma and related conditions in animal workers, in *Occupational Asthma* (eds E.J. Bardana, A. Montanaro and M.T. O'Hollaren), Hanley & Belfus Inc., Philadelphia, pp. 225–35.

Bardy, J.D., Malo, J.L. and Dolovich, J. (1987) Occupational asthma and IgE sensitisation in a pharmaceutical company processing psyllium. *Am. Rev. Respir. Dis.*, **135**, 1033–8.

Baur, X. and Fruhmann, G. (1979) Papain-induced asthma: diagnosis by skin test, RAST and bronchial provocation test. *Clin. Allergy*, **9**, 75–81.

Baur, X. and Jaeger, D. (1990) Airborne antigens from latex gloves. *Lancet*, **i**, 912.

Beeson, M.F., Dewdney, J.M., Edwards, R.G., Lee, D. and Orr, R.G. (1983) Prevalence and diagnosis of laboratory animal allergy. *Clin. Allergy*, **13**, 433–42.

Bernstein, D.I., Gallagher, J.S. and Bernstein, I.L. (1983) Mealworm asthma: clinical and immunologic studies. *J. Allergy Clin. Immunol.*, **72**, 475–80.

Bernstein, D.I., Smith, A.B., Moller, D.R. *et al.* (1987) Clinical and immunologic studies among egg-processing workers with occupational asthma. *J. Allergy Clin. Immunol.*, **80**, 791–7.

Bernstein, I.L. (1972) Enzyme allergy in populations exposed to long-term, low level concentrations of household laundry products. *J. Allergy Clin. Immunol.*, **49**, 219–37.

Bernstein, J.A., Kraut, A., Bernstein, D.I. *et al.* (1993) Occupational asthma induced by inhaled egg lysozyme. *Chest*, **103**, 532–5.

Bernton, H.S. and Brown, H. (1970) Insect allergy: the allergenicity of the excrement of the cockroach. *Ann. Allergy*, **28**, 543–7.

Bernton, H.S., McMahon, T.F. and Brown, H. (1972) Cockroach Asthma. *Br. J. Dis. Chest*, **66**, 61–6.

Bhagat, R., Swystun, V.A. and Cockroft, D.W. (1995) Occupational asthma caused by pea flour. *Chest*, **107**, 1772.

Block, G., Tse, K.S., Kijek, K., Chan, H. and Chan-Yeung, M. (1983) Baker's asthma. Clinical and immunological studies. *Clin. Allergy*, **13**, 359–70.

Block, G., Tse, K.S., Kijek, K., Chan, H. and Chan-Yeung, M. (1984) Baker's asthma. Studies of the cross-antigenicity between different cereal grains. *Clin. Allergy*, **14**, 177–85.

Botham, P.A., Davies, G.E. and Teasdale, E.L. (1987) Allergy to laboratory animals: a prospective study of its incidence and of the influence of atopy on its development. *Br. J. Ind. Med.*, **44**, 627–32.

Brisman, J. and Belin, L. (1991) Clinical and immunological responses to occupational exposure to α-amylase in the baking industry. *Br. J. Ind. Med.*, **48**, 604–8.

Brugnami, G., Marabini, A., Siracusa, A. and Abbritti, G. (1995) Work-related late asthmatic response induced by latex allergy. *J. Allergy Clin. Immunol.*, **96**, 457–64.

Bryant, D.H., Bocato, L.M., Mboloi, P.N. and Stuart, M.C. (1995) Allergy to laboratory animals among animal handlers. *Med. J. Aust.*, **163**, 415–18.

Burdorf, A., Lillienberg, L. and Brisman, J. (1994) Characterization of exposure to inhalable flour dust in Swedish bakeries. *Ann. Occup. Hyg.*, **38**(1), 67–78.

Burge, P.S., Edge, G., O'Brien, I.M., Harries, M.G., Hawkins, R. and Pepys, J. (1980) Occupational asthma in a research centre breeding locusts. *Clin. Allergy*, **10**, 355–63.

Cartier, A., Malo, J.L., Forest, F. *et al.* (1984) Occupational asthma in snow crab-processing workers. *J. Allergy Clin. Immunol.*, **74**, 261–9.

Cartier, A., Malo, J.L., Ghezzo, H., McCants, M. and Lehrer, S.B. (1986) IgE sensitisation in snow crab-processing workers. *J. Allergy Clin. Immunol.*, **78**, 344–8.

Chan-Yeung, M. and Malo, J.L. (1995) Current concepts. Occupational asthma. *N. Engl. J. Med.*, **333**, 107–12.

Chan-Yeung, M., Schulzer, M., MacLean, L., Dorken, E. and Grzybowski, S. (1980) Epidemiologic health survey of grain elevator workers in British Columbia. *Am. Rev. Respir. Dis.*, **121**, 329–38.

Charpin, C., Mata, P., Charpin, D., Lavaut, M.N., Allasia, C. and Vervloet, D. (1991) *Fel d* 1 allergen distribution in cat fur and skin. *J. Allergy Clin. Immunol.*, **88**, 77–82.

Cockcroft, A., Edwards, J., McCarthy, P. and Andersson, N. (1981) Allergy in laboratory animal workers. *Lancet*, **i**, 827–30.

Cohen, A.J., Forse, M.S. and Tarlo, S.M. (1993) Occupational asthma caused by pectin inhalation during the manufacture of jam. *Chest*, **103**, 309–11.

Colten, H.R., Polakoff, P.L., Weinstein, S.F. and Streider, D.J. (1975) Immediate hypersensitivity to hog trypsin resulting from industrial exposure. *N. Engl. J. Med.*, **292**, 1050–3.

Contreras, G.R., Rousseau, R. and Chan Yeung, M. (1994) Occupational respiratory diseases in British Columbia, Canada in 1991. *Occup. Environ. Med.*, **51**, 710–12.

Czuppon, A.B., Chen, Z., Rennert, S. *et al.* (1993) The rubber elongation factor of rubber trees (*Hevea brasiliensis*) is the major allergen in latex. *J. Allergy Clin. Immunol.*, **92**, 690–7.

Darke, C.S., Knowelden, J., Lacey, J. and Milford Ward, A. (1976) Respiratory disease of workers harvesting grain. *Thorax*, **31**, 294–302.

Davies, G.E. and McArdle, L.A. (1981) Allergy to laboratory animals: a survey by questionnaire. *Int. Archs Allergy Appl. Immunol.*, **64**, 302–7.

Davison, A.G., Britton, M.G., Forrester, J.A., Davies, R.J. and Hughes, D.T.D. (1983) Asthma in merchant seamen and laboratory workers caused by allergy to castor beans: analysis of allergens. *Clin. Allergy*, **13**, 553–61.

De Zotti, R., Larese, F., Bovenzi, M., Negro, C. and Molinari, S. (1994) Allergic airway disease in Italian bakers and pastry makers. *Occup. Environ. Med.*, **51**, 548–52.

De Zotti, R., Molinari, S., Larese, F. and Bovenzi, M. (1995) Pre-employment screening among trainee bakers. *Occup. Environ. Med.*, **52**, 279–83.

Delgado, J., Gomez, E., Palma, J.L. *et al.* (1994) Occupational rhinoconjunctivitis and asthma caused by *Tetranychus urticae* (red spider mite). A case report. *Clin. Exp. Allergy*, **24**, 477–80.

Douglas, J.D.M., McSharry, C., Baikie, L., Morrow, T., Miles, S. and Franklin, D. (1995) Occupational asthma caused by automated salmon processing. *Lancet*, **346**, 737–40.

Figley, K.D. and Elrod, R.H. (1928) Endemic asthma due to castor bean dust. *JAMA*, **90**, 79–82.

Figley, K.D. and Rawling, F.F.A. (1950) Castor bean: an industrial hazard as a contaminant of green coffee dust and used burlap bags. *J. Allergy*, **21**, 545–53.

Flindt, M.L.H. (1969) Pulmonary disease due to inhalation of derivatives of *Bacillus subtilis* containing proteolytic enzyme. *Lancet*, **i**, 1177–80.

Fowler, P.B.S. (1952) Printer's asthma. *Lancet*, **ii**, 755–7.

Franken, J., Stephan, U., Meyer, H.E. and Konig, W. (1994) Identification of alpha amylase inhibitor as a major allergen of wheat flour. *Int. Arch. Allergy Immunol.*, **104**, 171–4.

Frankland, A.W. (1953) Locust sensitivity. *Ann. Allergy*, **11**, 445.

Frankland, A.W. and Lunn, J.A. (1965) Asthma caused by the grain weevil. *Br. J. Ind. Med.*, **22**, 157–9.

Gaddie, J., Legge, J.S., Friend, J.A.R. and Reid, T.M.S. (1980) Pulmonary hypersensitivity in prawn workers. *Lancet*, **ii**, 1350–3.

Gailhofer, G., Wilders-Truschnig, M., Smolle, J. and Ludvan, M. (1988) Asthma caused by bromelain: an occupational allergy. *Clin. Allergy*, **18**, 445–50.

Galleguillos, F. and Rodriguez, J.C. (1978) Asthma caused by bromelin inhalation. *Clin. Allergy*, **8**, 21–4.

Gluck, J.C. and Pacin, M.P. (1986) Asthma from mosquito bites: a case report. *Ann. Allergy*, **56**, 492–3.

Gold, B.L., Mathews, K.P. and Burge, H.A. (1985) Occupational asthma caused by sewer flies. *Am. Rev. Respir. Dis.*, **131**, 949–52.

Gordon, S., Tee, R.D., Lowson, D., Wallace, J. and Newman Taylor, A.J. (1992) Reduction of airborne allergenic urinary proteins from laboratory rats. *Br. J. Ind. Med.*, **49**, 416–22.

Gordon, S., Tee, R.D., Nieuwenhuijsen, M.J., Lowson, D., Harris, J. and Newman Taylor, A.J. (1994) Measurement of airborne rat urinary allergen in an epidemiological study. *Clin. Exp. Allergy*, **24**, 1070–7.

Harindranath, N., Prakash, O.M. and Subba Rao, P.V. (1985) Prevalence of occupational asthma in silk filatures. *Ann. Allergy*, **55**, 511–15.

Hayes, J.P. and Newman Taylor, A.J. (1991) Bronchial asthma in a paediatric nurse caused by inhaled pancreatic extracts. *Br. J. Ind. Med.*, **48**, 355–6.

Hendrick, D.J., Davies, R.J. and Pepys, J. (1976) Bakers' asthma. *Clin. Allergy*, **6**, 241–50.

Home Office (1995) *Statistics of scientific procedures on living animals, Great Britain 1994*, HMSO, London. Cmd 3012. ISBN 0101301227.

Hudson, P., Cartier, A., Pineau, L. *et al.* (1985) Follow-up of occupational asthma caused by crab and various agents. *J. Allergy Clin. Immunol.*, **76**, 682–8.

Jaeger, D., Kleinhans, D., Czuppon, A.B. and Baur, X. (1992) Latex-specific proteins causing immediate-type cutaneous, nasal, bronchial, and systemic reactions. *J. Allergy Clin. Immunol.*, **89**, 759–68.

James, A.L., Zimmerman, M.J., Ee, H., Ryan, G. and Musk, A.W. (1990) Exposure to grain dust and changes in lung function. *Br. J. Ind. Med.*, **47**, 466–72.

Jimenez-Diaz, C., Lahoz, C. and Canto, G. (1947) The allergens of mill dust. Asthma in millers, farmers, and others. *Ann. Allergy*, **5**, 519–25.

Jones, R.N., Hughes, J.M., Lehrer, S.B. *et al.* (1982) Lung function consequences of exposure and hypersensitivity in workers who process green coffee beans. *Am. Rev. Respir. Dis.*, **125**, 199–202.

Juniper, C.P., How, M.J., Goodwin, B.F.J. and Kinshott, A.K. (1977) *Bacillus subtilis* enzymes: a 7 year clinical, epidemiological and immunological study of an industrial allergen. *J. Soc. Occup. Med.*, **27**, 3–12.

Jyo, T., Katasuya, T., Inoko, Y. *et al.* (1964) Studies on 'oystershucker's asthma', an asthma-like disease affecting oyster shuckers in Hiroshima prefecture (I). *Jpn. J. Allergol.*, **13**, 88–99.

Karr, R.M., Lehrer, S.B., Butcher, B.T. and Salvaggio, J.E. (1978) Coffee worker's asthma: a clinical appraisal using the radioallergosorbent test. *J. Allergy Clin. Immunol.*, **62**, 143–8.

Kaufman, G.L., Baldo, B.A., Tovey, E.R., Bellas, T.E. and Gandevia, B.H. (1986) Inhalant allergy following occupational exposure to blowflies. *Clin. Allergy*, **16**, 65–71.

Kay, A.B., MacLean, C.M.U., Wilkinson, A.H. and Gad EI Rab, M.O. (1983) The prevalence of asthma and rhinitis in a Sudanese community seasonally exposed to a potent airborne allergen (the 'green nimitti' midge, *Cladotanytarsus lewisi*). *J. Allergy Clin. Immunol.*, **71**, 345–52.

Kino, T. and Oshima, S. (1979) Allergy to insects in Japan. II. The reaginic sensitivity to silkworm moth in patients with bronchial asthma. *J. Allergy Clin. Immunol.*, **64**, 131–8.

Klaustermeyer, W.B., Bardana, E.J., Jr. and Hale, F.C. (1977) Pulmonary hypersensitivity to *Alternaria* and *Aspergillus* in baker's asthma. *Clin. Allergy*, **7**, 227–33.

Kolopp-Sarda, M.N., Massin, N., Gobert, B. *et al.* (1994) Humoral immune responses of workers occupationally exposed to wheat flour. *Am. J. Ind. Med.*, **26**, 671–9.

Lagier, F., Cartier, A., Somer, J., Dolovich, J. and Malo, J.-L. (1990) Occupational asthma caused by guar gum. *J. Allergy Clin. Immunol.*, **85**, 785–90.

Lagier, F., Vervloet, D., Lhermet, I., Poyen, D. and Charpin, D. (1992) Prevalence of latex allergy in operating room nurses. *J. Allergy Clin. Immunol.*, **90**, 319–22.

Lan, J.L., Lee, D.T., Wu, C.H., Chang, C.P. and Yeh, C.L. (1988) Cockroach hypersensitivity: preliminary study of allergic cockroach asthma in Taiwan. *J. Allergy Clin. Immunol.*, **82**, 736–40.

Lin, F.J., Kennedy, S.M. and Chan-Yeung, M. (1994) Effects of grain exposure and smoking on the longitudinal changes in immediate skin reactivity. *Clin. Exp. Allergy*, **24**, 949–54.

Licoln, T.A., Bolton, N.E. and Garret, A.S., Jr (1974) Occupational allergy to animal dander and sera. *J. Occup. Med.*, **16**, 465–9.

Longbottom, J.L. and Austwick, P.K.C. (1987) Allergy to rats: quantitative immunoelectrophoretic studies of rat dust as a source of inhalant allergen. *J. Allergy Clin. Immunol.*, **80**, 243–51.

Longbottom, J.L. and Price, J.A. (1987) Allergy to laboratory animals: characterization and source of two major mouse allergens, Ag1 and Ag3. *Int. Archs Allergy Appl. Immunol.*, **82**, 450–2.

Lunn, J.A. (1966) Millworkers' asthma: allergic responses to the grain weevil. *Br. J. Ind. Med.*, **23**, 149–52.

Lutsky, I.I. and Neuman, I. (1975) Laboratory animal dander allergy: I. an occupational disease. *Ann. Allergy*, **35**, 201–5.

Lutsky, I., Teichtahl, H. and Bar-Sela, S. (1984) Occupational asthma due to poultry mites. *J. Allergy Clin. Immunol.*, **73**, 56–60.

Malo, J.-L., Cartier, A., L'Archeveque, J. *et al.* (1990a) Prevalence of occupational asthma and immunologic sensitisation to guar gum among employees at a carpet-manufacturing plant. *J. Allergy Clin. Immunol.*, **86**, 562–9.

Malo, J.-L., Cartier, A., L'Archeveque, J. *et al.* (1990b) Prevalence of occupational asthma and immunologic sensitisation to psyllium among health personnel in chronic care hospitals. *Am. Rev. Respir. Dis.*, **142**, 1359–66.

Meredith, S.K. and McDonald, J.C. (1994) Work related respiratory disease in the United Kingdom, 1989–1992: report on the SWORD project. *Occup. Med.*, **44**, 183–9.

Milne, J. and Brand, S. (1975) Occupational asthma after inhalation of dust of the proteolytic enzyme, papain. *Br. J. Ind. Med.*, **32**, 302–7.

Mitchell, C.A. and Gandevia, B. (1971) Respiratory symptoms and skin reactivity in workers exposed to proteolytic enzymes in the detergent industry. *Am. Rev. Respir. Dis.*, **104**, 1–12.

Moneret-Vautrin, D.-A., Beaudouin, E., Widmer, S. *et al.* (1993) Prospective study of risk factors in natural rubber latex hypersensitivity. *J. Allergy Clin. Immunol.*, **92**, 668–77.

Musk, A.W., Venables, K.M., Crook, B. *et al.* (1989) Respiratory symptoms, lung function, and sensitisation to flour in a British bakery. *Br. J. Ind. Med.*, **46**, 636–42.

National Institute of Occupational Safety and Health (1984) Work related allergies in insect raising facilities. *MMWR*, **33**(31), 448–54.

Newhouse, M.L., Tagg, B., Pocock, S.J. and McEwan, A.C. (1970) An epidemiological study of workers producing enzyme washing powders. *Lancet*, **i**, 689–92.

Newman Taylor, A., Longbottom, J.L. and Pepys, J. (1977) Respiratory allergy to urine proteins of rats and mice. *Lancet*, **ii**, 847–9.

Newman Taylor, A.J., Gordon, S. and Tee, R.D. (1994) Influence of bedding, cage design, and stock density on rat urinary aeroallergen levels. *Am. J. Ind. Med.*, **25**, 89.

Nieuwenhuijsen, M.J., Sandiford, C.P., Lowson, D., Tee, R.D., Venables, K.M. and Newman Taylor, A.J. (1995) Peak exposure concentrations of dust and flour aeroallergen in flour mills and bakeries. *Ann. Occup. Hyg.*, **39**(2), 193–201.

Novey, H.S., Keenan, W.J., Fairshter, R.D., Wells, I.D., Wilson, A.F. and Culver, B.D. (1980) Pulmonary disease in workers exposed to papain: clinicophysiological and immunological studies. *Clin. Allergy*, **10**, 721–31.

Novey, H.S., Bernstein, I.L., Mihalas, L.S., Terr, A.I. and Yunginger, J.W. (1989) Guidelines for the clinical evaluation of occupational asthma due to high molecular weight (HMW) allergens. *J. Allergy Clin. Immunol.*, **84**, 829–33.

Ordman, D. (1946) Sewage filter flies (*Psychoda*) as a cause of bronchial asthma. *S Afr. Med.J.*, **20**, 32–5.

Orfan, N.A., Reed, R., Dykewicz, M.S., Ganz, M. and Kolski, G.B. (1994) Occupational asthma in a latex doll manufacturing plant. *J. Allergy Clin. Immunol.*, **94**, 826–30.

Osgood, H. (1957) Allergy to caddis fly (*Trichoptera*). II. Clinical aspects. *J. Allergy*, **28**, 292–300.

Ostrom, N.K., Swanson, M.C., Agarwal, M.K. and Yunginger, J.W. (1986) Occupational allergy to honeybee-body dust in a honey processing plant. *J. Allergy Clin. Immunol.*, **77**, 736–40.

Park, H.S., Kim, M.J. and Moon, H.-B. (1994) Occupational asthma caused by two herb materials, *Dioscorea batatas* and *Pinellia ternata*. *Clin. Exp. Allergy*, **24**, 575–81.

Pauwels, R., Devos, M., Callens, L. and van der Straeten, R. (1978) Respiratory hazards from proteolytic enzymes. *Lancet*, **i**, 669.

Pepys, J., Hargreave, F.E., Longbottom, J.L. and Faux, J. (1969) Allergic reactions of the lungs to enzymes of *Bacillus subtilis*. *Lancet*, **i**, 1181–4.

Pepys, J., Mitchell, J., Hawkins, R. and Malo, J.L. (1985) A longitudinal study of possible allergy to enzyme detergents. *Clin. Allergy*, **15**, 101–15.

Perlman, F., Press, E., Googins, J.A., Malley, A. and Poarea, H. (1976) Tussockosis: reactions to Douglas Fir Tussock moth. *Ann. Allergy*, **36**, 302–7.

Prichard, M.G., Ryan, G. and Musk, A.W. (1984) Wheat flour sensitisation and airways disease in urban bakers. *Br. J. Ind. Med.*, **41**, 450–4.

Quirce, S., Cuevas, M., Diez-Gomez, M.L. *et al.* (1992) Respiratory allergy to *Aspergillus*-derived enzymes in bakers' asthma. *J. Allergy Clin. Immunol.*, **90**(6), 970–8.

Rajka, G. (1961) Ten cases of occupational hypersensitivity to laboratory animals. *Acta Allergol.*, **16**, 168–76.

Ramazzini, B. (1713) Diseases of bakers and millers, in *de Morbis artificum*; translated by Wilmer Cave Wright (1940), University of Chicago Press, Chicago, pp. 235–52.

Renstrom, A., Malmberg, P., Larsson, K., Sundblad, B.-M. and Larsson, P.H. (1994) Prospective study of laboratory-animal allergy: factors predisposing to sensitisation and development of allergic symptoms. *Allergy*, **49**, 548–52.

Roberts, J.A. and Thomson, N.C. (1988) Tea-dust induced asthma. *Eur. Respir. J.*, **1**, 769–70.

Sallie, B.A., Ross, D.J., Meredith, S.K. and McDonald, J.C. (1994) SWORD '93: surveillance of work-related and occupational respiratory disease in the UK. *Occup Med.*, **44**, 177–82.

Salter, H.H. (1860) *On Asthma: its Pathology and Treatment*, John Churchill, London.

Sandiford, C.P., Tee, R.D. and Newman Taylor, A.J. (1994) The role of cereal and fungal amylases in cereal flour hypersensitivity. *Clin. Exp. Allergy*, **24**, 549–57.

Sandiford, C.P., Tee, R.D. and Newman Taylor, A.J. (1995) Identification of crossreacting wheat, rye, barley, and soya flour allergens using sera from individuals with wheat-induced asthma. *Clin. Exp. Allergy*, **25**, 340–9.

Seaton, A., Cherrie, B. and Turnbull, J. (1988) Rubber glove asthma. *Br. Med. J.*, **296**, 531–2.

Sheldon, J.M. and Johnston, J.H. (1941) Hypersensitivity to beetles (*Coleoptera*). Report of a case. *J. Allergy*, **12**, 493–4.

Sherson, D., Hansen, I. and Sigsgaard, T. (1989) Occupationally related respiratory symptoms in trout-processing workers. *Allergy*, **44**, 336–41.

Sjostedt, L. and Willers, S. (1989) Predisposing factors in laboratory animal allergy: a study of atopy and environmental factors. *Am. J. Ind. Med.*, **16**, 199–208.

Slavin, R.G. and Lewis, C.R. (1971) Sensitivity to enzyme additives in laundry detergent workers. *J. Allergy Clin. Immunol.*, **48**(5), 262–6.

Slovak, A.J.M. and Hill, R.N. (1981) Laboratory animal allergy: a clinical survey of an exposed population. *Br. J. Ind. Med.*, **38**, 38–41.

Smith, A.B., Bernstein, D.I., London, M.A. *et al.* (1990) Evaluation of occupational asthma from airborne egg protein exposure in multiple settings. *Chest*, **98**, 398–404.

Sorrell, A.H. and Gottesman, J. (1957) Mouse allergy. Case report. *Ann. Allergy*, **15**, 662–3.

Sotillos, M.M.G., Carmona, J.B. and Picon, S.J. (1991) Late asthma caused by inhalation of *Lepidoglyphus destructor*. *Ann. Allergy*, **67**, 126–8.

Spieksma, F.Th.M., Vooren, P.H., Kramps, J.A. and Dijkman, J.H. (1986) Respiratory allergy to laboratory fruit flies (*Drosophila melanogaster*). *J. Allergy Clin. Immunol.*, **77**, 108–113.

Stevenson, D.D. and Mathews, K.P. (1967) Occupational asthma following inhalation of moth particles. *J. Allergy Clin. Immunol.*, **39**, 274–83.

Subiza, J., Subiza, J.L., Escribano, P.M. *et al.* (1991) Occupational asthma caused by Brazil ginseng dust. *J. Allergy Clin. Immunol.*, **88**, 731–36.

Sutton, R., Skerritt, J.H., Baldo, B.A. and Wrigley, C.W. (1984) The diversity of allergens involved in bakers' asthma. *Clin. Allergy*, **14**, 93–107.

Sussman, G.L. and Beezhold, D.H. (1995) Allergy to latex rubber. *Ann. Intern. Med.*, **122**, 43–6.

Swanson, M.C., Campbell, A.R., O'Hollaren, M.T. and Reed, C.E. (1990) Role of ventilation, air filtration, and allergen production rate in determining concentrations of rat allergens in the air of animal quarters. *Am. Rev. Respir. Dis.*, **141**, 1578–81.

Swanson, M.C., Bubak, M.E., Hunt, L.W., Yunginger, J.W., Warner, M.A. and Reed, C.E. (1994) Quantification of occupational latex aeroallergens in a medical centre. *J. Allergy Clin. Immunol.*, **94**, 445–51.

Symington, I.S., Kerr, J.W. and McLean, D.A. (1981) Type I allergy in mushroom soup processors. *Clin. Allergy*, **11**, 43–7.

Tarlo, S.M., Wong, L., Roos, J. and Booth, N. (1990) Occupational asthma caused by latex in a surgical glove manufacturing plant. *J. Allergy Clin. Immunol.*, **85**, 626–31.

Tarlo, S.M., Sussman, G., Contala, A. and Swanson, M.C. (1994) Control of airborne latex by use of powder-free latex gloves. *J. Allergy Clin. Immunol.*, **93**, 985–9.

Tarlo, S.M., Liss, G., Corey, P. and Broder, I. (1995) A workers' compensation claim population for occupational asthma. Comparison of subgroups. *Chest*, **107**, 634–41.

Taylor, G., Davies, G.E., Altounyan, R.E.C. *et al.* (1976) Allergic reactions to laboratory animals. *Nature*, **260**, 280.

Tee, R.D., Gordon, D.J., Gordon, S. *et al.* (1992) Immune response to flour and dust mites in a United Kingdom bakery. *Br. J. Ind. Med.*, **49**, 581–7.

Uragoda, C.G. (1970) Tea maker's asthma. *Br. J. Ind. Med.*, **27**, 181–2.

Valdivieso, R., Subiza, J., Subiza, J.L., Hinojosa, M., de Carlos, E. and Subiza, E. (1994) Bakers' asthma caused by alpha amylase. *Ann. Allergy*, **73**, 337–42.

van Dishoeck, H.A.E. and Roux, D.J. (1939) Sensitisation to flour and flour illnesses amongst flour workers. *J. Hyg.*, **39**, 674–79.

van der Brempt, X., Ledent, C. and Mairesse, M. (1992) Rhinitis and asthma caused by occupational exposure to carob bean flour. *J. Allergy Clin. Immunol.*, **90**, 1008–10.

Vandenplas, O., Delwiche, J.-P., Evrard, G. *et al.* (1995) Prevalence of occupational asthma due to latex among hospital personnel. *Am. J. Respir. Crit. Care Med.*, **151**, 50–4.

Venables, K.M., Tee, R.D., Hawkins, E.R. *et al.* (1988a) Laboratory animal allergy in a pharmaceutical company. *Br. J. Ind. Med.*, **45**, 660–6.

Venables, K.M., Upton, J.L., Hawkins, E.R., Tee, R.D., Longbottom, J.L. and Newman Taylor, A.J. (1988b) Smoking, atopy, and laboratory animal allergy. *Br. J. Ind. Med.*, **45**, 667–71.

Walls, A.F. and Longbottom, J.L. (1983) Quantitative immunoelectrophoretic analysis of rat allergen extracts. *Allergy*, **38**, 501–12.

Walls, A.F., Newman Taylor, A.J. and Longbottom, J.L. (1985) Allergy to guinea-pigs: I allergenic activities of extracts derived from the pelt, saliva, urine and other sources. *Clin. Allergy*, **15**, 241–51.

Warren, C.P.W., Holford-Stevens, V. and Sinha, R.N. (1983) Sensitisation in a grain handler to the storage mite *Lepidoglyphus destructor* (Schrank). *Ann. Allergy*, **50**, 30–3.

Williams, P.B., Buhr, M., Weber, R.W., Volz, M.A., Koepke, J.W. and Selner, J.C. (1995) Latex allergen in respirable particulate air pollution. *J. Allergy Clin. Immunol.*, **95**, 88–95.

Wilson, J.A. (1971) Hamster-hair hypersensitivity in adults of low atopic status. *Br. Med. J.*, **4**, 341.

Wittich, F.W. (1940) Allergic rhinitis and asthma due to sensitisation to the Mexican bean weevil (*Zabrotes subfasciatus BOH*). *J. Allergy*, **12**, 42–45.

Wynn, S.R., Swanson, M.C., Reed, C.E., Penny, N.D., Showers, W.B. and Smith, J.M. (1988)

Immunochemical quantitation, size distribution, and cross reactivity of Lepidoptera (moth) aeroallergens in Southeastern Minnesota. *J. Allergy Clin. Immunol.*, **82**, 47–54.

Yassin, M., Lierl, M., Fischer, T., O'Brien, K., Cross, J. and Steinmetz, C. (1994) Latex allergy in hospital employees. *Ann. Allergy*, **72**, 245–9.

Yunginger, J.W. (1995) Natural rubber latex. *Immunol. Allergy Clin. North Am.*, **15**, 583–98.

Zweiman, B., Green, G., Maycock, R.L. and Hildreth, E.A. (1967) Inhalation sensitisation to trypsin. *J. Allergy*, **39**, 11–16.

26 Urban epidemic asthma

Josep M. Antó and Jordi Sunyer

INTRODUCTION

Episodes of asthma which affect less than a few thousand people over a brief period in a specific location are unique and offer a chance to better understand the pathogenesis of asthma. These outbreaks impose a substantial public health burden and major social concern in a community. When such epidemics occur, an explanation is necessary to alter the natural history of the current public health hazard and to prevent recurrence. Investigators of these local asthma outbreaks have parallels with those studying occupational asthma. Investigators in each field may find a previously unidentified agent which may be risk factor for asthma.

OUTBREAKS OF ASTHMA: A RECURRENT AVOIDABLE PROBLEM

Asthma outbreaks have been recognized since the beginning of this century (Table 26.1). Figley and Elrod (1928) described a time and space cluster of asthma affecting about 200 patients living within a 1-mile radius of a castor bean mill in Toledo, Ohio. The outbreak was attributed to inhalation of castor bean dust released from the mill and ceased after the factory stopped processing castor beans.

In August, 1952 in Bauru, Brazil – a city of 60 000 inhabitants – 150 people living near a castor bean mill suffered asthma crisis, leading investigators to suspect a point-source origin for the outbreak (Mendes and Ulhoa Cintra, 1954). Of these 150 cases, 28 among 30 tested had skin reactivity to castor bean extract and four had a positive bronchoprovocation test to castor bean extract. The mill was closed for a year so that changes to preclude the release of castor bean dust into the atmosphere could be introduced. A similar outbreak was reported in South Africa (Ordman, 1955). In both these outbreaks environmental intervention prevented their recurrence.

A different process occurred in New Orleans, Louisiana. Although asthma was frequent in New Orleans, an outbreak affecting about 350 patients who required therapy on a single day at one hospital's emergency room occurred in August, 1955. On this day, the asthma was severe in some individuals and two patients died. Lewis *et al.* (1962) advanced the hypothesis that the New Orleans asthma outbreaks were due to the inhalation of particles emanating from a waste dump site with spontaneous underground burning. However, since the outbreaks persisted years after the waste dump site had been removed this hypothesis was thought unlikely. In subsequent studies, investigators recognized that the asthma epidemic affected a subgroup of atopic asthmatics and although weather conditions may have affected the outbreak, neither atmospheric pollution nor chemical and industrial releases contributed to these illnesses. Although the causes were not clearly identified, researchers pointed out that a group of allergens, including spores of different fungal species, ambrosia pollen and vegetable detritus, explained this epidemic (Salvaggio *et al.*, 1971). Although excessive levels of these allergens were favored by meteorological conditions, the evidence was weak and a causal association was not proved.

Occupational Lung Disease. Edited by Daniel E. Banks and John E. Parker. Published in 1998 by Chapman & Hall. ISBN 0 412 73630 6

Table 26.1 Asthma outbreaks

Location (Reference)	*Year*	*Agent*	*Mechanism*	*People affected*	*Distribution of cases*
Ohio, USA (Figley and Elrod, 1928)	1928	Castor bean	Castor bean mill	200	Geographical cluster
Bauru, Brazil (Mendes and Ulhoa Cintra, 1954)	1952	Castor bean	Castor bean mill	150	Geographical cluster
South Africa (Ordman, 1955)	1955	Castor bean	Castor bean mill	–	Geographical cluster
New Orleans, USA (Salvaggio *et al.*, 1971; White *et al.*, 1993)	1953–77	Soybean?	Harbor operations?	>350	Time/geographical cluster
Brisbane, Australia (Morrison, 1960)	1960	?	?	90	Time cluster
New York, USA (Goldstein and Cuzick, 1978)	1969–71	?	Indoor exposure?	??	Time cluster
Birmingham, UK (Packe and Ayres, 1985)	1983	*D.exitalis*?	Thunderstorm?	26	Time cluster
Barcelona, Spain (Artó, 1989a)	1980–87	Soybean	Harbor silo	800	Time/geographical cluster
Cartagena, Spain (Navarro *et al.*, 1993)	1987	Soybean	Unloading transportation	30	Time/geographical cluster
Valencia, Spain (Alvarez-Dardet *et al.*, 1989)	1988	Soybean	Harbor unloading	–	Time cluster
Melbourne, Australia (Suphioglu *et al.*, 1992)	1990	Rye grass?	Thunderstorm?	1300	Time cluster
London, UK (Murray *et al.*, 1994)	1994	?	Thunderstorm?	1000	Time cluster
Tarragona, Spain (García-Ortega, 1995)*	1994	Soybean	Harbor activity	13	Time cluster

*Personal communication.

Years later, a re-analysis of the New Orleans data for the period 1969–1977, using a time-series approach, showed that the outbreaks had notably decreased. This decrease was attributed to an improvement in socioeconomic conditions of the susceptible population or to better medical care, rather than by a change in exposure or to an environmental agent (Goldstein and Cuzick, 1983; Goldstein and Salvaggio, 1984).

Unusual increases in the number of asthma emergency room visits in New York on a given day were first observed when assessing relationships among air pollution, meteorological conditions and respiratory morbidity in a retrospective time-series study. Goldstein and Rausch (1978) analyzed emergency room admissions of three hospitals in New York during 1969–1971. A number of unusual asthma days, based on departures from a Poisson distribution, were observed to occur simultaneously at the three hospitals. This led to the hypothesis that an environmental factor could have triggered such clusters. No relation was found with outdoor air pollution. Days of high asthma incidence were more likely on Sundays, Mondays, and Thanksgiving day, leading the authors to propose that exposure to indoor environmental agents might have been important (Goldstein and Cuzick, 1983). The explanation for these excessive asthma cases is not clear, but the high incidence of asthma on some days identified in New York probably resulted from a tendency for time clustering of asthma, rather than individual outbreaks.

SOYBEAN EPIDEMIC ASTHMA

Identification and descriptive study of asthma outbreaks in Barcelona

Asthma outbreaks in Barcelona were first identified in 1981 in the emergency room of the Hospital Clinic of

Barcelona following an abrupt and massive influx of patients with severe asthma. A retrospective study of six different outbreaks was carried out in 1983, comparing admissions during outbreaks with admissions on 12 randomly selected days (Ussetti *et al.*, 1983). All outbreaks showed similar features: most of the patients arrived at the hospital's emergency room within 4 hours of the onset of symptoms. The asthma attacks were unusually severe, with some patients needing mechanical ventilation. During the epidemics, the air had been still and atmospheric pressure high. No etiologic hypothesis was advanced on the basis of this study, although atmospheric pollution was considered a possible cause. The average 24-hour levels of sulfur dioxide (SO_2) and carbon particles were not unusually high, but short-term peaks in air pollution could not be excluded. During the seventh outbreak, high levels of nitric oxide (NO) were recorded during the hours at which most of asthmatics arrived at the emergency room. This coincidence between peak NO levels and peak emergency room assessments led investigators to suspect NO as an etiologic agent (Ussetti *et al.*, 1984).

In 1984, a Collaborative Asthma Group of epidemiologists, clinicians and other specialists from the main hospitals and research centers of Barcelona was established by the city's Department of Public Health. This group established a prospective monitoring system for respiratory emergency room visits in Barcelona. During 1985–1986, six outbreaks were identified by clinicians and all were confirmed by epidemiologic assessment. Retrospective data were also collected which allowed an evaluation of 12 outbreaks that had occurred since 1981. All of these outbreaks showed a simultaneous time and space cluster. District I, the area nearest the harbor, had the highest age- and sex-adjusted attack rate in each of the outbreaks. All outbreaks were also characterized by a statistically significant cluster of admissions occurring within a 4-hour or less period. The consistency of this feature was interpreted as strongly supporting the point-source hypothesis as well as confirming the potential location of the point source in the harbor or in the nearby industrial area. In all outbreaks, 23% to 56% of patients had been treated in an emergency room during a previous outbreak (Antó *et al.*, 1986a).

In the outbreaks that occurred after the monitoring system was put in place, patients were interviewed about possible triggers of their asthma, but no plausible hypothesis could be established. In addition, despite intensive environmental monitoring, unusual environmental events were not identified. As a consequence, even though an important number of asthma epidemics had been identified, no etiologic hypothesis could be ventured. However, the investigators hypothesized that the asthma epidemics were likely caused by a biologic or chemical agent spread from a point source, probably located in the harbor or in an important industrial area near the harbor (Antó *et al.*, 1986a). Air pollution was not shown to be a necessary cause of asthma outbreaks (Antó *et al.*, 1986b).

THE SOYBEAN HYPOTHESIS

In 1987, the Collaborative group recognized that at least three asthma epidemics had occurred on days when soybeans were unloaded in the harbor. However, complete data on ship unloadings were not available to screen for the relationship between soybean unloadings and the other outbreaks. Since asthma due to exposure to airborne soybean particulates had been reported on several occasions, and in the absence of previous reports of epidemic asthma due to soybean dust, we considered soybean unloadings an unlikely cause of these asthma outbreaks. Conversely, the only vegetable agent repeatedly reported as a cause of asthma epidemics was castor bean dust, a powerful agent in producing immediate allergy. Thus, the possibility of contamination of soybeans by castor bean dust during storage or shipping was considered (Antó *et al.*, 1990).

We tested the association between soybean unloadings and asthma outbreaks by determining the presence of specific IgE antibodies against 22 different commercial antigens, including both Barcelona airborne common antigens and the soybean and castor bean antigens. IgE reactivity was assessed with the RAST technique in serum samples from patients affected by the epidemics which had been stored since 1985. Results showed that 13/18 patients had specific IgE antibodies to soybean antigen, whereas only 4/18 reacted to the rest of the antigens. These results were considered to be strongly suggestive of a relationship between the unloading of soybeans in the Barcelona harbor and asthma outbreaks (Antó *et al.*, 1990).

Etiologic studies of soybean asthma epidemics

Two epidemiologic studies were designed to test the soybean hypothesis. First, the relationship between soybean unloadings in the harbor and the occurrence of asthma outbreaks was investigated in a time-ecological study. Second, the relationship between epidemic asthma and serum IgE against the soybean antigen was tested in a case-control study.

The retrospective study of asthma epidemics covered 1985–1986 (Antó *et al.*, 1989a), a period during which 13 asthma epidemic days had been identified. All products and the days on which each product was loaded or unloaded were identified. For soybeans, the use of two harbor silos (A and B) for soybean unloading was also examined. For each product unloaded, a risk ratio was calculated between the probability of an asthma epidemic on the days when the product was being unloaded and the probability of an asthma epidemic on days when it was not being unloaded. All 13 asthma epidemic days coincided with soybean unloadings. No asthma outbreak occurred in the absence of soybean unloading. Of the remaining products studied, only wheat showed a weak association to asthma although when adjusted for the unloading of soybeans it was no longer statistically significant.

When the site of soybean unloading (silo A or B) was taken into account, the association was equally strong in silo A but did not exist for silo B. This observation was consistent with the characteristics of the silos. Silo A was close to the city and did not have bag filters installed in the cyclone dust collection system, thereby allowing the release of high amounts of soybean dust.

A case-control study of the reactivity to soybean antigens was also conducted (Sunyer *et al.*, 1989). In September, 1987, two asthma epidemics occurred, resulting in 157 emergency room visits for asthma. These cases were compared with controls selected from emergency room visits for asthma on sporadic asthma days and when soybeans were unloaded under controlled conditions. Eighty-six controls were individually matched to cases by age, sex and area of residence. In 64/86 cases (74.4%) there was a reaction with commercial soybean antigen extracts, compared with only 4/86 (4.6%) controls (odds ratio = 61). The association was even stronger when the cases were tested with antigen prepared with soybean dust extract collected from a ship unloading on a day considered to be an epidemic day. No other serologic covariates confounded the association between serum anti-soybean IgE antibodies and epidemic asthma. These results strongly supported a causal relationship between the release of soybean dust and the asthma outbreaks. These results also suggested that the asthma outbreaks had an underlying allergic mechanism. Using RAST inhibition technique, highly statistically significant differences between soybean antigen concentrations were measured on the air of epidemic days in comparison with the air levels on non-epidemic days (7131.4 and 428 U/m^3, respectively) (Antó *et al.*, 1989b).

Intervention and follow-up

In September, 1987 soybean unloadings were stopped. Three months later, bag filters were installed at the top of silo A, and the unloading of soybeans was resumed under controlled conditions (Antó and Sunyer, 1990). The impact of this intervention was assessed comparing the concentration of the soybean allergen in 136 air samples collected for 9 months before and 24 months after the installation of filters. Emergency room visits for asthma before and after the intervention were also compared (Antó *et al.*, 1993).

Levels of airborne allergens on days when soybeans were unloaded fell from 324 to 25 U/m^3 (P<0.001) after the installation of filters. The number of days on which there were an unusually large number of visits to the emergency room for asthma and the number of epidemic days decreased from 29 to six and from 18 to zero, respectively. The mean daily number of emergency room and intensive care unit admissions for asthma on unloading days decreased from 8.3 to 5.4 and from 0.26 to 0.01, respectively (P <0.01). There was also a decrease in the levels of IgE to soybean in 38 asthmatics (P <0.001).

Thus, installation of bag filters at silo A greatly reduced soybean dust emissions during unloadings and eliminated the asthma outbreaks.

Other characteristics of soybean epidemic asthma

One of the most striking peculiarities of soybean epidemic asthma was its severity. Most patients experienced sudden asthma attacks and about 24 died during the outbreaks. Typically, most of the epidemic asthma patients experience asthma attacks with a sudden and severe onset and a rapid relief of symptoms after treatment at the emergency room (Ferrer *et al.*, 1990). A recent immunohistochemical study of soybean asthma death occurring in Barcelona during the outbreaks has suggested that an anaphylactic mechanism could also have been involved (Synek *et al.*, 1996). A similar course of illness has been reported by O'Hollaren *et al.* (1991) in selected asthmatics requiring admission to the intensive care unit. These cases were not related to soybean exposure, but the authors hypothesized that these severe attacks could have been caused by a large exposure to alternaria.

Age was also related to the risk of epidemic asthma. During the outbreaks people aged 65 years and over were eight times more likely to be admitted than those aged less than 14 years. In most of the outbreaks attack rates of asthma for those aged under 14 were not increased. This age pattern was confirmed in the case-control study and in the intervention study, which showed that the reduction in the mean number of emergency room admissions for asthma even after the installation of filters was higher in older groups. There is no clear explanation for this age pattern.

Atopy, defined as a positive reaction to skin tests or the presence of circulating IgE antibodies to common aeroallergens, was a risk factor for soybean epidemic asthma. In a case-control study (Sunyer *et al.*, 1989), specific IgE antibodies against *Dermatophagoides, Parietaria* and *Aspergillus* were identified. Asthmatics evaluated in the emergency room during an epidemic day were four times more likely to react to any of these allergens than the asthmatics treated on non-epidemic days (OR = 4.1); *Parietaria* showed the strongest association with epidemic asthma (OR = 9).

Another risk factor which proved to be associated with soybean epidemic asthma was smoking. Smokers were more likely to be epidemic asthmatics than non-smokers, regardless of the degree of soybean exposure or atopy. Atopic subjects who were smokers and residing near the point source were at seven-fold higher risk of epidemic asthma in comparison with those who lacked these three characteristics (Sunyer *et al.*, 1992).

Other asthma outbreaks due to soybean

Recognition of a causal association between asthma outbreaks and excessive soybean dust levels in Barcelona facilitated the identification of similar episodes elsewhere. In Cartagena, Spain, three asthma epidemic days were detected. Among 30 asthmatics treated on these epidemic days, 81% showed positive skin reactivity to the commercial soybean allergen of the cases compared with none of the controls (Hernando *et al.*, 1989). The mechanism of soybean contamination was likely due to uncovered trucks carrying soybeans (Navarro *et al.*, 1993).

More recently, more than 100 patients needed asthma therapy on a single day in Naples, Italy. The episode coincided with unloading soybeans from a ship in Naples's harbor. About half of the asthmatics showed serum IgE antibodies against soybeans (D'Amato *et al.*, 1994; Cocco *et al.*, 1995).

With regard to the New Orleans asthma outbreaks described above, a recent study has shown an association between the occurrence of such outbreaks and the presence of a ship carrying soybeans in the harbor. Thus, soybean dust inhalation may have been the cause of these asthma outbreaks (White *et al.*, 1993).

ASTHMA OUTBREAKS AND THUNDERSTORMS: A FIRM HYPOTHESIS

In 1983, Packe and Ayres (1985), at the East Birmingham Hospital, Birmingham, UK, on July 6th and 7th, reported a striking increase in the number of asthma emergency room visits admissions following a thunderstorm. Within 36 hours, 26 asthma cases were treated in the emergency room, compared with the usual two to three cases per day. A retrospective analysis of previous emergency room assessments showed that other asthma outbreaks had not occurred. This outbreak coincided with an important increase of several airborne spores, particularly *Sporobolomyces* and *Didymella exitialis*. However, when comparing epidemic patients with non-epidemic asthmatic controls, the prevalence of positive skin tests with numerous allergens, including *D. exitialis*, were no different (Packe and Ayres, 1986).

Substantial information linking asthma outbreaks with thunderstorms has been reported from Melbourne, Australia (Bellomo *et al.*, 1992). At least two large asthma outbreaks have been identified coinciding with thunderstorms leading to the hypothesis that pollen grains are ruptured in rainwater by osmotic shock. Each grain has the potential of releasing about 700 starch granules of respirable size. Suphioglu *et al.* (1992) showed that rye grass pollen grains contain a large quantity of starch granules containing Lol pIX and coated with Lol pl. Twelve patients involved in an asthma outbreak had positive skin reactions to both standard rye grass pollen and isolated starch granules compared with 7/13 outpatient controls ($P < 0.05$). Four of the epidemic cases showed early bronchial constriction after voluntary re-exposure to starch granules. The levels of chemical air pollution on epidemic days were below or similar to the levels in a control period.

While this is a relevant and well-established hypothesis, sufficient evidence to identify this as the etiology has not yet been provided. The association between thunderstorms and asthma in Melbourne, although likely to be real, does not totally allow a causal relationship to be established. Formal time series analysis with complete ascertainment of epidemic days is lacking and risk ratios for the relationship between asthma outbreaks and thunderstorms have not yet been provided. Thus, the possibility of asthma outbreaks on days without thunderstorms has not been assessed. In addition, the case control study included only 12 of the epidemic asthma cases from one single hospital out of the 300 or so patients affected.

Another outbreak of asthma following a thunderstorm has more recently been reported in London (Murray *et al.*, 1994). On the night of 24th–25th June, 1994, coinciding with a thunderstorm, there was a large increase in the number of visits for asthma in the emergency rooms of several hospitals of London and south-west England. In several hospitals the number of cases reported was around 100. This is likely to be the largest outbreak of asthma ever recorded. Taking into account the Melbourne experience, it may well be that a similar mechanism was involved, although other factors may have contributed. Several studies on the London's asthma outbreak are under development.

CONCLUDING REMARKS

We have recently compared soybean epidemic asthma with occupational asthma (Antó *et al.*, 1996). Risk factors, such as atopy and smoking which modify the effect of the causative agent occur in both groups. In both models, a given population is exposed in a given time to a novel allergen. This allows for identification of a possible causal relationship. Finally, a proportion of epidemic patients experienced severe and even near-fatal asthma attacks during outbreaks, even when they were located far from soybean point-source. This may be similar to the worker with occupational asthma who develops asthma without being directly exposed to the offending antigen.

In contrast, since the general population was exposed to soybean dust, the spectrum of the affected population in epidemic asthma cannot be compared with those who develop occupational asthma where important selection factors influence the exposed population. In addition, in contrast to many with occupational asthma, most of those affected had asthma before the soybean asthma attacks.

Like occupational asthma, outbreaks due to inhalation of point-source releases of dusts like castor bean and soybean should be considered an important local public health problem which is amenable to prevention. It is surprising that during the three decades which began with identification of the castor bean outbreaks to the time of the soybean outbreaks, other similar cases had not been reported. This might be explained by the fact that the scientific respiratory and public health community considered the castor bean episodes a local oddity. This was not the case for Dr Figley, who reported the first castor bean outbreak in 1928; in 1934, during a meeting at the Mayo Clinic, he made the following comment:

> *'Some years ago I reported quite a large number of cases of asthma which were occasioned by the grinding of castor bean dust in a castor bean mill in Toledo. There were some 200 of these patients living in a radius of a mile or so of this plant. The matter got into the courts, and this factory was prohibited from having anything to do with the handling of castor beans. They were not even allowed to import a single castor bean, much less grind any. Since that time, this castor bean division of the plant has been engaged in grinding soy beans and extracting the oil. Naturally, I have been on the lookout for cases of soy bean sensitivity, because they have not even troubled to use the dust collecting apparatus which was installed for the castor bean grinding operations. As a matter of fact, that dust collecting apparatus was packed up and shipped off to one of their plants, so they have taken no precaution against the dissemination in the air of soy bean dust from this plant. So far I have failed to see a single case of soy bean sensitivity in this vicinity in a matter of two years. I feel certain I will sooner or later, but I think it may be rather an interesting observation as to the necessity of time for many of these asthmatic individuals previously sensitized to castor bean dust, to develop sensitivity to this other substance.'*

Asthma outbreaks and thunderstorms may pose a quite different problem. If the thunderstorm hypothesis is right, asthma outbreaks would prove difficult if not impossible to prevent and people sensitized to seasonal pollens (up to one-third of the population) could be exposed sporadically to large amounts of allergen. In such circumstances outbreaks of asthma may overwhelm the capacity of health services, as was the case in London, and appropriate emergency plans should be made. From a scientific viewpoint, asthma outbreaks deserve further consideration because the causes of asthma have not been identified, and these particular outbreaks may provide a useful insight into the etiology of the condition.

REFERENCES

Alvarez-Dardet, C., Belda, J., Peña, M. and Nolasco, A. (1989) Outbreak of asthma associated with soybean dust. *N. Engl. J. Med.*, **321**, 1127–8.

Antó, J.M. and Sunyer, J. (1990) Epidemiological studies of Barcelona asthma epidemics. *Chest*, **98**, 185–90.

Antó, J.M., Sunyer, J. and the Barcelona Asthma Collaborative Group (1986a) A point-source asthma outbreak. *Lancet*, **i**, 900–3.

Antó, J.M., Sunyer, J. and Plasència, A. (1986b) Nitrogen dioxide and asthma outbreaks. *Lancet*, **ii**, 1096–7.

Antó, J.M., Sunyer, J., Rodríguez-Roisín, R., Suarez-Cervera, M., Vazquez, L. and the Toxicoepidemiological Committee (1989a) Community outbreaks of asthma associated with inhalation of soybean dust. *N. Engl. J. Med.*, **320**, 1097–1102.

Antó, J.M., Sunyer, J., Grimalt, J., Aceves, M. and Reed, C.E. (1989b) Outbreaks of asthma associated with soybean dust. *N. Engl. J. Med.*, **321**, 1128.

Antó, J.M., Sunyer, J., Reed, C.E., Sabrià, J., Martínez, F., Codina, R., Morell, F., Rodríguez-Roisín, R., Rodrigo, M.J., Roca, J. and Saez, M. (1993) Preventing asthma epidemics due to soybeans by dust control measures. *N. Engl. J. Med.*, **329**, 1760–3.

Antó, J.M., Sunyer, J. and Newman Taylor, A.J. (1996) Soybean epidemic asthma and occupational asthma compared: a review. *Thorax*, **51**, 743–9.

Bellomo, R., Gigliotto, P., Treloar, A., Holmes, P., Suphiolgu, C., Singh, M.B. and Kno, B. (1992) Two consecutive thunderstorm associated epidemics of asthma in the city of Melbourne. The possible role of rye grass pollen. *Med. J. Aust.*, **156**, 834–7.

Cocco, G., Schiano, M., Sacerdoti, G., Sagliocca, L. and Palumbo, F. (1995) Functional characteristics in soybean asthma. *Am. J. Respir. Crit. Care Med.*, **151**, A469.

D'Amato, G., Liccardi, G. and Cazzola, M. (1994) Environment and development of respiratory allergy. I. Outdoors. *Monaldi Arch. Chest Dis.*, **5**, 406–11.

Ferrer, A., Torres, A., Rodríguez-Roisín, R., Roca, J., Sunyer, J. and Antó, J.M. (1990) Characteristics of patients with soybean dust-induced acute severe asthma requiring mechanical ventilation. *Eur. Respir. J.*, **3**, 429–33.

Figley, K.D. (1934) Soy bean as a possible important source of allergy: discussion. *J. Allergy*, **5**, 322–3.

Figley, K.D. and Elrod, R.H. (1928) Endemic asthma due to castor bean dust. *JAMA*, **90**, 79–82.

Goldstein, I.F. and Cuzick, J. (1983) Daily patterns of asthma in New York and New Orleans: an epidemiologic investigation. *Environ. Res.*, **30**, 211–23.

Goldstein, I.F. and Rausch, L.E. (1978) Time series analysis of morbidity data for assessment of acute environmental health effects. *Environ. Res.*, **17**, 266–75.

Goldstein, I.F. and Salvaggio, J. (1984) The decline of New Orleans asthma epidemics. *Rev. Environ. Health*, **4**, 133–46.

Hernando, L., Navarro, C., Marquez, M., Zapatero, L. and Galvañ, F. (1989) Asthma epidemics and soybean in Cartagena (Spain). *Lancet*, **i**, 502.

Lewis, R., Gilkeson, M.M. and McCaldin, R.O. (1962) Air pollution and New Orleans asthma. *Public Health Rep.*, **77**, 947–54.

Mendes, E. and Ulhoa Cintra, A. (1954) Collective asthma, simulating an epidemic provoked by castor-bean dust. *J. Allergy Clin. Immunol.*, **25**, 253–9.

Morrison, I. (1960) It happened one night. *Med. J. Aust.*, **1**, 850–1.

Murray, V., Venables, K., Laing-Morton, T., Partridge, M. and Williams, D. (1994) Epidemic of asthma possibly related to thunderstorms. *Br. Med. J.*, **309**, 131–2.

Navarro, C., Márquez, M., Hernando, L., Galvañ, F., Zapatero, L. and Caravaca, F. (1993) Epidemic asthma in Cartagena, Spain, and its association with soybean sensitivity. *Epidemiology*, **4**, 76–9.

O'Hollaren, M.T., Yunginger, J.W., Offord, K.P. *et al.* (1991) Exposure to an aeroallergen as a possible precipitating factor in respiratory arrest in young patients with asthma. *N. Engl. J. Med.*, **324**, 359–63.

Ordman, D. (1955) An outbreak of bronchial asthma in South Africa, affecting more than 200 persons, caused by castor bean dust from an oil-processing factory. *Int. Arch. Allergy Appl. Immunol.*, **7**, 10–24.

Packe, G.E. and Ayres, J.G. (1985) Asthma outbreak during a thunderstorm. *Lancet*, **ii**, 199–204.

Packe, G.E. and Ayres, J.G. (1986) Aeroallergen skin sensitivity in patients with severe asthma during a thunderstorm. *Lancet*, **i**, 850–1.

Salvaggio, J., Seabury, J. and Schoenhardt, E.A. (1971) New Orleans asthma V. Relationship between asthma admission rates, semiquantitative pollen and fungal spore counts, and total particulate aerometric sampling data. *J. Allergy*, **48**, 96–114.

Sunyer, J., Antó, J.M., Rodrigo, M.J., Morell, F. and The Clinical and Toxicoepidemiological Committee (1989) Case-control study of serum Immunoglobulin-E antibodies reactive with soybean in epidemic asthma. *Lancet*, **i**, 179–82.

Sunyer, J., Antó, J.M., Sabrià, F., Rodrigo, M.J., Roca, J., Morell, F., Rodríguez-Roisín, R. and Codina, R. (1992) Risk factors of soybean epidemic asthma. The role of smoking and atopy. *Am. Rev. Respir. Dis.*, **145**, 1098–102.

Suphioglu, C., Singh, M.B., Taylor, P., Bellomo, R., Holmes, P., Puy, R. and Knox, R.B. (1992) Mechanism of grass-pollen-induced asthma. *Lancet*, **339**, 569–72.

Synek, M., Antó, J.M., Beasley, R., Frew, A.J., Holloway, L., Lampe, F.C., Lloreta, J.L., Sunyer, J., Thornton, A. and Holgate, S.T. (1996) Immunopathology of fatal soybean dust-induced asthma. *Eur. Respir. J.*, **9**, 54–7.

Ussetti, P., Roca, J., Agusti, A.G.N., Montserrat, J.M., Rodríguez-Roisín, R. and Agusti-Vidal, A. (1983) Asthma outbreaks in Barcelona. *Lancet*, **ii**, 280–1.

Ussetti, P., Roca, J., Agusti, A.G.N., Montserrat, J.M., Rodríguez-Roisín, R. and Agusti-Vidal, A. (1984) Another asthma outbreak in Barcelona. *Lancet*, **i**, 156.

White, M.C., Etzel, R.A. and Olson, D.R. (1993) Reexamination of epidemic asthma in New Orleans in relation to the presence of soy-carrying ships in the harbor. *Am. J. Epidemiol.*, **138**, 607.

27 The effects of cotton dust on the lung

Ragnar Rylander

This chapter is dedicated to the memory of R. S. F. Schilling, a pioneer in cotton dust research, an outstanding occupational physician, and a great humanitarian.

THE ORGANIC DUST ENVIRONMENT

Dusts of vegetable, animal and microbial origin have always been part of the human environment. Exposures to such organic dusts usually cause no specific respiratory problems, even though plants contain a number of extremely toxic substances (particularly those present in or produced by molds). A number of vegetable fibers such as flax, soft hemp and sisal appear to be the exception.

Cotton fibers were used in early civilizations to make string, cloth and insulation materials. The introduction of machine processing of large quantities of cotton in indoor environments, typically with poor ventilation, has resulted in high levels of airborne dust. As cotton was one of the first materials to be harvested and processed by machine, the first reports of disease caused by organic dusts are attributable to this fiber and to the particular disease it causes – byssinosis. The pathophysiology of byssinosis can be understood with modern epidemiologic and toxicologic methods. This chapter deals with the health effects associated with exposure to cotton dust, but the same effects are found following exposure to dusts from other vegetable fibers (Rylander, 1994).

During the growth of the cotton plant, there is an abundant increase of microbes on the plant. Molds are present, but the predominant microorganisms are Gram-negative bacteria, particularly *Pseudomonas* and *Enterobacter*. These are not human pathogens and form part of the normal plant environment. The number of microbes on the cotton plant appears to be determined by the climate; in rainy environments the contamination is large. Cotton grown in dry climates with irrigation is contaminated with relatively few microorganisms.

Cotton is harvested using machines. This approach results in cotton contaminated with branches, leaves, soil and other materials. In the gin, cotton is cleaned by air, mechanical and centrifugal forces, after which the fibers are densely packed into bales. In the cotton mill, the bales are opened and the cotton is fed into carding machines. The fibers are elongated and then drawn. Thereafter they are spun into yarn which is later woven. Each of these operations subjects the cotton fibers to large physical forces, creating aerosols of small dust particles consisting of cotton fiber and contaminants. The initial phases, such as opening and carding, are particularly dusty but the dustiness decreases as the fibers are mechanically cleaned.

EARLY REPORTS ON ADVERSE EFFECTS

One of the earliest records of adverse effects caused by cotton dust was used as evidence to reduce work hours for children. Visitors to Sunday schools in Manchester and Salford, England, noted that '. . . the boys and girls working in factories had a very peculiar hoarseness and hollowness of voice which seems to indicate that their

Occupational Lung Disease. Edited by Daniel E. Banks and John E. Parker. Published in 1998 by Chapman & Hall. ISBN 0 412 73630 6

lungs are effected' (Papers, 1819). Passier (1822) in France gave the first account of the clinical effects of inhaling cotton dust, i.e. cough and perpetual irritation of the lungs which forced the workers to change their employment.

Witnesses to the British Factory Commission on Child Labour in Factories (Factory Commission, 1831) noted that cotton dust had an adverse effect on breathing. At the same time, Kay (1831), a Manchester physician, described the effects on the lung as cotton spinner's phthisis (tuberculosis), characterized by a work-related cough which resolved on leaving work, but later became more severe and persistent. He described the symptoms as '. . . a short, dry, harassing cough which in the initial stages ceased immediately the workers left the mill . . .' and concluded that these complaints differed from ordinary chronic bronchitis with sputum production.

The first description that chest symptoms were more severe at the start of the working week was made in 1845 by two French physicians (Mareska and Heyman, 1845):

> *'All the workers have told us that the dust bothered them much less on the last days of the week than on Monday and Tuesday. The masters find cause of this increased sensitivity to be in excess of the Sunday, but the workers never fail to attribute it to the interruption of work which, they say, makes them lose, in part, their habituation to the dust'.*

High death rates from respiratory disease in English cotton towns led to investigations into the health of cotton workers some decades later. Greenhow (1861) described an asthma-like condition, worse at the beginning of the week. Proust (1877) suggested the term 'byssinosis' derived from the greek 'BΨΣΣΘΣ (byssos)', meaning linen or fine flax, as the name for the respiratory disease of textile workers.

The introduction of power-driven brushes in Lancashire mills in about 1890 to clean carding engines markedly worsened dust conditions throughout the card room (Home Office, 1932). Collis (1909) examined 126 workers from 31 Lancashire cotton mills. Nearly three-quarters had the characteristic asthmatic complaint on Mondays, with 51 severely affected.

Statistical evidence of increased illness among British cotton workers came from Bradford Hill's study of Trade Union records in the 1920s (Hill, 1930). For non-respiratory causes, the frequency of illness among male card room workers did not differ from other workers. For respiratory illness, card room workers aged 30 to 59 were absent three times as often as workers with little or no dust exposure.

Prausnitz (1936) related the development of symptoms to the duration of exposure:

> *'After working for years without any appreciable trouble except a little cough, cotton mill workers notice either a sudden aggravation of their cough, which becomes dry and exceedingly irritating. . . . These attacks usually occur on Mondays . . . but gradually the symptoms began to spread over the ensuing days of the week; in time the difference disappears and they suffer continuously'.*

Respiratory disease caused by cotton dust exposure has since been observed in practically every country where investigations have been undertaken (Schilling *et al.*, 1955; Lammers *et al.*, 1964; Belin *et al.*, 1965; Braun *et al.*, 1973; Haglind *et al.*, 1983; Christiani *et al.*, 1986; White, 1989). An unusually low prevalence of respiratory disease has been reported in the Australian cotton mills (Field and Owen, 1979; Gun *et al.*, 1983).

THE NATURE OF THE DISEASE

Numerous studies on cotton workers in many parts of the world show that exposure to cotton dust causes several different reactions and symptoms. By definition, byssinosis has been referred to as the 'Monday feeling of chest tightness' in the United Kingdom, whereas changes in respiratory function have been the dominant criterion in the United States. The other different adverse respiratory effects were defined at a meeting in Manchester as mill fever, pulmonary function changes, chest tightness, hyper-reactive airways and chronic bronchitis (Rylander *et al.*, 1987b). This range of effects is not unexpected, bearing in mind the various constituents of cotton dust, e.g. plant debris, cellulose fiber, bacteria and fungi, as well as the inevitable variations in an individual's susceptibility.

There is now increasing evidence that the symptoms are not unique to cotton dust but may be induced by a large variety of organic dusts. This suggests that the causative agent(s) may be common to several forms of organic dust. Table 27.1 reviews the different clinical effects found after exposure to cotton dust. The

Table 27.1 Effects after exposure to cotton dust

- Toxic pneumonitis
- Airways inflammation
- Chronic bronchitis
- Asthma
- Chest tightness

Modified after Rylander *et al.*, 1987a

different effects related to cotton dust exposure have been slightly modified after the Manchester working group report and are described in detail below.

Toxic pneumonitis, mill fever

An attack of fever may follow the first exposure to cotton or other organic dusts. This reaction, described by earlier writers (Arlidge, 1892; Trice, 1940; Schneiter *et al.*, 1942; Holness *et al.*, 1982) and on several occasions since, occurs after several hours of relatively high levels of dust exposure. Symptoms include shivering, muscle and joint pains, a dry cough and a feeling of tiredness. These complaints resemble the early stages of influenza (Von Essen *et al.*, 1990) and disappear the following day. There are no changes on the chest radiograph, but a slight leukocytosis and decreased spirometry values may be found. After repeated exposures, the reaction disappears (habituation) but recurs after exposure to unusually high levels or after a longer break in exposure such as that occurring after returning from work after a holiday.

Mill fever is rare in modern cotton mills with efficient dust control, but may occasionally occur in connection with dusty work such as the cleaning of carding machines. There is no reliable information on the relationship between susceptibility to developing toxic pneumonitis and chronic adverse effects on lung function following after a long-term exposure.

In toxic pneumonitis, the lung histology shows acute inflammation in the airways with cellular infiltration and edema in the parenchyma. This is a response to a toxic agent and not to an infection or an allergen. Toxic pneumonitis also occurs after exposure to other kinds of organic dusts and has also been named inhalation fever or organic dust toxic syndrome (ODTS).

Airways inflammation

There is experimental as well as epidemiologic evidence to join the effects of pulmonary function changes and hyper-reactive airways into a common syndrome, namely airways inflammation. This is the most widespread disease following exposure to cotton dust and can vary from a mild irritation in the nose and airways to a severe dry cough and difficulties in breathing. An important question in making the diagnosis of airways inflammation is: 'Do you often (more than a third of the time) have: (i) a dry cough; (ii) chest wheezing; (iii) nasal irritation; and (iv) throat irritation?'

Inflammation may cause airways constriction and a reduced FEV_1. The well-documented evidence of a gradual decrease in FEV_1 over the working day was first shown by Schilling *et al.* (1955). This decline in lung function is reversible with bronchodilator inhalation. Airways inflammation is accompanied by an increase in airway responsiveness as measured by a methacholine or histamine challenge test. There is considerable discussion whether airways inflammation should be accepted as a disease entity in itself or whether it merely represents a symptom. As the clinical findings of severe cough and airways narrowing can lead to a decrease in work ability, it is reasonable to regard this as an occupational disease.

Chronic bronchitis

Airways inflammation which persists for several years may be recognized as chronic bronchitis, an illness particularly common among the heavily exposed workers in the blowing and carding areas. Chronic bronchitis is usually defined as cough with phlegm more than 3 months per year for at least 2 years. Airways inflammation can progress to chronic bronchitis, determined by exposure level and the number of years of exposure.

The assumption that chronic bronchitis is the inevitable cause of disability in cotton workers is no longer tenable. Many workers have symptoms of cough without sputum or obstructive airway disease without sputum.

Occupational asthma

At the onset of asthma attributable to cotton dust exposures, symptoms of cough and difficulty in breathing tend to occur in the evenings and at night and improve during weekends and holidays. With continuous exposure, symptoms tend to occur at work and fail to improve during holidays. Many workers also have rhinitis, which may occur before the onset of asthma. Several studies have shown that many individuals with occupational asthma fail to recover even after years of removal from exposure. The most important prognostic factor appears to be the duration of exposure after the onset of symptoms. In general, the earlier the diagnosis and removal from exposure, the greater the chance of recovery.

A small proportion of cotton workers (less than 1%) develop decrements in lung function within 30 minutes of starting work. This increases in severity over the work week, indicative of occupational asthma. There are a variety of potential allergens in cotton dust, and the development of an IgE-mediated reaction is plausible; however, it may be difficult to distinguish this specific reaction from the more common non-specific airways inflammation. Those experiencing this type of reaction are generally forced to leave the industry. Therefor, this disease is usually not diagnosed in cross-sectional studies, as workers with asthma have already left the work place.

Chest tightness

A subjective feeling of chest tightness which gradually develops during the first day at work (Mondays in Western countries) and improves during the week is classically called 'byssinosis' (Schilling *et al.*, 1955; Schilling, 1956, 1960). Symptoms usually take years to develop, although they can be induced in previously unexposed subjects at high dose levels (Haglind and Rylander, 1984). With continued dust exposure, chest tightness extends to other days and may be accompanied by severe respiratory impairment. Symptoms of chest tightness and bronchoconstriction (in terms of a decrease in FEV_1) may present together, but can also occur independently. Chest tightness is not a feature unique to cotton dust exposure, as it also appears among persons working with other kinds of organic dusts (Donham *et al.*, 1989).

EPIDEMIOLOGIC STUDIES

Mortality

In 19th century England, mortality rates were calculated for cotton workers as a whole, making comparison between dusty and non-dusty occupations within the workplace impossible. For the years 1910–1912, the mortality rates for male strippers and grinders (working in card rooms in the mills) were calculated separately and showed a two-fold excess of respiratory and cardiovascular death compared to all men (Schilling and Goodman, 1951). There were similar excesses in these two groups in the years 1921–1923 and 1930–1932 compared with all men and to spinners and weavers working in the same industry but exposed to less dust. By the early 1950s, this excess in mortality among strippers and grinders had disappeared (Schilling, 1960), although there was further evidence of an increased mortality among English strippers and grinders in a follow-up study in the 1970s (Berry and Molyneux, 1981).

Other studies in the US (Merchant and Ortmeyer, 1981; Aldrich, 1982), and China (Li, 1991) recorded no excess of deaths from respiratory disease in workers exposed to cotton dust. A report from Finland followed a cohort of 1065 women and showed no increase in deaths from respiratory disease but an increased rate of workers applying for and receiving disability retirement among cotton workers compared to the general population (Koskela *et al.*, 1990).

Morbidity

The first systematic epidemiologic survey of some card room workers and controls was undertaken in Lancashire in the early 1950s (Schilling *et al.*, 1955) using questionnaires designed to evaluate the presence of chest tightness. A high prevalence of byssinosis was found overall (61%), and 11% of the card room workers had markedly reduced ventilatory capacity compared with 4.5% of the controls from the same area.

A scheme was developed for grading the severity of byssinosis based on the type and frequency of symptoms (Schilling, 1956; Mekky *et al.*, 1967). The duration of exposure was used as a measure of dose and related to the severity of the response. Based on interviews of large numbers of workers, this grading scheme was later modified to more accurately reflect the time interval for the development of symptoms (Berry *et al.*, 1973).

Mills using high-quality cotton to produce finer yarns had a lower prevalence of byssinosis than mills producing coarse yarns with a lower quality of cotton. Thus, in addition to exposure intensity and duration, both dose-related variables, the type of dust became an important variable for assessing exposure. Later, it was demonstrated that differences in the response of workers exposed to coarse and medium cottons were dependent not only on the type of cotton but on other variables, such as carding speed, humidification and ventilation, and manufacturing variables such as different yarn treatments (Berry *et al.*, 1973). In another study, a difference in the prevalence of byssinosis in mills processing different types of cotton was found (Jones *et al.*, 1979).

The next refinement in the relationship between exposure to cotton dust and response (either symptoms of chest tightness or a decline in pulmonary function) was in US studies comparing those who worked with 100% cotton to workers using the same cotton but in a 50 : 50 blend with synthetics and workers without exposure to cotton (Merchant *et al.*, 1973a). Workers exposed to 100% cotton had the highest prevalence of byssinosis independent of cigarette smoking, one of the confounders of the assessment in the change in lung function associated with exposure to cotton dust. This semi-quantitative relationship between dose and response to cotton dust was further refined in a group of textile workers stratified by sex, smoking category, work area and mill type (Merchant *et al.*, 1973b). A relationship was observed in each of these categories between byssinosis prevalence and/or change in FEV_1 and dust concentration, but only in the lower dust ranges.

A comprehensive survey of 4656 Lancashire cotton workers in the 1980s revealed a significant decline in the prevalence of byssinosis in all card room workers between the 1950s (51%) and the 1980s (14%) (Cinkotai *et al.*, 1988). The authors concluded that this downward trend was brought about by more effective dust control.

Prospective studies

Prospective studies (Kamat *et al.*, 1981; Beck *et al.*, 1982) showed accelerated annual declines in lung function (usually FEV_1) in cotton workers compared with controls with decreases in lung function related to the level and duration of dust exposure. A 5-year longitudinal study was performed in the US to identify whether current workplace standards for cotton dust prevented chronic adverse health effects (Glindmeyer *et al.*, 1991). The OSHA exposure limits for respirable cotton dust, averaged over 8 hours, are set at 200 $\mu g/m^3$ in yarn manufacturing and 750 $\mu g/m^3$ in slashing/weaving. Dust measurement records in these workplaces showed 5-year averages of 196 $\mu g/m^3$ in yarn manufacturing and 455 $\mu g/m^3$ in slashing/weaving.

The study comprised 1353 cotton and 464 synthetic workers exclusively in yarn manufacturing, in slashing/weaving, or in synthetic textile production. The prevalence of byssinosis was minimal; 1.3% for cotton workers and 0.4% for synthetic workers. Despite having lower overall dust exposures, yarn workers exhibited steeper annual declines in lung function than did slashers and weavers. This difference persisted within each smoking category. Among the yarn manufacturers, there was a synergistic effect of smoking and dust exposure on lung function. A significant dose–response relationship was seen in annual declines in FEV_1 forced vital capacity (FVC) and FEF_{25-75} and average exposure by mill, as shown in Figure 27.1.

Although compliance with current exposure limits in combination with health surveillance virtually eliminated byssinosis, there was evidence of chronic effects, measured by a rise in the rate of annual decline in lung function in yarn workers exposed to dust levels close to the limit. This supports the hypothesis that there can be long-term pulmonary effects without the classic symptoms of Monday-related chest tightness. Interestingly, in mills with synthetic fibers, workers had larger average declines over the years than cotton workers. Unrecognized and unmeasured causative exposures or selection bias could explain this result.

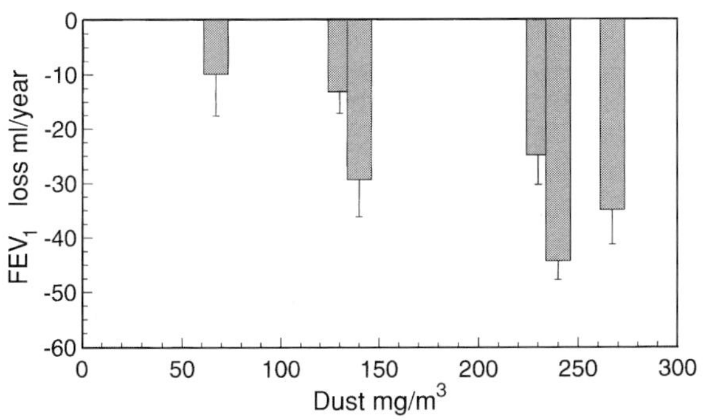

Figure 27.1 Relationship between exposure (dust mg/m^3) and predicted annual decline in FEV_1. (After Glindmeyer *et al.*, 1991.)

In a recent study in China, workers were examined before they started work in the mill and then at 10 weeks and 2 years thereafter (Li *et al.*, 1995). Persons with an atopic reaction to an extract of cotton dust before work showed a larger across-shift decrease in FEV_1, and an increased airway responsiveness when compared with workers with no atopic reaction. This suggests that atopy is a risk factor in cotton dust-related effects.

PULMONARY PATHOLOGY AND PATHOGENESIS

Important findings in autopsy studies include smooth muscle hypertrophy and mucus gland enlargement suggestive of chronic airways inflammation. Early autopsy studies of the lungs of cotton workers with byssinosis revealed no fibrosis. Emphysema was found (Edwards *et al.*, 1975), but more recent investigations indicate that emphysema is caused by cigarette smoking rather than by cotton dust (Pratt *et al.*, 1980; Moran, 1983; Honeybourne and Pickering, 1986). Yellow inclusion bodies (brown lung), described in some early studies, are not a general feature.

Recent developments in pulmonary toxicology have demonstrated that several of the symptoms seen after cotton dust exposure can be accounted for by inflammation (Ellakkani *et al.*, 1987; Holt, 1987; Ryan and Karol, 1991). Macrophages (and probably epithelial cells and platelets) become activated by cotton dust and secrete inflammatory mediators. These attract other cells types to the site of action.

The cellular hallmark of cotton dust exposure is a migration of neutrophils into lung tissue and later into airways. This inflammation can explain the appearance of toxic pneumonitis after an acute exposure at high levels as well as the subjective feeling of irritation, dry cough and objective features of a decreased FEV_1, and increased airway responsiveness indicative of airways inflammation. There are data suggesting that individuals with pre-existing inflammation (i.e. atopy) are at greater risk of responding to cotton dust, measured as the acute decrease in FEV_1 over a working shift (Sepulveda *et al.*, 1984; Jacobs, 1993).

Regarding occupational asthma, it was originally hypothesized that specific antibodies against one or several components in cotton dust were responsible for the effects, but efforts to detect IgG or IgE antibodies against cotton dust components have not been successful (Kutz *et al.*, 1980; Petronio and Bovenzi, 1983;

Rylander *et al.*, 1982). Asthma occurring after cotton dust exposure could be caused by T cell-mediated reactions, but investigations into this possible mechanism have not yet been reported.

The pathologic features explaining chest tightness have not yet been described. It has been suggested that chest tightness is caused by an increased adhesiveness of platelets which accumulate in the lung capillaries and increase the pulmonary artery pressure (Rylander, 1987). It is likely that chest tightness involves some kind of cell sensitization, as it takes repeated exposures over several years for the symptom to develop. An increased ability to produce procoagulant factor, indicative of blood monocyte activation, was found among cotton workers as compared with controls (Beijer *et al.*, 1990).

Early investigators suggested that bronchoconstriction after cotton dust exposure was caused by histamine, either in the dust itself or liberated by mast cells in contact with a toxic agent in the dust. These conclusions were based on experience from *in vitro* preparations, where extracts of cotton dust had been found to liberate histamine (Hayworth and MacDonald, 1937; Nicholls *et al.*, 1967). Histamine metabolites have been found in the urine of workers exposed to cotton dust, but anti-histamine drugs do not inhibit the exposure effects. It is likely that histamine in the urine is the late result of a general inflammatory response, activating a large number of mediators which are not directly involved in the pathogenesis of the disease.

CAUSATIVE AGENTS

That the cleanliness of cotton was related to disease was first observed by Leach (1863). Now it is generally agreed that the cotton fiber does not cause the various effects induced by cotton dust exposure. This was demonstrated in a study on 30 individuals in an experimental card room, comparing the effects of washed and unwashed cotton at the same dust level (Petsonk *et al.*, 1986). At dust levels around 0.6 mg/m^3, washed cotton produced a significantly lower decrease in FEV_1 than unwashed cotton.

In numerous epidemiologic studies, dose– response relationships have been found between the extent of chest tightness and other symptoms and the amount of airborne dust. These studies were generally made in industries using similar kinds of cotton, which implies that the contamination level varied with the dust level. When studies were undertaken in model card rooms using different kinds of cotton, no dose–response relationships for the amount of airborne dust were found, either for cotton workers or for previously unexposed subjects (Rylander *et al.*, 1985; Rylander and Haglind, 1986; Castellan *et al.*, 1987). This illustrates the fact that contamination of cotton lint, rather than the cotton itself, causes the biologic effects, and that the amount of this/these agents in cotton dust may vary due to different sources of the cotton fiber.

Much of the early work on causative agents concentrated on the bracts of the cotton plant. Various extracts from these bracts were found to cause bronchoconstriction (Schachter *et al.*, 1981, 1985). The concentration of the extract required to cause this effect was quite high, and bronchoconstriction occurred rapidly compared with the slow onset of bronchoconstriction after cotton dust exposure. In an experiment with carefully controlled exposure conditions, test persons were exposed to dust from ordinary cotton and to dust from cotton from which the bract leaves had been painstakingly removed (Castellan *et al.*, 1986). No effect of the removal of the bract contamination on lung function could be measured. This suggests that bract is not a unique source of the causative agent in cotton dust.

There is considerable information regarding the microbiologic contamination of cotton (Figure 27.2). Detailed measurements of microbes in cotton mills have been performed (Lacey and Lacey, 1987). Schneiter *et al.* (1942) concluded that an observed case of toxic pneumonitis was caused by inhalation of Gram-negative bacteria or an associated product contaminating the cotton. This report also associated the response with a specific type of cotton classed as 'graded' or 'tinged' – a phenomenon caused by microbial growth.

The association between bacteria and the acute response to cotton dust led to studies investigating the microflora of cotton. Clark *et al.* (1947) reported that contamination of cotton in the field with Gram-negative bacteria was dependent on environmental factors that influence the stage of boll maturity at the time of frost. Later studies demonstrated that the bacterial contamination of cotton was higher when grown in humid and rainy areas (Morey *et al.*, 1983) and that the bolls became contaminated when opened during the last part of the growing season (Millner *et al.*, 1982). The bract part of the cotton is particularly contaminated with bacteria (Rylander, *et al.*, 1979). A recent series of studies demonstrated that the lint becomes contaminated with Gram-negative bacteria within a few days of boll cracking.

Microorganisms contain a number of substances with strong biologic activity. The major known compounds are endotoxin from Gram-negative bacteria, peptidoglucans from Gram-positive bacteria, and $(1\rightarrow3)$-β-D-glucans from molds. Of these substances, endotoxins have been most extensively studied

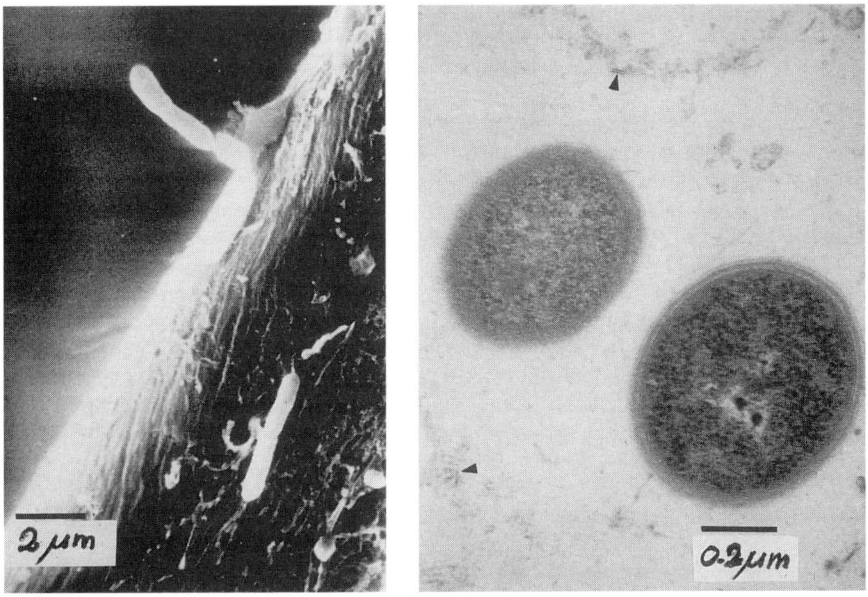

Figure 27.2 Gram-negative bacteria on cotton fiber (left) and in water with free endotoxin indicated by arrowheads (right).

(Figure 27.3). In the 1960s, Italian researchers measured endotoxin in cotton dust and exposed humans and animals to a crude preparation of endotoxin obtained from cotton. They observed bronchoconstriction and chest tightness among humans. Autopsy studies of the lungs of these animals showed features of chronic inflammation (Pernis *et al.*, 1961; Cavagna *et al.*, 1969).

A later study in Lancashire cotton mills demonstrated a significant relationship between the number of airborne Gram-negative bacteria and the prevalence of byssinosis, evaluated as chest tightness (Cinkotai *et al.*,

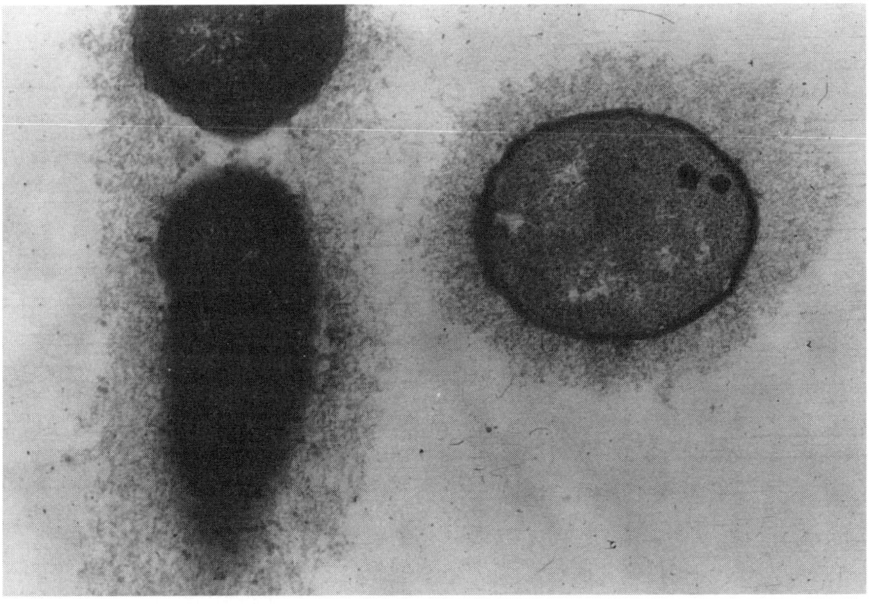

Figure 27.3 Gram-negative bacteria with endotoxin visible in the shaded area surrounding the bacterium.

1977). A similar relationship was found between the number of bacteria in cotton lint and lung function changes (Rylander and Lundholm, 1978; Rylander *et al.*, 1979).

Recent experimental studies have made it possible to assess the importance of different constituents of the dust. In a model card room study, dust from cotton grown in different locations was found to have a different potential for causing an effect on workers in a cotton mill (Rylander, and Haglind, 1986). In an experimental study, 15 card room workers were exposed to cotton from three geographic areas. There were no significant correlations between the decrease in FEV_1 and dust levels. Figure 27.4 shows the relation between airborne endotoxin levels and FEV_1 decrease over the shift. There was also a relation to subjective symptoms of byssinosis and an increase of neutrophils in blood.

Similar observations were reported from a study of previously unexposed subjects selected for increased airway responsiveness to cotton dust (Castellan *et al.*, 1987). Vertically elutriated gravimetric dust levels of 0.51 mg/m^3 of a cotton grown in California did not cause a decline in FEV_1. However, cottons grown in Mississippi, a more humid area, caused a mean decline of 6% in FEV_1 and 8% in FVC in 30 subjects exposed to dust levels of 0.25 mg/m^3.

These data show that the simple measurement of gravimetric dust is not an adequate measure of the biologic activity of the cotton dust. The content of active agents in the dust varies according to where the fiber is grown, the local cultivation practices, and the manufacturing process. A gravimetric measure of dust may be an appropriate measure of exposure if a single source of cotton is used for all processes in a group of textile mills. Under these conditions, the concentration of active agents in the dust will correlate closely with gravimetric dust levels.

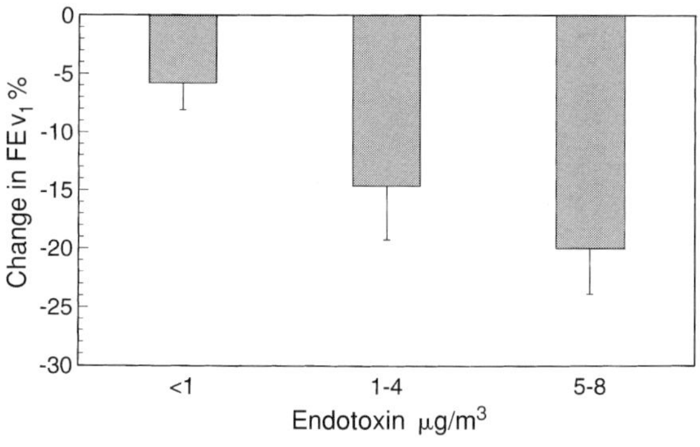

Figure 27.4 Relationship between decrease in FEV_1 during an experimental 6-hour exposure and airborne endotoxin to different cotton dusts. (After Rylander and Haglind, 1986.)

Proteases have been detected in both respirable and non-respirable dust from US and UK cotton mills (Tuma *et al.*, 1973; Cinkotai, 1976; Milton and Chawla, 1986; White, 1989). Some researchers have suggested they originate from bacteria and fungi which contaminate raw cotton and that they are unlikely to contribute to any disease process since they were sequestered within the organisms *per se*. However, data from one of the recent studies (Milton and Chawla, 1986), that took into account microbial contamination indicated that both coarse and respirable cotton dust contained a variety of preformed proteases, including both serine and metaloproteinases. These authors also demonstrated the presence of an elastolytic enzyme which was incompletely inhibited by α_1-antiprotease inhibitor.

DIAGNOSIS

The 'gold standard' for the classic diagnosis of byssinosis relies on the worker's description of chest tightness and its presence on the first day(s) of the working week. Roach and Schilling (1960) devised a scheme for grading the clinical stages of byssinosis. This was later modified by substituting 'difficulty in breathing' for 'breathlessness' and adding a third grade (Schilling *et al.*, 1963) (Table 27.2). Nowadays, with cleaner conditions in cotton mills, symptoms of chest tightness may be present on days with an unusually heavy exposure, such as when the machines are cleaned.

Spirometry before and after the work shift is often used to determine exposure-related declines. A rapid decline that does not improve over the working week is suggestive of occupational asthma, whereas the late onset of bronchoconstriction and no changes over the work week are suggestive of airways inflammation.

Table 27.2 Grading of byssinosis induced by cotton, flax, hemp or sisal dust

Grade 0:	Normal – no symptoms of chest tightness or cough
Grade 1/2:	Occasional chest tightness, or cough, or both, on first day of the work week
Grade 1:	Chest tightness on every first day of the work week
Grade 2:	Chest tightness on every first day and other days of the work week
Grade 3:	Grade 2 symptoms, accompanied by evidence of permanent incapacity from reduced ventilatory capacity

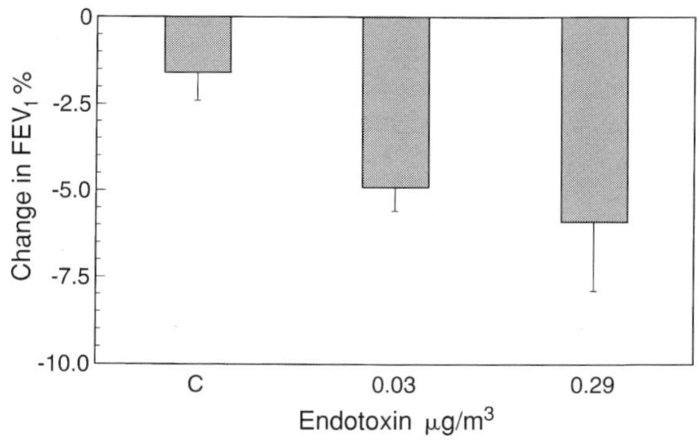

Figure 27.5 Decrease in FEV_1 among cotton workers after inhalation of 1.2 mg methacholine. The workers were employed in areas where different levels of airborne endotoxin are present. (After Rylander and Bergstrom, 1993.)

Table 27.3 Proposed guidelines for a 'no effect' threshold for endotoxin in organic dust

Disease	*Guidelines (ng/m^3)*
Toxic pneumonitis	200
Airways inflammation	10

From Rylander and Jacobs 1996.

Serial spirometry will give an indication of the annual fall in FEV_1 and whether this is larger than expected compared with the rate of decline of non-exposed workers.

Because airways inflammation is an important and widespread effect, measurements of airway responsiveness are increasingly used in epidemiologic investigations and will probably become useful for diagnostic purposes. Figure 27.5 shows the relation between airway responsiveness and the amount of airborne endotoxin in a cotton mill.

Questionnaires are a useful tool for diagnosing toxic pneumonitis, airways inflammation, chronic bronchitis and the few cases of occupational asthma that appear. For example, toxic pneumonitis is diagnosed from the clinical picture – fever, fatigue and shivering. The diagnosis can be made with the question: 'Have you, during the last 6 months in connection with dusty work, had episodes of influenza-like symptoms such as fever, chills, malaise, cough, muscle or joint pain, and felt completely well the following day?' These complaints usually occur only in the beginning of work in a mill, but may recur after a heavy exposure.

PREVENTION

Measures to control the adverse effects of exposure to cotton dust have depended on the reduction of the trash content of raw cotton and dust control in the workplace, together with health surveillance of exposed workers to identify susceptible individuals and remove them from exposure. In the Western world, dust control has significantly reduced the prevalence and severity of work-related disease (Jacobs, 1987; Glindmeyer *et al.*, 1991). However, as described above, simple reductions of dust levels will not always ensure prevention, and measurement and control of endotoxin would give better protection. A proposal for guidelines was presented by Rylander and Haglind (1986) and later by the Committee on Organic Dusts at the International Commission on Occupational Health (Rylander, and Jacobs, 1997) (Table 27.3). The 'no reaction' endotoxin threshold for toxic pneumonitis is 200 ng/m^3, for airways inflammation, exposures should remain below 10 ng/m^3.

REFERENCES

Aldrich, M. (1982) Mortality from byssinosis among New England cotton mill workers 1905–1912. *J. Occup. Med.*, **24**, 977–80.

Arlidge, J.T. (1892) *The hygiene, Disease and Mortality of Occupations*, London.

Beck, G.J., Schachter, E.N., Maunder, L.R. and Schilling, R.S.F. (1982) A prospective study of chronic lung disease in cotton textile workers. *Ann. Intern. Med.*, **97**, 645–51.

Beijer, L., Carvalheiro, M., Holt, P. and Rylander, R. (1990) Increased blood monocyte procoagulant activity in cotton mill workers. *J. Clin. Lab. Immunol.*, **33**, 125–27.

Belin, L., Bouhuys, A., Hoekstra, W., Johansson, M.-B., Lindell, S.-E. and Pool, J. (1965) Byssinosis in card room workers in Swedish cotton mills. *Br. J. Ind. Med.*, **22**, 101–8.

Berry, G. and Molyneux, M.K.B. (1981) A mortality study of workers in Lancashire cotton mills. *Chest*, **9S**, 115–55.

Berry, G., McKerrow, C.B., Molyneux, M.K.B., Rossiter, C.E. and Tonbleson, J.B.L. (1973) A study of the acute and chronic changes in ventilatory capacity of workers in Lancashire cotton mills. *Br. J. Ind. Med.*, **30**, 25–36.

Braun, D.C., Jurgiel, K.A., Kaschak, M. and Babyak, M.A. (1973) Prevalence or respiratory signs and symptoms among U.S. cotton workers. *J. Occup. Med.*, **15**, 414–19.

Castellan, R.M., Brashears, A.D., Cocke, J.B. and Olenchock, S.A. (1986) Removal of bract prior to harvest: Effects on card room dust levels and acute human ventilation responses, in *Proc. 10th Cotton Dust Res. Conf. Natl Cotton Council*, (eds R.R. Jacobs and P.J. Wakelyn), Memphis, TN, pp. 79–83.

Castellan, R.M., Olenchock, S.A., Kinsley, K.B. and Hankinson, J.L. (1987) Inhaled endotoxin and decreased spirometric values: an exposure – response relation for cotton dust. *New Engl. J. Med.*, **317**, 605–10.

Cavagna, G., Foa, V. and Vigliani, E.C. (1969) Effects in man and rabbits of inhalation of cotton dusts or extracts and purified endotoxins. *Br. J. Ind. Med.*, **26**, 314–21.

Christiani, D.C., Eisen, E.A., Wegman, D.H., Ye, T., Lu, P., Gong, Z. and Dai, H. (1986) Respiratory disease in cotton textile workers in the People's Republic of China: I. Respiratory symptoms; II. Pulmonary function results. *Scand. J. Work Environ. Health*, **12**, 404–5.

Cinkotai, F.F. (1976) The size-distribution and protease content of airborne particles in textile mill card rooms. *Am. Ind. Hyg. Assoc. J.*, **37**, 234–8.

Cinkotai, F.F., Lockwood, M.G. and Rylander, R. (1977) Airborne microorganisms and prevalence of byssinotic symptoms in cotton mills. *Am. Ind. Hyg. Assoc. J.*, **38**, 55–59.

Cinkotai, F.F., Rigby, A., Pickering, C.A.C., Seaborn, D. and Faragher, E. (1988) Recent trends in the prevalence of byssinotic symptoms in the Lancashire textile industry. *Br. J. Ind. Med.*, **45**, 782–9.

Clark, F.E., Hervey, R.J. and Bland, L.M. (1947) Occurrence of cotton fiber contaminated by *Aerobacter cloacae*. US Dept Agr. Tech. Bull., **935**, 1–28.

Collis, E.L. (1909) *Annual report of chief inspector of factories for 1908*, HMSO, London.

Donham, K., Haglind, P., Peterson, Y. and Rylander, R. (1989) Environmental and health studies of farm workers in Swedish swine confinement buildings. *Br. J. Ind. Med.*, **46**, 31–7.

Edwards, C., McCartney, J., Rooke, G. and Ward, F. (1975) The pathology of the lung in byssinotics. *Thorax*, **30**, 612–23.

Ellakkani, M.A., Alarie, Y., Weyel, D. and Karol, M.H. (1987) Chronic pulmonary effects in guinea pigs from prolonged inhalation of cotton dust. *Toxicol. Appl. Pharmacol.*, **54**, 354–69.

Factory Commission (1831/1832) Child Labour in factories, xv para, 10663.

Field, G.B. and Owen, P. (1979) Respiratory function in an Australian cotton mill. *Bull. Eur. Physiopathol. Respir.* **15**, 455–68.

Glindmeyer, H.W., Lefante, J.J., Jones, R.N., Rando, R.J., Abdel Kader, H.M. and Weill, H. (1991) Exposure-related declines in the lung function of cotton textile workers. Relationship to current workplace standards. *Am. Rev. Respir. Dis.*, **144**, 675–83.

Greenhow, H. (1861) Third report of the medical officer of the Privy Council. Sir John Simon, p. 152.

Gun, R.T., Janckewicz, T., Esterman, A., Roder, D., Antic, R., McEvoy, R.D. and Thornton, A. (1983) Byssinosis: a cross-sectional study in an Australian textile factory. *J. Soc. Occup. Med.* **33**, 119–25.

Haglind, P. and Rylander, R. (1984) Exposure to cotton dust in an experimental card room. *Br. J. Ind. Med.* **41**, 340–5.

Haglind, P., Bake, B. and Belin, L. (1983) Is mild byssinosis associated with small airways disease? *Eur. J. Respir. Dis.*, **64**, 449–59.

Hayworth, E. and MacDonald, A.D. (1937) On histamine in cotton dust, and in the blood of cotton workers. *J. Hyg.*, **37**, 232–42.

Hill, A.B. (1930) Sickness amongst operatives in Lancashire cotton spinning mills (with special reference to the card room), Ind. Hlth Res. Bd Rep. No. 59, HMSO, London.

Holness, D.L., Taraschuk, I.G. and Goldstein, M.B. (1982) Acute exposure to cotton dust. A case of mill fever. *JAMA*, **247**, 1602–3.

Holt, P.G. (1987) Current trends in research on the etiology and pathogenesis of byssinosis. *Am. J. Ind. Med.* **12**, 711–16.

Home Office (1932) *Report of departmental committee on dust in card rooms of the cotton industry*, London.

Honeybourne, D. and Pickering, C.A.C. (1986) Physiological evidence that emphysema is not a feature of byssinosis. *Thorax*, **41**, 6–11.

Jacobs, M., Boehlecke, B., van Hage-Hamsten, M. and Rylander, R. (1992) Bronchial reactivity, atopy and airway response to cotton dust. *Am. Rev. Respir. Dis.*, **148**, 19–24.

Jacobs, R.R. (1987) Strategies for prevention of byssinosis. *Am. J. Ind. Med.*, **12**, 717–30.

Jones, R.N., Diem, J.E., Glindmeyer, H.W., Dharmarajan, V., Hammad, Y.Y., Carr, J. and Weill, H. (1979) Mill effect and dose-response relationships in byssinosis. *Br. J. Ind. Med.*, **36**, 305–13.

Kamat, S.R., Kamat, G.R., Salpekar, V.Y. and Lobo, E. (1981) Distinguishing byssinosis from chronic obstructive pulmonary disease: results of a prospective five-year study of cotton workers in India. *Am. Rev. Respir. Dis.*, **124**, 31–40.

Kay, J.P. (1831) Observations and experiments concerning molecular irritation of the lungs as one source of tubercular consumption; and on spinner's phthisis. *North Engl. Med. Surg. J.*, **1**, 1348–63.

Koskela, R.S., Klockars, M. and Jarvinen, E. (1990) Mortality and disability among cotton mill workers. *Br. J. Ind. Med.*, **47**, 384–91.

Kutz, S.A., Olenchock, S.A., Elliot, J.A., Pearson, D.J. and Major, P.C. (1980) Precipitation of serum proteins by extracts of cotton dust and stems. *Environ. Res.* **22**, 476–84.

Lacey, J. and Lacey, M.E. (1987) Micro-organisms in the air of cotton mills. *Ann. Occup. Hyg.* **31**, 1–19.

Lammers, B., Schilling, R.S.F., Walford, J., Meadows, R., Roach, S.A., Van den Hoven, D., Van Genderen, D., Van der Veen, Y.G. and Wood, C.H. (1964) A study of byssinosis, chronic respiratory symptoms and ventilatory capacity in English and Dutch cotton mills, with special reference to atmospheric pollution. *Br. J. Ind. Med.*, **21**, 124–34.

Leach, J. (1863) Surat cotton, as it bodily affects operatives in cotton mills. *Lancet*, **ii**, 64–8.

Li, D. (1991) Preliminary data on mortality among cotton and flax workers in China, in *Health Effects of Grain and Other Organic Dusts*, (eds R. Rylander and A. William Musk), Committee on Organic Dusts, Report 2/91, pp. 16–17.

Li, D., Zhong, Y.N., Rylander, R., Ma, Q.Y. and Zhou, X.Y. (1995) Longitudinal study of the health of cotton workers. *J. Occup. Environ. Med.*, **52**, 328–31.

Mareska, J. and Heyman, J. (1845) Enquete sur le travail et la condition physique et morale des ouvriers employes dans les manufacturers de coton. *Ann. Soc. Med. Gand.*, **16**, 11:5.

Mekky, S., Roach, S.A. and Schilling, R.S.F. (1967) Byssinosis among winders in the industry. *Br. J. Occup. Med.*, **24**, 123–32.

Merchant, J.A. and Ortmeyer, C. (1981) Mortality of employees in two cotton mills in North Carolina. *Chest*, **79S**, 6–11.

Merchant, J.A., Lumsden, J.C., Kilburn, K.H., O'Fallon, W.M., Ujda, J.R., Germino, V.H. and Hamilton, J.D. (1973a) An industrial study of the biological effects of cotton dust and cigarette smoke exposure. *J. Occup. Med.*, **15**, 12–222.

Merchant, J.A., Lumsden, J.C., Kilburn, K.H., O'Fallon, W.M., Ujda, J.R., Germino, V.H. and Hamilton, J.D. (1973b) Dose response studies in cotton textile workers. *J. Occup.Med.*, **15**, 222–30.

Millner, P.D., Ericson, K.E. and Marsh, P.B. (1982) Bacteria on closed-boll and commercially harvested cotton. *Appl. Environ. Microbiol.* **44**, 335–62.

Milton, D.K. and Chawla, R.H. (1986) Cotton dust contains proteolytic and elastolytic enzymes not inhibited by alpha-1-proteinase inhibitor. *Am. J. Ind. Med.*, **9**, 247–60.

Moran, T.J. (1983) Emphysema and other chronic lung diseases in textile workers: an 18 year autopsy study. *Arch. Environ. Health.*, **38**, 267–76.

Morey, P., Fischer, J.J. and Rylander, R. (1983) Gram-negative bacteria on cotton with particular reference to climatic conditions. *Am. Ind. Hyg. Assoc J.*, **44**, 100–104.

Nicholls, P.J., Nicholls, G.R. and Bouhuys, A. (1967) Histamine release by compounds 45/80 and textile dusts from lung tissue *in vitro* in *Inhaled Particles and Vapours II*, (ed. C.N. Davies), Pergamon Press, London, pp. 69–74.

Papers on evidence for and against Sir Robert Peel's bill to reduce hours of work of children in cotton mills (1819) Manchester Central Library, Local History Department.

Passier, P. (1822) Traite'des maladies des artisans, Paris, p. 245.

Pernis, B., Vigliani, E.C., Cavagna, C. and Finulli, M. (1961) The role of bacterial endotoxins in occupational diseases caused by inhaling vegetable dusts. *Br. J. Ind. Med.*, **18**, 120–9.

Petsonk, E.L., Olenchock, S.A., Castellan, R.M., Banks, D.E., Mull, J.C., Hankinson, J.L., Bragg, K.C., Perkins, H.H. and Cocke, J.B. (1986) Human ventilatory response to washed and unwashed cottons from different growing areas. *Br. J. Ind. Med.*, **43**, 182–7.

Petronio, L. and Bovenzi, M. (1983) Byssinosis and serum IgE concentrations in textile workers in an Italian cotton mill. *Br. J. Ind. Med.*, **40**, 39–44.

Pratt, P.C., Wollmer, R.T. and Miller, J.A. (1980) Epidemiology of pulmonary lesions in nontextile and cotton textile workers: a retrospective autopsy analysis. *Arch. Environ. Health*, **35**, 133–8.

Prausnitz, C. (1936) *Investigations on respiratory dust disease in operatives in cotton industry*, Medical Research Council, special report series, No. 212. His Majesty' Stationery Office, London p.60.

Proust, A. (1877) *Traité d'Hygiène Publique et Privée*, G. Masson, Paris, **7**, 171.

Roach, S.A. and Schilling, R.S.F. (1960) A clinical and environmental study of byssinosis in the Lancashire cotton industry. *Br. J. Ind. Med.*, **17**, 1–9.

Ryan, L.K. and Karol, M.H. (1991) Release of tumor necrosis factor in guinea pigs upon acute inhalation of cotton dust. *Am. J. Respir. Cell Mol. Biol*, **5**, 93–8.

Rylander, R. (1994) Organic dust – from knowledge to prevention. *Scand. J. Work Environ. Health*, **20**, 116–22.

Rylander, R. (1987) The role of endotoxin for reactions after exposure to cotton dust. *Am. J. Ind. Med.*, **12**, 687–97.

Rylander, R. and Bergström, R. (1993) Bronchial reactivity among cotton workers. *Ann. Occup. Hyg*, **37**, 57–63.

Rylander, R. and Haglind, P. (1986) Exposure of cotton workers in an experimental cardroom with reference to airborne endotoxins. *Environ Health Perspect*, **66**, 83–6.

Rylander, R. and Jacobs, M. (eds) (1997) Endotoxins in the Environment – A criteria document. *Int. J. Occup. & Environ. Health*, **3**, S1–S48.

Rylander, R. and Lundholm, M. (1978) Bacterial contamination of cotton and cotton dust, and effects on the lung. *Br. J. Ind. Med.*, **35**, 204–7.

Rylander, R., Imbus, H.R. and Suh, M.W. (1979) Bacterial contamination of cotton as an indicator of respiratory effects among cardroom workers. *Br. J. Ind. Med.*, **36**, 299–304.

Rylander, R., Wold, A. and Haglind, P. (1982) Nasal antibodies against Gram-negative bacteria in cotton mill workers. *Int. Arch. Allergy Appl. Immunol.*, **69**, 330–34.

Rylander, R., Haglind, P. and Lundholm, M. (1985) Endotoxin in cotton dust and respiratory function decrement among cotton workers. *Am. Rev. Respir. Dis.* **131**, 209–13.

Rylander, R., Peterson, Y., Ong, S.G. and Liao, S. (1987a) Byssinosis in the Far East. *Am. J. Ind. Med.*, **12**, 647–802.

Rylander, R., Schilling, R.S.F., Pickering, C.A.C., Rooke, G.B., Dempsey, A.N. and Jacobs, R.R. (1987b) Effects after acute and chronic exposure to cotton dust – the Manchester criteria. *Br. J. Ind. Med.*, **44**, 557–79.

Schachter, E.N., Brown, S., Zuskin, E., Buck, M.G., Kolack, B. and Bouhuys, A. (1981) Airway reactivity and cotton bract induced bronchospasm. *Am. Rev. Respir. Dis.*, **123**, 272–6.

Schachter, E.N., Zuskin, E., Buck, M.G., Witek, T.J., Jr, Beck, G.J. and Tyler, D. (1985) Airway reactivity and cotton bract induced bronchial obstruction. *Chest*, **87**, 51–5.

Schilling, R.S.F. (1956) Byssinosis in cotton and other textile workers. *Lancet*, **2**, 261–5.

Schilling, R.S.F. (1960) The epidemiology of byssinosis, in *Symposium on Industrial Pulmonary Disease* (eds E.J. King and C.M. Fletcher), Churchill-Livingstone, London.

Schilling, R.S.F. and Goodman, N. (1951) Cardiovascular disease in cotton workers: Part I. *Br. J. Ind. Med.*, **7**, 77–90.

Schilling, R.S.F., Hughes, J.P.W., Dingwall-Fordyce, I. and Gilson, J.C. (1955) An epidemiological study of byssinosis among Lancashire cotton workers. *Br. J. Ind. Med.*, **12**, 217–27.

Schilling, R.S.F., Vigliani, E.C., Lammers, B., Valic, F. and Gilson, J.C. (1963) A report on a conference on byssinosis. Int. Cong. Series, No. 2, Excerpta Medica, Amsterdam, pp. 137–44.

Schneiter, R., Neal, P.A. and Caminita, B.H. (1942) Aetiology of acute illness among workers using low grade stained cotton. *Am. J. Public Health*, **32**, 1345–52.

Sepuldelva, M.J., Castellan, R.M., Hankinson, J.L. and Cocke, J.B. (1984) Acute lung function response to cotton dust in atopic and non/atopic individuals. *Br. J. Ind. Med.*, **41**, 487–91.

Trice, M.F. (1940) Card-room fever. *Textile World*, **90**, 68.

Tuma, J., Parker, I. and Braun, D. (1973) The proteolytic enzymes and the prevalence of signs and symptoms in U.S. cotton textile mills. *J. Occup. Med.*, **15**, 409–13.

Von Essen, S., Robbins, R.A., Thompson, A.B. and Rennard, S.I. (1990) Organic dust toxic syndrome; An acute febrile reaction to organic dust exposure distinct from hypersensitivity pneumonitis. *J. Toxicol. Clin. Toxicol.*, **28**, 389–420.

White, N.W. (1989) Byssinosis in South Africa. *S. Afr. Med. J.*, **75**, 435–42.

28 Byssinosis in India and the developing world

Rohini V. Chowgule

THE COTTON TEXTILE INDUSTRY IN INDIA

In the context of this chapter, byssinosis applies to the classical symptoms of chest tightness and shortness of breath on returning to cotton dust exposure after an absence (the acute response) and to the progressive chronic respiratory illness associated with permanent dyspnea and impaired lung function which may develop after years of cotton dust exposure.

Although world-wide recognition of byssinosis came only in the 1950s (Schilling *et al.*, 1955; Schilling, 1956), the relationship between respiratory impairment and exposure to cotton and textile dust had been described earlier (Ramazzini, 1713). Interestingly, Leach (1863), in response to the high mortality rates seen in some Lancashire mills, attributed byssinosis mainly to the use of coarse cotton from India (Surat).

India is the fourth major consumer of cotton (1969 million tons), after China, the US and the Confederation of Independent States (CIS) (Review of the World, Government of India, May–June 1994) (Table 28.1). The cotton textile industry is one of the largest organized industries in India and employs over seven million workers. Production was 78 million bales (one bale = 170 kg) in 1980–1981 and has increased to 122 million bales in 1994 (Cotton Advisory Board, Government of India). The primary product of the cotton textile industry is spun yarn, the raw material for the big composite mills, as well as for the handloom and powerloom sectors.

In 1994 there were 1175 registered mills of essentially two types – spinning mills (909) and composite mills (266). Steps in the processing of cotton in spinning mills include ginning, grading, baling, opening, picking, farfara, carding, combing, framing, roving and spinning. Composite mills have, in addition, weaving, dyeing, printing and folding sections.

Table 28.1 World raw cotton consumption (1993–1994) (total 19.430 million tons)

	Million tons
India	1.969
Pakistan	1.550
China	4.790
USA	2.240
Others	8.881

Source: Compendium of Textile Statistics, Government of India.

The work pattern in Indian mills is unlike that found elsewhere in the world. Mills run continuously for seven days a week in three shifts of 8 hours each. Workers get a weekly day off by rotation, and a 40-hour break on changing shifts. All workers start as temporary employees at the early age of 16–18 years. Boys as young as 10 years old were employed in the mills until legislation was passed [Child Labor and Prohibition and Regulation Act 1986 (Act 61)] banning child labor.

Occupational Lung Disease. Edited by Daniel E. Banks and John E. Parker. Published in 1998 by Chapman & Hall. ISBN 0 412 73630 6

COTTON EXPOSURE IN THE INDIAN MILLS

In order to minimize respiratory illness, the American Conference of Government Hygienists (ACGIH, 1983) recommended a dust content threshold exposure limit of 0.2 mg/m^3. The British Occupational Hygiene Society's recommendation for total dust level in recirculated air is 0.5 mg/m^3. Reducing exposures to these levels over the past 30 years in these countries has led to a dramatic decline in and near eradication of byssinosis, an illness which still occurs in 3–10% of cotton-exposed workers. Attaining these levels in the Indian mills has been difficult.

Damodaran *et al.* (1962) used a midget impinger and Chakravarti *et al.* (1971) used both an electrostatic precipitator and a millipore filter to determine dust levels in Indian mills. Chakravarti (using a millipore filter) found dust levels in the card room to be 2.0 mg/m^3; in the spinning area 0.9 mg/m^3; in the winding room 0.9 mg/m^3; and in weaving 1.2 mg/m^3 (Table 28.2). Dust levels in the 1980s (Kamat 1981; Kamat *et al.*, 1987; Parikh *et al.*, 1989) showed card room exposures to range from 2.9 to 4.6 mg/m^3, blow room exposures to range from 1.2 to 8.7 mg/m^3, spinning exposures to range from 1.2 to 1.4 mg/m^3 winding exposures to range from 1.7 to 3.3 mg/m^3, and weaving exposures to range from 1.2 to 3.3 mg/m^3 (Table 28.2). The lowest levels were from modern mills with air-conditioning, while highest levels were from the older mills. Dust levels in these Indian mills were thus 6 to 40 times higher than the recommended threshold values.

Work in other countries has proven a dose–response relationship between cotton dust levels (a surrogate for cotton and contaminant measurements in the workplace) and the acute symptoms of byssinosis. For example, at respirable dust concentration levels ranging from 0.4 mg/m^3 to 0.5 mg/m^3, the prevalence of byssinosis among card room workers increased from 20% to 50% (Roach and Schilling, 1960). Furthermore, a direct correlation has been shown between dust-years (the product of measured dust concentration and years of exposure) and the prevalence of byssinosis (Fox *et al.*, 1973). For exposures of less than 10 mg-years/m^3 the prevalence was 4%; for 10.1–20.0 mg-years/m^3, it was 10%; for 20.1–30.0 mg-years/m^3, it was 12%; and for more than 30 mg-years/m^3, the value was 22%.

Exposure to endotoxins present on Gram-negative bacteria

Cotton dust *per se*, does not cause the typical Monday sickness accompanied by chronic decrement in lung function. Endotoxins from Gram-negative bacteria (GNB) on raw baled cotton have been shown to have a strong association with byssinosis (Cinkotal *et al.*, 1977; Rylander *et al.*, 1979). Although the role of endotoxins as promoters of airway inflammation has been established, their role in the development of chronic disease is not clear.

Gokani *et al.* (1987) found the highest concentrations of total dust, respirable dust and endotoxin in the card room. The investigators showed GNB content in 55 airborne samples taken in Indian cardrooms to be variable, but with mean values to be as high as in US cottons. The endotoxin concentration in cardroom dust was recognized to be significantly greater ($P < 0.01$) than in the research laboratory of the same cotton mill as well as in the card room of a synthetic mill.

Bhagia *et al.* (1995) studied the amount of bract, the leaf-like structure which encloses the cotton boll, and which may contain agent(s) responsible for bronchoconstriction and symptoms of byssinosis, in two short staple varieties of cotton. Harvesting cotton by picking it directly rather than by picking the entire boll reduced bract content from 7% to 1% in the V_{797} variety and

Table 28.2 Airborne dust levels in Indian cotton mills*

Reference	*City*	*Blow room*	*Card room*	*Spinning*	*Weaving*	*Winding*	*Folding*
Chakravarti, (1971)	Madras	–	2.0	0.9	1.2	0.9	–
Kamat, (1987)	Madras	8.7	2.9	1.4	1.2	–	–
Kamat *et al.* (1981)	Bombay						
Old		7.6	4.6	1.3	1.4	1.7	1.1
Semi-modern		7.9	3.3	1.2	3.8	3.3	0.5
Modern		1.2	3.0	Trace	–	–	–
Parikh *et al.* (1989)	Ahmedabad	3.8	4.0	–	0.6	1.2	1.8

* American Conference of Governmental Industrial Hygienists – TLV for cotton dust = 0.2 mg/m^3.
British Occupational Hygiene Society TLV for cotton dust = 0.5 mg/m^3.

from 8% to 1% in the G_{13} variety. Using a pre-cleaning machine reduced the bract content from 7% to 1% for the V_{797} variety and from 8% to 2% for the G_{13} variety of cotton.

Histamines and immunoglobulins

Cotton dust extract induces the release of histamine in the human lung. Histamine may be involved in the etiology of byssinosis. On average, workers with byssinosis have a higher level of blood histamine than mill workers without respiratory symptoms. Parikh *et al.* (1987) correlated blood histamine levels with the symptoms of byssinosis by the day of the week. The highest levels were seen on Mondays, when workers experienced the most severe symptoms.

The symptoms of byssinosis may also be related to an interaction between cotton dust antigen and antibody in the serum. Continuing exposure to cotton dust resulted in increased titers of immunoglobulin G (IgG) (Taylor *et al.*, 1971). Indian investigators (Kamat, *et al.*, 1979; Karnik, *et al.*, 1987) showed a significant rise in IgG in cotton mill workers with work-related symptoms, including work-related cough with or without phlegm. A comparison group without cotton exposure but with bronchitis failed to show a rise in IgG. The increased amount of immunoglobulin level acts as a marker of exposure to cotton dust but, as is the case with increased serum histamine levels, is not diagnostic of any illness.

THE PREVALENCE OF BYSSINOSIS IN INDIA

Classic byssinosis

Byssinosis was first documented in India in 1950. Eight cross-sectional studies from 1950 to 1969 were reviewed by Gupta (1969) with 5804 workers in 24 of the 575 mills which existed in India at the time being studied. The prevalence of byssinosis ranged from 3% to 15% (Table 28.3).

Table 28.3 Prevalence of byssinosis in Indian cotton mill workers

Reference	*City*	*No. of mills*	*No. of workers*	*Byssinotics*	*Byssinosis grade (Schilling)*				*Indeterminate or By A**
					½	*1*	*2*	*3*	
Gupta (1969)	Ahmedabad	13	253	16 (7%)	–	6 (37%)	8 (50%)	2 (13%)	-
Raghavan *et al.* (1964)	Bombay	1	485	37 (8%)	–	25 (68%)	12 (68%)	–	–
Siddhu *et al.* (1966)	Kanpur	3	656	21 (3%)	–	11 (53%)	10 (47%)	–	34 (5%)
Vishwanathan, *et al.* (1971)	Delhi	1	950	80 (8%)	5 (6%)	25 (31%)	50 (63%)	–	150 (16%)
Thiruvengadam *et al.* (1970)	Madras	1	1,571	66 (4%)	10 (15%)	3 (5%)	42 (64%)	11 (17%)	33 (2%)
Kamat *et al.* (1981)	Bombay	3	814	30 (4%)	–	–	–	–	38 (5%)
Rastogi *et al.* (1985)	Lucknow	1	559	18 (3%)	–	10 (55%)	8 (45%)	–	51 (9%)
Gupta and Gupta (1986)	Delhi	1	277	56 (20%)	–	–	–	–	–
Parikh *et al.* (1989)	Ahmedabad	3	929	59 (6%)	–	13 (22%)	42 (71%)	4 (7%)	–
Bgarjatiya *et al.* (1990)	Rajasthan	1	616	149 (24%)	37 (25%)	78 (52%)	25 (17%)	9 (6%)	–

* By A = Work-related cough with or without phlegm. Chest tightness, wheeze and dyspnea are also absent.

In more recent studies from Lucknow (Rastogi *et al.*, 1985), Delhi (Gupta and Gupta, 1986), Ahmedabad (Parikh *et al.*, 1989), and Rajasthan (Bgarjatiya *et al.*, 1990), as well as a 5-year prospective study done in Bombay (Kamat *et al.*, 1981), the prevalence ranged from 3% to 24% (Table 28.3). Despite this relatively large percentage of workers with disease, the prevalence in India during these years was similar to that occurring in other countries during the same general period.

Indeterminate byssinosis

In India, despite the relatively high dust levels, the prevalence of classical byssinosis has remained comparable with the level of byssinosis in other countries. This prevalence of classical byssinosis may be due to open ventilation in worksheds, or alternatively, this may be explained by the fact that mill workers comprise a survivor population. Many workers leave their jobs upon developing symptoms or develop symptoms after retirement and go unreported.

Yet, all investigators have invariably documented another symptom complex, that is, work-related cough, with or without phlegm. Dyspnea and chest tightness were not present. Siddhu *et al.* (1966), Thiruvengadam *et al.* (1970), and Vishwanathan *et al.* (1967) termed this symptom complex 'indeterminate byssinosis', while Kamat *et al.* (1981) termed this 'By A' (byssinosis A). Similar symptoms have been reported in Britain by Roach and Schilling (1960). Valic and Zuskin (1972) and Fox *et al.* (1973) did not differentiate these symptoms from those of chronic bronchitis.

The prevalence of 'By A' was 5% (Kamat *et al.* 1981) and the prevalence of indeterminate byssinosis ranged from 5% to 16% in a series of reports (Siddhu *et al.* 1966; Vishwanathan *et al.*, 1967; Thiruvengadam *et al.*, 1970). Indeterminate byssinosis appeared more common than classical byssinosis in these same reports.

Factors affecting prevalence

In the cotton processing industry, the two dustiest work areas are the blow room and card room. Internationally, the prevalence of byssinosis is highest among card room workers. In Britain, the prevalence was 60% (Roach and Schilling, 1960); in Egypt, 27% (El Batawi, 1962); in Sweden, the prevalence ranged from 25% to 60% (Belin *et al.*, 1965; Haglind *et al.*, 1981); in Yugoslavia, from 21% to 88% (Valic and Zuskin, 1972; Zuskin *et al.*, 1975); and in the US from 26% to 38% (Bouhuys *et al.*, 1967; Merchant *et al.*, 1973).

In blow room workers in India, the prevalence of classical byssinosis ranged between 4% and 37%, in the card room between 7% and 47%, while less dusty areas like winding, weaving and other departments showed a prevalence of 1% to 31% (Table 28.4). Generally, the dustiest areas of mills showed a higher prevalence. Other studies, however, have shown a high prevalence of byssinosis even in the weaving (21%), spinning (31%), and winding (25%) sections, with an excess prevalence in the other departments also (Gupta and Gupta, 1986; Bgarjatiya *et al.*, 1990). The authors attributed these high figures to the use of coarse cotton and the lack of adequate ventilation.

World-wide, the severity of byssinosis increases with age and with years of exposure. In India, the prevalence of classical byssinosis (15%) is highest between 49 and 60 years (Siddhu *et al.*, 1966; Thiruvengadam *et al.*, 1970). The highest prevalence (13%) of indeterminate byssinosis occurred between 30 and 34 years.

Table 28.4 Prevalence of byssinosis among Indian cotton workers by work area

Reference	*City*	*Blow room*	*Card room*	*Reeling & spinning*	*Winding*	*Weaving*	*Waste mill*
Gupta (1969)	Ahmedabad	3 (4%)	13 (7%)	–	–	–	–
Siddhu *et al.* (1966)	Kanpur	11 (9%)*	–	3 (2%)	3 (5%)	4 (2%)	–
Vishwanathan, (1971)	Delhi	19 (12%)	14 (19%)	37 (10%)	–	2 (1%)	15 (15%)
Thiruvengadam *et al.* (1970)	Madras	–	57 (8%)	1 (1%)	–	3 (3%)	–
Kamat *et al.* (1976)	Bombay	–	14 (42%)	38 (10%)	22 (11%)	3 (3%)	–
Rastogi *et al.* (1985)	Lucknow	8 (6%)*	–	6 (2%)	4 (3%)	–	–
Gupta and Gupta (1986)	Delhi	11 (37%)	14 (47%)	25 (17%)	–	5 (21%)	–
Parikh *et al.* (1990)	Ahmedabad	16 (30%)	28 (37%)	–	–	–	–
Bgarjatiya *et al.* (1990)	Rajasthan	28 (28%)	21 (31%)	11 (31%)	63 (25%)	–	–

* Combined prevalence of byssinosis from blow and card room.

Two studies compared the peak age-related prevalence of byssinosis with Schilling's grades of byssinosis. The first report (Thiruvengadam *et al.*, 1970) showed a pattern similar to that seen elsewhere. The maximum number of workers with grade 1/2 byssinosis were aged between 20 and 30 years; those with grade 1 byssinosis were mostly aged 31–41 years; and those with grades 2 or 3 byssinosis were mostly aged between 51 and 60 years. The second report (Kamat *et al.*, 1976) showed the maximum prevalence of grade 1 byssinosis at 44 years of age. Grades 2 and 3, however, were most often seen at 39 and 40 years, respectively. The differences in ages may be attributable to differing dust exposures.

Roach and Schilling (1960) showed that the maximum prevalence of byssinosis to occur after 25 years of service, though when service exceeded 30 years, prevalence increased to 71%. Most Indian data (Figure 28.1) show a prevalence rate of 1 to 6% in the first 5 years of exposure. In general, with increasing years of employment, the percent of workers with byssinosis increased. The range was 5% to 56% in those working for 6 to 16 years and 25% to 80% in those with 16–25 years of service. Only one study was done of workers in service for over 36 years (Siddhu *et al.*, 1966) and showed a prevalence of 14% (Figure 28.1).

For the most part, the maximum number of Indian mill workers with byssinosis have 16 to 25 years of service (Figure 28.1). Indeterminate byssinosis is most common in those working for 6–10 years. Thus, compared with cotton workers in other countries, Indian textile workers, on the whole, develop byssinosis at an earlier age. This could be due to the work pattern of Indian mills, the coarse cotton used, and the high levels of dust in these mills.

The role of smoking in the development of byssinosis is not clearly established. Two perspectives exist. In the first view, smoking accelerates both the development of byssinosis and the decline in ventilatory capacity. Smokers have one to four times more byssinosis than non-smokers. In the second view, the effects of cotton dust and smoking on lung function are only additive. That is, the amount of lung function loss due to one factor is not changed by the presence or absence of the other.

In the past, the high prevalence of byssinosis in countries like Britain, US and Sweden was attributed to their cold, damp climates. The prevalence in India is perhaps also affected by local climatic conditions, with summer dust levels higher than during the monsoon season. However, Egypt has a similar climate to India but a higher prevalence of byssinosis (27%).

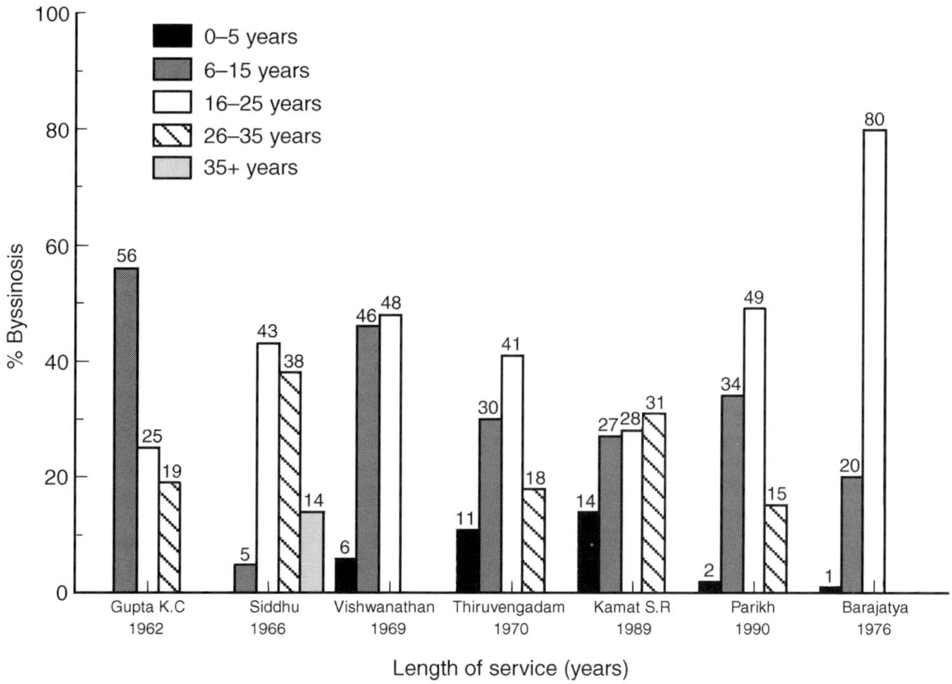

Figure 28.1 Relationship of prevalence of byssinosis to length of service of cotton workers.

CLINICAL MANIFESTATIONS

The natural history of classical byssinosis is well described in Schilling's (1956) classification. In India, work-related cough with or without phlegm associated with a temporary and reversible decline in lung function occurs more often than chest tightness. This led Siddhu *et al.* (1966) and Thiruvengadam *et al.* (1970) to define 'Indeterminate byssinosis'.

Kamat *et al.* (1981), on the basis of a 5-year prospective study, suggested the following categories to describe comprehensively the pattern seen in India.

1. Byssinosis – those with work-related dyspnea, chest tightness or wheeze.
2. 'By A' – work-related cough along with, or without, the above symptoms.
3. Byssinosis and 'By A' – symptoms of the above two categories plus symptoms of bronchitis.

A study group of the World Health Organization (1983), combined both the clinical manifestations and the objective measurements of lung functions into one classification (Table 28.5). This is the most comprehensive classification. It eliminates Schilling's grade 1/2 byssinosis and expands on his grade 1. It further allows for the categorization of secondary symptoms, i.e. dry cough related to exposure to dust with or without persistent phlegm and chest tightness as respiratory tract irritation (RTI). The classification further extends to the objective measurements of byssinotic lung function and defines lung function abnormalities as either acute or chronic on the basis of specified parameters.

The frequency of byssinotic symptoms in India is different from that seen elsewhere (Table 28.6). Chest tightness is traditionally diagnostic of byssinosis, ranging between 49% and 75% in most Indian studies. Most investigators have attributed this anomaly (as compared with chest tightness which occurs in 100% of workers reported to have byssinosis in Western studies) to linguistic difficulties that may have resulted in an inaccurate recording of symptoms (Table 28.6).

Table 28.5 Classification of clinical manifestations and lung function changes in respiratory disorders from exposure to vegetable dusts causing byssinosis or respiratory tract irritation

Classification	*Symptoms*
Grade O	No symptoms
Byssinosis	
Grade B1	Chest tightness and/or shortness of breath on most of first days back at work
Grade B2	Chest tightness and/or shortness of breath on the first and other days of the working week
Respiratory tract irritation (RTI)	
Grade RTI 1	Cough associated with dust exposure
Grade RTI 2	Persistent phlegm (i.e. on most days during 3 months of the year) initiated or exacerbated by dust exposure
Grade RTI 3	Persistent phlegm initiated or made worse by dust exposure either with exacerbations of chest illness or persisting for 2 years or more
Lung function	
1. *Acute changes*	
No effect	A consistent[a] decline in FEV_1 of less than 5% or an increase in FEV_1 during the work shift
Mild effect	A consistent[a] decline of between 5–10% in FEV_1 during the work shift
Moderate effect	A consistent[a] decline of between 10–20% in FEV_1 during the work shift
Severe effect	A decline of 20% or more in FEV_1 during the work shift
2. *Chronic changes*	
No effect	FEV_1[b]–80% of predicted value[c]
Mild to moderate effect	FEV_1[b]–60–79% of predicted value[c]
Severe effect	FEV_1[b] less than 60% of predicted value[c]

[a] A decline occurring in at least three consective tests made after an absence from dust exposure for 2 days or more.
[b] Predicted values should be based on data obtained from local populations or similar ethnic and social class groups.
[c] By a pre-shift test after an absence from dust exposure of 2 days or more.

Adapted from World Health Organization study group (1983) Recommended health-based occupational exposure limits from elected vegetable dust. WHO, Geneva.

Table 28.6 Frequency of chest symptoms in workers with byssinosis

Reference	*Country/ City*	*Byssinosis cases*	*Chest tightness (%)*	*Breath-lessness (%)*	*Monday sickness (%)*	*Dry cough (%)*	*Cough with expectoration (%)*	*Wheeze (%)*	*Cough with cold (%)*
Gupta (1969)	Ahmedabad	16	75	94	–	44	–	–	–
Kamat *et al.* (1976)	Bombay	102	49	64	22	–	–	25	92
Parikh *et al.* (1989)	Ahmedabad	59	93	80	98	68	37	–	–
Schrag and Coullet (1970)	US	63	100	6	100	59	54	39	–
Fishwick *et al.* (1994)	UK	23	100	17	100	30	–	26	–

Other commonly seen symptoms are breathlessness (64–94%), dry cough (44–68%), and cough with expectoration (37%). Wheezing (25%) and cough with cold (92%) were studied in one report (Kamat *et al.*, 1976). Monday sickness, defined in two Indian studies as chest tightness with breathlessness, was reported to be 22% by Kamat *et al.* (1976), but 98% by Parikh *et al.* (1989). Mill fever occurs in only 40% of recently employed workers in Indian mills.

Chronic bronchitis varied from 7% to 24%. Vishwanathan *et al.* (1967) showed that 13% to 21% of these workers developed byssinosis. Pulmonary tuberculosis was found in 8–19% of cotton mill workers (Kamat *et al.*, 1981; Gupta and Gupta, 1986), a level higher than that in the general population. Bronchial asthma was recognized in 2% of workers with classical byssinosis and in 24% of those with indeterminate byssinosis (Table 28.7).

DIAGNOSIS AND THERAPY OF BYSSINOSIS

Byssinosis is diagnosed on the basis of work-related symptoms. There are no specific laboratory or radiologic tests. Lung function data may help in the diagnosis and in defining the extent of impairment.

Lung function tests have been used to detect pulmonary changes associated with byssinosis ever since 1958. Like symptoms, lung function abnormalities depend upon the dose of cotton dust and the type of job in the mill. A typical annual rate of decline in forced vital capacity (FVC) among British textile workers was 54 ml/year and 32 ml/year for workers not exposed to textile dusts (Berry *et al.*, 1973).

A 5-year prospective study of lung function and chest symptoms in cotton workers (Kamat *et al.*, 1981) (Table 28.8) showed that the rate of FVC and FEV_1 decline was statistically significantly greater in byssinotic compared with that in asymptomatic workers. Surprisingly, workers unexposed to cotton dust but with chest complaints of dyspnea or cough unrelated to cotton mill work had a greater fall in FVC and in FEV_1 compared with workers with chest complaints unrelated to work and without cotton dust exposure. There was no obvious explanation for findings in the latter group; however, the authors thought that some 'unexposed' group numbers who worked in administrative offices close to the carding and blowing sections may in fact have been uncountingly exposed to cotton dust.

Table 28.7 Respiratory diseases in cotton workers

Reference	*City*	*No. of workers*	*Chronic bronchitis (%)*	*Tuberculosis (%)*	*Asthma (%)*
Gupta (1969)	Ahmedabad	253	–	19	23
Siddhu *et al.* (1966)	Kanpur	656	13	–	–
Kamat *et al.* (1981)	Bombay	899	7	10	–
Gupta and Gupta (1986)	Delhi	277	24	8	2

Table 28.8 Pulmonary function changes from first to fifth year by symptoms in fifth year

Symptom category	*No.of subjects*	*Mean decrease (±SEM) (%)* FVC (ml/yr)	*FEV$_1$ (ml/yr)*
Asymptomatic			
DE	398	112 ± 23	76 ± 33
C	155	111 ± 30	68 ± 27
Chest symptoms unrelated to work			
DE	113	118 ± 42*	88 ± 37†
C	45	164 ± 70*	110 ± 54†
Work-related chest symptoms			
DE	41	144 ± 74	114 ± 67†
C	5‡	–	

* P <0.05; †P <0.05.

‡ Data omitted because based number of subject too small.

DE, work areas exposed to cotton dust; C, control work areas not exposed to cotton dust.

Adapted with permission from Kamat, S.R. *et al.* (1981) Distinguishing byssinosis from chronic obstructive pulmonary disease. *Am. Rev. Respir. Dis.*, **124**, 31–40.

Pulmonary function tests aid in the diagnosis of byssinosis by measuring the deterioration in lung function seen across a Monday workshift. The mean across shift decline in FEV$_1$ for byssinotic smokers and non-smokers (Parikh *et al.*, 1990) was 260 ml and 230 ml (Table 28.9). These figures are similar to Schilling's (1981) findings, a loss of 200 ml of FEV$_1$ being significant. For non-byssinotic smokers, non-smokers and controls, the mean declines were 50 ml, 80 ml and 10 ml, respectively.

This same study related the levels of mean FEV$_1$ decline across-shift with the grade of byssinosis (Table 28.10). A statistically significant increase in mean FEV$_1$ decline correlated with an increase in the grade of byssinosis. Grade 1 showed a 150-ml decline; grade 2 a 250-ml decline; and grade 3 a 580-ml decline. Pre-shift FEV$_1$ values in those with byssinosis were less than predicted and attributed to chronic changes in lung function.

The treatment of byssinosis in India, like elsewhere, focuses on providing relief of chest symptoms. Kamat *et al.* (1975) used beta-agonist aerosols. With inhalation of the relatively long-lasting beta-agonist aerosols (albuterol, orciprenaline), the FEV$_1$ showed a sustained improvement lasting approximately 4 hours of 10–15%. Raghavan *et al.* (1964) used intravenous aminophylline and subcutaneous injections of adrenaline and with this approach, improved the mean FEV$_1$ 23% from baseline.

PREVENTION, SURVEILLANCE, DUST CONTROL AND COMPENSATION

Prevention of byssinosis has not been suitably emphasized in India. Since early studies on cotton workers showed a low prevalence of byssinosis, the government agencies have not recognized the need to strictly enforce preventive measures.

The most vital preventive measures are dust control and the treatment of raw cotton to eliminate endotoxins. Universally recommended methods for reducing dust include the use of air-exhaust devices, automation of blow room and mixing procedures (dust-suppressant additives are available), and oiling coarse cotton. Endotoxin content can be reduced by bleaching, steaming and washing cotton.

Table 28.9 Relationship between grades of byssinosis and lung function*

Grade	*No. of workers*	*Mean age (years)*	*Mean height (cm)*	*Mean FEV$_1$ Decline during the shift (l)*	*Mean FEV$_1$ pre-shift value (l)*	*Mean FEV$_1$ predicted value (l)*	*Difference (l)*
I	12	39.0 ± 7.5	163.8 ± 5.5	–0.2 ± 0.1	2.41 ± 0.79	2.52 ± 0.24	0.11
II	41	41.3 ± 8.4	162.8 ± 6.5	–0.3 ± 0.2	1.94 ± 0.81	2.45 ± 0.34	0.51
III	3	47.3 ± 2.5	163.3 ± 3.5	–0.6 ± 0.2	1.45 ± 0.13	2.33 ± 0.10	0.88

* Values are mean ± SD.

Adapted with permission from Parikh, J.R. *et al.* (1990) Acute and chronic changes in pulmonary function among Indian textile workers. *J. Soc. Occup. Med.* **40**, 71–74.

Table 28.10 Mean change in FEV_1 during the shift for control, non-byssinotic and byssinotic workers (three mills)

	Change in FEV_1 during the shift					
	*Smokers**			*Non-smokers**		
Group	*FEV_1 at beginning of shift*	*FEV_1 at end of shift*	*Difference*	*FEV_1 at beginning of shift*	*FEV_1 at end of shift*	*Difference*
Control (n = 184)	2.46 ± 0.68 (84)	2.47 ± 0.67	+0.01	2.66 ± 0.69 (100)	2.65 ± 0.65	–0.01
Non-byssinotic (n = 158)	2.38 ± 0.69 (90)	2.33 ± 0.66	–0.05	2.41 ± 0.61 (68)	2.33 ± 0.59	–0.08
Byssinotic (n = 56)	2.00 ± 0.86 (37)	1.74 ± 0.84	–0.26	2.05 ± 0.86 (19)	1.82 ± 0.98	–0.23

Figures in parentheses indicate number of workers.

* Values are means ± SD.

Adapted with permission from Parikh, J.R. *et al.* (1990). Acute and chronic changes in pulmonary function among Indian textile workers. *J. Soc. Occup. Med.*, **40**, 71–4.

Byssinosis was officially recognized as a disabling occupational lung disease in Britain in 1940. In India, byssinosis was mentioned as one of 29 notifiable diseases only in the 1984 amendment of the third schedule of the Factories Act 1948, sections 89 and 90. This act also mandates measurements of cotton dust in the work environment.

Recommendation of acceptable dust levels have been made by the Central Labor Institute, in Bombay. Raw cotton dust exposure levels, as required by the Factories Act 1948, are 0.2 mg/m^3 (weighted concentration) over 8 hours and 0.6 mg/m^3 for a short-term, 15-minute exposure. These levels are taken from existing guidelines in Britain and the US. There has been no work done to determine acceptable cotton dust levels in Indian workplaces. Inspectors of the Industrial Safety and Health Department of the Indian Government are responsible for the enforcement of dust control and other preventive measures.

BYSSINOSIS IN OTHER DEVELOPING COUNTRIES

China

China is the largest consumer of cotton in the world (4.79 million tons; Table 28.1) with a cotton industry which employs 15 million people. Byssinosis was first reported in 1964 and several studies have shown byssinosis prevalence to range from 2% to 10% Liu, 1987; Lu *et al.*, 1987; Christiani *et al.*, 1994a; Jiang *et al.*, 1995) (Table 28.11). Most studies report measured dust levels from 3.0 to 19.9 mg/m^3.

A 5-year prospective study (Christiani *et al.*, 1994b) measured lung function changes in cotton workers, using silk workers as controls. The mean decline in FEV_1 was 40 ml/year for the former group and 31 ml/year for the latter. Workers with an acute response (a 5% or greater decline in FEV_1 across the shift recognized at the first survey) showed a greater decline in FEV_1 (57 ml/year) than those with a less acute response (35 ml/year). The prevalence of byssinosis was 7% at the start of the survey.

The prevalence of symptoms studied as recently as 1995 (Jiang *et al.*, 1995), showed typical Monday sickness in 9%, cough with phlegm in 18%, and chronic bronchitis in 11% of workers. After the work shift, 17% of workers showed a fall in FEV_1 of >5%, and 4% showed a fall in FEV_1 of >10%.

Hong Kong

The overall prevalence of classic byssinosis in three Hong Kong surveys was 2% in a preliminary survey (Morgan and Ong, 1981), 3% in the second survey (increasing to 6% in the blowing and carding sections) (Ong *et al.*, 1985) and 2% in the third report (Ong *et al.*, 1987) (Table 28.11). Although 20% of workers examined had work-related, non-specific respiratory symptoms with obstructive ventilatory function, the authors did not include them in any diagnostic group.

Table 28.11 Prevalence of byssinosis in other developing countries

Country/Reference	*Byssinosis (%)*	*Ginnery (%)*	*Blow room (%)*	*Farfara (%)*	*Card room (%)*	*Simplex (%)*	*Ring frame (%)*	*Weaving (%)*
Hong Kong								
Morgan and Ong (1981)	2	–	–	–	–	–	–	–
Ong *et al.* (1985)	3	–	–	–	–	–	–	–
Ong *et al.* (1987)	2	–	–	–	–	–	–	–
China								
Lu *et al.* (1987)	6	–	–	–	–	–	–	–
Liu (1987)	4	–	–	–	–	–	–	–
Jiang *et al.* (1995)	2	–	–	–	–	–	–	–
Christiani *et al.* (1994a)	7–10	–	–	–	–	–	–	–
Eygpt								
El Batawi (1962)	–	38	–	–	27	–	–	–
El Batawi *et al.* (1964)	30	33	18	50	43	–	–	–
Noweir *et al.* (1984)	–	–	–	–	13	–	–	–
Sudan								
Khogali (1969)	–	20	–	48	–	–	–	–
Khogali (1976)	21	–	–	–	–	–	–	–
Awaad *et al.* (1986)	–	–	67	–	40	42	37	–
Awaad *et al.* (1987)	–	–	–	–	–	–	–	–
Ethiopia								
Woldeyohannes *et al.* (1991)	–	–	43	–	38	–	17	4
Seboxa and Abebe (1994)	46	–	–	–	58	–	57	37
Cameroon								
Takam and Nemery (1988)	18	–	–	–	28*	–	–	–

* Combined for opening, carding, spinning.

The low prevalence of classic symptoms was attributed, by these authors, to three factors. Firstly, a weekly off-system did not exist. Secondly, the term 'chest tightness' could not be communicated effectively. Finally, they postulated a genetic resistance to the causative agents of byssinosis, though this hypothesis was ruled out in another report (Li, 1982) which showed a high prevalence (59% and 74%) among carders.

Egypt

One of the first studies on byssinosis in Africa was conducted in Egypt (El Batawi, 1962). This showed a byssinosis prevalence of 38% among ginning workers and 27% in those employed in the carding sections. A later study (El Batawi *et al.*, 1964) showed an overall prevalence of 30% with 33% in the ginning, 50% in farfara, 18% in the blowing, and 43% in the carding sections. A significant decline in the indirect maximum breathing capacity (IMBC) of dust-exposed workers was also shown, their capacity being 10.1 l/min less than for normal cotton workers and 19.6 l/min less than for power station workers (controls).

Another study (Noweir *et al.*, 1984) showed the concentration of cotton dust to be four to 20 times higher than the TLV defined by the ACGIH (1983). The prevalence of byssinosis was 21% in the opening and clearing sections, and 13% in the carding and combing sections.

Sudan

In Sudan (Khogali, 1969), the prevalence of byssinosis was 20% in ginnery and 48% in farfara workers. In a follow-up study, Khogali (1976) showed an overall byssinosis prevalence of 21%. The level of chronic bronchitis was 29% in a non-smoker population with cotton dust exposure.

Awaad *et al.* (1986) related the prevalence of byssinosis with the type of cotton used. In a mill using coarse cotton, the prevalence was 67% in the blowing, 42% in the simplex, 40% in the carding and drawframe, and 37% in the ring frame sections. The prevalence of chronic bronchitis ranged from 29% to 47%. In second survey, Awaad *et al.* (1987) showed that when two mills using different types of cotton – coarse and fine – were compared, the prevalence of byssinosis was 37% and 1% and the prevalence of chronic bronchitis was 29% and 2%, respectively.

Ethiopia

A study in Ethiopia (Woldeyohannes, 1991) found that the concentration of air dust was four to 17 times that of the WHO recommended exposure level of 0.2 mg/ m^3. This resulted in high prevalence rates of byssinosis: 43% in the blowing, 38% in the carding, 24% in the drawing, 24% in the simplex, 17% in the ringframe, 11% in the preparatory and 4% in the weaving sections, with no relationship to smoking being seen. The incidence of chronic bronchitis was high, ranging from 48% in the dustiest to 18% in the least dusty sections of mills, and was shown to be associated with smoking. The prevalence of asthma, showed a significant relationship to cumulative cotton dust exposure, and ranged from 7% to 21%.

Overall prevalence of byssinosis in a 1994 study (Seboxa and Abebe, 1994) was 46%, and was 58% in the carding, 57% in the ringframe sections, while lowest prevalence was in the weaving section (37%). The prevalence of bronchitis was 49%, asthma 12–29% and tuberculosis 5%.

Classic byssinosis (Takam and Nemery, 1988) was seen to occur mainly after a weekly break in 18% of the workers. Section-wise prevalence of byssinosis in the opening, carding and spinning departments was 28%.

CONCLUSION

Although the Indian textile industry – one of the largest organized industries in India – has involved a large population of workers for many years, the health hazards of these mills have been largely ignored. This is despite the recognition that cotton dust exposure is considered an important cause of respiratory disability and mortality.

Byssinosis was for a long time not considered life-threatening because the early studies in India showed a relatively low prevalence of the disease, despite high dust counts. This prevalence could be due to the work-pattern of mill workers who work for several months and then go home to their villages for long spells. A second reason could be that symptomatic workers leave their jobs for other employment. Local climatic conditions and open worksheds which allow for cross-ventilation in dusty areas could also contribute to the realtively low prevalence levels (in relation to dust levels) seen in Indian mills.

The symptom complex as described by Schilling and which is traditionally diagnostic of byssinosis differs greatly from that reported in India. There is confusion among Indian surveys on reclassifying byssinosis, although the commonly seen symptoms in India are typically work-related cough with or without phlegm, with Monday sickness and chest tightness less frequent. Symptoms of classic byssinosis are seen with less frequency, perhaps because workers have a 40-hour break between shifts unlike the usual 2 days off each week. Often, poor Indian workers do not take a weekly break at all and do not report symptoms for fear of unemployment.

There are two unexpected differences on comparing with studies from the rest of the world. Firstly, the severity of byssinosis in India occurs more in younger age groups, likely the result of cotton workers starting work at an early age. Secondly, the number of Indian workers with byssinosis after 30 years of service is low, likely because many develop respiratory complaints and ventilatory impairment from byssinosis earlier in their career, take early retirement, and are not accounted for in any studies.

Lung function studies in India differ from those elsewhere. Deterioration of lung function occurs at a more rapid rate and even controls have lower lung capacity values. This is because several factory workers of lower socioeconomic groups show a higher prevalence in other associated respiratory diseases. Workers do not use suitable protective measures, and many are smokers. Ambient pollution and overcrowding in the work area also contribute to reduced ventilatory capacity.

The picture of byssinosis in India as gleaned from existing literature is incomplete. A need exists for systematic, multi-centered prospective epidemiologic studies, to form a clearer understanding of the importance of this disease in the country as a whole. This approach can only lead to better control of dust exposures.

REFERENCES

American Conference of Governmental Industrial Hygienists (1983) *Threshold limit values and biological exposure indices.*

Awaad Elkarim, M.A., Osman, Y., El Haimi, Y.A.A. *et al.* (1986) Byssinosis: environmental and respiratory symptoms among textiles workers in Sudan. *Int. Arch. Occup. Environ. Health*, **57**, 101–8.

Awaad Elkarim, M.A., Onsman, Y., and Sulieman, H. (1987) Prevalence of byssinosis and respiratory symptoms among spinners in Sudanese cotton mills. *Am. J. Ind. Med.*, **12**, 281–9.

Belin, L., Bouhuys, A., Hoekstra, W. *et al.* (1965) Byssinosis in cardroom workers in Swedish cotton mills. *Br. J. Ind. Med.*, **22**, 101–3.

Berry, G., McKerrow, C.B., Molyneux, M.K.B. *et al.* (1973) A study of the acute and chronic changes in ventilatory capacity of workers in Lancashire cotton mills. *Br. J. Ind. Med.*, **30**, 25–36.

Bhagia, L.J., Parikh, J.R. and Kashyap, S.K. (1995) Bractless cotton in prevention of byssinosis. *Indian J. Ind. Med.*, **41**(2), 39–41.

Bgarjatiya, M.K., Mathur, R.N., Swaroop, A. *et al.* (1990) Byssinosis in cotton textile workers of Kishangarh. *Indian J. Chest Dis. Allied Science*, **32**, 215–23.

Bouhuys, A., Heaphy, L.J., Jr, Schilling, R.S.F. *et al.* (1967) Byssinosis in the United States. *N. Engl. J. Med.*, **227**, 170–5.

Chakravarti, S., Arunachalam, K., Pauldas, A. *et al.* (1971) Studies on byssinosis- Part III. *J. Industr. Assoc. Occup. Health*, **17**, 129–41.

Christiani, D.C., Ye, T.T., Wegman, D.H. *et al.* (1994a) Cotton dust exposure, across-shift drop in FEV_1, and five year change in lung function. *Am. J. Respir. Crit. Care Med.*, **150**, 1250–5.

Christiani, D.C., Ye, T.T., Wegman, D.H. *et al.* (1994b) Variability in symptom reporting, across-shift drop in FEV_1 and longitudinal change in pulmonary function among cotton textile workers. *Chest*, **105**, 1713–23.

Cinkotai, F.F., Lockwood, M.G., Rylander, R. *et al.* (1977) Airborne microorganisms and prevalence of byssinotic symptoms in cotton mills. *Am. Ind. Hyg. Assoc. J.*, **38**, 554–9.

Compendium of textile statistcs (1994) The Textile Commissioner, Government of India, Bombay.

Damodaran, V.N., Gupta, S.N. and Viswanathan, R. (1962) Byssinosis in cotton textile workers. *Indian J. Chest Dis.*, **4**, 36–40.

El Batawi, M.A. (1962) Byssinosis in the cotton industry in Egypt. *Br. J. Ind. Med.*, **19**, 126–30.

El Batawi, M.A., Schilling, R.S.F., Valic, F. *et al.* (1964) Byssinosis in the Egyptian cotton industry changes in ventilatory capacity during the day. *Br. J. Ind. Med.*, **21**, 13–19.

Fishwick, D., Fletcher, A.M., Pickering, C.A. *et al.* (1994) Respiratory symptoms and dust exposure in Lancashire cotton and man-made fiber mill operatives. *Am. J. Respir. Crit. Care Med.*, **150**, 441–7.

Fox, A.J., Tombleson, J.B.L., Watt, A. *et al.* (1973) A survey of respiratory disease in cotton operatives. Part II: symptoms, dust estimations and the effect of smoking habit. *Br. J. Ind. Med.*, **30**, 48–53.

Gokani, V.N., Doctor, P.B. and Ghosh, S.K. (1987) Isolation and identification of gram negative bacteria from raw baled cotton and synthetic textile fibers with special reference to environmental GNB and endotoxin concentrations of textile mill. *Am. Ind. Hyg. Assoc. J.*, **48**, 511–14.

Gupta, M.N. (1969) Review of byssinosis in India. *Indian J. Med. Res.*, **57**, 1776–89.

Gupta Saroj and Gupta, B.K. (1986) A study of byssinosis and associated respiratory disorders in cotton mill workers. *Indian J. Chest Dis. Allied Science*, **28**(4), 183–8.

Haglind, P., Lundholm, M., Rylander, R. *et al.* (1981) Prevalence of byssinosis in Swedish cotton mills. *Br. J. Ind. Med.*, **38**, 138–43.

Jiang, C.Q., Lam, T.H., Kong, C. *et al.* (1995) Byssinosis in Guangzhou, China. *Occup. Environ. Med.*, **52**, 268–72.

Kamat, S.R. (1987) Evaluation of health hazards in textile industry. *The Indian Practioner*, **XL**, 303–10.

Kamat, S.R., Store, S.D., Karandikar, K.N. *et al.* (1975) Bronchodilator trials in byssinosis. *Indian J. Chest Dis.*, **XVII**, 151–7.

Kamat, S.R., Kamat, G.R., Kale, S.C. *et al.* (1976) Prospective 5 year study of cotton textile mill workers in Bombay, India. Report to Indian Council of Medical Research.

Kamat, S.R., Taskar, S.P., Iyer, E.R. *et al.* (1979) Discrimination between byssinosis and chronic bronchitis in cotton mill workers by serum immunoglobulin patterns. *J. Soc. Occup. Med.*, **29**, 102–6.

Kamat, S.R., Kamat, G.R., Salpekar, V.Y. *et al.* (1981) Distinguishing byssinosis from chronic obstructive pulmonary diseases. *Am. Rev. Respir. Dis.*, **124**, 31–40.

Karnik, A.B., Parikh, J.R., Suthar, A.M. *et al.* (1987) Immunoglobulins, serum proteins and lactate dehydrogenase levels in workers exposed to cotton dust. *Indian J. Med. Res.*, **85**, 222–6.

Khogali, M. (1969) A population study in cotton ginnery workers in the Sudan. *Br. J. Ind. Med.*, **26**, 308–13.

Khogali, M. (1976) A follow-up study of cotton ginnery workers in Sudan. *Br. J. Ind. Med.*, **33**, 166–74.

Leach, J. (1863) Surat cotton, as it bodily affects operatives in cotton mills. *Lancet*, **2**, 648–9.

Li, D.H. (1982) Byssinosis. *Chinese J. Prev. Med.*, **14**, 377–80.

Liu, M.Z. (1987) The health investigation of cotton testile workers in Beijing. *Am. J. Ind. Med.*, **12**, 759–64.

Lu, P.L., Christiani, D.C., Ye, T.T. *et al.* (1987) The study of byssinosis in China: a comprehensive report. *Am. J. Ind. Med.*, **12**, 743–53.

Merchant, J.A., Lumsden, J.C., Kilburn, K.H. *et al.* (1973) Dose–response studies in cotton textile workers. *J. Occup. Med.*, **15**, 222–30.

Morgan, P.C.M. and Ong, S.G. (1981) First report of byssinosis in Hong Kong. *Br. J. Ind. Med.*, **38**, 290–2.

Noweir, M.H., Noweir, K.H., Osman, H.A. *et al.* (1984) An environmental and medical study of byssinosis and other respiratory conditions in the cotton textile industry in Egypt. *Am. J. Ind. Med.*, **6**, 173–83.

Ong, S.G., Jam, T.H., Wong, C.M. *et al.* (1985) Byssinosis in Hong Kong. *Br. J. Ind. Med.*, **42**, 499–502.

Ong, S.G., Jam, T.H., Wong, C.M. *et al.* (1987) Byssinosis and other respirators' problems in the cotton industry of Hong Kong. *Am. J. Ind. Med.*, **12**, 773–7.

Parikh, J.R., Chatterjee, B.B., Rao, N.M. *et al.* (1986) The clinical manifestations of byssinosis in Indian textile workers. *J. Soc. Occup. Med.*, **36**, 24–8.

Parikh, J.R., Venkatakrishna, X., Bhatt, H. *et al.* (1987) Blood histamine levels in cotton dust exposed workers in a textile mill of Ahmedabad. *Am. J. Ind. Med.*, **12**(4), 439–43.

Parikh, J.R., Bhagia, L.J., Majumdar, P.K. *et al.* (1989) Prevalence of byssinosis in textile mills at Ahmedabad, India. *Br. J. Ind. Med.*, **46**, 787–90.

Parikh, J.R., Majumdar, P.K., Shah, A.R. *et al.* (1990) Acute and chronic changes in pulmonary functions among Indian textile workers. *J. Soc. Occup. Med.*, **40**, 71–4.

Raghavan, P., Nagendra, A.S., Thacker, P.V. *et al.* (1964) Respiratory function studies in byssinosis. *Indian J. Chest. Dis.*, **6**, 38–47.

Ramazzini, B. (1713) Diseases of workers. Translation of the Latin text *De Morbis Artificum* by W.C. Wright, Hafner Publishing Co., New York, NY, 1964.

Rastogi, S.K., Clerk, S.K., Chandra, H.T. *et al.* (1985) Byssinosis prevalence in a cotton spinning mill. *Indian J. Occup. Health*, **28**, 87–95.

Roach, S.A. and Schilling, R.S.F. (1960) A clinical and environmental study of byssinosis in the Lancashire cotton industry. *Br. J. Ind. Med.*, **17**, 1–9.

Rylander, R., Imbus, H.R. and Suh, M.W. (1979) Bacterial contamination of cotton as an indicator of respiratory effects among cardroom workers. *Br. J. Ind. Med.*, **36**, 299–304.

Schilling, R.S.F. (1956) Byssinosis in cotton and other textile workers. *Lancet*, 261–267, 319–323.

Schilling, R.S.F., Hughes, J.P.W., Dingwall-Fordyce, I. *et al.* (1955) An epidemiological survey of byssinosis amongst cotton workers. *Br. J. Ind. Med.*, **12**, 217–27.

Schrag, P.E. and Coullett, A.D. (1970) Byssinosis in cotton textile mills. *Am. Rev. Respir. Dis.*, **101**, 497–503.

Seboxa, T. and Abebe, Y. (1994) Byssinosis and tuberculosis among textile mill workers in Bahar Dar, Ethiopia. *Tropical Geography Medicine*, **46**, 180–3.

Siddhu, C.M.S., Kedar Nath, Mehrotra, P.K. *et al.* (1966) Byssinosis among cotton and jute workers in Kanpur. *Indian J. Med. Res.*, **54**, 980–94.

Takam, J. and Nemery, G. (1988) Byssinosis in a textile factory in Cameroon: a preliminary study. *Br. J. Ind. Med.*, **35**, 803–9.

Taylor, G., Massoud, A. and Lucas, F. (1971) Studies on the aetiology of byssinosis. *Br. J. Ind. Med.*, **28**, 145–51.

Thiruvengadam, K.C. (1967) A study of byssinosis. Occupational Health Expert Group Meeting. Indian Council of Medical Research, New Delhi.

Thiruvengadam, K.V., Kamat, S.R., Arunachalam, K. *et al.* (1970) Studies of byssinosis. Part I. *Indian J. Ind. Med.*, **16**, 149–65.

Valic, F. and Zuskin, E. (1972) Effects of different vegetable dust exposures. *Br. J. Ind. Med.*, **29**, 293–7.

Vishwanathan, R., Damodaran, V.N., Subramanian, P.A.V. *et al.* (1967) Report of the inquiry on byssinosis in cotton textile workers. *Indian J. Med. Res.*, **55**, 185–92.

Woldeyohannes, M., Bergevin, Y., Mgeni, A.Y. *et al.* (1991) Respiratory problems among cotton textile mill workers in Ethiopia. *Br. J. Ind. Med.*, **48**, 110–15.

World Health Organization Study Group (1983) Recommended health based occupational exposure limits for selected vegetable dusts, Tech. Rep. Series No. 684, WHO, Geneva.

Zuskin, E., Valic, F., Butkovic, D. and Bouhuys, A. (1975) Lung function in textile workers. *Br. J. Ind. Med.*, **32**, 283–8.

29 Hypersensitivity pneumonitis

John E. Salvaggio, Ann E. Vockroth and Manuel Lopez

INTRODUCTION

Hypersensitivity pneumonitis (HP), or extrinsic allergic alveolitis (AA), represents a group of immunologic pulmonary diseases associated with intense and/or repeated exposure to finely dispersed organic dusts that affect the distal portion of the lung (Pepys, 1969; Schatz and Patterson, 1983; Fink, 1984; Salvaggio, 1987). The clinical and pathologic findings are similar regardless of the type of inhaled organic dust. In the classic acute phase of the disease affected patients have episodes of fever, cough and dyspnea 4–6 hours after exposure to the appropriate organic dust (e.g. hay, pigeon droppings). Symptoms are frequently mistaken for those of bacterial or viral pneumonia. In the most insidious cases associated with prolonged exposure to smaller quantities of the relevant antigen, an afebrile chronic form of the disease may occur. This chronic form is associated with dyspnea, malaise, weakness and weight loss. Pulmonary function abnormalities range from diffusion defects with restrictive dysfunction to varying degrees of obstructive dysfunction. Severe, chronic disease is indistinguishable from other fibrotic lung diseases.

ETIOLOGY

Occupational exposures

There are numerous causes of hypersensitivity pneumonitis (Table 29.1). In the US, the majority of recognized etiologic agents are derived from occupational exposures such as farming, sugar cane harvesting, working with cereal grains or wood products, and packing mushrooms. Agricultural workers are frequently exposed to actinomycetes which are ubiquitous in soil, compost and decaying vegetable matter. Farmer's lung is the name given to HP caused by exposure to moldy hay in the late winter and early spring. Working with certain birds, either as an occupation or as a hobby, puts an individual at risk for pigeon breeder's or bird fancier's disease, and similar diseases caused by antigens derived from sensitizing proteins in bird droppings and serum. Laboratory worker's hypersensitivity pneumonitis often results from exposure to rat urinary proteins.

Both the outdoor and the indoor working environments can provide exposure to HP antigens. Heating, cooling and ventilation systems can become contaminated with microbial antigens. Microorganisms in contaminated water, either of bacterial, fungal, nematode or parasitic origin, can cause the disease. Patients with these forms of HP are usually exposed to aerosols, mists or steams derived from saunas, or contaminated air conditioning, heating and humidification systems.

Of increasing interest are continued reports of HP in workers exposed to reactive simple chemicals such as aerosols of isocyanates and anhydrides. Such inorganic chemicals in the form of vapors can also induce rhinitis and asthma and in higher concentrations can act as potent respiratory tract and skin irritants.

In contrast to these occupational exposures, Japanese summer-type HP is caused by exposure within the home environment to an imperfect yeast, *Trichosporon*

Occupational Lung Disease. Edited by Daniel E. Banks and John E. Parker. Published in 1998 by Chapman & Hall. ISBN 0 412 73630 6

Table 29.1 Causes of hypersensitivity pneumonitis

Disease	*Source of antigen*	*Probable antigen*
Vegetable products		
Farmer's lung disease	Moldy hay	Thermophilic actinomycetes, *Micropolyspora faeni. Thermoactinomyces vulgaris, Aspergillus* spp.
Bagassosis	Moldy pressed sugar cane (bagasse)	Thermophilic actinomycetes, *T. sacchari. T. vulgaris*
Mushroom worker's disease	Moldy compost	Thermophilic actinomycetes, *M. faeni, T. vulgaris*
Suberosis	Moldy cork	*Penicillium* spp.
Malt worker's lung	Contaminated barley	*Aspergillus clavatus*
Maple bark disease	Contaminated maple logs	*Cryptostroma corticale*
Sequoiosis	Contaminated wood dust	*Graphium* spp., *Pullularia* spp.
Wood pulp worker's disease	Contaminated wood pulp	*Alternaria* spp.
Humidifier lung	Contaminated humidifiers, dehumidifiers, air conditioners	Thermophilic actinomycetes, *T. candidus, T. vulgaris, Penicillium* spp., *Cephalosporium* spp., amoebas
Familial hypersensitivity pneumonitis	Contaminated wood dust in walls	*Bacillus subtilis*
Cheesewasher's disease	Cheese casings	*Penicillium* spp.
Wood trimmer's disease	Contaminated wood trimmings	*Rhizopus* spp., *Mucor* spp.
Thatched roof disease	Dried grasses and leaves	*Sacchoromonospora viridis*
Tea grower's disease	Tea plants	Unknown
Coffee worker's lung	Green coffee	Unknown
Streptomyces hypersensitivity pneumonia	Contaminated fertilizer	*Streptomyces albus*
Cephalosporium hypersensitivity pneumonitis	Contaminated basement (sewage)	*Cephalosporium*
Sauna taker's disease	Sauna water	*Pullularia* spp.
Detergent worker's disease	Detergent	*B. subtilis* enzymes
Japanese summer-type HP	Contaminated home environment	*Trichosporon cutaneum*
Paprika splitter's lung	Paprika dust	*Mucor stolonifer*
Animal products		
Pigeon breeder's disease	Pigeon droppings	Altered pigeon serum (probably IgA)
Duck fever	Duck feathers	Duck proteins
Turkey handler's disease	Turkey products	Turkey proteins
Laboratory worker's hypersensitivity pneumonitis	Rat fur	Male rat urine
Pituitary snuff taker's disease	Pituitary powder	Bovine and porcine proteins
Insect products		
Miller's lung	Wheat weevils	*Sitophilus granarius*
Reactive simple chemicals		
TDI hypersensitivity pneumonitis	Toluene diisocyanate	Altered proteins (albumin + others)
TMA hypersensitivity pneumonitis	Trimellitic anhydride	Altered proteins
MDI hypersensitivity pneumonitis	Diphenylmethane diisocyanate	Altered proteins
Epoxy resin lung	Heated epoxy resin	Phthalic anhydride

cutaneum. This mold grows on the damp walls of older houses in southern Japan, mainly in the hot, humid summer months.

Specific antigens

Because antigens of actinomycete and avian (pigeon) and more recently *Trichosporon* origin have been the most extensively studied and characterized, they require more detailed discussion.

(a) Actinomycete antigens
Actinomycetes are ubiquitous organisms which may be the most important sources of antigen in diverse forms of HP such as those associated with home central air conditioning, farming, sugar cane processing, mushroom growing and contaminated humidification systems. *Actinomycetes* are classified as bacteria based on their cell wall structure and susceptibility to antibacterial drugs; however, they share many features with fungi, including hyphae formation, and are often mistaken and identified as such (Figure 29.1).

Pathogenic species include various thermophilic and mesophilic actinomycetes, such as *Faeni rectivirgula* (formerly *Micropolyspora faeni*) and the *Thermoactinomyces* species *T. vulgaris, T. sacchari* and *T. candidus*. They grow best in decaying organic matter such as hay and bagasse (compressed sugar cane fiber), under optimal conditions of high humidity at temperatures between 37°C and 60°C. High numbers of spores are present in contaminated material. Lacey and Lacey have reported studies showing the presence of up to 1.6×10^9 actinomycete spores in the air after disturbing moldy hay. Because particle sizes are smaller than 6 μm, it has been estimated that a farmer working in this environment might inhale and retain in the lung 750 000 spores per minute (Lacey and Lacey, 1964).

The antigenic components of *F. rectivirgula* (previously *M. faeni*), the most important source of antigen in moldy hay, have been studied by Edwards (1978), Roberts (1978) and others. Roberts (1978), using a discontinuous gel gradient polyacrylamide technique, has identified over 30 different protein bands with different electrophoretic patterns for each of several actinomycete species tested. Several of these proteins demonstrated enzymatic activity. The molecular nature and biologic activity of actinomycete antigens are current areas of intense investigation.

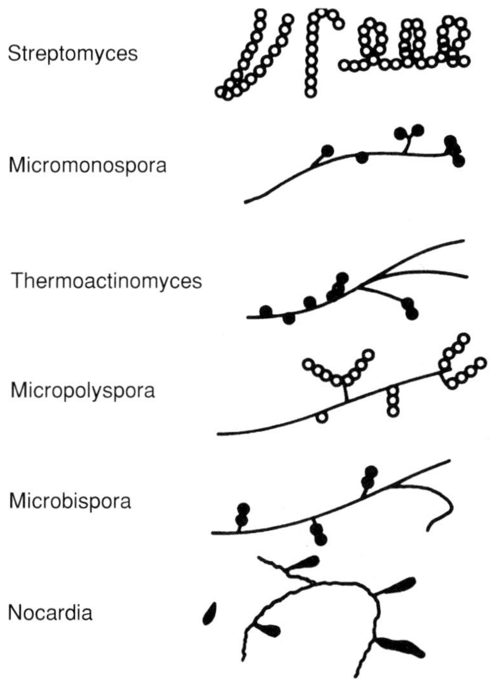

Figure 29.1 Schematic representation of various genera of the Actinomycetales order.

(b) Avian (pigeon) antigens
Proteins derived from feathers, serum and excrement of several avian species have been shown to produce hypersensitivity pneumonitis. An extensive study by Fredericks (1978) using a soluble, undialyzed pigeon dropping extract yielded four major antigens on immunoelectrophoresis. One of these antigens, a glycoprotein with a molecular weight of 200 kDa, was shown likely to be a pigeon IgA molecule derivative. Another of these antigens was susceptible to treatment with mercaptoethano lurea and appears to be a mucopolysaccharide protein conjugate. As in the case of actinomycete antigens, intensive studies are in progress to elucidate the molecular and biologic properties of these potent avian antigens.

(c) Summer HP antigens
Trichosporon cutaneum, an imperfect mold, is the causative agent in most cases of Japanese summer-type HP. High titers of antibodies to *T. cutaneum* are found in more than 95% of patients but less than 10% of controls (Shimazu *et al.*, 1994). There are at least two pathogenic serotypes. The causative antigen in serotype II is felt to be an acidic polysaccharide composed of mannose, xylose and glucuronic acid (Mizobe *et al.*, 1995).

CLINICAL AND HISTOPATHOLOGIC FEATURES

Factors influencing the clinical features of HP are (i) the immunogenicity of the inhaled antigen; (ii) the pattern of exposure; and (iii) the susceptibility of the host. Of these, exposure intensity and frequency are the most important factors determining the clinical pattern. As a result, despite the broad spectrum of disease and the large number of inciting agents, HP can be grouped into acute and chronic forms. Intermittent exposure of short duration produces an acute form that is usually reversible, as seen in farmers who work occasionally with moldy hay, or in sugar cane workers who handle moldy bagasse. More continuous exposure to relatively smaller amounts of antigen, as seen in bird handlers, may lead to a chronic, minimally reversible disease. A subacute form lies between these two extremes. Entities within each form share a similar course and histopathologic features, independent of the specific etiologic agent.

Acute form

Symptoms in the acute form are often explosive in nature, and there is usually a clear temporal relation to exposure. Typically, the patient complains of fever (as high as 104°F), chills, malaise and dyspnea. Cough and myalgia are not uncommon. Symptoms usually begin 4–6 hours after an intense exposure and last 18–24 hours or longer after the exposure ceases. On physical examination, the patient may appear toxic, even cyanotic, and bibasilar rales are usually prominent. Wheezing is rare. In some cases the only physical abnormality is fever. Not surprisingly, viral or bacterial pneumonia is often mistakenly diagnosed.

The tissue reaction accompanying the acute symptoms is that of alveolar and interstitial inflammation, with marked prominence of lymphocytes. A large percentage of the lymphocytes accumulating in the alveoli are CD8+ T cells with cytolytic activity (Figure 29.2). This is in contrast to the CD4+ lymphocytic infiltrate seen in sarcoidosis. In addition to lymphocytes, there are increased numbers of plasma cells and 'activated' alveolar macrophages. Giant cells and large macrophages with foamy cytoplasm are often found (Figure 29.3). Intraalveolar exudate may be present. Antigen is present in the interstitium and in the alveolar septae, as demonstrated by immunofluorescence. Immunoglobulin or complement has only rarely been demonstrated (Wenzel *et al.*, 1971). Obstructive bronchiolitis has been reported and may account for the obstructive pattern seen in some patients. The histologic changes begin to return to normal as the symptoms resolve.

Radiographic and spirometric abnormalities parallel the time course of symptoms in the acute form. When the patient is acutely ill, bilateral patchy infiltrates are typically present. Characteristically, the infiltrates are of an interstitial or alveolar nodular pattern (Figure 29.4). They are usually bilateral and equally distributed but may spare the apices. Less often there are linear infiltrates or a 'hilar haze' pattern. The latter is frequently confused with acute pulmonary edema. In some cases, the chest X-ray may be normal, particularly in the early stages.

In most cases, spirometry shows a restrictive pattern accompanying the acute symptoms which reverts to normal when the patient is asymptomatic. Vital capacity and other lung volumes are decreased. Diffusing capacity and lung compliance are reduced. Airway resistance usually remains normal. Blood gases show

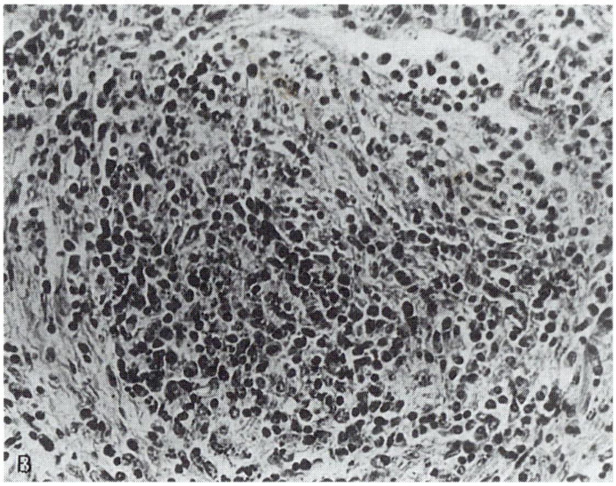

Figure 29.2 Loosely organized granuloma containing many lymphocytes. This specimen was obtained from a case of bagassosis and is typical of pulmonary granulomas.

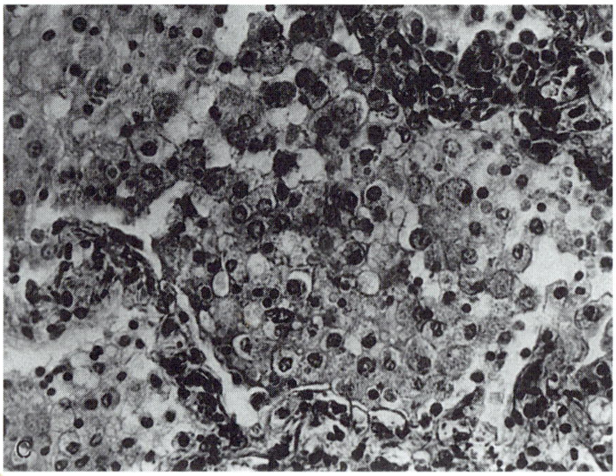

Figure 29.3 Marked parenchymal infiltrate with foamy macrophages. Note the finely stippled cytoplasm producing the characteristic foamy appearance.

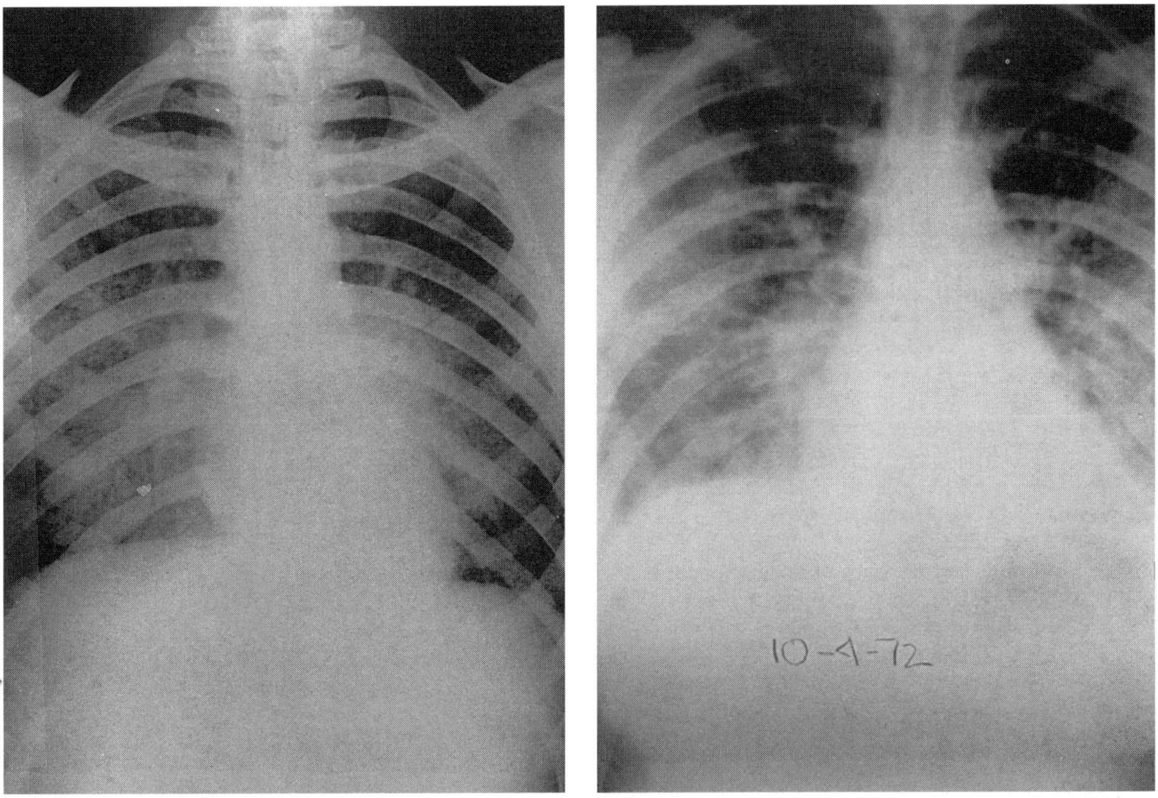

(a) (b)

Figure 29.4 (a) Diffuse micronodular pattern of acute bagossosis. (b) Bilateral patchy and diffuse pneumonitis in a case of humidifier lung.

arterial oxygen desaturation, worse with exercise, and a mild reduction in P_{CO_2}. A minority of patients may display an obstructive pattern early after allergen exposure. There is also often a persistent response or a late-phase response in hypersensitivity pneumonitis, consisting of a recurrence of the restrictive defect after several additional hours. This is in contrast to the obstructive defect seen in the late phase of an asthmatic response.

A period for induction or sensitization precedes the onset of symptoms. This can range from months to even several years of work with the relevant inhaled organic dust. The host factors which make some exposed persons susceptible to disease and others not have yet to be determined, but once an individual develops symptomatic disease, attacks will recur each time the patient inhales a sufficient amount of antigen. Because the diagnosis is frequently not recognized, the typical patient returns repeatedly to the source of exposure, with repeated exacerbations of disease.

Subacute form

A more insidious onset of cough, sputum, malaise and dyspnea marks the subacute form. Anorexia, easy fatiguability and weight loss may also occur. Diffuse rales are the major physical finding, and fever is generally absent. Histologically, the subacute form is characterized by non-caseating granulomata similar to those found in sarcoidosis, but less organized. Subacute disease is diagnosed less often and, therefore, less information about its course and prognosis is available.

Chronic form

The chronic, irreversible form may occur either in the presence or in the absence of pre-existing explosive attacks seen in the acute form. In other words, fibrosis can result either from repeated exposure to large amounts of antigen or from prolonged exposure to small amounts of antigen. The latter is common in bird fancier's disease, ventilation pneumonitis, and other forms of the disease where there is almost continuous, year-round exposure to relevant antigens in the indoor environment.

Presenting symptoms are progressive dyspnea plus fatigue, anorexia and weight loss. These symptoms occur late in the course of the disease, often after irreversible interstitial fibrosis has occurred. In the absence of acute attacks, there are no early signs or symptoms of chronic HP except for possible chronic cough. At the time of presentation, the physical examination shows diffuse rales. In advanced cases, there may also be signs of respiratory failure or cor pulmonale.

Histologic sections from peripheral lung tissue obtained from patients with chronic HP show a variable degree of interstitial fibrosis with the upper zones often affected. Fibrosis may be focal with adjacent areas of emphysema or diffuse, causing microcysts similar to those in fibrosing alveolitis. The granulomata seen in earlier stages may either persist or disappear. Immunofluorescent studies typically fail to demonstrate antigen (Barrowcliff *et al.*, 1968; Wenzel *et al.*, 1971), unless the specimens are taken early after antigen exposure (Barrowcliff *et al.*, 1968; Ghose *et al.*, 1974).

Radiographically and functionally, chronic HP cannot be distinguished from any other advanced interstitial fibrosis. Chest X-ray shows diffuse interstitial fibrosis (Figure 29.5). The periphery of the lung may be more involved, creating the so-called 'white wall' appearance. With increasing fibrosis, loss of lung volume and cardiac enlargement are seen. As with acute disease, chronic disease is associated with a restrictive pattern, characterized by decreased lung volumes and diffusing capacity. In a few cases there may also be an element of obstruction.

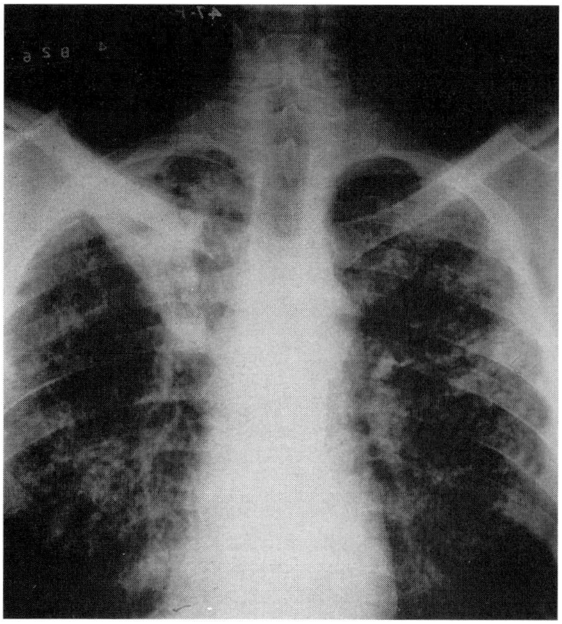

Figure 29.5 Extensive fibrosis in chronic bagassosis.

PATHOGENESIS

A complex relationship between environmental and host-related factors influences the manner by which inhaled organic dust induces lesions of HP (Reynolds, 1986; Salvaggio and de Shazo, 1986). No single immune mechanism can explain all the features. Although the pathogenesis of many immune-mediated diseases can be explained mainly in terms of the Gel and Coombs classification, it is naive to presume that only one immunopathogenic mechanism is involved. Several types of immunologically induced tissue injury, as well as non-immunologic factors, would appear to be operative in this disease, either simultaneously or in different stages of disease activity.

Host factors

There appears to be a host susceptibility factor, although its precise nature has yet to be determined. Only a small proportion of exposed workers in any given industry will become symptomatic. The percentage of affected workers varies according to the industry, but frequently ranges from 5% to 20%. A recent questionnaire found that 8% of pigeon breeders in the Canary Islands had classic HP symptoms (de Castro *et al.*, 1993). This is similar to prevalence rates of 7% for farmers in the British Isles (Boyd, 1971; Grant *et al.*, 1972) and 15% of office workers exposed to a contaminated air conditioning system (Banaszak *et al.*, 1970). Genetic factors remain elusive. HP occurs most often in adults, but children occasionally develop the disease. Atopic status does not appear to be important as a disease determinant. Several studies have failed to demonstrate any association between specific type I HLA antigens and susceptibility (Rodey *et al.*, 1979; Flaherty *et al.*, 1980; Muers *et al.*, 1982). Among HLA type II antigens, HLA DR3, DR7 and DQw3 have been loosely associated with certain etiologic agents in certain populations (Rittner *et al.*, 1983; Selman *et al.*, 1987; Ando *et al.*, 1989), but no clear linkage has been established.

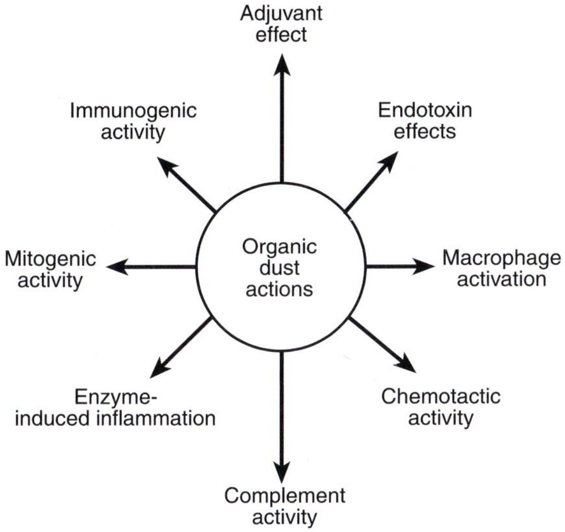

Figure 29.6 Variety of biologic effects induced by organic dust.

Effects of organic dust

Organic dusts produce a variety of biologic effects and may be directly or indirectly toxic to the pulmonary epithelium, often containing enzymes and histamine releasers. Directly toxic substances include compounds with enzymatic activity (Schorlemmer *et al.*, 1977), endotoxin (Rylander *et al.*, 1978), substances that induce non-specific precipitins, and histamine releasers (Burrell and Pokorney, 1977). Certain organic dusts may also directly activate the alternative complement pathway (Edwards *et al.*, 1974), contributing to increased vascular permeability and chemotactic migration of polymorphonuclear leukocytes to the lung (Figure 29.6).

Organic dusts activate alveolar macrophages (Stankus *et al.*, 1978), cells which are key to the early inflammatory response of many granulomatous lung diseases, including HP. Inflammatory cytokines, lipoxygenase products, oxidants and proteases are among the important mediators released from alveolar macrophages. Following activation, alveolar macrophages produce large amounts of proinflammatory cytokines, including interleukin-1 (IL-1), tumor necrosis factor alpha (TNFα) and PDGF. High levels of these substances are found in lavage fluid in experimental animals undergoing challenge and also in the lavage fluid from patients (Denis *et al.*, 1993). Indeed, Denis and co-workers have shown that release of IL-1 and TNFα from macrophages can be stimulated by *M. faeni*. TNFα is likely to be directly pathogenic because recombinant IL-1 and TNF administered to animals can induce granulomata (Kasohara *et al.*, 1989). Similarly, neutralizing TNF by infusion of an anti-TNF antibody significantly abrogates experimental HP (Denis *et al.*, 1991a).

The inflammatory consequences of these direct injurious effects and those modulated by complement and macrophages could be important factors in the pathogenesis of HP. Many other cell types may also be stimulated, directly or indirectly, by these antigens (Figure 29.7 (a,b)).

Immunopathogenesis

There is some evidence to suggest that mast cells may play a role in the pathogenesis of HP. Several authors have reported increased mast cells in the bronchoalveolar lavage fluid (BAL) of farmer's lung patients and a correlation between disease activity and mast cells in lung tissue sections (Kopp *et al.*, 1985; Bjermer *et al.*, 1988; Laviolefte *et al.*, 1991; Pesci *et al.*, 1991). Histopathologic sections show that these mast cells are in the degranulated state (Miadonna *et al.*, 1994). The exact role of the mast cell in pathogenesis is not known, but in experimental HP, mast cell-deficient mice demonstrate a significant reduction in HP-like lesions compared with normal mice. In humans, symptomatic farmers show increased numbers of mast cells in lung tissue, but so do exposed, asymptomatic farmers (Scarier *et al.*, 1982). Although the increase in mast cells is of unknown significance, it is possible that they are stimulated by IL-3 and modulate the inflammatory response.

Although increased numbers of bronchoalveolar mast cells have been detected in farmer's lung, there is little evidence for participation of IgE-mediated (type I) reactions in these diseases. Patients have normal IgE levels, eosinophilia is usually absent, and there is no higher incidence of HP in atopic patients. It has been noted, however, that a substantial number of patients with pigeon breeder's disease demonstrate immediate-type wheal and flare skin reactions to pigeon antigens, and about 10% of these patients have asthma-like symptoms on inhalation challenge (Schorlemmer *et al.*, 1977). Muers and colleagues have suggested that these type I reactions may be mediated by short-term or 'latent' sensitizing antibody of the IgG_4 subtype (Muers *et al.*, 1982).

The demonstration by Wenzel and co-workers (1971) of *F. rectivirgula* antigen, immunoglobulin and complement in the bronchial walls and mononuclear cells in lung biopsy specimens from cases of farmers lung has suggested a type II cytotoxic mechanism. These investigators speculate that antigens may be

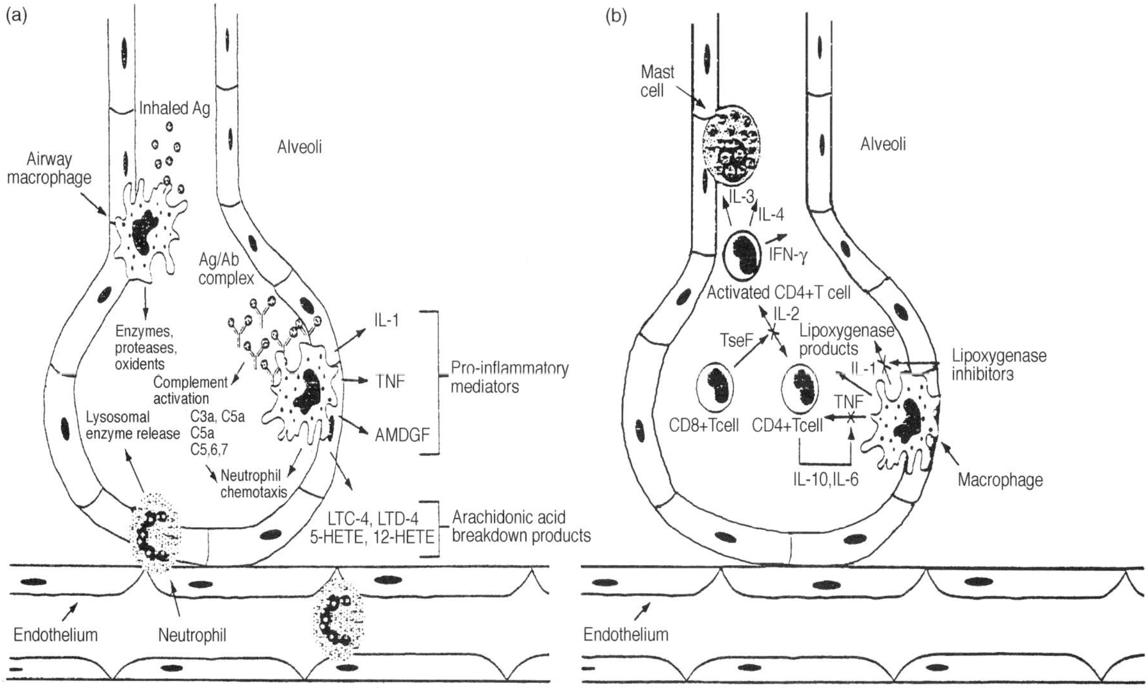

Figure 29.7 (a) Early inflammatory events in acute hypersensitivity pneumonitis. (b) Proposed immunoregulatory mechanisms in chronic hypersensitivity pneumonitis.

adsorbed to cells and in the presence of complement react with the specific antibodies of the IgG and IgM class to produce complement-mediated cytotoxic damage. There is no other substantial evidence to support this view.

Immune complex-induced tissue injury (type III) has been thought to play a major role in the pathogenesis of these diseases. The time course of the symptoms after exposure, the presence of precipitating antibodies against the specific etiologic agent, and the detection of antigen, immunoglobulin and complement components in lung lesions are among the arguments in favor of this hypothesis. Recent observations have also indicated that immune complexes *per se* can both release proinflammatory cytokines, including IL-1 and TNFα from BAL cells and induce pulmonary granuloma formation (Warreu *et al.*, 1989; Piguet *et al.*, 1990; Denis *et al.*, 1991b). Immune complexes are also known to activate alveolar macrophages and be a prerequisite to cell-mediated lung injury. However, there are difficulties with this hypothesis. First, precipitating antibodies are related to environmental exposure and not to disease. Indeed, most patients with precipitating antibodies and continuing exposure do not develop the disease. Second, serum complement levels are not decreased in patients after inhalation challenge with antigen (Fink, 1973). Third, pulmonary histologic abnormalities are not those of classic antigen–antibody complex-mediated vasculitis.

Although histopathologic findings may be related to the timing of the biopsy in relation to the evolution of the disease (Schuyler *et al.*, 1978), evidence for a significant role for both delayed (type IV) and immune complex (type III) hypersensitivity in these diseases is strong. In this regard, it is known that agents or procedures capable of inhibiting cell-mediated hypersensitivity such as cortisone acetate, neonatal thymectomy, antimacrophage sera, specific antisera against certain proinflammatory cytokines and cyclosporine have either markedly diminished or prevented development of experimental forms of granulomatous pneumonitis in animal model systems (Hunninghake *et al.*, 1979). Suppression of granulomatous lesions and anamnestic antibody response to ovalbumin (OA) in rabbits by cyclosporine is illustrated in Figure 29.8.

In addition to the histopathologic findings (granuloma formation), there is infiltration of the alveolar and interstitial tissue with macrophages, lymphocytes. Patients exhibit many of the phenomena associated with cell-mediated (T cell-dependent) and humoral (B cell-dependent) hypersensitivity such as antigen-induced lymphocyte proliferation, lymphokine produc-

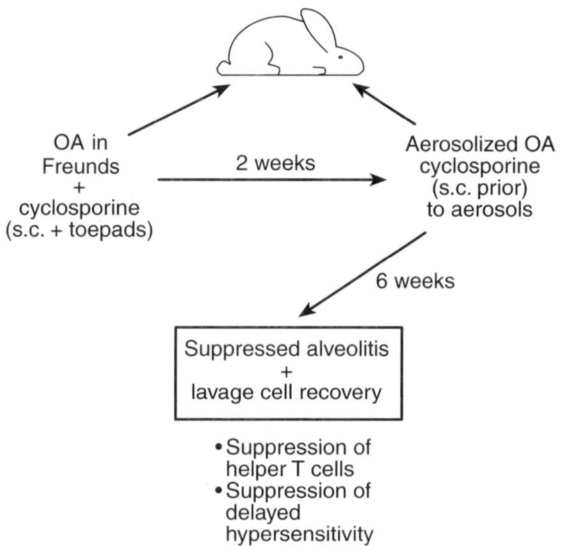

Figure 29.8 Cyclosporine effect in experimental hypersensitivity pneumonitis.

tion, specific precipitating antibody and increased immunoglobulin levels (Denis *et al.*, 1993). Sensitized lymphocytes occur in both blood and lung but may be detected at times only locally in the lung (Schuyler *et al.*, 1978; Hunninghake *et al.*, 1979; Keller *et al.*, 1984). A significant increase in the percentage of T cells is seen in BAL fluid from patients with chronic HP, although asymptomatic, exposed individuals also have high percentages of lymphocytes (Keller *et al.*, 1984). Immunoregulatory events have been studied in patients with HP as well as in animal models of the disease. Using peripheral blood cells of asymptomatic pigeon breeders, Keller and co-workers (Keller *et al.*, 1984) have demonstrated spontaneous and pigeon serum-induced T-suppressor cells that were not present in symptomatic pigeon breeders. In further studies by this group using BAL, there was a reduction in T-suppressor cell activity in the lymphocytes from patients with pigeon breeder's disease, in comparison with lymphocytes from asymptomatic but similarly exposed individuals (Yamasaki *et al.*, 1991). These studies suggest immunoregulatory abnormalities in patients with pigeon breeder's disease. A recent study on the responsiveness of BAL T lymphocytes to mitogens in patients with summer-type HP (Yamasaki *et al.*, 1991) demonstrated that their proliferative response was lower than the response of peripheral blood T cells of the same patients. The decrease response was not due to the presence of suppressor T cells or suppressor factors. There was, however, a decreased capacity of the BAL T lymphocytes to produce IL-2. The significance of these observations is not clear at the present; like the increased numbers of mast cells in this disease (Hasiam *et al.*, 1987; Soler *et al.*, 1987).

Another interesting experimental finding in HP is that appropriately sensitized animals with chronic immunospecific granulomatous pulmonary inflammation associated with T cell-dependent hypersensitivity can become 'desensitized' after a series of antigen challenges, either by intravenous or aerosolized route (Figure 29.9) (Unger *et al.*, 1968; Hargreave *et al.*, 1972; Allen *et al.*, 1981). This desensitization has been shown to be associated with the gradual resolution and disappearance (rather than progression) of characteristic pulmonary mononuclear cell infiltrates. Animals challenged repeatedly with unrelated antigens are refractory to development of pulmonary inflammatory responses for more than 2 months (Parraft *et al.*, 1975). After the refractory period, an anamnestic response is induced with repeated challenge. Evidence suggests that in experimental HP, the negative modulatory effect can be transferred to normal syngeneic recipients with non-antigen-specific suppressor macrophages, lymphocytes or NK cells. These cells have been thought to be Ia^+ antigen-specific suppressor T cells that are H_2-IJ restricted. Overall, these studies provide evidence that pulmonary granulomatous lesions of experimental HP are modulated after repeated antigen exposure and that suppressor factors ultimately dampen or modulate the granulomatous response.

These studies suggest the possibility that HP develops as a result of complex events involving an initial immunization, development of granulomatous inflammation, and the ultimate modulation of the pulmonary inflammatory response by immunoregulatory processes.

DIAGNOSIS

When a patient presents acutely, there is often a clear temporal relationship between exposure to the offending agent and development of the clinical syndrome. When this is the case, no further testing is necessary, and the diagnosis is confirmed by demonstrating remission of symptoms following extended removal from the source of antigen. However, when it is not clear if an agent in the patient's environment is related to symptoms, inhalation challenge can be used to establish that relationship. Laboratory tests are helpful to document exposure but are hampered by the frequent finding of similar abnormalities in exposed but asymptomatic individuals.

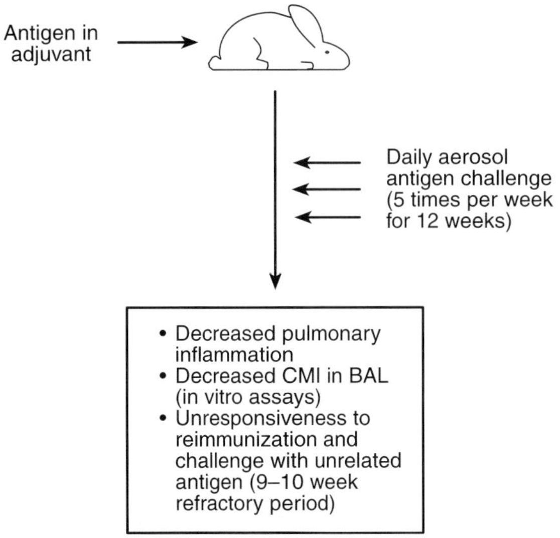

Figure 29.9 Schematic representation of animal model of desensitization.

HP should be suspected in patients with respiratory symptoms and signs of restrictive disease. To prevent fibrosis, it is important to make an early diagnosis. A carefully taken clinical history may suggest a possible temporal relationship between symptoms and certain activities, such as working with hay, working in a building, or a particular hobby. For example, the exposed individual may manifest increased symptoms several hours after beginning work or shortly after returning home. If exposure occurs at home, as in the case of contaminated air-conditioning systems or in association with certain hobbies, symptoms may occur during weekends and periods at home. Complicating the clinical picture is the fact that, in many cases, the levels of environmental antigens may fluctuate widely and the patient may have a combination of a low-grade exposure with occasional acute exacerbations. The chronic form, although less dramatic and more difficult to relate to environmental exposure, may be more common and certainly more incapacitating than the acute and subacute forms. For this reason, physicians should be keenly aware of the possibility of HP in patients presenting with respiratory problems, particularly with symptoms of bronchitis and dyspnea without obvious causes such as cigarette smoking.

When the history is not conclusive, a challenge procedure may be required. Certain laboratory procedures, discussed in the following paragraphs, greatly aid in the diagnosis of HP.

Immunologic studies

(a) Antigen-specific IgG

Patients with HP invariably have circulating levels of IgG directed against the causative antigen; however, a large number of exposed but asymptomatic workers will also have high levels of specific IgG. In farmer's lung, the relatively insensitive technique of Ouchterlony double gel diffusion will identify specific IgG-precipitating antibodies in all ill patients and in approximately 50% of similarly exposed asymptomatic individuals (Bamdad, 1980). More sensitive ELISA and RIA techniques demonstrate presence of some antigen-specific IgG in all exposed individuals (Parraft *et al.*, 1975; Bamdad, 1980). It remains to be seen if the level of IgG detected has any predictive value, for instance in comparing the level in an acutely ill patient relative to a similarly exposed but healthy individual. (This would be analogous to the comparison of anti-*Aspergillus* antibody levels in suspected ABPA patients relative to mold-sensitive asthmatics as controls, where a ratio above a certain value supports the diagnosis.) At the present time, a positive result is valuable only to document exposure. A negative result, on the other hand, is highly significant in acute cases. Assuming the correct antigen is being investigated, lack of IgG antibodies to that antigen essentially rules out the disease.

(b) Skin tests

Skin tests are not helpful in the diagnosis of HP. Many patterns of skin test reactivity are possible, including immediate, late and dual phase reactions. Patients may have false positive reactions due to irritation, or the test may be positive due to an immediate wheal and flare IgE-type allergic response (asthma or allergic rhinitis) which is unrelated to HP. Skin tests in patients with pigeon breeder's disease may evoke a late or dual-phase IgE-dependent reaction. The dual response includes an immediate wheal and flare reaction followed in 4–8 hours by a large area of dermal and subcutaneous edema. In the past, it was hypothesized that late skin test reactions may reflect the presence of IgG antibodies, but this had not been conclusively shown. Additionally, up to 40–60% of exposed asymptomatic individuals will also demonstrate similar patterns of skin reactivity (Schatz *et al.*, 1976). The use of isolated delayed hypersensitivity (48-hour) types of skin tests has not been described in human HP.

(c) T-cell function

A number of *in vitro* tests to demonstrate cellular immune responses have been employed in the evaluation of HP. They include tests for lymphocyte transformation (blastogenic factor production) and the production of lymphokines such as migration inhibitory factor (MIF). These assays have been of some use in evaluating the immunologic mechanisms in HP, but none is of practical use in routine clinical practice. Some studies have shown that in patients with pigeon breeder's disease, avian antigens stimulate peripheral blood lymphocytes primarily from symptomatic subjects but not from asymptomatic exposed individuals (Fink *et al.*, 1974). However, other studies have shown that lymphocytes from some exposed asymptomatic subjects will also undergo blastogenesis *in vitro* upon exposure to antigen (Reynolds *et al.*, 1977). With the use of direct and indirect standard capillary tube assays, MIF production in response to avian antigens has, however, generally been shown to occur in symptomatic patients but not in asymptomatic exposed subjects (Semenzato *et al.*, 1986). Although these cellular studies have been helpful in understanding the mechanisms of hypersensitivity pneumonitis, they are not used routinely in clinical practice for evaluation of patients suspected of having the disease.

BAL fluid analysis

BAL lavage can be extremely useful to help elucidate the pathogenesis of HP. It is most useful to compare groups of patients, as there is much overlap among individuals. At present, BA lavage contributes little to the diagnosis of an individual patient.

In the normal lung, most of the inflammatory and immune effector cells recovered in the lavage fluid are macrophages. Typically, in normal non-smokers, macrophages represent about 90% of the cells; a small percentage (6–8%) are lymphocytes (Weinberger *et al.*, 1978). In HP, sarcoidosis and other respiratory disorders, there is an increased percentage of lymphocytes (Mornex *et al.*, 1984). The lymphocyte subpopulations are similar to those seen in peripheral blood. High numbers of CD8+ T cells in the alveolar sacs are the distinguishing characteristic of HP (Figure 29.8) (Bascon *et al.*, 1985; Tentin *et al.*, 1988). In patients with pigeon breeder's diseases lavage fluid contains a large proportion of lymphocytes. However, several pigeon breeders without clinical manifestations of the disease also have high percentages of lymphocytes. T cells may be more highly activated in HP. There are two lines of evidence supporting this. First, lymphocytes from patients with farmer's lung show enhanced lymphocyte proliferation to mitogen stimulation with phytohemagglutinin, compared with asymptomatic farmers. Second, there is higher density of some cell surface markers in HP than in sarcoidosis (Satake *et al.*, 1993). Interesting is the finding by Denis and colleagues (1993) that BAL macrophages secreted significantly higher amounts of TNFα spontaneously, but not in response to LPS (lipopolysaccharide).

Acute onset of symptoms is highly correlated with a CD8+ lymphocytic infiltrate. Murayama *et al.* (1993) compared BAL fluids from acute, summer-type HP patients with that from chronic, fibrotic HP patients (mostly bird fanciers). They found a significantly greater number of CD8+ cells in the acute group, whereas CD4+ lymphocytes were increased to the same level in both groups. Yoshizawa *et al.* (1995), performing serial brouchoalveolar lavages, found that CD8+ numbers decreased more slowly in summer-type HP than in bird fanciers. This is surprising, given the more rapid symptom resolution in the former.

The time interval between antigen exposure and BAL assessment is critical, and has been a confounding variable in many studies. Lavage fluids collected early in the disease course show increased numbers of neutrophils and complement components. Miadonna found a negative correlation between histamine levels and CD8+ lymphocytes, which may suggest that mast cells are activated earlier in the process than are lymphocytes (Miadonna *et al.*, 1994). A group of Japanese investigators found greater CD8+ infiltrate in a non-fibrosis group (mostly summer type HP) compared with a fibrosis group (bird fanciers and HP of unknown etiology) (Yoshizawa *et al.*, 1988).

Overall, BAL cell profiles distinguish diseased from healthy, unexposed individuals, but the cell profile does not distinguish ill from asymptomatic/exposed individuals. Functional studies, such as mitogen proliferation, and analysis of cell surface markers performed on cells recovered from BAL may be more helpful to distinguish disease from exposed but asymptomatic groups. As mentioned above, mitogen responsiveness is increased in farmer's lung patients compared with healthy farmers. Ohtsuka *et al.* (1994) found increased lymphocyte migration in patients with HP as well as pulmonary sarcoidosis, compared with normal controls.

Longitudinal studies of BAL cell populations in farmers with HP have revealed that individuals who were initially evaluated from 2 weeks to 4 months after an acute episode and continued to work regularly on farms, exhibited increased NK cell numbers and function together with high numbers of CD8 suppressor cells and reversal of CD4, CD8 ratios (Trentin *et al.*, 1988). Conversely, individuals who discontinued exposure to specific antigens in the work area but continued to live in normal agricultural environments demonstrated a decrease in CD8 suppressor cells and recovery of the CD4 helper cell populations within a periods of 6 months. CD8+ cellular infiltrates persisted in those individuals with continued exposure to antigens in contrast to recovery of the CD4 helper cell population within 6 months in individuals who discontinued work. These data suggest that lymphocytic alveolitis is a chronic event in patients with HP and that its intensity might be modulated by the extent and degree of exposure to inhaled organic dust antigen. Other studies of BAL in patients with HP have shown increased levels of albumin, fibronectin and angiotensin converting enzyme (ACE), even in asymptomatic healthy farmers regularly exposed to organic dust antigen (Larsson *et al.*, 1992). The BAL fluid from asymptomatic farmers demonstrated normal levels of hyaluronic acid and procollagen III N-terminal peptide, two markers of fibroblast activation, whereas these two markers were significantly increased in symptomatic farmers (Yoshida et al., 1989). This difference may reflect the intensity of the inflammatory response. If these findings are confirmed they may be used in the evaluation and follow-up of patients with HP. Taken together, these findings suggest that healthy farmers exposed to organic dust antigens can exhibit subclinical lymphocytic alveolitis, albumin leakage and alveolar macrophage activation. The significance of increased fibronectin and ACE levels is unknown. Fibronectin has chemotactic and stimulatory effects on fibroblasts, but the individuals under study were asymptomatic and had no demonstrable pulmonary fibrosis. It is of interest that in these studies, procollagen III N-terminal peptide and hyaluronan were normal in BAL effluences of these subjects, suggesting that no fibroblast activation occurred.

Inhalation challenge

In many instances of diagnostic uncertainty or dispute, or for medicolegal reasons, it may be important to perform inhalation challenge testing to establish the diagnosis of HP. There are two methods of challenge.

(a) 'Natural' or 'biologic' challenge
This method is accomplished by re-exposing the patient to the suspected environment. After at least 72 hours away from the suspected environment, the patient returns briefly, and clinical and laboratory parameters, including pulmonary function studies, are monitored for 24 hours. There is minimal danger in this type of challenge, because the degree of exposure has been encountered safely many times previously in this 'natural' setting. Thus, there is no need for expensive hospital admission. On the other hand, this challenge does not help in the identification of specific antigen responsible for the symptoms. For this purpose, a more controlled laboratory challenge is required.

(b) Laboratory inhalation challenge
This type challenge is carried out in a specialized laboratory. After the patient undergoes measurement of baseline pulmonary function, an extract of the suspected antigen is inhaled via a mechanical nebulizer. Symptoms, signs, pulmonary function and white blood count are monitored over the next 24 hours. The most common positive response occurs 4–6 hours after challenge and is characterized clinically by chills, fever, cough, dyspnea and crepitant rales on auscultation. Pulmonary function tests show a restrictive defect with a decrease in forced vital capacity (FVC) and diffusing capacity, (D_{L}co). Less commonly, this reaction may be preceded by acute bronchospasm. This controlled laboratory challenge has been used frequently to identify etiologic antigens and is the only method to prove a direct relationship between a suspected offending antigen and the disease process. In sensitized individuals, the symptoms and physiologic abnormalities can be reproduced accurately in the laboratory; nevertheless, results of challenge studies must be interpreted critically, because endotoxin and other irritants found in crude extracts may induce febrile reactions in certain normal individuals. Another potential risk involves the possibility of producing a severe pneumonitic episode that might ultimately contribute to permanent lung damage. In some cases, the benefits of early recognition and consequent avoidance of antigen outweigh the potential risks of challenge testing.

Samples used in the laboratory are usually collected from the workplace or home, except in the case of potential sensitizing chemicals. Challenge should be undertaken with caution, and the patient needs to be monitored for at least several hours. A typical regimen is 15 minutes of antigen exposure by nebulization followed by 12–24 hours of monitoring. Monitoring typically includes FEV_1 and FVC, D_{L}co (diffusing capacity), ABG (arterial blood gases), temperature and leukocyte count. The expected reaction is reproduction of symptoms and signs of HP, with symptoms onset 4–6 hours after challenge (Figure 29.10)

TREATMENT AND PROGNOSIS

Allergen avoidance is of the utmost priority in HP, as it is in other allergic respiratory diseases. Once the antigen has been identified, a variety of precautions may be attempted. The best long-term management depends on complete elimination of antigen from the patient's environment. When elimination of antigen is not feasible or the etiologic agent is not identified, the best approach is to remove the patient from the environment. Air-filtering systems, protective face

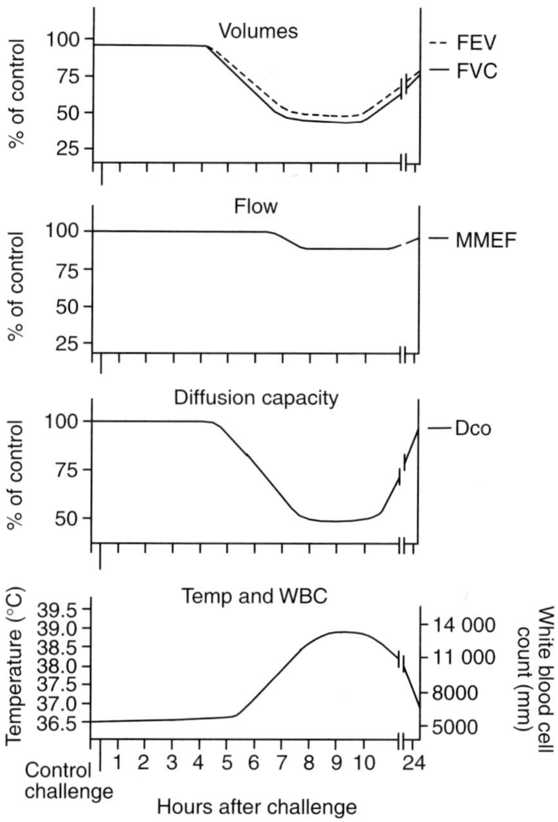

Figure 29.10 Response to laboratory inhalation challenge.

masks and changes in ventilation systems can decrease the amount of antigen inhaled and should be used when complete avoidance is not feasible. Drug treatment is indicated when the antigen cannot be avoided or if the disease is progressing despite avoidance.

Environmental manipulation

Elimination of antigen from the patient's environment not only will help to control the disease but also will prevent sensitization in other exposed individuals. Changes in the handing of occupational agents, as has occurred with sugar cane following the discovery of bagassosis, can control the patient's disease and also prevent sensitization of other individuals. Maple bark disease and bagassosis are now quite rare in the locations where they were first described, owing to simple alterations of the handling of allergenic materials.

In farming and other occupations involving compost, dispersion of actinomycete spores can be minimized by keeping the compost wet. Propionic acid 1% spray can suppress growth of actinomycete spores. Certain organic dusts can be rendered sterile by pasteurization that exposes fibers to very high temperatures. In the case of humidifier fever, frequent cleaning and changing of water should prevent microbial growth. In the summer-type HP in Japan, cleaning and disinfecting areas of the house contaminated with *Trichosporon cutaneum* has been shown to be useful in prevention of symptoms (Yoshida *et al.*, 1989).

When elimination of the antigen is not feasible and the patient has no other reasonable alternatives than to continue working despite exposure, a protective face mask may help to diminish the amount of antigen inhaled. Limited information indicates that some protective masks can protect farmers from developing farmer's lung (Gourley and Braidwood, 1971) but it is unlikely that commonly used masks can reduce exposure adequately.

Drug therapy

Pharmacologic treatment should be used when avoidance is impossible or inadequate. Most mild, acute episodes will resolve spontaneously provided the patient is removed from the environment. More severe attacks require treatment with steroids (e.g. prednisone 60 mg daily) plus oxygen and other supportive measures. Steroid dose should be maintained until there is significant clinical improvement, a clearing chest X-ray and normal pulmonary function tests. This may require as long as 2–3 weeks in severe cases. Antibiotics are not indicated, and bronchodilators and cromolyn are not of proven benefit. Immunotherapy is not helpful and may be contraindicated because of the danger of inducing immune complex-mediated vasculitis related to the high levels of IgG antibody already present.

Steroids hasten clinical improvement, most likely by their effects on cytokine production. Treatment with either steroids or avoidance was shown by Denis and colleagues (1993) to be associated with a decline in spontaneous and LPS-induced IL-1 from BAL macrophages. This same team demonstrated that treatment produced a modest increase in NK cell activity, compared with pretreatment values.

Course and prognosis

The rate of clinical improvement depends on the timeliness of diagnosis and implementation of treatment. Mild, acute episodes often resolve in 1–3 days without further sequelae. Several authors have found no long-term beneficial or deleterious effects from steroids. (Mönkäre and Haahtela, 1987; Kokkarinen *et al.*, 1992). Indeed, Mönkäre and Haahtela (1987) followed patients with farmers' lung for 5 years, concluding that steroid treatment had no long-term effect on symptoms or pulmonary function.

The majority of researchers find that pulmonary function tests return to normal as the symptoms resolve. Kokkarinen *et al.* (1992 found that improvement in FVC and D_{L}co can continue for 1–2 years following a first diagnosed attack of farmers lung. Also of note, 7% (7/101) of patients in their sample were diagnosed with asthma during the 5-year follow-up period.

Less is known about natural history and prognosis with chronic exposure, but outcomes appear to be highly variable. Braun and co-workers (1979) cited a 30% 5-year morbidity rate in patients with farmer's lung. Other studies, however, suggest that continued allergen exposure does not uniformly lead to clinical deterioration in all patients (Bourke *et al.*, 1989; Cuthbert and Gordon, 1983). The factors portending a more benign course remain to be determined.

SUMMARY

Like extrinsic asthma and some forms of interstitial lung disease, HP is an immune-mediated lung disease caused by the inhalation of antigen. Unlike asthma, it is not mediated by IgE, and, unlike other interstitial lung diseases, it involves the alveoli as well as the interstitium. In the US, occupational exposures yield the greatest percentage of inciting antigens, including the prototypical farmer's lung and bagassosis as well as bird breeder's lung, malt worker's lung and many others. In Japan, summer-type HP results most often from household exposures. All share a similar presentation and course whose manifestations depend most on exposure frequency and intensity. Chronic, subacute and acute exposures lead to distinct presentations, varying from insidious dyspnea to an explosive febrile illness. The diagnosis requires a high degree of suspicion plus careful attention to the environmental history, often supplemented by immunologic studies, inhalation challenge and analysis of bronchoalveolar lavage fluid. Because these diseases may have a tremendous socioeconomic impact, with significant disability, physicians should be thoroughly familiar with their diagnosis and treatment.

REFERENCES

Allen, E.M., Sternick, J.L., Scarier, D.J. and Moore, V.L. (1981) BCG-induced chronic pulmonary inflammation of splenomegaly in mice: suppression of PHA-induced proliferation, delayed hypersensitivity to sheep erythrocytes and chronic pulmonary inflammation by soluble factors from adherent spleen cells. *Cell. Immunol.*, **58**, 61.

Ando, M., Hirayama, K., Okubo, R., Araki, S. and Sasazuki, T. (1989) HLA-DQw3 in Japanese summer-type hypersensitivity pneumonitis induced by *Trichosporon cutaneum. Am. Rev. Respir. Dis.*, **140**, 948–50.

Bamdad, S. (1980) Enzyme-linked immunosorbent assay (ELISA) for IgG antibodies in farmer's lung disease. *Clin. Allergy*, **10**, 161–71.

Banaszak, E., Thiede, W. and Fink, J. (1970) Hypersensitivity pneumonitis due to contamination of an air conditioner. *N. Engl. J. Med.*, **283**, 271–6.

Barrowcliff, D.F. and Arblaster, P.G. (1968) Farmer's lung: a study of an early fatal case. *Thorax*, **23**, 490.

Bascon, R., Kennedy, T.P., Levitz, D. and Zeiss, C.R. (1985) Specific bronchoalveolar lavage IgG antibody in hypersensitivity pneumonitis from diphenylmethane diisocyanate. *Am. Rev. Respir. Dis.*, **131**, 463.

Bjermer, L., Engstrom-Laurent, A., Lundgren, R., Rosenhall, L. and Haligren, R. (1988) Bronchoalveolar mastocytosis in farmer's lung is related to the disease activity. *Arch. Intern. Med.*, **148**, 1362–65.

Bourke, S.J., Banham, S.W., Carter, R., Lynch, P. and Boyd, G. (1989) Longitudinal course of extrinsic allergic alveolitis in pigeon breeders. *Thorax*, **44**, 415–18.

Boyd, D. (1971) The incidence of farmer's lung in Caithness. *Scott. Med. J.*, **16**, 261–2.

Braun, S.R., do Pico, G.A., Tsiatis, A., Dickie, H.A. and Rankin, J. (1979) Farmer's lung disease: long term clinical and physiologic outcome. *Am. Rev. Respir. Dis.*, **117**, 185–91.

Burrell, R. and Pokorney, H. (1977) Mediators of experimental hypersensitivity pneumonitis. *Int. Arch. Allergy Appl. Immunol.*, **55**, 161.

Cuthbert, O.D. and Gordon, M.F. (1983) Ten year follow-up of farmers with farmer's lung. *Br. J. Ind. Med.*, **40**, 173–6.

de Castro, R., Carillo, T., Castillo, R., Blanco, C., Diaz, F. and Uevas, M. (1993) Relationships between characteristics of exposure to pigeon antigens: clinical manifestations and humoral immune response. *Chest*, **103**, 1059–63.

Denis, M., Bedard, M., Laviolette, M. and Cormier, Y. (1993) A study of monokine release and natural killer activity in bronchoalveolar lavage of subjects with farmer's lung. *Am. Rev. Respir. Dis.*, **147**, 934–9.

Denis, M., Cormier, Y. and Tardif, E. (1991a) Tumor necrosis factor plays as essential role in determining hypersensitivity pneumonitis in a mouse model. *Am. J. Respir. Cell. Mol. Biol.*, **5**, 477–83.

Denis, M., Cormier, Y., Tardif, J., Ghadirian, E. and Laviolette, M. (1991b) Hypersensitivity pneumonitis: whole *Micropolyspora faeni* or antigens thereof stimulate the release of proinflammatory cytokines from macrophages. *Am. J. Respir. Cell Mol. Biol.*, **5**, 198–203.

Edwards, J.H. (1978) Methodology involved in fractionating *Micropolyspora faeni* antigens, in *Proceedings of NIAID Workshop on Antigens in Hypersensitivity Pneumonitis*, (eds J. Fink and J. Salvaggio), *J. Allergy Clin. Immunol.*, **61**, 199.

Edwards, J.H., Baker, J.T. and Davies, B.H. (1974) Precipitin test negative farmer's lung. Activation of the alternative pathway of complement by moldy hay dust. *Clin. Allergy*, **4**, 379.

Fink, J. (1984) Hypersensitivity pneumonitis. *J. Allergy Clin. Immunol.*, **74**, 1–9.

Fink, J., Moore, V. and Barboriak, J. (1974) Cell mediated hypersensitivity in pigeon breeders. *Int. Arch. Allerg. Appl. Immunol.*, **47**, 498.

Fink, J.N. (1973) Hypersensitivity pneumonitis. *J. Allergy Clin. Immunol.*, **52**, 309.

Flaherty, D.K., Brau, S.R., Mark, J.J. *et al.* (1980) Serologically detectable HLA-AB, and C loci antigens in Farmer's lung disease. *Am. Rev. Respir. Dis.*, **122**, 437.

Fredericks, W. (1978) Antigens in pigeon droppings extracts, in *Proceedings of NIAID Workshop on Antigens in Hypersensitivity Pneumonitis*, (eds J. Fink and J. Salvaggio), *J. Allergy Clin. Immunol.*, **61**, 199.

Ghose, T., Landrigan, D., Killeen, R. and Dill, J. (1974) Immunopathological studies in patients with farmer's lung. *Clin. Allergy*, **4**, 119–29.

Gourley, C. and Braidwood, G. (1971) The use of dust respirators in the prevention of recurrence of farmer's lung. *Trans. Soc. Occup. Med.*, **21**, 93–5.

Grant, I., Blyth, W., Vardrop, V. *et al.* (1972) Prevalence of farmer's lung in Scotland: a pilot survey. *Br. Med. J.*, **1**, 530–4.

Hargreave, F., Hinson, K.F., Reid, L., Simon, G. and McCarthy, D.S. (1972) The radiologic appearance of allergic alveolitis due to bird sensitivity (bird fancier's lung). *Clin. Radiol.*, **23**, 1–10.

Haslam, P.L., Deward, A., Butcher, P., Primeft, Z.S., Newman-Taylor, A. and Turner-Warwick, M. (1987) Mast cells, atypical lymphocytes, and neutrophils in bronchoalveolar lavage in extrinsic allergic alveolitis: comparison with other interstitial lung diseases. *Am. Rev. Respir. Dis.*, **135**, 35–47.

Hunninghake, G., Gadek, J., Kawanami, O., Ferrens, V. and Crystal, R. (1979) Inflammatory and immune processes in human lung in health and disease: evaluation by bronchoalveolar lavage. *Am. J. Pathol.*, **97**, 149.

Kasohara, K., Kobayashi, K., Shimak, Y. *et al.* (1989) The role of monokines on granuloma formation in mice: the ability of interleukin-1 and tumor necrosis factor-alpha to induce lung granulomas. *Clin. Immunol. Immunopathol.*, **51**, 419–25.

Keller, R., Swartz, S., Schlueter, P., Shlomo, B. and Fink, J. (1984) Immunoregulation in hypersensitivity pneumonitis: phenotypic and functional studies in bronchoalveolar lavage lymphocytes. *Am. Rev. Respir. Dis.*, **130**, 766–71.

Kokkarinen, J.I., Tukiainen, H.O. and Terho, E.O. (1992) Effect of corticosteroid treatment on the recovery of pulmonary function in farmer's lung. *Am. Rev. Respir. Dis.*, **145**, 3–5.

Kokkarien, J., Tukianinen, H. and Terho, E. (1993) Recovery of function in farmer's lung. A five-year follow-up study. *Am. Rev. Respir. Dis.*, **147**, 793–6.

Kopp, W.C., Dierks, S.E., Butler, J.E., Upadrasta, B.S. and Richerson, H.B. (1985) Cyclosporine immunomodulation in rabbit model of chronic hypersensitivity pneumonitis. *Am. Rev. Respir. Dis.*, **132**, 1027.

Lacey, J. and Lacey, M. (1964) Spore concentration in the air of farm buildings. *Trans. Br. Mycol. Soc.*, **47**, 547.

Larsson, K., Eklund, A., Maimberg, P., Bjermer, L., Lundgren, R. and Belin, L. (1992) Hyaluronic acid (Hyaluronan) in BAL fluid distinguishes farmers with allergic alveolitis from farmers with asymptomatic alveolitis. *Chest*, **101**, 109–14.

Laviolefte, M., Cormier, Y., Loiseau, A., Soler, P., Leblanc, P. and Hance, A.J. (1991) Bronchoalveolar mast cells in normal farmers and subjects with farmer's lung. *Am. Rev. Respir. Dis.*, **144**, 855–60.

Marayama, J., Yoshizowa, Y., Ohtsaka, M. and Hasegawa, (1993) Lung fibrosis in hypersensitivity pneumonitis. Association with $CD4^+$ but not CD8+ cell common alveolitis and insidious onset. *Chest*, **104**, 38–43.

Miadonna, A., Pesci, A., Tedeschi, A. *et al.* (1994) Mast cell and histamine involvement in farmer's lung disease. *Chest*, **195**, 1184–9.

Mizobe, T., Ando, M., Yamasaki, H. and Misaki, A. (1995) Purification and characterization of the serotype-specific polysaccharide antigen of *Trichosporon cutaneum* II: A disease-related antigen of Japanese summer-type hypersensitivity pneumonitis. *Clin. Exp. Allergy*, **25**, 265–72.

Mönkäre, S. and Haahtela, T. (1987) Farmer's lung: a 5-year follow-up of eighty-six patients. *Clin. Allergy*, **17**, 143–51.

Mornex, J., Cordier, G., Pages, J. *et al.* (1984) Activated lung lymphocytes in hypersensitivity pneumonitis. *J. Allergy Clin. Immunol.*, **74**, 719.

Muers, M.F., Faux, J.A., Ting, A. *et al.* (1982) HLA-

A,B,C, and HLA-DR antigens in extrinsic allergic alveolitis (budgerigar fancier's lung disease). *Clin. Allergy*, **12**, 47.

Ohtsuka, M., Yoshizawa, Y., Naitou, T. *et al.* (1994) The motility of lung lymphocytes in hypersensitivity pneumonitis and sarcoidosis. *Am. J. Respir. Crit. Care. Med.*, **149**, 455–9.

Parraft, D., Nielsen, K. and Boyd, G. (1975) The quantitation of antibody in farmer's lung syndrome using a radioimmunoassay. *Clin. Exp. Immunol.*, **20**, 217–25.

Pepys, J. (1969) Hypersensitivity disease of the lung due to fungi and other organic dusts. *Monogr. Allergy*, **4**, 44.

Pesci, A, Bertorelli, G. and Olivieri, D. (1991) Mast cells in bronchoalveolar lavage fluid and in transbronchial biopsy specimens of patients with farmer's lung disease. *Chest*, **100**, 1197–202.

Piguet, P.F., Collart, M.A., Grau, G.E., Sappino A.P. and Vassaiii, P. (1990) Requirement of tumour necrosis factor for development of silica-induced pulmonary fibrosis. *Nature*, **344**, 245–7.

Reynolds, H.Y. (1986) Concepts of the pathogenesis and lung reactivity in hypersensitivity pneumonitis. *Ann. N.Y. Acad. Sci.*, **465**, 287.

Reynolds, H.Y., Fulmer, J.D., Kazmierowski, J.A., Roberts, W.C., Frank, M.M. and Crystal, R.G. (1977) Analysis of cellular and protein content of bronchoalveolar lavage fluid from patients with idiopathic pulmonary fibrosis and chronic hypersensitivity pneumonitis. *J. Clin. Invest.*, **59**, 165–75.

Rittner, C., Sennekamp, F., Molienhauer, E., *et al.* (1983) Pigeon breeders lung: association with HLA-DR3. *Tissue Antigens*, **21**, 374–9.

Roberts, R. (1978) Fractionation and characterization of thermophilic actinomycete, in *Proceedings of NIAID Workshop in Antigens in Hypersensitivity Pneumonitis*, (eds J. Fink and M. Salvaggio) *J. Allergy Clin. Immunol.*, **61**, 199.

Rodey, G.E., Fink, J. and Koethe, S. (1979) A study of HLA-A,B,C, and DR specificities in pigeon breeder's disease. *Am. Rev. Respir. Dis.*, **119**, 755.

Rylander, R., Haglind, P., Landholn, M., Maftsby, I. and Stengrist, I. (1978) Humidifier fever and endotoxin exposure. *Clin. Allergy*, **8**, 561.

Salvaggio, J.E. (1987) Hypersensitivity pneumonitis. *J. Allergy Clin. Immunol.*, **79**, 558.

Salvaggio, J.E. and deShazo, R.D. (1986) Pathogenesis of hypersensitivity pneumonitis. *Chest*, **89**, 1905.

Satake, N., Nagai, S., Kawatani, A. *et al.* (1993) Density of phenotypic markers on BAL T-lymophcytes in hypersensitivity pneumonitis, pulmonary sarcoidosis and bronchiolitis obliterans with organizing pneumonia. *Eur. Respir. J.*, **6**, 477–82.

Scarier, D., Sternick, J., Allen, E. and Moore, V. (1982) Immunogenetics of BCG induced anergy in mice: control by genes linked to the IgH complex. *J. Immunol.*, **128**, 1466.

Schatz, M. and Patterson, R. (1983) Hypersensitivity pneumonitis: general considerations. *Clin. Rev. Allergy*, **1**, 451–67.

Schatz, M., Patterson, R., Fink, J. and Moore, F. (1976) Pigeon breeder's disease. II. Pigeon antigen induced proliferation of lymphocytes from symptomatic subjects. *Clin. Allergy*, **6**, 7–17.

Schorlemmer, H.U., Edwards, J.H., Davies, P. and Allison, A.C. (1977) Macrophage responses to moldy hay dust. *Micropolyspora faeni* and zymosan, activators of complement by the alternative pathway. *Clin. Exp. Immunol.*, **27**, 198.

Schuyler, M., Thigpen, T. and Salvaggio, J. (1978) Local pulmonary immunity in pigeon breeder's disease. *Ann. Intern. Med.*, **88**, 355–8.

Selman, M., Teran, L., Mendoza, A. *et al.* (1987) Increase of HLA-DR7 in pigeon breeder's lung in a Mexican population. *Clin. Immunol. Immunopathol.*, **44**, 63–70.

Semenzato, G.C., Agostini, R., Zambello, R. *et al.* (1986) Lung T cells in hypersensitivity pneumonitis: phenotypic and functional analysis. *J. Immunol.*, **137**, 1164–72.

Shimazu, K., Ando, M., Sakata, T., Yoshida, K. and Araki, S. (1994) Hypersensitivity pneumonitis induced by *Trichosporon cutaneum*. *Am. Rev. Respir. Dis.*, **130**, 407–11.

Soler, P., Nioche, S., Basset, F. *et al.* (1987) Role of mast cells in the pathogenesis of hypersensitivity pneumonitis. *Thorax*, **42**, 565–72.

Stankus, R.P., Cashner, F.M. and Salvaggio, J.E. (1978) Bronchopulmonary macrophage activation in the pathogenesis of hypersensitivity pneumonitis. *J. Immunol.*, **120**, 685.

Trentin, L., Marcer, G., Chilosi, M. *et al.* (1988) Longitudinal study of alveolitis in hypersensitivity pneumonitis patients. An immunologic evaluation. *J. Allergy Clin. Immunol.*, **10**, 577–85.

Unger, J.D., Fink, J.N. and Unger, G.E. (1968) Pigeon breeder's disease: roentgenographic lung findings in hypersensitivity pneumonitis. *Radiology*, **90**, 683–7.

Warren, J., Yabross, K. and Remick, D. (1989) Tumour necrosis factor participates in the pathogenesis of acute immune complex alveolitis in the rat. *J. Clin. Invest.*, **84**, 1873–82.

Weinberger, E.S., Weinberger, S.E., Kelman, J.A., Elson, N.A., Young, R.C., Jr., Reynolds, H.Y., Fulmer, J.D. and Crystal, R.G. (1978) Bronchoalveolar lavage in interstitial lung disease. *Ann. Intern. Med.*, **89**, 459–66.

Wenzel, F.J., Emanuel, D.A. and Gray, R.L. (1971) Immunofluorescent studies in patients with farmers lung. *J. Allergy*, **48**, 224.

Yamasaki, H., Kinoshita, T., Ohmura, T., Ando, M., Soda, K., Sakata, T., Araki, S. and Onoue, K. (1991) Lowered responsiveness of bronchoalveolar ravage T lymphocytes in hypersensitivity pneumonitis. *Am. J. Respir. Cell Mol. Biol.*, **4**, 417–25.

Yoshida, K., Ando, M., Sakata, T. and Araki, S. (1989) Prevention of summer-type hypersensitivity pneumonitis: effect of elimination of *Trichosporon cutaneum* from the patient's homes. *Arch. Environ. Health*, **44**, 317–22.

Yoshizawa, Y., Miyake, S., Sumi, Y., Hisauch, K., Sato, T. and Kurup, VP. (1995) A follow-up study of pulmonary, function tests, bronchoalveolar lavage cells, and humoral and cellular immunity in bird feeder's lung. *J. Allergy Clin. Immunol.*, **96**, 122–9.

Yoshizawa, Y., Nomura, A., Ohdama, S. *et al.* (1988) The significance of complement activation in the pathogenesis of hypersensitivity pneumonitis. Sequential changes of complement components and chemotactic activities of bronchoalveolar lavage fluids. *Int. Arch. Allergy Appl. Immunol.*, **87**, 417–23.

30 Acute toxic chemical exposures: mechanisms and outcomes

Paul Cullinan and
A.J. Newman Taylor

INTRODUCTION

Toxic chemicals are those which, when inhaled, have the potential to cause damage to the respiratory tree. Given that most agents, including oxygen, have such a capacity at high enough concentration this is a widely encompassing definition. In this chapter we examine the pulmonary pathophysiology and outcome of acute inhalation of chemicals in toxic concentrations. We have specifically excluded discussion of microbial toxins and those agents which cause respiratory hypersensitivity. Toxic inhalations can result in damage to other organ systems (such as the neurologic effects of solvent inhalation); conversely, administration by other routes can cause pulmonary damage, as after paraquat ingestion. These situations are not discussed.

Much of the pathophysiological literature is based on the results of experimental challenges in animals – as are most industrial safety standards. Because of important anatomic, metabolic and immunologic inter-species variations these findings should be extrapolated to humans with caution. On the other hand, it is difficult to study the outcome of acute, high-intensity toxic exposures directly in humans: such events are, by nature, unexpected though often occurring on a background of chronic, low-level exposure. Precise details of exposure levels and measurements of pre-exposure morbidity, crucial to the attribution of cause and effect, are frequently unavailable. Most events (with the important exception of chemical warfare) occur in close association with industrial processes and in particular during manufacture, storage or transport of chemicals. Those exposed are usually employees, but others in the vicinity can also be at risk; the opportunity to study the effects of exposure may be very different between these two groups. Recently an interest in domestic, high-concentration exposures has also developed (Centers for Disease Control, 1991). Paradoxically, there is generally a mismatch between the probability of an accidental exposure occurring and the facilities to assess its effects; situations where industrial safety concerns are of low priority seldom have the toxicologic, epidemiologic and clinical resources to examine the outcomes of accidental exposures.

MECHANISMS

In the absence of a unifying classification of toxic exposures it is difficult to predict the pulmonary

Occupational Lung Disease. Edited by Daniel E. Banks and John E. Parker. Published in 1998 by Chapman & Hall.
ISBN 0 412 73630 6

response to inhaled toxic chemicals. Broadly, however, several inter-related factors govern the response: the dose and site of deposition within the lung, the extent – if any – of pre-existing respiratory disease, and the function of pulmonary protective and repair processes.

Toxic chemicals may be inhaled as gases, vapors, fumes or aerosols which are relatively stable suspensions of solid or liquid particles in a gas, frequently air. Most toxic exposures resulting in acute pulmonary damage are gases or vapors. The dose of chemical is that quantity delivered directly to the target organ and is dependent on the concentration of exposure, its physicochemical properties, the environmental circumstances of the exposure, the total ventilation of the subject, and the efficacy of pulmonary and behavioral protective mechanisms. The experience of the Bhopal disaster in 1984 provides a good example of the complex interactions between these factors. In the early hours of a December morning an explosion at the Union Carbide (India) plant in the city (Plate 13) resulted in the release of about 27 tons of toxic materials. Most of this was probably methylisocyanate which, with a low boiling point (39°C) and high vapor pressure (348 mmHg at 20°C) was readily vaporized by the heat of the explosion, condensed in the cool night air, and was inhaled in aerosol form. The heat of the respiratory tract encouraged revaporization and hence deeper penetration into the lungs. Most of those exposed were initially asleep but were woken by the intensely irritating nature of the fume; those who subsequently fled into the path of the plume would have further increased their exposure by raising their minute ventilation. Others were able to reduce their pulmonary dose by covering their faces with wet cloths and thus dissolving the highly soluble isocyanate.

The site of deposition and the clearance of inhaled gases are primarily dependent on their reactivity and water solubility; high proportions of inhaled gaseous sulfur dioxide or ammonia for example – both readily soluble in water – are dissolved in the upper respiratory tract with consequently low concentrations reaching the distal airways (Dalhamn and Strandberg, 1960). Less-soluble gases, including ozone, oxides of nitrogen and metal fumes, tend to penetrate deeper into the lung and cause inflammation predominantly in alveoli. Gases of intermediate solubility, notably chlorine, may distribute throughout the respiratory tree, penetrating increasingly with increasing intensity of exposure. Similarly, aerosols of >5 μm mass mean diameter tend to impact in the nasopharynx while smaller particles penetrate into the lower and distal airways. Such topographic differentiation may readily be obscured in very high concentration exposures. Other factors which determine the site of deposition of particles include their density and aerodynamic diameter and the anatomy of the upper and lower respiratory tree.

Many of the acute effects of high-dose, toxic inhalations can be considered to be exaggerated physiologic responses which protect the lower airways: these include cough, sneeze, bronchoconstriction and increased mucus production. Since many concepts of respiratory disease are couched in these terms there is an indistinct boundary between the induction of protective mechanisms and pathologic outcomes. While this is seldom a problem in dramatic exposures, it may lead to confusion when interpreting the outcomes of low-intensity exposures, or of animal experiments.

On contact with airway epithelium, irritant chemicals, at sufficient dose, cause inflammation, edema, muscle contraction and neurohumoral stimulation. Certain cell types – ciliated epithelial cells and Type I alveolar pneumocytes – are especially sensitive to toxic exposures (Man *et al.*, 1986). Cilial loss, epithelial inflammation, desquamation and necrosis and a pulmonary exudate may all occur. Epithelial damage following high intensity exposures may be focal rather than regional (Boatman and Frank, 1974), perhaps reflecting local cytoprotective mechanisms. A common feature after toxic inhalations is disruption of the airway epithelium. While at high doses this may be the result of cell death and necrosis, at lower doses there is an increase in paracellular permeability (Gordon *et al.*, 1983) which allows greater access of large molecules to the neurohumoral and inflammatory constituents of the submucosa. Acute, high-intensity exposures to SO_2 also cause goblet cell and submucosal gland hypertrophy and hyperplasia with an increase in basal mucus secretion (Seltzer *et al.*, 1984). These changes occur in central airways and only overwhelming exposures cause damage to the peripheral bronchial tree. Similar responses may follow ozone and nitrogen dioxide exposures but since these agents are less soluble than SO_2, the effects tend to be more peripheral and affect the epithelium of small bronchioles and alveolar cells. The molecular mechanisms responsible for cellular injury are poorly understood but probably reflect in part the nature and dose of the toxic insult. Oxidative gases, including SO_2, O_3 and oxides of nitrogen, primarily act through the formation of free radicals which, among other effects, alter the permeability of cellular plasma membranes.

With highly soluble or reactive agents the inflammatory processes tend to be confined to the nasal epithelium and upper airways, causing immediate symptoms and early recognition by those exposed.

Irritant-provoked engorgement of nasal, vascular erectile tissue can increase the epithelial surface area and increase further nasopharyngeal scrubbing. In contrast, high exposures to insoluble compounds may continue for long periods before the subject is alerted; such compounds are said to have poor warning levels.

Direct stimulation of effector cells or of sensory afferent nerves in the submucosa may lead to sensations of irritation or breathlessness, cough or bronchial hyper-reactivity, a process first described by Hogg (1981). Normal, physiologic pulmonary responses including cough and acute bronchoconstriction, may be the result of direct afferent nerve stimulation in the larger airways. These mechanisms tend to reduce further the dose to the lower respiratory tree, although they may be overwhelmed in the circumstances of very high exposure.

It has been suggested that a degree of tolerance through repeated, low-level exposures may also lead to some threshold adaptation; with the result that workers exposed to chlorine vapor for example may tolerate more severe exposures after many years of employment because of an induced reduction in olfactory sensitivity (Laciak and Sipa, 1958).

Acute pulmonary responses to toxic inhalations are followed by a slow process of repair which may include squamous metaplasia, cellular proliferation and transformation (Witschi, 1990) and persistent changes in basement membrane architecture. Occasionally a fibrotic or obliterative bronchiolitis may develop, most clearly described following high-dose exposure to oxides of nitrogen (McAdams, 1955). The nature and outcome of repair processes probably depends on the dose and site(s) of deposition of the initiating exposure and also on whether or not there is continuing (or prior) low-level exposure to the agent. Thus, the prognosis of those with acute, high-dose, occupational exposures – where continuing exposure is probable – may be different from that of cases with single, non-occupational exposures. The direct consequences of irritant exposure need to be differentiated from those attributable to subsequent infection. In a guinea pig model, exposure to chlorine vapor has been suggested to increase susceptibility to *Mycobacterium tuberculosis* infection (Arloing *et al.*, 1940). The apparently high incidence of infection in the lungs of those experiencing high-dose irritant exposures may result from direct injury to pulmonary inflammatory cells or, in the longer term, from anatomic and functional distortion.

Persistent airway hyper-reactivity after acute, high-dose exposures to respiratory irritants may be the result of continuing epithelial inflammation and a consequent increase in epithelial permeability or reduction in afferent receptor conduction thresholds (Blanc and Schwartz, 1994). The stimulus for this remains unclear but may reflect continual exposure to the wide variety of non-specific irritants encountered in ambient air. There have been a small number of reports of the pathologic changes in the airways of patients with irritant-induced asthma. In early cases, epithelial cell injury with desquamation and bronchial wall inflammation with plasma cell and lymphocyte (but not eosinophil) infiltration was reported (Brooks *et al.*, 1985). A later study of bronchial biopsies in four men with irritant-induced asthma, between 1 and 32 months after exposure to ammonia fumes, described denuded epithelium, chronic submucosal inflammation and focal thickening of the basement membrane (Bernstein *et al.*, 1989).

FREQUENCY OF ACUTE INHALATION EPISODES

Reliable information on the frequency of acute inhalation accidents at work is scarce. In the United Kingdom, the Surveillance of Work and Occupational Respiratory Diseases scheme (SWORD), set up in 1989, provides what is probably the most complete, systematically collected source of information. Respiratory and occupational physicians in the UK are invited to report new cases of occupationally related lung disease on a monthly basis, from which data incidence rates by age, sex and occupational group can be estimated. Between 1990 and 1993, 1180 inhalation accidents were reported (Sallie and McDonald, 1996), some 10% of all reported cases and the fifth most common disease category. The majority of cases followed volatile chemicals – most frequently chlorine (12% of all reports). Occupational groups with the highest rates were chemical processors and engineers/electricians. The incidence was five times higher in men and, surprisingly, greater in those aged over 30 years than in younger men. The importance of chlorine exposures has also been emphasized in the United States where it is estimated that 191 000 workers are at risk of acute inhalation injury (National Institute for Occupational Safety and Health, 1976) – though in that country it has been suggested that domestic or environmental exposures may pose a more frequent risk than occupational exposures (Blanc *et al.*, 1991). A surveillance programme for occupational respiratory disease in two states in the United States (SENSOR) identified, between 1988 and 1992, 42 cases of irritant-induced asthma though the lack of any information on the size of the population at-risk makes this figure difficult to interpret (Reilly *et al.*, 1994).

PERSISTENT EFFECTS OF ACUTE TOXIC INHALATION

Two types of persistent effect will be examined here; chronic, non-reversible airflow limitation and irritant-induced asthma. On occasion the distinction between these is difficult to determine clinically or functionally.

In order to establish a causal link between persistent respiratory disease and a particular toxic exposure a number of conditions need to be satisfied. The first, and a necessary condition, is temporality; that is the exposure must be demonstrated to have preceded the onset of respiratory symptoms or measured defect in pulmonary function. Other features of a relationship which suggest causality – rather than mere association – are its strength and specificity and the demonstration of an exposure–response relationship. The accidental and unpredictable nature of most peacetime toxic exposures make it difficult to ensure that these conditions are satisfied; in other words to estimate, as a basis for comparison, what would have been the outcome in the absence of such exposure. In very few cases is there an unbiased measurement of symptoms or lung function before the exposure. The limited range of pulmonary responses to a wide variety of environmental insults means that many proposed outcomes of toxic exposure – with the arguable exception of bronchiolitis obliterans (which can also follow viral infections) – are non-specific; one result, together with the generally small number of individuals examined, has been that causal inferences from most reported relationship have been relatively weak.

In the absence of prior information about lung function, or of a suitable referent group, useful causal information may be gained from an examination of exposure–response relationships within the exposed group. It is usually very difficult to estimate, after the event, an individual's exposure to a toxic release, let alone the dose to his or her target organ(s). Attempts to model prior exposures with any precision have generally been unsuccessful; Jones *et al.* (1986) used meterologic and topographic data to model the spread of chlorine vapour from a derailed freight train in the United States. They were unable, however, to associate the severity of acute respiratory symptoms or subsequent changes in lung function among those exposed with estimates of individual exposure intensity derived from two different mathematical predictions. On the other hand, a simple measurement of distance from the spill at the time of the accident correlated closely with early triage status. Similarly, sophisticated attempts to model exposures to the Bhopal gas leak in 1984, using admittedly scanty meteorologic information from the time of the disaster (Singh and Ghosh, 1987), have been less useful in assessing dose–response relationships than crude measures based on location (Cullinan *et al.*, 1996) or on local death rates (Andersson *et al.*, 1990). Retrospective assessments of local conditions and of personal and behavioral variations are inevitably difficult and sophisticated modeling may lead to greater misclassification of individual exposures and a weakening of observed exposure–response relationships (Armstrong *et al.*, 1994).

Chronic, non-reversible airflow limitation

Most reported assessments of long-term respiratory damage have followed exposures to chlorine or its derivatives. Chlorine – in addition to phosgene and sulfur mustard – was used extensively as a chemical weapon in Europe during the First World War. Clinical examination of 2000 soldiers, three months after 'gassing' by a variety of agents, revealed 'bronchitis' in 30% and evidence compatible with 'emphysema' in 20% (Berghoff, 1919). Of 96 men exposed mainly to chlorine, assessed between 8 and 10 years later, nine (9.4%) had symptomatic and radiologic evidence consistent with chronic airflow obstruction which the authors attributed to one or more gassing incidents (Gilchrist and Matz, 1933). Penington (1954) concluded that although immediate morbidity and mortality following military exposures was high, permanent pulmonary damage was uncommon. None of these early series, in common with several others, used a comparison group or attempted any dose–response assessment and it remains difficult to determine to what extent their findings are attributable to the gassing incidents rather than to the high contemporary rates of childhood and adult respiratory infection – including the 1919 pandemic of influenza – and the emergence of widespread cigarette smoking among European men at the same time.

More recently, the effects of liquid chlorine spills have been studied after two, non-occupational incidents. In the first, a cloud of chlorine vapor from a de-railed train was dispersed over several square miles, with concentrations of at least 400 ppm measurable seven hours after the release (Weill *et al.*, 1969). Seventeen people were hospitalized - the majority with pulmonary edema - and one (an infant) died shortly after admission. The lung function of 11 of the 16 survivors and one other person reporting severe immediate reactions was assessed 3 or 7 (or in three cases both) years after the incident. Reduced vital capacity was detected in one child and a low FEV_1 in an elderly woman; other reductions could be explained by cigarette smoking or obesity. Only one subject (a 53-year-old with chronic bronchitis) was symptomatic. No attempt to correlate pulmonary function with estimated exposure was made in this small case series.

The second study – reported by the same group – also examined the long-term effects of heavy chlorine exposure after a rail accident (Jones *et al.*, 1986). Eight of those exposed died as a direct result of acute, chlorine-induced pulmonary damage; 23 were hospitalized with respiratory injuries. Over the next 6 years, at approximately annual intervals, lung function was measured in 113 survivors, categorized by an estimate of acute exposure. No changes in flow rates, lung volumes or gas transfer which could not be accounted for by smoking were found, nor was there any relationship between functional decline and estimated chlorine exposure. No measurements of lung function prior to the accident were available and the authors point out that a single, lasting reduction in lung function could not be excluded. Changes (not statistically significant) in mean FEV_1 decline at a 2-year assessment were greater than those at the final (6-year) examination; this feature, which the authors attribute to random fluctuation about a mean, emphasizes the importance of a sufficiently long follow-up period in studies of this sort.

Other, smaller studies also suggest that single, high-intensity exposures to chlorine fumes in the community do not result, among survivors, in persistent, measurable respiratory disease (Ploysonang *et al.*, 1982; Hasan *et al.*, 1983). Similarly, persistent morbidity (lasting 14 days or more) is rare in those experiencing acute irritant exposures in the home, many of which are due to chlorine-containing bleaching products (Blanc *et al.*, 1991) though in these cases, where detailed information on exposure and prior lung function are absent, it is difficult to attribute any causal role to the irritant exposures.

Repeated, high-concentration exposures – generally on a background of chronic, lower level exposures – may occur in chlorine-exposed workforces. Kennedy and others examined 321 pulpmill workers, 189 of whom reported personal exposure to one or more (chlorine or ClO_2) gassing incidents. Internal comparisons, after stratification by three categories of cigarette smoking, demonstrated an increased frequency of some respiratory symptoms and reduced mid-maximal flows and FEV_1/FVC ratios among 'gassed' workers. The authors suggest that the explanation for the contrast between these findings and those from the single-episode community studies lies in the repeated nature of occupational exposures whereby inflammatory responses to initial exposure(s) do not recover before subsequent incidents – or are perpetuated by continuous low-level exposures. A second study of 58 'gassed' workers demonstrated a similar reduction in airflow at low lung volumes compared with 81 non-exposed co-workers (Chester *et al.*, 1969). Given the cross-sectional nature of these two studies, and their reliance on subjective reporting of past exposures, it is not possible to exclude an increased tendency by those with respiratory disease to recall gassing incidents.

The 1984 Bhopal gas leak resulted in approximately a quarter of a million people being exposed to varying concentrations of an irritant mix, most of which was probably methyl isocyanate. About 2500 inhabitants died within 1 month; the long-term experience of the survivors arguably represents the largest series of irritant-exposed subjects. The acute effects of heavy exposure were consistent with inhalation of a volatile (and deeply penetrating), highly irritative aerosol; large numbers of people died rapidly with bronchial necrosis or pulmonary edema (Indian Council of Medical Research, 1985). Within months of the disaster, clinical, radiologic and pathologic evidence suggested persistent airflow obstruction associated with an obliterative bronchiolitis among survivors (Weill, 1987). Subsequent case reports, of small numbers, proposed a variety of other pulmonary outcomes including interstitial fibrosis (Vijayan *et al.*, 1995). Systematic attempts to examine the chronic effects of exposure have generally been hampered by the study of highly selected groups of subjects and the difficulty of assigning individual exposures and thereby assessing exposure–response relationships (Vijayan *et al.*, 1989; Kamat *et al.*, 1992; Rastogi *et al.*, 1988). Andersson *et al.*, (1990) however, in a study conducted 3 years after the disaster, demonstrated that respiratory symptoms were more frequently reported by those in crudely designated high-exposure categories and that this relationship was not accounted for by smoking. In a further follow-up survey of current, adult inhabitants a consistent gradient of obstructive symptoms across six categories of estimated exposure was observed and confirmed by simple spirometry (Cullinan *et al.*, 1996); as in all such epidemiologic studies it was not possible to account entirely for potentially confounding factors (in particular socioeconomic status) although the findings were consistent with the persistence of small airways obstruction in survivors of what was an unusually high-intensity exposure.

In one small study, information on lung function before the event was available and used to examine the effects of (mainly) sulfur dioxide exposure on subsequent lung function (Harkonen *et al.*, 1983). Seven Finnish pyrite miners were exposed to levels of SO_2 estimated at between 300 and 1600 ppm for between 20 and 45 minutes following an underground

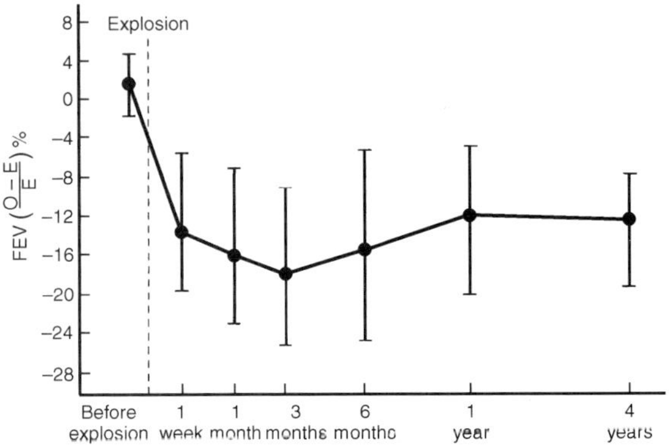

Figure 30.1 Mean FEV_1 values in seven pyrite miners during 4 years after acute exposure to high concentrations of SO_2. (Adapted with permission from Harkonen *et al.*, 1983.)

dust explosion; one died in the accident but most of the others were followed periodically for 4 years. All had had repeated medical examinations, including lung function measurements, before the event. Unfortunately, the authors did not present individual results, but across the survivor group average values of FEV_1, FVC and maximal mid-expiratory flow were reduced throughout the period of follow-up. The maximum reduction occurred 1 week after the accident; subsequent improvements plateaued after 3 months, leaving the mean FEV_1 approximately 13% below predicted value, with a smaller proportional reduction in FVC (Figure 30.1). Four subjects, at the end of the period of study, had demonstrable bronchial hyper-reactivity on histamine challenge, although measurements of this index before the accident were not available. Airflow obstruction after acute SO_2 inhalation has been reported elsewhere (Charan *et al.*, 1979) and, when fatal, may be associated with obliterative bronchiolitis (Galea, 1964).

Smoke from fires contains a wide variety of particulate matter in addition to potentially toxic fumes which commonly include carbon monoxide, hydrogen chloride, hydrogen cyanide, oxides of nitrogen, styrene and several aldehydes (Loke *et al.*, 1988). Thermal injury to the upper respiratory tract may also occur leading, acutely, to laryngeal edema and obstruction. Tracheal and upper bronchial stenosis may result (Colice *et al.*, 1986), as may endobronchial polyposis (Williams *et al.*, 1983).

Lower airway function in smoke victims has been more difficult to assess. In 13 consecutive patients treated in a Scottish hospital for acute smoke inhalation, spirometry and histamine challenge testing was conducted 3 days and 12 weeks after the incident (Kinsella *et al.*, 1991). Measurement of serum carboxyhemoglobin at presentation provided an estimate of inhaled smoke dose, but lung function values before the fire were not available. After 3 days, mean FEV_1 was 76% of predicted values and did not increase thereafter. Bronchial hyper-reactivity was demonstrated in all but one subject at the first assessment (mean histamine PC_{20} 1.65 mg/ml); by 3 months this had improved in 10 subjects, worsened in two and remained unchanged in one patient with a mean value of 5.43 mg/ml. There were no relationships between exposure carboxyhemoglobin and initial FEV_1 or bronchial reactivity. Similarly, 14 survivors of an underground railway fire in London were examined 6 months and (seven subjects) 24 months after the incident; all had inhaled high concentrations of smoke and 10 had suffered cutaneous burns (Fogarty *et al.*, 1991). At both assessments mean maximum airflow at 25% of vital capacity was at the lower limit of the predicted 95% confidence interval and residual volume at the upper, suggesting a persistent defect in small airway flow; at least one of these defects was present in all seven patients studied at 2 years.

In 1985, an air crash in England resulted in the immediate deaths of 52 of the 137 passengers and crew; eight survivors required treatment in intensive care, four with assisted ventilation. These and seven less severely affected passengers were followed for a period of 3 months (O'Hickey, 1987) with assessment of respiratory symptoms and measurements of lung flows and bronchial hyper-responsiveness. At the final assessment all eight of those more severely affected (in contrast to the remaining seven) continued to report wheezing and breathlessness. In most subjects lung function, initially depressed in those with the greater initial problems, had returned to normal, although two asthmatic patients retained markedly increased airway hyper-responsiveness and required more treatment than before the incident.

As with the non-occupational chlorine exposures discussed above, these case series describe the experience of those acutely exposed to high concentrations of an unfamiliar mixture. The experience of firefighters provides a useful means of examining more chronic or repeated high-concentration exposures. In general, studies of this group have not demonstrated any persistent deficit in lung function attributable to smoke inhalation; given the probable, overall high level of fitness in firefighters however, results based on 'predicted' values of lung function (derived from general populations) need to be interpreted with care. Using these indices, Musk and others (1982) were unable to detect any acceleration in decline in FEV_1

or FVC over 6 years in a large group of Boston firefighters; or, during 5 years of retirement, in a similar population (Musk *et al.*, 1977). Smaller studies of firefighters have, in general, reported similar findings (Loke *et al.*, 1980; Douglas *et al.*, 1985) but it should be noted that they were both cross-sectional in design and unable to take account of the possible effects of survival bias. One, large study however – again in Boston – demonstrated an accelerated decline in lung function over 1 year; by an average of 68 ml in FEV_1 (against an expected decline of 25 ml) and 77 ml (versus 30 ml) in FVC, differences not wholly accounted for by reported smoking (Peters *et al.*, 1974). Interestingly, the rates of decline were greater in those who had at some stage been hospitalized for smoke inhalation – a crude index of accumulated exposure. It is not clear to what extent, if any, the contrast between the results of this early study and those subsequent to it are attributable to the increased efficiency of personal protective equipment in the intervening years.

Table 30.1 Criteria for the diagnosis of irritant-induced asthma

1. Documented absence of preceding respiratory complaints.
2. Other types of pulmonary diseases were excluded.
3. Onset of symptoms occurred after a specific exposure incident or accident.
4. Exposure was to a gas, smoke, fume or vapor that was present in high concentrations and had irritant qualities.
5. Onset of symptoms occurred within 24 hours of the exposure and persisted for a least 3 months.
6. Symptoms were clinically consistent with asthma, with cough, wheezing and dyspnea predominating.
7. Appropriate challenge testing demonstrated increasing airway responsiveness.
8. Pulmonary function tests may show airflow obstruction.

Irritant-induced asthma

It is well established that exposure at work to aeroallergens can cause asthma and airway hyper-responsiveness. In 1985, Brooks *et al.* reported the first cases of airway hyper-responsiveness attributed to irritant exposure; the authors labeled this syndrome RADS (reactive airways dysfunction syndrome) although recently the term 'irritant-induced asthma' has become more widely used. For these purposes a useful definition of an irritant agent is one able to cause pulmonary tissue damage through non-immunologic mechanisms (Kennedy, 1992), although it is important to note that some agents (such as some diisocyanates) may, depending on the inhaled concentration cause irritant or hypersensitivity asthma. Little is known about the pathophysiology of irritant-induced asthma (see above) since opportunities to study human patients in detail are rare and few convincing animal models have yet to be developed. However, Sheppard and others (1988) have demonstrated bronchial hyper-responsiveness to substance P in a guinea pig model exposed to toluene diisocyanate.

Brooks' original report described 10 patients who developed persistent asthma following single exposures to a variety of respiratory irritants, including acid fumes, smoke, uranium hexachloride and ammonia fumes. In three subjects symptoms started immediately after exposure – among the remainder the average latency was 9 hours. At final assessment (between 1 and 12 years after the initiating incident) all had increased airway responsiveness to inhaled metacholine.

None of these cases reported pre-existing respiratory disease, but a major limitation to the understanding of irritant-induced asthma is the almost universal lack of information on respiratory function, particularly airway responsiveness, before the incident. In order to overcome this a conservative set of criteria for the diagnosis of irritant-induced asthma are generally used (Table 30.1), although it should be recognized that these are somewhat self-fulfilling and will inevitably exclude those with pre-existing asthma whose disease is made worse by high-concentration irritant exposure(s). Furthermore, such a definition complicates the study of those factors which determine susceptibility to the development of irritant-induced asthma following exposure – among which baseline bronchial reactivity may be important. Indeed, restriction of cases to those with no evidence of pre-existing airway hyper-reactivity may result in an apparent increase in risk among non-asthmatics or non-atopics (Tarlo and Broder, 1989).

Much of the literature on irritant-induced asthma, as with other persistent effects of toxic exposures, is in the form of case reports or small case series. The list of responsible agents reflects those which are generally associated with high-dose irritant exposures and it is not possible at this stage to assign particular hazards to any group of compounds. Chlorine (Donnelly and Fitzgerald, 1990; Moore and Sherman, 1991), ammonia vapour (Bernstein *et al.*, 1989), smoke (Moisan, 1991) and toluene diisocyanate (Luo *et al.*, 1988) have all been implicated. In a larger series, Cone *et al.* (1994) reported the experience of a group of subjects exposed to irritant fumes produced by the derailment of a train carrying 19 000 gallons of metam sodium pesticide. Of an estimated 3000 people living in the exposed zone, some 700 were seen in local medical facilities for spill-associated symptoms and the subsequent course of 197 (selected by medical practice) was reported. After 3 months 20 (10%) – without prior histories of respiratory disease – had developed symptoms and measurable airway hyper-responsiveness consistent with a diagnosis of irritant-induced asthma. Unfortunately, the authors considered that any attempt to model the relationship between exposure and outcome to be impractical. Those who developed irritant-induced asthma did not differ importantly in smoking habit or reported allergies from those without respiratory symptoms.

An estimate of the incidence of irritant-induced asthma in the United Kingdom can be made from SWORD data (Sallie and McDonald, 1996). Occupational physicians representing 383 patients reported to have been involved in inhalation accidents were asked for details of their subsequent health and returns to work. Some 70% had recovered within a week (and 78% returned to work) but 47 (12%) cases had persistent ill-health, of whom 11 (3% of the original total) had developed airway hyper-reactivity. The agents associated with this continuing ill-health were those commonly causing inhalation accidents – chlorine, sulfur dioxide, oxides of nitrogen, ammonia. Blanc and others (1991) have estimated, from outcome information from 323 (47%) of 683 consecutive referrals to a poison information center, that persisting symptoms (lasting 2 or more weeks) were present in 6% of those exposed to toxic inhalation incidents, although it is not clear from these figures to what extent they describe exacerbation of pre-existing disease.

A further estimate of incidence can be derived from a study of hospital workers following a single, brief, high-intensity exposure to glacial (100%) acetic acid after a laboratory accident (Kern, 1991). Of 56 employees in the laboratory area during the 2½ hours after the incident, 51 were studied 8 months later by questionnaire and measurement of non-specific bronchial reactivity. Exposure was estimated, indirectly and retrospectively, by an industrial hygienist. An unusual feature of this study was the existence of (symptomatic) pre-exposure health information, and the inclusion of the entire exposed cohort. At final assessment eight subjects reported symptoms consistent with irritant-induced asthma; 24 agreed to undergo metacholine challenge, after which a diagnosis of irritant-induced asthma was made in four, an incidence rate of approximately 8%. A clear relationship between estimated category of exposure and subsequent risk of asthma development was shown, even after exclusion of those with pre-existing asthmatic symptoms.

CONCLUSIONS

The potential for high-dose exposure to respiratory irritants is high in societies where industrial chemical use is frequent, and arguably increasing. Although much of the risk is to those with occupational exposure, there are also risks to the wider community during transport or accidental release from factories. The effects may differ between those experiencing single exposures and those whose exposure is on a background of frequent, low-level contact. The available literature suggests that both reversible and non-reversible airflow obstruction may be attributable to single, high-dose toxic exposures, although these are probably uncommon outcomes in those who survive the initial episode.

The dose to the respiratory tree of a toxic exposure depends on a complex mixture of host, agent and environmental factors; these are incompletely understood and the possibilities for controlled research in humans are very limited. It is questionable whether the results of animal dosing models are appropriate to human exposures, but they are frequently used and in the absence of any alternative are a source of potentially useful information. A better understanding of the chronic effects of high-intensity toxic exposures will only follow more systematic analysis of such events. Ironically, the unpredictable nature of chemical accidents, and their relative rarity, tend to preclude any systematic study of either their frequency or outcome. Thus, much of the available literature is in the form of case reports or small series. Many of the larger studies have been devoted to highly selected groups of subjects, usually those presenting in the immediate aftermath of the incident to one or more medical facilities. Unbiased information on respiratory symptoms or on the degree of lung function before the incident is almost invariably lacking, making it difficult to interpret any but the most severe of subsequent deficiencies. Regular monitoring of respiratory symptoms and lung function (including bronchial reactivity) before a potential event is only feasible in occupational settings – and requires considerable investment given the rarity of these events in those situations where such surveillance is likely to be introduced. An alternative strategy for determining causality, the establishment of a relationship between estimated exposure and outcome, is similarly difficult given the problems with establishing, in retrospect, a scale of exposure among those affected. Those (too) few studies which have attempted this latter approach have been successful where they have used crude measures of exposure, reducing the possibilities of misclassification.

REFERENCES

Andersson, N., Ajwani, M.K., Mahashabde, S. *et al.* (1990) Delayed eye and other consequences from exposure to methyl isocyanate: 93% follow-up of exposed and unexposed cohorts in Bhopal. *Br. J. Ind. Med.*, **47**, 553–8.

Arloing, F., Berthet, E. and Viallier, J. (1940) Action of chronic intoxication by low concentration chlorine fumes on experimental guinea pigs. *Presse Med.*, **48**, 361.

Armstrong, B.K., White, E. and Saracci, R. (1994) Principles of exposure measurement in epidemiology, *Monographs in Epidemiology and Biostatistics, vol. 21*, Oxford University Press, Oxford.

Berghoff, R.S. (1919) The more common gases; their effects on the respiratory tract. *Arch. Intern. Med.*, **24**, 678–84.

Bernstein, I.L., Bernstein, D.L., Weiss, M. and Cambell, G.O.P. (1989) Reactive airways dysfunction syndrome (RADS) after toxic exposure to ammonia fumes. *J. Allergy Clin. Immunol.*, **83**, 173.

Blanc, P.D. and Schwartz, D.A. (1994) Acute pulmonary responses to toxic exposures, in *Textbook of Respiratory Medicine* (eds J.F. Murray and J.A. Nadel), W.B. Saunders Co., Philadelphia, pp. 2050–61.

Blanc, P.D., Galbo, M., Hiatt, P. and Olsen, K.R. (1991) Morbidity following acute irritant inhalation in a population-based study. *JAMA*, **266**, 664–9.

Boatman, E.S. and Frank, R. (1974) Morphologic and ultrastructural changes in the lungs of animals during acute exposure to ozone. *Chest*, **65**, 9S–11S

Brooks, S.M., Weiss, M.A. and Bernstein, I.L. (1985) Reactive airways dysfunction syndrome (RADS): persistent asthma after high-level irritant exposures. *Chest*, **88**, 376–84.

Centers for Disease Control (1991) Chlorine gas toxicity from mixture of bleach with other cleaning products. *MMWR*, **40**, 619–29.

Charan, N.B., Myers, C.G., Lakshminarayan, S. and Spencer-Tracy, M. (1979) Pulmonary injuries associated with acute sulphur dioxide inhalation. *Am Rev Respir Dis.*, **119**, 555–60.

Chester, E.H., Gillespie, D.G. and Krause, F.D. (1969) The prevalence of chronic obstructive pulmonary disease in chlorine workers. *Am. Rev. Respir. Dis.*, **99**, 365–73.

Colice, G.L., Munster, A.M. and Haponik, E.F. (1986) Tracheal stenosis complicating cutaneous burns: an underestimated problem. *Am. Rev. Respir. Dis.*, **134**, 1315–18.

Cone, J.E., Wugofski, L., Balmes, J.R., Das, R., Bowler, R., Alexeef, G. and Shusterman, D. (1994) Persistent respiratory health effects after a metam sodium pesticide spill. *Chest*, **106**, 500–8.

Cullinan, P., Acquilla, S.D. and Dhara, V.R. (1996) Long term morbidity in survivors of the 1984 Bhopal gas leak. *Nat. Med. J. India*, **9**, 5–10.

Dalhamn, T. and Strandberg, L. (1960) Acute effects on rate of ciliary beat in rabbit trachea (*vivo* and *vitro*) and absorption capacity of nasal cavity. *Int. J. Air Water Poll.*, **4**, 154.

Donnelly, S.C. and Fitzgerald, M.X. (1990) Reactive airways dysfunction syndrome (RADS) due to chlorine gas exposure. *Irish J. Med. Sci.*, **159**, 275–7.

Douglas, D.B., Douglas, R.B., Oakes, D. and Scott, G. (1985) Pulmonary function of London firemen. *Br. J. Ind. Med.*, **42**, 55–8.

Fogarty, F.W., George, P.J.M., Solomon, M., Spiro, S.G. and Armstrong, R.F. (1991) Long term effects of smoke inhalation in survivors of the King's Cross underground station fire. *Thorax*, **46**, 914–18.

Galea, M. (1964) Fatal sulphur dioxide inhalation. *Can. Med. Assoc. J.*, **91**, 345–7.

Gilchrist, H.L. and Matz, P.B. (1933) The resident effects of warfare gases: the use of chlorine gas with reports of cases. *Med. Bull. Vet. Admins.*, **9**, 229–70.

Gordon, R.E., Case, B.W. and Kleinerman, J. (1983) Acute NO_2 effects on penetration and transport of horseradish peroxidase in hamster respiratory epithelium. *Am. Rev. Respir. Dis.*, **128**, 528–33.

Harkonen, H., Nordman, H., Koshonen, O. and Vinblad, I. (1983) Long term effects of exposure to sulphur dioxide. Lung function 4 years after a pyrite dust explosion. *Am. Rev. Respir. Dis.*, **126**, 890–3.

Hasan, F.M., Gehshan, A. and Fuleihan, F.J.D. (1983) Resolution of pulmonary dysfunction following acute chlorine exposure. *Arch. Environ. Health*, **38**, 76–80.

Hogg, J.C. (1981) Bronchial mucosal permeability and its relationship to airways hyperreactivity. *J. Allergy Clin. Immunol.*, **67**, 421–5.

Indian Council of Medical Research (1985) Health effects of exposure to toxic gas at Bhopal: an update on ICMR sponsored researches. New Delhi, India

Jones, R.N., Hughes, J.M., Glindmeyer, H. and Weill, H. (1986) Lung function after acute chlorine exposure. *Am. Rev. Respir. Dis.*, **134**, 1190–5.

Kamat, S.R., Patel, M.H., Pradhan, P.V. *et al.* (1992) Sequential respiratory, psychologic and immunologic studies in relation to methyl isocyanate exposure over two years with model development. *Environ. Health Perspect.*, **97**, 241–53.

Kennedy, S.M. (1992) Acquired airway hyperresponsiveness from nonimmunogenic irritant exposure. *Occup. Med.*, **7**, 287–300.

Kern, D.G. (1991) Outbreak of the reactive airways dysfunction syndrome after a spill of glacial acetic acid. *Am. Rev. Respir. Dis.*, **144**, 1058–64

Kinsella, Carter, R, Reid, W.H., Cambell, D. and Clark, C.J. (1991) Increased airway reactivity after smoke inhalation. *Lancet*, **337**, 595–7.

Laciak, J. and Sipa, K. (1958) The importance of the sense of smell in workers in some branches of the chemical industry. *Med. Prac.*, **9**, 85–90.

Loke, J., Farmer, W., Matthay, R.A., Putman, C.E. and Walker-Smith, G.J. (1980) Acute and chronic effects of fire fighting on pulmonary function. *Chest*, **77**, 369–73.

Loke, J., Matthay, R.A. and Walker-Smith, G.J. (1988) The toxic environment and its medical implications with special emphasis on smoke inhalation, in *Pathophysiology and Treatment of Inhalation Injuries* (ed. J. Loke), Lung Biology in Health and Disease, vol. 34, Marcel Dekker Inc., New York.

Luo, J.C.J., Nelson, K. and Fishbein, A. (1988) Persistent reactive airways dysfunction after exposure to toluene diisocyanate. *Br. J. Ind. Med.*, **47**, 239–41.

Man, S.F., Hulbet, W.C., Mok, K., Ryan, T. and Thomson, A.B. (1986) Effects of sulfur dioxide on pore populations of canine tracheal inflammation. *J. Appl. Physiol.*, **60**, 416–26.

McAdams, A.J. (1955) Bronchiolitis obliterans. *Am. J. Med.*, **19**, 314–22.

Moisan, T. (1991) Prolonged asthma after smoke inhalation; a report of three cases in a review of previous reports. *J. Occup. Med.*, **33**, 458–61.

Moore, B.B. and Shereman, M. (1991) Chronic reactive airway disease following acute chlorine gas exposure in an asymptomatic atopic patient. *Chest*, 855–6.

Musk, A.W., Peters, J.M. and Wegman, D.H. (1977) Lung function in fire fighters II. A five year follow-up of retirees. *Am. J. Public Health*, **67**, 630–3.

Musk, A.W., Peters, K.M., Bernstein, L., Rubin, C. and Monroe, C.B. (1982) Pulmonary function in firefighters; a six year follow-up in the Boston Fire Department. *Am. J. Ind. Med.*, **3**, 3–9.

National Insitute for Occupational Safety and Health (1976) NIOSH criteria for a recommended standard: occupational exposure to chlorine. Publication no. 76–170 NIOSH, Cincinnatti.

O'Hickey, S.P. (1987) Manchester air disaster. *Br Med. J.*, **294**, 1663–7.

Penington, A.H. (1954) War gases and chronic lung disease. *Med. J. Aust.*, **1**, 510.

Peters, J.M., Theriault, G.P., Fine, L.J. and Wegman, D.H. (1974) Chronic effect of fire fighting on pulmonary function. *N. Engl. J. Med.*, **291**, 1320–2.

Ploysonang, Y., Beach, B.C. and DiLisio, R.E. (1982) Pulmonary function changes after acute inhalation of chlorine gas. *South. Med. J.*, **75**, 23–6.

Rastogi, S.K., Gupta, B.N., Husain, T., Kumar, A., Chandra, S. and Ray, P.K. (1988) Effect of exposure to toxic gas on the population of Bhopal: Part II. Respiratory impairment. *Ind. J. Exp. Biol.*, **26**, 161–4.

Reilly, M.J., Rosenman, K.D., Watt, F.C. *et al.* (1994) Surveillance for occupational asthma – Michigan and New Jersey 1988–1992. *MMWR CDC Surveillance Summaries*, **194**(43),9–17.

Sallie, B. and McDonald, J.C. (1996) Inhalation accidents reported to the SWORD surveillance project 1990–1993. *Ann. Occup. Hyg.*, (in press).

Seltzer, H.H., Scanlon, P.D., Drazen, J.M., Ingram, R.H. and Reid, L. (1984) Morphologic correlation of physiologic changes caused by SO_2-induced bronchitis in dogs. *Am. Rev. Respir. Dis.*, **129**, 790–7.

Sheppard, D., Thompson, J.E., Scypinski, L., Dusser, D., Nadel, J.A. and Borson, D.B. (1988) Toluene diisocyanate increases airway responsiveness to substance P and decreases airway neutral endopeptidase. *J. Clin. Invest.*, **81**, 111–15.

Singh, M.P. and Ghosh, S. (1987) Bhopal gas tragedy: model simulation of the dispersion scenario. *J. Hazardous Mat.*, **17**, 1–21.

Tarlo, S.M. and Broder, I. (1989) Irritant-induced occupational asthma. *Chest*, **96**, 297–300.

Vijayan, V.K., Pandey, V.P., Sankaran, K., Mehrotra, Y., Darbari, B.S. and Misra, N.P. (1989) Bronchoalveolar lavage study in victims of toxic gas leak at Bhopal. *Indian J. Med. Res.*, **90**, 407–14.

Vijayan, V.K., Sankaran, K., Sharma, S.K. and Misra, N.P. (1995) Chronic lung inflammation in victims of the toxic gas leak at Bhopal. *Respir. Med.*, **89**, 105–11.

Weill, H. (1987) Disaster at Bhopal: the accident, early findings and respiratory health outlook in those injured. *Bull. Eur. Physiopathologie Respiratoire*, **23**, 587–90.

Weill, H., George, R., Schwarz, M. and Ziskind, M. (1969) Late evaluation of pulmonary function after acute exposure to chlorine gas. *Am. Rev. Respir. Dis.*, **99**, 374–9.

Williams, D.O., Vanecko, R.M. and Glassroth, J. (1983) Endobronchial polyposis following smoke inhalation. *Chest*, **84**, 774–6.

Witschi, H. (1990) Responses of the lung to toxic injury. *Environ. Health Perspect.*, **85**, 5–13.

Part Four Pulmonary carcinogenesis

31 Lung carcinogenesis: molecular and cellular mechanisms

Heather H. Nelson, Xi Wang and Karl T. Kelsey

INTRODUCTION

Carcinogenesis is a complex process driven by the accumulation of DNA changes within a population of cells, and is classically broken into three stages: initiation, promotion, and progression. These heritable genomic alterations, called mutations, result in phenotypic changes that manifest as neoplastic growth. Initiation refers to the first genetic change occurring within one cell that confers a growth advantage. This initiated cell then undergoes clonal expansion during the promotion stage. During this period there may be further accumulation of mutations and the growth of the emerging cell population becomes unregulated. The final phase of cancer development is termed progression, and involves additional genetic changes that render the tumor aggressive leading to the metastatic state. Histologic changes, from preneoplasia to carcinoma, accompany the genetic changes that occur during the promotion and progression phases.

Lung cancer is a particularly complex disease, involving mutations in at least 10 to 20 genes (Harlow 1994). Not all of the genes involved in lung cancer development have been identified, nor is it clear whether mutations must occur in a specific order for carcinoma to develop. However, it is clear that cancer is a mutation-driven process, and that lung cancer in particular is effected by environmental mutagen exposure. The recent use of molecular methods in cancer epidemiology has led to the recognition that histologically similar cancers can possess different patterns of genetic changes, indicative of specific mutational events (Bishop, 1991; Weinberg, 1991). The molecular characterization of these etiologically different cancers will aid in the further subclassification of lung disease, and may prove useful in identifying intervention strategies and treatment alternatives. Additionally, the field of molecular epidemiology aims to identify host factors that confer increased risk for development of exposure-related lung cancer to aid in prevention and control strategies.

A MOLECULAR MODEL FOR CARCINOGENESIS

The classic initiation, promotion, progression model described above derives from toxicology studies involving the application of coal tar and croton oil to animal skin (Rous and Kidd, 1941; Berenblum, 1941). When coal tar or corton oil are independently applied to skin, no tumor formation is observed. However, if coal tar exposure precedes application of croton oil the animals develop tumors at the site of application. When

Occupational Lung Disease. Edited by Daniel E. Banks and John E. Parker. Published in 1998 by Chapman & Hall. ISBN 0 412 73630 6

exposed in the reverse order no tumors develop. More refined experiments were subsequently conducted in development of a mouse skin model of carcinogenesis (Berenblum and Shubik, 1947; Boutwell, 1964). Application of the potent carcinogen DMBA (7,12-dimethylbenzanthracene) followed by the phorbol ester TPA (12-*O*-tetradecnoyl-phorbol-13 acetate) to mouse skin results in benign tumors, as seen with the experiments with coal tar and croton oil. With continued TPA exposure some tumors progress to carcinomas. The resultant hypothesis is that cancer development has multiple, distinct phases. Continued experiments using DMBA and TPA application to mouse skin have defined a molecular basis for the classic carcinogenic model involving genetic damage and clonal growth as driving forces in the development of cancer (Bizub *et al.*, 1986; Quintanilla *et al.*, 1986).

Mutagenesis

Initiation refers to the first somatic DNA change in a cancer-related gene that renders the cell prone to deregulation. In the mouse skin model this occurs with exposure to DMBA, a highly mutagenic polyaromatic hydrocarbon. Carcinogens primarily exert their effects by inducing heritable genetic changes in the DNA sequence, a process termed mutagenesis. Carcinogens may act as mutagens by either directly interacting with DNA or by inducing the production of highly reactive radical species in the DNA environment. The chemical modification of DNA will result in a mutation when it is not properly recognized and repaired by the cell before DNA replication and cell division.

Mutagens may induce either large or small chromosomal changes with both types leading to a potentially cancerous cell. Large chromosome events are characterized by changes in chromosome number or changes in chromosome structure. Changes in chromosome structure primarily result from attempted repair of DNA double strand breaks. These aberrations may be lethal at cell division, such as dicentric and acentric chromosomes that are unable to segregate properly at mitosis. Other large chromosome rearrangements are viable and induce mutations, such as chromosome translocations, large deletions of DNA or insertions of additional bases into the normal sequence of DNA. Small chromosome changes are equally important in mutagenesis. They typically involve single base substitutions, and small DNA deletions or insertions. These mutations arise from many different processes including single strand breaks, misreplication of adducted DNA (DNA with exogenous material covalently bound), and misrepair of other DNA lesions.

Carcinogen exposure, metabolism and mutagenesis

Classic carcinogens have most often been discovered and confirmed as human cancer-causing agents by epidemiologic methods. Cancer epidemiology is a well-developed discipline that has sought to define associations between population exposure to possible carcinogens and subsequent disease, explore dose–response relationships and model the magnitude of the risk associated with exposure to these compounds. These data have then been the basis of subsequent experimental work that has attempted to understand the mechanism of action of carcinogens, including their metabolism and the DNA structural alteration associated with carcinogen–DNA binding.

Carcinogen metabolism has been widely studied and includes a series of seminal experiments that sought to define the conversion of procarcinogens to ultimate carcinogens using animal tissues to activate chemical carcinogens to reactive, electron-deficient intermediates (Miller, 1981). The production of electrophilic metabolic intermediates of polyaromatic compounds that are found in cigarette smoke has been used as a paradigm for investigation of carcinogen metabolism in human systems. Activation of these compounds via the multiple forms of the cytochrome P450 enzymes leads to DNA adduct-forming intermediates (such as epoxides) that are further detoxified by additional steps in metabolism (Law, 1990). The conjugation of these reactive compounds, such as epoxides, quinones and diols is completed by enzymes systems such as glutathione transferases, sulfotransferases and UDP glucuronyl transferases, producing conjugates that are water-soluble, excretable and hence no longer toxic.

This model of carcinogen metabolism has led to the supposition that understanding the precise balance between activation of procarcinogens and the detoxification of reactive intermediates is critical to estimation of individual risk of cancer from exposure to agents subject to this process. This has proven complex, however, since these enzyme systems are known to have overlapping substrate specificity and many are inducible *in vivo* by a wide variety of exposures. That is, prior exposure to a variety of classes of compounds will induce the P450 enzymes, potentially altering the metabolism of procarcinogens. Common inducers include medications (barbiturates, rifampin, clofibrate, phenytoin), ethanol, food additives (e.g. BHA and BHT), insecticides (e.g. DDT), known toxins (e.g. dioxins) and some solvents (including toluene and xylene). The kinetics of enzyme induction are variable and consequently, systematic understanding of biologically relevant enzyme levels is problematic.

In addition to the polyaromatic hydrocarbons, *N*-nitroso compounds have been shown to be potent lung carcinogens. Nitrosamides are direct-acting carcinogens that do not require activation, while nitrosamines are P450-activated to become carcinogens. Tobacco-specific nitrosamines, derived from nicotine and nicotine-like tobacco constituents, are particularly important mutagens found in cigarettes (Hoffman and Hecht, 1985). These have been intensively studied and are likely very important chemical carcinogens involved in the induction of lung cancer.

As the genes that code for these metabolic enzymes are cloned, large studies have been initiated to investigate whether functional polymorphisms in these genes are associated with cancer risk. Lung cancer has been widely studied, with investigators hypothesizing that variation in metabolism of tobacco-related carcinogens will be associated with variation in lung cancer occurrence in smokers. The results of initial molecular epidemiologic studies indicate that inherited variation in some of these enzymes contribute to a modest increase in risk of lung cancer (Devereux *et al.*, 1996). Thus, these studies are beginning to elucidate both the magnitude of the role of metabolism of tobacco-products in lung cancer and will also allow for the further understanding of individual susceptibility to lung cancer.

As reactive metabolic intermediates are generated and bind to DNA, the production of clinically relevant mutations necessarily involves processing the lesion. When the cell encounters DNA lesions, including covalently bound adducts generated by carcinogen metabolism, they are generally repaired in cells where mitosis must occur. In cells that do not divide, however, DNA-based lesions may go unrepaired. One example of a tissue where cell division does not occur is the neuron in the central nervous system (CNS). These cells are terminally differentiated and do not divide. Consequently, DNA-based lesions are not processed in the CNS. Because of this, neurons are relatively resistant to the effects of carcinogens, including DNA alkylating agents and radiation. In addition, cancers of the CNS are extraordinary rare. This is contrasted with tissues where active cellular growth is necessary for normal function. Epithelial tissue, such as that found throughout the respiratory tract, is actively replicating. Thus, lesions of the DNA are subject to enzymatic DNA repair that occurs with remarkable fidelity before replication. This process of DNA repair is highly complex but recently has become more amenable to mechanistic study at the molecular level.

Several human clinical conditions, such xeroderma pigmentosum, Bloom's syndrome and ataxia telanectasia are well-known examples of diseases that arise as a result of DNA repair defects (Gatti, 1995; German, 1995; Lambert *et al.*, 1995). The genes that are responsible for these and other similar diseases are rapidly being cloned and the disease-associated genetic alterations are being described (Ellis *et al.*, 1995; Savitsky *et al.*, 1995). This, along with animal and *in vitro* work, has aided in discovery of multiple classes of enzymes that catalyze the repair of specific types of DNA lesions. The genetics and biochemistry of enzymes such as glycosylases, helicases and exo- and endonucleases (Hanawalt, 1994; Bohr, 1995) are coming under intense scrutiny and will certainly result in rapid advances of this field in the near future. As these systems are described, it will provide additional information about the nature of mutagenic and carcinogenic lesions, including identification of those that are the most difficult to repair, the most associated with error-prone repair, and hence most carcinogenic. This will help to more completely elucidate the picture of chemical carcinogenesis.

Endogenous mutagens

Endogenous exposures have recently begun to receive more attention as an event contributing to the development of cancer. In particular, reactive oxygen species are believed to be important in mutagenesis. In test systems, reactive oxygen has been shown to act as a mutagen, directly damaging DNA (Reid and Loeb, 1993; Du *et al.*, 1994; Kato *et al.*, 1994). Oxygen metabolism results in the production of superoxide anion, a compound that is quickly detoxified to hydrogen peroxide by the enzyme SOD (super oxide dismutase). However, in the presence of certain metals, hydrogen peroxide may react to form hydroxy radicals (Halliwell, 1987; Kasprzak, 1995), considered to be the most mutagenic of the reactive oxygen species.

Endogenous exposure to active oxygen species may occur from outside or within a target cell. The inflammatory response results in a hostile cell environment, with reactive oxygen species being released into the cell microenvironment (Allen, 1979). Exposure to reactive oxygen may also occur within a cell through lysosome release and membrane lipid peroxidation (Marnett, 1987). Either type of exposure may result in oxygen radicals that are highly DNA reactive and may potentially induce cancerous mutations.

Another endogenous mutagen source may be the cell signaling molecule, nitric oxide (Ohshima and Bartsch, 1994). Nitric oxide, like oxygen, has free radical properties and may be present as several different chemical species. It is produced by activated phagocytes, and can form carcinogenic nitrosamines. Additionally, nitric oxide has been shown to be genotoxic in several different *in vitro* test systems (Wink *et al.*, 1991; Arroyo 1991, Nguyen *et al.*, 1992; Maragos *et al.*, 1993). However, the role of nitric oxide in human carcinogenesis has not been clearly demonstrated.

The carcinogenic potential of reactive oxygen and nitric oxide species lies in their ability to induce DNA strand breaks and DNA base deamination. This mechanism of mutation may be particularly important in lung cancer development given the inflammation that accompanies many exogenous exposures. A role for reactive oxygen species in occupational carcinogenesis is strongest for asbestos and metal exposure (Mossman and Marsh, 1989; Kasprzak, 1995).

Non-genotoxic carcinogens

Carcinogens are not always mutagens, as demonstrated in the mouse model of carcinogenesis. Phorbol esters are not DNA reactive, and have not been shown to induce mutations. Rather, they have been shown to interact with protein kinase C (PKC) to alter cell growth (Blumberg, 1988). Similarly, dioxin has been hypothesized to act as a carcinogen by inducing the aryl hydrocarbon (Ah) receptor (Nebert *et al.*, 1993). These agents are considered to be carcinogenic without being DNA reactive because of their ability to alter cell homeostasis, often through promotion of rapid, unscheduled clonal growth. Mutations are known to occur with regularity in normal cells (approximately one in one million cells will have a spontaneously occurring mutation at a given gene locus), due to endogenous mutagen exposure and failed DNA repair. Rapid cell growth increases the likelihood that a cell will accumulate mutations, particularly in the clonal expansion of cells carrying an initial deregulating mutation. In the mouse skin model some papillomas progress to carcinomas without additional exposure to exogenous mutagens, evidence that rapid cell growth contributes to carcinogenesis.

Non-genotoxic compounds, such as the phorbol esters, may act as carcinogens by promoting cell growth. Other exposures, such as nickel, may act by suppressing immune function (Haley *et al.*, 1990; Smialowicz *et al.*, 1984, 1985, 1987). The immune system is believed to be important in the recognition and removal of abnormal, precancerous cells. Exposures that suppress this immunologic function may be considered carcinogenic since they promote the continued growth of potentially cancerous cells.

LUNG CANCER GENES

Mutations contribute to cancer development when they occur in genetic sequence critical to maintaining controlled growth and cell differentiation. Some genes are believed to be important in many different cancers, such as *p53* and the Retinoblastoma gene (*Rb*), while others are critical for a particular tissue, such as *BRCA*1 and breast cancer. Historically, any gene found mutated within a tumor was termed an oncogene. However, we are now able to further classify cancer-related genes based on protein function and type of mutation required for cancer development.

ONCOGENES

Proto-oncogenes encode proteins involved in cell signaling, and are important in receiving and processing extracellular information for growth control. These proteins are positive regulators of cell proliferation and, when not mutated, integrate extracellular signals for cell growth; examples include growth factors, growth factor receptors, kinase signaling proteins and transcription factors. These genes require a single mutation, usually a small chromosome change in one allele, to contribute to the cancer cell phenotype. However, other mechanisms such as gene overamplification or altered gene transcription that result in overexpressed protein may contribute to cell autonomy. Once alterations occur in these genes they are referred to as oncogenes, and cells that harbor these changes may no longer properly process extracellular signals. Oncogene alterations are activating events, and the result is a continuous signal for the cell to keep dividing and an autonomous growth phenotype.

Ras

The *ras* gene was the first human oncogene to be isolated and studied in human cancers. Oncogenic *ras* was discovered based upon its ability to transform cultured fibroblast cells into hyperproliferative foci. Upon cloning, human *ras* was found to have homology to the transforming elements of murine sarcoma viruses (Parada *et al.*, 1982). There are three highly homologous human forms of *ras*. Their nomenclature is based on homology to the Harvey (H-*ras*) and Kirsten (K-*ras*) murine sarcoma viruses and the transforming gene found in neuroblastoma (N-*ras*). All three forms are expressed in human tissues and have been shown as targets for mutation in a variety of cancers.

The *ras* genes encode small (21 kDa) guanine-binding proteins with intrinsic GTPase activity. Ras proteins have been shown to play a critical role in several different signaling pathways of growth and differentiation (Marshall, 1995). The most well-understood function of *ras* is the integration of extracellular stimulation into a cascade of kinase signaling, resulting in the transcription of genes involved in cell growth (Lowery and Willumsen, 1993; Magnuson *et al.*, 1994;

Marshall, 1995). The signaling process begins with the ligand activation of growth receptors located at the cell surface. Adapter proteins, such as SOS, recognize and bind to the activated receptor. Ras proteins then interact with the src homology domains of the adaptor proteins, resulting in membrane localization of ras. Hence, ras serves as a docking protein to link extracellular events to further cell signaling and acts as a signaling switch. The kinase signaling pathway downstream of ras involves multiple genes termed raf, MAPKK, MAPK, and transcription factors including the AP-1 family.

There are a limited number of oncogenic mutations seen in the *ras* genes: they occur exclusively at codons 12, 13 and 61. The phenotypic change is always one of activation such that the protein remains predominantly in the active, GTP-bound state. There are very few reports of mutations at H- or N-*ras* in primary lung cancers: 90% of all observed *ras* mutations occur at K-*ras*, the remaining 10% in H- and N- *ras* (Kern and Filderman, 1993). K-*ras* mutation is not observed in small cell lung cancer (SCLC) but is estimated at 20–30% in non-small cell lung cancer (NSCLC) (Rodenhuis *et al.*, 1988; Slebos *et al.*, 1990; Suzuki *et al.*, 1990; Capella *et al.*, 1991; Mitsudomi *et al.*, 1991a,b; Rodenhuis and Slebos, 1992; Husgafvel-Pursiainen *et al.*, 1993; Minna, 1993; Bongiorno *et al.*, 1994; Isobe *et al.*, 1994; Li *et al.*, 1994; Silini *et al.*, 1994; Ridanpaa *et al.*, 1994). In lung cancer, K-*ras* mutation is strongly associated with adenocarcinoma histology and mutations occur almost exclusively in codon 12 (Bos, 1989; Kobayashi *et al.*, 1990; Slebos *et al.*, 1990; Mitsudomi *et al.*, 1991a,b; Rodenhuis and Slebos, 1992; Sugio *et al.*, 1992; Minna, 1993; Gazdar *et al.*, 1994; Mills *et al.*, 1995).

K-*ras* mutation in human NSCLC has been associated with smoking (Kobayashi *et al.*, 1990; Bongiorno 1991, Slebos *et al.*, 1990; Rodenhuis and Slebos, 1992; Sugio *et al.*, 1992; Husgafvel-Pursiainen *et al.*, 1993), and asbestos exposure (Husgafvel-Pusriainen *et al.*, 1993; Vainio *et al.*, 1993). Animal models have also demonstrated an association of PAH with the occurrence of K-*ras* mutation in mice (Mass *et al.*, 1993; You *et al.*, 1994; Nesnow *et al.*, 1995).

K-*ras* shows promise as an oncogene with clinical implications in the diagnosis of NSCLC (Table 31.1). It is the only oncogene shown to predict lung cancer occurrence when positively assayed in patient sputum (Mao *et al.*, 1994). This may prove to be a useful tool in screening and early detection of lung cancer (Table 31.2). Further, K-*ras* mutation is associated with decreased survival time, showing a stronger association with patient outcome than any other lung cancer-related gene studied to date (Slebos *et al.*, 1990, Mitsudomi *et al.*, 1991a,b; Rodenhuis and Slebos, 1992; Sugio *et al.*, 1992; Rosell *et al.*, 1993; Kern *et al.*, 1994; Silini *et al.*, 1994; Kashii *et al.*, 1995).

Table 31.1 Summary of K-*ras* codon 12 mutations in non-small cell lung cancer (NSCLC)

	All histologies	*Adenocarcinoma*	*Reference*
	17/85 (20%)	–	Capella *et al.* (1991)
	14/48 (29%)	12/21 (57%)	Husgafvel-Pursiainen *et al.* (1993)
		8/41 (20%)	Li *et al.* (1994)
		6/63 (9.5%)	Lung *et al.* (1992)
	8/15 (53%)	7/14 (50%)	Mao *et al.* (1994)
		13/28 (46%)	Mills *et al.* (1995)
		10/54 (18.5%)	Slebos *et al.* (1991)
		15/115 (13%)	Sugio *et al.* (1992)
		18/57 (31.5%)	Westra *et al.* (1993)
	45/280 (16%)	41/141 (29%)	Rodenhuis and Slebos (1992)
	9/66 (14%)	2/22 (9%)	Rosell *et al.* (1993)
		2/21 (9.5%)	Lima *et al.* (1994)
		16/42 (38%)	Kern *et al.* (1994)
		6/30 (20%)	Isobe *et al.* (1994)
		7/29 (24%)	Bongiorno *et al.* (1994)
		33/109 (30%)	Silini *et al.* (1994)
	19/99 (19%)	7/36 (19%)	Ridanpaa *et al.* (1994)
	4/57 (7%)	4/31 (13%)	Kawajiri *et al.* (1996)
		10/67 (15%)	Kobayashi *et al.* (1990)
	14/141 (10%)	7/19 (37%)	Vachtenheim *et al.* (1995)
Total	130/791 (16%)	224/940 (24%)	

Table 31.2 K-*ras* mutations and lung cancer recurrence

Reference	*Total Patients*	*Percent Adeno-carcinoma*	*Enrolment Criteria*	*Prevalence codon 12 K-ras mutation in adenocarcinoma*	*Indep. association with recurrence*	*Comments*
Kashii *et al* (1995)	97	72%	<45 or >50 years old	10% (7)	+	limited power
Isobe *et al* (1994)	30	100%	30 of 31 consec. stag I surg, pts.	23% (7)	–	limited power
Silini *et al* (1994)	109	100%	Surg. pts. 1984–91	70% (76)	+	
Rosell *et al* (1993)	66	33%	Surg. pts. 1988–90	14% (3)	+	limited power
Sugio *et al* (1994)	115	100%	Surg. pts. 1975–91 curative resection on 89	16% (18)	+	29 fresh DNA, 86 paraffin DNA
Rodenhuis *et al* (1988)	280	50%	Surg. series	30% (41)	+	38 newly added from prev. report
Harada *et al* (1992)	116	44%	Surg. series	72% (39)	+	Antibody IHC only
Mitsudomi *et al* (1991a)	66	61%	estab. cell lines	29% (19)	+	cell line analysis only
Kern *et al* (1990)	44	100%	44 of 46 surg. pts.	36% (16)	+	paraffin-derived DNA
Slebos *et al* (1990)	69	100%	Surg. series	28% (19)	+	paraffin-derived DNA
Nishio *et al* (1992)	147	56%	Surg. series	80% (66)	+	Antibody IHC only

Myc

There are three forms of the human *myc* gene, c-, N- and L-*myc*. The proteins encoded by *myc* genes are transcription factors that are activated by growth stimuli. The *myc* oncogenes, unlike the *ras* genes, are overexpressed in small cell lung cancers either by gene amplification or altered transcription regulation (Little *et al.*, 1983; Wong *et al.*, 1986; Krystal *et al.*, 1988; Takahashi *et al.*, 1989; Saksela, 1990; Prins *et al.*, 1993). Approximately one-fourth of primary SCLC tumors have DNA overamplification of one of the *myc* genes (Johnson *et al.*, 1992; Makela *et al.*, 1992). Protein overexpression is observed in 80–90% of these tumors, whereas no oncogenic *myc* is seen in NSCLC (Slebos *et al.*, 1986). *Myc* overexpression, like K-*ras* mutation, is associated with poorer lung cancer survival (Johnson 1979, Brennan *et al.*, 1991; Johnson *et al.*, 1992).

Growth factor receptors

Two growth factor receptors, *c-erbB*-1 and *c-erbB*-2, are implicated in lung tumorigenesis. *c-erbB*-1 is the gene that encodes for the epidermal growth factor receptor (EGFR), an important protein for receiving signals for epithelial cell proliferation. The EGFR is a tyrosine kinase receptor that dimerizes upon ligand binding, initiating a cascade of phosphorylation signaling that results in the transcriptional activation of genes involved in cell growth. The oncogenic form of *c-erbB*-1 is overexpressed in NSCLC, either by amplification or altered transcription, but is not overexpressed in SCLC (Damstrup *et al.*, 1992). *c-erbB*-1 overexpression is observed in 15–40% of adenocarcinoma, 25–40% of large cell carcinoma, and 65–90% of squamous cell carcinoma (Cerny *et al.*, 1986). Overexpression of the EGFR has been correlated with poor prognosis in patients with NSCLC (Veale *et al.*, 1993).

c-erbB-2 (also called HER2 or neu) is a tyrosine kinase receptor homologous to the EGFR. This receptor is believed to be important in epithelial cell growth and differentiation, yet its ligand remains unknown. As with *c-erbB*-1 and *myc*, the oncogenic phenotype is conferred by protein overexpression. Approximately 30% of tumours examined abnormally express this protein (Schneider *et al.*, 1989; Kern *et al.*, 1990; Weiner *et al.*, 1990; Tateishi *et al.*, 1991; Shi *et al.*,

1992). Overexpression has also been associated with shortened survival, most notably in adenocarcinoma (Kern *et al.*, 1990; Tateishi 1990).

TUMOR SUPPRESSOR GENES

Tumor suppressor genes, sometimes referred to as anti-oncogenes, comprise the second class of cancer-related genes. Proteins encoded by tumor suppressor genes are typically involved in monitoring intracellular signals for cell replication and division. Conceptually they are considered negative regulators of cell growth, and are responsible for pausing DNA synthesis and cell division when the cell is not stimulated to grow or when DNA damage is present. Mutations in these genes, like proto-oncogenes, result in an uncontrolled cell growth phenotype. Tumor suppressor genes are distinct from proto-oncogenes in that they require inactivating mutations that involve both alleles (Knudson, 1971). This may occur by two separate inactivating mutations, loss of one allele and mutation of the other, or a dominant mutation that results in a gene product that functionally inactivates the non-mutated gene product.

P53

The human *p53* gene, located on the short arm of chromosome 17, encodes a nuclear phosphoprotein that is involved in the signal transduction pathways that regulate DNA synthesis and repair, genomic plasticity and programmed cell death (Hollstein *et al.*, 1991; Levine *et al.*, 1991; Montenarch, 1992; Vogelstein and Kinzler, 1992; Harris, 1993; Harris and Hollstein, 1993). Mutations in this gene are probably the most common genetic alterations known in human tumors (Harris and Hollstein, 1993). The *p53* gene has 11 exons that code for a 393 amino acid protein. A large number of sites in this gene are targets for mutation (Vogelstein and Kinzler, 1992), although most occur in the highly conserved region of exons 5–9 (Greenblatt *et al.*, 1994). Conclusions regarding the potential causative factors in human carcinogenesis have been reached by analyzing differences in mutational spectra within the *p53* gene (Sommer, 1990; Hollstein *et al.*, 1991; Harris, 1993).

Greenblatt *et al.* (1994) have recently reviewed the literature and presented wide-ranging evidence suggesting that differences in the patterns of mutations in the *p53* gene likely reflect different etiologies for a variety of cancers. Analysis of gene mutations has been done comparing characteristics of patients with and without mutations in a number of different types of tumors. In addition, the types of mutations (transition, transversion, deletion), location and sequence have also been examined for relationship with parameters of interest for a variety of cancers. An additional example of the use of *p53* mutation data in an epidemiologic context is illustrated by examining the large body of *in vitro* data (in bacteria, yeast and mammalian cells) showing that mutations induced by exogenous carcinogens exhibit marked strand bias, occurring preferentially from damage to the non-transcribed DNA strand (Bohr *et al.*, 1987; Ruggeri *et al.*, 1993). This is thought to occur because organisms preferentially repair the transcribed DNA strand in an effort to preserve the essential functions of short-lived proteins. The majority of recent epidemiologic data addressing this question, particularly in studies of *p53* mutation, have been consistent with the *in vitro* observations (reviewed in Greenblatt *et al.*, 1994). Hence, exposure-induced mutations in the *p53* gene arise from damage to the non-transcribed DNA strand.

NSCLC recurrence and altered p53

A significant number of investigators have reported that p53 alterations in NSC lung tumors predict cancer recurrence (Table 31.3). Quinlan *et al.* (1992) studied 114 cases of stage I and II NSCLC using IHC (anti-p53 monoclonal antibody Pab-1801) and reported that recurrence was significantly associated with p53 accumulation. Horio *et al.* (1993) found that *p53* mutations in NSCLC detected by SSCP analysis were independently associated with shortened survival in 71 patients with stage I–IIIA disease, including a negative impact on recurrence in very early stage disease. Mitsudomi *et al.* (1993) used SSCP to study mutations in exons 5–8 of 120 NSCLC patients and reported that *p53* mutations in tumors were significantly associated with disease recurrence, although this was not true for early-stage disease. Marchetti *et al.* (1993) used Pab-1801 and Pab-240 to stain for persistent p53 protein in 53 NSCLCs. In this dataset, positive staining was associated with metastatic recurrence. In a small study of 36 NSCLC patients, Lee *et al.* (1994) reported that mutations detected in the *p53* gene (particularly in exon 8) using reverse transcription and PCR amplification with subsequent sequencing, were associated with nodal metastasis. Isobe *et al.* (1994) also reported results from a study of 30 NSCLC patients. This study is one of the few where both sequencing and IHC were used to detect *p53* abnormalities. They found both sequence-positive mutations as well as IHC to be associated with recurrence. Based upon these limited data, the authors suggested that IHC-positive status was more highly correlated with recurrence. Carbone *et al.* (1994) similarly used both IHC and PCR/SSCP mutation detection to study this question in 75 NSC lung cancer patients undergoing curative resection of

Table 31.3 Mutations in *p53* and lung cancer recurrence

Reference	*Total no. of patients*	*Method*	*Distribution of cell type*	*Enrolment criteria*	*Prevalence of p53 abnormalities (%)*	*Indep. Assoc. w recur.*	*Comments*
Quinlan *et al.* (1992)	114	IHC (PAb1801)	48 Ad, 66Sq	Surg pts 1974–88	43	+	No control for confounders
Horio *et al.* (1993)	71	PCR/SSCP exon 5–8	34Ad, 33Sq	Surg pts 1982–91	49	+	Control for tumor size, nodes, hist, stage
Mitsudomi *et al.* (1993)	120	PCR/SSCP exon 5–8	47Ad, 65Sq, 8Lg	Surg pts 1986–89	43	+	Control for stage, gender
Marchetti *et al.* (1993)	53	IHC (PAb1801) (Pab240) PCR/SSCP ex 4–9	15Ad, 27Sq, 11 other	Consec. surg pts 1991	58 IHC 47 PCR/SSCP	(+)	Associated with nodal inv
Lee *et al.* (1994)	36	Rt-PCR sequencing	24Ad, 11Sq, 1 other	Surg pts	42	(+)	Assoc with nodal inv
Isobe *et al.* (1994)	30	PCR/DGGE ex5–9 IHC (DO-7)	30Ad	30/31 consec Ad pts	37(IHC) 37(PCR)	+	Control for smoking, tumor size, *ras*, gender
Carbone *et al.* (1994)	85	PCR/SSCP ex 4–9 IHC (BP53–12)	43 Ad, 34 Sq, 8 other	Surg pts 1987–88	64(IHC) 53(PCR)	-PCR (+)IHC	Control for age, sex, hist, stage, *ras*. Marginal + IHC
Ebina *et al.* (1994)	123	IHC(DO-7)	75 Ad, 13 Sq	Prospec clin trial	39	+	Recur. after curative resection only
Harpole *et al.* (1995)	271	IHC (PAb1801)	100Ad, 95 Sq, 23 Lg, 37 other	Consec surg pts 1980–88	38	+	Control for sympt, stage, sex, tumor size, *erbB*-2 age, vasc inv
Volm *et al.* (1992)	81	IHC (PAb1801)	81 Sq	Surg pts	43	–	No multivariate analysis
Morkve *et al.* (1993)	112	IHC(PAb1801)	49 Ad, 50 Sq, 4 Lg, 9 other	Consec surg pts w paraffin	77	–	Trend toward + assoc w recur.
Casson *et al.* (1994)	19	IHC(DO-7) PCR/SSCP ex5–9	5 Ad, 14 Sq	Surg pts 1988	26(IHC), 26(PCR)	(+)	3/5 + had recur.
Passlick *et al.* (1995)	73	IHC(PAb1801)	35Ad, 30 Sq 6 Lg, 2 other	Surg pts	45	–	*p53* + assoc w better prognosis
Volm *et al.* (1994)	209	IHC (PAb1801)	NR	Surg pts	51	–	*p53* + assoc w better prognosis
Kashii *et al.* (1995)	97	PCR/SSCP exons 5–9	70 Ad, 12 Sq, 7Lg, 8 SC	Surg pts <45 or >50yo, '80–92	34	–	No multivar. analysis
Top *et al.* (1995)	54	IHC(DO-7) (BP53–12) PCR/DGGEex 5–8	32 Ad, 11 Sq, 9 Lg, 2 other	Surg pts	69 (IHC), 69(PCR)	–	*p53*+assoc w better prognosis excl ras
Fontanini *et al.* (1993)	101	IHC(PAb1801)	64 Sq, 37 Non-Sq	Surg pt 1991–92	68	(+)	Assoc w nodal inv& mets, no assoc w recur
Hiyoshi *et al.*	88	IHC(PAb1801)	46 Ad, 26 Sq, 5 LG, 11 other	Surg pts	49	(+)	Assoc w nodal inv stage, mets

Ad, adenocarcinoma; Sq, squamous

their tumor. In this study IHC detection of *p53* abnormalities was associated with tumor recurrence, while no association with recurrence was observed when using only PCR/SSCP detection of p53 alterations. Ebina and colleagues (1994) also studied this in 123 NSCLC, including 71 primary and 52 metastatic tumors, where 63 of the patients were candidates for surgical cure. In this smaller group, detection of *p53* alteration by IHC (using monoclonal mouse anti-human antibody DO-7) was associated with recurrence. Finally, in a recently published large series Harpole *et al.* (1995) reported on *p53* and recurrence in a series of 271 consecutive patients with stage I NSCLC that were surgically treated with curative intent. Using IHC (Pab-1801), detection of p53 protein was marginally associated with recurrence, controlling for symptoms, tumor size, differentiation and the presence of vascular invasion.

At the same time, a number of investigators have reported *p53* abnormalities to be either not associated with recurrence or associated with disease-free survival. Volm *et al.* (1992) used IHC (Pab-1801) in studying 81 surgically treated patients with squamous cell lung cancer and found no association of *p53* alteration with recurrence. Morkve *et al.* (1993) studied 112 patients with NSCLC and found that patients with either no or extremely high levels of *p53* staining had a better prognosis than the other groups. In a small study of 19 patients, both IHC detection of *p53* alterations and PCR/SSCP methods did not correlate with recurrence (Casson *et al.*, 1994). Passlick *et al.* (1995) reported on 73 NSCLC patients and saw a favorable outcome in the IHC (Pab-1801) positive patients. Volm *et al.* (1994), in an enlarged series of 209 patients, also reported IHC positivity (Pab-1801) to be associated with no recurrence of disease. Kashii *et al.* (1995) studied 97 patients and found PCR/SSCP determined mutation in *p53* to not be associated with recurrence. Finally, Top *et al.* (1994) and Fontanini *et al.* (1993) reported no association between *p53* alteration and recurrence. Top *et al.* (1994) used both DGGE and PCR sequencing as well as IHC (PabDO7 and BP53–12) detection of *p53* alteration in 54 primary NSCLC patients and Fontanini *et al.* (1993) used IHC (Pab-1801) detection alone in 101 cases of NSCLC.

Hence, a clear understanding of the implications of *p53* alterations in surgically resectable NSCLC remains elusive. The literature, although by now quite large, is difficult to synthesize due to the many complexities involved in comparing studies. Interpretation of these reports is limited by a number of factors, including the methods used to determine if *p53* is altered [IHC (including different antibodies for staining) vs. PCR/SSCP and PCR/DGGE-based sequencing], the selection criteria for inclusion of patients, size of the study, and control for potential confounders (e.g. stage and tumor size as well as smoking, age, gender, occupation, presence of symptoms, etc.) and follow-up time period.

Mutation spectra

A combination of numerous oncogene and tumor suppressor gene alterations are required for the development of clinically evident lung cancer. However, there currently is no clear consensus as to whether specific genes must be mutated in a defined order to develop disease. Analysis of mutational patterns of specific genes may prove useful in defining disease pathways and etiologies. Despite the apparent heterogeneity in observed mutation patterns in lung tumours (Mulshine *et al.*, 1993; Minna, 1993; Gazdar *et al.*, 1994 Husgafvel-Pursiainen *et al.*, 1995), some general patterns have been observed with regard to the patterns of *p53* and K-*ras* mutations by histology. Additionally, some conclusions regarding the potential causative factors in human carcinogenesis have been reached by analyzing differences in mutation spectra withing the *p53* gene of various cancers, including lung cancer (Harris *et al.*, 1969; Sommer, 1990; Hollstein *et al.*, 1991; Somers *et al.*, 1992). This is particularly useful in workplace-related disease, where exogenous exposures are most well defined.

Small genetic mutations within gene coding regions fall into three basic categories: insertions, deletions, and base substitutions. This third type of mutation, along with large deletions, are the most common genetic events in primary tumor analysis. Base substitutions may be further categorized as transitions (purine to purine or pyrimidine to pyrimidine) or transversions (purine to pyrimidine and vice versa). In analysis of lung tumors there is is a propensity of mutations at guanine residues, particularly for the G:C to T:A transversion (Greenblatt *et al.*, 1994).

The over-representation of G to T transversions in lung cancer may be explained using several different mechanistic models. These models rely on the 'A rule', whereby default DNA polymerase inserts adenine when the correct base pair is not recognized (Larson and Strauss, 1987). There are two types of lesions that best explain the G to T transversions seen in lung cancer. Polyaromatic hydrocarbon metabolites have been shown *in vivo* to react with the N^2 position of guanine. These types of lesions are bulky and helix-distorting; if they remain adducted at DNA replication the result is insertion of adenine and a G to T transversion. Other carcinogenic exposures are believed to result in DNA oxidation, either by components of the carcinogen or and inflammatory response to exposure. One well-studied oxidative lesion is 8-oxo-guanine. This lesion is not helix-distorting, but is rather unstable and *in vitro* has been shown to result in apurination. DNA polymerase inserts adenine opposite apurinic sites leading to the observed G to T transversion.

Correlations of mutational spectra with exposures in other tumor types have also led to speculation concerning etiologic agents involved in the genesis of specific lesions in *p53*. The most dramatic examples include tandem dipyrimidine mutations in skin cancer (Brash *et al.*, 1991), G:C to T:A transversions in the third base of codon 249 in aflatoxin B-exposed patients with hepatocellular carcinoma (Bressac *et al.*, 1991; Hsu *et al.*, 1991), and G:C to T:A transversions in the second base of codon 249 in radon-exposed uranium miners with primarily squamous cell lung cancer (Taylor *et al.*, 1994). Specific mutational spectra have also been found in many tumor types including head and neck tumors (related to cigarette smoking and alcohol ingestion), bladder cancer (related to smoking), and colon cancers (transitions at CpG sites presumably associated with endogenous mutagens) (Greenblatt *et al.*, 1994). The colon cancer model, which is particularly well developed, has been interpreted as reflective of endogenous processes as a result of this over-representation of transition events at sites (CpG dinucleotides) where cytosine can be methylated and subject to deamination, resulting in misrepair of a G:T mismatch, leading to transition mutations (Tornaletti and Pfeifer, 1995).

Another interesting observation is that tumors that occur in the absence of known carcinogen exposure (e.g. lung cancer in non-smoking, non-asbestos-exposed patients) most often are found not to have *p53* mutations (Kondo *et al.*, 1992; Suzuki *et al.*, 1992; Takeshima *et al.*, 1993; Ryberg *et al.*, 1994). Hence, endogenous DNA alterations that give rise to cancers may occur through pathways that are different either in the sequence of gene alterations or in the specific pattern of genes altered. The pattern of mutation in genes will likely differ based upon carcinogen, DNA target size for gene alteration, and other factors such as age, gender, family history as a surrogate for unmeasured factors involved in susceptibility to disease, as well as other characteristics that are risk factors for the disease of interest.

It has been estimated that approximately 70% of SCLC have *p53* mutations (Greenblatt *et al.*, 1994). Mutations in NSCLC are less common, with adenocarcinoma having the lowest prevalence at approximately 30% (Greenblatt *et al.*, 1994). Review of the mutational spectra in lung cancer has shown that G:C to T:A transversion mutations are unusually common (with the mutation evident most often on the non-transcribed strand) and that mutations at CpG sites occur with a prevalence lower than most other cancers (reviewed in Greenblatt *et al.*, 1994). The G:C to T:A transversions have been associated with polyaromatic hydrocarbon exposure (Ruggeri *et al.*, 1993) and are presumably the result of deoxyguanine adducts occurring after enzymatic activation of cigarette smoke-associated carcinogens. Multiple studies have demonstrated that the occurrence of G:C to T:A transversion mutations is associated with lifetime cigarette consumption.

Other work has suggested that the mutational spectra of *p53* mutations differs by histology. Where smoking is most strongly associated with this disease (small cell and squamous cell) there is a high prevalence of *p53* alterations, with G:C toT:A transversions being common. In comparing spectra, Greenblatt *et al.* (1994) suggested that SCLC has an excess of A:T to G:C lesions with a corresponding increase in insertions and deletions in NSCLC. In adenocarcinoma, a higher proportion of G:C to A:T lesions at non-CpG sites has been seen and this has been interpreted as evidence that other carcinogens are etiologically important in this disease. Guinee *et al.* (1995) have suggested that an excess of transversion mutations in *p53* may also occur more often in women and, further, that this be associated with the changing nature of the lung cancer epidemic.

As high-quality epidemiolgic data have become available for use in studies of *p53* spectra in lung cancer, other very important observations have emerged. One important example is the observation that asbestos exposure is associated with *p53* mutations and, further, that the predominant lesion in these cases is the G:C to T:A transversion (Wang *et al.*, 1995). This observation may have important implications for the mechanism of action of asbestos in lung cancer. These data suggest that asbestos may act by effectively increasing the transport of polyaromatic carcinogens to the lung epithelium, as has been suggested by several investigators using *in vitro* and *in vivo* models (Lakowicz *et al.*, 1980; Brown *et al.*, 1983; Menard *et al.*, 1986). It is also possible that oxidative mechanisms are involved, as it has been demonstrated that base substitutions are produced by oxygen free radicals such as those associated with asbestos exposure (Mossman and Marsh, 1989). Hence, as databases are enlarged, more detailed comparisons of *p53* status and carcinogen exposure, diet, gender, family history, etc. will be possible. These analyses will no doubt delineate important associations that reveal the distinctive patterns of genes involved in the genesis of lung cancer.

The retinoblastoma gene

Rb, the gene associated with development of retinoblastoma, was the first tumor suppressor gene described (Cavenee *et al.*, 1983). *Rb* encodes a 105-kDa nuclear phosphoprotein critical to negative regulation of cell cycle progression and DNA synthesis. Rb protein is regulated in a cell-cycle dependent manner by cyclin-dependent kinases. When hypophosphorylated, Rb binds to the E2F transcription factors, inhibiting their ability to stimulate gene transcription. Before DNA synthesis the Rb protein is hyperphosphorylated, releasing E2F, resulting in transcription of genes involved in DNA replication. Mutated, or dysregulated Rb is impaired in its ability to block entry into the DNA synthesis phase of cell growth.

Nearly 100% of SCLC and 10–20% of NSCLC demonstrate altered *Rb* expression (Harbour *et al.*, 1988; Yokota *et al.*, 1988; Horowitz *et al.*, 1989; Hensel *et al.*, 1990; Reissmann *et al.*, 1993). Very few primary NSCLC tumors have *Rb* gene mutations; however, 13% of SCLC tumors have detectable mutations (Harbour *et al.*, 1988).

Chromosome 3P

In the early 1980s a large number of SCLC cell lines were analyzed karyotypically with the unexpected result that a great majority of them had deletions in the short arm of chromosome 3 (Whang-Peng *et al.*, 1982; Yunis, 1983). There are three to four regions on chromosome 3p that are believed to harbor tumor suppressor genes important in the development of lung cancer (Brauch 1990; Hibi *et al.*, 1992). These include 3p25–26, 3p21 and 3p14.2.

Chromosome 3p is lost or displays loss of heterozygosity (LOH) in a high percentage of lung tumors [60% in NSCLC and nearly 100% in SCLC (Thiberville *et al.*, 1995)], as do each of the chromosome

Table 31.4 Chromosome 3p loss of heterozygosity (LOH) in non-small cell lung cancer (NSCLC)

Pathology	*No. of patients*	*No. of markers tested*	*LOH (%)*	*Reference*
Resected tumors				
	7		85.7 (6/7)	Hung *et al.* (1995)
	71	5	48.6 (34/70)	Horio *et al.* (1993)
	48	13	79.5 (31/39)	Hibi *et al.* (1992)
	38	12	63.2 (24/38)	Shiseki *et al.* (1994)
	37	1	54.1 (20/37)	Tsuchiya *et al.* (1992)
	30		66.7 (20/30)	Testa and Siegfried (1992)
	48	7	72.7 (32/44)	Rabbitts *et al.* (1989)
	18	2	25.0 (2/8)	Becker *et al.* 1988
	36	2	62.5 (5/8)	Yokota *et al.* (1987)
	54	3	51.6 (16/31)	Weston *et al.* (1989)
Squamous cell carcinoma				
	6	4	100 (6/6)	Sundaresan *et al.* (1992)
	6	12	83.3 (5/6)	Shiseki *et al.* (1994)
	7	1	100 (7/7)	Tsuchiya *et al.* (1992)
	29	7	79.3 (23/29)	Rabbitts *et al.* (1989)
	20	13	70 (14/20)	Hibi *et al.* (1992)
Adenocarcinoma				
	28	12	60.1 (17/28)	Shiseki *et al.* (1994)
	26	1	35 (9/26)	Tsuchiya *et al.* (1992)
	10	7	60 (6/10)	Rabbitts *et al.* (1989)
	18	2	83.3 (5/6)	Yokota *et al.* (1987)
	15	13	87 (13/15)	Hibi *et al.* (1992)
Large cell carcinoma				
	4	12	50.0 (2/4)	Shiseki *et al.* (1994)
	2	7	100 (2/2)	Rabbitts *et al.* (1989)
	4	13	100 (4/4)	Hibi *et al.* (1992)
Hyperplasia	17*		76.5 (13/17)	Hung *et al.* (1995)*
Dysplasia	7*		42.9 (3/7)	Hung *et al.* (1995)*
	9	4	100 (9/9)	Sundaresan *et al.* (1992)
Non-invasive tumor	4		100 (4/4)	Hung *et al.* (1995)
Stage I tumors	18	12	44.4 (8/18)	Shiseki *et al.* (1994)
Brain metastases	22	12	81.2 (18/22)	Shiseki *et al.* (1994)

* No. of lesions

regions described above, a characteristic of tumor suppressor gene loci (Table 31.4). Additionally, analysis of preneoplastic lesions indicates that 3p alteration may be an early event in lung cancer development (Sundaresan *et al.*, 1992; Shiseki *et al.*, 1994; Hung *et al.*, 1995). Of prognostic significance, LOH at 3p has been weakly associated with decreased lung cancer survival time (Horio *et al.*, 1993).

One lung cancer gene candidate is the von-Hippel Lindau gene located at chromosome 3p25–25. However, the absence of mutations in this gene indicates it probably is not a critical gene for lung cancer development and that a separate gene is involved (Sekido *et al.*, 1994). The 3p21 region appears to be particularly important in lung cancer, as the majority of tumors of all histologies show LOH at this locus (Brauch *et al.*, 1987; Kok *et al.*, 1987; Naylor *et al.*, 1987). Chromosome deletion or LOH in this region approaches 100% in SCLC. To date, no candidate gene has been cloned at this site.

Yunis (1983), in an examination of chromosome fragile sites found that the most common fragile site in human chromosomes is 3p14.2. Interestingly, Yunis (1988) also showed substantial evidence of an effect of carcinogens on the expression of this fragile site. These experiments involved exposure of cultured cells to chemical carcinogens; BPDE (benezo[a]pyrene diol epoxide), benzene and nitrosamines all induced the expression of the fragile site at 3p14.2. This fragile site (FRA3B) is classified as a common fragile site, and has been linked to several cancers, particularly renal cell carcinoma and lung cancer (Murata *et al.*, 1992; Druck *et al.*, 1995). The current consensus is that FRA3B represents a region of fragility, and not a single site. Further, there is no evidence that the FRA3B region contains trinucleotide repeats similar to those seen with fragile X. These observations have made the study of this fragile site complicated and the cloning of cancer-related genes in this region difficult.

CONTROVERSIES IN OCCUPATIONAL CARCINOGENESIS – THE CASE OF ASBESTOS

It is instructive to examine a current controversy in carcinogenesis to illustrate not only our progress, but also to reveal the limitations of the data with respect to truly understanding the complex mechanisms of carcinogenesis. There have been many attempts to understand the mechanisms of asbestos carcinogenesis over the years. Current consensus on a likely mechanism revolves around the ability of asbestos to induce oxidative damage (Mossman and Marsh, 1989). In this scheme, oxidative damage is responsible for the production of DNA damage, resulting in cell death, fibrosis and also the induction of neoplasia. However, some have advanced the notion that asbestos can only contribute to the induction of lung cancer when interstitial lung disease is also present (Hughes and Weill, 1991).

In this proposed scheme, asbestos acts to 'promote' lung cancer; being much more potent in smokers, but being operative only 'when and if the exposure to the mineral is of sufficient duration and intensity to cause asbestosis' (Craighead, 1995).

The problem that is inherent in this hypothesis is readily apparent. Both fibrosis of the lung and cancer of the lung are dose-related occurrences. In addition, smoking and asbestos exposure clearly interact in a synergistic fashion to produce lung cancer (Selikoff *et al.*, 1968), and this increase in risk must necessarily also depend upon the exposure of an individual to each compound. Consequently, in epidemiologic study the overwhelming majority of cancers will occur in those people who have the highest exposure. These people, of course, will be most likely to have asbestosis, regardless of whether the process that produces lung cancer has anything to do with fibrosis.

Only if the biologic process that gives rise to fibrosis itself also directly induces genetic changes important for the production of lung cancer (or creates conditions that enhance the likelihood of these mutations in relevant cells) can it be *necessary* for interstitial lung disease to be present in order for asbestos to cause lung cancer. While it is certainly possible that the process of inducing fibrosis elaborates growth factors that stimulate clonal expansion of previously mutant cells, little direct evidence that this occurs has been presented to date. Thus, it can be said that, to date, there is no direct evidence that there is any necessity for asbestosis to be present for a lung cancer to be caused by asbestosis. Arguments will persist that are based upon the epidemiology and politics of the disease, but the defining end to this argument will be the more profound mechanistic understanding of the induction of fibrosis and the induction of malignant disease.

Acknowledgements

These studies were supported by ES-00002, ES-06717, ES/CA-06409 and ES-08357.

REFERENCES

Allen, R.C. (1979) Reduced, radical, and excited state oxygen in leukocyte microbicidal activity, in *Lysosomes in Applied Biology and Therapeutics* (eds J.T. Dingle, P.L.J. Jacques, and I.I.H. Shaw), North-Holland, Amsterdam, p.197.

Arroyo, P.L. and Hatch-Pigott, H.F. (1992) Mutagenicity of nitric oxide and its inhibition by antioxidants. *Mutat. Res.*, **281**, 193–202.

Berenblum, I. (1941) The cocarcinogenic action of croton resin. *Cancer Res.*, **1**, 44–50.

Berenblum, I. and Shubik, P. (1947) The role of croton oil applications, associated with a single painting of a carcinogen, in tumor induction of the mouse skin. *Br. J. Cancer*, **1**, 379–82.

Bishop, J. (1991) Molecular themes in oncogenesis. *Cell*, **64**, 235–48.

Bizub, D., Wood, A.W. and Skalka A.M. (1986) Mutagenesis of Ha-ras oncogene in mouse skin tumors induced by polycyclic aromatic hydrocarbons. *Proc. Natl. Acad. Sci. USA*, **83**, 6048–52.

Blumberg, P.M. (1988) Protein kinase C as the receptor for the phorbol ester tumor promoters: sixth Rhoads memorial award lecture. *Cancer Res.*, **48**, 1.

Bohr, V.A. (1995) DNA repair fine structure and its relations to genomic instability. *Carcinogenesis*, **16**, 2885–92.

Bohr, V.A., Phillips, D.H. and Hanawalt, P.C. (1987) Heterogeneous DNA damage and repair in the mammalian genome. *Cancer Res.*, **47**, 6426–36.

Bongiorno, P.F., Whyte, R.I., Lesser, E.J., Moore, J.H., Orringer, M.B. and Beer, D.G. (1994) Alterations of K-ras, p53, and erbB-2/neu in human lung adenocarcinomas. *J. Thorac. Cardiovasc. Surg.*, **107**, 590–5.

Bos, J.L. (1989) Ras oncogenes in human cancer: a review. *Cancer Res.*, **49**, 4682–9.

Boutwell, R.K. (1964) Some biological aspects of skin carcinogenesis. *Prog. Exp. Tumor Res.*, **4**, 207–50.

Brash, D.E., Rudolph, J.A. and Simon, J.A. (1991) A role for sunlight in skin cancer: UV-induced p53 mutations in squamous cell carcinoma. *Proc. Natl Acad. Sci. USA*, **88**, 10124–8.

Brauch, H., Johnson, B. Hovis, J. *et al.* (1987) Molecular analysis of the short arm of chromosome 3 in small-cell and non-small-cell carcinoma of the lung. *N. Engl. J. Med.*, **317**, 1109–13.

Brennan, J., O'Connor, T. and Makuch, R.W. (1991) myc family DNA amplification in 107 tumor and tumor cell lines from patients with small cell lung cancer treated with different combination chemotherapy regimens. *Cancer Res.*, **51**, 1708–12.

Bressac, B., Kew M. and Wands, J. (1991) Selective G to T mutations of p53 gene in hepatocellular carcinoma from southern Africa. *Nature*, **350**, 429–31.

Brown, R.C., Gleming, G.T. and Knight, A.I. (1983) Asbestos affects in the vitro uptake and detoxification of aromatic compounds. *Environ. Health Perspect.*, **51**, 315–18.

Capella, G., Cronauer-Mitra, S., Peinado, M.A. and Perucho, M. (1991) Frequency and spectrum of mutations at codons 12 and 13 of the C-K-ras gene in human tumors. *Environ. Health Perspect.*, **93**, 125–31.

Carbone, D.P., Mitsudomi, T. Chiba, I. *et al.* (1994) p53 immunostaining positivity is associated with reduced survival and is imperfectly correlated with gene mutations in resected non-small cell lung cancer. *Chest*, **106**, 337S–381S.

Casson, A.G., McCuaig, S., Craig, I., Ayed, A., Inculet, R., Kerkvliet, N. and O'Malley, F. (1994) Prognostic value and clinicopathologic correlation of p53 gene mutations and nuclear DNA content in human lung cancer: a prospective study. *J. Surg. Oncol.*, **56**, 13–20.

Cavenee, W.K., Dryja, T.P., Phillips, R.A., *et al.* (1983) *Nature*, **305**, 779–84.

Cerny, T., Barnes, D.M. and Hasleton, P. (1986) Expression of epidermal growth factor receptor (EGF-R) in human lung tumors. *Br. J. Cancer*, **54**, 265–9.

Craighead, J.E. (1995) Carcinogenesis, in *Pathology of Environmental and Occupational* (ed. J.E. Craighead) Mosby, pp. 411–36.

Damstrup, L., Rygaard, K. and Spang-Thomsen, M. (1992) Expression of the epidermal growth factor receptor in human small cell lung cancer cell lines. *Cancer Res.*, **52**, 3089–93.

Devereux, T.R., Taylor, J.A. and Barrett, J.C. (1996) Molecular mechansism of lung cancer. Interaction of environmental and genetic factors. Giles F. Filley Lecture. *Chest*, **109**, 14S–19S.

Druck, T., Kastury, K. Hadaczek, P. *et al.* (1995) Loss of heterozygosity at the familial RCC t(3;8) locus in most clear cell renal carcinomas. *Cancer Res.*, **55**, 5348–55.

Du, M.Q., Carmichael, P.L. and Phillips, D.H. (1994) Induction of activating mutations in human c-Ha-ras-1 proto-oncogene by oxygen free radicals. *Mol. Carcinogen.*, **11**, 170–5.

Ebina, M., Steinberg, S.M., Mulshine, J.L. and Linnoila R.I. (1994) Relationship of p53 overexpression and up-regulation of proliferating cell nuclear antigen with the clinical course of non-small cell lung cancer. *Cancer Res.*, **54**, 2496–503.

Ellis, N.A., Groden, J., Ye, T.Z., Straughen, J., Lennon, D.J., Ciocci, S., Proytcheva, M. and German, J. (1995) The Bloom's syndrome gene product is homologous to RecQ helicases. *Cell*, **83**, 566–666.

Fontanini, G., Bigini, D. and Vignati, S. (1993) p53 expression in non-small cell lung cancer: clinical and biological correlations. *Anticancer Res.*, **13**, 737–42.

Gatti, R.A. (1995) Ataxia-telangiectasia. *Dermatol. Clin.*, **13**, 1–6.

Gazdar, A.F., Bader, S., Hung, J. *et al.* (1994) Molecular genetic changes found in human lung cancer and its precursor lesions. Cold Spring Harbor *Symp. Quant. Biol.*, **LIX**, 00.

German, J. (1995) Bloom's syndrome. *Dermatol. Clin.*, **13**, 7–18.

Greenblatt, M.S., Bennett, W.P., Hollstein, M. and Harris, C.C. (1994) Mutations in the p53 tumor suppressor gene: clues to cancer etiology and molecular pathogenesis. *Cancer Res.*, **54**, 4855–78.

Guinee, D.G., Travis, W.D., Travers, G.E. *et al.* (1995) Gender comparisons in human lung cancer; analysis of p53 mutations, anti-p53 serum antibodies and c-erbB-2 expression. *Carcinogenesis*, **16**, 993–1002.

Haley, P.J., Shopp, G.M., Benson, J.M. *et al.* (1990) The immunotoxicity of three nickel compounds following 13-week inhalation exposure in the mouse. *Fund. Appl. Toxicol.*, **15**, 476–87.

Halliwell, B. (1987) Oxidants and human disease; some new concepts. *FASEB J.*, **1**, 358.

Hanawalt, P.C. (1994) Transcription-coupled repair and human disease. *Science*, **266**, 1957–8.

Harada, M., Dosaka-Akita, H., Miyamoto, H., Kuzumaki, N. and Kawakami, Y. (1992) Prognostic significance of the expression of ras oncogene product in non-small cell lung cancer. *Cancer*, **69**, 72–7.

Harbour, J.W., Sali, S.L. and Whang-Peng, J. (1988) Abnormalities in structure and expression of the human retinoblastoma gene in SCLC. *Science*, **241**, 353–7.

Harlow, E. (1994) An introduction to the puzzle. *Cold Spring Harbor Symp. Quant. Biol.*, **LIX**, 709–23.

Harpole, D.H., Herndon, J.E., Wolfe, W.G., Iglehart, J.D. and Marks, J.R. (1995) A prognostic model of recurrence and death in stage I non-small cell lung cancer utilizing presentation, histopathology, and oncoprotein expression. *Cancer Res.*, **55**, 51–6.

Harris, C.C. (1993) At the crossroads of molecular carcinogenesis and risk assessment. *Science*, **262**, 1318–27.

Harris, C.C. and Hollstein, M. (1993) Clinical implications of the p53 tumor-suppressor gene. *N. Engl. J. Med.*, **329**, 1318–26.

Harris, H., Miller, O.J., Klein, G., Worst, P. and Tachibana, T. (1969) Suppression of malignancy by cell fusion. *Nature*, **222**, 368.

Hensel, C.H., Hsieh, C.L. Gazdar, A.F. *et al.* (1990) Altered structure and expression of the human retinoblastoma susceptibility gene in small cell lung cancer. *Cancer Res.*, **50**, 3067–72.

Hibi, K., Takahashi, T., Yamakawa, K. *et al.* (1992) Three distinct regions involved in 3p deletion in human lung cancer. *Oncogene*, **7**, 445–9.

Hoffmann, D. and Hecht, S.S. (1985) Nicotine-derived N-nitrosamines and tobacco-related cancer: current status and future directions. *Cancer Res.*, **45**, 935.

Hollstein, M., Sidransky, D., Vogelstein, B. and Harris, C.C. (1991) p53 mutations in human cancers. *Science*, **253**, 49–53.

Horio, Y., Takahashi, T., Kuroishi, T., *et al.* (1993) Prognostic significance of p53 mutations and 3p deletions in primary resected non-small cell lung cancer. *Cancer Res.*, **53**, 1–4.

Horowitz, J.M., Yandell, D.W. and Park, S.H. (1989) Point mutational inactivation of the retinoblastoma antioncogene. *Science*, **243**, 937–40.

Hsu, I.C., Metcalf, R.A. and Sun, T. (1991) Mutational hotspot in the p53 gene in human hepatocellular carcinomas. *Nature*, **350**, 427–8.

Hughes, J.M. and Weill, H. (1991) Asbestosis as a precursor of asbestos related lung cancer: results of a prospective mortality study. *Br. J. Ind. Med.*, **48**, 229–33.

Hung, J., Kishimoto, Y., Sugio, K., Virmani, A., McIntire, D.D., Minna, J. and Gazdar, A.F. (1995) Allele-specific chromosome 3p deletions occur at an early stage in the pathogenesis of lung carcinoma. *JAMA*, **273**, 558–63.

Husgafvel-Pursiainen, K., Hackman, P., Ridanpaa, M. *et al.* (1993) K-ras mutations in human adenocarcinoma of the lung: association with smoking and occupational exposure to asbestos. *Int. J. Cancer*, **53**, 250–6.

Husgafvel-Pursiainen, K., Ridanpaa, M., Anttila, S. and Vainio, H. (1995) p53 and ras gene mutations in lung cancer: implications for smoking and occupational exposures. *J. Environ. Med.*, **37**, 69–76.

Isobe, T., Hiyama, K., Yoshida, Y., Fujiwara, Y. and Yamakido, M. (1994) Prognostic significance of p53 and ras gene abnormalities in lung adenocarcinoma patients with stage I disease after curative resection. *Jpn. J. Cancer Res.*, **85**, 1240–6.

Johnson, B.E., Ihde, D.C. and Makuch, R.W. (1987) myc family oncogene amplification in tumor cell lines established from small cell lung cancer patients and its relationship to clinical course. *J. Clin. Invest.*, **79**, 1629–34.

Johnson, B.E., Brennan, J.F., Ihde, D.C. and Gazdar, A.F. (1992) myc family DNA amplification in tumors and tumor cell lines from patients with small cell lung cancer. *J. Natl Cancer Inst. Monogr.*, **13**, 39.

Kashii, T., Mizushima, Y., Lima, C.E. and Noto, H. (1995) Studies on clinicopathological features of lung cancer patients with K-ras/p53 gene alterations: comparison between younger and older groups. *Oncology*, **52**, 219–25.

Kasprzak, K.S. (1995) Possible role of oxidative

damage in metal-induced carcinogenesis. *Cancer Invest.*, **13**, 411–30.

Kato, T., Watanabe, M. and Ohta, T. (1994) Induction of the SOS response and mutations by reactive oxygen-generating compounds in various escherichia coli mutants defective in the mutM, mutY or soxRS loci. *Mutagenesis*, **9**, 245–51.

Kawajiri, K., Eguchi, H., Nakachi, T., Sekiya, M., Yamamoto M. (1996) Association of CYP1A1 germ line polymorphisms with mutations of the p53 gene in lung cancer. Cancer Res **56**: 72–76.

Kern, J.A. and Filderman A.E. (1993) Oncogenes and growth factors in human lung cancer. *Lung Cancer*, 31–41.

Kern, J.A., Schwartz, D.A. and Nordberg, J.E. (1990) p185neu expression in human lung adenocarcinomas predicts shortened survival. *Cancer Res.*, **50**, 5184–91.

Kern, J.A., Slebos, R.J.C., Top, B. *et al.* (1994) C-erbB-2 expression and codon 12 k-ras mutations both predict shortened survival for patients with pulmonary adenocarcinomas. *J. Clin. Invest.*, **93**, 516–20.

Knudson, A.G. (1971) Mutation and cancer: statistical study of retinoblastoma. *Proc. Natl Acad. Sci. USA*, **68**, 820–3.

Kobayashi, T., Tsuda, H., Noguchi, M., Hirohashi, S., Shimosato, Y., Goya, T. and Hayata, Y. (1990) Association of point mutation in c-Ki-ras oncogene in lung adenocarcinoma with particular reference to cytologic subtypes. *Cancer*, **66**, 289–94.

Kok, K., Osinga, J., Carritt, B. *et al.* (1987) Deletion of a DNA sequence at the chromosome region 3p21 in all major types of lung cancer. *Nature*, **330**, 578–81.

Kondo, K., Umemoto, A., Akimoto, S., Uyama, T., Hayashi, K., Ohnishi, Y. and Monden, Y. (1992) Mutations in the p53 tumor suppressor gene in primary lung cancer in Japan. *Biochim. Biophys. Acta*, **183**, 1139–46.

Krystal, G., Birrer, M. and Way, J. (1988) Multiple mechanisms for transcriptional regulation of the myc gene family in small-cell lung cancer. *Mol. Cell. Biol.*, **8**, 3373–81.

Lakowicz, J.R., Bevan, D.R. and Riemer, S.C. (1980) Transport of a carcinogen, benzo(a)pyrene, from particulates to lipid bilayers: a model for the fate of particle-adsorbed polynuclear aromatic hydrocarbons which are retained in the lungs. *Biochim. Biophys. Acta*, **629**, 243–58.

Lambert, W.C., Kuo, H.R. and Lambert, M.W. (1995) Xeroderma pigmentosum. *Dermatol. Clin.*, **13**, 169–209.

Larson, K.L. and Strauss, B.S. (1987) Influence of template strandedness on *in vitro* replication of mutagen-damaged DNA. *Biochemistry*, **26**, 2471–9.

Law, M.R. (1990) Genetic predisposition to lung cancer. *Br. J. Cancer*, **61**, 195.

Lee, L.N., Shew, J.Y., Sheu, J.C. *et al.* (1994) Exon 8 mutation of p53 gene associated with nodal metastasis in non-small-cell lung cancer. *Am. J. Respir. Crit. Care. Med.*, **150**, 1667–71.

Levine, A.J., Momand, J. and Finlay, C.A. (1991) The p53 tumor suppressor gene. *Nature*, **351**, 453–6.

Li, Z-H., Zhang, J., Weiss, L.M. and Shibata, D. (1994) c-K-ras and p53 mutations occur very early in adenocarcinoma of the lung. *Am. J. Pathol.*, **114**, 303–9.

Lima, C.E.Q., Mizushima, Y., Kashii, T. *et al.* (1994) Sex differences in clinicopathological features of peripheral T1 adenocarcinoma of the lung. *Anticancer Res.*, **14**, 721–4.

Little, C., Nau, M. and Carney, D. (1983) Amplification and expression of the c-myc oncogene in human lung cancer cell lines. *Nature*, **306**, 194–6.

Lowery, D.R. and Willumsen, B.M. (1983) Function and regulation of ras. *Annu. Rev. Biochem.*, **62**, 851–91.

Lung, M.L., Wong, M., Lam, W.K., Lau, K.S., Kwan, S., Fu, K., Cheung, H. and Yew, W.W. (1992) Incidence of ras oncogene activation in lung carcinomas in Hong Kong. *Cancer*, **70**, 760–3.

Magnuson, N.S., Beck, T., Vahidi, H., Hahn, H., Smola, U. and Rapp, U.R. (1994) The raf-1 serine/threonine protein kinase. *Semin. Cancer Biol.*, **5**, 247–53.

Makela, T.P., Shiraishi, M., Borrello, M.G., Sekiya, T. and Alitalo, K. (1992) Rearrangement and co-amplification of L-myc and rlf in primary lung cancer. *Oncogene*, **7**, 405–9.

Mao, L., Jruban, R.J., Boyle, J.O., Tockman, M. and Sidransky, D. (1994) Detection of oncogene mutations in sputum precedes diagnosis of lung cancer. *Cancer Res.*, **54**, 1634–7.

Maragos, C.M., Andrews, A.W., Keefer, L.K. and Elespuru, R.K. (1993) Mutagenicity of glyceryl trinitrate (nitroglycerin) in *Salmonella typhimurium*. *Mutat. Res.*, **298**, 187–95.

Marchetti, A., Buttitta, F., Merlo, G. *et al.* (1993) *Cancer Res.*, **53**, 2846–51.

Marnett, L.J. (1987) Peroxyl free radicals: potential mediators of tumor initiation and promotion. *Carcinogenesis*, **8**, 1365.

Marshall, M.S. (1995) Ras target proteins in eukaryotic cells. *FASEB J.*, **9**, 1311–18.

Mass, M.J., Jeffers, A.J. and Ross, J.A. (1993) K-ras oncogene mutations in tumours and DNA adducts formed by benz[j]aceanthrylene and benzo[a]pyrene in the lungs of strain A/J mice. *Mol. Carcinogen.*, **8**, 186–92.

Menard, H.N., Khorami, J., Jouve, J.L. and Dunnigan, J. (1986) The adsorption of polyaromatic compounds on natural and chemically modified asbestos fibers. *Environ. Res.*, **40**, 84–91.

Miller, C.W., Simon, K., Aslo, A. *et al.* (1992) Mutations in human lung tumors. *Cancer Res.*, **52**, 1695–8.

Mills, N.E., Fishman, C.L., Rom, W.N., Dubin, N. and Jacobson, D.R. (1995) Increased prevalence of K-ras oncogene mutations in lung adenocarcinoma. *Cancer Res.*, **55**, 1444–7.

Minna, J.D. (1993) The molecular biology of lung cancer pathogenesis. *Chest*, **103**, 449–56.

Mitsudomi, T., Steinberg, S.M., Oie, H.K. *et al.* (1991a) ras gene mutations in non-small cell lung cancers are associated with shortened survival irrespective of treatment intent. *Cancer Res.*, **51**, 4999–5002.

Mitsudomi, T., Viallet, J. and Mulshine, J.L. (1991b) Mutations of ras genes distinguish a subset of non-small-cell-lung cancer cell lines from small-cell lung cancer lines. *Oncogene*, **5**, 1353–62.

Mitsudomi, T., Oyama, T., Kusano, T., Osaki, T., Nakanishi, R. and Shirakusa, T. (1993) Mutations of the p53 gene as a predictor of poor prognosis in patients with non-small-cell lung cancer. *J. Natl. Cancer Inst.*, **85**, 2018–23.

Montenarch, M. (1992) Functional implications of the growth-suppressor/oncoprotein p53. *Int. J. Oncol.*, **1**, 37–45.

Morkve, O., Halvorsen, O.J., Skajervan, R., Sangeland, L., Gulsvik, A. and Laerum, O.D. (1993) Prognostic significance of p53 protein expression and DNA ploidy in surgically treated non-small cell lung carcinomas. *Anticancer Res.*, **13**, 571–8.

Mossman, B.T. and Marsh, J.P. (1989) Evidence supporting a role for oxygen species in asbestos-induced toxicity and lung disease. *Environ. Health Perspect.*, **81**, 91–4.

Mulshine, J.L., Treston, A.M., Brown, P.H., Birrer, M.J. and Shaw, G.L. (1993) Initiators and promoters of lung cancer. *Chest*, **103**, 4S–11S.

Murata, M., Otsuka, M., Hayakawa, Y., Takahashi, E., Tsuji, H. and Hori, T. (1992) Expressivity of a common fragile site, fra(3)(p 14.2), in patients with cancer and other diseases. *Jpn. J. Hum. Genet.*, **37**, 205–13.

Naylor, S.L., Johnson, B.E., Minna, J.D. and Sakaguchi, A.Y. (1987) Loss of heterozygosity of chromosome 3p markers in small cell lung cancer. *Nature*, **32**, 451–4.

Nebert, D.W., Pugu, A. and Vasiliou, V. (1993) Role of the Ah receptor and the dioxin-inducible [Ah] gene battery in toxicity, cancer and signal transduction. *Ann. N.Y. Acad. Sci.*, **685**, 624–40.

Nesnow, S., Ross, J.A., Stoner, G.D. and Mass, M.J. (1995) Mechanistic linkage between DNA adducts, mutations in oncogenes and tumorigenesis of carcinogenic environmental polycyclic aromatic hydrocarbons in strain A/J mice. *Toxicology*, **105**, 403–13.

Nguyen, T., Brunson, D., Crespi, C.L., Penman, B.W., Wishnok, J.S. and Tannenbaum, S.R. (1992) DNA damage and mutation in human cells exposed to nitric oxide *in vitro*. *Proc. Natl. Acad. Sci. USA*, **89**, 3030–4.

Nishio, H., Nakamura, S.-I., Horai, T., Ikegami, H. and Matsuda, M. (1992) Clinical and histopathologic evaluation of thc expression of H-ras and fes oncogene products in lung cancer. *Cancer*, **69**, 1130–6.

Ohshima, H. and Bartsch, H. (1994) Chronic infections and inflammatory processes as cancer risk factors: possible role of nitric oxide in carcinogenesis. *Mutat. Res.*, **305**, 253–64.

Parada, L.F., Tabin, C.J. and Shih, C. (1982) Human EJ bladder caricnoma oncogene is homologue of Harvey sarcoma virus ras gene. *Nature*, **297**, 474–8.

Passlick, B., Izbicki, J.R., Haussinger, K., Thetter, O. and Pantel, K. (1995) Immunohistochemical detection of p53 protein is not associated with a poor prognosis in non-small-cell lung cancer. *J. Cardiovasc. Surgery*, **109**, 1205–11.

Prins, J., DeVries, E.G.E. and Mulder, N.H. (1993) The myc family of oncogenes and their presence and importance in small-cell lung carcinoma and other tumor types. *Anticancer Res.*, **13**, 1373–86.

Quinlan, D.C., Davidson, A.G., Summers, C.L., Warden, H.E. and Doshi, H.M. (1992) Accumulation of p53 protein correlates with a poor prognosis in human lung cancer. *Cancer Res.*, **52**, 4828–31.

Quintanilla, M., Brown, K., Ramsden, M. and Balmain, A. (1986) Carcinogen-specific mutation and amplification of Ha-ras during mouse skin carcinogenesis. *Nature*, **322**, 78–80.

Rabbitts, P., Douglas, J., Daly, M. *et al.* (1989) Frequency and extent of allelic loss in the short arm of chromosome 3 in non-small-cell lung cancer. *Genes Chromosom Cancer*, **1**, 95–105.

Reid, T.M. and Loeb, L.A. (1993) Effect of DNA-repair enzymes on mutagenesis by oxygen free radicals. *Mutat. Res.*, **289**, 181–6.

Reissmann, P.T., Koga, H., Takahashi, R. *et al.* (1993) Inactivation of the retinoblastoma susceptibility gene in non-small-cell lung cancer. The Lung Cancer Study Group. *Oncogene*, **8**, 1913–19.

Ridanpaa, M., Karjlainen, A. and Antilla, S. (1994) Genetic alterations in p53 and k-ras in lung cancer in relation to histopathology of the tumor and smoking history of the patient. *Int. J. Oncol.*, **5**, 1109–17.

Rodenhuis, S. and Slebos, R.J. (1992) Clinical significance of ras oncogene activation in human lung cancer. *Cancer Res.*, **52**, 2665s–2669s.

Rodenhuis, S., Slebos, R.J.C. and Boot, A.J.M. (1988) Incidence and possible clinical significance of K-ras oncogene activation in adenocarcinoma of the human lung. *Cancer Res.*, **48**, 5738–41.

Rosell, R., Li, S., Skacel, Z. *et al.* (1993) Prognostic impact of mutated K-ras gene in surgically resected non-small cell lung cancer patients. *Oncogene*, **8**, 2407–12.

Rous, P. and Kidd, J.G. (1941) Conditional neoplasms and subthreshold neoplastic states: a study of the tar tumours of rabbits. *J. Exp. Med.*, **73**, 369–90.

Ruggeri, B., DiRado, M., Zhang, S.Y., Bauer, B., Goodrow, T. and Klein-Szanto, A.J.P. (1993) Benzo[a]pyrene-induced murine skin tumours exhibit frequent and characteristic G to T mutations in the p53 gene. *Proc. Natl. Acad. Sci. USA*, **90**, 1013–17.

Ryberg, D., Kure, E., Lystad, S. *et al.* (1994) p53 mutations in lung tumors: relationship to putative susceptibility markers for cancer. *Cancer Res.*, **54**, 1551–5.

Saksela, K. (1990) myc genes and their deregulation in lung cancer. *J. Cell Biochem.*, **42**, 153–80.

Savitsky, K., Barshira, A., Gilad, S. *et al.* (1995) A single ataxia telangiectasia gene with a product similar to PI-3 kinase. *Science*, **268**, 1749–53.

Schneider, P.M., Hung, M.-C. and Chiocca, S.M. (1989) Differential expression of the c-erbB-2 gene in human small cell and non-small cell lung cancer. *Cancer Res*, **49**, 4968–71.

Sekido, Y., Bader, S., Latif, F. *et al.* (1994) Molecular analysis of the von Hippel-Lindau disease tumor suppressor gene in human lung cancer cell lines. *Oncogene*, 1599–1604.

Selikoff, I.J., Hammond, E.C. and Churg, J. (1968) Asbestos exposure, smoking and neoplasia. *JAMA*, **204**, 106–110.

Shi, D., He, G. and Co, S. (1992) Overexpression of the c-erbB-2/neu-encoded p 185 protein in primary lung cancer. *Mol. Carinogen.*, **5**, 213–18.

Shiseki, M., Kohno, T., Nishikawa, R., Sameshima, Y., Mizoguchi, H. and Yokota, J. (1994) Frequent allelic losses on chromosomes 2q, 18q, and 22q in advanced non-small cell lung carcinoma. *Cancer Res.*, **54**, 5643–48.

Silini, E.M., Bosi, F., Pellegata, N.S. *et al.* (1994) K-ras gene mutations: an unfavorable prognostic marker in stage I lung adenocarcinoma. *Virchows Archiv.*, **424**, 367–73.

Slebos, R., Evers, S. and Wagenaar, S. (1986) Cellular protooncogenes are infrequently amplified in untreated non-small-cell lung cancer. *Br. J. Cancer*, **59**, 76–80.

Slebos, R.J.C., Hruban, R.H., Dalesio, O., Mooi, W.J., Offerhaus, J.G.A. and Rodenhuis, S. (1990) Relationship between K-ras oncogene activation and smoking in adenocarcinoma of the human lung. *J. Natl. Cancer Inst.*, **83**, 1024–7.

Smialowicz, R.J., Rogers, R.R., Riddle, M.M. and Scott, G.A. (1984) Immunologic effects of nickel: I. Suppression of cellular and humoral immunity. *Environ. Res*, **33**, 413–27.

Smialowicz, R.J., Rogers, R.R., Riddle, M.M., Garner, R.J., Rowe, D.G. and Luebke, R.W. (1985) Immunologic effects of nickel: II. Suppression of natural killer cell activity. *Environ. Res.*, **36**, 56–66.

Smialowicz, R.J., Rogers, R.R., Rowe, D.G., Riddle, M.M. and Luebke, R.W. (1987) The effects of nickel on immune function in the rat. *Toxicology*, **44**, 271–81.

Somers, K.D., Merrick, M.A., Lopez, M.E., Incognito, L.S., Schecter, G.L. and Casey, G. (1992) Frequent p53 mutations in head and neck cancer. *Cancer Res.*, **52**, 5997–6000.

Sommer, S.S. (1990) Mutagen test. *Nature*, **346**, 22–3.

Sugio, K., Kishimoto, Y., Virmani, A.K., Hung, J.Y. and Gazdar, A.F. (1994) K-ras mutations are a relatively late event in the pathogenesis of lung carcinomas. *Cancer Res.*, **54**, 5811–15.

Sugio, K., Ishida, T., Yokoyama, H., Inoue, T., Sugimachi, K. and Sasazuki, T. (1992) ras gene mutations as a prognostic marker in adenocarcinoma of the human lung without lymph node metastasis. *Cancer Res.*, **52**, 2903–6.

Sundaresan, V., Ganly, P., Hasleton, P., Rudd, R., Sinha, G., Bleehen, N.M. and Rabbitts, P. (1992) p53 and chromosome 3 abnormalities, characteristic of malignant lung tumours, are detectable in preinvasive lesions of the bronchus. *Oncogene*, **7**, 1989–97.

Suzuki, H., Takahashi, T., Kuroishi, T., Suyama, M., Ariyoshi, Y. and Ueda, R. (1992) p53 mutations in non-small cell lung cancer in Japan: association between mutations and smoking. *Cancer Res.*, **52**, 734–6.

Suzuki, Y., Orita, M. and Shiraishi, M. (1990) Detection of ras gene mutations in human lung cancers by single-strand conformation polymorphism analysis of polymerase chain reaction products. *Oncogene*, **5**, 1037–43.

Takahashi, T., Obata, Y. and Sekido, Y. (1989) Expression and amplification of myc gene family in small cell lung cancer and its relation to biological characteristics. *Cancer Res.*, **49**, 2683–8.

Takeshima, Y., Seyama, T., Bennett, W.P. *et al.* (1993) p53 mutations in lung cancers from non-smoking atomic bomb survivors. *Lancet*, **342**, 1520–1.

Tateishi, M., Ishida, T. and Mitsudomi, T. (1991) Prognostic value of c-erbB-2 protein expression in human lung adenocarcinoma and squamous cell carcinoma. *Eur. J. Cancer*, **27**, 1372–5.

Taylor, J.A., Watson, M.A. and Devereux, T.R. (1994) p53 mutation hotspot in radon-associated lung cancer. *Lancet*, **343**, 86–7.

Testa, J.R. and Siegfried, J.M. (1992) Chromosome abnormalities in human non-small cell lung cancer. *Cancer Res.*, **52**, 2702s–2706s.

Thiberville, L., Bourguignon, J., Metayer, J. *et al.* (1995) Frequency and prognostic evaluation of 3p21–22 allelic losses in non-small-cell lung cancer. *Int. J. Cancer*, **64**, 371–7.

Top, B., Moi, W.J., Klaver, S.G. and Boerrigter, L. (1995) Comparative analysis of p53 gene mutations and protein accumulation in human non-small-cell lung cancer. *Int. J. Cancer*, **64**, 83–91.

Tornaletti, S. and Pfeifer, G. (1995) Complete and tissue-independent methylation of CpG sites in the p53 gene: implications for mutations in human cancers. *Oncogene*, **10**, 1493–9.

Tsuchiya, E., Nakamura, Y., Weng, S-Y, Nakagawa, K., Tsuchiya, S., Sugano, H. and Kitagawa, T. (1992) Allelotype of non-small cell lung carcinoma–comparison between loss of heterozygosity in squamous cell carcinoma and adenocarcinoma. *Cancer Res.*, **52**, 2478–81.

Vachtenheim, J., Horakova, I., Novotna, H., Opalka, P. and Roubkova, H. (1995) Mutations of k-ras oncogene and absence of H-ras mutations in squamous cell carcinomas of the lung. *Clin. Cancer Res.*, **1**, 359–65.

Vainio, H., Husgafvel-Pursiainen, K., Anttila, S., Karjalainen, A., Hackman, P. and Partanen, T. (1993) Interaction between smoking and asbestos in human lung adenocarcinoma: role of K-ras mutations. *Environ. Health Perspect.*, **101**(S3), 189–92.

Veale, D., Kerr, N. and Gibson, G.J. (1993) The relationship of quantitative epidermal growth factor receptor expression in non-small cell lung cancer to long-term survival. *Br. J. Cancer*, **68**, 162–5.

Vogelstein, B. and Kinzler, K.W. (1992) p53 function and dysfunction. *Cell*, **70**, 523–6.

Volm, M., Efferth, T. and Mattern, J. (1992) Oncoprotein (c-myc, c-erbB1, c-erbB2, c-fos) and suppressor gene product (p53) expression in squamous cell carcinomas of the lung. Clinical and biological correlations. *Anticancer Res.*, **12**, 11–20.

Volm, M., van Kaick, G. and Mattern, J. (1994) Analysis of c-fos, c-jun, c-erbB1, c-erbB2 and c-myc in primary lung carcinomas and their lymph node metastases. *Clin. Exp. Metast.*, **12**, 329–34.

Wang, X., Christiani, D.C., Wiencke, J.K. *et al.* (1995) Mutations in the p53 gene in lung cancer are associated with cigarette smoking and asbestos expsoure. *Cancer Epidemiol. Biomark. Prevent.*, **4**, 543–8.

Weiner, D.B., Nordberg, J. and Robinson, R. (1990) Expression of the neu gene-encoded protein (p 185) in human non-small cell lung cancer. *Cancer Res.*, **50**, 5184–91.

Weinberg, R.A. (1991) Tumor suppressor genes. *Science*, **254**, 1138–46.

Weston, A., Willey, J.C., Modali, R. *et al.* (1989) Differential DNA sequence deletions from chromosomes 3, 11, 13, and 17 in squamous-cell carcinoma, large cell carcinoma, and adenocarcinoma of the human lung. *Proc. Natl Acad. Sci. USA*, **86**, 5099–103.

Westra, W.H., Slebos, R.J.C., Offerhaus, G.J.A. *et al.* (1993) K-ras oncogene activation in lung adenocarcinomas from former smokers. *Cancer*, **72**, 432–8.

Whang-Peng, J.T., Knusten, A., Gazdar, S.M. *et al.* (1991) *Genes Chrom Cancer*, **3**, 168–88.

Wink, D.A., Kasprzak, K.S., Maragos, C.M. *et al.* (1991) DNA deaminating ability and genotoxicity of nitric oxide and its progenitors. *Science*, **254**, 1001–3.

Wong, A.J., Ruppert, J.M. and Eggleston, J. (1986) Gene amplification of c-myc and N-myc in small cell carcinoma of the lung. *Science*, **223**, 461–4.

Yokota, J., Wada, M., Shimosato, Y., Terada, M. and Sugimura, T. (1987) Loss of heterozygosity on chromosomes 3, 13, and 17 in small-cell carcinoma and on chromsome 3 in adenocarcinoma of the lung. *Proc. Natl Acad. Sci. USA*, **84**, 9252–6.

Yokota, J., Akiyama, T., Fung, Y.-K. *et al.* (1988) Altered expression of the retinoblastoma (RB) gene in small cell carcinoma of the lung. *Oncogene*, **3**, 471–5.

You, L., Wang, D., Galati, A.J. *et al.* (1994) Tumor multiplicity, DNA adducts and K-ras mutation pattern of 5-methylchrysene in strain A/J mouse lung. *Carcinogenesis*, **15**, 2613–18.

Yunis, J.J. (1983) The chromosomal basis of human neoplasia. *Science*, **221**, 227–36.

Yunis, J.J. (1988) Fragile sites, mutagens and genomic rearrangements in cancer, in *Phenotypic Variations in Populations*, (eds A.D. Woodhead, M.A. Bender and R.C. Leonard), Plenum Publishing, New York, pp. 11–21.

32 Occupationally induced pulmonary malignancy (excluding asbestos-related cancers)

David C. Christiani

INTRODUCTION

Most of the exposures that have been judged to cause cancer in humans have occurred in the workplace (Tomatis, 1988). A large proportion of these occupational exposures are lung carcinogens, the most important of which is asbestos. The public health impact of asbestos is large enough and the human data on asbestos health effects are substantial enough that a separate chapter in this volume is devoted to it. Hence, in this chapter, workplace lung cancer, *except* for asbestos-related disease, will be reviewed.

Clearly, tobacco smoke is known to cause the highest proportion of lung cancer. Next in order of importance, a variety of workplace exposures have been identified to have caused lung cancer. Tobacco and many workplace exposures are relatively strong causes of cancer; the cancer rate among exposed groups has been reported to be 10 to 100 times the rate in unexposed groups. Workers are likely to be exposed to a wide range of potent carcinogens and therefore epidemiologic studies are more likely to identify a strong carcinogen than a weak one. Other causes of cancer in humans, while possibly present among a larger proportion of the population, are likely to be weak and therefore difficult to identify.

Several methodologic problems may influence epidemiologic investigations of occupational exposures and lung cancer (Doll, 1985). A detailed discussion of these problems is beyond the scope of this chapter, but will be described briefly.

The retrospective cohort design is most often used to assess occupational exposures and lung cancer. The design incidence of diseases in a historic cohort of workers with the expected occurrence in a reference population. The ratio of observed deaths in the worker cohort to the expected number of deaths is highly dependent upon the reference population chosen for comparison. Moreover, the parameter derived from these studies, the standardized mortality ratio (SMR), from different worker cohorts are not directly comparable (Hennekens and Buring, 1987).

Another critical aspect of study design is estimation of *exposure*. Exposure estimates are usually based on job title (a surrogate for exposure) or on other methods of job categorization which may be extracted from personnel records, from subjective evaluation of specific jobs, or from extrapolations from measurements

Occupational Lung Disease. Edited by Daniel E. Banks and John E. Parker. Published in 1998 by Chapman & Hall. ISBN 0 412 73630 6

of air contaminants in work areas in which specific jobs are performed. Cumulative exposures are often estimated by simply multiplying the job-specific figure by years exposed. These various methods for estimating intensity and duration of exposures have great potential for misclassification of an individual worker's actual personal exposure (Smith, 1987).

Proper adjustment for confounders and effect modifiers, especially cigarette smoking, is essential. Workers are often exposed to multiple agents (in addition to tobacco) which are possible causes of lung cancer, but information on these exposures may be lacking. Lack of information on smoking or other exposures does not necessarily nullify findings of a lung cancer association in an occupational cohort, but the results should be interpreted cautiously when moderate elevations of risk are detected (Siemiatycki *et al.*, 1988). Degrees of confounding bias related to smoking, ethnic group and socioeconomic status needs to be adjusted in estimates of the associations between occupation and cancer.

As suggested by this discussion, the designation of a substance as a human carcinogen is a judgment. The International Agency for Research on Cancer (IARC) reviews the animal and human data available on potential carcinogens. Since 1971, IARC has developed a program to provide a continuing evaluation of the scientific literature relating to causes of cancer. Each year one or more working groups meet in Lyon, France to consider the scientific information available on one or more substances to which there is potential human exposure. Beginning in 1977 and continuing to the present, these working groups designate a substance as to its potential for carcinogenicity in humans (IARC, 1979). These designations are:

- Group 1: The agent is carcinogenic to humans.
- Group 2A: The agent is probably carcinogenic to humans.
- Group 2B: The agent is possibly carcinogenic to humans.
- Group 3: The agent is not classifiable as to its carcinogenicity to humans.
- Group 4: The agent is probably not carcinogenic to humans.

Each of the above designations takes into account information from epidemiologic studies in humans and, to a lesser extent, from experimental studies in laboratory animals. In order for an agent to be judged as Group 1, there must be sufficient evidence of information in humans based on epidemiologic studies. Recently, IARC added mechanism of action at the cellular level to be considered as a factor in judging potential human carcinogenicity. Group 2A has limited evidence for humans and sufficient evidence for animals. Group 2B has either inadequate evidence in humans together with sufficient evidence in both humans and animals. Group 4 requires suggestive lack of carcinogenicity in both humans and animals. Finally, other agents fall into Group 3.

In 1979, the first comprehensive listing of agents judged to be causes of human cancer was published by IARC (1979). A number of these agents are lung carcinogens (Table 32.1). One of these agents had been found to be carcinogens before 1900; and only one (bis-choloromethyl ether or BCME) initially identified after 1970. It should also be noted that the year an agent or process was first thought to be a carcinogen is arbitrary and other dates have been suggested by researchers in the field (Vincis *et al.*, 1995). A characteristic of many of the agents listed in Table 32.1 is their strong association with lung cancer; often a rate ratio of 10 or more, depending upon intensity and duration of exposure to a working population.

Table 32.1 Agents and processes judged by IARC in 1979 to be lung carcinogens

Substance or process	*Year suggested to be a carcinogen*
Arsenic and certain arsenic compounds	(1822)*
Asbestos	1935
Bis(chloromethyl) ether and chloromethyl ether	1973
Chromium and certain chromium compounds	1948
Underground hematite mining	1956
Mustard gas	1955

* Refers to work environment rather than to specific compound.

Table 32.2 Agents judged by IARC from 1980–1987 to be human lung carcinogens

Substance or process	*Year suggested to be a carcinogen*
Aluminum production	1981
Coal gassification	1936
Coke production	1971
Iron and steel founding	1977
Soot	1985
Talc containing asbestiform fibers	1979

Table 32.3 Agents judged by IARC from 1988–1995 as being lung carcinogens

Substance or process	*Year suggested to be a carcinogen*
Beryllium and beryllium compounds	1979
Cadmium and cadmium compounds	1976
Nickel	–
Spray-painting	1976
Radon	(1879)*
Sulfuric acid mist	1952

* Refers to work environment rather than specific compound.

Table 32.4 Agents judged by IARC from 1979–1995 as probable human lung carcinogens

Substance or process	*Year suggested to be a carcinogen*
Acrylonitrile	1978
Diesel particulate	1983
Formaldehyde	–
Glass manufacture	1987
Non-arsenical pesticides (spraying)	1979
Silica	1986

Between 1980 and 1987 the occupational exposures deemed to be lung carcinogens was expanded to include several more categories (Table 32.2). Since 1987 the agents listed in Table 32.3 have been categorized as workplace lung carcinogens by IARC.

Table 32.4 lists occupational agents that, to date, have been classified by IARC as Group 2A: probable human carcinogens. These agents represent a mixture of chemical compounds and manufacturing processes or general exposures. They have been more recently suggested as possible lung carcinogens. In this group there is limited epidemiologic information on which to provide an unambiguous judgment of causality. In parts of the industrializing world, however, there may be sufficiently large working populations with a wide enough range of exposure to yield such unambiguous information in the future. Agents deemed possible lung carcinogens we listed in Table 32.5. At present, there is insufficient or unconvincing data to consider these compounds as probable lung carcinogens.

Table 32.5 Agents judged by IARC to be possible human lung carcinogens

- Acetaldehyde
- Very fine vitreous fibers
- Welding fumes

DATA ON SPECIFIC LUNG CARCINOGENS AND PROCESS

There is a large body of information on IARC Group I lung carcinogens and certain occupations with mixed exposures from toxicologic and epidemiologic studies in the industrialized world.

With few exceptions, only sparse epidemiologic information is available on the occurrence of occupational lung cancer from industrializing countries. Reviewed below are those for which English-language published studies from both the industrialized and industrializing world are available.

KNOWN CARCINOGENS

Asbestos

Relatively reliable information is available on cancer risks from occupational exposure to asbestos and asbestos products in developing countries. Asbestos-related malignancies are described in detail in Chapters 15–18.

Arsenic

Although arsenic has generally not been found to induce cancer in experimental animals, epidemiologic studies from around the world clearly show that inorganic arsenic is a human lung carcinogen. This evidence emanates mainly from observations among cooper smelter workers who are exposed to arsenic trioxide. Inorganic arsenic is a constituent of most copper ores and, to a lesser extent, of lead and zinc ores. Arsenic is removed during the smelting process and the rest emitted as industrial effluent.

In 1948 Hill and Fanning (1948) reported a five-fold increase in lung cancer deaths among workers manufacturing arsenical pesticides used as sheep dip. Studies of larger cohorts in the 1960s and 1970s reported increased standardized mortality ratios (SMRs) for lung cancer among Montana copper smelter workers, with the first large-scale epidemiologic study reported in 1969 (Lee and Fraumeni, 1969). Extended follow-up of the cohort confirmed

Table 32.6 SMR for lung cancer according to cumulative exposure to airborne arsenic among 8000 Montana copper smelter workers

Cumulative arsenic exposure ($mg/m^3 \times$ years)	*No. of lung cancer deaths*	*SMR*
<0–9	20	168
0.9–1.9	33	256
2.0–8.3	43	167
8.4–41.6	115	315
41.7–208.3	54	429
208.4–416.6	20	571
>416.7	17	680

Source: adapted from Lee-Feldstein (1986).

excess lung cancer risk in smelter workers and clarified the exposure–response relationship between cumulative exposure and risk (Table 32.6) (Lee-Feldstein, 1986).

Similar observations have come from follow-up studies of workers in copper smelters elsewhere in the US (Enterline *et al.*, 1987), as well as in Japan (Tokudome, 1976), Sweden (Jarup *et al.*, 1989) and The People's Republic of China (Xu *et al.*, 1991).

In some of the epidemiologic studies done to date, the potential interaction between cigarette smoking and arsenic has been evaluated. In the US Montana smelter study, although most of the men smoked, the risk of lung cancer increased with rising arsenic exposure among both smokers and non-smokers (Welch *et al.*, 1982). The association appeared more additive than multiplicative. Similar findings were reported in smelter workers in the south-west China province of Yunnan (Taylor *et al.*, 1989).

Another major source of occupational arsenical exposure around the world has been the production of pesticides, fungicides and herbicides. The earliest report of arsenic-related lung cancer came from the study of workers making arsenic trioxide pesticides which were used to control insects infesting sheep. Epidemiologic studies of American workers involved in arsenical pesticide production revealed excess mortality from lung cancer, with as much as a 10-fold increase for workers with light exposures for 15 or more years (Mabuchi *et al.*, 1979).

Although systematic studies of cancer risk have not been performed on workers exposed to arsenic in other settings, it is important to note that airborne arsenic exposure can take place in a variety of industries including pressure-treated wood and glass production.

Coke-oven emissions, coal gasification

Polycyclic aromatic compounds are formed mainly from pyrolysis or incomplete combustion of organic matters: polycyclic hydrocarbons (PAH) comprise the major subgroup of the family of compounds. A large fraction of PAH is adsorbed to particles produced by combustion. Occupational settings where exposure may occur to PAH include aluminum production, iron and steel founding, coke production and coal gassification. Compounds containing PAH are used in a variety of settings and include coal tar, coal tar pitches and untreated and mildly treated mineral oils. Ash from fuel combustion and diesel exhaust also contains PAH. Hence, workplace exposures to PAH are common, though few data are available on exposures to specific compounds or on detailed human exposure characterization. In the mid-1980s, IARC estimated that worldwide, over 2 million workers were exposed to PAH in the basic processes alone of aluminum production, iron and steel founding, coke production and coal gassification (IARC, 1984). The strongest epidemiologic evidence for human lung cancer risk after exposure to PAH comes from studies of coke oven workers. The highest risk for lung cancer occurs with work on the topside of coke ovens, and the risk increases with increasing intensity and duration of exposure. In one study of steel workers working at the coke ovens, the relative risks for lung cancer for topside workers (Plate 14) compared with non-oven workers increased to 16 with 15 or more years of employment (Redmond, 1983). Lifetime estimates of lung cancer risk associated with work on a coke oven have shown that after 40 years of work, beginning at age 20, there is a 40% chance of developing lung cancer by age 85; in other words, 15 times that for non-exposed men (Dong *et al.*, 1988).

A few large-scale studies have been done in the industrializing world. Several cohort studies of workers exposed to polyaromatic hydrocarbons in coking and coal-gas plants in China have been performed (Wu, 1988). In these studies the referent cohort consisted of workers at a primary steel rolling mill. A cohort of 21 995 workers at 19 coke plants had a Standardized Incidence Ratio (SIR) for lung cancer of 2.55 (90% CI; 2.13–3.03). A cohort of 3107 workers from six coal-gas plants had an SIR for lung cancer of 3.66 (90% CI; 2.36–5.43). In a cohort of 5122 workers in two coke plants, the SIR for lung cancer was 2.6 (90% CI; 1.79–3.60).

Similarly, from China have come reports of cohort studies of workers at shale-refining and fuel-oil producing plants. In a cohort of 6285 workers at a large shale-refining plant, increased SMR's for lung cancer (240; 90% CI: 185–307) and esophageal cancer (244;

90% CI: 157–311) were observed (Wu, 1988). In a cohort of 12 422 workers at an oil-refining plant, no elevated risk of lung cancer was found, though the SMR for liver cancer was 1.99 (90% CI: 1.27–2.99). In a cohort of 3774 workers at a synthetic oil plant, an excess risk for lung (SMR 2.02; 90% CI: 1.34–2.93) and digestive tract (SMR 1.56; 90% CI: 1.13–2.09) cancers were found.

Coal tars or asphalts arc by-products from the distillation of coal for gassification and coke production. Coal tars are used in road-paving, roofing and other industries. There have been a few epidemiologic studies of workers exposed to these forms of PAH. Mortality studies of roofers revealed a small excess in lung cancer mortality, but smoking was not adjusted for (IARC, 1987). A case-control study at an automobile metal stamping plant assessed cancer risk and specific jobs, and found an elevated odds ratio (13.2, 95% CI 1.1–154.9) for lung cancer among maintenance welders and millwrights at the plant. Workers in these jobs were exposed to coal tar volatiles and welding fumes (Silverstein *et al.*, 1985).

Diesel exhaust

Diesel exhaust is a complex mixture of substances, characterized by PAHs surrounding an elemental carbon core. The gas phase includes carbon monoxide and oxides of nitrogen. Diesel exhaust has been shown to cause lung cancer in animals (Ishinishi *et al.*, 1986) and the particulate, rather than gaseous, phase appears to be the cause.

Several large epidemiologic investigations of occupational exposure to diesel exhaust have been conducted in North American railroad workers. Howe *et al.* (1983) performed a mortality study of 43 826 retired workers at the Canadian National Railway Company. They reported relative risks for lung cancer among workers probably or possibly exposed to diesel exhaust. The estimates were 1.35 and 1.2 respectively, compared with unexposed workers. A study of 55 407 US railroad workers by Garshick *et al.* (1988) reported an elevated risk of lung cancer among workers with the longest possible duration of exposure to diesel exhaust with a relative risk of 1.45 (95% CI, 1.11–1.89). An earlier case-control study of 1256 lung cancer deaths among the same cohort found an odds ratio of 1.41 (1.06–1.88) for subjects working in a job exposed to diesel exhaust for 20 or more years after adjusting for smoking and asbestos exposure.

Several other epidemiologic studies of exposed workers have confirmed the elevated risk of lung cancer and support the IARC 1989 judgment that occupational exposure to diesel exhaust is a probable human lung carcinogen (Boffetta *et al.*, 1988, 1990; Gustavsson *et al.*, 1986, 1990; Steenland *et al.*, 1990; Emmelin *et al.*, 1993).

MINING AND OCCUPATIONAL PULMONARY MALIGNANCY

Workers mining copper and certain other ores may be exposed to arsenic, radon, and free silica. Studies of hard-rock miners in the USA (Wagoner *et al.*, 1964) (Plate 15) and Great Britain (Hodgson and Jones, 1990) have reported increased lung cancer risks.

Studies on occupational lung cancer among miners have been published from the People's Republic of China (coal, tin, hematite) and South Africa (gold mining).

In a large cohort of coal miners in Anhui, China, Yang *et al.* (1985) reported an unadjusted proportion of deaths from malignancies to increase from 0.3% in 1953 to 5.6% in 1980. The study did not control for cigarette smoking.

An early case series of 22 lung-cancer deaths in Zimbabwe reported that 13 had worked in mines (12 gold, one asbestos) and silica and arsenic were suggested as possible agents (Osburn, 1969). However, no exposure assessment was done and smoking was not adjusted for, though smoking prevalence was in general quite low at the time in this population.

There have been several studies of South African *gold miners*. Some data have suggested an association between work in these mines and lung cancer (Wyndam *et al.*, 1986), while other work has not (Hessel *et al.*, 1990).

Investigators in China reported an extremely high incidence of lung cancer among tin miners in Yunnan Province, where the highest lung cancer mortality rate in men among 2392 counties or cities in the country during a nation-wide survey conducted in 1973–75 was reported. Among male miners aged 15 or over who had underground exposure, the crude annual incidence for 1983–1985 was 585/100 000 (Sun, 1987, 1989; Xuan *et al.*, 1993). Exposures in the tin mines are likely complex and may include arsenic, radon decay products and free silica. Qiao *et al.* (1989) reported the results of a case-control study of 107 prevalent cases of lung cancer among tin miners exposed to radon decay products in Yunnan. Subjects in the highest quartile of exposure had an odds ratio of 9.5 (95% CI; 2.7–33.1) compared with those without exposure to radon and after adjusting for arsenic exposure, age, year starting work and tobacco use. Risk increased with exposure duration. Another report from the same research group assessed arsenic exposure in the same subjects (Taylor *et al.*, 1988). The results showed that subjects in the highest quartile of cumulative arsenic exposure had a relative risk (RR) of 22.6 compared with subjects without exposure after adjusting for tobacco and radon exposure. Moreover, a positive exposure–response relationship was observed. The median duration of exposure was 28 years and exposure duration more strongly predicted risk than intensity, although misclassification of exposure intensity is common in such studies.

Radon exposure among Chinese tin miners was also examined. A case-control study of 69 cases and 138 controls in the Dachang tin area of Guangxi Province was reported by Wu *et al.*, (1989). The authors reported that the main risk factors for lung cancer in this group were time of exposure to smelting, time of exposure to underground mining, and age at which last occupation began: all factors related to occupational exposures. Cigarette smoking was found to be synergistic with these factors. A recent study from a mine in the same area measured low radon levels and found elevated lung cancer mortality among the miners as compared with community-based controls (Fu *et al.*, 1994).

A recent study from China of hematite miners revealed an excess risk for lung cancer (Chen *et al.*, 1990), with an SMR of 370 compared with nation-wide male population rates. The SMR was 480 for miners who had first been employed before exhaust ventilation and wet-drilling were introduced.

Several case-control studies of non-ferrous miners in Southern Africa have been reported, mainly from the Balawayo (Zimbabwe) Cancer Registry which operated during 1963–1977 in an area where farming and mining were the two major economic activities (Parkin *et al.*, 1994). There was an association between mining and lung cancer (odds ratio 3.1; 95% CI: 1.07–1.60), but not with other types of cancer. The risk for lung cancer was particularly elevated among copper, gold and nickel miners, while it was not elevated among coal and chromium miners. Farmers were at a reduced risk for lung, esophageal and liver cancer. A limitation of this study was the absence of occupational information on 50% of the cases.

NON-MINING METAL EXPOSURES: CHROMIUM, CADMIUM, NICKEL

(a) Chromium

Chromium is a metal that can exist in any oxidation state between −2 and +6. Most chromium compounds contain the metal in the +3 (trivalent) or +6 (hexavalent) state, and most naturally occurring chromium is found as oxides in chromite ore. Cr (VI) causes cancer in animal studies and is the compound currently regulated (IARC, 1990). Most chromium consumed in the USA is in the production of stainless (rust-resistant) steel. Workers may be exposed in the production of stainless steel, other chrome alloys, chrome-containing pigments, chrome plating and stainless steel welding.

There have been over 50 epidemiologic studies of chromium exposure and lung cancer (IARC, 1990). The largest investigations have been performed on chromium production workers, chromate paint makers and chrome platers. Steeland *et al.* (1996) estimated overall relative risk of lung cancer in these studies is 2.78 (2.47–3.52) and the studies which were able to address confounders (tobacco, asbestos, nickel) rejected these confounders as the major source of the observed elevated risk.

A retrospective cohort study of 2982 workers in Chinese chromate manufacturing plants revealed an increased risk for lung cancer (RR = 2.7; 90% CI: 1.3–5.5) (Wu, 1988). Risk was related to exposure latency, extent of exposure and smoking habits.

(b) Nickel

The principal uses (currently) of nickel and nickel salts are in the production of stainless steel, non-ferrous alloys, electroplating and battery manufacture (IARC, 1990). IARC considered the evidence sufficient for the carcinogenicity of nickel sulfate and for the 'combinations of nickel sulfides and oxides encountered in the nickel refining industry'. However, the IARC considered the evidence in humans for the carcinogenicity of metallic nickel and nickel alloys inadequate.

The most current review of cancer among nickel-exposed workers is the 1990 report of The International Committee on Nickel Carcinogenesis in Man (ICNCM, 1990). This review of nine cohort and one case-control study reports that lung cancer and nasal cancer were consistently and significantly increased among exposed workers at nickel refineries. Steenland *et al.* used data from this report to estimate a combined relative risk of 1.56 (95% CI 1.41–1.73). Interestingly, studies of workers engaged in other nickel-related industries such as nickel alloys and stainless steel welders, have not shown consistently elevated lung cancer deaths.

(c) Cadmium

Cadmium is principally used in electroplating, in compounds that serve as stabilizers for plastics and as pigments, in electrodes in batteries and in alloys (IARC, 1976). In experimental animals, cadmium induces tumors of the lung (Heinrich, 1992) as well as other sites including prostate (Waalkes *et al.*, 1992a) and lymphatics (Waalkes *et al.*, 1992b). Cadmium exposure often takes place in the context of other carcinogen exposure such as nickel and arsenic. This results in confounding in epidemiologic studies (Boffetta, 1992). However, a USA study of metal waste reclamation workers revealed an excess risk of lung cancer often excluding those with arsenic exposure (Thun *et al.*, 1986; Stayner *et al.*, 1992). The RR for lung cancer was 1.49 (95% CI 0.96–2.22) with an increasing relative risk with increasing cumulative exposure to cadmium with an RR of 2.72 (95% CI 1.24–5.18) for workers in the highest exposure category.

(d) Beryllium

Beryllium exposure occurs mainly in mining, refining, and in the manufacture of ceramics, electronic and aerospace equipment. In 1980, IARC concluded that there was sufficient animal evidence for the carcinogenicity of beryllium metal and several beryllium compounds (IARC, 1980). Since that time two cohort mortality studies have been reported and in both an excess of lung cancer was found. One study was based on the US Beryllium Case Registry Mortality Study of 689 women and men. The lung cancer SMR was 2.00 (95% CI 1.33–2.89). Ward *et al.* (1992) reported the results of a cohort mortality study of 9225 males from seven US beryllium plants. The overall lung cancer SMR was 1.24 (95% CI 1.10–1.39). Lung cancer SMRs increased with latency and with work in factories with higher pneumoconiosis SMR, and smoking was adjusted for (based on limited data). In a lung cancer incidence study of patients with beryllium disease, Steenland and Ward, (1991) reported an SMR of 2.00 (95% CI, 1.33–2.89). The authors reported that it was unlikely that smoking was a major confounder in this study. In 1993, the IARC concluded that the evidence was now sufficient to conclude that beryllium is carcinogenic to humans (IARC, 1993).

(e) Iron

A proportionate mortality study of 578 deceased South African iron molders was reported by Sitas *et al.*, (1989). Non-whites were excluded due to small sample size and lack of population comparison at that time. An increased proportionate mortality ratio (PMR) was found for lung cancer (PMR = 1.7, $p < 0.05$).

A population-based case-control study of 1249 lung cancer patients and 1345 controls in Shenyang, China was reported by Xu *et al.* (1991). Employment in the non-ferrous metal smelting and refining industry was associated with a smoking-adjusted increased risk for lung cancer among males (RR = 3.5; 95% CI: 1.6–1.8). Most of the males had worked in the smelting industry for at least 20 years.

A retrospective cohort study from China of 18 795 workers at copper and arsenic smelters and an ore mine found an SMR for lung cancer of 678 (90% CI; 511–582) (Wu, 1988). In this study there was a significant association between lung cancer and airborne arsenic levels.

SILICA AND OCCUPATIONAL PULMONARY CANCER

The IARC reviewed the evidence for carcinogenicity of crystalline silica in 1987 and concluded that there was sufficient evidence for carcinogenicity in animals and limited evidence for carcinogenicity in humans (IARC, 1987).

In rats, inhaled silica causes both fibrosis and lung cancer in relatively low doses (1 mg/m^3) (Muhle *et al.*, 1989). The problems of confounding lung carcinogens such as smoking and radon exposure and of selection bias in the detection of pneumoconiosis cases, complicate the epidemiologic assessment of silica exposure (McDonald *et al.*, 1989). The balance of evidence indicates that silicotic patients have increased risk for lung cancer (Goldsmith, 1994). It is less clear whether silica exposure in the absence of silicosis carries increased risk for lung cancer.

The association between human exposure to silica and risk for bronchogenic carcinoma has been examined in autopsy series, case-control series drawn from workers with silicosis or from patients with lung cancer, and in population-based groups of silica-exposed workers. An elevated SMR, approximately 1.5-fold the expected number of cancer deaths, was detected in large population based studies in Massachusetts (Dubrow and Wegman, 1984) and Canada (Siemiatycki *et al.*, 1990). Workers exposed to silica in Italy (Mastrangelo *et al.*, 1988), the Netherlands (Meijers *et al.*, 1990), and in some, but not all, Nordic countries showed three- to five-fold increased cancer risk (Lynge *et al.*, 1986).

Reports from many countries have identified lung cancer with increased frequency among workers compensated for silicosis (Finkelstein *et al.*, 1987; Infante *et al.*, 1989; Merlo *et al.*, 1990; Ng *et al.*, 1990; Carta *et al.*, 1991; Partanen *et al.*, 1994; Finkelstein, 1995), with a relative risk for lung cancer compared with the general population of 1.3 to 6.9. A slight excess cancer mortality was found after adjusting for the effects of smoking, and cancers were found in never-smokers in two series (Zambon *et al.*, 1987; Chiyootani *et al.*, 1990).

Metal ore miners with silica exposure in the United States (Amandus and Costello, 1991), the United Kingdom (Hodgson and Jones, 1990), China (Chen *et al.*, 1990, 1992; McLaughlin *et al.*, 1992), and South Africa (Hessel *et al.*, 1986; Hnizdo and Sluis Cremer, 1991) experienced a significant increase in mortality due to lung cancer, with two- to five-fold increases in risk. Several of these studies adjusted for the effects of smoking ((Chen *et al.*, 1990; Hessel *et al.*, 1990; Amandus and Costello, 1991; Hnizdo and Sluis Cremer *et al.*, 1991; McLaughlin *et al.*, 1992). Miners may also be exposed in the underground work environment to radon, arsenic and diesel exhaust (Carta *et al.*, 1994; Samet *et al.*, 1994).

In addition, an increased lung cancer risk (SMR 1.5–2.0) has been demonstrated in a variety of other non-mining dusty trades including granite workers in Vermont (Costello and Graham, 1988) and China (Chia *et al.*, 1991), foundry workers (Sherson *et al.*, 1991), German slate workers (Mehnert *et al.*, 1990), North Carolina workers in dusty trades (Amandus *et al.*, 1992), and California diatomaceous earth workers (exposed to amorphous silica and cristobalite) (Checkoway *et al.*, 1993). Ceramic or pottery workers demonstrated increased cancer risk in Sweden (Tornling *et al.*, 1990), and Italy (Lagorio *et al.*, 1990), but not in China. Workers in foundries, quarries (Plate 16), and manufacturing industries do not experience radon exposure, although some may be exposed to PAHs or other lung carcinogens.

The available data support the conclusion that silicosis produces increased risk for bronchogenic carcinoma. The cancer risk may also be increased by smoking and other carcinogens in the workplace. The increased risk associated with silica exposure appears to be less than the risk conferred by asbestos exposure. Epidemiologic studies provide convincing evidence for increased cancer risk among tobacco smokers with silicosis. Less information is available for never-smokers and for workers exposed to silica but who do not have silicosis. For workers with silicosis, the risks for lung cancer are relatively high and consistent among various countries and investigators. Silicosis should be considered a condition which predisposes workers to an increased risk of lung cancer. This concern about cancer influences decisions about permissible exposure limits.

SUSCEPTIBILITY

Individual variability in carcinogen metabolism may explain differential susceptibility to lung cancer. In recent years molecular techniques in the epidemiologic study of cancer have been applied in three areas: determination of internal and biologically effective dose, detection of early biologic effects, particularly mutations and cytogenic changes, and assessment of variations in individual susceptibility to carcinogens, mainly via metabolic polymorphisms.

Among the 3800 chemicals that have been identified in tobacco smoke, a large number are biologically active compounds. The most important families of carcinogens are polycyclic aromatic hydrocarbons (PAHs), aromatic amines, nitroso compounds, volatile organic compounds (e.g., benzene, formaldehyde), and radioactive elements such as polonium–210. Chemical compounds derived from tobacco smoke have been measured in biological specimens of smokers and non-smokers. DNA adducts are formed when chemical carcinogens react with DNA. Several methods of detecting adducts in lung and surrogate tissue, including the sensitive ^{32}P-postlabeling procedure for PAHs, have been described in detail. Among the class of pulmonary carcinogens known as PAHs, benzo-α-pyrene forms DNA adducts in the lung that are associated with smoking. Recent research indicates that surrogate tissue (blood mononuclear cells) adducts in cancer patients (Wiencke *et al.*, 1995). Such validation will permit assessment of interventions such as smoking cessation, chemoprevention, and elimination of passive-smoking exposures in reducing PAH-DNA adducts in peripheral-blood mononuclear cells. Use of blood adducts will also permit epidemiologic study of large populations exposed to PAHs in air pollution.

Although the presence of PAH-DNA adducts has not been definitely associated with lung cancer, it is clear that these adducts appear to be important in the pathway leading to cancer (Wogan and Gorelick, 1985), and interventions aimed at reducing their presence in high-risk individuals are on the horizon. Focusing on this 'upstream' marker of dose and early effect is particularly important in lung-cancer prevention, a condition for which there is no effective early detection marker and for which treatment is an ineffective means of disease control.

In recent years mutations in specific genes can be used as 'fingerprints' of specific exposures, as surrogate and points of cancer, or for further subtyping of cancer to clarify causal associations (McMichael, 1994). Epidemiologic evidence suggests the association between chemical exposure tumor-suppressor gene (*p*53) and oncogene (*k*-ras) mutation and cancer occurrence (Hollstein *et al.*, 1991; Slebos *et al.*, 1991; Suzuki *et al.*, 1992; Vahakangas *et al.*, 1992; Husgafvel-Pursiainen *et al.*, 1993). Investigations concerning the *ras* oncogene family or the *p*53 tumor-suppressor gene have considered the association between lung cancer and tobacco smoking. Results of these studies suggest that the mutational spectrum of lung cancer depends upon whether the person was a smoker and whether the person was exposed to asbestos (Wang *et al.*, 1995).

Genetic susceptibility to lung cancer has been examined in a number of epidemiologic studies that incorporate molecular markers of susceptibility to carcinogens. It has been known for over 30 years that smoking and family history are independent risk factors for lung cancer (Sellers *et al.*, 1990). Additional evidence has been provided by the study of metabolic polymorphisms, that is, the different ability to metabolize chemical carcinogens. Traditional monogenic susceptibility to cancer is related to rare diseases (e.g., xeroderma pigmentosa, Li-Fraumeni syndrome) or small subgroups of common diseases (e.g., *BRCA1* in familial breast cancer) and involves rare mutations that can be identified through linkage analysis. Polygenic susceptibility involves frequently-occurring genetic polymorphisms, confers a low to medium elevation of risk for frequent diseases and is identified through epidemiologic (usually case-control) studies. Examples of this include the *CYP2D6* polymorphism phenotype and lung-cancer risk (Amos *et al.*, 1992) in Whites and the *CYP1A1* and *GSTM1* polymorphisms in lung and bladder cancer, respectively, among Japanese (Nakachi *et al.*, 1993).

Recent work in the US suggests that GSTM1 persons with low dietary intake of antioxidants are at high risk of squamous-cell carcinoma of the lung, in contrast to persons with null genotype and high antioxidant intake (Garcia-Closas *et al.*, 1995), after controlling for smoking intensity, duration, age, and time since quitting smoking. This kind of investigation may lead to antioxidant chemoprevention in ex-smokers and illustrates a dual clinical approach to disease control: exposure cessation combined with cumulative risk modification on the secondary level. If genotype-lung cancer risk is modified by a factor amenable to intervention, such as dietary intake of antioxidants (Ames *et al.*, 1993), then dietary changes and/or supplementation may be a useful intervention strategy for individuals who are attempting to quit or have already quit smoking.

Tobacco smoke contains significant amounts of another class of carcinogens, aromatic amines (Patrianakos *et al.*, 1979). Those compounds have been shown to be associated with bladder cancer in smokers (Vineis, 1992). Epidemiologic studies have suggested that the type of tobacco associated with the highest risk of bladder cancer (air-cured tobacco) is also richer in arylamines and that smokers of air-cured tobacco have higher levels of 4-aminobiphenyl-hemoglobin adducts in their blood compared with smokers of flue-cured tobacco (Bartsch *et al.*, 1990). Some arylamines (including 4-aminobiphenyl and β-2-naphthylamine) are among the most potent human carcinogens. Measurement of 4-aminobiphenyl-hemoglobin adducts provides evidence of another biologically-effective dose marker used in epidemiologic investigation.

Another common polymorphic trait in humans, the *N*-acetylation phenotype, modulates the concentration of 4-aminobiphenyl-hemoglobin adducts in both smokers and non-smokers (Talaska *et al.*, 1990). The *N*-acetyltransferase (NAT) polymorphism has been associated with bladder cancer in epidemiologic studies. NAT is a non-inducible enzyme that deactivates aromatic amines (Heim, 1988). The slow acetylator genotype prevalence is about 50% in the White population. Slow acetylators exposed to aromatic amines have been reported to have an increased risk of bladder cancer, as well as higher levels of 4-aminobiphenyl adducts (Vineis *et al.*, 1990, 1994).

The importance of their work in the molecular epidemiology of lung cancer promises to lead to the development of exciting opportunities in environmental health research, which will add greatly to our understanding of the human-health consequences of toxic exposures (Christiani, 1996).

POPULATION ATTRIBUTABLE RISK

Lung cancer is the most common malignancy in the US and its incidence is increasing in many industrializing countries. In the US lung cancer ranks second only to bladder cancer in the proportion of cases thought to be due to occupational exposure. It is logical to assume that this epidemic pattern will be reproduced in many of the world's industrializing nations as both tobacco and occupational exposures increase.

In the US each year there are approximately 90 000 lung-cancer deaths among males, and approximately 100 000 newly-diagnosed cases. In females, the comparable estimates are 45 000 and 60 000 respectively. Several attempts have been made to estimate the proportion of lung cancers due to occupational exposure. From large case-control studies, Morabia *et al.* (1992) estimated 9% of male lung cancers were attributable to occupational exposures. Using similar methodology, Vineis *et al.* (1988) estimated that between 3% and 17% were attributable to occupational exposures. Doll and Peto (1981) estimated that 15% of male and 5% of female lung cancers were due to occupational exposures, without any formal methodology. Steenland *et al.* (1996) based estimates on the proportion of the population exposed to specific occupational carcinogens and used estimated relative risks for these carcinogens, using a formula for calculating attributable risk (or etiologic fraction) from Kleinbaum *et al.* (1982). Excluding the contribution of radon at work, approximately 9000 annual lung cancer cases among US males can be attributed to occupational exposures (about 9%) and among females, about 900 cases (2%) annually. Asbestos exposure (see chapters 15–18) accounts for the majority of the attributable cases. If radon exposure at work is considered, the authors estimated 10 000 cases among males and 1900 cases among females annually. Of course, these estimates apply to the USA only and Steenland *et al.* omitted from their analysis workplace carcinogens for which relatively few workers in the US are exposed (>30 000). These carcinogens included bis-chloromethyl ether, coke and coal gasifications, and soot. In some parts of the world (e.g., East Asia), considerably larger numbers of workers may be exposed to these and other lung carcinogens. Also, with rapid changes in demographic patterns, smoking habits, and industrial development for much of the industrializing world, we are likely to witness recurrent lung cancer epidemics.

CLINICAL PRESENTATION AND MANAGEMENT

Occupational lung cancer displays no unique clinical features with which to inform the clinician regarding causation. Most patients who develop bronchogenic carcinoma do not consult a physician until symptoms of the disease become manifest. The initial symptoms may vary with the patient population and the cell type of the tumor. A detailed discussion of clinical diagnosis and management is beyond the scope of this chapter and the reader is referred to the many textbooks of pulmonary medicine which are now available. However, a few salient points should be noted by clinical and preventive-medicine practitioners alike.

First, there is a tendency among clinicians to discard the diagnosis of lung cancer in lifelong non-smokers. However, in an assessment of the reasons for missed diagnosis in patients with lung carcinoma found at autopsy, one group concluded that reliance upon the well-accepted association between smoking and lung cancer was a significant cause for misdiagnosis among non-smokers (McFarlane *et al.*, 1986a,b). The complete epidemiologic setting in which patients are evaluated must be considered in clinical practice. Similarly, in preventive medicine, recognition of a population exposure profile is important in order to take steps to reduce workers' risk of exposure to lung carcinogens.

Second, lung cancer is one of the most important human neoplasms, not only because of its incidence but also because of its dismal prognosis. Overall, only about 15% of patients survive five years or longer.

Third, the vast majority of lung cancer is preventable. The critical preventive intervention is elimination or reduction in exposure to lung carcinogens.

In many parts of the industrializing world, worker exposure to lung carcinogens is increasing. Some countries, such as China, are undergoing societal transformations at an unprecedented rate (Christiani, 1984, 1988). In order to prevent epidemics of occupational lung cancer, preventive measures must be instituted now, and not when the human and economic toll of disease beckons too loudly to be ignored.

Acknowledgements

These studies were supported by NIH grants ES/CA 06409 and ES00002 and NIOSH grant OH02421.

REFERENCES

Amandus, H. and Costello, J. (1991) Silicosis and lung cancer in U.S. metal miners. *Arch. Environ. Health*, **46**, 82–9.

Amandus, H.E., Castellan, R.M., Shy, C., Heineman, E.F. and Blair A. (1992) Reevaluation of silicosis and lung cancer in North Carolina dusty trades workers. *Am. J. Ind. Med.*, **22**, 147–53.

Ames, B.N., Shinenaga, M.K. and Hagen, T.M. (1993) Oxidants, antioxidants and the degenerative diseases of aging. *Proc. Natl. Acad. Sci. USA*, **90**, 7915–22.

Amos, C.I., Caporaso, N.E. and Weston, A. (1993) Host factors in lung cancer risk: a review of interdisciplinary studies. *Cancer Epidem. Biomarkers Prev.*, **1**, 505–13.

Bartsch, H., Caporaso, N., Coda, M., Kadlubar, F., Malaveille, C., *et al.* (1990) Carcinogen hemoglobin adducts, urinary mutagenicity, and metabolic phenotype in active and passive smokers. *J. Natl. Cancer Inst.*, **82**, 1826–31.

Boffetta, P. (1992) Methodological aspects of the epidemiological association between cadmium and cancer in humans. In *Cadmium in the Human Environment: Toxicity and Carcinogenicity.* (eds., G. Nordberg, R. Herber, L. Alessio). Lyon, France, IARC, pp. 425–434.

Boffetta, P., Harris, R. and Wynder E. (1990) Case-control study on occupational exposure to diesel exhaust and lung cancer risk. *Am. J. Ind. Med.*, **17**, 577–91.

Boffetta, P., Stellman, S. and Garfinkel, L. (1988) Diesel exhaust exposure and mortality among males in the American Cancer Society Prospective Study. *Am. J. Ind. Med.*, **14**, 403–15.

Carta, P., Cocco, P. and Picchiri, G. (1994) Lung cancer mortality and airways obstruction among metal miners exposed to silica and low levels of radon daughters. *Am. J. Ind. Med.*, **25**, 489–506.

Carta, P., Cocco, P.L. and Casula, D. (1991) Mortality from lung cancer among Sardinian patients with silicosis. *Br. J. Ind. Med.*, **48**, 122–9.

Checkoway, H., Heyer, N.J., Demers, P.A. and Breslow N.E. (1993) Mortality among workers in the diatomaceous earth industry. *Br. J. Ind. Med.*, **50**, 586–97.

Chen, J., McLaughlin, J.K. and Zhang J.Y., (1992) Mortality among dust-exposed Chinese mine and pottery workers. *J. Occup. Med.*, **34**, 311–6.

Chen, S.Y., Hayes, R.B., Liang, S.R., Li, Q.G., Stewart, P.A. and Blair, A. (1990) Mortality experience of haematite mine workers in China. *Br. J. Ind. Med.*, **47**, 175–81.

Chia, S.E., Chia, K.S., Phoon, W.H. and Lee, H.P. (1991) Silicosis and lung cancer among Chinese granite workers. *Scand. J. Work Environ. Health*, **17**, 170–4.

Chiyotani, K., Saito, K., Okubo, T. and Takahashi, K. (1990) Lung cancer risk among pneumoconiosis patients in Japan, with special reference to silicotics. IARC Sci. Publ. pp. 95–104.

Christiani, D.C. (1984) Occupational health in the People's Republic of China. *Am. J. Public Health*, **74**, 58–64.

Christiani, D.C. (1988) Modernization and occupational cancer. *J. Occup. Med.*, **30**, 975–6.

Christiani, D.C. (1996) Utilization of biomarker data for clinical and environmental intervention. *Environ. Health Perspect.*, **104**(Suppl 5): 921–5.

Costello, J. and Graham, W.G. (1988) Vermont granite workers' mortality study. *Am. J. Ind. Med.*, **13**, 483–97.

Doll, R. (1985) Occupational cancer: A hazard for epidemiologists. *Int. J. Epidemiol.*, **14**, 22–31.

Doll, R. and Peto R. (1981) The causes of cancer: Quantitative estimates of avoidable risks of cancer in the U.S. today. *J. Natl Cancer Inst.*, **66**, 1191–308.

Dong, M.H., Redmond, C.K., Mazumdar, S. and Constantino, J.P. (1988) A multi-stage approach to the cohort analysis of lifetime lung cancer risk among steelworkers exposed to coke-oven emissions. *Am. J. Epidemiol.*, **128**, 860–73.

Dubrow, R. and Wegman, D.H. (1984) Cancer and occupation in Massachusetts: a death certificate study. *Am. J. Ind. Med.*, **6**, 207–30.

Emmelin, A., Nystrom, L. and Wall, S. (1993) Diesel exhaust and smoking: a case-referent study. *Epidemiology*, **4**, 237–44.

Enterline, P.E., Marsh, G.M., Esmen, N.A., Henderson, V.L., Callahan, C.M. and Paik, M. (1987) Some effects of cigarette smoking, arsenic and SO_2 on mortality among US copper smelter workers. *J. Occup. Med.*, **29**, 831–8.

Finkelstein, M., Liss, G.M., Krammer, F. and Kusiak, R.A. (1987) Mortality among workers receiving compensation awards for silicosis in Ontario 1940–85. *Br. J. Ind. Med.*, **44**, 588–94.

Finkelstein, M.M. (1995) Radiographic abnormalities and the risk of lung cancer among workers exposed to silica dust in Ontario. *Can. Med. Assoc. J.*, **152**, 37–43.

Fu, H., Gu, X., Jon, X., Yu, S., Wu, K. and Guidotti, T.L. (1994) Lung cancer among tin miners in southeast China: silica exposure, silicosis and cigarette smoking. *Am. J. Ind. Med.*, **26**, 373–81.

Garcia-Closas, M., Kelsey, K.T., Wiencke, J. and Christiani, D.C. (1995) Nutrient intake as a modifier of the association between lung cancer and glutathione S-transferase μ deletion. In: Proceedings of the Eighty-sixth Annual Meeting of the American Association for Cancer Research, Toronto, Canada, 281.

Garshick, E., Schenker, M.B., Munoz, A., *et al.* (1988) A retrospective cohort study of lung cancer and diesel exhaust in railroad workers. *Am. Rev. Respir. Dis.*, **137**, 820–25.

Goldsmith, D. (1994) Silica exposure and pulmonary cancer, in *Epidemiology of Lung Cancer*, Marcel Dekker, New York, pp. 245–98.

Gustafsson, L., Wall, S., Larsson, L. and Skog, B. (1986) Mortality and cancer incidence among Swedish dock workers: A retrospective cohort study. *Scan. J. Work Environ. Health*, **12**, 22–6.

Gustavsson, P. and Reuterwall, C. (1990) Mortality and incidence of cancer among Swedish gas workers, *Br. J. Ind. Med.*, **47**, 169–74.

Heim, D.W. (1988) Acetylator genotype and arylamine–induced carcinogenesis, *Biochim. Biophys. Acta*, **948**, 36–66.

Heinrich U. (1992) Pulmonary carcinogenicity of cadmium by inhalation in animals, in *Cadmium in the Human Environment: Toxicity and carcinogenicity*, (eds G. Nordberg, R. Herber and L. Alessio), IARC, Lyon, France, pp. 405–14.

Hennekens, C.H. and Buring, J.E. (1987) *Epidemiology in Medicine*. Little Brown and Company, Boston.

Hessel, P.A., Sluis-Cremer, G.K. and Hnizdo E. (1986) Case-control study of silicosis, silica exposure, and lung cancer in White South African gold miners, *Am. J. Ind. Med.*, **10**, 57–62.

Hessel, P.A., Sluis-Cremer, G.K. and Hnizdo E. (1990) Silica exposure, silicosis, and lung cancer: a necropsy study, *Br. J. Ind. Med.*, **47**, 4–9.

Hill, A.B. and Fanning, E.L. (1948) Studies in the incidence of cancer in a factory handling inorganic compounds of arsenic. I. Mortality Experience in the Factory. *Br. J. Ind. Med.*, **5**, 1–6.

Hnizdo, E. and Sluis Cremer, G.K. (1991) Silica exposure, silicosis, and lung cancer: a mortality study of South African gold miners, *Br. J. Ind. Med.*, **48**, 53–60.

Hodgson, J.T. and Jones R.D. (1990) Mortality of a cohort of tin miners 1941–86, *Br. J. Ind. Med.*, **47**, 665–76.

Hollstein, M., Sidransky, D., Vogelstein, B. and Harris C.C. (1991) *p53* mutations in human cancers, *Science*, **253**, 49–53.

Howe, G.R., Fraser, D., Lindsay, J., Presnal, G. and Yu S.Z. (1983) Cancer mortality (1965–1977) in relation to diesel fume and coal exposure in a cohort of retired railway workers. *J. Natl. Cancer Inst.* **70**, 1015–1019.

Husgafvel–Pursiainen, K., Hackman, P., Ridanpaa, M., *et al.*, (1993) K-*ras* mutations in human adenocarcinoma of the lung: association with smoking and occupational exposure to asbestos, *Int. J. Cancer*, **53**, 250–6.

IARC (1976) *Cadmium, Nickel, Some Epoxides, Miscellaneous Industrial Chemicals and General Considerations on Volatile Anaesthetics, Monograph 11.* IARC Lyon, France.

IARC (1979) Monographs on the Evaluation of Carcinogenic Risks to Human. Chemicals and Industrial Processes Associated with Cancer in Humans, IARC, Lyon, France.

IARC (1980) Some metals and metallic compounds, Monograph 23. IARC, Lyon, France.

IARC (1984) Monographs on the evaluation of the carcinogenic risk of chemicals to humans. Polynuclear Aromatic Compounds, Part 3: Industrial Exposures in Aluminum Production, Coal Gasification, Coke Production and Iron and Steel Founding. Vol. 34, IARC Lyon, France.

IARC (1987) Monographs on the evaluation of the carcinogenic risk of chemical to humans. Polynuclear Aromatic Compounds, Part 4: Bitumens, Coal–Tars and Derived Products, Shale Oils and Soots. Vol. 35, IARC, Lyon, France.

IARC (1987) An evaluation of carcinogenic risk to humans. Overall evaluation of carcinogenicity: An updating of IARC. Monograph Vols 1–42, Suppl 7, IARC, Lyon, France.

IARC (1990) *Chromium, Nickel, and Welding, Monograph 49.* IARC, Lyon, France.

IARC (1993) *Beryllium, Cadmium, Mercury and Exposures in the Glass Manufacturing Industry, Monograph 58*, IARC, Lyon, France.

ICNCM (International Committee on Nickel Carcinogenesis in Man) (1990) Report of the International Committee on Nickel Carcinogenesis in Man, *Scand. J. Work Environ. Health*, **16**, 1–84.

Infante Rivard, C., Armstrong, B., Petitclerc, M., Cloutier, L.G. and Theriault, G. (1989) Lung cancer mortality and silicosis in Quebec, 1938–85. *Lancet*, **2**, 1504–7.

International Agency for Research on Cancer (1987) Silica. I.A.R.C. Monogr. Eval. Carcinog. Risk Chem. Hum., **42**, 39–143.

Ishinishi, N., Koizumni, A., McClellan, R. and Stober, W. (eds) (1986) *Carcinogenic and Mutagenic Effects of Diesel Engine Exhaust*, Elsevier, New York.

Jarup, L., Pershugen, G. and Wall, S. (1989) Cumulative arsenic exposure and lung cancer in smelter workers: a dose-response study, *Am. J. Ind. Med.*, **15**, 31–41.

Kleinbaum, D., Kupper, L. and Morganstern, H. (1982) *Epidemiologic Research*, Lifetime Learning, Belmont CA.

Lagorio, S., Forastiere, F., Michelozzi, P., *et al.* (1990) A case-referent study on lung cancer mortality among ceramic workers, IARC Sci. Publ., pp. 21–8.

Lee–Feldstein, A. (1986) Cumulative exposure to arsenic and its relationship to respiratory cancer among upper smelter employees, *J. Occup. Med.*, **28**, 296–302.

Lynge, E., Kurppa, K., Kristofersen, L., Malker, H. and Sauli, H. (1986) Silica dust and lung cancer: results from the Nordic occupational mortality and cancer incidence registers, *J. Natl Cancer Inst.*, **77**, 883–9.

Mabuchi, K., Lilienfeld, A.M. and Snell, L.M. (1979) Lung cancer among pesticide workers exposed to inorganic arsenicals, *Arch. Environ. Health*, **34**, 312–20.

Mastrangelo, G., Zambon, P., Simonato, L. and Rizzi, P. (1988) A case-referent study investigating the relationship between exposure to silica dust and lung cancer, *Int. Arch. Occup. Environ. Health*, **60**, 299–302.

McDonald, J.C. (1989) Silica, silicosis, and lung cancer (editorial). *Br. J. Ind. Med.*, **46**, 289–91.

McFarlane, M.J., Feinstein, A.R. and Wells, C.K. (1986a) Clinical features of lung cancers discovered as a postmortem 'surprise', *Chest*, **90**, 520.

McFarlane, M.J., Feinstein, A.R. and Wells, C.K. (1986b) Necropsy evidence of detection bias in the diagnosis of lung cancer, *Arch. Intern. Med.*, **146**, 1695.

McLaughlin, J., Chen, J., Mustafa, D., *et al.*, (1992) A nested case-control study of lung cancer among silica-exposed workers in China, *Br. J. Ind. Med.*, **49**, 167–71.

McMichael, A.J. (1994) Molecular epidemiology: new pathway or new traveling companion? *Am. J. Epidem.*, **40**, 1–11.

Mehnert, W.H., Staneczek, W., Mohner, M., *et al.* (1990) A mortality study of a cohort of slate quarry workers in the German Democratic Republic. IARC Sci. Publ., pp. 55–64.

Meijers, J.M., Swaen, G.M., van Vliet, K. and Borm, P.J. (1990) Epidemiologic studies of inorganic dust-related lung diseases in The Netherlands, *Exp. Lung Res.*, **16**, 15–23.

Merlo, F., Doria, M., Fontana, L., Ceppi, M., Chesi, E. and Santi L. (1990) Mortality from specific causes among silicotic subjects: a historical prospective study. IARC Sci. Publ., pp. 105–11.

Morabia A., Markowitz S., Garibaldi K. and Wynder E. (1992) Lung cancer and occupation: Results of a multicentre case-control study. *Br. J. Ind. Med.*, **49**, 721–7.

Muhle, H., Takenadka, S., Mohr, U., Dasenbroack, C. and Mermelstein, R. (1989) Lung tumor induction upon long-term low-level inhalation of crystalline silica. *Am. J. Ind. Med.*, **15**, 343–6.

Nakachi, K., Imai, K., Hayashi, S. and Kawajiri, K. (1993) Polymorphisms of the CYP1A1 and glutathione S-transferase genes associated with susceptibility to lung cancer in relation to cigarette dose in a Japanese population. *Cancer Res.*, **53**, 2994–9.

National Research Council (1990) Health Effects of Exposure to Low Levels of Ionizing Radiation. BEIR V. National Academy Press, Washington.

Ng., T.P., Chan, S.L. and Lee, J. (1990) Mortality of a cohort of men in a silicosis register: further evidence of an association with lung cancer. *Am. J. Ind. Med.*, **17**, 163–71.

Osburn, H.S. (1969) Lung cancer in a mining district in Rhodesia. *S. Afr. Med. J.*, **43**, 1307–12.

Parkin, D.M., Vizcaino, A.P., Skinner, M.E. and Ndhlovu, A.C. (1994) Cancer patterns and risk factors in the African population of southwestern Zimbabwe, 1963–1977. *Cancer Epidem. Biomarkers Prev.*, **3**, 537–47.

Partanen, T., Pukkala, E., Vainio, H., Kurppa, K. and Koskinen, H. (1994) Increased incidence of lung and skin cancer in Finnish silicotic patients. *J. Occup. Med.*, **36**, 616–22.

Patrianakos, C. and Hoffman, D. (1979) Chemical studies of tobacco smoke, LXIV. On the analysis of aromatic amines in cigarette smoke. *J. Anal. Chem.*, **3**, 150–4.

Qiao, Y.L., Taylor, P.R., Yao, S.X., *et al.* (1989) Relation of radon exposure and tobacco use to lung cancer among tin miners in Yunnan Province, China. *Am. J. Ind. Med.*, **16**, 511–21.

Redmond, C.K. (1983) Cancer mortality among coke-oven workers. *Environ. Health Perspect.*, **52**, 67–73.

Samet, J.M., Pathak, D.R., Morgan, M.V. *et al.* (1994) Silicosis and lung cancer risk in underground uranium miners. *Health Phys.*, **66**, 450–3.

Sellers, T.A., Bailey-Wilson, J.E., Elston, R.C., Wilson, A.E., Elston, R.C. and Rotschild, H. (1990) Evidence for Mendelian inheritance in the pathogenesis of lung cancer. *J. Natl. Cancer Inst.*, **82**, 1272–9.

Sherson, D., Svane, O. and Lynge, E. (1991) Cancer incidence among foundry workers in Denmark. *Arch. Environ. Health*, **46**, 75–81.

Siemiatycki, J., Gerin, M., Dewar, R. and Lakhani, R., (1990) Silica and cancer associations from a multicenter occupational case-referent study. *IARC Scientific Publications*, **97**, 29–42.

Siemiatycki, J., Wacholder, S., Dewar, R., Cardis, E., Greenwood, C. and Richardson L. (1988) Degree of confounding bias related to smoking, ethnic group, and socio-economic status in estimates of the association between occupation and cancer. *J. Occup. Med.*, **30**, 617–25.

Silverstein, M., Maizlish, N., Park, R. and Mirer, F. (1985) Mortality among workers exposed to coal-tar pitch volatiles and welding emissions. *Am. J. Public Health*, **75**, 1283–7.

Sitas, F., Douglas, A.J. and Webster, E.C. (1989) Respiratory disease mortality patterns among South African iron moulders. *Br. J. Ind. Med.*, **46**, 310–15.

Slebos, R.J.C., Hruban, R.H., Dalesio, O., Mooi, W.J., Offerhaus, J.A. and Rodenhuis S. (1991) Relationship between K-*ras* oncogene activation and smoking in adenocarcinoma of the human lung. *J. Natl Cancer Inst.*, **83**, 1024–7.

Smith, T.J. (1987) Exposure assessment for occupational epidemiology. *Am. J. Ind. Med.*, **12**, 249–68.

Stayner, L., Smith, R., Thun, M., Schnorr, T. and Lemen, R. (1992) A dose-response analysis and quantitative assessment of lung cancer risk and occupational cadmium exposure. *Ann. Epidemiol.*, **2**, 177–94.

Steenland, K., Silverman, D. and Hornung, R. (1990) Case-control study of lung cancer and truck driving in the Teamsters Union. *Am. J. Public Health*, **80**, 670–4.

Steenland, K., Loomis, D., Shy, C. and Simonsen, N. (1996) Review of occupational lung carcinogens. *Am. J. Ind. Med.*, **29**, 474–90.

Steenland, K. and Ward, E. (1991) Lung cancer incidence among patients with beryllium disease: A cohort mortality study. *J. Natl. Cancer Inst.*, **83**, 1380–5.

Sun, S.Q. (1987) Etiology of lung cancer at the Gejiu tin mine, China. Princess Takamatsu Symp. **18**, 103–15.

Sun, S.Q., You, Z.Y., Tan, S.Y., Chen, Y.W. and Zhao, G.F. (1989) Pathohistogenetic approach on the etiology of Yunnan tin miner's lung cancer. *Chin. Med. J.*, **102**(5), 347–55.

Suzuki, H., Takahashi, T., Kuroishi, T. *et al.* and (1992) Mutations in non-small cell lung cancer in Japan: association between mutations and smoking. *Cancer Res.*, **52**, 734–6.

Talaska, G., Dooley, K.L. and Kadlubar, F.F. (1990) Detection and characterization of carcinogen-DNA adducts in exfoliated urothelial cells from 4-aminobiphenyl-treated dogs. *Carcinogenesis*, **11**, 639–46.

Taylor, P.R., Qiao, Y.L., Schatzkin, A., *et al.* Relation of arsenic exposure to lung cancer among tin miners in Yunnan Province, China. *Br. J. Ind. Med.*, **46**, 881–886.

Thun, M., Schnort, T. and Halperin, W. (1986) Mortality from lung and prostatic cancer in U.S. cadmium workers. Presented at the Workshop on Cadmium and Cancer, Oxford, England, September 29-October 1, 1986.

Tokudome, S. and Kuratsune, M. (1976) A cohort study on mortality from cancer and other causes among workers at a metal refinery. *Int. J. Cancer*, **17**, 310–17.

Tomatis, L. (1988) The contribution of the IARC monographs to the identification of cancer risk factors. *Ann. N.Y. Acad. Sci.*, **534**, 31–8.

Tornling, G., Hogstedt, C. and Westerholm, P. (1990) Lung cancer incidence among Swedish ceramic workers with silicosis. IARC Sci. Publ., 113–9.

Vahakangas, K.H., Samet, J.M., Metcalf, R.A. *et al.* (1992) Mutations of p53 and *ras* genes in radon-associated lung cancer from uranium miners. *Lancet*, **339**, 576–80.

Vineis, P. (1992) Epidemiological models of carcinogens: the example of bladder cancer. *Cancer Epidem. Biomarkers Prev.*, **1**, 149–53.

Vineis, P., Bartsch, H., Caporaso, N., *et al.* (1994) Genetically-based *N*-acetyltransferase metabolic polymorphism and low-level environmental exposure to carcinogens. *Nature*, **369**, 154–6.

Vineis, P., Cantor, K., Gonzales, C., Lynge, E. and Vallyathan, V. (1995) Occupational cancer in developed and developing countries. *Int. J. Cancer*, **15**, 655–60.

Vineis, P., Caporaso, N., Tannenbaum, S.R., *et al.* (1990) Acetylation phenotype, carcinogen hemoglobin adducts and cigarette smoking.

Vineis, P., Thomas, T., Hayes, R., *et al.* (1988) Proportion of lung cancers in males due to occupation in different areas of the U.S. *Int. J. Can.*, **42**, 851–6.

Waalkes, M., Rehm, S., Perantoni, A. and Coogan, T. (1992a) Cadmium exposure in rats and tumors of the prostate, in *Cadmium in the Human Environment: Toxicity and Carcinogenicity*, (eds. G. Norberg, R. Herber, L. Alessio), Lyon, France, IARC, pp. 391–400.

Waalkes, M., Rehm, S., Sass, B. and Ward, J. (1992b) Induction of tumors of the hematopoietic system by cadmium in rats, *Cadmium in the Human Environment: Toxicity and Carcinogenicity*, (eds. G. Norberg, R. Herber, L. Alessio), Lyon, France, IARC, pp. 401–404.

Wagoner, J.K., Archer, V.E., Carroll, B., Holaday, D.A. and Lawrence, P.A. (1964) Cancer mortality patterns among US uranium miners and millers, 1950–1962. *J. Natl Cancer Inst.*, **32**, 787–801.

Wang, X., Christiani, D.C., Wiencke, J.K., *et al.* (1995) Mutations in the *p*53 gene in lung cancer are associated with cigarette smoking and asbestos exposure. *Cancer Epidem. Biomarkers Prev.*, **4**, 543–8.

Ward, E., Okan, A., Ruder, A., Fingerhut, M. and Steenland K. (1992) A mortality study of workers at seven beryllium processing plants. *Am. J. Ind. Med.*, **22**, 885–904.

Welch, K., Higgins, I., Oh, M. and Burchfiel, C. (1982) Arsenic exposure, smoking and respiratory cancer in copper smelter workers. *Arch. Environ. Health*, **37**, 325–35.

Wiencke, J., Varkonyi, A., Semey, K., *et al.* (1995) Validation of blood mononuclear cell DNA-adducts as a marker of DNA damage in human lung. *Cancer Res.*, **55**, 4910–15.

Wogan, G.N. and Gorelick, N.S. (1985) An overview of chemical and biochemical dosimetry of exposure to genotoxic chemicals. *Environ. Health Perspect.*, **62**, 5–18.

Wu, K.G., Fu, H., Mo, C.Z. and Yu, L.Z. (1989) Smelting, underground mining, smoking, and lung cancer: a case-control in a tin mine area. *Biomed. Environ. Sci.*, **2**, 98–105.

Wu, W. (1988) Occupational cancer in the People's Republic of China. *J. Occup. Med.*, **30**, 968–74.

Wyndham, C.H., Bezuidenhout, B.N., Greenacre, M.J. and Sleeis-Cremer, G.K. (1986) *Br. J. Ind. Med.*, **43**, 677–684.

Xu, Z.Y., Blot, W.J., Li, G. *et al.* (1991) Environmental determinants of lung cancer in Shenyang, China, *IARC Sci. Publ.*, **105**, 460–5.

Xuan, X., Lubin, J., Li, J., *et al.* (1993) A cohort study in southern China of tin miners exposed to radon decay products. *Health Phys.*, **64**, 120–31.

Yang, M.D., Wang, J.D., Wang, Y.L. *et al.* (1985) Changes in health conditions in the Huainan coal mine in the past three decades. *Scand. J. Work Environ. Health*, **11(4)**, 64–7.

Zambon, P., Simonato, L. and Mastrangelo, G. *et al.* (1987) Mortality of workers compensated for silicosis during the period 1959–1963 in the Veneto region of Italy. *Scand. J. Work Environ. Health*, **13**, 118–23.

Index

Numbers in bold refer to illustrations; numbers in italic refer to tables

Abeitic acid 349, 354, 355
Accident fund 86
Accident/incident investigations, gases and vapors 61–2
Acetic acid exposure 460
Acrylates 349
Actinomycetes 435, 437
Active sampling 65–6
Acute
hypersensitivity pneumonitis 438–9
inhalation injury, metallic compounds 285–6
sequelae 286
toxic/irritant exposures 453–61
asthma 336–7, 458–60
frequency of episodes 455
mechanisms 453–5
persistant effects of 456–60
Adenocarcinoma
chromosome 3p loss of heterozygosity *477*
histochemistry 235
distinguishing from mesothelioma 242
K-*ras* codon 12 mutations *471*
p53 mutations 476
recurrent disease *474*
Adhesion molecule expression, asthma 334
Adult respiratory distress syndrome (ARDS), metallic compound inhalation 285, 286
Aerodynamic diameters, particle 41
Aerosols 53
Agricultural workers 435
see also Farmer's lung
Air stagnation, passive samplers 66–7
Air-purifying respirators 60–1
Airflow limitation, acute toxic exposures 456–9, 460
Airway responsiveness, assessment 309–23
airway resistance 20
medical history 310–11
methacholine/histamine challenge 311, 314, 315–18, 328
factors affecting results 317
methods 315–16
standards for testing 24
peak flow monitoring 21, 25, 310, 311, 312–15, 318
specific occupational challenge tests 318–22
methodologic issues 320–1
spirometry 311–12, 318
see also Spirometry
Airways obstruction 69–79
asbestosis 242
asthma 340–1
atopy, family history 76
childhood illnesses 77
clinically important 75–6
effect of dust on 70–2
quantitative estimates 72–5
host susceptibility 76
obesity 77
silicosis 205
socioeconomic status 78
see also Chronic obstructive pulmonary disease (COPD)
'Alcalase' 390
Aldehydes 349
asthma induction 339, 349, 357–60
sampling methods *65*
n-Alkanes, gas/particle partitioning in tobacco smoke 56, 57
Allergen exposure, concurrent 317
α_1-antitrypsin phenotypes 249
α-amylase 389
Aluminum
inhalation therapy 127–8, 194
toxicity *47*, *282*
COPD 288
corundum plant workers 111
parenchymal lung disease 294–5
potroom asthma 290–1
Alveolar macrophages 139, **141**, 143
hypersensitivity pneumonitis 438, 441, 442
macrophage-derived inflammatory mediators 143
pneumoconiosis 174, 175
acute toxic chemical exposures 454
asbestosis 223, 224
clearance of particles 154
silicosis 194, **195**
Alveolar-arterial oxygen difference (P(A-a)O_2 26
Alveoli, dust deposition 41, 139
Ambient sampling, PM_{10} particles 42
American College of Pathologists, asbestosis grading *226*

American Conference of Government Industrial Hygienists (ACGIH)
  respirable dust
    definition 39
    size-selective sampling criteria 40
  threshold limit value (TLV)
    cotton dust 422
    crystalline silica 43, 44, 200–1
American Medical Association
  dyspnea scale *27*
  guidelines, disability evaluation 29, 242
    exercise testing 31
American Thoracic Society
  guidelines, disability evaluation 29
  spirometry values, interpretation 21, *22*, 31
Amines 349
  aromatic 493
  asthma induction 356–7
Ammonia 460
  exposure
    persistent airway hyper-reactivity 455
    site of deposition 454
  sampling methods *65*
Amosite *46*, 220, 255, 257
Amphiboles 107, 109, 111, 220, 221, 223, 241
  mesothelioma 234
Amylase protease 390
Anaerobic threshold (AT) 25, 26
Anaphylaxis, soybean asthma epidemic 404
Anatomy, respiratory system 17, **18**
'Angel's wings' **199**
Anhydrides 349
  occupational asthma 335, 338, 348, 360–3
Anhydrite *46*
Animal studies
  man made vitreous fibers (MMVF) 266–7
  silicosis, data interpretation 194
Anthophyllite 221, 255
  toxicity *46*
    lung cancer 256, 257, 258
    mesothelioma 258
Anti-oxidant defenses 145, 146
Anti-TNF antibody, influence on fibrosis **151**
Antibiotic-induced asthma
  ceftazidime inhalation challenge **331**
  related manuscripts *346*
Antimony, toxicity *282*
  acute inhalation 286
  antimoniosis 292–3
  lung cancer 297
Antinuclear antibodies (ANA) 175, 176, 206
Antioxidant chemoprevention 493
Arachidonic acid derivatives, leukocyte-derived 153, 154
Arithmetic mean 54, 55, 63
Aromatic amines 493
Arsenic toxicity *47*, *282*
  acute inhalation 286
  carcinogenicity 487–8
    lung cancer 297, 487–8, 489–90
Arterial blood gases 23
Asbestos 219, 239
  carcinogenicity 250, 251, 476, 478, 485, 487
  deposits, major global sites *221*
  fibers 35
    counting 36, 241
    deposition and lung clearance 222
    in environment and tissues 221–2, 240
    size and shape 146, **147**
    specificities 222–3
    width 241
    worker susceptibility to 224
  industry *219*, 221
    Brazil 107, 109–11
    Finland 255–6
    Russia 247, 251
    South Africa 99–100
  mineralogy 220–1
  prevention of related diseases 236
  in public buildings 240
Asbestosis 121, 122, **123**, 223–9, 236, 239, 256
  clinical findings 226–7
    Russian observations 248
  diagnostic criteria 223–5, 241
  dose-response relationship 225
  early detection 228–9
  latency period 122, 224, 249
  pathology 225–6
  prevalence *8*
    Brazil 110
    Finland 257
    Russia 248–50, 252
  radiology
    chest CT 125, 126, 227–8, 229, 248
    chest radiograph 223, 227, 229
  treatment and follow-up 229, 242, 243–4
    long-term 242, 243
  workers' compensation 89–90, 229
Ash, coal 161
Assessment, pulmonary impairment 26–30
  approach to evaluation 30–1
  respiratory impairment evaluation report 31–2
  see also Pulmonary function tests
Asthma
  urban epidemics 401–6
  see also Occupational asthma
Asthmatic predisposition 76
Atelectasis, rounded 232, 233, 236
Atmospheric pollution, asthma 403
Atopy 327
  in asthma 334, 338, 394
    grain and flour workers 389
    laboratory animal asthma 384
    platinum sensitivity 353
    soybean epidemic 404–5
  family history 76
Attribution, respiratory impairment evaluation report 32
Australian sheep blowfly, asthma induction 387
Autoantibodies in CWP 176
Avian (pigeon) antigens 437
  see also Pigeon breeder's disease
Azo dyes, asthma induction 349, 355

B lymphocytes 332
Bagassosis *436*, **438**, 447, 448
  chest X-ray **439**, **440**

Bakers, asthma 387, 388, 389
Barcelona, soybean epidemic asthma 402–5
Barium toxicity *282*
  baritosis 292
Base substitutions, gene mutations 475–6
Basement membrane thickening, asthma 329, 330, 340, 352
Basophils, immediate asthmatic reactions 333
Bauxite *46*
Beetle larvae, asthma induction 387
Berkson-type grouping, exposure measurements 55
Beryllium lymphocyte transformation test (LTT) 293–4
Beryllium toxicity *47*, *282*, 286, 291
  berylliosis 112, 281, 293–4
  COPD 288
  lung cancer 297, 491
Bhopal disaster 56, 454, 456, 457
Biopersistence of MMVF 265
Biopsies
  asthma 340
  open-lung, asbestos-related disease 241, 244
  silicosis 199, 200
Biotite *46*
Bird fancier's disease 435, 440
  see also Pigeon breeder's disease
Bituminous coal mining, air-purifying respirators 60
'Black Box Syndrome' 67
Black coal 161
Black lung *see* Coal workers' pneumoconiosis (CWP)
Black Lung Act 30
Blowfly, asthma induction 387
Blue asbestos *see* Crocidolite
Brazil 105–13
  asbestos-related diseases 109–11
  coal workers' pneumoconiosis (CWP) 107–8, 184, 186
  disease control strategies 112–13
  dust exposure legislation 106–7
  organic dust exposure 112
  population distribution 105
  silicosis 108–9
    epidemics **4**, *5*
Breath-holding method, diffusion capacity measurement 22–3
British Medical Research Council (BMRC), MRE horizontal elutriator 39–40
British National Coal Board (NCB), dust sampling 44
British Occupational Hygiene Society, cotton dust exposure 422
Bromelain, asthma induction 391
Bronchial provocation testing 24–5, 311, 328–9
  methacholine/histamine challenge 311, 314, 315–18, 328
    factors affecting results 317
    methods 315–16
  specific occupational challenge tests 25, 289, 290, 318–22, 331
    false positive/negative chamber challenges 319–20
    frequency of monitoring 322
    hospital admission 322
    hypersensitivity pneumonitis 446, **447**
    indications *319*
    low-molecular weight compounds 348, 358
    methodologic issues 320–2
Bronchiogenic carcinoma *see* Lung cancer
Bronchitis 79, 112
  coal dust 167, 168, 171
  cotton workers 411, 427, 431
  silica dust 70, 71, 207, 215
  socioeconomic status 78
Broncho-alveolar lavage (BAL)
  asbestos workers **222**
    asbestosis 224, 229
    ferruginous bodies 241
  coal workers' pneumoconiosis (CWP) 175
  hard metal lung disease 296
  hypersensitivity pneumonitis 441, 443, 445–6
  occupational asthma 330, 334, 351
  plicatic acid exposure 333
  silicosis 128, 194, 195, 196, 199
Bronchodilator therapy, CWP 176
Brown coal 161
Butterfly allergens, asthma 385
Byssinosis 409–17, 421–31
  causative agents 414–16
  diagnosis 416–17, 427–8
  epidemiology *8*, *112*, 412–13
    prevalence, developing world *423*-5
  nature of disease 410–12
    clinical manifestations, developing world 426–7
  pathology and pathogenesis 413–14
  prevention 417
    in developing world 428, 429
  worker's compensation 87–8, 429

Cadmium
  toxicity *47*, *282*
    acute inhalation 285
    chronic inhalation 280
    emphysema 287
    lung cancer 297, 298, 490–1
  workers, half-mask air-purifying respirators 60
Calcification
  asbestos-related pleural plaques 231
  silicosis opacities **199**, 202, **203**
Cameroon, byssinosis *430*
Canada
  asbestos in public buildings 240
  MMVF study, lung cancer 271
  organic dust asthma 381
  silicosis
    epidemics *5*
    surveillance 11
  workers' compensation
    malignancies 90
    occupational asthma 87
Canadian Medical Association's Task Force on Occupational Respiratory Disease, disability evaluation 29–30
Caplan's syndrome 169, 199, 204
Carbon dioxide, partial pressure ($PaCO_2$) 23
Carbon pneumoconiosis **9–10**

Carcinoembryonic antigen (CEA) 235
Carcinogenesis 467
  molecular model 467–70
Carcinogens *486–7*, 487–9
  metabolism and mutagenesis 468–9, 492
  non-genotoxic 470
  specific agents
    arsenic 487–8
    asbestos 250, 251, 476, 478, 485, 487
      see also Lung cancer; Mesothelioma
    coke-oven emissions 488–9
    crystalline silica 171, 192, 193, 208–9, 491–2
    diesel exhaust 489
    metals 280, 490–1
Cardiovascular limitation, estimating 26
Carmine, asthma induction 356
Carob bean flour, asthma induction 389–90
Castor beans, asthma induction 392, 401, 406
Cat allergens, asthma induction 383
CD4+ helper cells
  asthma 330, 331, 333, 334, 350
    cytokine production 349–50
CD8+ suppressor cells
  asthma 334, 350, 351, 376
  hypersensitivity pneumonitis 438, 443, 445
CD25+ cells, asthma 350
Cedar workers, asthma 328, 335, 339, 340, 350
  plicatic acid challenge 317, 331
Cellular transformation tests, MMVF 266
Cement asthma 290
Central Federation of Industrial Professional Associations, Germany 85
Central Limit Theorem 63
Centriacinar emphysema 71
Centrilobar emphysema 71
Ceramic workers, lung cancer 492
Cerium, toxicity *283*
  interstitial fibrosis 294
Chalcedony *46*
Chamber of Mines, South Africa 96, 100
Charcoal tube sampling 66
Chemical species, particles 36
Chest radiography
  hypersensitivity pneumonitis 438, **439**, **440**
  in impairment evaluation 30
  lung volume measurements 18, 19
  pneumoconiosis 119, 123–4
    1986 diagnostic criteria 188
    asbestosis 223, 227, 229, 248
    benign pleurisy 230
    berylliosis 293
    coal workers' pneumoconiosis (CWP) 168, 169
    diffuse pleural thickening 232
    mesothelioma 235
    pleural plaques 231
    rounded atelectasis 232–3
    silicosis 98, **122**, **124**, 192, **198**, **199**, 202–5
    silicotuberculosis **216**
Chest tightness, byssinosis 412, 426
Childhood illnesses, COPD susceptibility 77
Chile, CWP prevalence 186
China 494
  byssinosis 429, *430*
  coal workers' pneumoconiosis (CWP) 183, 184–6
  dust sampling 11
    total dust sampler 38, **39**
  lung cancer 489
    coke and coal-gas workers 488–9
    hematite miners 490
    tin miners 490
  silicotuberculosis 217
Chlorine 53
  chlorine dioxide, sampling methods *65*
  exposure 456, 457, 460
    chloramine-T, asthma induction 357
    domestic bleaching products 457
    respiratory protection 61
    site of deposition 454
    tolerance to 455
    tuberculosis susceptibility 455
    vapor spread, derailed train 456–7
Chromium toxicity *47*, *282*
  asthma 290
  contact dermatitis 289
  lung cancer 297, 490
Chromosome
  chromosome 3p 477–8
  damage tests, glass microfibers 266
  rearrangements 468
Chronic hypersensitivity pneumonitis 439, 440
Chronic obstructive pulmonary disease (COPD) 69, 70, 286–7
  childhood illnesses 77
  host susceptibility factors 76
  metal workers 287–8
  MMVF exposure 269
  prevalence, South Africa 100
  see also Airways obstruction
Chrysotile *46*, 109, 219, 240, 247, 255
  fibers 220, 223, 241
    lung clearance 222
    mesothelioma 234
  threshold limit values (TLV) 107
Classical byssinosis 423, 425
Closed-circuit
  method, lung volume measurement 18
  particle generation systems 321
Coal *46*, 162
  characteristics 161, *162*
  dust 174, 175
    composition 161
    effect on airways obstruction 71–2
    effect on lung function 73
    sampling 44–5, 48
  gasification, lung cancer 488–9
  macules 71, **172**, **173**
  mining
    early 20th century **163**
    today 163–7
  tar 162, 489
Coal workers' pneumoconiosis (CWP) 6, 71, 85, 121, 122, 167, 168, 178
  chest radiograph 168, 169

development of, cellular and immunologic factors 174–6
early descriptions, 1831: 162
epidemiology and prevalence *8*, 9, 10–11, 169–72
Brazil 107–8, 184
China 183, 184–6
India **186**
South Africa 100
management 176–7
pathology 172–4
prevention 177
in China 187
treatment, experimental approaches
aluminum inhalation *127*
whole lung lavage (WLL) 132–3
workers' compensation 88–9
Cobalt toxicity *47*, *282*, 286, 291
asthma 290, 351
COPD 288
fibrosing alveolitis 281
hard metal lung disease 100, 295–7
lung cancer 297
Cockroaches, asthma induction 387
Coffee beans, asthma induction 392
Coke 162
coke-oven emissions, lung cancer 488–9
Collagen mRNA probes 133
Colophony 349
asthma 354–5
Columbia, CWP prevalence 186
Compensation 3–4, 83–92
asbestos-related diseases 242
asbestosis 89–90, 229
byssinosis 87–8
coal workers pneumoconiosis (CWP) 88–9
entitlement programs 29–30
in Germany 91
beginnings of workers compensation 83–4
modern approaches 84–5
malignancies 90–1, 234
occupational asthma 13–14, 86–7, 91
radiation exposure 90–1
silicosis 88
workers' claims database 9
tuberculosis, South Africa 97
in United States 85–6, 91
Compensation authorities
Germany 84, 85
United States 86
Complement activation 143
Computed tomography (CT), pneumoconiosis 124–6
asbestosis 224, 227–8, 229, 242
diffuse pleural thickening 232
pleural plaques 231, 242
rounded atelectasis 232-**3**
silicosis 204–5
Contact dermatitis, chromium-induced 289
Continuous mining 165
Continuous-filament glass fibers 263, 264
Conventional mining 165
Copper
smelting
air-purifying respirators 60
lung cancer 297, 487–8, 489
toxicity *282*
Cor pulmonale, compensation 91
Cornstarch glove powder 393
Correction factors, dust sample measurements 40
Corticosteroids
hypersensitivity pneumonitis 447
isocyanate-induced asthma 377
pneumoconiosis 128, **129**, 207
Cotton
dust, effects on lung 409–17
early reports 409–10
epidemiologic studies 412–13
see also Byssinosis
industry, India 421
cotton exposure 422–3
Cough 168, 207, 226
acute toxic inhalation 455
Counting dust particles 36–7
Crackles, in asbestosis 226, 228
Cricket allergens, asthma induction 386
Cristobalite *46*
threshold limit value (TLV) 201
Crocidolite *46*, 219, 220, 223, 257, 292
Crystalline silica 35, 43–4, 191, 192
ambient levels 193, 202
analysis, particle size dependence 44
carcinogenicity 171, 192, 193, 208–9, 491–2
coal dust sampling 44–5
exposure limits 192, 193
development of exposure standard 200–1
see also Silicosis
Cyanoacrylates 349
Cycle industry, abrasive wheel polish workers 214, 215
Cyclone respirable dust sampler **39**, 40
Cytokine secretion
asthma 332–3, 334, 349–50, 351
hypersensitivity pneumonitis 441, 442
pneumoconiosis 150–3, **154**
asbestosis 223–4
silicosis 195

Death certificate data 8, 9, 11, 12, 170, 171
Dental technicians, pneumoconiosis 111, 281, 297
Desensitization, hypersensitivity pneumonitis 443, **444**
Desquamative interstitial pneumonitis (DIP) 296
Detergent products, asthma induction 390–1
Diamond polishers, pneumoconiosis 281
hard metal lung disease 295, 296
Diatomaceous earth *46*, 191
exposure limits 201
Diborane, acute inhalation 286
4,4'-Dicyclohexylmethane-diisocyanate (HMDI) *56*, 375
Diesel emissions
carcinogenicity 489
sampling 45
Diffuse nodular silicosis **202**
Diffusing capacity (transfer factor) test 21, 22–3, 31
clinical use 23

1.4-Diiosocyanato-benzene (PPDI) 375
1.6-Diiosocyanato-hexane *see* Hexamethylene diphenyl diisocyante (HDI)
1.5-Diiosocyanato-naphthalene (NDI) 375
Dioxin 470
4,4'-Diphenylmethane-diisocyanate *see* Methylene diphenyl diisocyante (MDI)
Direct-reading devices 67
Directly Observed Therapy, Short Course Treatment (DOTS) 97
Disability 27
  insurance 86
  malingering 28
  respiratory impairment
    evaluation report 31–2
    pulmonary function guidelines 29, 242, *243*
  see also Compensation
Disease schedules 86
Disincentives, physician motivation 91
DNA
  damage tests, MMVF 266
  direct interaction of metals 280
  modification 468, 470
  repair defects 469, 470
Dog allergens, asthma induction 383
Dolomite *46*
Douglas fir tussock moth 385
Dust
  control 48–9
  exposure, measuring 5, 11, 35, 36–43, 120–1
    dust deposition 41–2
    international comparisons 45, 48
  generation, exposure chambers 321
  particle composition 43–5
    analysis **9–10**
    mixed dust exposures 45
Dyes, asthma induction 349, 355–6
Dyspnea 26, 30
  asbestosis 226, 248
  byssinosis 427
  diffuse pleural thickening 231
  grading 27, **28**
  silicosis 207, 215
Earning power, reduction in 85
Effective protection factor (EPF) 60
Egg protein, asthma induction 392
Egypt, byssinosis 425, *430*
Electric power generation 162
Electrocardiography 30–1
Electronic
  peak flow devices 315
  spirometers 312
Electronics factory workers, solder flux 355
Emphysema 79
  cadmium inhalation 287
  coal dust 71
  host susceptibility factors 76
  silica dust 70–1
  workers' compensation 90
Endotoxins, Gram-negative bacteria 414–15, 422, 423
  guidelines for 'no effect' threshold *417*
  reducing content in cotton 428
Environmental
  exposures 77–8
    ambient sampling PM_{10} sampler 42
    asbestos fibers 221, 239–40, 247
    Sahara desert 214
  tobacco smoke, vapor/particle interactions 56, 57
Environmental Protection Agency (EPA) 42
  ambient quartz levels 193, 202
Enzymes, asthma induction 390–1
Eosinophils, asthma 329, 330, 333–4, 349, 351, 377
  late asthmatic reaction 330
  sputum 352–3
Epidemics
  occupational illnesses 4
    silicosis *5*, 6–7
  urban asthma 401–6
Epidermal growth factor receptor (EGFR) 472
Epithelial
  cells, effect of activation *141*
    proliferation 155
  damage, acute toxic chemical exposures 454
  disruption, asthma 329, 334, 336
Epoxy resins, asthma induction 361
Ethiopia, byssinosis *430*, 431
Ethylenediamine (EDA), asthma induction 356–7
European MMVF study, lung cancer 270, *271*
European Respiratory Society, lung function test standardization 21
European Society for Clinical Respiratory Physiology 31
  rating procedure for respiratory disability 29
Exercise testing 25–6, 31, *311*
Expiratory reserve volume (IRV) 18
Exposure
  chambers, bronchial challenge tests 320–1
    aldehydes 359
    exposure conditions and incremental increases 321–2
  distributions 53–5
  estimates 485–6
  factors affecting 55–61
  limits
    asbestos fibers 236
    cotton dust 413, 422, 429
    crystalline silica 43, 44, 107, 192, 193, 200–1
    environmental asbestos 221
    see also Maximum allowable concentration (MAC); Threshold limit value (TLV)
  metrics 36–7
Extrinsic allergic alveolitis *see* Hypersensitivity pneumonitis
Extrinsic factors, occupational exposures **55**, 59–61

Farmer's lung 435, *436*, 447
  BAL fluid 441, 445, 446
  IgG antibodies 444
  prognosis 448
Federal Coal Mine Health and Safety Act, 1969, US 44, 89
Feldspar *46*
Ferrous metals, definition 279
Ferruginous bodies, lung tissue 221–2, 241, 257

Fibers
asbestos 35
counting 36, 241
deposition and clearance 222
in environment and tissues 221–2, 240
size and length 146, **147**, 223
specificities 222–3
width 241
man-made vitreous (MMVF)
classification and composition 263–4
properties 264–5
Fibroblasts 139, 144
effect of activation *141*
proliferation 155
Fibronectin 223, 224, 446
Fibrosis 133
asbestos-related 125, 222, 223, 244
compensation 90, 91
hypersensitivity pneumonitis 440
metal exposure 291, 292
MMVF exposure 268–9
see also Progressive massive fibrosis (PMF)
Filter cassettes, total dust 38
Finger clubbing, asbestosis 226, 248
Finland
asbestos-related disease
asbestosis 257
lung cancer 257–8
mesothelioma 258
risk reduction, national program 258–9
history of asbestos 255–6
silicosis epidemics *5*
Firefighters, toxic chemical exposures 458–9
Fish processing workers, asthma 392
Flaviatase, asthma induction 391
Fletcher's dyspnea scale *27*
Flint *46*
Flour, asthma induction 339, 387–90
Flow spirometers 21
Fluorite *46*
Food coloring agents, asthma 356
Forced Expiratory Flow ($FEF_{200-1200}/FEF_{25-75\%}$) 20
Forced Expiratory Volume (FEV_1) 20
childhood illnesses 77
clinically important airways obstruction 75–6
comparibility of baseline 317
degree of impairment, clinically important 69
host susceptibility factors 76
obesity 77
predicted decline, multiple regression analysis 72–5, 78, 79
longitudinal studies *74–5*
Forced vital capacity (FVC) maneuver 20
Forestry workers, asthma 385
Formaldehyde 339, 349
asthma induction 357–8
sampling methods *65*
France, silicosis epidemics *5*
Fraud, coal mine sampling 120
Free radicals 35, 144, **145**, 147–9, 174, 224, 280
Friction material production 110, 251
Fruit flies, asthma induction 387
Fruits, allergy to 393
Fullers earth *46*
Fume generation systems 321
Functional residual capacity (FRC) 17, 18
significance of changes in 19
Fungal allergens
asthma in bakers 389
in atmosphere 401
thunderstorms 405
grain harvesting 388

Gallium scans, asbestosis 125, 224, 229
Gas transfer 21, 22–4
Gases 53–67, 454
exposure assessment strategies 61–4
exposure distributions 53–5
factors affecting occupational exposures 55–61
sampling 64–7
site of deposition and clearance 454
Gastric cancer
asbestos workers 249, 250, 251
coal miners 171, 172
Gauley Bridge tunnel construction 6, 88
Gemstone workers, Hong Kong 215
see also Diamond polishers, pneumoconiosis
Gene expression, cytokine 150–3
General Insurance Act of the German Reich 84
General Mining Act, 1865, Germany 84
Generic passive sampler 66
Genetic factors
asthma development 338, 339
lung cancer 470–8, 492–3
Geometric mean 54, 55
calculating 54
HDI-based spray paint manaufacture, hygiene survey 60
Geometric standard deviation 55, 63
calculating 54
German Liability Insurance Act, 1871: 84
Germany
beginnings of workers' compensation 83–4
modern German compensation approaches 84–5
Giant cell interstitial pneumonitis (GIP) 296, 297
Giant cells, hypersensitivity pneumonitis 438
Ginseng, asthma induction 392
Glass
fibers, cytotoxicity 266
wool 263, 264
lung cancer 270
lung fibrosis 269
Glutaraldehyde 349
asthma induction 358–9, 360
Gold miners
Brazil, silicosis 108
South Africa 95–9
airways obstruction 70, 71, 77
HIV 98–9
lung cancer 489
silicosis and tuberculosis 95–8
Gold, toxicity *282*
Government policies 3–4, 192
Grain and flour, asthma induction 387–90

Gram-negative bacteria, cotton plant 409, 414–15, 422, 423
Granite *46*
  see also Vermont granite industry
Granuloma formation
  hypersensitivity pneumonitis **438**, 440, 442, 443
  metal exposure 293–4
  silicosis 195, 196
Graphite *46*
Gravimetric analysis 37
Greenburg-Smith impinger 37
Growth factor
  receptors, lung tumorigenesis 472
  release 224
GSTM1 polymorphisms, lung cancer 493
Guinea pig allergens, asthma induction 383, 384
Gums, asthma induction 394
Gypsum *46*

Half-mask air-purifying respirators, protection evaluation 60
Hamster allergens, asthma induction 383
Hard metal lung disease 100, 295–7
Health care workers
  acetic acid, high intensity exposure 460
  asthma
    aldehyde-induced 358, 359, 360
    latex-induced 393
Heart, effect of pneumoconiosis 168–9, 248
Hematite miners, lung cancer 490
Hemoptysis 226
Henna dye, asthma induction 356
Hepatoglobin (Hp I) 249
Herbal products, asthma induction 392
Hewett's procedure 63
Hexachlorophene, asthma induction 357
Hexahydrophthalic anhydride **360**
  asthma induction 361–2
Hexamethylene diisocyante (HDI) 13, 375
  asthma 376, 377
  polyurethane paint manufacture, hygiene survey 59–60
  vapor pressure/saturated vapor concentration *56*
High-molecular weight compounds, asthma 328, 331, 338, 347, 381–94
  enzymes 390–1
  generation, exposure chambers 321
  grain and flour 387–90
  gums and latex 393–4
  IgE antibodies 335
  laboratory animals and insects 382–7
  seafoods 391–2
High-resolution computed tomography (HRCT), pneumoconiosis 125, 126
  asbestosis 224, 227–8
Histamine
  challenge test 24, 311, 315–18, 328
  release, byssinosis 414, 423
History
  medical 30, 86–7
  occupational 30, 120, 310
  smoking 30
HIV, South African gold miners 96, 97, 98–9, 100
HLA antigens 339
Homogenous exposure groups (HEG) 62, 63
Honeybees, asthma induction 387
Hong Kong
  byssinosis 429, 430
  gemstone workers 215
Horizontal elutriator, MRE 39–40, 44
Host susceptibility factors, COPD 76
Human immunodeficiency virus *see* HIV
Humidifier lung *436*, 447
  chest X-ray **439**
Hydrochloric acid vapor, vapor/particle interactions 57
Hydrogen peroxide 145, 194, 469
Hydrogen selenide, acute inhalation 286
Hydroxyl radical 144, 149, 194
Hypersensitivity
  type I 332–3, 334–6
  type III and IV 442
Hypersensitivity pneumonitis 35, 291, 435–48
  clinical and histopathologic features 437–40
  diagnosis 443–6
    immunologic studies 444–5
  etiology 435–7
    beryllium workers 293
    glass filament workers 269
    specific antigens 437
    vapor/particle interactions 57
    zirconium workers 294
  pathogenesis 440–3
  prognosis 447–8
  treatment 446–7
Hypoxemia, asbestosis 248

Idiopathic lung disease 292
IgE antibodies 76
  occupational asthma 318, 327, 328, 333, 338, 350
    formaldehyde 358
    high-molecular weight allergens 335
    platinum salts 288, 340
    snow-crab asthma 340
IgG antibodies
  antigen-specific, hypersensitivity pneumonitis 444
  byssinosis 423
  coating of particles, pneumoconiosis 142
  occupational asthma 347, 350, 358
Iglau Mining Rules 83
IL-8 production, asbestos exposure 152, **153**
Immediate asthmatic reactions 330, 333, 348
Immune system involvement
  hypersensitivity pneumonitis 441–3
  occupational asthma 331–4, **336**
    allergic and non-allergic asthma 334
    development of immune response 331–2
    low-molecular weight compounds 349–53
    type I hypersensitivity 332–3, 334–6
  pneumoconiosis 142–4, 175–6

Immunization 177
Immunoglobulins
in CWP and PMF 176
see also IgE antibodies; IgG antibodies
Impactors 36
Impinger
count data 37
sampler 37–8
solutions 36, 37
Income, loss of 87
Indeterminate byssinosis 424, 425, 426
India
Bhopal disaster 56, 454, 456, 457
byssinosis 421–9, 431
CWP prevalence **186**
silicosis 213–15
epidemics *5*
epidemiology 213–15
silicotuberculosis 215–16
Indigotine, asthma induction 356
Industrial Accidents Act, 1884, Germany 84
Industrial hygienists 35, 37, 61, 236
activities 53
see also American Conference of Government Industrial Hygienists (ACGIH)
Inflammation of airways
asthma, immunologic mechanisms 333–4
low-molecular weight compounds 351–3
byssinosis, clinical features 411
hypersensitivity pneumonitis 441, **442**
Inflammatory changes, hypersensitivity pneumonitis 441, **442**
Infrared spectroscopy, crystalline silica analysis 44
Inhalable (inspirable) fraction 40–1
dust sampling 42
equipment **38**
Inhalation
fever *see* Toxic pneumonitis
studies, MMVF 267–8
Insects
asthma induction 385–7, 387, 388, 389
prevalence surveys *386*
hypersensitivity pneumonitis *436*
Inspiratory reserve volume (IRV) 18
Institute of Medicine (IOM), training occupational medicine 91
Insulation wools 263, 264, 265, 266
animal inhalation studies 268
lung cancer 270, 272
lung fibrosis 269
Interleukins, asthma
IL-4 334
IL-5 333, 334, 350–1
IL-8 350
International Association for Research on Cancer (IARC), carcinogens *486–7*
beryllium 491
crystalline silica 208, 491
nickel 490
polycyclic aromatic hydrocarbons 488
The International Committee on Nickel Carcinogenesis in Man (ICNCM) 490
International Labor Office (ILO) classification, chest radiograph 123–4, 202, 227
International size-selective dust definition 40, **41**
Interstitial
lung disorders, metallic compounds 291–7
macrophages 139, 143, 144
Intracavitary studies, MMVF 267
Intracellular signaling pathway activation, pneumoconiotic particles **148**-9
Intrinsic physical properties, toxicants 55–9
Invalidity and Old Age Insurance Act, 1889, Germany 84
Iron
role of particle-derived 149
toxicity *47*, *283*
COPD 287
lung cancer 297, 298, 491
siderosis 111, 292
silicosis 214
Irritant-induced asthma *see* Reactive airways dysfunction syndrome (RADS)
Isocyanate-induced asthma 4, 12, 13–14, 340, 348, 375–8, 459
diagnosis 377
HLA antigens 339
IgE antibodies 335, 376
isocynate sensitization 13
mechanisms of 376–7
cell types found 330–31, 334, 350, 351, 352, 353
treatment 377
Isocyanate-induced hypersensitivity pneumonitis 57
Isocyanates 12–13
hazards and safe practices 375–6
reactivity 58–9
sampling methods *65*
vapor pressures/saturated vapor concentrations 13, *56*
vapor/particle interactions 57
see also Bhopal disaster
Isoniazid 207, 216
prophylaxis 207–8, 217
Isophorone diioscyanate (IPDI) 56
Italy, silicosis epidemics *5*

Japanese summer-type HP 435, 437, 447, 448
Johns-Manville Corporation 89

K-*ras* 470, 471, 492
codon 12 mutations in NSCLC *471*
mutations and lung cancer recurrence *472*
Kaolin *46*
Korea, CWP prevalence 186
Kyanite *46*

Laboratory
animal workers
asthma 338, 339, 382–5
hypersensitivity pneumonitis 435
studies, respiratory impairment 30
Lanthanides, toxicity *283*
interstitial fibrosis 294

Large cell carcinoma, chromosome 3p loss of heterozygosity *477*
Late
  asthmatic reactions 320, 322, 330–1, 333, 348
  response, hypersensitivity pneumonitis 439
Latency periods 4, 6–7, 8
  asbestos-related pleural plaques 230
  asbestosis 121, 122, 224, 249
  coal workers' pneumoconiosis (CWP) 8, 185–6
  malignancies 90
    mesothelioma 122, 240
  occupational asthma 345
Latex, asthma induction 393–4
Lead
  smelter workers, air-purifying respirators 60–1
  toxicity *47*, *283*
    chronic inhalation 280
Leon Commission of Enquiry into the Health and Safety of mineworkers 97, 100–1
Leukocyte-derived
  arachidonic acid derivatives 153, 154
  cytokines 150–3
  oxidants 144–6
  proteases 149, 150
Leukotriene B_4 levels, isocyanate challenge 351, 377
Ligand-metal complexes 348–9
Limestone *46*
Lithium, toxicity *283*
  acute inhalation 286
Living conditions
  Indian silica workers 216
  South African gold miners 95, 96
Locust allergens, asthma induction 386
Loft MMVF insulation, airborne fiber levels 266
Log-normal distributions 53, **54**
  arithmetic mean, geometric mean and geometric standard deviation comparison 55
  probability density function 54
London asthma outbreak, 1994: 405
Longwall mining **7**, 165, 167
Low-molecular weight compounds, asthma 322, 328, 331, 333, 338, 340, 345–63
  agents causing asthma *347*, 353–63
  biochemical mechanisms 348–9
  immunologic mechanisms 349–53
    IgE antibodies 335
  related manuscripts *346–7*
Lumsden-Lynch vertical elutriator 42, **43**
Lung
  lining fluid, particle interaction 142
  transplant, silicosis 208
  volumes 17–19
    significance of changes in 19, 69–70
    timed 20
Lung cancer 467, 485–94
  asbestos workers 122, 222, 233, 234, 236, 240
    Brazil 110
    Finland 256, 257–8
    p53 mutations 476
    Russia 250, 251, 252
    treatment 243
  chest CT 126
  clinical presentation and management 494
  coal miners 171, 488
  crystalline silica workers 204, 207, 208–9
  genes
    oncogenes 470–2
    tumor suppressor genes 473–8
  individual susceptibility 492–3
  man made vitreous fiber (MMVF) exposure 270–3
    animal studies 268
  metal workers 297–8
  population at risk 493–4
  uranium miners 85, 90
  workers' compensation 90, 234, 241
Lymph nodes
  hilar enlargement 203, 204
  particle transport to 154–5, 195
Lymphocytes
  function in CWP and PMF 175
  hypersensitivity pneumonitis 438, 442, 443, 445
  occupational asthma 35, 331, 332, 333, 350
  see also T lymphocytes

Macrophage scavenger receptor 142
Macrophages 139, **141**, 413, 445
  attraction to particle deposition site 143–4
  clearance of pneumocytic particles 142–3, 154, 155, 174, 195
    incomplete phagocytosis 146, **147**
  effect of activation *141*, 175, 194
    cytokine production 223–4
  multiple overlapping roles 142
  see also Alveolar macrophages
Magnesium toxicity *283*
Magnetic resonance imaging (MRI), pneumoconiosis 125
Major histocompatibility complex (MHC) 331
Malawi, HIV prevalence 99
Maleic anhydride, asthma induction 360, 362
Malignancies *see* Gastric cancer; Lung cancer; Mesothelioma; Uterine cancer
Malingering 28
Man-made vitreous fibers (MMVF) 263–73
  biometrology 265
  classification and composition 263–4
  experimental data 266–8
  human data 268–73
  industrial and environmental hygiene data 265–6
  properties 264–5
Manganese toxicity *47*, *283*
  acute inhalation 286
  chronic inhalation 280
  COPD 288
Mannequin studies, total dust 38
Maple bark disease 447
Marble *46*
Mass, dust particles 36, 37

Mast cells
  asthma 329, 332–3, 352
  byssinosis 414
  hypersensitivity pneumonitis 441, 445
'Maxatase', asthma induction 391
Maximal oxygen uptake (VO_{2max}) 26
Maximum allowable concentration (MAC) 192
  asbestos levels, Russia 247
  crystalline silica 200–1
Mealworms, asthma induction 387
Mean concentration, vapors and gases 64
Medical
  history 30, 86–7
  surveillance 176, 177, 189
    Brazil, pneumonconiosis 107
    Finland 258–60
    South Africa, tuberculosis 97
    US, coal worker's surveillance program 170
Melanoptysis 174
Mercury toxicity *47*, *283*
  acute inhalation 285–6
  chronic inhalation 280
Mesothelioma 231, 234–6, 239–40, 241
  clinical findings 235
  histochemistry 234–5
  latency period 122, 240
  MMVF-induced 267, 271
  pathology 234
  prevalence
    Finland *256*, 257, 258
    Russia 251
    South Africa 99–100
  radiology 235
  treatment 244
  workers' compensation 90
Metal
  fume fever 281, 285
  speciation 280
Metals
  definitions 279
  exposure to 281, 285–98
    acute toxic effects 281, 285–6
    carcinogenicity 280, 490–1
    chronic obstructive pulmonary disease (COPD) 286–8
    intestitial lung disorders 291–7
    macro/microanalysis of 298
    occupational asthma 288–91
  general toxicity 279–80
    summary *282–4*
Metam sodium pesticide exposure 460
Methacholine challenge test 311, 314, 315–18, 328
  sources of variability *317*
  standards for testing 24
Methylene diphenyl diisocyante (MDI) 13, 375
  asthma 376, 377
  hypersensitivity pneumonitis 57
  vapor pressure/saturated vapor pressure *56*
Methylisocyanate (MIC) 56
  sampling methods *65*
  see also Bhopal disaster
Mexican bean weevil, asthma induction 387
Mica mining, India 214, 216
Microorganisms
  contaminated water 435
  cotton plant 409, 414–15, 416
Microvascular leakage, asthma 331, 351–2
Midget impinger 37
Migrant laborers 95, 98
Migration inhibitory factor (MIF) 444–5
Mill fever 411
Mill workers
  asthma induction 387–8, 389
  see also Cotton dust, effect on lung
'Mine effect' 77
Mine Safety and Health Administration (MSHA)
  ambient sampling, quartz 193
  respirable dust standard (RDS) 45
Miners' Compensation Authority, Germany 84
Mines Phthisis Act, 1916, South Africa 97
Mini-Finland health survey 259
Mining 3
  asbestos 109, 221
  coal 163–7, 489
    longwall mines **7**, 165, 167
    roof bolters 120
    surface mining 120–1, 164–5
  gold 95–9, 108, 489
  mica 214
  tin 489
  see also Asbestosis; Coal workers' pneumoconiosis (CWP); Lung cancer; Mesothelioma; Silicosis
Minute ventilation, maximal exercise 26
MIP-1α/MIP-2 upregulation 152, **153**
Moisture in coal 161
Moldy hay 437
Monday sickness 427
Moth allergens, asthma 385, 386
Mouse
  allergens, asthma induction 383, 384
  skin model, carcinogenesis 468
MRE horizontal elutriator 39–40, 44
Mucosal edema, asthma 351–2
Multiple regression analysis, lung function decline 72–5, 78, 79
Muscovite *46*
Mushroom powder, asthma induction 392
Mutagenesis 468, 469
Mutagens, endogenous 469–70
Mutations 467, 493
  mutation spectra 475–6
Myc oncogenes 472
Mycobacterial infection
  association with silicosis 7, 11, **204**, 206, 207
  coal miners 177
  see also Silicotuberculosis; Tuberculosis
Myocardial dysfunction, asbestosis 248
N-acetyltransferase (NAT) polymorphism 492

NADPH oxidase 141, 144
Nasal cancer 35
National Institute for Occupational Safety and Health (NIOSH)
  abrasive blasting 7
  Agricultural Research Service, insect-induced asthma 385
  crystalline silica exposure limits 192, 201, 206
  flow rate, respirable dust sampling 40

National Institute for Occupational Safety and Health – *continued*
  National Study of Coal Workers' Pneumoconiosis 9, 10–11
  spirometry testing, standardization 21
Nepheline *46*
Netherlands, silicosis epidemics *5*
Neurons, resistance to carcinogens 469
Neutrophils 144–6
  asthma 330, 336, 377
  byssinosis 413
NF-kB transcription factor activation 148
Nickel toxicity *47*, *283*
  acute inhalation 286
  asthma 289–90, 351
  lung cancer 297, 470, 490
Nicotine, vapor/condensed phase partitioning 57
Nitric oxide (NO) 403, 469–70
Nitrogen dioxide 281
  fibrotic bronchiolitis 455
  sampling methods *65*
  site of deposition 454
Nitrosamides 469
Nitrous oxide (NO·) 144
Nodules, Caplan's syndrome 169, 199, 204
Non-small cell lung cancer (NSCLC)
  chromosome 3p LOH *477*
  K-*ras* codon 12 mutations *471*
  recurrence, altered *p53* 473, *474*, 475
Northern fowl mites, asthma induction 386
'Nuisance dust' 45

Obesity, airways obstruction 77
Obstructive impairment 26
Occupational asthma 12, 112, 309
  acute toxic chemical inhalation 336–7, 458–60
  airway responsiveness 309–23, 329
    bronchial provocation testing 24–5, 315–22
    history 310–11
  cotton dust exposure 411, 413–14, 427, 431
  high-molecular weight compounds 321, 328, 331, 335, 338, 347, 381–94
    enzymes 390–1
    grain and flour 387–90
    laboratory animals and insects 382–7
    seafoods 391–2
  impairment and disability, evaluation 29
  isocyanate-induced 4, 12, 13–14, 375–8, 458, 459
    see also Isocyanate-induced asthma
  low-molecular weight compounds 322, 328, 331, 333, 338, 340, 345–63
    biochemical mechanisms 348–9
    immunologic mechanisms 349–53
    specific agents 353–63
  mechanisms of 327–41
    acute toxic/irritant exposures 336–7
    asthmatic reactions 330–1
    immune hypersensitivity 334–6
    immunologic mechanisms 331–4
    pathologic features 329–30
    population changes in airway responsiveness 337
    repeated submassive inflammatory exposures 337
  metal-induced 288–91, 295
    potroom asthma 290–1
  natural history 339–41
  prevalence 12, 345
  risk factors 337–9
  worker's compensation 86–7, 91, 309
  see also Urban epidemic asthma
Occupational Disease in Mines and Works Act, 1994, South Africa 97
Occupational Health and Safety Act, 1970, US 86, 200
Occupational history 30, 120, 310
Occupational Safety and Health Administration (OSHA) 53
  compliance sampling 64
  cotton dust exposure limits 413
  crystalline silica exposure limits 192, 200
Olivine *46*
Oncogenes 470–2, 492
Opacities, radiographic
  ILO classification 123–4
  silicosis 202–3
Open-circuit method, lung volume measurement 18
Organic dust *see* Cotton dust; High-molecular weight compounds; Hypersensitivity pneumonitis
Organic dust toxic syndrome (ODTS) *see* Toxic pneumonitis
Organometallic compounds 280
Osmium, toxicity *283*, 286
Oxidative stress 144, 146
  and the epithelium 146–7
Oxygen
  diffusing capacity 21, 22
  maximal oxygen uptake (VO_{2max}) 26
  partial pressure (PaO_2) 23, 26
  saturation of hemoglobin (SaO_2) 23, 26
Ozone 281, 317
  site of deposition and clearance 454
p53 gene 473, 492
  altered and NSCLC recurrence 473, *474*, 475, 476

Pancreatic extract, asthma induction 391
Papain, asthma induction 391
Paraphenylethylenediamine (PD), asthma induction 356
Parenchymal abnormalities
  chest CT 125
  ILO classification 123–4
Partial pressures 23
Partial-period sequential samples 48
Particle
  count 36–7
  deposition 139, **140**
    attraction of macrophages 143–4
  interaction with lung lining fluid 142
  marcophage response to pneumoconiotic 142–3
  mass 36, 37
  shape 146

size 35, 36, 37–43, 146
crystalline silica 44, 194
surface chemistry 147
transport to lymph nodes 154–5
see also Vapor/particle interactions
Passive
sampling 66–7
smoking 57, 77
Peak flow (PF)
measurements at workplace 25, 310, 311, 312–15, 318
intepretation of readings 313–14
meters 21
Peat 161
Pectin, asthma induction 392
Peripheral airways obstruction, asbestos workers 72
Permissable exposure limit (PEL) 192
Pesticides
arsenical 487, 488
metam sodium 460
Phagocytosis **141**, 143, 154, 195, 266, 331
generation of oxidants **144**
incomplete 146, **147**, 223
Pharmaceutical-induced asthma, related manuscripts *346*
Phlogopite *46*
Phorbal esters 470
Phosgene 53, 456
Phosphate rock miners 111
Phosphine, acute inhalation 286
Phosphorus *47*
Phthalic anhydride, asthma induction 360, 361
Physical examination 30
Pigeon breeder's disease 435, *436*, 440, 441, 443
BAL fluid 445
skin tests 444
T-cell function 444–5
Pit diggers, Brazil **4**, 108, *109*
Placebo, occupational agent challenges 322
Plasminogen activator 224
Platinum
platinum salt asthma 100, 288–9, 338, 353–4
IgE antibodies 335, 340
skin tests 349
toxicity *47*, *283*
Plethysmography
airway resistance measurements 20
lung volume measurements 18, 19
Pleural abnormalities
asbestos-related
diffuse pleural thickening 231–2, **233**, 236, 244
pleural effusion 122, 230
pleural plaques 230–1, 236, 240, 241, 243, 259
rounded atelectasis 232, 233, 236
chest CT 125
ILO classification 124
MMVF exposure 269
see also Mesothelioma
Pleurisy, asbestos-related 229–30, 236
Plicatic acid 328
antibodies to 335, 350
challenge tests 317, 330
exposure 333
PM_{10} sampler **42**, 43
Pneumoconiosis 4, 35, 69–79
definition 119
diagnosis 119–26
dust exposure history 119–21
pathologic criteria 189
mechanisms of 139–56
free radical activity 147–9
see also Free radicals
future research 155–6
immune system involvement 142–4
neutrophils, role of 144–6
oxidative stress 146–7
site of deposition and target cells 139–41
prevalence 7–12
treatment 188–9
experimental approaches 126–33
workers' compensation 88–90, 91
see also Asbestosis; Coal workers' pneumonconiosis (CWP); Man-made vitreous fibers (MMVF); Metals: exposure to; Silicosis
Pneumonia 100
Pneumothorax 177
Poland, CWP prevalence 11
Pollen grains, rupture in thunderstorms 405
Polyaromatic hydrocarbons (PAH)
lung cancer 488–9
PAH-DNA adducts 492
sampling 45
tobacco smoke
electrophilic metabolic intermediates 468
gas/particle partitioning 56, 57
Polyurethane
foam production
TDI reactivity 58–9
workers, homogenous exposure groups (HEGs) *62*
paint manufacture, hygiene survey 59–60
Potroom asthma 290–1
Poultry mites, asthma induction 386
Prawns, asthma induction 391–2
Prednisolone, silicosis 128
Prednisone
hypersensitivity pneumonitis 447
silicosis 128, **129**
Prevention 7, 8, 12–14
asbestos-related diseases 236
coal workers' pneumonconiosis (CWP) 177
context of 91–2
dust control 48–9
measuring effectiveness 4, 5
respiratory protective equipment 4, 7, 49, 60–1, 120, 239
US model 92
Primary prevention 7, 8, 177
Printer's asthma 394
Procollagen 3 levels 223
Progressive massive fibrosis (PMF) 168, 169, **173**, 176, 178, 198, 199
in asbestosis 227
immunization 177
low flow oxygen 176–7
pathology 173–4
respiratory failure 177
superoxide release 175
Proteases
in cotton dust 416
leukocyte-derived 149, 150

Proto-oncogenes 141, 470
Protoporphyrin, intracavitary photoactivated 244
'Pseudoplaques' 205
Psyllium workers, asthma 338, 394
Public buildings, asbestos 240
Pulmonary
- alveolar proteinosis (PAP) 132
- gas exchange limitations, estimating 26
- impairment, assessing 26–30
  - approach to evaluation 30–1
  - optimal time for evaluation 28, 29

Pulmonary function tests 17, 20–3, 31
- acute toxic chemical exposures
  - smoke inhalation 458
  - sulfur dioxide inhalation 457–8
- asbestos dust exposure 72
  - asbestosis 226, 229
  - guidelines to permanent respiratory impairment 242, *243*
- byssinosis 427, *428*, *429*, 431
- coal dust exposure 71, 72, *187*
- hypersensitivity pneumonitis 448
- occupational asthma 311–15
- silica dust exposure 70, 71, 205–6
- *see also* Spirometry

Pulse oximetry 24
Pumice *46*
PVNO therapy, silicosis 128, 130, 134
Pyrazinamide (PZA) 207, 216
Pyrite miners, sulfur dioxide exposure 457–8

Quartz *see* Crystalline silica
Quartzite *46*

Rabbit allergens, asthma induction 383, 384
Radiation exposure
- radon gas 90, 297, 493
  - domestic 297
  - tin miners, China 490
- uranium *48*, 85, *284*
- workers' compensation 90–1

Rank of coal 161
Ras gene 470-*2*, 492
Rat allergens, asthma induction 383, 384
Re-breathing method, diffusion capacity measurement 23
Reactive airways dysfunction syndrome (RADS) 286, 327–8, 336–7
- acute toxic chemical exposures 459–60
- basement membrane thickening 330
- isocyanate-induced asthma 376

Reactive oxygen species (ROS) 144, **145**, 174
- importance in mutagenesis 469, 470

Reactivity 58–9
Record-keeping 4, 12
Red spider mite, asthma induction 386
Refractory ceramic fibers (RCF) 263, 264, 269
- animal inhalation studies 268
- biopersistence 265

Relative standard deviation, vapors and gases 64
Residual volume (RV) 17
- significance of changes in 19

Respirable dust 39–40
- fraction, definition 41
- Respirable Dust Standard (RDS) 45
- sampling 42
  - coal dust 44–5

Respirators 49, 60–1, 120, 177, 239
- respirator protection factor (RPF) 60

Respiratory
- impairment evaluation report 31–2
- infections 77, 317
- protective equipment 4, 7, 49
  - types 49
  - *see also* Respirators
- system, functional anatomy 17, **18**
- tract irritation (RTI) 426

Restrictive impairment 27
Retinoblastoma gene 470, 476, 477
Retreat mining 165, 167
Retrospective cohort study design 485
Rh blood grouping 249
Rheumatoid
- arthritis 206
- factor 169, 175–6, 206

Rhinitis 335, 355, 356
Ricin, asthma induction 401, 403, 406
Rifampicin 207, 216
- prophylaxis 217

Right ventricular hypertrophy (RVH) 169
Rock drilling, overburden removal **164**-5
Rock wool 263, 264
- biopersistence 265
- lung cancer 270, 272, 273
- lung fibrosis 268, 269

Roof bolters 120
Room and pillar mines 165, **166**
Rounded atelectasis 232, 233, 236
Rubber elongation factor (REF) 393
Russia
- asbestos-related diseases 247–52
- chrysotile production 219

Rye
- flour, asthma induction 388
- grass pollen 405

Sample
- duration 45
- size, gases and vapors 62–3, *64*

Sampling 120
- frequency 48
- location 48
- technique
  - dust particles 36–7
  - vapors and gases 64–7

Sandblasting 202
Sandstone *46*
Scleroderma 100
- fibrogenic dust exposure 169

Screening *see* Medical: surveillance
Seafoods, asthma induction 391–2
Secondary prevention 8, 177
Selenium toxicity *47*, *283*
Sensitization 86
Settling velocity, dust particles 38

Sewage filter flies, asthma induction 387
Shape, dust particles 146
Shaver's disease 295
Shipyard workers, asbestos-exposure 72
Shortwall mining 167
Sickness fund 86
Siderosis 111, 292
Silica 5, *46*
 dust 120, 174
  effect on airways obstruction 70–1
  effect on lung function 73
 flour 202
  analysis 44
  mill workers 214
 see also Crystalline silica
Silicon carbide, interstitial lung disease 295
Silicoproteinosis 192, 196, 204, 205, 206
Silicosis 4, 85, 121, **122**, 191–209
 in Brazil 108–9
 clinical considerations 206–8, 215–17
  airways obstruction 70
 diagnosis 199–200
 epidemics *5*, 6–7
 lung cancer risk 491, 492
 massive fibrotic lesions **174**
 pathogenesis 194–6
  role of TNFα 151
 pathology 196–8, 199
 prevalence 7, *8*, 11
  China **186**
  India 213–15
 pulmonary function 205–6
 radiology
  chest CT 204–5
  chest radiographs **6**, **122**, **124**, **129**, **198**, **199**, 202–4
 risk occupations 202
 in South African gold miners 96, 97–8, 100
 treatment, experimental approaches 126–33
 workers' compensation 88
Silicotic nodules 196, **197**, 198, 199
Silicotuberculosis 108, 207
 chest radiograph **216**
 China 217
 India 215–16
Silkworm moth, asthma induction 385, 386, 387
Silver toxicity *47*, *283*
Single-breath technique 23
Sintering 295
Size
 dust particles 35, 36, 37–43, 146
 man-made vitreous fibers (MMVF) 264
Skin testing
 hypersensitivity pneumonitis 444
 occupational asthma 318, 350
  grain and flour workers 389
  nickel hypersensitivity 290
  platinum salt hypersensitivity 288–9, 349, 354
 tuberculin skin test 207
Slag wool 263, 264, 267
 lung cancer 270, 272, 273
 lung fibrosis 269
Slate *46*
 pencil industry workers 214
Small cell lung cancer (SCLC) 471
 p53 mutations 476
Smoke inhalation, acute lung injury 336, 337, 458–9
Smoking
 cessation 176
 decline in FEV_1 70, 73
 and effect of dust 75
 effect on immune system 339
 history 30
 passive smoking 57, 77
 see also Tobacco smoke
Smoking, role in
 asthma 394
  grain and flour workers 389
  laboratory animal asthma 385
  platinum salts 348, 354
 bronchitis 168
 byssinosis 425
 COPD 70, 72, 77, 269, 286–7
 emphysema 71
 lung cancer 90, 486
  asbestos exposure 233, 240
  MMVF exposure 272
  silica exposure 209
 metal exposure lung disease 280
 silicosis 205, 215
Snow-crab asthma 87, 340, 392
Social security system, Germany 83, 84
Socioeconomic status, respiratory disease 78
 South African gold mine workers 99
Soldering fluxes, asthma induction 356
 colophony 354, 355
Solubility, vapor-phase toxicants 57–8
Source characteristics, toxicants 59–60
South Africa
 asbestos industry 99–100, 220, *221*
 coal mining 100
 cotton workers, byssinosis compensation 87–8
 gold miners 95–9
  airways obstruction
   dust effects 70, 71
   effects of obesity 77
  HIV 96, 97, 98–9, 100
  lung cancer 489
  silicosis 11, 97–8, 100
  social context of disease 99
  tuberculosis 96, 97, 100
 iron workers, lung cancer 491
Soybean epidemic asthma 402–5, 406
 etiologic studies 403–4
 intervention and follow-up 404
Spectrophotometers 23
Spirometry 17, 20–1, 31, 310
 American Thoracic Society interpretation 21, *22*
 byssinosis 416, 417
 hypersensitivity penumonitis 438
 occupational asthma 310, 311–12, 318
 sources of variation 21
Spray painting 12, 57
 asthma 357
 hygiene survey 59–60
Sputum production
 coal worker's pneumoconiosis 168
 isocyanate-induced asthma 352–3
Squamous cell carcinoma 476
 chromosome 3p loss of heterozygosity *477*
 GSTM1 polymorphisms 493
 recurrence and *p53* mutations *474*, 475

Stability of sample, vapors and gases 64
Standard methods, vapor and gas sampling 64, *65*
Standard Mortality Ratio (SMR) 485
- coal miners, all causes of death 170–1, 172
- lung cancer
  - asbestos workers 250, 251
  - copper smelter workers, airborne arsenic *488*
  - crystalline silica workers 491, 492
  - MMVF workers 270

Standard Reference Material (NBS-SRM#1878) 44
Stannosis 292
Steady-state method, diffusion capacity measurement 23
Steel workers
- COPD 287
- lung cancer 297
- lung function decline 76
  - obesity 77
- silicosis 214

Stellite 295
Stone quarry workers 215
Stonemasons, silicosis 7, 108, *109*, 194
Storage mites, asthma induction 388, 389
Streptomycin 207, 216
Styrene 349
Subacute hypersensitivity pneumonitis 439
Sudan, byssinosis 430, 431
Sulfur 161
- sulfur dioxide 281
  - lung function measurements 457–8
  - sampling methods *65*
  - site of deposition and clearance 454
  - solubility 57
  - vapor/particle interactions 57
- sulfur mustard 456

Surface mining
- dust sampling 120–1
- rock drilling, overburden removal **164**-5

Surveillance 7–12
- SENSOR program 455
- SWORD data 12, 381, 455, 460
- *see also* Medical: surveillance

Systemic toxicity 35, 39
Systems of Health Insurance Act, 1883, Germany 84

T lymphocytes
- asthma 329, 330, 351, 376
  - cytokine secretion 332, 349–50
- hypersensitivity pneumonitis 438, 443, 444–5

Talc *46*
Talcosis 111
Talvite method, crystalline silica analysis 44
Tea dust, asthma induction 392
Tellurium *48*
Terminal bronchiolar epithelium 139, 142
Tertiary prevention 8
Tetrachlorophthalic anhydride, asthma induction 360, 361
Tetrandrine, silicosis treatment 130–1, 134
Textile Occupational Disease Panel, North Carolina Industrial Commission 87
Thermal Insulation Manufacturers' Association (TIMA), MMVF classification 263
Thermal precipitators 44
Thoracic
- dust sampling 42
- gas volume (TGV) 18
- size fraction 41

Threshold limit values (TLV)
- asbestos fibers 236
- cotton dust 413, 422
- crystalline silica 43, 44, 107, 192, 193, 200–1
- isocyanates 375
- respirable chrysotile 107

Thunderstorms, asthma outbreaks 405, 406
Tidal volume (TV) 18
Tin
- mining, lung cancer 489
- toxicity *48*, *283*
  - stannosis 292

Tires
- latex antigens, asthma 394
- repair workers, silicosis 108, 109

Titanium *48*, *283*
- granuloma formation 294
- titanium dioxide (TiO_2) 146
  - COPD 288

'TMA flu' 362
TNFα, role in pneumoconiosis 150, 151, **152**, 153
Tobacco smoke 485, 492
- aromatic amines 493
- nitrosamines 469
- polyaromatic hydrocarbons (PAH)
  - electrophilic metabolic intermediates 468
  - gas/particle partitioning 56, 57
  - PAH-DNA adducts 492
- *see also* Smoking

Toluene diisocyanate (TDI) 12, 13, 375
- hypersensitivity pneumonitis 57
- occupational asthma 350, 351, 376, 377, 460
  - inflammatory cells 352, 376
- reactivity 58–9
- sampling methods *65*
- vapor pressure/saturated vapor concentration *56*

Tool sharpening 295
Total dust 37–9
Total lung capacity (TLC) 17
- body plethysmograph 18
- significance of changes in 19

Toxic pneumonitis 285, 411, 413, 414
Training, occupational medicine 91
Transcription factor activation 148
Transition metals 280
Transmission electron microscopy (TEM), fiber counts 221
Tremolite 240
Trichosporon cutaneum 435, 437, 447
Tridymite *46*, 191
- threshold limit value (TLV) 201

Trimellitic anhydride (TMA), asthma induction 362–3
Trypsin, asthma induction 391
Tuberculin skin test 207
Tuberculosis
- asbestos workers, Russia 248
- chlorine vapor exposure 455
- coal miners 177, 187
  - China 184
- cotton workers, India 427

crystalline silica workers 192, 204, 206, 207, 208
India 213, 215–17
gold miners, South Africa 95, 97, 100
association with HIV 99
history 96
see also Silicotuberculosis
Tumor suppressor genes 473–8, 492
Tungsten toxicity *48*, *283*

Uganda, HIV prevalence 99
Ultrafine particles 146, 147
Union Carbide plant explosion *see* Bhopal disaster
United Kingdom
coal mining
coal workers' pneumoconiosis 168, 169, 170, 171
gastric cancer 171
lung disease, history 162–3
compensation
asbestosis 90
byssinosis 87
occupational asthma 87
London asthma outbreak, 1994: 405
United States
asbestos building materials 240
coal mining **164**-7
coal workers' pneumoconiosis (CWP) 169–70, 171–2
malignancies 171
US coal workers' surveillance program 170
compensation 85–6
asbestosis 89–90
byssinosis 87
coal workers' pneumoconiosis 88–9
compensation fund 91
radiation exposure 90–1
silicosis 88
cotton workers 413
lung cancer
MMVF study 270–1
population at risk 493–4
model of prevention 92
SENSOR surveillance program 455
silicosis 193–4
epidemics *5*
Uranium
mining
East Germany 85
United States, workers' compensation 90–1
toxicity *48*, *284*
lung cancer 297
see also Radiation exposure: radon gas
Urban epidemic asthma 401–6
Uterine cancer, asbestos workers 250

Vanadium toxicity *48*, *284*
acute inhalation 286
asthma 291
Vapors 53, 454
mercury inhalation 285–6
sampling 64–7
solubility 57–8
vapor pressure 13, 55, 56
vapor/particle interactions 56, 57
Ventilation systems 49
Ventilatory limitation, estimating 26
Vermont granite industry 4
crystalline silica exposure standard 200, 201
lung cancer 209
lung function decline 70, 205–6
Vertical elutriator, cotton dust sampling 42, **43**
Victoria, Princess Royal 84
Vital capacity (VC) 17, 18
Vitallium 297
Vitreous fibers *see* Man-made vitreous fibers (MMVF)
Volume spirometers 21
Von-Hippel Lindau gene 478

Wales, coal mining
coal workers' pneumoconiosis (CWP) 169
gastric cancer 171
Water use, dust suppression 49
Weevils, asthma induction 387, 388
Weight gain, effect on lung function 77
Welding
occupational asthma 291
'welder's pneumoconiosis' 292
Welfare and Retirement Fund, US 89
Western red cedar *see* Cedar workers
Wet-chemistry method, crystalline silica analysis 44
Wheat flour, asthma induction 387, 388
immunotherapy 388
Whole-lung lavage (WLL) 131–3, 134, 208
Wood dust
challenge tests 322
inhalable dust sampling 42
see also Cedar workers, asthma
Wool dyes, asthma induction 356
Work-related lung diseases surveillance report 7, 8, 14
Working Level Months (WLM) 85, 90
exposure requirements for compensation 91
Workmen's Compensation Act, South Africa 87–8
World Health Organization (WHO)
classification, byssinosis and respiratory tract irritation *426*
CWP in developing nations 184
tuberculosis 100
World-wide problem 3
surveillance and recognition 7–12

X-ray diffraction, dust particle analysis **10**
crystalline silica 44, 196
Xinin, silicosis treatment 131

Yan method 315

Zimbabwe, mining activities
CWP prevalence 186
lung cancer 489, 490
Zinc toxicity *48*, *284*
asthma 291
zinc chloride inhalation 286
zinc oxide, metal fume fever 281, 285
Zirconium toxicity *284*
granuloma formation 294